Problems and Solutions

PHYSICAL CHEMISTRY

C Kalidas
Professor (Retd), Department of Chemistry
Indian Institute of Technology Madras

M V Sangaranarayanan
Professor, Department of Chemistry
Indian Institute of Technology Madras

Universities Press

PROBLEMS AND SOLUTIONS: PHYSICAL CHEMISTRY

UNIVERSITIES PRESS (INDIA) PRIVATE LIMITED

Registered office
3-6-747/1/A & 3-6-754/1, Himayatnagar, Hyderabad 500 029, Telangana, India
info@universitiespress.com; www.universitiespress.com

Distributed by
Orient Blackswan Private limited

Registered office
3-6-752 Himayatnagar, Hyderabad 500 029, Telangana, India

Other offices
Bengaluru, Bhopal, Chennai, Guwahati, Hyderabad, Jaipur, Kolkata,
Lucknow, Mumbai, New Delhi, Noida, Patna, Visakhapatnam

First published 2020

ISBN: 978-93-89211-18-4

Typeset in Times New Roman 10/12 by SRS Technologies, Puducherry

Printed at
Akash Press, Delhi

Published by
Universities Press (India) Private Limited
3-6-747/1/A & 3-6-754/1, Himayatnagar, Hyderabad 500 029, Telangana, India

Preface

Problems and Solutions: Physical Chemistry presents an extensive collection of diverse types of worked out examples, multiple choice questions and practice problems. A large number of examples on all the major topics covered in physical chemistry courses find a place. In order to make the book student-friendly, detailed steps have been provided in the worked out problems in all chapters. Numerical and conceptual questions are illustrated in detail for easy comprehension of students. Since uniform coverage for each topic is difficult, the level of questions ranges from the fundamental to the advanced level with the hope that the book will be found useful by a wide spectrum of students at all levels. The theoretical aspects required to solve the examples are also provided, although the book is primarily intended as a companion to textbooks in physical chemistry.

This book is divided into 15 chapters. Chapter 1 provides an elaborate collection of problems on the gaseous state with particular emphasis on various equations of state and critical phenomena. In Chapter 2, examples spanning various topics such as laws of thermodynamics, concept of fugacity and activity in gases and liquids, as well as the application of Maxwell equations and calculation of absolute entropies have been discussed. Chapter 3 on electrochemistry deals with examples from ionic solutions and the Debye–Hückel–Onsager theory to kinetics of electrochemical processes. Since it is not easy to delineate the ionics from Nernst equations, examples complimentary to both are provided in an overlapping manner. Different worked out examples in chemical kinetics have been dealt with in Chapter 4, wherein, apart from the conventional questions on determination of order and rate constants, questions on Marcus theory and relaxation methods have been provided, thus making this chapter unique. The field of 'liquid state' has been covered in Chapter 5 wherein the subtle features of refractive index, polarisability and their relevance in structure elucidation (through parachor) has been outlined. Chapter 6 deals with surface chemistry, colloids and interfaces. Apart from adsorption isotherms, electrokinetic and electrocapillary phenomena, colloidal electrolytes have also been discussed with numerical examples. In Chapter 7, solid state chemistry examples are provided wherein the Miller indices, defects in solids and lattice types are illustrated with adequate examples. In Chapter 8, various topics in photochemistry such as actinometry, kinetics of photochemical reactions, photosynthesis and radiation chemistry are discussed and numerical examples are provided. The importance of the Jablonski diagrams, Stern–Volmer equation and flash photolysis techniques have been emphasised.

An extremely important topic in classical physical chemistry is solutions. The treatment of solutions using Henry's law and Raoult's law is done with various examples. Furthermore, numerical examples involving the van't Hoff factors and osmotic pressure of electrolytes are also illustrated. Non-ideal behaviour of electrolyte solutions is discussed in terms of activity coefficients from the Debye–Hückel limiting law. The phase rule and phase equilibrium is discussed in Chapter 10. The method of constructing phase diagrams for binary and ternary systems is highlighted and will be found especially useful for chemical engineers and metallurgists. Chapter 11 deals with a few important topics in polymer chemistry such as molecular weight determinations, glass transition temperature and viscosity relations. In Chapter 12, quantum chemistry examples are covered with illustrative problems covering wave functions, particle in a box, harmonic oscillator and rigid rotor. Chapter 13 discusses acid–base equilibria in an extensive manner on account of its importance in analytical and industrial chemistry. Several intricate examples for solubility products, buffers and ion selective electrodes are provided. The concept of acidity function, essential for comparing acidities in different solvents has been illustrated. In Chapter 14, a variety of examples on nuclear fission and fusion reactions as well as isotopes and nuclear reaction cross-section rates have been provided. Chapter 15 provides a brief overview of applications of statistical thermodynamics; classical and quantum distribution laws are illustrated with typical examples. The different types of partition functions have been given along with their use in computing various thermodynamic quantities.

In each chapter, a large number of worked out examples have been provided. The answers for the practice problems will aid the student to independently verify the calculations involved.

The essential objective of the book is to provide in a comprehensive manner, a variety of worked out examples in all the topics of physical chemistry for students of various degree courses, especially at the undergraduate and post-graduate levels. It is hoped that the book will be valuable to sharpen the problem-solving skills of students.

Prof. Kalidas thanks his family for support and encouragement during all phases of the writing of the book. Prof. Sangaranarayanan thanks his family for the continuous enthusiastic support as well as his research scholars S Harinipriya, V Sudha, R Ramya, V Divya, S Preethi and P Sivasakthi for their help during the writing of various chapters. Prof. S Mahalakshmi, Head, Department of Chemistry, VIT Chennai, has provided valuable suggestions for the improvement of the book. Special thanks are due to Dr Gita S Dattatri, Universities Press, for the efficient and immaculate handling of the 'chaotic material' in a patient manner.

C Kalidas and M V Sangaranarayanan
IIT Madras, 2020

Contents

Fundamental Constants

Fundamental constant	Numerical value	Symbol
Avogadro's number	6.023×10^{23} mol^{-1}	N or N_0
Boltzmann's constant	$\begin{cases} 1.381 \times 10^{-23}\ \text{J K}^{-1} \\ 8.626 \times 10^{-23}\ \text{eV K}^{-1} \\ 0.695\ \text{cm}^{-1}\ \text{K}^{-1} \end{cases}$	k or k_B
Universal gas constant	$\begin{cases} 8.314\ \text{J mol}^{-1}\ \text{K}^{-1} \\ 1.99\ \text{cal mol}^{-1}\ \text{K}^{-1} \end{cases}$	R
Molar volume at STP	22.4 L	
Planck's constant	6.626×10^{-34} J s	h
Velocity of light	3×10^{10} cm s^{-1}	c
Charge of the electron	1.60×10^{-19} C	e
Bohr radius	0.529×10^{-8} cm	a_0
Mass of an electron	9.11×10^{-28} g	m_e
Mass of a proton	1.67×10^{-24} g	m_p
Atomic mass unit	1.661×10^{-24} g	amu
Faraday constant	96500 C equiv^{-1}	F
Mass of neutron	1.675×10^{-24} g	m_n

Other useful data

1 faraday = 26.8 ampere hours

$RT/F = 0.0257$ V at 298 K

CHAPTER 1

GASEOUS STATE

CHAPTER OUTLINE

1.1 Introduction

Among the three states of matter – solids, liquids and gases – gases are the simplest to deal with. The complexity in their properties increases in the order gases < liquids < solids.

A gas is defined as a homogeneous species whose volume increases continuously with a decrease of pressure at constant temperature. When different gases in any proportion are brought in contact, they diffuse rapidly into each other and form a homogeneous mixture. Gases are highly compressible compared to solids and liquids.

1.2 Gas Laws

1.2.1 Boyle's Law

At constant temperature, the volume of a given mass of gas is inversely proportional to the pressure. That is,

$$P \propto \frac{1}{V} \text{ or } PV = \text{Constant} \tag{1.1}$$

We may also write Eq. 1.1 as

$$P_1V_1 = P_2V_2 \tag{1.2}$$

where P_1, V_1 and P_2, V_2 are the pressures and volumes at a given temperature for a given mass of gas. The plot of P vs $1/V$ is linear (at constant temperature) passing through the origin.

1.2.2 Charles' Law

At constant pressure, the volume of a given mass of gas expands by the same fraction of its volume at 0°C per one degree rise in temperature. If V_0 and V_t are the volumes of a gas at 0°C and t°C, and α is the coefficient of expansion,

$$V_t = V_0\left(1+\alpha t\right) \tag{1.3}$$

α is the same for all gasses and is equal to 1/273.16. Thus,

$$\frac{V_t}{V_0} = \left(1+\frac{1}{273.16}t\right) = \frac{273.16+t}{273.16} \tag{1.4}$$

A plot of V_t/V_0 against t (in °C) is linear with a slope of 1/273.16. The intercept of the graph is 1.0 at t = 0°C. An alternate form of Eq. 1.4 is

$$\frac{V_t}{V_0} = \frac{T}{T_0} \left(\text{where } T = 273.16+t\right) \text{ and } T_0 = 273.16 \tag{1.5}$$

T_0 and T are temperatures on the absolute temperature scale (K). An alternate form of Eq. 1.5 is

$$\frac{V_1}{V_2} = \frac{T_1}{T_2} \text{ or } V \propto T\left(T \text{ being in K}\right) \tag{1.6}$$

1.2.3 Combined Form of Boyle's and Charles' Law

It is given by

$$\frac{PV}{T} = \text{Constant} \tag{1.7}$$

Equation 1.7 is known as the *Equation of State*, and for a given state, say at STP, the constant can be evaluated. Using $P = 1$ atm, $V = 22.414$ litres (for one mole of gas and $T = 273.16$ K, Eq. 1.7 may be written as

$$\frac{1 \times 22.414}{273.16} = R = 0.0821 \text{ L atm K}^{-1} \text{ mol}^{-1} \tag{1.8}$$

For n moles of gas, Eq. 1.7 is given as

$$PV = nRT \tag{1.9}$$

where V = volume of n moles of gas under a pressure P and temperature T.

1.2.4 Units of Gas Constant *R*

- R has the units of energy or work deg^{-1}mol^{-1} and its numerical value varies with the unit chosen. If the value of V is chosen in cm^3,

$$R = 0.0821 \times 1000 = 82.1 \text{ cm}^3 \text{ atm K}^{-1} \text{ mol}^{-1} \tag{1.10}$$

- R in CGS units

$$R = 8.316 \times 10^7 \text{ ergs K}^{-1} \text{ mol}^{-1} = 8.316 \text{ J K}^{-1} \text{ mol}^{-1}$$

$$= \frac{8.316}{4.184} = 1.987 \text{ cal K}^{-1} \text{ mol}^{-1} \tag{1.11}$$

- ***R* in SI units:** In SI units, pressure is given in pascals (Pa) and volume is expressed in m^3.

$$R = \frac{PV}{T} = \frac{101325 \text{ N m}^{-2} \times 22.4 \times 10^{-3} \text{ m}^3 \text{ mol}^{-1}}{273.16 \text{ K}}$$

$$= 8.316 \text{ N m K}^{-1} \text{ mol}^{-1} = 8.316 \text{ J K}^{-1} \text{ mol}^{-1} \tag{1.12}$$

$$\left(\text{since N m} = 1 \text{ J}\right)$$

- A different form of ideal gas equation: If W g of a gas of molar mass M is present in a volume V,

$$PV = \frac{W}{M} RT \tag{1.13}$$

or

$$P = \frac{W}{V} \times \frac{RT}{M} = d \times \frac{RT}{M} \tag{1.14}$$

where d is the density of the gas in g L^{-1} or kg m^{-3}. It is seen that P is directly proportional to d at a temperature T.

Equation 1.14 may be used for determining the molar mass, M, of the gas. A plot of d/P against P is made and the linear graph is extrapolated to $P = 0$, that is, $\left(\frac{d}{P}\right)_{p \to 0} = \frac{M}{RT}$. Using known R and T, M can be calculated.

1.2.5 Absolute and Relative Density of a Gas

The absolute density of any substance is defined as mass per unit volume, that is, in g L^{-1} or kg m^{-3}. The relative density of a gas is defined as the ratio of the absolute density to that of a standard substance, which is usually hydrogen gas. Thus,

The relative density (also known as vapour density)

$$= \frac{\text{Weight of a known volume of gas at a given temperature and pressure}}{\text{Weight of the same volume of H}_2 \text{ gas at the same temperature and pressure}}$$

The weight of 11.2 L (or dm^3) of hydrogen gas at STP is close to unity and hence this volume is chosen to define relative density. Thus, relative density of gas

$$= \frac{\text{Weight of 11.2 dm}^3 \text{ of a gas at given } T \text{ and } P}{\text{Weight of 11.2 dm}^3 \text{ of H}_2 \text{ at the same } T \text{ and } P} \tag{1.15}$$

The denominator is unity since it refers to weight of 11.2 dm^3 of H_2. Thus,

$$\text{Absolute density of a gas} \left(\text{in g L}^{-1} \text{ or in kg m}^{-3}\right)$$

$$= \frac{\text{Relative density}}{11.2} \tag{1.16}$$

Also,

$$\text{Molecular weight of the gas} = \text{Relative density} \times 2 \tag{1.17}$$

1.2.6 Abnormal Vapour Density

In some cases, the experimental vapour density of a gas is much less (or sometimes greater) than the theoretical vapour density [= (molecular weight)/2]. For example, for NH_4Cl the experimental vapour density is 14.6 at 673 K whereas the theoretical value is 26.75. This abnormal value is explained on the basis of dissociation of NH_4Cl.

Degree of dissociation of a gas or vapour from vapour density: The degree of dissociation of a gas (or vapour) is the fraction of the total number of molecules that has undergone dissociation. When a gas dissociates, more molecules are formed and at constant temperature and pressure, the volume of the system increases. So, the density, which is inversely proportional to volume, decreases.

If we start with one mole of a gas and α is the degree of dissociation, the amount of gas remaining undissociated = $(1 - \alpha)$

Total amount of gas pesent at equilibrium

$$= 1 - \alpha + \left(1 + \sum_i \gamma_i\right)\alpha = \left(1 + \alpha \sum_i \gamma_i\right) \tag{1.18}$$

γ_i is the stoichiometric number and is positive for products and negative for reactants. It is the difference in the total number of products to reactants.

The density of an ideal gas at constant P and T is inversely proportional to the number of moles for a definite weight of the gas. The ratio of the density of undissociated gas, d_0, to the density of the partially dissociated gas, d, is given by

$$\frac{d_0}{d} = \left(1 + \alpha \sum_i \gamma_i\right) \quad (1.18a)$$

or

$$\alpha = \frac{d_0 - d}{d \sum_i \gamma_i} \quad (1.19)$$

For example, the dissociation of

$$PCl_5(g) \rightleftharpoons PCl_3(g) + Cl_2(g)$$

$$\sum_i \gamma_i = \sum_i \text{Products} - \sum_i \text{Reactants} = 2 - 1 = 1$$

$$\alpha = \frac{d_0 - d}{d} \quad (1.20)$$

Note that d_0 and d are measured at the same pressure and are in the same units. The same relation for the calculation of α will be valid even if the total pressure is allowed to vary keeping the volume of the system constant.

1.2.7 Dalton's Law of Partial Pressures

Dalton's law states that at any given temperature, the total pressure exerted by a gas mixture is equal to the sum of the partial pressures of the constituent gases. Hence, the total pressure P, is given by,

$$P = p_1 + p_2 + p_3 + \ldots \quad (1.21)$$

where p_1, p_2 and p_3 are the partial pressures of the gases 1, 2 and 3. The partial pressure p_i of the species i in the mixture is defined as the pressure the gas would exert if it alone occupied the total volume of the mixture V at the given temperature. If n_1, n_2 and n_3 represent the the number of moles of each of the gases 1, 2 and 3 in the mixture in the total volume V, Eq. 1.21 may be written as

$$P = \left(\frac{n_1}{V} + \frac{n_2}{V} + \frac{n_3}{V}\right) RT \quad (1.22)$$

or

$$(p_1 + p_2 + p_3)V = (n_1 + n_2 + n_3)RT = nRT \quad (1.23)$$

where $n = n_1 + n_2 + n_3$

$$\frac{p_1}{P} = \frac{n_1}{n} = X_1; \frac{p_2}{P} = \frac{n_2}{n} = X_2; \frac{p_3}{P} = \frac{n_3}{n} = X_3 \quad (1.24)$$

where, X_1, X_2 and X_3 are the mole fractions of the different gases. Thus we may write,

$$p_1 = X_1P, \; p_2 = X_2P, \; p_3 = X_3P \quad (1.25)$$

1.2.8 Graham's Law of Diffusion

Gases diffuse at different rates depending on their densities. Graham's law states that the rate of diffusion of a gas is inversely proportional to the square root of its density. If D_1 and D_2 are the rates of diffusion of two gases whose densities are d_1 and d_2,

$$\frac{D_1}{D_2} = \sqrt{\frac{d_2}{d_1}} = \sqrt{\frac{M_2}{M_1}} \quad (1.26)$$

where M_1 and M_2 are molecular weights of gases 1 and 2. This law is used in the separation of the isotope of U_{235} from U_{238}. It is also useful in the determination of gas densities. For example, if t_1 and t_2 are the times required for the same volume V of gases 1 and 2 to effuse through a pin hole,

$$\frac{D_1}{D_2} = \frac{V}{t_1} \times \frac{t_2}{V} = \sqrt{\frac{d_2}{d_1}} = \sqrt{\frac{M_2}{M_1}} \quad (1.27)$$

NOTE

- D_1 and D_2 are the rates of diffusion of gases 1 and 2.

As another possibility, if two gases having volumes V_1 and V_2 pass through a pin hole in the same time interval t, the rates of diffusion

$$\frac{D_1}{D_2} = \frac{V_1}{t} \times \frac{t}{V_2} = \sqrt{\frac{d_2}{d_1}} = \sqrt{\frac{M_2}{M_1}} \quad (1.28)$$

1.3 Kinetic Theory of Gases

The gas laws discussed so far are based on experimental observation. A theoretical basis to explain the laws comes from the *kinetic theory of gases*. This theory is based on several postulates or assumptions which are given below.

- The gas consists of a large number of small particles (atoms or molecules) moving about in various directions.
- During continuous movement, the molecules collide with each other frequently. In this process, there is no loss of translational energy and hence the collisions are said to be *elastic*.
- The elastic collisions of the gas molecules on the walls of the vessel results in their pressure exerted by the gas.
- There are no forces of attraction or repulsion between gas molecules.
- The gas molecules are very small in size such that their actual volume is negligible in comparison to the total volume of gas.

1.3.1 Expression for the Pressure of a Gas and the Kinetic Energy

The pressure P of the gas on the basis of the model is

$$P = \frac{1}{3} \times \frac{mn\overline{C^2}}{V} \quad (1.29)$$

In Eq. 1.29, m = mass of the gas molecule (in g or kg), n = number of gas molecules, $\overline{C^2}$ = mean square velocity of the gas molecules (cm^2 s^{-2} or m^2 s^{-2}), V = volume of the gas (in cm^3 or m^3) and P is expressed as dynes cm^{-2} or N m^{-2} (in SI unit).

If the total kinetic energy (KE) of all the gas molecules in one mole of gas is given by

$$\text{KE} = E_k = \frac{1}{2} \times mN\overline{C^2}$$

Equation 1.29 may be related to E_k according to

$$PV = \frac{1}{3} mN\overline{C^2} = \frac{2}{3} \times \frac{1}{2} \times mN\overline{C^2} = \frac{2}{3} E_k \quad (1.30)$$

or $$E_k = \frac{3}{2}PV = \frac{3}{2}RT \text{ (for 1 mol of gas)} \tag{1.31}$$

NOTE

- N is Avogadro's number in Eq. 1.30.

The kinetic energy per molecule of gas (E_k) is equal to

$$E_k = \frac{3}{2} \times \frac{RT}{N} = \frac{3}{2}kT \tag{1.32}$$

where k is the Boltzmann constant and T is in K. Since $E_k \propto T$ from Eq. 1.31, all gases have the same kinetic energy (per mol of gas) at a given temperature T. All gas laws may be derived from the kinetic theory (Eq. 1.29).

1.3.2 Different Types of Velocities of Gases

Root mean square velocity (C_{rms}): The root mean square velocity of a gas is defined by

$$C_{\text{rms}} = \sqrt{\overline{C^2}} = \sqrt{\frac{\overline{C_1^2} + \overline{C_2^2} + \overline{C_3^2} + \ldots + \overline{C_n^2}}{n}} \tag{1.33}$$

where $\overline{C_1^2}, \overline{C_2^2}, \ldots \overline{C_n^2}$ are the mean square velocities of gaseous molecules 1, 2, 3 and so on.

From Eq. 1.30,

$$PV = \frac{2}{3} \times \frac{1}{2} \times mN\overline{C^2} = \frac{2}{3} \times \frac{1}{2} \times M\overline{C^2}$$

$$\overline{C^2} = \frac{3PV}{M} = \frac{3RT}{M}$$

$$\therefore \quad \sqrt{\overline{C^2}} = \sqrt{\frac{3RT}{M}} \tag{1.34}$$

When considering n moles of a gas, $\frac{PV}{n} = RT$ and $\frac{nM}{V} = d$, d being the density of the gas at temperature T and pressure P,

$$C_{rms} = \sqrt{\frac{3PV}{nM}} = \sqrt{\frac{3P}{d}} \tag{1.34a}$$

Maxwell's law of distribution of molecular velocities: As a result of the continuous collisions among themselves and also with the walls of the container, the molecules of a gas do not move with the same velocity. There is a continuous interchange of momentum and according to Maxwell, the distribution of molecular velocities among various gas molecules is given by

$$\frac{1}{n} \times \frac{dn}{dc} = 4\pi \left(\frac{M}{2\pi RT} \right)^{3/2} e^{\left(\frac{-MC^2}{2RT} \right)} C^2 \tag{1.35}$$

In Eq. 1.35, R, M and T have their usual significance and C is the velocity of the gas. The LHS of the equation gives the fraction of the molecules having speeds in the range C and $C + dC$ for a unit interval of speed. It is thus the probability of finding gas molecules in the velocity range of C+dC and it depends on the molar mass and temperature of the gas. A plot of LHS of Eq. 1.35 against C is shown in Fig. 1.1 at various temperatures T_1 and T_2.

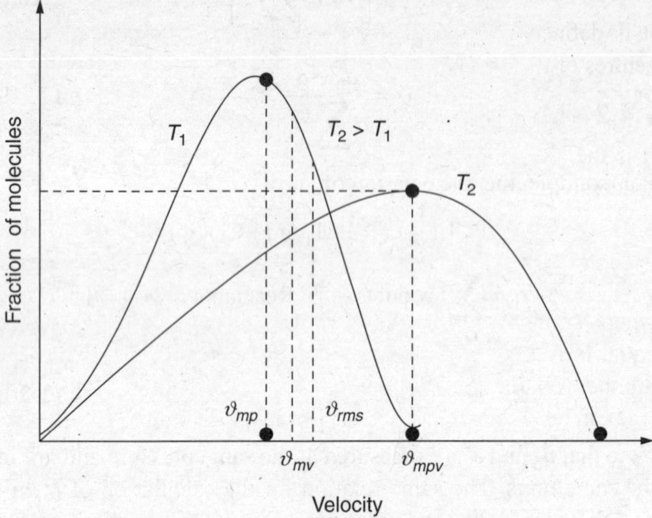

Fig. 1.1 Dependence of the fractional occupancy on velocities at two different temperatures

In the diagram ϑ_{mv} and ϑ_{mpv} are known as *mean velocity* and *most probable velocity* respectively and are defined as follows.

Mean velocity (ϑ_{mv}): The mean velocity is defined by

$$\vartheta_{mv} = \bar{\vartheta} = \frac{\vartheta_1 + \vartheta_2 + \vartheta_3 + \ldots + \vartheta_n}{n} \tag{1.36}$$

where ϑ_1, ϑ_2 and ϑ_3 are the velocities possessed by gas molecules 1, 2 and 3 and n is their total number. This velocity is also equal to

$$\vartheta_{mv} = \sqrt{\frac{8RT}{\pi M}} \tag{1.37}$$

Most probable velocity (ϑ_{mpv}): Most probable velocity is the velocity possessed by most molecules in a gas and is given by

$$\vartheta_{mpv} = \sqrt{\frac{2RT}{M}} \tag{1.38}$$

ϑ_{mpv}, ϑ_{mv} and ϑ_{rms} differ slightly from each other and at a given temperature are in the ratio

$$\vartheta_{mpv} : \vartheta_{mv} : \vartheta_{rms} = 1.0 : 0.92 : 0.82$$

and thus $\vartheta_{rms} > \vartheta_{mv} > \vartheta_{mpv}$.

All these velocities are directly proportional to the square root of the temperature and inversely proportional to the square root of the molar mass of the gas. Two significant features of the curves in Fig. 1.1 are, (i) the maximum of the curve is seen to shift to the right indicating an increase of molecular velocities with temperature, and (ii) the maximum also becomes flatter at higher temperatures indicating a wider distribution of velocities. The significant effect of temperature on molecular velocities arises due to the exponential term of Eq. 1.35.

1.4 Other Terms Relevant to Kinetic Energy

1.4.1 Collision Diameter (σ)

It is defined as the distance of closest approach between the centres of two gas molecules.

1.4.2 Mean Free Path (l)

It is the average distance travelled by a gas molecule between successive collisions and is given by

$$l = \frac{1}{\sqrt{2}\pi\sigma^2 n} = \frac{V}{\sqrt{2}\pi\sigma^2 N} \tag{1.39}$$

where n is the number of gas molecules per unit volume. It is also equal to N/V where V is the volume occupied by an Avogadro number (N) of gas molecules.

l can be calculated at STP in the following manner.

$$n = \frac{6.023\times10^{23}}{22.4\times10^{-3}} = 2.69\times10^{25} \text{ molecules m}^{-3}$$

Assuming a collision diameter $\sigma = 2\times10^{-10}$ m,

$$l = \frac{1}{\sqrt{2}\times\frac{22}{7}\times\left(2\times10^{-10}\text{ m}\right)^2\times2.69\times10^{25}\text{ m}^{-3}} = 2.09\times10^{-7}\text{ m}$$

Effect of temperature and pressure on l

$$n = \frac{N}{V} = \frac{N\times P}{RT};\, l = \frac{RT}{\sqrt{2}\pi\sigma^2 PN}\,(\text{from Eq. 1.39}) \tag{1.40}$$

$\therefore l \propto \frac{1}{P}$ at constant T. Also, $l \propto T$ at constant P. Thus, l increases as P decreases at a given temperature and l increases with an increase in T at a given pressure. Thus, it is possible to write that

$$l \propto \frac{T}{P} \tag{1.41}$$

1.4.3 Viscosity of a Gas (η)

It is defined as the force per unit area necessary to maintain a velocity difference of unity between two parallel gas layers which are one cm apart. Thus,

$$\eta = \frac{\text{Tangential force}}{\text{Area}\times\text{Gradient of velocity}}$$

$$= \frac{\text{N}}{\text{m}^2}\times\frac{1}{\text{m}\times\text{s}^{-1}\text{m}^{-1}} = \text{N m}^{-2}\text{ s}^{1} = \text{pascal second}$$

$$= \text{kg}\times\text{m}\times\text{s}^{-2}\times\text{m}^{-2}\times\text{s} = \text{kg m}^{-1}\text{ s}^{-1}$$

Considering gas molecules as elastic spheres and the distribution of molecular velocities, an accurate relation for η is given by

$$\eta = \frac{1}{2}m\times n\times C_{mv}\times l \tag{1.42}$$

In Eq. 1.42, m = mass of the gas molecule and n = number of gas molecules per unit volume. Thus

$$\eta = \frac{1}{2}d\times C_{mv}\times l \tag{1.43}$$

where d is the density of the gas in kg m^{-3}, C_{mv} is the mean velocity and l is the mean free path. l can thus be calculated if η, d and C_{mv} are known. From Eq. 1.42, it is seen that η depends on $n\times l$ and hence is independent of pressure (because $n \propto P$ and $l \propto 1/P$).

Using C_{mv} from Eq. 1.37 and l from Eq. 1.39 in Eq. 1.42 for η, we get

$$\eta = \frac{1}{2}m\sqrt{\frac{8kT}{\pi m}}\times n\times\frac{1}{\sqrt{2}\pi\sigma^2 n}$$

$$= \frac{\sqrt{kTm}}{\pi^{3/2}\sigma^2} = \frac{\sqrt{\frac{R}{N}\times\frac{M}{N}\times T}}{\pi^{3/2}\sigma^2} = \frac{\sqrt{RTM}}{\pi^{3/2}N\sigma^2} \tag{1.44}$$

Using measured values of η and known values of N, the collision diameter σ can be calculated from Eq. 1.44.

1.4.4 Collision Number (Z)

Z_{11} is defined as the number of collisions occurring between the molecules of a gas per unit volume per unit time and is given by,

$$Z_{11} = \frac{1}{2}\sqrt{2}\pi\sigma^2 n^2 C_{mv} \tag{1.45}$$

where all the terms have their usual significance. But the collisions undergone by a single molecule with the other molecules is given by

$$Z_1 = \sqrt{2}\pi\sigma^2 n C_{mv} \tag{1.46}$$

1.5 Principle of Equipartition of Energy

Principle of equipartition of energy states that each form of energy associated with a molecule contributes ½ R to the heat capacity of the gas (or ½ k per molecule). In other words, it contributes ½ RT to the average energy per mole of the gas. A molecule possesses three kinds of energy—translational, rotational and vibrational energies. We prefer to use E as the symbol for internal energy, instead of U.

(i) For a mono-atomic gas like He, Ar possessing only *translational energy*, which can be split into x, y, z components, each contributes ½ RT per mole, that is, $3\times1/2\,RT = (3/2)\,RT$

$$E_K = \frac{3}{2}RT \tag{1.47}$$

(ii) For a diatomic molecule (which is linear), the molecule can rotate about two axes perpendicular to the line connecting the two atoms. Contribution due to rotational energy

$$= 2\times\frac{1}{2}\times R = R \text{ per mole}$$

Contribution to the average energy of the gas at temp T

$$= 2\times\frac{1}{2}\times RT = RT \text{ per mole} \tag{1.48}$$

(iii) For a linear molecule having n atoms,

$$\text{no. of vibrational modes} = (3n-5) \tag{1.49}$$

For a non-linear molecule having n atoms,

$$\text{no. of vibrational modes} = (3n-6) \tag{1.50}$$

Each mode of vibration contributes RT to the energy per mole. This is because the *vibrational energy* is partly kinetic in nature and partly potential.

1.5.1 Contribution of Various Forms of Energy to the Total Energy in Gases

(i) Diatomic linear molecules like H_2, O_2, N_2, CO, HCl

Contribution due to translational energy

$$= 3\times\frac{1}{2}\times RT = \frac{3}{2}RT \tag{1.51}$$

Contribution due to rotational energy

$$= 2\times\frac{1}{2}\times RT = RT \tag{1.52}$$

Contribution due to vibrational energy

$$= (3n-5) = (3\times 2-5)RT = RT \tag{1.53}$$

Total contribution due to vaious forms of energy

$$= \frac{3}{2}RT + RT + RT = \frac{7}{2}RT \tag{1.54}$$

(ii) Polyatomic linear molecules like CO_2, CSO (n = 3 as an example)

Total contribution due to various forms of energy

$$= 3\times\frac{1}{2}RT(\text{translational}) + 2\times\frac{1}{2}\times RT(\text{rotational})$$

$$+ (3\times 3-5)\,\text{RT(vibrational)} = \frac{13}{2}RT \tag{1.55}$$

(iii) Polyatomic nonlinear molecule like NH_3, O_3, NO_2, CH_4

Total contribution due to various forms of energy

$$= 3\times\frac{1}{2}RT(\text{translational}) + 3\times\frac{1}{2}\times RT(\text{rotational})$$

$$+ (3\times 3-6)\,\text{RT(vibrational)} = 6RT \tag{1.56}$$

1.5.2 Heat Capacity of Gases

(i) ***Heat capacity at constant volume (C_V):*** The heat taken by a system at constant volume (q_V) is utilised in increasing the kinetic energy of the gas molecules.

$$q_V = dE \tag{1.57}$$

where dE is the change in the internal energy.

The heat capacity at constant volume (c_V) is given by

$$C_V = \frac{q_V}{dT} = \left(\frac{\partial E}{\partial T}\right)_V \tag{1.58}$$

The kinetic energy of translation per mole of an ideal gas E_K is given by Eq. 1.47 and therefore,

$$C_V = \left(\frac{\partial E_K}{\partial T}\right)_V = \left(\frac{\partial\left(\frac{3}{2}RT\right)}{\partial T}\right)_V$$

$$= \frac{3}{2}R = \frac{3}{2}\times 8.314 = 12.47\ \text{J K}^{-1}\ \text{mol}^{-1} \tag{1.59}$$

This value is true in the case of monoatomic gases. However, in the case of diatomic or polyatomic gases, some of the supplied heat is utilised in causing rotation and vibration of molecules in addition to translation. Therefore, in such cases, more heat is required to raise the temperature by 1° and thus the molar heat capacity will be greater than 12.47 J K^{-1} mol^{-1}.

(ii) ***Heat capacity at constant pressure (C_p):*** For one mole of an ideal gas,

$$C_P - C_V = R \quad \text{or} \quad C_P = \frac{3}{2}R + R = \frac{5}{2}R \tag{1.60}$$

or

$$\frac{C_P}{C_V} = \gamma = \frac{5}{3} = 1.667 \tag{1.61}$$

In the case of monatomic gases, the values of C_P, C_V and hence γ are in agreement with the theoretical values. For diatomic and polyatomic molecules, the experimentally determined heat capacity values are always higher than that predicted by theory (assuming that no rotational and vibrational modes exist) and also increase with increase in temperature. Correspondingly, γ values are also observed to be less than the theoretical value of 1.667. It is therefore important in the case of polyatomic molecules that the rotational and vibrational contributions to heat capacity are taken into consideration.

1.5.3 Real Gases

The equation of state (Eq. 1.9) for an ideal gas is not exactly obeyed by most gases over all ranges of temperature and pressures. Boyle's and Charles' laws are found to be applicable to many gases at higher temperatures and low pressures. Deviations begin to show up from these laws at higher pressures and lower temperatures. Thus, in general, all gases are real gases.

(i) ***Compressibility factor (Z):*** This quantity best represents the deviation of gases from ideal behaviour and is given by

$$Z = \frac{V_{\text{actual}}}{V_{\text{ideal}}} = \frac{PV}{RT} \tag{1.62}$$

where $V_{\text{ideal}} = RT/P$. For an ideal gas, Z = 1 and is independent of pressure and temperature whereas for real gases, Z is a function of both temperature and pressure. Equation 1.62 for n moles of a gas may be written as

$$Z = \frac{PV}{nRT} = \frac{(PV)_{\text{actual}}}{(PV)_{\text{ideal}}} \tag{1.63}$$

(ii) ***Variation of Z with pressure at a given temperature:*** For an ideal gas, Z = 1 at all pressures and a plot of Z against pressure is a straight line parallel to the pressure axis. For hydrogen and helium, $Z > 1$ at all pressures at 273 K. For nitrogen and methane, for example, $Z < 1$ at low pressures (≈ 100 atm) but is greater than 1 at higher pressures. For carbondioxide, an easily liquefiable gas, Z decreases sharply below a pressure of around 100 atm but increases subsequently.

The effect of temperature is revealed from the plot of PV vs P at various temperatures. For example, in the case of nitrogen gas, this curve passes through a minimum in the temperature range 200–250 K and the minimum disappears as the temperature is raised. At higher temperatures, a continuous increase of PV is observed with increasing pressure.

(iii) ***Causes for deviation from ideal behaviour:*** The deviation of gases from ideal behaviour arises mainly due to the invalidity of the two assumptions (*vide infra*). This is demonstrated by the following experimental observations.

- A gas can be liquefied by lowering the temperature and increasing the pressure. It can also be solidified. This shows that molecules of a gas do occupy definite volume.
- Decrease in the compressibility factor with pressure and the Joule–Thompson effect are further evidences for the existence of intermolecular forces.

(iv) ***Correction due to a finite volume of gas:*** Because of the finite size of the gas molecules, the actual space available for the movement of the gas molecules is less than V, the measured volume of the gas (at a given temperature and pressure). A correction term, b, known as effective volume per mole of the gas (also known as co-volume), must be subtracted from the total volume to give the ideal volume, that is, the actual volume available for their movement. It has the units dm^3 mol^{-1} (CGS units) or m^3 mol^{-1} (in SI units).

$$\therefore V_{\text{ideal}} = V - b;\ \text{For } n \text{ moles of gas, } V_{\text{ideal}} = V - nb \qquad (1.64)$$

b is also referred to as excluded volume and is a constant, it is characteristic of each gas. The excluded volume per molecule of the gas is $4 \times 4/3 \times \pi r^3$, where r is the radius of the gas molecule.

$$\text{The excluded volume per molecule of gas } (b) = 4 \times \text{volume occupied by the gas molecule} \qquad (1.65)$$

(v) ***Correction for molecular attraction among gas molecules:*** For n moles of a gas occupying a total volume V, the correction factor to be added to the observed pressure P is,

$$\frac{n^2 a}{V^2}$$

For one mole of the gas, this correction term is a/V^2. a is also a constant for the gas and independent of temperature. It has the units,

$$a = \frac{\text{pressure} \times \text{volume}^2}{\text{mol}^2} = \frac{\text{atm} \times \left(\text{dm}^3\right)^2}{\text{mol}^2}$$

$$= \text{atm} \times \text{dm}^6 \times \text{mol}^{-2} \text{ in CGS units}$$

$$a = \frac{\text{N} \times \text{m}^{-2}\left(\text{m}^3\right)^2}{\text{mol}^2} = \text{N m}^4 \text{ mol}^{-2} \text{ in SI units}$$

The corrected equation incorporating the above corrections may be written as follows.

For n moles of the gas,

$$P_{\text{ideal}} V_{\text{ideal}} = \left(P + \frac{n^2 a}{V^2}\right)\left(V - nb\right) = nRT \qquad (1.66)$$

For one mole of an ideal gas

$$\left(P + \frac{a}{V^2}\right)\left(V - b\right) = RT \qquad (1.67)$$

The larger the value of a, the greater are the intermolecular forces among the gas molecules. For example, SO_2 has $a = 0.7$ N m^4 mol^{-2}, CO_2 has $a = 3.6$ N m^4 mol^{-2}. They are easily liquefiable. Equations 1.66 and 1.67 are known as van der Waals equations.

Units of a and b: To convert a in atm lit^2 mol^{-2} (or atm dm^6 mol^{-2}) into N m^4 mol^{-2}, multiply the former by 0.101325. To convert b in L mol^{-1} (or dm^3 mol^{-1}) into m^3 mol^{-1} multiply the former by 10^{-3}.

(vi) ***Some aspects of the van der Waals equation:*** Although the van der Waals equation qualitatively explains the deviations of real gases from ideal behaviour, it fails at high pressures and at low temperatures, especially near the critical temperature.

At low pressures: Under these conditions Eq. 1.67 may be written as

$$PV + \frac{a}{V} = RT \qquad (\text{since } b \ll V)$$

or

$$\frac{PV}{RT} = 1 - \frac{a}{VRT} \qquad (1.68)$$

Thus at low pressures $Z < 1$. It is seen that as pressure increases, $a/(VRT)$ increases and hence Z decreases with increase of pressure.

At high pressures: In this case, b cannot be neglected compared to V, in the second term of Eq. 1.67. However, a/V^2 may be neglected in comparison to P. Equation 1.67 thus reduces to,

$$P\left(V - b\right) = RT \quad \text{or} \quad PV = RT + bP$$

or

$$\frac{PV}{RT} = Z = 1 + \frac{bP}{RT} \qquad (1.69)$$

Thus Z is greater than 1 and increases with pressure.

At very low pressures and high temperatures: Under these conditions $b << V$, and $a/V^2 << P$. Equation 1.67 thus reduces to the ideal gas equation. For gases like hydrogen and helium, at ordinary temperature, a/V^2 is more predominant at low pressures whereas b is more predominant at high pressures. But a for these gases is so low that b predominates even at low pressures. The result is that no dip is seen for these gases in the PV–P graph at ordinary temperatures.

(vii) ***Definition of Boyle temperature, T_B:*** It is the temperature at which a real gas behaves ideally over a range of pressures. At this temperature, the effects of appreciable size of gas molecules and the intermolecular attraction between them roughly compensate each other. Above the Boyle temperature, the slope of Z vs P is positive. In this situation the size effect predominates (thus, $b > a/RT$). Below the Boyle temperature, the slope of Z vs P is negative; thus $b < a/RT$ and the intermolecular forces predominate over the effects of size.

(viii) *Relation of inversion temperature (T_i) to Boyle temperature (T_B):* It may be recalled that inversion temperature T_i is the temperature above which a gas shows a heating effect on Joule–Thompson expansion. It is related to T_B through the relation

$$T_i = \frac{2a}{Rb} = 2T_B \tag{1.70}$$

1.5.4 Critical Constants of a Gas

(i) **Critical temperature, T_c:** The maximum temperature at which a gas can be liquefied is referred to as *critical temperature*. Above this temperature, the liquid phase cannot exist irrespective of the pressure applied.

(ii) **Critical pressure, P_c:** It is the pressure required to cause liquefaction of a gas at the critical temperature T_c.

(iii) **Critical volume, V_c:** It is the volume occupied by one mole of a gas at the critical temperature T_c and critical pressure P_c.

1.5.5 Principle of Continuity of States

Andrew's experiments on the liquefaction of carbon dioxide, which involve the study of P–V isotherms at various temperatures revealed the following. Above a certain temperature, known as critical temperature, a gas cannot be liquefied however high the applied pressure is. At this temperature (T_c), a gas becomes indistinguishable from its liquid state indicating that its transition from gas to liquid state or vice verse is a continuous process. This indistinguishability between the two states (the liquid and gaseous states) and its continuous transition from one to the other is referred to as the principle of *continuity of states*. The critical temperature for CO_2 is 304.1 K.

Differences in the use of 'gas' and 'vapour': The term 'vapour' is employed when a gas is below its critical temperature (T_c). Above the critical temperature, the substance is known as 'gas'.

1.5.6 Relation Between van der Waals Equation and Critical Constants

The van der Waals equation is a cubic equation in V and hence for any given value of P and T, it will have three values for V, all of which or some of which, may be real. However at the critical temperature, the values of V approach each other and become identical and real too. Comparing the coefficients of the van der Waals equation and the coefficients of cubic equation derived in terms of V_c, the critical constants P_c, V_c, T_c may be expressed in terms of a and b as shown below:

$$P_C = \frac{a}{27b^2}; V_C = 3b; T_C = \frac{8a}{27Rb} \tag{1.71}$$

The constants a, b and R may also be expressed in terms of critical constants according to

$$a = 3P_C V_C^2; b = \frac{V_C}{3}; R = \frac{8a}{27bT_C} = \frac{8P_C V_C}{3T_C} \tag{1.72}$$

It is also possible to express a and b in terms of P_c and T_c. Substituting the value of V_c from the last term of Eq. 1.72 in the terms for a and b, we get

$$a = \frac{3\times P_C \times 9(RT_c)^2}{64P_c^2} = \frac{27(RT_c)^2}{64P_c} \tag{1.73}$$

and

$$b = \frac{3RT_c}{8P_c} \times \frac{1}{3} = \frac{RT_c}{8P_c} \tag{1.74}$$

a and b can be evaluated from Eqs 1.73 and 1.74.

1.5.7 Law of Corresponding States

Using the values of a, b and R from Eqs 1.73, 1.74 and 1.72 and substituting them in the van der Waals equation, the following relation

$$\frac{P}{P_C} = \frac{8(T/T_c)}{3(V/V_c)-1} - \frac{3}{(V/V_c)^2} \tag{1.75}$$

is obtained. Employing the symbols

$$\frac{P}{P_c} = \alpha; \frac{V}{V_c} = \beta; \frac{T}{T_c} = \gamma$$

(known as reduced variables of state)

Eq. 1.75 may be written as

$$\alpha = \frac{8\gamma}{3\beta-1} - \frac{3}{\beta^2} \tag{1.76a}$$

or

$$\left(\alpha + \frac{3}{\beta^2}\right)(3\beta-1) = 8\gamma \tag{1.76b}$$

Equations 1.76 do not contain any constants specific to any gas and hence must apply to all gases. Further, one reduced variable α, has been expressed as a function of two other reduced variables β and γ.

From this equation, note that two gases having the same reduced temperature (γ) and reduced pressure (α) must occupy the same reduced volume (β). This statement is known as the *law of corresponding states*. Gases under these conditions are said to be in corresponding states. This law is not however, exact but is applied in chemical engineering calculations.

1.5.8 Other Equations of State

Apart from the van der Waals equation, other equations of state were proposed by Dieterici, Berthelot, and so on.

(i) ***Dieterici equation:*** The relation proposed by Dieterici is

$$P = \frac{RTe^{\frac{-a}{VRT}}}{(V-b)} \tag{1.77}$$

This equation gives results in agreement with the van der Waals equation up to moderate pressures.

(ii) ***Berthelot equation:*** This equation is given by

$$P = \frac{RT}{(V-b)} - \frac{a}{TV^2} \tag{1.78}$$

A modified form of Berthelot equation has the form

$$P = \frac{RT}{V}\left(1 + \left[\frac{9}{128\gamma} - \frac{27}{64\gamma^3}\right]\alpha\right) \tag{1.79}$$

where α and γ have the same significance as explained earlier.

(iii) *Virial equation:* This equation expressing the variation of pressure, volume and temperature is given by

$$PV = RT\left(1+\frac{B}{V}+\frac{C}{V^2}+\frac{D}{V^3}+\ldots\right) \quad (1.80)$$

where, B, C and D are known as second, third and fourth *virial coefficients* and V is the molar volume of gas. The virial coefficients are functions of temperature.

(iv) *Caraahan-Starling equation of state:* $\frac{PV}{RT} = \frac{(1+q+q^2-q^3)}{(1-q)^3}$ where q denotes the packing fraction given by $q = \frac{nN\sigma^3}{6V}$ and σ is the hard sphere diameter.

1.6 Liquefaction of Gases

Gases can be liquefied after cooling them below their critical temperature and then subjecting them to high pressures. There are several methods available for the liquefaction of gases but only a few of them are listed here.

(i) *Employing freezing mixtures:* Some of the freezing mixtures used for the liquefaction of gases are

- ice–salt mixtures (such as NaCl), and
- mixture of ether and solid CO_2.

When a salt like NaCl or $CaCl_2$ is added to ice, it dissolves in the thin film of water always present with ice. The salt solution has a lower freezing point than the ice which is at a higher temperature. This results in the melting of the ice. Both processes – melting of ice and formation of salt solution – are endothermic in nature and hence the temperature of the salt mixture decreases.

(ii) *Application of Joule-Thomson effect:* Expansion of a compressed gas into a region of low pressure or vacuum under adiabatic conditions results in lowering of temperature. In this process, work is done by the gas molecules in overcoming intermolecular forces. This principle is the basis for liquefaction of air by *Linde's process*.

(iii) *Principle of adiabatic demagnetisation:* This method was developed by Giauque and Debye independently. A paramagnetic salt like cerium fluoride or another rare earth salt is initially cooled with liquid helium in the presence of a strong magnetic field. The liquid helium is then pumped away keeping the magnetised salt at liquid helium temperature under adiabatic conditions. The magnetic field is next switched off resulting in the salt undergoing adiabatic transformation. This transformation requires energy which comes from the crystal lattice and thereby resulting in a fall of temperature. Temperatures as low as 10^{-3} K were achieved by this method.

Worked Out Problems

Densities of Gases and Diffusion

Problem 1.1

(a) A gas is enclosed in a 0.5 m^3 vessel at a pressure of 202.65 kPa and a temperature of 400 K. What is the change in temperature required if it occupies a volume of 0.2 m^3 under a pressure of 303.98 kPa?

(b) The density of a gas is 100 kg m^{-3} at 300 K under a pressure of 101.325 kPa. What will be its density at 400 K at the same pressure?

Solution

(a) $\frac{P_1V_1}{T_1} = \frac{P_2V_2}{T_2}; \frac{202.65\times0.5}{400} = \frac{303.98\times0.2}{T_2}$

$$T_2 = \frac{303.98\times0.2\times400}{202.65\times0.5} = 240 \text{ K}$$

It is necessary to bring the temperature down to 240 K.

The change in temperature required = 400 – 240 = 160 K.

(b) $PV = \frac{W}{M}RT$ or $P = \frac{W}{V}\times\frac{RT}{M} = \frac{dRT}{M}$

$d = \frac{MP}{RT}$, that is, $d \propto \frac{1}{T}$

$\frac{d_1}{d_2} = \frac{T_2}{T_1}; \frac{10}{d_2} = \frac{400}{300}; d_2 = \frac{300\times10}{400} = 7.5 \text{ kg m}^{-3}$

Problem 1.2

(a) The relative density of chlorine at STP is 35.48. What is its absolute density at 27°C and 2 atm pressure?

(b) The density of a mixture of nitrogen and oxygen at 400 K and 1 atm pressure (101.325 kPa) is 0.920 g L^{-1}. Calculate the mole fraction of the gases in the mixture.

Solution

(a) Relative density of $Cl_2 = \frac{\text{Weight of 11.2 dm}^3\ Cl_2 \text{ at STP}}{\text{Weight of 11.2 dm}^3\ H_2 \text{ at STP}}$

$$35.48 = \frac{\text{Weight of 11.2 dm}^3\ Cl_2 \text{ at STP}}{1}$$

or, absolute density of Cl_2 at STP $= \frac{35.48}{12} = 3.168 \text{ kg m}^{-3}$

but $d = \frac{MP}{RT}$, that is, $d \propto \frac{P}{T}; \frac{d_1}{d_2} = \frac{P_1}{T_1}\times\frac{T_2}{P_2}$

$\frac{3.168}{d_2} = \frac{1}{2}\times\frac{300}{273}; d_2 = \frac{3.168\times2\times273}{1\times300} = 5.766 \text{ kg m}^{-3}$

(b) Density of mixture of N_2 and O_2 = 0.920 kg m^{-3}

$$0.920 = \frac{M_{N_2}P_{N_2}}{RT} + \frac{M_{O_2}P_{O_2}}{RT}$$

$$= \frac{0.028\times P_{N_2}}{8.314\times400} + \frac{0.032(101.325 - p_{N_2})}{8.314\times400}$$

$$0.920\times 8.314\times 400 = 3059.55$$
$$= 0.028P_{N_2} + 3242.4 - 0.032p_{N_2}$$
$$-0.004P_{N_2} = -182.85$$
$$\text{or } P_{N_2} = \frac{182.85}{0.004} = 45712.5 \text{ Pa} = 45.71 \text{ kPa}$$
$$P_{O_2} = 101.325 - 45.71 = 55.615 \text{ kPa}$$
$$\text{Mole fraction of N}_2 = \frac{45.71}{101.325} = 0.451$$
$$\text{Mole fraction of O}_2 = \frac{55.615}{101.325} = 0.549$$

Problem 1.3

(a) 12.0 g of helium and 80.0 g of argon are kept in a vessel of 5 dm^3 capacity at 127°C. Calculate the total pressure of the system in atm. Also calculate the partial pressures of the gases in units of atm and kPa.

(b) The density of partially dissociated PCl_5 gas at 250°C is 17.0 kg m^{-3} at a pressure of 506.625 kPa. What is the degree of dissociation of the gas?

Solution

(a) $$\text{Moles of He} = \frac{12.0}{4.0} = 3.0$$
$$\text{Moles of argon} = \frac{80.0}{40.0} = 2.0$$
Total moles = 3.0 + 2.0 = 5.0

Partial pressure of He = 3.5 × 32.82 = 19.69 atm

Partial pressure of Ar = 2.5 × 32.82 = 13.13 atm

Total pressure
$$= \frac{5.0(\text{mol})\times 8.314(\text{N m mol}^{-1}\text{K}^{-1})\times 400(\text{K})}{5\times 10^{-3}(\text{m}^3)}$$
$$= 3325600 \text{ Pa} = 3325.6 \text{ kPa} = \frac{3325.6}{101.325} = 32.82 \text{ atm}$$

(b) Density of partially dissociated gas (d) = 17.0 kg m^{-3}

Density of gas if it was not dissociated
$$= \frac{208.5\times 10^{-3}\times 5\times 101.325}{8.314\times 523} = 24.29 \text{ kg m}^{-3}$$
$$\alpha = \frac{d_0 - d}{d(\gamma-1)} = \frac{24.29-17.0}{17.0(2-1)} = \frac{7.9}{17.0} = 0.465$$

Graham's Law of Diffusion

Problem 1.4

(a) A mixture of helium and oxygen in the ratio of 5:2 diffuses through a pinhole at certain temperature. What is the initial ratio of helium to oxygen leaking through?

(b) A certain volume of each of the gases, nitric oxide and an unknown gas, diffuse through a pinhole in the ratio of time intervals 56.3:82.2. What is the molar mass of the unknown gas?

Solution

(a) $$\frac{r_{He}}{r_{O_2}} = \sqrt{\frac{M_{O_2}}{M_{He}}} = \sqrt{\frac{32}{4}} = 2.83$$

Initial ratio of He: O_2 leaking through = 2.83×5:1×2 = 14.15:2 = 7.08:1

(b) Let M_1 = mol. wt of NO and M_2 = mol. wt of unknown gas

For the same volume of gas to pass through the pin hole, the relation $\frac{D_1}{D_2} = \sqrt{\frac{M_2}{M_1}}$ must hold, where, D_1 and D_2 are the rates of diffusion.

$$\frac{V_1}{56.3}\times\frac{82.2}{V_1} = \sqrt{\frac{M_2}{30}};\ \sqrt{\frac{M_2}{30}} = 1.46$$
$$\sqrt{M_2} = 5.477\times 1.46 = 8.0;\ M_2 = 64.0$$

Maxwell's Law of Distribution of Molecular Velocities

Problem 1.5

(a) Ammonia gas at a pressure of 152.0 kPa and HCl gas at a pressure of 304 kPa were released at the same time from the ends of a 5 m long tube. Calculate the distance from the HCl gas end where the dense fumes due to the formation of ammonium chloride appear.

(b) Calculate the most probable, rms and average speeds of carbon dioxide gas molecules which have a density of 3.57 kg m^{-3} at 300 K and 202.35 kPa pressure.

Solution

(a) Let x m be the distance from the HCl gas end where the dense fumes of ammonium chloride vapours appear.

Distance from ammonia gas end = $(5 - x)$ m

$$\frac{(5-x)/t}{x/t} = \frac{r_{NH_3}}{r_{HCl}} = \frac{p_{NH_3}}{\sqrt{M_{NH_3}}}\times\frac{\sqrt{M_{HCl}}}{p_{HCl}} = \frac{152}{304}\times\frac{\sqrt{36.5}}{17}$$
$$\text{or } \frac{(5-x)}{x} = 0.5\times\frac{6.04}{4.12} = 0.733$$
$$5 - x = 0.733x;\ x = \frac{5}{1.733} = 2.88 \text{ m}$$

Fumes of ammonium chloride will appear at a distance of 2.88 m from the HCl gas end.

(b) $$\vartheta_{rms} = \sqrt{\frac{3RT}{M}} = \sqrt{\frac{3PV}{M}} = \sqrt{\frac{3P}{d}}$$
$$= \sqrt{\frac{3\times 202350(\text{kg m}^{-1}\text{ s}^{-2})}{3.57(\text{kg m}^{-3})}} = \sqrt{170042.017} = 412.4 \text{ m s}^{-1}$$

$$\vartheta_{mpv} = \sqrt{\frac{2RT}{M}} = \sqrt{\frac{2PV}{M}} = \sqrt{\frac{2P}{d}} = \sqrt{\frac{2\times202350}{3.57}}$$
$$= \sqrt{113361.34} = 336.7 \text{ m s}^{-1}$$

$$\vartheta_{Ave} = \sqrt{\frac{8RT}{\pi M}} = \sqrt{\frac{8P}{\pi d}} = \sqrt{\frac{8\times202350}{3.1416\times3.57}}$$
$$= \sqrt{144335.80} = 379.9 \text{ m s}^{-1}$$

Problem 1.6

(a) At what temperature will the rms velocity of sulfur dioxide molecules at 400 K be the same as the most probable velocity of helium atoms? What is the kinetic energy of 5 moles of helium at this temperature?

(b) The critical constants P_c and T_c of a certain gas are 5097 kPa and 154.5 K respectively. Calculate its van der Waals constants a and b in SI units. Also find its Boyle temperature from these data.

Solution

(a) C_{rms} of $SO_2 = \sqrt{\frac{3RT}{M_{SO_2}}} = \sqrt{\frac{3\times8.314\times400}{0.064}}$
$$= \sqrt{155887.5} = 394.8 \text{ m s}^{-1}$$

$$C_{mpv} \text{ of He} = \sqrt{\frac{2RT}{M_{He}}}$$

$$\text{or } 394.8 = \sqrt{\frac{2\times8.314\times T}{0.004}} = 64.47\times\sqrt{T}$$

$$\sqrt{T} = \frac{394.8}{64.47} = 6.124 \text{ or } T = 37.5 \text{ K}$$

$$\text{Kinetic energy of 5 mol of He} = \frac{3}{2}nRT$$
$$= \frac{3}{2}\times5\times8.314\times37.5 = 2338.3 \text{ J}$$

(b) $P_c = 5097$ kPa; $T_c = 154.5$ K

$$\frac{P_cV_c}{RT_c} = \frac{3}{8} \text{ or } V_c = \frac{3}{8}\times\frac{8.314\times154.5}{5097\times1000} = 9.45\times10^{-5} \text{ m}^3 \text{ mol}^{-1}$$

$$b = \frac{V_c}{3} = \frac{=9.45\times10^{-5}}{3} = 3.15\times10^{-5} \text{ m}^3 \text{ mol}^{-1}$$

$$P_C = \frac{a}{27b^2} \text{ or } a = P_c\times27b^2$$
$$= 5097\times10^3\times27\times\left(3.15\times10^{-5}\right)^2$$
$$= 5097\times10^3\times9.92\times10^{-7} = 0.1365 \text{ N m}^4 \text{ mol}^{-2}$$

Problem 1.7

10^{22} xenon gas atoms are enclosed in a vessel of 1 dm^3 capacity.

(a) If the average velocity of its atoms is 1000 m s^{-1}, calculate the pressure exerted by the gas in (i) kPa, (ii) atm (atomic mass of Xe = 131.3).

(b) What is the temperature of the gas?

Solution

(a) Mass of the gas atom $= \frac{131.3}{6.023\times10^{23}}$
$$= 2.18\times10^{-22} \text{ g} = 2.18\times10^{-25} \text{ kg}$$

Pressure, $P = \frac{1}{3}\frac{mn\bar{C}^2}{V}$ where $\bar{C}^2$
= mean of the squares of the velocity

$$\text{Average velocity, } \vartheta_{ave} = \sqrt{\frac{8RT}{\pi M}} = 1000 \text{ m s}^{-1}$$

$$\vartheta_{rms} = \sqrt{\frac{3RT}{M}} = \sqrt{\frac{8RT}{\pi M}\times\frac{3\pi}{8}}$$
$$= 1000\times\sqrt{1.178} = 1000\times1.085 = 1085 \text{ m s}^{-1}$$

$$\text{or, } P = \frac{1}{3}\times\frac{10^{22}\times2.18\times10^{-25}\times(1085)^2 \text{ kg m}^2 \text{ s}^{-2}}{1\times10^{-3} \text{ m}^3}$$
$$= \frac{0.727\times10^{-3}(1085)^2 \text{ kg m}^2 \text{ s}^{-2}}{10^{-3}}$$
$$\text{or, } P\left(\text{kg m s}^{-2} \text{ m}^{-2} = \text{kg m}^{-1} \text{ s}^{-2}\right)$$
$$= 855.84 \text{ kPa} = \frac{855.84}{101.325} = 8.45 \text{ atm}$$

(b) The total kinetic energy $= \frac{1}{2}\times m\times n\times\bar{C}^2$
$$= \frac{1}{2}\times2.18\times10^{-25}\times(1085)^2\times10^{22}$$
$$\text{or, } \frac{3}{2}kT = \frac{1}{2}\times2.18\times10^{-25}\times(1085)^2$$
(where k = Boltzmann constant)

$$T = \frac{1}{2}\times2.18\times10^{-25}\times(1085)^2\times\frac{2}{3}\times\frac{1}{1.38\times10^{-23}}$$
$$= \frac{0.727\times10^{-2}\times1177225}{1.38} = 6201.7 \text{ K} = 5928.7°\text{C}$$

Mean Free Path of Gases

Problem 1.8

(a) The collision diameter of nitrogen gas molecules is 3.72×10^{-10} m. Calculate the mean free path of the gas in metres at 350 K and 202.35 kPa.

(b) What will be the pressure of the gas in atm at 350 K when the mean free path of the gas is twice that obtained in (a)?

Solution

(a) Mean free path $= l = \frac{1}{\sqrt{2}\pi n\sigma^2}$

(n = number of molecules per unit volume)

To calculate n at 350 K and 202.65 kPa

$$n = \frac{N}{V} = x \times N_{Avo} \text{ where } x = \text{no. of moles of gas}$$

N_{Avo} = Avogadro's number

$$\text{but, } \frac{x}{V} = \frac{P}{RT} \text{ or } n = \frac{P \times N_{Avo}}{RT}$$

$$= \frac{202.65 \times 6.023 \times 10^{23} \times 10^3}{8.314 \times 350}$$

$$= 0.419 \times 10^{26} \text{ molecules m}^{-3}$$

$$l = \frac{1}{1.414 \times 3.1416 \times 4.19 \times 10^{25} \times (3.72 \times 10^{-10})^2}$$

$$= \frac{1}{257.57 \times 10^5} = 3.88 \times 10^{-8} \text{ m}$$

(b) When mean free path = 7.76×10^{-8} m

Let us first calculate n, that is, $n = \dfrac{P \times N_{Avo}}{RT}$

$$\text{but } n = \frac{1}{\sqrt{2} \times \pi \times 7.76 \times 10^{-8} \times (3.72 \times 10^{-10})^2}$$

$$= \frac{1}{477.03 \times 10^{-28}} = 2.1 \times 10^{25} \text{ molecules m}^{-3}$$

$$\text{now, } P = \frac{2.1 \times 10^{25} \times 8.314 \times 350}{6.032 \times 10^{23}}$$

$$= 101457.6 \text{ Pa} = 101.4576 \text{ kPa} = 1.001 \text{ atm}$$

Equipartition Principle

Problem 1.9

(a) Calculate the heat capacity of an organic molecule (gas) having five atoms on the basis of the principle of equipartition of energy. If the experimental C_p of the gas at 400 K is 66.9 J K^{-1} mol^{-1}, how many vibrational degrees of freedom are not effective?

(b) Calculate the value of $\gamma (C_p/C_V)$ for the non-linear ammonia molecule. The experimental value of γ for the gas is 1.30 at 298 K. Account for the discrepancy. Also compare the experimental value of C_V = 27.3 J K^{-1} mol^{-1} with the theoretical value from the Solution of (a).

Solution

(a) For a gaseous organic molecule with $n = 5$

Contribution due to translational degrees of freedom

$$= 3 \times \frac{1}{2} \times RT = \frac{3}{2} RT$$

Contribution due to rotational degrees of freedom

$$= 3 \times \frac{1}{2} \times RT = \frac{3}{2} RT$$

Contribution due to vibrational degrees of freedom

$$= (3n - 6) \times RT = (3 \times 5 - 6) RT = 9RT$$

$$\text{Total energy} = \frac{3}{2} RT + \frac{3}{2} RT + 9RT = 12RT$$

$$\left(\frac{\partial E}{\partial T}\right)_V = C_V = \frac{\partial (12RT)}{\partial T}$$

$$= 12R = 12 \times 8.314 = 99.8 \text{ J K}^{-1} \text{ mol}^{-1}$$

Experimental C_P = 66.9 J K^{-1} mol^{-1}

or C_V = 66.9 − 8.3 = 58.6 J K^{-1} mol^{-1}

The molecule thus has 41.2 J K^{-1} mol^{-1} less of C_V than the predicted value, based on the equipartition principle. Since each vibrational degree of freedom contributes R = 8.37 J, the number of vibrational degrees of freedom not contributing to the total energy are 41.2/8.37 ≈ 5.

(b) For ammonia, NH_3, $n = 4$

Contribution due to translational degrees of freedom

$$= 3 \times \frac{1}{2} \times RT = \frac{3}{2} RT$$

Contribution due to rotational degrees of freedom

$$= 3 \times \frac{1}{2} \times RT = \frac{3}{2} RT$$

Contribution due to vibrational degrees of freedom

$$= (3n - 6) \times RT = (3 \times 4 - 6) RT = 6RT$$

$$\text{Total energy} = \frac{3}{2} RT + \frac{3}{2} RT + 6RT = 9RT$$

$$\left(\frac{\partial E}{\partial T}\right)_V = C_V = \frac{\partial (9RT)}{\partial T}$$

$$= 9R = 9 \times 8.314 = 74.8 \text{ J K}^{-1} \text{ mol}^{-1}$$

$$C_P = C_V + R = 74.8 + 8.3 = 83.1 \text{ J K}^{-1} \text{ mol}^{-1}$$

$$\gamma = \frac{C_P}{C_V} = \frac{83.1}{74.8} = 1.11; \text{experimental} = 1.30$$

The experimental value of γ is slightly higher than that predicted by the principle of equipartition of energy. Thus, the vibrational degrees of freedom in the case of ammonia seems to match with the theoretical value despite the small difference observed. However, a comparison of experimental value of C_V = 27.3 J K^{-1} mol^{-1} with the theoretical value C_V = 74.8 J K^{-1} mol^{-1} shows that the experimental value is far lower indicating that not all vibrational degrees of freedom are fully utilised at this temperature.

Problem 1.10

(a) One mole of a gas at 298 K requires 484 J of heat to raise its temperature by 20 degrees under constant volume conditions. If the vibrational degrees of freedom do not contribute to the heat capacity, what conclusion can be reached about the structure of the molecule?

(b) Estimate the translational, vibrational and rotational energy contributions to the average energy per mole of the gas which is tetraatomic. Assume that one vibrational mode contributes to the energy at 298 K.

Solution

(a) $Q_V = nC_V\Delta T$, that is, $484 = 1\times C_V \times 20$

or $C_V = \dfrac{484}{20} = 24.2\ \text{J K}^{-1}\ \text{mol}^{-1}$

Contribution due to three translational degrees of freedom to heat capacity $= 3\times\dfrac{1}{2}\times R = \dfrac{3}{2}8.314 = 12.45$ J

Difference between C_V and translational contribution

$= 24.2 - 12.4 = 11.8 \approx 12$ J

Contribution of the rotational degree of freedom to heat capacity $= \dfrac{1}{2}\times R = 4.15\ \text{J K}^{-1}\ \text{mol}^{-1}$

Thus, the total rotational degrees of freedom contributed by the gas $= \dfrac{12}{4.15} \approx 3$

Therefore the molecule must be non-linear.

(b) The molecule has four atoms

Contribution due to translational degrees of freedom

$= \dfrac{3}{2}RT = \dfrac{3}{2}\times 8.314\times 298 = 3710$ J

Contribution due to rotational degrees of freedom

$= \dfrac{3}{2}RT = \dfrac{3}{2}\times 8.314\times 298 = 3710$ J

Contribution from one vibrational degree of freedom

$= RT = 8.314\times 298 = 2473.4$ J

Total energy $= 3710 + 3710 + 2473.4 = 9893.4$ J

Problem 1.11

(a) A mercury diffusion pump is attached to a cold trap maintained at 153 K. Given that the vapour pressure of mercury at this temperature is 1.33×10^{-14} Pa, calculate the number of Hg vapour molecules per m^3 in the cold trap.

(b) The viscosity of nitrogen gas is 1.67×10^{-5} Pa s or kg m^{-1} s^{-1} at STP. Calculate its mean free path in metres and its collision diameter.

Solution

(a) $n = \dfrac{PV}{RT} = \dfrac{1.33\times10^{-14}\times 1}{8.314\times 153}\left(\dfrac{\text{N m}^{-2}\ \text{m}^3}{\text{N m K}^{-1}\ \text{mol}^{-1}\ \text{K}} = \text{mol}\right)$

$= 1.33\times 7.86\times10^{-14}\times10^{-14} = 1.045\times10^{-17}$ mol

Number of molecules $= 1.045\times10^{-17}\times 6.023\times10^{23}$

$= 6.3\times10^{6}$ molecules m^{-3}

(b) $\eta = \dfrac{1}{3}\times d\times C_{\text{ave}}\times l$

C_{ave} of the gas has to be calculated first.

$C_{\text{ave}} = \sqrt{\dfrac{8RT}{\pi M}} = \sqrt{\dfrac{8\times 8.314\times 273}{3.1416\times 0.028}} = \sqrt{\dfrac{18157.78}{0.088}}$

$= \sqrt{206338.4} = 454.24\ \text{m s}^{-1}$

Density of the gas $= \dfrac{MP}{RT} = \dfrac{0.028\times101325}{8.314\times 273}$

$= \dfrac{2837.1}{2269.7} = 1.249\ \text{kg m}^{-3}$

$1.67\times10^{-5} = \eta = \dfrac{1}{3}\times 1.249\times 454.24\times l$

$l = \dfrac{3\times 1.67\times10^{-5}}{1.249\times 454.24}\left(\dfrac{\text{kg m}^{-1}\ \text{s}^{-1}}{\text{kg m}^{-3}\ \text{m s}^{-1}} = \text{m}\right)$

$l = 8.83\times10^{-8}$ m

Collision diameter σ is related to l according to l

$= \dfrac{1}{\sqrt{2}\pi\sigma^2 n} = \dfrac{V}{\sqrt{2}\pi\sigma^2 N}$

n = no. of molecules per unit volume

$= \dfrac{x\times N_{\text{Avo}}}{V} = \dfrac{P\times N_{\text{Avo}}}{RT}$

$= \dfrac{101.325\times10^{3}\times 6.023\times10^{23}}{8.314\times 273}$

$= 2.69\times10^{25}$ molecules m^{-3}

where x = no. of moles

$\sigma^2 = \dfrac{1}{1.414\times 3.1416\times 8.83\times10^{-8}\times 2.69\times10^{25}}$

$= 9.48\times10^{-3}\times10^{-17} = 9.48\times10^{-20}$

$\sigma = \sqrt{9.48\times10^{-20}} = 3.08\times10^{-10}$ m

NOTE

We can also calculate σ^2 directly from the equation,

$$\sigma^2 = \frac{2d}{3\pi n\eta}\sqrt{\frac{RT}{\pi M}}$$

where $n = \dfrac{P\times N_{\text{Avo}}}{RT}$, η = viscosity, d = density and M = molar mass.

Problem 1.12

(a) The mean free path of oxygen gas molecules at 298 K and 101.325 kPa is 7.0×10^{-8} m. What is the mean free path of the gas if the gas is compressed to 5066.25 kPa and heated to 348 K.

(b) Using the equipartition principle, calculate the change in internal energy of 2 moles of SO_3 gas on heating it from 300 K to 500 K.

Solution

(a) Mean free path,

$$l \propto \frac{T}{P}$$

$\dfrac{l_1}{l_2} = \dfrac{T_1}{P_1}\dfrac{P_2}{T_2}$, that is, $\dfrac{7\times10^{-8}}{l_2} = \dfrac{298}{101.325}\times\dfrac{5066.25}{348}$

$$\frac{7\times10^{-8}}{l_2}=\frac{50\times298}{348}=42.82$$

$$l_2=\frac{7\times10^{-8}}{42.82}=1.63\times10^{-9}\ \text{m}$$

(b) Contribution due to translational degrees of freedom

$$=3\times\frac{1}{2}\times R=\frac{3}{2}R$$

Contribution due to rotational degrees of freedom

$$=3\times\frac{1}{2}\times R=\frac{3}{2}R$$

Contribution due to vibrational degrees of freedom

$$=(3n-6)\times R=(3\times4-6)R=6R$$

Total contribution to

$$C_V=\frac{3}{2}R+\frac{3}{2}R+6R=9\times8.3=74.7\ \text{J K}^{-1}\ \text{mol}^{-1}$$

Change in energy on heating from 300 K to 500 K

$$=Q=nC_V\Delta T=2\times74.7\times200=29880\ \text{J}=29.88\ \text{kJ}$$

van der Waals Equation of State

Problem 1.13

(a) The van der Waals constants a and b for methane are 0.2280 N m^4 mol^{-2} and 4.28×10^{-5} m^3 mol^{-1} respectively. Calculate the critical constants for the gas.

(b) Using the same values of a and b, calculate the pressure exerted by 20 moles of methane in a vessel of 5 dm^3 capacity at 27°C from the ideal gas equation and from the van der Waals equation.

Solution

(a) For CH_4, $a = 0.228$ N m^4 mol^{-2}; $b = 4.28\times10^{-5}$ m^3 mol^{-1}

$$P_c=\frac{a}{27b^2}=\frac{0.228}{27\times\left(4.28\times10^{-5}\right)^2}\frac{\text{N m}^4\ \text{mol}^{-2}}{\text{m}^6\ \text{mol}^{-2}}$$

$$=\frac{0.228}{494.6\times10^{-10}}$$

$$=4.61\times10^{-4}\times10^{10}=4.61\times10^6\ \text{Pa}=4.61\times10^3\ \text{kPa}$$

$$V_c=3b=3\times4.28\times10^{-5}=12.84\times10^{-5}\ \text{m}^3\ \text{mol}^{-1}$$

$$T_c=\frac{8a}{27Rb}$$

$$=\frac{8\times0.228}{27\times8.314\times4.28\times10^{-5}}\frac{\text{N m}^4\ \text{mol}^{-2}}{\text{N m K}^{-1}\ \text{mol}^{-1}\ \text{m}^3\ \text{mol}^{-1}}$$

$$=\frac{1.824}{0.00961}=189.8\text{K}$$

(b) $$P=\frac{nRT}{V}=\frac{20\times0.0821\times300}{5}$$

$$=98.52\ \text{atm}=9.982\times10^3\ \text{kPa}$$

Using van der Waals equation

$$\left(P+\frac{n^2a}{V^2}\right)(V-nb)=nRT$$

or $$P=\frac{nRT}{V-nb}-\frac{n^2a}{V^2}$$

$$=\frac{20\times8.314\times300}{5\times10^{-3}-(20\times4.28\times10^{-5})}-\frac{400\times0.228}{\left(5\times10^{-3}\right)^2}$$

$$=\frac{49884}{5\times10^{-3}-85.6\times10^{-5}}-\frac{91.2}{25\times10^{-6}}$$

$$=12037644.8-3.65\times10^6$$

$$=12.04\times10^6-3.65\times10^6=8.39\times10^3\ \text{kPa}$$

Problem 1.14

(a) The critical temperature and critical pressure of a certain gas are 154 K and 5096.6 kPa respectively. What are its van der Waals constants a and b in SI units?

(b) Calculate the Boyle temperature and inversion temperature of the gas.

Solution

(a) $$a=\frac{27\left(RT_c\right)^2}{64P_C}=\frac{27\times(8.314\times154)^2}{64\times5096.6\times10^3}$$

$$=\frac{27\times(1280.36)^2}{326182.4\times1000}=0.1357\ \text{N m}^4\ \text{mol}^{-2}$$

$$b=\frac{RT_c}{8P_c}=\frac{8.314\times154}{8\times5096.6\times10^3}=\frac{1280.36}{40772.8\times10^3}$$

$$=\frac{1.2804}{40772.8}=3.14\times10^{-5}\ \text{m}^3\ \text{mol}^{-1}$$

(b) Boyle temperature

$$T_B=\frac{a}{Rb}=\frac{0.1357}{8.314\times3.14\times10^{-5}}=\frac{0.1356}{2.611\times10^{-4}}=519.34\ \text{K}$$

Inversion temperature, $T_i=2T_B=2\times519.34=1038.68$ K

Problem 1.15

(a) The critical temperature and critical density of a certain gas are 647 K and 315.8 kg m^{-3}. Given its mass as 18, calculate its van der Waals constants.

(b) What is the radius of the gas molecule?

Solution

(a) $T_c=647\text{K};\ d_c=315.8\ \text{kgm}^{-3}$

V_c is given by $$\frac{\text{molar mass}}{\text{density}}=\frac{0.018}{315.8}=5.7\times10^{-5}\ \text{m}^3\ \text{mol}^{-1}$$

Since $V_c=3b$; $$b=\frac{V_c}{3}=\frac{5.7\times10^{-5}}{3}=1.9\times10^{-5}\ \text{m}^3\ \text{mol}^{-1}$$

But $b = 4\times$ volume occupied by a mol of gas molecules

$$V=\frac{b}{4}=\frac{1.9\times10^{-5}}{4}=0.475\times10^{-5}\ \text{m}^3\ \text{mol}^{-1}$$

Volume accupied by a molecule

$$=\frac{0.475\times10^{-5}}{6.023\times10^{23}}$$

$= 7.88\times10^{-30}\ m^3\ (molecule)^{-1}$

$$a=\frac{T_c\times 27Rb}{8}$$

$$a=\frac{647\times27\times1.9\times10^{-5}\times8.314}{8}=0.34494\ N\ m^4\ mol^{-2}$$

(b) Assuming the molecule to be spherical,

$$\frac{4}{3}\pi r^3 = 7.88\times10^{-30};\ r^3=\frac{7.88\times10^{-30}\times3}{4\times3.1416}=1.88\times10^{-30}$$

$$r=\sqrt[3]{1.88\times10^{-30}}=1.23\times10^{-10}\ m$$

Problem 1.16

(a) A spherical balloon having a diameter of 2.0 m weighs 2.0 kg. It is filled with helium (at. mass = 4.0) at a pressure of 1.5 atm and 298 K. (i) How many moles of helium are in the balloon? (ii) If it rises to an altitude where the temperature is 200 K and pressure 0.2 atm, what is the volume of the balloon? (iii) What is the pay load of the balloon? (density of air = 1.30 kg m^{-3}) [payload = mass of displaced air – (mass of balloon + mass of helium)]

(b) In a gaseous diffusion experiment which was used to determine the degree of dissociation of an oxide of nitrogen, it was found that helium diffused four times faster than the partially dissociated gas under similar conditions of temperature and pressure. If the molar mass of oxide of nitrogen is 92, what is its degree of dissociation? The dissociation reaction is

$$X_2O_4(g) \rightleftharpoons 2XO_2(g)$$

Solution

(a) (i) Moles of helium in the balloon

$$=\frac{1.5\times101.325\times10^3\times4.19}{8.314\times298}=257$$

$$\left(\text{Volume of balloon}=\frac{4}{3}\pi r^3 =\frac{4}{3}\times3.1416\times(1.0)^3=4.19\ m^3\right)$$

(ii) $\dfrac{P_1V_1}{T_1}=\dfrac{P_2V_2}{T_2}$, that is, $\dfrac{1.5\times101.325\times10^3\times4.19}{298}$

$$=\frac{0.2\times101.325\times10^3\times V_2}{200}$$

$$V_2=\frac{1.5\times101.325\times10^3\times4.19\times200}{0.2\times101.325\times10^3\times298}=21.1\ m^3$$

(iii) Mass of balloon + mass of He $= 2.0+(257\times0.004)$

$= 2.0+1.028 = 3.028$ kg

Mass of displaced air $= 4.19\times1.3 = 5.45$ kg

Payload $= 5.45-3.028 = 2.42$ kg

(b) $$\frac{r_{He}}{r_{gas\ mix}}=\sqrt{\frac{M_{gas\ mix}}{M_{He}}}\ \text{or}\ \frac{M_{gas\ mix}}{M_{He}}=(4.0)^2$$

$$M_{gas\ mix}=4.0\times(4.0)^2=64.0$$

If α is the degree of dissociation, we have 64.0

$$=\frac{(1-\alpha)M_{undiss\ gas}+2\times\alpha\times46}{1+\alpha}=\frac{(1-\alpha)92+2\times\alpha\times46}{1+\alpha}$$

$$92-92\alpha+92\alpha=64.0+64\alpha$$

$$64\alpha=92-64=28;\ \alpha=\frac{28}{64}=0.44$$

NOTE

- Molar mass of $XO_2 = 46$

Problem 1.17

(a) 0.874×10^{-3} kg of the vapour of a halogen displaced 0.137 dm^3 of air in a Victor Meyer apparatus at 294.5 K and 104.4 kPa pressure. Calculate the vapour density of the halogen. If the molar mass of the pure halogen is 254, what is its degree of dissociation?

(b) Hydrogen gas has a compressibility factor, Z value, of 1.05 at 273 K and 10132.5 kPa. Calculate b of van der Waals equation. Given that the T_c of the gas is 33.2 K, calculate a of van der Waals equation.

Solution

(a) Volume of vapour at STP

$$\frac{P_1V_1}{T_1}=\frac{P_2V_2}{T_2};\ \frac{101.325\times V_1}{273}=\frac{104.4\times0.137\times10^{-3}}{294.5}$$

$$V_1=\frac{104.4\times0.137\times10^{-3}}{294.5}\times\frac{273}{101.325}=1.308\times10^{-4}\ m^3$$

$$\text{Density of vapour}=\frac{0.874\times10^{-3}}{1.308\times10^{-4}}=6.682\ kg\ m^{-3}$$

Weight of 11.2×10^{-3} m^3 of vapour

$= 6.682\times0.0112 = 0.07484$ kg m^{-3}

Weight of 11.2×10^{-3} m^3 of $H_2 = 1\times10^{-3}$ kg

$$\text{Relative density of vapour}=\frac{0.07484}{0.001}=74.84\ kg\ m^{-3}$$

Molar mass $= 2\times74.84 = 149.68$

$$\alpha=\frac{254-149.68}{149.68}=0.697$$

(b) $Z = 1.05$ for H_2 at 273 K and 10132.5 kPa

$$Z=1.05=1+\frac{bP}{RT}$$

$$\text{or } Z=1.05-1.0=0.05=\frac{b\times10132.5\times1000}{8.314\times273}$$

$$b = \frac{0.05 \times 8.314 \times 273}{10132.5 \times 1000}$$

$$= \frac{113.49}{10132.5 \times 10^3} = 1.12 \times 10^{-5}\ \text{m}^3\ \text{mol}^{-1}$$

$T_c = 33.2$ K; but $T_c = \dfrac{8a}{27Rb}$

$$33.2 = \frac{8 \times a}{27 \times 8.314 \times 1.12 \times 10^{-5}}$$

$$a = \frac{33.2\ \text{K} \times 27 \times 8.314\ \text{N m K}^{-1}\ \text{mol}^{-1} \times 1.12 \times 10^{-5}\ \text{m}^3\ \text{mol}^{-1}}{8}$$

$$= 0.0104\ \text{N m}^4\ \text{mol}^{-2}$$

Problem 1.18

(a) The Boyle temperature of methane is 500 K. The compressibility factor Z is 1.21 at 1000 K and 60795 kPa. Calculate the van der Waals constants a and b for the gas.

(b) Calculate the critical parameters P_c, V_c and T_c of the gas and also its critical compressibility factor.

Solution

(a) Boyle temp $T_B = \dfrac{a}{bR}$

$$\frac{a}{b} = 500 \times 8.314 = 4157\ \text{N m mol}^{-1}$$

$$Z = 1 + \frac{1}{RT}\left(b - \frac{a}{RT}\right)P$$

$$1.21 - 1 = 0.21 = \frac{60795 \times 1000}{8.314 \times 1000}\left(b - \frac{4157b}{8.314 \times 1000}\right)$$

$$= 7312.4(b - 0.5b) = 3656.2b$$

$$b = \frac{0.21}{3656.2} = 5.744 \times 10^{-5}\ \text{m}^3\ \text{mol}^{-1}$$

Now, $\dfrac{a}{b} = 4157$ or $\dfrac{a}{5.744 \times 10^{-5}} = 4157$

$$a = 4157 \times 5.744 \times 10^{-5} = 0.2388\ \text{N m}^4\ \text{mol}^{-2}$$

(b)

$$P_c = \frac{a}{27b^2} = \frac{0.2388\ \text{N m}^4\ \text{mol}^{-2}}{27\left(5.744 \times 10^{-5}\right)^2\ \text{m}^6\text{mol}^{-2}}$$

$$= \frac{0.2388}{890.82 \times 10^{-10} \times 1000} = 2680.7\ \text{kPa}$$

$$= \frac{2680.7}{101.325} = 26.46\ \text{atm}$$

$V_c = 3b$; or $V_c = 3 \times 5.744 \times 10^{-5}\ \text{m}^3\ \text{mol}^{-1}$

$$= 17.23 \times 10^{-5}\ \text{m}^3\ \text{mol}^{-1}$$

$$T_c = \frac{8a}{27Rb} = \frac{8 \times 0.2388}{27 \times 8.314 \times 5.744 \times 10^{-5}}$$

$$= \frac{1.9104}{0.012894} = 148.2\ \text{K}$$

Problem 1.19

(a) The critical pressure of mercury vapour is 3.6×10^{-5} kPa and its critical temperature is 1900 K. (i) Calculate the van der Waals constants a and b. (ii) Calculate V_c using P_c and T_c data employing the ideal gas equation and also the equation connecting P_c, V_c and T_c by van der Waals equation. Why do the values of V_c in (ii) differ from each other? Give reasons.

(b) Calculate the reduced pressure of oxygen gas which has a reduced volume of 8.67 and reduced temperature of 2.6 using the law of corresponding states.

Solution

(a) (i) van der Waals constant a

$$= \frac{27(RT_c)^2}{64P_c} = \frac{27(8.314 \times 1900)^2}{64 \times 3.6 \times 10^5 \times 10^3}$$

$$= \frac{27 \times (1.5797 \times 10^4)^2}{230.4 \times 10^8} = \frac{67.38}{230.4} = 0.2924\ \text{N m}^4\ \text{mol}^{-2}$$

$$b = \frac{RT_c}{8P_c} = \frac{8.314 \times 1900}{8 \times 3.6 \times 10^5 \times 10^3}$$

$$= \frac{157.97 \times 10^2}{28.8 \times 10^8} = 5.485 \times 10^{-6}\ \text{m}^3\ \text{mol}^{-1}$$

(ii) $V_C = \dfrac{RT_c}{P_c} = \dfrac{8.314 \times 1900}{8 \times 3.6 \times 10^5 \times 10^3}$

$$= \frac{157.97 \times 10^2}{3.6 \times 10^8} = 43.88 \times 10^{-6}\ \text{m}^3\ \text{mol}^{-1}$$

NOTE

- $V_C = 3b$

$$V_C = \frac{3RT_c}{8P_c} = \frac{3}{8} \times 43.88 \times 10^{-6} = 16.455 \times 10^{-6}\ \text{m}^3\ \text{mol}^{-1}$$

The large difference between the values of V_c is due to the fact that the van der Waals equation is not accurate near the critical state.

(b) $\dfrac{V}{V_c} = 8.67$; $\dfrac{T}{T_c} = 2.6$

$$\left(\alpha + \frac{3}{(8.67)^2}\right)([3 \times 8.67] - 1) = 8 \times 2.6$$

$$(\alpha + 0.04)25.01 = 20.8$$

$$25.01\alpha + 1.0004 = 20.8$$

$$\alpha = \frac{19.8}{25.01} = 0.792$$

Problem 1.20

(a) A certain gas has a molar volume 24.4% greater than that predicted by the ideal gas law at 500 K and 6.0×10^4 kPa. (i) Calculate Z and (ii) the molar volume of the gas. Which factor, a or b, predominates under these conditions?

(b) What is the value of the second virial coefficient B from the equation

$$Z = 1 + \frac{Bp}{RT} + \frac{CP^2}{(RT)^2} \cdots$$

Solution

(a) $V_{act} = V_{ideal} + 0.244V_{ideal} = 1.244V_{ideal}$

$$Z = \frac{V_{act}}{V_{ideal}} = \frac{1.244V_{ideal}}{V_{ideal}} = 1.244$$

$$V_{act} = \frac{1.244 \times 8.314 \times 500}{6.0 \times 10^4 \times 10^3} = 8.619 \times 10^{-5}\ \text{m}^3\ \text{mol}^{-1}$$

$$= 8.619 \times 10^{-2}\ \text{dm}^3\ \text{mol}^{-1}$$

Since $Z > 1$, the Z vs P plot has a positive slope. Hence $b > \frac{a}{RT}$. Thus, the size effect is predominant.

(b) Neglecting the terms beyond the second one,

$$Z = 1 + \frac{Bp}{RT} \text{ or } B = \frac{(1.244 - 1) \times 8.314 \times 500}{6.0 \times 10^4 \times 10^3}$$

$$= 1.691 \times 10^{-5}\ \text{m}^3\ \text{mol}^{-1}$$

Problem 1.21

(a) The second virial coefficient for a ketone is -1.6×10^{-3} m^3 mol^{-1} at 393 K and 101.325 kPa. Calculate its compressibility factor. Does this value indicate the predominance of the b or a factor of the van der Waals equation for this vapour?

(b) Calculate the temperature and pressure at which one mole of hydrogen and helium will be in a corresponding state with one mole of O_2 at 101.325 kPa and 300 K. (P_c of O_2 = 5096.6 kPa; T_c of O_2 = 154 K)

Solution

(a) $Z = 1 + \frac{Bp}{RT}$

$$\text{or } Z = 1 - \frac{1.6 \times 10^{-3} \times 101.325 \times 10^3}{8.314 \times 393} = 1 - 0.05 = 0.95$$

The fact that $Z < 1$ indicates that the plot of Z vs P gives a negative slope. Thus $b < a/RT$ in van der Waals equation, and hence the term a predominates over b.

(b) P_{red} of $O_2 = \frac{101.325}{5096.6} = 0.02$ $\quad T_{red}$ of $O_2 = \frac{300}{154} = 1.95$

For H_2, $P_{red} = \frac{P}{p_c}$ or $P = 0.02 \times 1297 = 25.94$ kPa

$T_{red} = \frac{T}{T_c}$ or $T = 1.95 \times 33.2 = 64.7$ K

For He, $P_{red} = \frac{P}{p_c}$ or $P = 0.02 \times 229 = 4.58$ kPa

$T_{red} = \frac{T}{T_c}$ or $T = 1.95 \times 5.25 = 10.24$ K

NOTE

- For H_2, $P_c = 1297$ kPa, $T_c = 33.2$ K
- For He, $P_c = 229$ kPa, $T_c = 5.25$ K

Kinetic Theory of Gases

Problem 1.22

(a) A cylinder of 0.1 m^3 capacity can withstand a maximum pressure of 3.04×10^4 kPa. Suppose the cylinder contains 13.2 kg of a gas (molar mass 0.044), calculate the lowest temperature at which it may explode. The van der Waals constants of the gas are a = 0.360 N m^4 mol^{-2}, and b = 4.30×10^{-5} m^3 mol^{-1}.

(b) Which of the following statements pertaining to the kinetic theory are correct?

(i) Attractive and repulsive forces exist among the gas molecules.

(ii) The volume occupied by the gas molecules is negligible in comparison to total volume of the gas.

(iii) Pressure exerted by the gas molecules is due to their collisions on the walls of the container.

(iv) Collisions among the gas molecules results in the loss of translational energy.

Solution

(a) $$\left(P + \frac{n^2 a}{V^2}\right)(V - nb) = nRT$$

$$\left(3.04 \times 10^4 \times 10^3 + \frac{(300)^2 \times 0.36}{(0.1)^2}\right)(0.1 - 300 \times 4.30 \times 10^{-5})$$

$$= 300 \times 8.314 \times T$$

$$(3.04 \times 10^7 + 3.24 \times 10^6)(0.1 - 0.01290) = 2494.2T$$

$$T = \frac{33.64 \times 10^6 \times 0.0871}{2494.2} = \frac{2.93 \times 10^6}{2494.2} = 1175\ \text{K} = 902°\text{C}$$

(b) Statements (ii) and (iii) are correct.

Problem 1.23

Calculate the number of collisions an oxygen molecule undergoes in one second at a pressure of 101.325 kPa and 300 K. Also, calculate the total number of collisions among its molecules per m^3 per second. The collision diameter of the oxygen molecules is 3.57×10^{-10} m.

Solution

$$Z_1 = \sqrt{2} \times \pi \times \sigma^2 \times nC_{mv}$$

$$n = \frac{NP}{RT} = \frac{6.023 \times 10^{23} \times 101.325 \times 10^3}{8.314 \times 300}$$

$$= \frac{610.28 \times 10^{26}}{2494.2} = 2.45 \times 10^{25}\ \text{molecules m}^{-3}$$

$$C_{mv} = \sqrt{\frac{8RT}{\pi M}} = \sqrt{\frac{8\times 8.314\times 300}{3.1416\times 32\times 10^{-3}}}$$

$$= \sqrt{198.48\times 10^3} = 445.5 \text{ m s}^{-1}$$

$$\sigma = 3.57\times 10^{-10} \text{ m}$$

$$\therefore Z_1 = 1.414\times 3.1416\times\left(3.57\times 10^{-10}\right)^2\times 2.45\times 10^{25}\times 445.5$$

$$= 61794.6\times 10^{-20}\times 10^{25} = 6.18\times 10^9 \text{ collisions s}^{-1}$$

$$Z_{11} = \frac{1}{2}nZ_1 = \frac{1}{2}\times 2.45\times 10^{25}\times 6.18\times 10^9$$

$$= 7.57\times 10^{34} \text{ collisions m}^{-3}\text{ s}^{-1}$$

Problem 1.24

A container of 1 dm^3 capacity contains 3×10^{22} gas molecules each of mass 5.0×10^{-26} kg which are moving with a rms velocity of 500 m s^{-1}. Given that the collision diameter of the gas molecule is 2×10^{-10} m, calculate (i) the molar mass of the gas (ii) the pressure of the gas in kPa (iii) the kinetic energy per mole of the gas (iv) the temperature of the gas and (v) the mean free path of the gas molecules in metres.

Solution

(i) $C_{rms} = \sqrt{\frac{3PV}{M}}$ or $500 = \sqrt{\frac{3.1.25\times 10^5\times 10^{-3}}{M}}$

$$(500)^2 = 25\times 10^4 = \frac{3.1.25\times 10^2}{M}$$

$$M = \frac{3.1.25\times 10^2}{25\times 10^4} = 0.0015 \text{ kg} = 1.5 \text{ g}$$

(ii) $P = \frac{1}{3}\frac{mn\bar{C}^2}{V} = \frac{\frac{1}{3}\times 5.0\times 10^{-26}\text{ kg}\times 3\times 10^{22}\times (500)^2\text{ m}^2\text{s}^{-2}}{1\times 10^{-3}\text{ m}^3}$

$$= \frac{5\times 25\times 10^{-4}\times 10^4}{10^{-3}}$$

$$= 1.25\times 10^5 \text{ kg m}^{-1}\text{ s}^{-2} = 1.25\times 10^5 \text{ Pa} = 1.25\times 10^2 \text{ kPa}$$

(iii) KE per mole $= \frac{3}{2}PV = \frac{3}{2}\times 1.25\times 10^5\times 10^{-3}$

$$= 1.875\times 10^2 \text{ J mol}^{-1}$$

(iv) KE per mole $= \frac{3}{2}RT;\ T = \frac{187.5\times 2}{3\times 8.314} = 15.03$ K

(v) $l = \frac{1}{\sqrt{2}\pi\sigma^2 n}$

$$= \frac{1}{1.414\times 3.1416\times\left(2\times 10^{-10}\right)^2\times 3\times 10^{22}\times 10^3}$$

$$= \frac{1}{53.31\times 10^5} = 0.0187\times 10^{-5} = 1.87\times 10^{-7} \text{ m}$$

Problem 1.25

Two moles of a gas obeying the van der Waals equation has P = 950 kPa, $V = 5.0\times 10^{-3}$ m^3 mol^{-1}, $T = 310$ K and $a = 0.680$ N m^4 mol^{-2}. Calculate b and its compressibility factor. What is its virial coefficient B? What is the radius of the molecule?

Solution

$$\left(P + \frac{n^2 a}{V^2}\right)(V - nb) = nRT$$

$$(V - nb) = \frac{nRT}{\left(P + \frac{n^2 a}{V^2}\right)};\ b = \left(V - \frac{nRT}{\left(P + \frac{n^2 a}{V^2}\right)}\right)\frac{1}{n}$$

$$b = \left\{5\times 10^{-3} - \frac{2\times 8.314\times 310}{950\times 10^3 + \frac{4\times 0.680}{\left(5\times 10^{-3}\right)^2}}\right\}\frac{1}{2}$$

$$\frac{0.005 - 0.00487}{2} = 0.000065 = 6.5\times 10^{-5} \text{ m}^3\text{ mol}^{-1}$$

$$Z = \frac{PV}{nRT} = \frac{950\times 10^3\times 5\times 10^{-3}}{2\times 8.314\times 310} = 0.921$$

$$Z = 1 + \frac{Bp}{RT};\ B = \frac{-0.079\times 8.314\times 310}{950\times 10^3} = -2.14\times 10^{-3} \text{ m}^3\text{ mol}^{-1}$$

$$\text{Excluded volume} = V - nb = 5\times 10^{-3} - 2\times 6.5\times 10^{-5}$$

$$= 0.00487 \text{ m}^3\text{ mol}^{-1}$$

$$\text{Excluded volume per molecule} = \frac{4.87\times 10^{-3}}{6.023\times 10^{23}} = 8.086\times 10^{-27}$$

$$r^3 = \frac{8.086\times 10^{-27}\times 3}{4\times 3.1416} = 1.93\times 10^{-27} \text{ m}^3$$

$$r = \sqrt[3]{1.93\times 10^{-27}} = 1.245\times 10^{-9} \text{ m} = 1.245\times 10^{-7} \text{ cm}$$

Problem 1.26

(a) Draw the probability distribution curve, based on the Maxwell–Boltzmann equation, against C, the velocity of gas molecules at two temperatures T_1 and T_2 ($T_2 > T_1$). Mark the root mean square (rms), average (mv) and most probable (mpv) velocities on the curve. Explain the salient features of the curve.

(b) The density and viscosity of helium at STP is 0.18 kg m^{-3} and 1.86×10^{-5} Pa s respectively. What is the mean free path of the gas molecules?

Solution

(a)

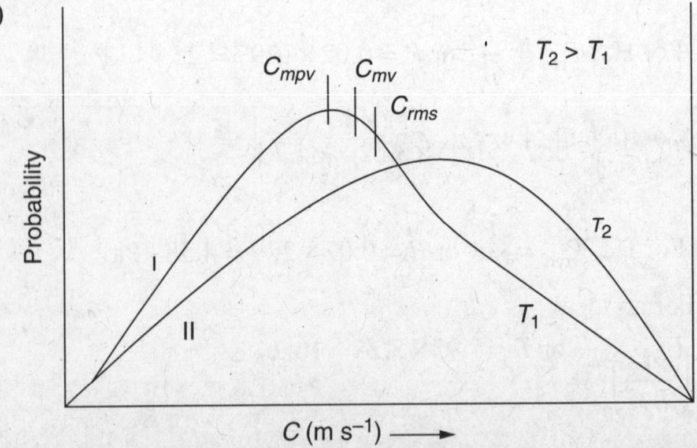

Important features of the curve:

(i) it is seen that $C_{rms} > C_{ave} > C_{mpv}$,

(ii) at higher temperatures, the maximum shifts to the right indicating increase in velocity as the temperature goes up, and

(iii) the maximum becomes flatter at higher temperatures indicating wider distribution of velocities.

(b) $\eta = \frac{1}{3} \times m \times n \times C_{ave} \times l$

where m = mass of gas molecule, n = no. of molecules per unit volume and l = mean free path.

$$(m \times n = d) \quad l = \frac{3\eta}{dC_{ave}} = \frac{3 \times 1.86 \times 10^{-5}}{0.180 \times C_{ave}}$$

$$C_{ave} = \sqrt{\frac{8RT}{\pi M}} = \sqrt{\frac{8 \times 8.314 \times 273}{3.1416 \times 0.004}}$$

$$= \sqrt{144946.5} = 1202.06 \text{ m s}^{-1}$$

$$l = \frac{3 \times 1.86 \times 10^{-5}}{0.180 \times 1202.06} = 2.58 \times 10^{-7} \text{ m}$$

Isothermal Expansion and Coefficient of Compressibility

Problem 1.27

(a) Calculate the pressure of air on the top of Mount Everest which is 8850 m above sea level. Assume that the atmosphere has a uniform temperature of 17°C and that the pressure at sea level is 101.325 kPa. The mean molar mass of air may be taken as 29.0 g mol^{-1}.

(b) Define the terms coefficient of isothermal expansion and the coefficient of compressibility. Derive their values for an ideal gas.

Solution

(a) The pressure relating to height is given by $\frac{P}{P_0} = e^{-MgZ/RT}$

where P and P_0 refer to the heights at height Z and that at sea level, M = molar mass of air, g = acceleration due to gravity, Z = required height, R = gas constant and T = temperature in K. Substituting the relevant data,

$$\ln \frac{P}{P_0} = -\frac{MgZ}{RT}; \; 2.303 \log \frac{P}{P_0} = -\frac{MgZ}{RT}$$

$$\log \frac{P}{101.325} = -\frac{29 \times 10^{-3} \times 9.81 \times 8850}{8.314 \times 300 \times 2.303} = -0.4383$$

$$\frac{P}{101.325} = 3.645 \times 10^{-1} \text{ or } P = 36.93 \text{ kPa (or 0.365 atm)}$$

(b) Coefficient of expansion $= \frac{1}{V}\left(\frac{\partial V}{\partial T}\right)_P$

Coefficient of compressibility $= \frac{-1}{V}\left(\frac{\partial V}{\partial P}\right)_T$

For an ideal gas $PV = RT$,

$$\left(\frac{\partial V}{\partial T}\right)_P = \frac{R}{P}; \frac{1}{V}\left(\frac{\partial V}{\partial T}\right)_P = \frac{R}{PV} = \frac{R}{RT} = \frac{1}{T}$$

$$PdV + VdP = RdT \text{ at constant } T, \, dT = 0$$

$$\left(\frac{\partial V}{\partial P}\right)_T = -\frac{V}{P}; \frac{-1}{V}\left(\frac{\partial V}{\partial P}\right)_T = \frac{V}{P} \times \frac{1}{V}$$

$$-\frac{1}{V}\left(\frac{\partial V}{\partial P}\right)_T = \frac{1}{P}$$

Viscosity of Gases

Problem 1.28

The viscosity of ethylene gas is 1.008×10^{-5} Pa s at 293 K. Calculate the number of collisions a molecule of ethylene undergoes in 1 s at a pressure of 101.325 kPa and 293 K. Also, calculate the total number of collisions per unit volume per unit s under these conditions.

Solution

$$\sigma^2 = \frac{2d}{3\pi n\eta}\sqrt{\frac{RT}{\pi M}}; \; Z_1 = \sqrt{2}\pi\sigma^2 nC_{ave}$$

$$= \sqrt{2} \times \pi \times \frac{2d}{3\pi n\eta} \times \sqrt{\frac{RT}{\pi M}} \times n \times C_{ave}$$

$$d = \frac{MP}{RT} = \frac{28 \times 10^{-3} \times 101.325 \times 10^3}{8.314 \times 293} = 1.165 \text{ kg m}^{-3}$$

$$Z_1 = \frac{\sqrt{2} \times 2 \times 1.165 \times \sqrt{\frac{8RT}{\pi M}}\sqrt{\frac{RT}{\pi M}}}{3 \times 1.008 \times 10^{-5}}$$

$$= \frac{8 \times 1.165 \times 8.314 \times 293}{3 \times 1.008 \times 10^{-5} \times 3.1416 \times 0.028}$$

$$= 8.535 \times 10^9 \text{ collisions s}^{-1}$$

$$Z_{11} = \frac{1}{2} nZ_1 = \frac{1}{2} \times \frac{NP}{RT} \times 8.535 \times 10^9$$

$$= \frac{1}{2} \times \frac{6.023 \times 10^{23} \times 101.325 \times 10^3}{8.314 \times 293} \times 8.535 \times 10^9$$

$$= 4.268 \times 10^9 \times 0.25 \times 10^{26} = 1.07 \times 10^{35} \text{ collisions m}^{-3} \text{ s}^{-1}$$

Redlich–Wong Equation of State

Problem 1.29

(a) A gas contains molecules with speeds as follows. 10 molecules have a speed of 4 m s^{-1}, 15 molecules have a speed of 6 m s^{-1} and 5 molecules have a speed of 5 m s^{-1}. Calculate their rms average and most probable velocities. (Use the rms velocity calculated by the definition for calculating the average and most probable velocities.)

(b) The Redlich–Wong equation of state is given by $P = \frac{RT}{V-b} - \frac{a}{\sqrt{T}V(V+b)}$. Calculate P using the above equation given that, T = 293 K, $V = 0.23 \times 10^{-3}$ m^3 mol^{-1}, $a = 0.366$ N m^4 mol^{-2}, $b = 4.28 \times 10^{-5}$ m^3 mol^{-1}. Compare this value with that obtained from the van der Waals equation. Also calculate Z and Z_c given that P_c = 7366.3 kPa, V_c = 95×10^{-6} m^3 mol^{-1} and T_c = 304.2 K.

Solution

(a) $$\bar{C}^2 = \frac{10\times4^2+15\times6^2+5\times5^2}{10+15+5}$$

$$= \frac{160+540+125}{30} = \frac{825}{30} = 27.5$$

$$C_{rms} = \sqrt{27.5} = 5.24 \text{ m s}^{-1}$$

$$\frac{C_{rms}}{C_{mpv}} = \frac{1}{0.82} \Rightarrow C_{mpv} = 5.244\times0.82 = 4.300 \text{ m s}^{-1}$$

$$\frac{C_{rms}}{C_{ave}} = \frac{1}{0.92} \Rightarrow C_{ave} = 5.244\times0.92 = 4.824 \text{ m s}^{-1}$$

(b) $$P = \frac{8.314\times293}{0.23\times10^{-3}-4.28\times10^{-5}} - \frac{0.366}{\sqrt{293}\times0.23\times10^{-3}\times\left(0.23\times10^{-3}+4.28\times10^{-5}\right)}$$

$$= \frac{2436}{0.1872\times10^{-3}} - \frac{0.366}{17.12\times0.0627\times10^{-6}}$$

$$= 13012.8\times10^3 - \frac{0.366}{1.073\times10^{-6}}$$

$$= 13012.8\times10^3 - 0.341\times10^6$$

$$= 1.3013\times10^7 - 0.0341\times10^7$$

$$= 1.267\times10^7 \text{ Pa} = 1.267\times10^4 \text{ kPa} = 125.04 \text{ atm}$$

P from van der Waals equation

$$P = \frac{RT}{V-b} - \frac{a}{V^2}$$

$$= \frac{8.314\times293}{0.23\times10^{-3}-4.28\times10^{-5}} - \frac{0.366}{\left(0.23\times10^{-3}\right)^2}$$

$$= 13012.8\times10^3 - 6.919\times10^6$$

$$= 13.013\times10^6 - 6.919\times10^6$$

$$= 6.094\times10^6 \text{ Pa} = 6094 \text{ kPa} = 60.1 \text{ atm}$$

Z from Redlich – Wong equation

$$= \frac{PV}{RT} = \frac{1.267\times10^4\times10^3\times0.23\times10^{-3}}{8.314\times293}$$

$$= \frac{0.2914\times10^4}{2436} = 1.2$$

Z from van der Waals equation $= \dfrac{6094\times10^3\times0.23\times10^{-3}}{8.314\times293}$

$$= \frac{1401.6}{2436} = 0.575$$

$$Z_C = \frac{P_C V_C}{RT_C} = \frac{7366.3\times10^3\times95\times10^{-6}}{8.314\times304.2} = 0.277$$

> **NOTE**
>
> The wide divergence between the values of P, Z and Z_C calculated from the Redlich–Wong and van der Waals equations is perhaps due to the closeness of the temperature to the T_c of CO_2.

Compressibility of Gases

Problem 1.30

(a) A diver swims up to the surface of water very fast from a depth of 3.0 m (about 10 ft) without exhalation of air from his lungs. Given that for a normal metabolic function a partial pressure of 20.5 kPa of O_2 is necessary, what is the excess volume per cent of oxygen that must be removed from the body at that depth so that the diver does not suffer from excess O_2 in his lungs?

> **NOTE**
>
> • A pressure of 101.325 kPa is equivalent to 10.3 m height of water.

(b) The compressibility factor, Z, for propane from the compressibility reduced P vs reduced T plot is 0.82 at T_c = 96.7°C and P_c = 41.95 atm. Given that, T_r and P_r are 1.50 and 4.0 respectively, calculate the molar volume of the gas in m³ and compare it with the molar volume from ideal gas equation. Which factor, a and b of van der Waals equation, predominates under these conditions?

Solution

(a) Pressure at 3 m depth $= \dfrac{3.0}{10.3} = 0.291$ atm

$= 0.291\times101.325 = 29.48$ kPa

This is over and above the atmospheric pressure.

∴ Pressure at 3 m depth $= 101.325+29.48 = 130.81$ kPa

Partial pressure of O_2 = mole fraction $\times P_{Total}$

$$= \frac{V_{O_2}}{V_{O_2}+V_{N_2}} P_{Total}$$

Since air contains 0.21 mole fraction of O_2,

$$0.21\times101.325 = \frac{V_{O_2}}{V_{O_2}+V_{N_2}}\times130.81$$

or, $$\frac{V_{O_2}}{V_{O_2}+V_{N_2}} = \frac{0.21\times101.325}{130.81} = 0.163$$

Hence, the Vol% of O_2 (≈ mol% of O_2) must be reduced from 21 ro 16.3 = 3.7. If it is not reduced, the partial pressure of O_2 in the body will be high and injurious to health.

(b) T_r of propane = 1.5; $T_c = 273+96.7$

$= 369.7$ K; $T = 1.5\times369.7 = 554.55$ K

P_r of propane = 4.0; P_c = 41.95 atm, or $p = 41.95\times4$

$= 167.80 \text{ atm} = 167.80\times101.325 = 17002.3$ kPa

$$\therefore 0.82 = \frac{PV}{RT}, \text{ or, } V = \frac{8.314\times554.55\times0.82}{17002.3\times1000}$$

$$= \frac{3780.6}{17002.3\times1000} = 2.22\times10^{-4} \text{ m}^3 \text{ mol}^{-1}$$

From the ideal gas equation, $nV = \dfrac{RT}{P} = \dfrac{8.314\times554.55}{17002.3\times1000}$

$$= \frac{4610.53}{17002.3\times1000} = 2.71\times10^{-4} \text{ m}^3 \text{ mol}^{-1}$$

Since the value of Z is less than 1, the limiting slope of Z vs P plot is negative. Therefore b is the predominant factor and that the molecular size overrides the intermolecular attraction.

EXERCISES

Practice Problems

1. (a) A steel tank has a volume capacity of 0.45 m^3 and is filled with 1.0 kg of oxygen gas. What is its pressure in kPa and atm at 27°C?
   (b) It is later observed that the tank developed a leak and the pressure drops to 101.325 kPa. How much oxygen remained in the flask? How much leaked out?
2. (a) An empty flask of volume 0.15×10^{-3} m^3 weighs 0.0685 kg. The flask was next filled with a volatile liquid and placed in a bath maintained at 368 K and the liquid inside evaporates. After cooling and the vapour again condensed, the flask weighs 0.0687 kg. The barometric pressure under these conditions is 98.7 kPa. Use this data to calculate the molar mass of the liquid.
   (b) A mixture of noble gases consisting of 6×10^{-3} kg of helium, 0.01 kg of neon and 0.04 kg of krypton is placed in a flask at STP. Calculate the mole fraction and partial pressure of each of the constituents. (Molar mass of He = 4×10^{-3} kg; Ne = 20.2×10^{-3} kg; Kr = 83.8×10^{-3} kg)
3. (a) Methane and ethane are present in a vessel at mole fractions of 0.748 and 0.252 respectively at 300 K and 202.65 kPa. What is the density of the mixture in kg m^{-3}?
   (b) The average molar mass of selenium vapour dissociating according to

$$Se_6(g) \rightleftharpoons 3Se_2(g)$$

   is 215.0 at 970 K. what is its degree of dissociation? (Atomic mass of Se = 79.0)
4. (a) In a diffusion study, hydrogen effuses through a pinhole in 300 s. The pressure dropped in the system from 106.7 kPa to 100 kPa. The effusion of an unknown gas required 1900 s under the same pressure drop conditions. Calculate the molar mass of the gas.
   (b) An oxygen cylinder contains 0.16 kg of O_2 under a pressure of 2.03×10^4 kPa and 300 K. On heating to 350 K, half the amount of O_2 was lost. What is the pressure of the gas remaining in atm?
5. (a) At what temperature will the root mean square velocity of SO_2 gas molecules be equal to the average velocity of O_2 molecules at 400 K? What will be the most probable velocity of SO_2 gas molecules under the same conditions?
   (b) If the pressures of SO_2 and O_2 are 202.63 kPa and 303.375 kPa respectively, at the temperature calculated above, what is the ratio of d_{SO2}/d_{O2}?
6. (a) The critical temperature and critical pressure of NH_3 are 405.5 K and 1.135×10^4 kPa respectively. Calculate the van der Waals parameters a and b. What is its critical volume? Explain the difference.
   (b) Starting from the van der Waals equation, derive a relation between Z, the compressibility factor, the coefficients of van der Waals equation and the pressure of the gas. How do you get the limiting slope $(\partial Z/\partial P)_T$ at $P = 0$?
7. (a) The second virial coefficient B for n-butane at 293 K is -0.778×10^{-3} m^3 mol^{-1}. Calculate Z for this gas at 1013.25 kPa. Also calculate Z for this gas at 273 K and 506.6 kPa given B as -0.918×10^{-4} m^3 mol^{-1}. Are the Z values in the correct order to be expected under these conditions?
   (b) The van der Waals constants a and b for SO_2 are 0.6780 N m^4 mol^{-2} and 5.64×10^{-5} m^3 mol^{-1} respectively. Calculate the critical parameters, P_c, V_c and T_c, and calculate the critical compressibility factor.
8. (a) Calculate the pressure of a gas obeying the van der Waals equation and having $V = 2.0$ dm^3, $T = 300$ K, $n = 20$, $a = 0.1505$ N m^4 mol^{-2} and $b = 3.98\times10^{-5}$ m^3 mol^{-1}. What is its value in atm? Compare this value with that obtained from the ideal gas equation.
   (b) Calculate the radius of the molecule. What are the Boyle and inversion temperatures of the gas?
9. (a) In a vessel of 2 dm^3 capacity there are 3×10^{22} gas molecules whose rms velocity is 475 m s^{-1}. (i) Use this data, to calculate the pressure of the gas in kPa and atm. (ii) What is the kinetic energy of the gas per mole? How much is the kinetic energy per molecule of the gas in J? (iii) What is the temperature of the system and what is the molar mass of the gas? The mass of each gas molecule may be taken as 3.32×10^{-24} g.
   (b) The heat capacity C_P of ethane at 400 K is 65 J K^{-1} mol^{-1}. How many degrees of vibrational freedom are not effective at this temperature?
10. (a) The Boyle temperature of isopentane is 1555 K. The compressibility factor $Z = 0.43$ at $P = 8721.2$ kPa and $T = 507.1$ K. Calculate the van der Waals constants a and b from these data.
    (b) Calculate the change in internal energy of 2 moles of arsine (AsH_3) when it is heated from 300 K to 400 K using the principle of equipartition of energy.
11. (a) The viscosity of H_2S is 1.17×10^{-5} Pa s at STP. Calculate the mean free path of the gas molecules in metres. What will be the value of l if the temperature is raised to 400 K and the pressure is reduced to 50.66 kPa. Also calculate the collision diameter from the data of η at STP.
    (b) Calculate the number of collisions a molecule of H_2S undergoes in one second at STP. What are the total number of collisions among its molecules per m^3 per s?

12. (a) Calculate the pressure of air on the top of the mountain Kanchenjunga located on the India–Nepal border which is 28170 ft above sea level. Assume that air has a uniform temperature of 20°C and that pressure at sea level is 1 atm. Air has a molar mass of 29 g mol^{-1}.

(b) Ethylene has a molar volume 10% less than that predicted by the ideal gas equation. Given that its critical temperature and critical pressure are 283 K and 5.12 MPa, calculate Z, and the van der Waals constants a and b. What is its second virial coefficient at 300 K and 101.325 kPa?

13. (a) Uranium hexafluoride has a high vapour pressure of 11 kPa at 293 K. Its critical pressure and critical temperature are 4.6 MPa and 503 K. Calculate the pressure and temperature of one mole of sulfur hexafluoride which has the same reduced pressure and reduced temperature as UF_6? (SF_6 has a $T_c = 319$ K and $P_c = 3.76$ MPa)

(b) For the gas n-butane, Z is 0.425 at a pressure of 3.8×10^3 kPa and 425 K. Calculate the volume occupied by 5.0 mmol of the gas under these conditions. What is its second virial coefficient?

14. (a) If hydrogen selenide (H_2Se) is a rigid non-linear molecule, what is its C_P value on classical considerations? How does this value change if vibrational degrees of freedom are taken into account?

(b) 2×10^{-3} kg of N_2O_4 gas gives a total pressure of 130.1 kPa when it is partially dissociated in a 5×10^{-4} m^3 vessel at 300 K. Calculate the molar mass of the partially dissociated gas and its degree of dissociation.

15. (a) The collision diameter of SO_2 gas molecules is 4.3×10^{-10} m. Calculate the viscosity of the gas at 300 K. What is the mean free path of the gas molecules at a pressure 202.65 kPa and the temperature given?

(b) At what temperature will the most probable velocity of Cl_2 gas molecules be the same as the average velocity of NH_3 gas molecules at 500 K?

16. State whether the following statements are true or false. Give the correct statements where the statements are considered as false.

(a) The time required for a given volume of gas to pass through an orifice is inversely proportional to the rate of diffusion but directly proportional to the molar mass of gas.

(b) The kinetic energy of sulfur dioxide gas molecules is twice that of molecules of oxygen at 300 K.

(c) According to the kinetic theory of gases, most molecules possess the velocity given by the expression $\sqrt{\dfrac{2RT}{M}}$ than by the mean velocity equation.

(d) The distribution of velocities of gaseous species among the three categories is not affected by increasing the temperature of the system.

(e) The mean free path of gas molecules decreases with pressure but increases with temperature.

(f) The viscosity of any given gas is directly proportional to the square root of the temperature and inversely proportional to the square of the collision diameter.

(g) The gases A, B and C have the following values of van der Waals constants a.

Gas	a (N m^4 mol^{-2})
A	0.658
B	0.422
C	0.678

The ease of liquefaction of the gases increases in the order C < B < A.

(h) At Boyle temperature, a real gas behaves ideally over a range of pressures.

(i) The compressibility factor, Z, of a gas is independent of pressure at a given temperature.

(j) The Joule–Thomson expansion effect of a gas is reversed above the inversion temperature.

(k) The heat capacities of gases approach values calculated from the principle of equipartiton of energy at higher temperatures.

(l) The plot of V against absolute temperature (K) at a given pressure is linear with an intercept.

17. (a) The vapour pressure of water at 293 K is 2333 Pa. The relative humidity at a certain place at the same temperature is 75%. Calculate the mass of water in one litre of air under these conditions.

(b) If the density of the dry air at this temperature (293 K) is 1.180 kg m^{-3}, what is the density of humid air?

(c) Nitric oxide has critical temperature T_c = 180 K and critical pressure P_c = 6.48×10^6 Pa. Calculate the critical volume V_c using (i) the ideal gas equation, (ii) the van der Waals equation, and (iii) V of the gas using the modified Berthelot equation

$$P = \frac{RT}{V}\left(1+\left[\frac{9}{128\gamma}-\frac{27}{64\gamma^3}\right]\alpha\right)$$

at a pressure of P = 202.65 kPa and a temperature of 400 K.

18. (a) Calculate the total number of collisions occurring between the hydrogen and iodine molecules according to the reaction

$$H_2(g)+I_2(g)\rightleftharpoons 2HI(g)$$

at 600 K in unit volume per second. The average collision diameter of the reactants may be taken as 3.5×10^{-10} m.

(b) Calculate the value of α for oxygen at β = 133.3 and γ = 2.42 using the law of corresponding states and modified Berthelot equation.

19. (a) 100 ml of SO_2 and 55.4 ml of an unknown gas pass through an orifice in the same time under similar conditions. Calculate the molar mass of the unknown gas.
(b) A gas X has a viscosity 1.5 times that of another gas Y. Calculate the ratio of their mean free paths and their collision diameters (molar mass of X = 28; molar mass of Y = 16). The temperature may be taken as 280 K.

20. (a) A mixture of propane and n-butane is kept in a vessel at a temperature of 400 K with their partial pressures $p_{\text{propane}} = 101.325$ kPa and $p_{n\text{-butane}} = 303.975$ kPa.
(i) Calculate the density of the mixture.
(ii) What is the most probable velocity of the molecules in the mixture?
(iii) Suppose the vessel develops a leak and the gases leak out, what is the ratio of rates of effusion of the gases?
(iv) Given that the collision diameters of propane and n-butane are 5.118×10^{-10} m and 4.687×10^{-10} m, calculate the viscosities of these gases and the viscosity of the mixture.

21. Show that the ratio between coefficient of thermal expansion and compressibility for a van der Waals gas is $R/(V - b)$.

22. Derive the Joule–Thomson coefficient for gases obeying the van der Waals equation of state.

Multiple Choice Questions

1. The plot of PV (Y-axis) against P (X-axis) at constant temperature gives **(a)** rectangular hyperbola **(b)** a straight line passing through the origin **(c)** a horizontal line parallel to the pressure axis (X-axis) **(d)** a straight line with a negative slope.

2. The density of a gas at 57°C and 1 atm pressure is 1.1 g L^{-1}. Its density in kg m^{-3} at 27°C under the same pressure is **(a)** 1.56 **(b)** 1.26 **(c)** 1.84 **(d)** 2.11.

3. A spherical balloon has a radius of 1.0 m and weighs 1 kg. If it is to be filled with helium at 1 atm and 300 K (density of air = 1.2 kg m^{-3}), the payload (in kg) is **(a)** 3.36 **(b)** 5.50 **(c)** 6.33 **(d)** 4.85.

4. Helium at a pressure of 506.6 kPa and hydrogen at a pressure of 810.6 kPa diffuse through a pinhole at a given temperature. Their relative rates of diffusion ($r\text{He}/r\text{H}_2$) are **(a)** 0.495 **(b)** 0.638 **(c)** 0.526 **(d)** 0.442.

5. A mixture of gases consisting of 60 mol% methane and the rest propane is to be kept in a 0.5 m^3 gas cylinder. If the methane is to be pumped into the evacuated cylinder at a temperature of 300 K and 101.325 kPa pressure, the total pressure of the system (in kPa) after propane also is pumped into the cylinder is **(a)** 198.6 **(b)** 174.5 **(c)** 189.6 **(d)** 168.9.

6. The relative density of N_2O_4 under certain experimental conditions of temperature is 62 and on dissociation, the relative density is 45.6. Hence the degree of dissociation of the gas is **(a)** 0.63 **(b)** 0.45 **(c)** 0.80 **(d)** 0.36.

7. The relative density of xenon at STP is 66.0 (molar mass of Xe = 131.3 g mol^{-1}). The pressure of the gas at 300 K (in atm) is **(a)** 1.053 **(b)** 1.005 **(c)** 1.024 **(d)** 1.009.

8. In a comparative study of diffusion rates between argon and bromine vapour, it was observed that argon diffused 1.5 times faster than partially dissociated bromine vapour under certain temperature and pressure conditions. From these data, the degree of dissociation of bromine vapour is **(a)** 0.66 **(b)** 0.87 **(c)** 0.81 **(d)** 0.78.

9. The van der Waals equation for n moles of a gas is **(a)** $(P + n\,a^2/V^2)(V - b) = nRT$ **(b)** $(P + n^2a/V^2)(V - nb) = nRT$ **(c)** $(P + na/V^2)(V - nb) = RT$ **(d)** $(P + n^2a/V)(V - nb) = nRT$.

10. The van der Waals constant a (in L^2 atm mol^{-2}) for three gases X, Y, Z is 3.610, 6.493, 4.170 respectively. The ease of liquefaction of these gases increases in the order **(a)** Z < X < Y **(b)** Y < X < Z **(c)** X < Z < Y **(d)** Y < Z < X.

11. The number of collisions between the molecules of a gas in a vessel at constant temperature **(a)** increases linearly with increase of pressure and becomes zero as the pressure becomes zero **(b)** decreases linearly with increase of pressure **(c)** is unaffected by pressure **(d)** decreases exponentially with pressure.

12. The compressibility factor, Z, for hydrogen at a given temperature **(a)** is equal to unity at all pressures **(b)** is less than unity at low pressures but increases as the pressure is increased passing through a minimum **(c)** continuously increases beyond unity as the pressure is increased **(d)** decreases continuously from unity with increasing pressure.

13. The root mean square velocity (rmsv), mean velocity (mv), and most probable velocities (mpv) of a gas differ from each other at a given temperature. Which of the following ratios among the following is correct? **(a)** $C(\text{rmsv})/C(\text{mv}) = 1.205$ **(b)** $C(\text{mv})/C(\text{mpv}) = 1.122$ **(c)** $C(\text{rmsv})/C(\text{mpv}) = 1.150$ **(d)** $C(\text{mv})/C(\text{rmsv}) = 0.980$.

14. The temperature at which the most probable velocity of oxygen gas molecules is equal to the root mean square velocity of SO_2 gas molecules at 600 K is (in K) **(a)** 385 **(b)** 400 **(c)** 540 **(d)** 450.

15. The average velocity of helium (molar mass = 4.0) at 400 K is equal to the most probable velocity of H_2 at (in K) **(a)** 455.2 **(b)** 254.5 **(c)** 525.4 **(d)** 369.2.

16. Boyle temperature indicates the **(a)** temperature above which a liquid phase does not exist **(b)** temperature at which the Joule–Thomson coefficient, $\mu = 0$ **(c)** temperature at which a real gas behaves ideally over a range of pressures **(d)** temperature at which Boyle's law is obeyed by all gases.

17. The van der Waals constants a and b for SO_2 are 0.6780 N m^4 mol^{-2} and 5.64×10^{-5} m^3 mol^{-1} respectively. The inversion temperature of the gas is **(a)** 2890 **(b)** 3485 **(c)** 4264 **(d)** 2980.

18. The equation of state obeyed by all gases and involving the critical constants P_c, V_c, T_c of a gas is given by **(a)** $(\alpha+3/\beta)(\beta-1)=8/3\times\gamma$ **(b)** $(\alpha+3/\beta^2)(3\beta-1)=8\gamma$ **(c)** $(\alpha\times\beta+3/\beta)(2\times\beta-1)=3\times\gamma$ **(d)** $(\alpha\times\beta^2+3)(\beta-1)=3/8\times\gamma$

(**Note:** $P/P_c = \alpha$, $V/V_c = \beta$, $T/T_c = \gamma$).

19. The van der Waals constants a and b for methane gas are 0.2280 N m^4 mol^{-2} and 4.28×10^{-5} m^3 mol^{-2} respectively. Its P_c, V_c, T_c in that order (in atm, dm^3, K) are **(a)** 94.5, 0.182, 198.8 **(b)** 59.4, 0.154, 245.6 **(c)** 66.7, 0.243, 224.3 **(d)** 45.9, 0.128, 189.8.

20. Pick out the correct statement among the postulates of kinetic theory of gases given by **(a)** attractive and repulsive forces exist among gas molecules **(b)** the collisions among gas molecules are elastic **(c)** the volume occupied by gas molecules is finite and cannot be neglected in comparison to the total gas volume **(d)** the random collisions among the gas molecules leads to the pressure exerted by the gas.

21. The density of carbon monoxide gas at STP is 1.25 kg m^{-3}. The rms velocity of the gas at 500 K is (in m s^{-1}) **(a)** 287 **(b)** 325 **(c)** 211 **(d)** 422.

22. The correct statement among the following about the mean free path of a gas, l, is **(a)** it does not depend on the temperature and pressure of the gas **(b)** it is proportional to the pressure of the gas at constant temperature **(c)** it is inversely proportional to the temperature of the gas at constant pressure **(d)** it is directly proportional to temperature and inversely proportional to the pressure of the gas.

23. The root mean square velocity of argon (molar mass = 40) at a certain temperature at which its average kinetic energy is 2 kJ mol^{-1} is (in m s^{-1}) **(a)** 30 **(b)** 20 **(c)** 10 **(d)** 40.

24. The root mean square velocity of nitrogen gas molecules at 273 K is 493.5 m s^{-1}. The density of the gas is 1.25 kg m^{-3} under the same temperature and a pressure of 101.325 kPa. If the mean free path of the gas molecules is 2×10^{-7} m, the viscosity in Pa s is **(a)** 3.78×10^{-5} **(b)** 2.4×10^{-3} **(c)** 1.5×10^{-6} **(d)** 3.2×10^{-4}.

25. The distribution of molecular velocities curves, $(1/n^*)$ dn/dC vs C reveals that the **(a)** maximum becomes very sharp as the temperature is increased **(b)** maximum shifts to the right indicating that there is an increase of molecular velocities with temperature **(c)** rms velocity to be at the top of the maximum in the curves **(d)** maximum in the curve remains unchanged with temperature.

26. The total number of collisions occurring between the molecules of a gas per unit volume per unit time is **(a)** proportional to the number of molecules of the gas at constant T **(b)** proportional to the temperature at constant pressure of the gas **(c)** proportional to $T^{1/2}$ keeping other conditions constant **(d)** is independent of temperature under all conditions.

27. According to the principle of equipartition of energy, for a molecule containing X_a atoms, the number of vibrational degrees of freedom for a gaseous polyatomic molecule is **(a)** 3 **(b)** 2 **(c)** $3X_a - 5$ **(d)** $3X_a - 6$.

28. For the non-linear nitrogen dioxide molecule, the heat capacity, C_v, on the basis of equipartition principle is (in J K^{-1} mol^{-1}) **(a)** 36.5 **(b)** 49.9 **(c)** 41.8 **(d)** 59.6.

29. If the experimental value of C_v for NO_2 is 31.6 J K^{-1} mol^{-1}, the number of ineffective vibrational modes are **(a)** 2 **(b)** 1 **(c)** 4 **(d)** 3.

30. For a multi-component gaseous system with the mole fraction of the ith component as x_i, the van der Waals equation of state is given by (where g_i refers to the interaction constant of the ith component)

(a) $\dfrac{P}{RT} = \Sigma_{i=1}^{2} \dfrac{1}{V - x_i b_i} - \dfrac{1}{V^2}\Sigma_{i=1}^{2} g_i x_i^2$

(b) $\dfrac{P}{RT} = \Sigma_{i=1}^{2} \dfrac{x_i}{V - x_i b_i} - \dfrac{1}{V^2}\Sigma_{i=1}^{2} g_i x_i^2$

(c) $\dfrac{P}{RT} = \Sigma_{i=1}^{2} \dfrac{x_i}{V - x_i b_i} - \Sigma_{i=1}^{2} g_i^2 \dfrac{x_i^2}{v_i^2}$

(d) $\dfrac{P}{RT} = \Sigma_{i=1}^{2} \dfrac{b_i x_i}{V - x_i b_i} - \Sigma_{i=1}^{2} g_i x_i^2$

31. For a gas obeying the van der Waals equation of state, C_v is **(a)** independent of temperature **(b)** dependent on temperature as well as volume **(c)** dependent only on temperature **(d)** dependent only on volume.

32. For gases obeying the Dieterici equation of state, the ratio $P_c V_c / R T_c$ is nearly **(a)** 3.69 **(b)** 0.27 **(c)** 0.33 **(d)** 1.0

33. Assuming that the van der Waals constant b for O_2 is 0.0319 L mol^{-1}, the approximate mean free path of O_2 (in 10^{-5} cm) at 27°C and 1 atm pressure is **(a)** 0.54 **(b)** 1.36 **(c)** 2.14 **(d)** 1.62

Answer Keys

Practice Problems

1. (a) $P(\text{in kPa}) = 173.21$; $P(\text{in atm}) = 1.71$
   (b) O_2 remaining in flask = 18.28 moles = 0.585 kg; O_2 leaked out = 0.415 kg
2. (a) Molar mass of liquid = 41.0 g
   (b) $X_{He} = 0.607$; $X_{Ne} = 0.200$; $X_{Kr} = 0.193$; $p_{He} = 61.5$ kPa; $p_{Ne} = 20.26$ kPa; $p_{Kr} = 19.55$ kPa
3. (a) Density of mixture = 1.59 kg m^{-3}
   (b) $\alpha = 0.600$
4. (a) Molar mass of unknown gas = 80.14
   (b) Pressure of remaining gas = 117.1 atm
5. (a) $T = 679.06$ K
   (b) $\dfrac{d_{SO_2}}{d_{O_2}} = 0.787$
6. (a) $a = 0.423$ N m^4 mol^{-2}; $b = 3.713\times10^{-5}$ m^3 mol^{-1};
   $V_C = 11.13\times10^{-5}$ m^3 mol^{-1};
   V_C from $Z_C = 7.22\times10^{-5}$ m^3 mol^{-1}.
   van der Waals equation is not an accurate description of the behaviour of gases near the critical state.
   (b) Theoretical question
7. (a) Z at 101.325 kPa and 293 K = 0.676; Z at 506.6 kPa and 273 K = 0.795
   Z is in the expected order, that is, closer to 1.0 at lower temperature and lower pressure.
   (b) $P_c = 7.89\times10^3$ kPa; $V_C = 16.92\times10^{-5}$ m^3 mol^{-1}
   $T_C = 428.4$ K; $Z_c = \dfrac{P_C V_C}{RT_C}$
   $= \dfrac{7.89\times10^6 \times 16.92\times10^{-5}}{8.314\times428.4} = 0.375$
8. (a) $P = 2.629\times10^4$ kPa = 259.4 atm
   P from ideal gas equaiton $= 2.494\times10^4$ kPa = 246.1 atm
   (b) Radius of the molecule $= 1.767\times10^{-10}$ m
9. (a) (i) $P = 3.745$ kPa = 0.037 atm;
   (ii) Kinetic energy per molecule $= 3.75\times10^{-22}$ J
   KE per mole = 226 J mol^{-1}; T of the gas = 18.1 K
   Molar mass of the gas = 2 g
   (b) Vibrational degrees of freedom not contributing to the energy = 14
10. (a) $b = 1.33\times10^{-4}$ m^3 mol^{-1}; $a = 1.719$ N m^4 mol^{-2}
    (b) ΔE for two moles of gas = 14.97 kJ
11. (a) $l = 5.6\times10^{-8}$ m; l at 400 K and 50.66 kPa $= 16.42\times10^{-8}$ m
    $\sigma = 4.73\times10^{-10}$ m
    (b) Z_1 of $H_2S = 1.102\times10^{10}$ collisions s^{-1}
    $Z_{11} = 1.48\times10^{35}$ m^{-3} s^{-1}
12. (a) Pressure at height of 8586.2 m = 37.18 kPa
    (b) $Z = 0.9$; $a = 0.4561$ N m^4 mol^{-2}
    $b = 5.744\times10^{-5}$ m^3 mol^{-1}
    Second virial coefficient $B = -2.46\times10^{-3}$ m^3 mol^{-1}
13. (a) Pressure of SF_6 = 8.99 kPa
    Temperature of SF_6 = 186 K
    Volume accupied by 5 millimoles of gas
    (b) $= 1.967\times10^{-6}$ m^3; $B = -5.35\times10^{-4}$ m^3 mol^{-1}
14. (a) C_P value for classical consideration = 33.2 J mol^{-1} K^{-1}
    C_P value if vibrational modes are also taken into account = 58.1 J mol^{-1}K^{-1}
    (b) Molar mass of partially dissociated gas = 0.0767 kg; degree of dissociation, $\alpha = 0.200$
15. (a) η of SO_2 at 300 K $= 2.038\times10^{-5}$ Pa s
    Mean free path, $l = 2.488\times10^{-8}$ m
    (b) $T = 2658.7$ K
16. Theoretical question
17. (a) Mass of water in 1 litre of air $= 1.293\times10^{-2}$ g
    (b) Density of moist air = 1.193 kg m^{-3}
    (c) (i) V_C from ideal gas equation $= 230.94\times10^{-6}$ m^3
    (ii) V_C from van der Waals' gas equation $= 86.6\times10^{-6}$ m^3
    (iii) V_C from modified Berthelot gas equation $= 16.4\times10^{-5}$ m^3

18. (a) Total number of collisions at 600 K

$= 1.457 \times 10^{29}$ m^{-3} s^{-1}

(b) α from law of corresponding states $= 4.836 \times 10^{-2}$

α from modified Berthelot equation $= 6.455 \times 10^{-2}$

19. (a) Molar mass of unknown gas = 208.5

(b) $\dfrac{l_x}{l_y} = 1.134$

(c) $\dfrac{\sigma_x}{\sigma_y} = \dfrac{1}{1.065} = 0.939$

20. (a) $d_{\text{mixture}} = 6.642$ kg m^{-3}

(b) $C_{MPV} = 349.34$ m s^{-1}

(c) Ratio of rates of effusion of propane to butane

$= 0.383$

(d) $\eta_{\text{propane}} = 1.38 \times 10^{-5}$ Pa s; $\eta_{\text{butane}} = 1.885 \times 10^{-5}$ Pa s

$\eta_{\text{mix}} = 1.759 \times 10^{-5}$ Pa s

21. Theoretical question

22. $$\left(\frac{\partial T}{\partial P}\right)_H = \frac{\left(\dfrac{2a}{RT} - b\right)}{C_p}$$

Multiple Choice Questions

1. (c)	**2.** (b)	**3.** (a)	**4.** (d)	**5.** (d)	**6.** (d)	**7.** (b)	**8.** (d)	**9.** (b)	**10.** (c)
11. (a)	**12.** (c)	**13.** (b)	**14.** (d)	**15.** (b)	**16.** (c)	**17.** (a)	**18.** (b)	**19.** (d)	**20.** (b)
21. (c)	**22.** (d)	**23.** (b)	**24.** (a)	**25.** (b)	**26.** (c)	**27.** (d)	**28.** (b)	**29.** (a)	**30.** (b)
31. (c)	**32.** (b)	**33.** (b)							

CHAPTER
2 THERMODYNAMICS

CHAPTER OUTLINE

Important Concepts and Formulae

Thermochemistry
- Conventions Followed in Considering Thermochemical Equations
- Different Enthalpy Changes
- Heats of Formation of Compounds, ΔH_f^0
- Hess' Law of Constant Heat Summation
- Bond Enthalpies
- Variation of Heat of Reaction with Temperature

Second Law of Thermodynamics
- Carnot's Cycle
- Entropy Changes in Various Processes
- Entropy Change in the Free Expansion of an Ideal Gas
- Helmholtz Free Energy (*A*) and Gibbs Free Energy (*G*) and their Variation with *T*, *P* and *V*
- ΔG as a Criterion of Spontaneity of a Process

Third Law of Thermodynamics
- Evaluation of Entropy Changes in Chemical Reactions
- Partial Molar Properties

2.1 Important Concepts and Formulae

In the study of thermodynamics, it is necessary to understand the significance of the various terms that are used.

System: A system, in the thermodynamic sense, is any part of the physical universe separated from the rest by a definite boundary. For example, a gas enclosed in a vessel at a definite temperature and pressure. A system may be *isolated*, *closed* or *open* depending upon whether mass or energy or both are transferred across the boundary.

Isolated system: A system that does not exchange mass or energy across its boundary is an isolated system. No visible effects are observed in the surroundings in transformations involving such systems.

Open system: When transfer of mass and energy across a system occurs, it is known as an open system. For example, a reaction vessel containing sodium iodide solution to which lead nitrate is continuously added to obtain lead iodide upon heating.

Closed system: No exchange of mass between system and surroundings is permissible in such a system.

Surroundings: The rest of the universe excluding the system is considered for this purpose. For example, a thermostat surrounding a gas enclosed in a vessel to supply heat.

State of a system: The thermodynamic state of a system is defined by fixing some variables such as temperature and pressure. These variables are chosen based on experimental requirements.

Change of state: This is defined by specifying the initial and final states of the system.

Path of transformation: This implies the various stages that the system goes through, from a specified initial to a final chosen state. For example, the isothermal expansion of a gas from an initial volume V_1 to a final volume V_2.

Energy: We define it as the capacity of a system to do work. It may take several forms such as mechanical work, electrical work or heat.

Work: It is a quantity that passes across the boundary of a system during a change of state. Its effect is observable on the surroundings. Examples are mechanical work, electrical work, surface work, and so on.

Intensive property: It is the property that is not dependent on the amount or mass of the material in the system. It has the same value at any point in the system. Examples are temperature, density, refractive index, and so on.

Extensive property: This property is characterised by the fact that it is an addition of values of that property in various parts of the system. Typical examples are mass, volume, and so on.

Cyclic process: After undergoing a series of changes, when a system returns to its original state, it is said to be a cyclic process.

Thermodynamic equilibrium: A system in which thermal equilibrium, mechanical equilibrium (implying no pressure imbalances within the system), chemical equilibrium (in case of chemical reaction) and absence of movement of significant amount of material in the system exists is said to be in thermodynamic equilibrium. Systems which are in thermodynamic equilibrium are of great importance.

Reversible and irreversible processes: A given change of state may be carried out in a reversible or irreversible manner in the thermodynamic sense. The condition for reversibility is that after a given change of state, the system and surroundings must not differ by more than an infinitesimal amount in variables like T or P. In other words, the change of state of the system must be carried out so slowly that this condition is met. An illustrative example is the reversible adiabatic expansion of gases. Since reversible processes need to be carried out infinitesimally slowly, they do not have practical significance. In reality, irreversible processes, that is, those processes that occur at a finite rate are valuable but the difference between them and reversible processes is that the latter give information on the maximum work realisable for a given change state of the system.

Some common examples of reversible processes are, frictionless motion, elastic stretching, electric circuits with no resistance, and so on. The mixing of different gases is an example of an irreversible process.

Zeroth law of thermodynamics: This law is stated as 'two systems which are both in thermal equilibrium with a third system, are also in thermal equilibrium with each other'. The importance of this law lies in the fact that it leads to the definition of temperature.

First law of thermodynamics: It is a statement of law of conservation of energy which may be expressed as 'energy cannot be created or destroyed but may be transformed from one form to the other'. Mathematically, it is expressed as

$$\Delta E = Q + W \tag{2.1}$$

where ΔE = change in internal energy of the system (the symbol U is also employed for the internal energy), Q = heat absorbed by the system and W = work done on the system. For an infinitesimal change, Eq. 2.1 is written as

$$dE = q + w \tag{2.2}$$

(i) The following conventions are adopted for Q and W as per IUPAC:
- W is positive when work is done on the system,
- W is negative when work is done by the system,
- Q is positive when heat is absorbed by the system, and
- Q is negative when heat is given out by the system.

(ii) *State functions and path functions*
- E is a property depending only on the state of the system. Thus for a change of the system from state 1 to state 2,

$$\Delta E = E_2 - E_1 \tag{2.3}$$

where E_2 and E_1 are the energies of the system in states 2 and 1. Thus E is a state function. E is independent of how the change of state is brought about. Mathematically, one may write for such a change of state

$$\int_1^2 dE = E_2 - E_1 = \Delta E \tag{2.4}$$

The cyclic integral of such a change, $\oint dE = 0$

- Q and W are dependent on the path of transformation. For a small change in either of them (denoted as A),

$$\int_1^2 dA = A \tag{2.5}$$

- E is an extensive property, that is, its value varies depending on whether one considers per mole or per gram of the substance. For an ideal gas, E depends only on the temperature. Thus for an isothermal process involving an ideal gas, $\Delta E_T = 0$.

Enthalpy, H: It is defined by the relation

$$H = E + PV \tag{2.6}$$

where P and V refer to the pressure and volume of the system. H is an extensive property and a state function. For a finite process occurring at constant pressure P,

$$\Delta H = Q_P \tag{2.7}$$

Similarly, it may be shown that $\Delta E = Q_V$ (2.8)

For an ideal gas, $\Delta H_T = 0$ (at constant temperature)

Some typical processes involving an ideal gas or a liquid

Isothermal reversible expansion or compression of a gas: When a gas expands against an opposing pressure P_{op} or (P_{ext}) and the volume changes by an infitesimal amount dV, the work done W is given by

$$W = -P_{op} dV \tag{2.9}$$

The negative sign indicates that there is a decrease in the internal energy of the gas. W is positive for the reverse process of compression. For a finite process, Eq. 2.9 is written as,

$$W = -P_{op} \Delta V \tag{2.10}$$

where $\Delta V = V_2 - V_1$, V_2 and V_1 are the final and initial volumes. Note that P_{op} is constant in this case.

For n moles of an ideal gas undergoing a reversible expansion at constant temperature T, the work of expansion W for a change of state from an initial volume V_1 to a final volume V_2, is given by

$$W = -\int_{V_1}^{V_2} P_{ex} dV = -\int_{V_1}^{V_2} \frac{nRT}{V} dV = -nRT \ln \frac{V_2}{V_1} \tag{2.11}$$

Note that W is negative in this case since $V_2 > V_1$. For the reverse process of compression, W is positive because, $V_2 < V_1$. Equation 2.11 may be written in terms of pressure as

$$W = -nRT \ln \frac{P_1}{P_2} \tag{2.12}$$

where P_1 and P_2 are the initial and final pressures of the gas.

Since T is constant for this process, $\Delta E_T = 0$, $\Delta H_T = 0$

Reversible isothermal vaporisation of a liquid (say water or methanol): The same Eq. 2.10 applies, but $P_{op} \approx P_{vap}$ with P_{vap} being the vapour pressure of the liquid at the given temperature. For vapourisation of n moles of liquid, say at its boiling point,

$$W = -P_{b.p.}\left(V_v - V_l\right) \approx -P_{b.p.}V_v = -nRT \qquad (2.13)$$

assuming the vapour to be ideal. For this type of phase change,

$$\Delta H = Q_P \text{ and } \Delta E = \Delta H - P\Delta V = Q_P - nRT \qquad (2.14)$$

Heat capacity (C): The heat capacity at constant volume C_V is defined by,

$$C_V = \left(\frac{\partial E}{\partial T}\right)_V \qquad (2.15a)$$

Similarly, the heat capacity at constant volume C_P is defined by,

$$C_P = \left(\frac{\partial H}{\partial T}\right)_P \qquad (2.15b)$$

For a finite change in ΔH or ΔE of the system,

$$\Delta H = C_P\Delta T \text{ and } \Delta H = C_V\Delta T \qquad (2.16)$$

Note that C_P and C_V are extensive in nature and the unit is J K^{-1} mol^{-1}. For an ideal gas,

$$C_P - C_V = nR \quad (n = \text{number of moles of gas}) \qquad (2.17)$$

Adiabatic reversible expansion or compression in ideal gases: By definition, for an adiabatic process, $q = 0$. For such a process involving an ideal gas, the following relations are valid.

$$\frac{T_2}{T_1} = \left(\frac{V_1}{V_2}\right)^{\gamma-1} \qquad (2.18)$$

$$\frac{P_2}{P_1} = \left(\frac{V_1}{V_2}\right)^{\gamma} \qquad (2.19)$$

where $\gamma = \dfrac{C_P}{C_V}$

$$\frac{T_2}{T_1} = \left(\frac{P_2}{P_1}\right)^{\frac{\gamma-1}{\gamma}} \qquad (2.20)$$

Note that Eqs 2.18 to 2.20 are valid only when the process is carried under thermodynamically reversible conditions. Hence we can summarise the equations for adiabatic reversible processes, connecting P, V and T as follows:

$$PV^{\gamma} = \text{Constant}$$

$$TV^{\gamma-1} = \text{Constant}$$

$$TP^{\frac{1-\gamma}{\gamma}} = \text{Constant}$$

Joule–Thomson coefficient and its significance: The Joule–Thomson experiment is carried out under isenthalpic conditions and μ_{JT}, the *Joule–Thomson coefficient* of a gas, is given by

$$\mu_{JT} = \left(\frac{\partial T}{\partial P}\right)_H \qquad (2.21)$$

where ∂T is the change of temperature for a given change in pressure ∂P. We may also write,

$$\left(\frac{\partial H}{\partial P}\right)_T = -C_P \times \mu_{JT} \qquad (2.22)$$

which is used for calculating the LHS of Eq. 2.22 knowing C_P and μ_{JT}. μ_{JT} is positive for all gases, that is, there is a fall of temperature on *J–T* expansion at or below room temperature whereas for hydrogen and helium, μ_{JT} is negative, that is, there is an increase in temperature on *J–T* expansion.

Inversion temperature is the temperature above which μ_{JT} is negative. The inversion temperature of most gases is much higher than the room temperature. Joule–Thompson expansion is used in the production of liquefied gases. For a gas obeying van der Waals equation, μ_{JT} is given by

$$\mu_{JT} = \frac{\frac{2a}{RT} - b}{C_P} \qquad (2.23)$$

where a and b are the van der Waals constants for the gas and C_P is the heat capacity at constant pressure.

2.2 Thermochemistry

The subject deals with heat changes associated with chemical reactions under conditions of constant temperature and pressure. Reactions may be exothermic or endothermic. In the former case, heat is given out by the system and Q_P is negative. For endothermic reactions, heat is absorbed by the system and Q_P (or ΔH) is positive.

2.2.1 Conventions Followed in Considering Thermochemical Equations

The following example is considered for this purpose

$$C_6H_{12}O_6(s) + 6O_2(g) \rightarrow$$
$$6H_2O(l) + 6CO_2(g) \quad \Delta H^0 = -2860\,\text{kJ} \qquad (2.24)$$

(i) The state of the substance solid(s), liquid(l), or gas(g) has to be indicated.

(ii) ΔH^0 indicates the enthalpy change under standard state conditions, that is, that all substances are taken in their standard states (298 K and 1 atm). Also, the substances must be in their most stable form under the given conditions.

(iii) The enthalpies of all elements in their most stable forms in their standard states are assigned a value of zero. For example, for carbon, graphite is the most stable form and not diamond. $H^0_{298} = 0$ for graphite at 1 atm.

(iv) The enthalpy change of reaction 2.24 may be clearly shown as,

$$\Delta H^0 = \left[6H^0_{H_2O(l)} + 6H^0_{CO_2(g)} - H^0_{C_6H_{12}O_6(s)} - 6H^0_{O_2(g)}\right]$$
$$= -2860\,\text{kJ} \quad (2.25)$$

where H^0 refers to standard molar enthalpies of the various species in the states shown in brackets. Note that for the reverse reaction of Eq. 2.25, ΔH^0 is positive, that is, it is endothermic.

2.2.2 Different Enthalpy Changes

Standard enthalpy of combustion, ΔH^0_C: ΔH^0_C of an organic compound is the heat change associated with the combustion of one mole of the compound to give CO_2 gas and liquid water under standard state conditions.

Standard enthalpy of hydrogenation, ΔH^0_{hy}: It is defined in similar way as above for an organic unsaturated compound.

Enthalpy of solution, ΔH^0_{soln}: It is the enthalpy change associated with the addition of one mole of the solute to a specified amount of the solvent (say water) under standard state conditions.

Enthalpy change at infinite dilution, ΔH^0_{infdil}: It is the enthalpy change associated with the addition of one mole of the solute to a large quantity of the solvent such that further addition of the same results in no further evolution of heat.

Enthalpy or heat of neutralisation, ΔH^0_{neut}: It is the heat change associated with the neutralisation of one mole of a strong acid by a solution of a strong base both being in infinitely dilute solutions. This may be expressed as

$$H^+(aq) + OH^-(aq) \rightarrow$$
$$H_2O(aq) \quad \Delta H^0_{neut} = -57\,\text{kJ} \quad (2.26)$$

The heat of neutralisation of a weak acid such as acetic acid, by sodium hydroxide under similar conditions as the above, liberates a smaller amount of heat (= 55.0 kJ), the difference being due to the enthalpy required to dissociate the acid into ions.

For calculating enthalpies of formation of ions in aqueous solution, $\Delta H^0_i\left(i = OH^-, Cl^- \ldots\right)$, it is assumed by convention that $\Delta H^0_{H^+}$ is equal to zero. For example,

$$\Delta H^0(\text{HCl}, aq) = \Delta H^0_{H^+}(aq) + \Delta H^0_{Cl^-}(aq) \quad (2.27)$$

Knowing $\Delta H^0_{(\text{HCl},\,aq)}$ and making use of the assumption that $\Delta H^0_{H^+}(aq) = 0$, $\Delta H^0_{Cl^-}(aq)$ may be calculated. This can be extended to other ions.

2.2.3 Heats of Formation of Compounds, ΔH^0_f

The heat change associated with the formation of one mole of the compound from its elements in their most stable states under standard state conditions of 298 K and 1 atm is referred to as standard heat of formation. For example,

$$S(Rh) + O_2(g, 1\,\text{atm}) \rightarrow$$
$$SO_2(g, 1\,\text{atm}) \quad \Delta H^0_{f(298\,\text{K})} = -297.0\,\text{kJ} \quad (2.28)$$

Since $H^0_{S(Rh)}$ and $H^0_{O_2(g)}$ by definition have zero enthalpy values, $\Delta H^0_{SO_2(g)}$ is equal to –297.0 kJ. Using ΔH^0_f of different compounds, it is possible to determine the enthalpy changes of desired reactions. Consider the reaction,

$$Na_2CO_3(s) + H_2(g) \rightarrow 2NaHCO_3(s) \quad (2.29)$$

$$\Delta H^0_r = 2 \times \Delta H^0_{f(\text{NaHCO}_3)} - \Delta H^0_{f(\text{Na}_2\text{CO}_3)} - \Delta H^0_{f(\text{H}_2)}$$
$$= 2 \times (-951) - (-1131) - 0 = -771\,\text{kJ}$$

2.2.4 Hess's Law of Constant Heat Summation

It states that by combining (adding or subtracting) the enthalpy changes of two or more suitable reactions algebraically, it is possible to determine the enthalpy change of a desired reaction. Hess's law is very useful to determine the enthalpy changes of reactions not determinable by experimentation. Many examples are found in standard books.

2.2.5 Bond Enthalpies

The bond enthalpy of a bond is the enthalpy change associated with the breakage of a bond in the gas phase. For example, the breaking of a C–H bond in a gas molecule of C_2H_6 is given by

$$C_2H_6(g) \rightarrow C_2H_5(g) + H(g) \quad \Delta H^0 = +410\,\text{kJ} \quad (2.30)$$

Actually, the bond dissociation enthalpy is an average value of the values obtained on a number of similar compounds. The bond enthalpies of various bonds such as C–H, N–H, C=O, C=C, can be used to estimate the enthalpy changes of various reactions occurring in the gas phase. For example,

$$C_2H_4(g) + 3O_2(g) \rightarrow$$
$$2CO_2(g) + 2H_2O(g) \quad \Delta H^0 = X \quad (2.31)$$

ΔH^0 of the reaction may be computed as follows.

- Reactants side: breakage of 4 C–H bonds + breakage of 1 C=C bond + breakage of 3 O=O bonds requires $4 \times 413 + 1 \times 612 + 3 \times 497 = 3755$ kJ
- Products side: Formation of 4 C=O bonds + formation of 4 O–H bonds results in

$$-4 \times 738 + (-4 \times 463) = -4804\,\text{kJ}$$
$$\Delta H^0_{reac} = 3755 + (-4804) = -1049\,\text{kJ}$$

2.2.6 Variation of Heat of Reaction with Temperature

ΔH^0 of the reaction varies with temperature and it may be written as,

$$\left(\frac{\partial \Delta H^0}{\partial T}\right)_p = \left(\frac{\partial H^0_{\text{products}}}{\partial T}\right)_p - \left(\frac{\partial H^0_{\text{reactants}}}{\partial T}\right)_p \quad (2.32)$$

Table 2.1 Equations for Q, W, ΔE and ΔH for isothermal and adiabatic processes from the initial state (P_1, T_1, V_1) to the final state (P_2, V_2, T_2)

Free expansion under	Q	W	ΔE	ΔH
Isothermal irreversible	W	$P_{ext} \times [V_2 - V_1]$	0	0
Isothermal reversible	W	$nRT \ln \frac{V_2}{V_1}$	0	0
Adiabatic reversible	0	$-\Delta E$	$nC_V(T_2 - T_1)$	$nC_P(T_2 - T_1)$
Adiabatic irreversible	0	$nC_V(T_2 - T_1)$	$nC_V(T_2 - T_1)$	$nC_P(T_2 - T_1)$

$$= C^0_{p(\text{products})} - C^0_{P(\text{reactants})} = \Delta C^0_P \text{ or } \left(\frac{\partial \Delta H^0}{\partial T}\right)_p = \Delta C^0_P \quad (2.33)$$

Equation 2.33 is known as Kirchhoff's equation. Rearranging and integrating the equation between two temperature limits, we get

$$\left(\Delta H^0\right)_{T_2} = \left(\Delta H^0\right)_{T_1} + \Delta C^0_P (T_2 - T_1) \quad (2.34)$$

assuming that ΔC^0_P is independent of temperature in the given temperature range. Sometimes, it is necessary to consider the variation of heat capacity with temperature as well. Table 2.1 provides a summary of various thermodynamic quantities for isothermal and adiabatic processes.

Sign convention: There are two methods of writing ΔE, either as in Eq. 2.1, which is the IUPAC convention, or as $\Delta E = Q - W$ which is the sign convention of Clausius. This difference arises since, in the Clausius convention, work done by the system is considered to be positive. The entries in Table 2.1 employ the Clausius convention. It is essential to be aware of both conventions, although final interpretations and *magnitudes* will not differ for Q or W.

2.3 Second Law of Thermodynamics

This law aims at providing a criterion for the spontaneity of a process and it does so by introducing the concept of entropy by means of the *Carnot cycle*. Before stating the law, the importance of spontaneous processes will be mentioned.

Spontaneous processes: These are processes that occur of their own accord and finally reach a state of equilibrium. They are important because one can derive useful work by operating such processes. It is worthwhile, therefore, to consider whether any criterion can be used for the spontaneity of a process. No such information is obtained from the first law.

Statement of second law: The law may be stated in two ways.

(i) All spontaneous processes are thermodynamically irreversible.

(ii) It is impossible to construct a machine operating in a cycle and connected to a single heat reservoir to generate work in the surroundings

Concept of entropy (S): Although the determination of absolute entropy is possible for any substance through the third law (see later), an infinitesimal increase, dS, in a process is defined as

$$dS = \frac{dq_{rev}}{T} \quad (2.35)$$

where dq_{rev} is the quantity of heat absorbed in an infinitesimal stage of an appreciable process at the temperature T in kelvin. S is a state function and an extensive property. Note that dq_{rev}/T is an exact differential like E or H.

For a finite process involving a change of state from state 1 and state 2

$$\Delta S = \int_1^2 \frac{dq_{rev}}{T} = \frac{Q_{rev}}{T} = (S_2 - S_1) \quad (2.36)$$

Furthermore, the entropy change for the system and surroundings for reversible and irreversible processes are given by the following equations:

$$\Delta S_{sys} + \Delta S_{surr} = 0 \text{ (rev)}$$

$$\Delta S_{sys} + \Delta S_{surr} > 0 \text{ (irreversible)}$$

2.3.1 Carnot's Cycle

This is a reversible cycle and involves the use of one mole of ideal gas as working substance. There are four steps in the cycle:

(i) It is first subjected to an isothermal reversible expansion from an initial volume V_1 to a final volume V_2 by keeping the gas at a higher temperature reservoir T_1 (see later).

(ii) The gas is next subjected to an adiabatic reversible expansion from the volume V_2 to V_3 until the temperature falls to a value T_2, that is, the temperature of the sink.

(iii) In the next step, the gas is isothermally compressed at a temperature T_2 until its volume changes from V_3 to V_4.

(iv) In the final step, the gas is subjected to adiabatic reversible compression until its volume changes from V_4 to V_1 and the temperature rises to T_1 from T_2. By this process, the gas returns to its initial state and completes a cycle.

The Carnot engine operates in a cycle between two heat reservoirs one at a higher temperature T_1 and another at a lower temperature T_2. In one cycle, it absorbs a certain quantity of heat

Q_1 from the source at temperature T_1, converts part of it into work W and returns the rest to the lower temperature reservoir at the temperature T_2. The efficiency η, of the heat engine, defined by the ratio of work W, performed to the quantity of heat Q_1, taken from the higher temperature reservoir is given by

$$\eta = \frac{W}{Q_1} = \frac{T_1 - T_2}{T_1} \tag{2.37}$$

where T_1 = temperature of the higher temperature reservoir and T_2 = temperature of the lower temperature reservoir. More details can be obtained from standard books. There are two points of interest in Eq. 2.37.

(i) When $T_2 = T_1$, $\eta = 0$, that is, no work is produced from an isothermal process.
(ii) When $T_2 = 0$ K (absolute zero), $\eta = 1$ which is not observed. Therefore it is not possible to observe or attain absolute zero.

2.3.2 Entropy Changes in Various Processes

Phase equilibria: For a fusion process, such as the melting of ice to liquid water at the melting point, ΔS_{fusion} is given by,

$$\Delta S_{\text{fusion}} = S_{\text{liquid}} - S_{\text{solid}} = \frac{\Delta H_f}{T_f} \tag{2.38}$$

where ΔH_f = enthalpy of fusion of solid (ice) and T_f = melting point the solid (ice) in K. ΔS_{vap} for the vaporisation of a liquid (say water) into vapour at the boiling point T_b,

$$\Delta S_V = \frac{\Delta H_V}{T_b} \tag{2.39}$$

where ΔH_V = enthalpy of vaporisation of liquid and T_b = boiling point the liquid in K. Note that ΔH_f and ΔH_V are positive.

Therefore, $$S_V > S_L > S_s \tag{2.40}$$

Entropy may also be interpreted as a measure of disorder of a system on the basis of Eq. 2.40.

Entropy changes involving an ideal gas

(i) Isothermal reversible expansion of an ideal gas: For a finite change involving the expansion of n moles of an ideal gas from an initial volume V_1 to a final volume V_2 at temperature T,

$$\Delta S = nR \ln \frac{V_2}{V_1} = nR \ln \frac{P_1}{P_2} \tag{2.41}$$

where P_1 = initial pressure and P_2 = final pressure. From Eq. 2.41, it is noted that ΔS is positive (since $P_1 > P_2$ and $V_2 > V_1$).

(ii) Reversible process involving change of volume and temperature of *n* moles of an ideal gas: In this case,

$$\Delta S = n\left(C_V \ln \frac{T_2}{T_1} + R \ln \frac{V_2}{V_1} \right) \tag{2.41}$$

where T_1 and T_2 = initial and final temperatures, V_1 and V_2 = initial and final volumes and C_V = molar heat capacity of gas at constant volume. At constant volume, Eq. 2.41 simplifies to

$$\Delta S_V = nC_V \ln \frac{T_2}{T_1} \tag{2.42}$$

Equation 2.41 may also be expressed in terms of change in pressure and temperature as

$$\Delta S = n\left(C_P \ln \frac{T_2}{T_1} + R \ln \frac{P_1}{P_2} \right) \tag{2.43}$$

which simplifies at constant pressure to

$$\Delta S_P = nC_P \ln \frac{T_2}{T_1} \tag{2.44}$$

Entropy of mixing of a number of ideal gases: Consider n_A and n_B moles of two ideal gases enclosed in a rectangular box separated by a partition with the temperature and pressure of each being T and P. After removing the partition, the total number of moles is $n_A + n_B$ and the temperature and pressure are the same as before. The entropy changes associated with the mixing process, ΔS_{mix} is given by,

$$\Delta S_{mix} = -NR\left[x_A \ln x_A + x_B \ln x_B \right] \tag{2.45}$$

which may be also written as

$$\Delta S_{mix} = -NR\left[x_A \ln x_A + (1 - x_A)\ln(1 - x_A) \right] \tag{2.46}$$

In Eqs 2.45 and 2.46, N = total number of moles and x are the mole fractions of the components. From these equations, it is seen that ΔS_{mix} is positive because the xs are fractions. The positive ΔS_{mix} is explained as arising from the orderly arrangement of gases before mixing to the disorderly state of mixed gases after mixing.

Trouton's rule: For many liquids, especially the hydrocarbons, the entropy of vapourisation, ΔS_{vap} at the boiling point, T_b, is given by

$$\Delta S_{vap} = \frac{\Delta H_{vap}}{T_b} = 88 \text{ J K}^{-1} \text{ mol}^{-1} \tag{2.47}$$

This equation is useful to evaluate approximate values of ΔH_{vap} at the boiling point of a liquid and also the ebullioscopic boiling point constant of the liquid.

2.3.3 Entropy Change in the Free Expansion of an Ideal Gas

From Eq. 2.41, ΔS_{gas} (ΔS_{sys}) in the isothermal reversible expansion of an ideal gas is positive since $V_2 > V_1$. The entropy change of surroundings (heat reservoir), ΔS_{surr}, is equal to

$$\Delta S_{Surr} = -\frac{Q_{rev}}{T} = -nR \ln \frac{V_2}{V_1} \tag{2.48}$$

$$\Delta S_{Sys} + \Delta S_{Surr} = 0 \quad \text{(for a reversible process)} \tag{2.49}$$

Consider the same expansion of the gas into vacuum which is irreversible in a thermodynamic sense. In this case, ΔS_{sys} remains unchanged, and $\Delta S_{surr} = 0$, because no heat is taken by the system from the reservoir.

∴ The sum of $\Delta S_{sys} + \Delta S_{surr}$ is positive,

$$\Delta S_{sys} + \Delta S_{surr} > 0 \tag{2.50}$$

Combining Eqs 2.49 and 2.50, we may write

$$\Delta S_{sys} + \Delta S_{surr} \geq 0 \tag{2.51}$$

Generalising, for an infinitesimal change, we write

$$dS_{sys} + dS_{surr} \geq 0 \tag{2.52}$$

where the equality sign applies for a reversible process and the inequality sign applies for an irreversible process.

2.3.4 Helmholtz Free Energy (*A or F*) and Gibbs Free Energy (*G*) and their Variation with *T, P* and *V*

The Helmholtz free energy, A, is defined by

$$A = E - TS \tag{2.53}$$

For a finite process carried out reversibly under isothermal conditions, the change in A, ΔA_T, is given by

$$\Delta A_T = W_{rev} = W_{max} \tag{2.54}$$

That is, the work done on the system, W_{rev} is also the maximum work done under these conditions. (The symbol F is also used for Helmlholtz free energy.)

The Gibbs free energy, G, is defined by

$$G = H - TS = E + PV - TS \tag{2.55}$$

For a finite process carried at constant temperature T and pressure P under reversible conditions,

$$-\Delta G_{T,P} = W_{rev} - P\Delta V \tag{2.56}$$

The physical significance of Eq. 2.56 is that the decrease in free energy under constant temperature and pressure conditions is equal to all forms of work excluding work of expansion. Such work may include electrical work, surface work, and so on. From the relations,

$$\left(\frac{\partial A}{\partial V}\right)_T = -P \tag{2.57a}$$

$$\left(\frac{\partial A}{\partial T}\right)_V = -S \tag{2.57b}$$

$$\left(\frac{\partial G}{\partial P}\right)_T = V \tag{2.57c}$$

$$\left(\frac{\partial G}{\partial T}\right)_P = -S \tag{2.57d}$$

the relations,

$$\Delta A_T = nRT \ln \frac{V_1}{V_2} = nRT \ln \frac{P_2}{P_1} \tag{2.58}$$

and

$$\Delta G_T = nRT \ln \frac{P_2}{P_1} \tag{2.59}$$

for the isothermal reversible expansion of n moles of an ideal from an initial state V_1 to a final state V_2 are obtained. It is seen that for this process, $\Delta A_T = \Delta G_T$ and that both quantities are negative. Helmholtz free energy may also be denoted as F.

2.3.5 ΔG as a Criterion of Spontaneity of a Process

For a small change in G,

$$dG = dE + PdV + VdP - TdS - SdT \tag{2.60}$$

If $dE + PdV = TdS$ (combination of 1st and 2nd law) as in a reversible process,

$$dG = VdP - SdT \tag{2.61a}$$

But if $dE + PdV < TdS$ as in an irreversible process,

$$dG < VdP - SdT \tag{2.61b}$$

At constant T and P, Eq. 2.62 becomes $dG_{T,P} \leq 0$ which for a finite process may be written as,

$$\Delta G_{T,P} \leq 0 \tag{2.62}$$

A similar equation may be obtained for a process at equilibrium or for its spontaneity in terms of Helmholtz free energy as $dA_{V,T} \leq 0$ or $\Delta A_{V,T} \leq 0$. From Eq. 2.55, at constant T,

$$\Delta G = \Delta H - T\Delta S \tag{2.63}$$

The sign of ΔG depends on the terms ΔH and ΔS. If ΔH is negative (as in an exothermic reaction) and ΔS is positive, ΔG is negative and the reaction is feasible. If ΔH is positive and ΔS is largely positive such that $T\Delta S$ overwhelms ΔH, the reaction is still feasible. Under conditions when ΔH is positive and ΔS is negative, ΔG is positive and the reaction is not feasible. Equation 2.63 may also be written in the form

$$\Delta G = \Delta H + T\left(\frac{\partial(\Delta G)}{\partial T}\right)_P \tag{2.64}$$

Equations 2.63 and 2.64 are known as Gibbs–Helmholtz equations.

2.4 Third Law of Thermodynamics

The third law of thermodynamics deals with absolute entropies of pure elements and compounds. It states that *the entropy of any element or compound in a perfectly crystalline form is zero at absolute zero of temperature*. Its determination is based on the relation

$$\left(\frac{\partial S}{\partial T}\right)_P = \frac{C_P}{T} \tag{2.65}$$

Rearranging and integrating we get

$$S_T - S_0 = \int_0^T \frac{C_P}{T} dT$$

or, $$S_T = \int_0^T \frac{C_P}{T} dT \quad \text{(since } S_0 = 0 \text{ by the third law)} \tag{2.66}$$

In determining the third law entropy, entropy change due to allotropic modification (example rhombic to monoclinic sulfur), melting of solid to liquid at melting point, liquid to vapour at boiling point must be taken into account. The general equation in such cases may be written as

$$S_T = \int_0^{T'} \frac{C_P}{T} dT + \frac{\Delta H'}{T'} + \int_{T'}^{T''} \frac{C_P'}{T} dT \qquad (2.67)$$

where $\Delta H_T'$ = enthalpy change for transition from one solid to another solid form, T' = temperature of transition, T'' = melting point of new solid form, and so on. Depending upon the temperature up to which entropy is to be evaluated, further terms to Eq. 2.67 are added. There are exceptions to the third law, such as the gas N_2O, which is due to some disorder in the arrangement of the N_2O molecule. In this case, entropy does not approach zero as the temperature tends to absolute zero. This non-zero value of entropy at absolute zero is often called *residual entropy*. In the case of CO, the residual entropy per mole is R ln(2).

2.4.1 Evaluation of Entropy Changes in Chemical Reactions

Consider a general chemical reaction,

$$a\text{A} + b\text{B} + \ldots\ldots \rightleftharpoons x\text{X} + y\text{Y} + \ldots\ldots \qquad (2.68)$$

where a, b, x, y are the stoichiometric coefficients and A, B, X, Y represent the reactants and products. The entropy change associated with the reaction, ΔS^0, is given by

$$\Delta S^0 = \left[xS_X^0 + yS_Y^0 + \ldots\right] - \left[aS_A^0 + bS_B^0 + \ldots\right] \qquad (2.69)$$

Note that S^0 represents the entropies of the given species in the standard state (1 atm, 298 K). The entropy changes at any other temperature are obtained from the relation,

$$\left(\frac{\partial \Delta S^0}{\partial T}\right)_P = \left(\frac{\partial S^0_{\text{products}}}{\partial T}\right)_P - \left(\frac{\partial S^0_{\text{reactants}}}{\partial T}\right)_P$$

$$= \sum \frac{C^0_{P(\text{products})}}{T} - \sum \frac{C^0_{P(\text{reactants})}}{T} \qquad (2.70)$$

2.4.2 Partial Molar Properties

(i) Consider an open system in which the composition of the system may be altered by adding or removing amounts (as happens in a chemical reaction) of a substance. A change in an extensive parameter such as, E, H, S, A, G involving an open system must include another variable, namely, the composition of the system in addition to variables like T, P, V, n_1, n_2, ... The partial molar property $\overline{X_1}$ due to component 1 is defined by

$$\overline{X_1} = \left(\frac{\partial X}{\partial n_1}\right)_{T,P,n_2,\ldots} \qquad (2.71)$$

It is the increase in the property X when one mole of the component 1 is added to an infinitely large amount of solution that its composition remains constant. Using G, the Gibbs free energy as an example, we may write

$$dG = \left(\frac{\partial G}{\partial T}\right)_{P,n_1,n_2,\ldots} dT + \left(\frac{\partial G}{\partial P}\right)_{T,n_1,n_2,\ldots} dP$$

$$+ \left(\frac{\partial G}{\partial n_1}\right)_{T,P,n_2,\ldots} dn_1 + \left(\frac{\partial G}{\partial n_2}\right)_{T,P,n_1,\ldots} dn_2 \qquad (2.72)$$

$$= -SdT + VdP + \mu_1 dn_1 + \mu_2 dn_2 + \ldots\ldots$$

$$= -SdT + VdP + \sum_i \mu_i dn_i \qquad (2.73)$$

Note that the above equation takes into account the change in the number of moles of the constituents. At constant T and P, Eq. 2.73 simplifies to

$$dG_{T,P} = \sum_i \mu_i dn_i \qquad (2.74)$$

The quantity $\left(\frac{\partial G}{\partial n_i}\right)_{T,P,n_j,\ldots} = \mu_i$ is known as chemical potential (or the partial molar Gibbs free energy) of the component i and may be defined as above. It is an intensive property.

(ii) *Variation of μ_i with temperature and pressure:*

$$\left(\frac{\partial \mu_i}{\partial T}\right)_{P,n_j,\ldots} = \frac{\partial}{\partial T}\left(\frac{\partial G}{\partial n_i}\right)_{T,P,n_j,\ldots} = \frac{\partial}{\partial n_i}\left(\frac{\partial G}{\partial T}\right)_{P,n_j,\ldots}$$

$$= -\left(\frac{\partial S}{\partial n_i}\right)_{P,n_j,\ldots} = -\overline{S_i} \qquad (2.75)$$

where $\overline{S_i}$ is the partial molar entropy of species i. A similar relation for variation of μ_i with pressure is given by

$$\left(\frac{\partial \mu_i}{\partial P}\right)_{T,n_j,\ldots} = \left(\frac{\partial V}{\partial n_i}\right)_{P,n_j,\ldots} = \overline{V_i} \qquad (2.76)$$

where $\overline{V_i}$, is the partial molar volume of species i.

(iii) *Gibbs–Duhem equation:* For a system of two components 1 and 2 at constant T and P, Eq. 2.74 may be written as

$$dG = \mu_1 dn_1 + \mu_2 dn_2 \qquad (2.77)$$

On integration, Eq. 2.77 gives

$$G = \mu_1 \int_0^{n_1} dn_1 + \mu_2 \int_0^{n_2} dn_2 \qquad (2.77a)$$

(note that in the initial state, $n_1 = 0$ and $n_2 = 0$ and hence $G = 0$). In the above integration, the composition is kept constant throughout and thus μ_1 and μ_2 are taken out of the integral. Equation 2.77a becomes

$$G = \mu_1 n_1 + \mu_2 n_2$$

which is written in a general manner as

$$G = \sum_i \mu_i n_i \qquad (2.78)$$

A useful relation in dealing with a binary mixture of two components in solution is obtained by differentiating Eq. 2.78 as

$$dG = n_1 d\mu_1 + \mu_1 dn_1 + n_2 d\mu_2 + \mu_2 dn_2 \qquad (2.79)$$

Comparing Eq. 2.79 with Eq. 2.74 and simplifying, we get,

$$-SdT + VdP = \sum_i n_i d\mu_i \qquad (2.80)$$

Table 2.2 Entropy changes of a system under various transformations

Process	Entropy change (ΔS)
Heat transfer of the same material from T_h to T_l	$C_P(T-T_l)=C_P(T_h-T)$ Using energy balance, $T=\frac{(T_h+T_l)}{2}$ $\Delta S_{sys}=\int_{T_h}^{T} C_P\frac{dT}{T}+\int_{T_l}^{T} C_P\frac{dT}{T}$
Chemical reactions $aA+bB \rightarrow cC+dD$	$\Delta S^0_{298}=c\left[S^0_{298}\right]_C+d\left[S^0_{298}\right]_D-\left\{aS^0_{298,\,A}+bS^0_{298,\,B}\right\}$ $\Delta S^0_T=\Delta S^0_{298}+\int_{298^*}^{T}\Delta C_P\frac{dT}{T}\overrightarrow{AB}$ $\Delta C_P=c\left[C_P\right]_C+d\left[C_P\right]_D-a\left[C_P\right]_A-b\left[C_P\right]_B$
Irreversible adiabatic changes $P_1V_1T_1$ to $P_2V_2T_2$	$\Delta S=nC_P\ln\frac{V_2}{V_1}+nC_V\ln\frac{P_2}{P_1}$
N moles of solid heated from T_i to T_b	$\int_{T_i}^{T_f}\frac{nC_P(\text{solid})dT}{T}+\frac{nL_{\text{fusion}}}{T_f}+\int_{T_f}^{T_b}\frac{nC_P(\text{liquid})dT}{T}+\frac{nL_{vap}}{T_b}$
Mixing of two ideal gases	$\Delta S_{mix}=-R\sum x_i\ln x_i$ for one mole
Vaporisation	$\Delta S_{vap}=\frac{\Delta H_{Vap}}{T_b}\approx 85\text{ J K}^{-1}$ (Trouton rule)
Simple heating at constant pressure	$\Delta S=\int_{T_1}^{T_2} nC_P\frac{dT}{T}$
Reversible adiabatic process	$\Delta S_{sys}=0$
Reversible isothermal process from V_1 to V_2	$\Delta S_{sys}=nR\ln\frac{V_2}{V_1}$
Irreversible isothermal expansion from V_1 to V_2	$\Delta S_{sys}=nR\ln\frac{V_2}{V_1}$
Change in pressure, at constant temperature	$\Delta S_{sys}=nR\ln\frac{P_{\text{initial}}}{P_{\text{final}}}$

which is known as Gibbs–Duhem equation. From Eq. 2.80 at constant T and P, for a two component system, it is possible to write,

$$d\mu_2 = -\left(\frac{n_1}{n_2}\right)d\mu_1 \quad (2.81)$$

which shows the variation of chemical potential with composition.

(iv) *Duhem–Margules equation:* This equation may be considered as a special case of application of the Gibbs–Duhem equation to a two-component liquid mixture consisting of components A and B. For such a system, the Gibbs–Duhem equation may be written as,

$$x_A d\mu_A + x_B d\mu_B = 0 \quad (2.82)$$

where x_A and x_B are the mole fractions of A and B. Equation 2.82 may be simplified to,

$$\left(\frac{\partial\mu_A}{\partial \ln x_A}\right)_{T,P} dx_A + \left(\frac{\partial\mu_B}{\partial \ln x_B}\right)_{T,P} dx_B = 0 \quad (2.83)$$

Since $dx_A = -dx_B$ (since $x_A + x_B = 1$ and $dx_A + dx_B = 0$)

Equation 2.83 can be expressed as

$$\left[dx_A\left(\left(\frac{\partial\mu_A}{\partial \ln x_A}\right)_{T,P} - \left(\frac{\partial\mu_B}{\partial \ln x_B}\right)_{T,P}\right)\right] = 0$$

$$\text{or}\quad \left(\frac{\partial\mu_A}{\partial \ln x_A}\right)_{T,P} = \left(\frac{\partial\mu_B}{\partial \ln x_B}\right)_{T,P} \quad (\text{since } dx_A \neq 0) \quad (2.84)$$

But $\partial\mu_A \propto \partial \ln P_A$ and $\partial\mu_B \propto \partial \ln P_B$ which follow from the relation $\mu_i = \mu_i^0 + RT\ln p_i$

$$\left(\frac{\partial \ln p_A}{\partial \ln x_A}\right)_{T,P} = \left(\frac{\partial \ln p_B}{\partial \ln x_B}\right)_{T,P} \quad (2.85)$$

where p_A and p_B are the partial pressures of A and B. Equation 2.85 is known as Duhem–Margules equation.

(v) *Relation between chemical potential and pressure in an ideal gas or a mixture of ideal gases:* For an ideal gas, the relation

$$G = G^0_{(T)} + RT\ln\frac{P}{P_0} = G^0_{(T)} + RT\ln\frac{P}{1\text{ atm}}$$

$$= G^0_{(T)} + RT\ln P \quad (2.86)$$

is valid. $G^0_{(T)}$ is the free energy per mole of gas at 1 atm and the given temperature. Thus the free energy of an ideal gas is proportional to the pressure. In a mixture of ideal gases, the chemical potential of the component A is related to the partial pressure p_A according to

$$\mu = \mu^0_{A(T)} + RT\ln p_A \quad (2.87)$$

$\mu^0_{A(T)}$ is the chemical potential of pure A at 1 atm and a given temperature T. Using Dalton's law, $p_A = x_A P$ Equation 2.87 may be written as,

$$\mu = \mu^0_{A(T)} + RT\ln P + RT\ln x_A \quad (2.88)$$

$$\mu = \mu^{0'}_{A(T)} + RT\ln x_A$$

$$\left(\text{where } \mu^{0'}_{A(T)} = \mu^0_{A(T)} + RT\ln P\right) \quad (2.89)$$

x_A is the mole fraction of the gas and P is the total pressure. Note that when $x_A = 1$, that is, pure A, $\mu_A = \mu_A^{0'}$. An equation similar to entropy of mixing, ΔS_{mix} may be obtained for the free energy of mixing, ΔG_{mix}, on mixing a number of ideal gases A, B, C at given temperature T and pressure P as,

$$\Delta G_{mix} = NRT\left(x_A\ln x_A + x_B\ln x_B + x_C\ln x_C\right) \quad (2.90)$$

where N = total number of moles ($n_A + n_B + n_C$) and the xs are mole fractions.

(vi) *Thermodynamics of chemical equilibria in ideal gases:* Consider the general chemical reaction

$$l\text{L} + m\text{M} + \ldots\ldots \rightleftharpoons x\text{X} + y\text{Y} + \ldots \quad (2.91)$$

occurring in a mixture of ideal gases at a given temperature T. In Eq. 2.91, l, m, x, y, represent stoichiometric coefficients and L, M, X, Y represent the reactants and products. The change in Gibbs free energy of the reaction at any stage is given by

$$\Delta G = \sum G_{\text{products}} - \sum G_{\text{reactants}}$$

$$= \left(x\mu^0_{\text{X}} + y\mu^0_{\text{Y}} - l\mu^0_{\text{L}} - m\mu^0_{\text{M}}\right) + RT\ln\frac{p^x_{\text{X}}p^y_{\text{Y}}}{p^l_{\text{L}}p^m_{\text{M}}} \quad (2.92)$$

$$= \Delta G^0 + RT\ln Q_P \quad (2.93)$$

When equilibrium is attained in the system, $\Delta G = 0$, or,

$$\Delta G^0 = -RT\ln\left(Q_P\right)_{eqbm} = -RT\ln K_P \quad (2.93a)$$

where $K_P = \left(\dfrac{p^x_X \times p^y_Y}{p^l_L \times p^m_M}\right)_{eqbm}$ and it depends only on temperature. Equation 2.93 may be written as,

$$\Delta G = RT\ln\frac{Q_P}{K_P} \quad (2.94)$$

When $Q_P < K_P$, ΔG is negative and the reaction proceeds spontaneously from left to right. When $Q_P = K_P$, $\Delta G = 0$ and the reaction is at equilibrium. When $Q_P > K_P$, ΔG is positive and reaction 2.91 is non-spontaneous. If the mixture of gases is a real gas, the expression for the equilibrium constant contains a quantity fugacity instead of partial pressure and K_P is replaced by K_f given by

$$K_f = \frac{f^x_{\text{X}} \times f^y_{\text{Y}}}{f^l_{\text{L}} \times f^m_{\text{M}}} \quad (2.95)$$

where the fugacities take account of deviation of gases from ideal behaviour.

(vii) *Equilibrium constants K_C and K_F:* The equilibrium constant K_P may be expressed in terms of concentrations of various species (K_C) or mole fractions (K_F) as,

$$K_P = K_C (RT)^{\Delta n} \quad (2.96)$$

$$K_P = K_F (P)^{\Delta n} \quad (2.97)$$

where $\Delta n = (x + y + \ldots) - (l + m + \ldots)$ of Eq.2.91.

Using suitable data such as the degree of dissociation, it is possible to relate K in terms of α (degree of dissociation).

(viii) *Variation of equilibrium constants K_P and K_C with temperature:* Variation of K_P with temperature is given by

$$\left(\frac{\partial \ln K_P}{\partial T}\right) = \frac{\Delta H^0}{RT^2} \quad (2.98)$$

where ΔH^0 = standard enthalpy change of the reaction. If ΔH^0 is positive (reaction is endothermic), K_P increases with temperature. If ΔH^0 is negative (reaction is exothermic), K_P decreases with temperature. The variation of K_C with temperature is given by

$$\frac{d \ln K_C}{dT} = \frac{\Delta E^0}{RT^2} \quad (2.99)$$

where ΔE^0 = standard internal energy change of the reaction.

(ix) *Standard Gibbs free energies of formation and Gibbs energy changes of chemical reactions:* By assigning that free energies of all elements in their standard states [stable state at 1 atm ($\approx$ 1 bar)] and at all temperatures are zero, the standard Gibbs energies of formation of various compounds $\left(\Delta G_f^0\right)$ may be arrived at. It is common, however to list such values of ΔG_f^0 at 1 atm and 298 K. Using ΔG_f^0 data, the ΔG^0 of the desired reaction may be calculated.

(x) *Fugacity:* The term fugacity is applied in connection with real gases. It may be referred to as idealised pressure. It is related to the chemical potential of the gas according to

$$\mu = \mu_T^0 + RT \ln f \quad (2.100)$$

Using the relation, $d\mu_{real} = \overline{V_{real}}\, dP$ and $d\mu_{id} = \overline{V_{id}} dP$, we write

$$d\left(\mu_{real} - \mu_{id}\right) = \left(\overline{V_{real}} - \overline{V_{id}}\right) dP \quad (2.101)$$

Integrating between the limits of very low pressure P^l and P, we get

$$\left(\mu_{real} - \mu_{id}\right) - \left(\mu^l - \mu_{id}^l\right) = \int_{P^l}^{P} \left(\overline{V_{real}} - \overline{V_{id}}\right) dP \quad (2.102)$$

As $P^l \to 0$, properties of real gases approach that of ideal gases or $\mu^l \to \mu_{id}^l$. Thus,

$$\mu_{real} - \mu_{id} = \int_{P^l}^{P} \left(\overline{V_{real}} - \overline{V_{id}}\right) dP \quad (2.103)$$

Expressing μ_{real} and μ_{id} in terms of fugacity and pressure respectively and simplifying, we get,

$$\ln f = \ln P + \frac{1}{RT}\int_0^P \left(\overline{V_{real}} - \overline{V_{id}}\right) dP \quad (2.104)$$

We can also evaluate f by using the relation, $Z = \dfrac{\overline{V_{real}}}{\overline{V_{id}}}$ and $\overline{V_{id}} = \dfrac{RT}{P}$ and rewrite Eq. 2.104 as,

$$\ln f = \ln P + \int_0^P \frac{Z-1}{P} dP \quad (2.105)$$

(where Z is the compressiblity factor of the gas)

f can be obtained by plotting $\dfrac{Z-1}{P}$ as a function of P and integrating the curve so obtained from $P = 0$ to P.

(xi) *Activity and activity coefficient:* The activity concept is generally applied in the case of liquid systems especially when they deviate from ideal behaviour, that is, non-compliance of the mixture to Raoult's law or in cases where the solute is non-volatile and in dilute solution and such dilute solutions deviate from Henry's law. The activity may be defined in a similar way as fugacity and thus, one may write in the case of a liquid mixture containing a component b as,

$$\mu_b = \mu_b^0 + RT \ln a_b \quad (2.106)$$

In the description of rational activity we write that as $x_b \to 1$ (as the system tends to pure liquid b), $\mu_b \to \mu_b^0$ and ln $a_b \to 0$ and $a_b \to 1$ as $x_b \to 1$, that is, $a_b = 1$ for pure liquid b. Comparing Eq. 2.106 with that of an ideal liquid mixture,

$$\mu_{b(id)} = \mu_b^0 + RT \ln x_b \quad (2.107)$$

$$\mu_b - \mu_{b(id)} = RT \ln \frac{a_b}{x_b} \quad (2.108)$$

$\dfrac{a_b}{x_b} = \gamma_b$ rational activity coefficient (note x_b = mole fraction of b) and $\gamma_b \to 1$ as $x_b \to 1$.

Activities are measured from vapour pressure measurements and

$$a_b = \frac{p_b}{p_b^0} \quad (2.109)$$

p_b = partial vapour pressure of b, P_b^0 = vapour pressure of pure liquid b. If the liquid mixture is ideal, $a_b = x_b$ and $\gamma_b = 1$, for all compositions of the liquid mixture. If the component b of the liquid mixture shows a positive deviation from Raoult's law $\gamma_b > 1$. If it shows negative deviations from Raoult's law, $\gamma_b < 1$.

(xii) *Practical activity and activity coefficients:* These forms of activity and activity coefficients are used in the case of volatile and non-volatile solutes in dilute solutions. For a solvent that is present in a larger amount, however, the rational activity and rational activity coefficients are used. If y represents the solute in an ideal solution,

$$\mu_{y(id)} = \mu_y^{++} + RT \ln m_y \text{ (where } m_y = \text{molality)} \quad (2.110)$$

For the solute in an actual solution,

$$\mu_y = \mu_y^{++} + RT \ln a_y \text{ (where } a_y = \text{activity of } y) \quad (2.111)$$

$$\mu_y - \mu_{y(id)} = RT \ln \frac{a_y}{m_y}$$

where $\frac{a_y}{m_y} = \gamma_y$ and γ_y is the practical activity coefficient. When the solute concentration is very low $m_y \to 0, \gamma_y \to 1$ and $a_y \to m_y$. Note that μ_y^{++} is the chemical potential of a hypothetical ideal solution of unit molality.

In the case of a volatile solute, it is possible to relate its vapour pressure p_y to its activity through Henry's law constant K_y as

$$p_y = K_y a_y \quad (2.112)$$

This equation may be considered as an idealised form of Henry's law.

$$\frac{p_y}{m_y} = K_y \frac{a_y}{m_y} \quad (2.113)$$

By measuring $\frac{p_y}{m_y}$ as function of m_y, we can evaluate K_y by extrapolating the curve to $m_y = 0$. The extrapolated value is equal to K_y. Using the K_y, a_y may be evaluated using the measured p_y from Eq. 2.112.

Worked Out Problems

Calculation of Various Thermodynamic Quantities for Different Processes

Problem 2.1

Two moles of an ideal gas expand isothermally and reversibly at 27°C from an initial volume of 25 dm^3 to a final volume of 250 dm^3. Calculate (i) W (ii) ΔE (iii) ΔH (iv) Q (v) ΔS (vi) ΔG and (vii) ΔA for this process.

Solution

Since the process is isothermal, (ii) and (iii), that is, $\Delta E = 0$ and $\Delta H = 0$.

(i) $\Delta E = 0 = Q + W$ or $Q = -W = -nRT \ln \frac{V_2}{V_1}$

$$W = -Q = -nRT \ln \frac{V_2}{V_1}$$

$$= -2 \times 8.314 \times 300 \times 2.303 \times \log \frac{250}{25} = -11.49 \text{ kJ}$$

(iv) $Q = -W = 11.49$ kJ

(v) $\Delta S = \frac{Q_{rev}}{T} = \frac{11.49 \text{ kJ}}{300} = 38.3 \text{ J K}^{-1}$

(vi) $\Delta G = nRT \ln \frac{V_1}{V_2} = 2 \times 8.314 \times 300 \times 2.303 \times \log \frac{25}{250}$

$$= -11.49 \text{ kJ}$$

(vii) Since $\Delta G = \Delta A$ at constant T, $\Delta A = -11.49$ kJ

Problem 2.2

One mole of an ideal gas is compressed isothermally at 310 K against an external pressure of 20 bar from an initial volume of 100 dm^3 to a final volume of 50 dm^3. Is this process reversible or irreversible? Assume that 1 bar = 1 atm. Calculate (i) W (ii) Q (iii) ΔS (iv) ΔG (v) ΔA.

Solution

This is not a reversible process because the gas is compressed against a constant external pressure.

(i) $W = -P_{ext} \times \Delta V = -20(50 - 100) = 20 \times 50 \text{ L atm}$

$$= \frac{1000 \times 4.184}{0.0821} = 50.96 \text{ kJ}$$

(ii) $Q = -W = -50.96 \text{ kJ}$ $(\Delta E = 0 \text{ at constant } T)$

(iii) $\Delta S = nR \ln \frac{50}{100} = 1 \times 8.314 \times 2.303 \times \log \frac{50}{100}$

$$= -5.76 \text{ J K}^{-1}$$

(iv) $\Delta G = nRT \ln \frac{V_1}{V_2} = 1 \times \frac{8.314}{1000} \times 310 \times 2.303 \times \log \frac{100}{50}$

$$= 1.79 \text{ kJ}$$

(v) $\Delta G = \Delta A$ for this process as it is isothermal.

$\therefore \Delta A = 1.79$ kJ

Problem 2.3

One mole of liquid benzene evaporates to vapour at its boiling point 80.2°C under reversible conditions. What is the work done in this process? What are ΔE and ΔH? Calculate the entropy change of the system (benzene) and the surroundings (ΔH_{vap} of benzene = 30.68 kJ mol^{-1}).

Solution

The vapour pressure at this temperature is equal to 1 atm and this is also the external pressure.

(i) $W = -P_{ext} \times \Delta V = W = -P_{ext} \times (V_{Vap} - V_l) \approx -P_{ext} \times V_{Vap}$

$$= -nRT = -1 \times 8.314 \times 353.2 = -2937 \text{ J} = -2.94 \text{ kJ}$$

(ii) $\Delta H = Q_P = 30{,}680 \text{ J} = 30.68 \text{ kJ}$

(iii) $\Delta E = Q + W = 30.68 + (-2.94) = 27.74 \text{ kJ}$

$$\Delta S \text{ of system} = \frac{Q_{rev}}{T} = \frac{30680}{353.2} = 86.9 \text{ J K}^{-1}$$

$$\Delta S_{surr} = -\frac{30680}{353.2} = -86.9 \text{ J K}^{-1}$$

Problem 2.4

5 moles of an ideal gas (C_V = 12.47 J K^{-1} mol^{-1}) are compressed adiabatically and reversibly from an initial pressure of 1 bar (~1 atm) and 300 K to a final pressure of 20 bar. Calculate (i) the final temperature of the gas (ii) W (iii) ΔE (iv) ΔH (v) ΔS.

Solution

(i) $\dfrac{T_2}{T_1} = \left(\dfrac{P_2}{P_1}\right)^{\frac{\gamma-1}{\gamma}}; \dfrac{T_2}{300}$

$$= \left(\frac{20}{1}\right)^{\frac{1.64-1}{1.64}} \left(\gamma = \frac{C_P}{C_V} = \frac{20.78}{12.47} = 1.64\right)$$

or, $\dfrac{T_2}{300} = 20^{0.39} \Rightarrow T_2 = 963.8$ K

(ii) $\Delta E = Q + W = 0 + W$ or, $W = 41.39$ kJ

(iii) $\Delta E = nC_V\left[T_2 - T_1\right] = 5 \times 12.47 \times (963.8 - 300)$

$= 41388 \text{ J} = 41.39 \text{ kJ}$

(iv) $\Delta H = nC_P\left[T_2 - T_1\right] = 5 \times 20.78 \times 663.8$

$= 68969 \text{ J} = 68.97 \text{ kJ}$

(v) $\Delta S = n\left(C_P \ln\dfrac{T_2}{T_1} + R\ln\dfrac{P_1}{P_2}\right)$

$$= 5 \times \left(20.78 \times 2.303 \times \log\frac{963.8}{300} + 8.314 \times 2.303 \times \log\frac{1}{20}\right)$$

$$= 11.52\left(10.53 + (-10.82)\right) = -3.04 \text{ J K}^{-1}$$

Problem 2.5

Two moles of a diatomic gas (C_V = 20.8 J K^{-1} mol^{-1}) at 300 K and 20 bar pressure is adiabatically expanded until its pressure falls to 2 bar. Calculate (i) W (ii) Q (iii) ΔE (iv) ΔH and (v) ΔS.

Solution

The process is an adiabatic irreversible expansion. First, T_2 needs to be estimated.

(i) $W = -P_{ext} \times (V_2 - V_1) = -2\left(\dfrac{nRT}{P_2} - \dfrac{nRT}{P_1}\right)$

$$= -2 \times 0.0821 \times \left(\frac{T_2}{2} - \frac{T_1}{20}\right) = -0.164T_2 + 4.93$$

But $\Delta E = nC_V\left[T_2 - T_1\right] = 2 \times \dfrac{20.8 \times 0.0821}{8.314} \times (T_2 - 300)$

$\therefore 0.417T_2 - 125.1 = -0.164T_2 + 4.93$

$$T_2 = \frac{130.0}{0.581} = 223.7 \text{ K}$$

We can also calculate T_2 from the relation,

$$T_2 = \frac{T_1\left(\dfrac{P_2}{P_1} + \dfrac{C_V}{R}\right)}{\left(\dfrac{C_V}{R} + 1\right)}$$

(This equation gives a slightly different value for T_2 = 222.8 K.)

(ii) Since it is adiabatic expansion, $Q = 0$ and $\Delta E = W$

(iii) $\Delta E = nC_V\left(T_2 - T_1\right) = 2 \times 20.8 \times (223.7 - 300)$

$= -3174 \text{ J} = -3.17 \text{ kJ}$

(iv) $\Delta H = nC_P\left(T_2 - T_1\right) = 2 \times 29.1 \times (223.7 - 300)$

$= -4441 \text{ J} = -4.44 \text{ kJ}$

(v) $\Delta S = n\left(C_P \ln\dfrac{T_2}{T_1} + R\ln\dfrac{P_1}{P_2}\right)$

$$= 2 \times 2.303 \times \left(29.1 \times \log\frac{223.7}{300} + 8.314 \times \log\frac{20}{2}\right)$$

$$= 4.606\left(-3.707 + 8.314\right) = 21.2 \text{ J K}^{-1}$$

Joule–Thomson Coefficient and Expansion

Problem 2.6

The van der Waals constants a and b of a gas are 0.1378 N m^4 mol^{-2} and 3.18×10^{-5} m^{3}mol^{-1} respectively. The C_P of the gas is 29.1 J K^{-1}mol^{-1}. Calculate the Joule–Thomson coefficient at 300 K in units of K atm^{-1}. Also calculate the enthalpy change for the isothermal compression of 5 moles of the gas from 1 atm. to 300 atm. What is the inversion temperature of the gas?

Solution

$$\mu_{J-T} = \frac{\left(\dfrac{2a}{RT} - b\right)}{C_P}$$

$$= \frac{\dfrac{2 \times 0.1378 \text{ N} \times \text{m}^4 \times \text{mol}^{-2}}{8.314 \text{ N} \times \text{m} \times \text{K}^{-1}\text{mol}^{-1} \times 300 \text{ K}} - 3.15 \times 10^{-5} \text{m}^3 \times \text{mol}^{-1}}{29.1 \text{ N} \times \text{m} \times \text{K}^{-1} \times \text{mol}^{-1}}$$

$$= \frac{(1.105 \times 10^{-4} - 0.315 \times 10^{-4})\text{m}^3 \times \text{mol}^{-1}}{29.1 \text{ N} \times \text{m} \times \text{K}^{-1} \times \text{mol}^{-1}}$$

$$= 2.714 \times 10^{-6} \text{N}^{-1} \times \text{m}^2 \times \text{K}$$

or, $\left(\dfrac{\partial T}{\partial P}\right)_H = 2.714 \times 10^{-6} \text{ N}^{-1}\text{m}^2\text{K} \times 1.013 \times 10^{-5} \text{ N m}^{-2}\text{atm}^{-1}$

$= 0.275 \text{ K atm}^{-1}$

Inversion temperature

At inversion temperature, $\mu_{J-T} = 0; \dfrac{2a}{RT} = b$

or, $T = \dfrac{2a}{Rb} = \dfrac{2 \times 0.1378 \text{ N} \times \text{m}^4 \times \text{mol}^{-2}}{8.314 \text{ N} \times \text{m} \times \text{K}^{-1}\text{mol}^{-1} \times 3.15 \times 10^{-5} \text{ m}^3 \times \text{mol}^{-1}}$

$= 1052.3$ K

$\Delta H: \left(\dfrac{\partial T}{\partial P}\right)_H = -\dfrac{1}{C_P}\left(\dfrac{\partial H}{\partial P}\right)_T$

or, $\left(\dfrac{\partial H}{\partial P}\right)_T = -C_P\left(\dfrac{\partial T}{\partial P}\right)_H = -C_P \times \dfrac{\left(\dfrac{2a}{RT} - b\right)}{C_P}$

$$\left(\frac{\partial H}{\partial P}\right)_T = -0.79 \times 10^{-4} \text{ m}^3 \text{ mol}^{-1}$$

This may be written as, $\frac{\Delta H}{\Delta P} = -0.79\times10^{-4}$

or, $\Delta H = -0.79\times10^{-4}\ \text{m}^3\ \text{mol}^{-1}\times(300-1)\text{atm}$

$\times1.013\times10^5\ \text{N}\times\text{m}^{-2}\ \text{atm}^{-1}$

$= -0.79\times299\times10\times1.013 = -2392.8\ \text{J mol}^{-1}$

For 5 moles of gas, $\Delta H = -5\times2392.8 = -11964\ \text{J}$

Enthalpy Changes for Chemical Reactions

Problem 2.7

Calculate the enthalpy change of the reaction,

$$2P_4(s)+10O_2(g)\rightarrow 2P_4O_{10}(s)$$

given the following thermochemical data.

(i) $PCl_3(g)+Cl_2(g)\rightarrow PCl_5(g)\quad \Delta H = -84.0\ \text{kJ}$

(ii) $4PCl_3(g)\rightarrow P_4(s)+6Cl_2(g)\quad \Delta H = 1226\ \text{kJ}$

(iii) $2POCl_3(g)\rightarrow 2PCl_3(g)+O_2(g)\quad \Delta H = 572.0\ \text{kJ}$

(iv) $P_4O_{10}(s)+6PCl_5(g)\rightarrow 10POCl_3(g)\quad \Delta H = -610\ \text{kJ}$

Solution

Reverse Eq. (ii) and multiply by 2 to get,

(v) $2P_4(s)+12Cl_2(g)\rightarrow 8PCl_3(g)\quad \Delta H = -2452\ \text{kJ}$

Reverse Eq. (iii) and multiply by 10 to obtain,

(vi) $20PCl_3(g)+10O_2(g)\rightarrow 20POCl_3(g)\quad \Delta H = -5720\ \text{kJ}$

Reverse Eq. (iv) and multiply by 2 to get,

(vii) $20POCl_3(g)\rightarrow$

$2P_4O_{10}(s)+12PCl_5(g)\quad \Delta H = 1220\ \text{kJ}$

Add Eqs (v), (vi) and (vii) and rearrange to get,

(viii) $2P_4(s)+10O_2(g)+12PCl_3(g)+12Cl_2(g)\rightarrow$

$2P_4O_{10}(s)+12PCl_5(g)\quad \Delta H = -6952\ \text{kJ}$

Multiply Eq. (i) by 12 to get,

(ix) $12PCl_3(g)+12Cl_2(g)\rightarrow 12PCl_5(g)\quad \Delta H = -1008\ \text{kJ}$

Eqs (viii) – (ix) gives

$$2P_4(s)+10O_2(g)\rightarrow 2P_4O_{10}(s)$$

$$\Delta H = -6952+1008\ \text{kJ} = -5944\ \text{kJ}$$

Problem 2.8

(a) Calculate ΔH^0 at 298 K for the reaction,

$CaCl_2(aq)+2NaOH(aq)\rightarrow Ca(OH)_2+2NaCl(aq)$

given the following data.

$\Delta H^0_{f(\text{NaOH},\,aq)} = -470\ \text{kJ};\ \Delta H^0_{f(\text{CaCl2},\,aq)}$
$= -877\ \text{kJ};\ \Delta H^0_{f(\text{Ca2+},\,aq)} = -543\ \text{kJ};\ \Delta H^0_{f(\text{OH}^-,\,aq)}$
$= -230.0\ \text{kJ};\ \Delta H^0_{f(\text{Na+},\,aq)} = -240\ \text{kJ};\ \Delta H^0_{f(\text{Cl}-,\,aq)} = -167\ \text{kJ}$

(b) Calculate ΔH^0 at 298 K for the reaction,

$KCl(s)+200H_2O\rightarrow KCl(\text{in }200H_2O)$

given ΔH^0 of the reaction,

$HCl(\text{in }200H_2O)+KOH(\text{in }200H_2O)\rightarrow$

$KCl(\text{in }200H_2O)+H_2O(l)$ as – 57 kJ

and ΔH^0 for HCl (in 200 H_2O) and KOH (in 200 H_2O) as –166 kJ and –482 kJ respectively.

$\Delta H^0_{f(\text{H2O},\,l)} = -286\ \text{kJ}$

Solution

(a) $CaCl_2(aq)+2NaOH(aq)\rightarrow$

$Ca(OH)_2(aq)+2NaCl(aq)\Delta H^0 = ?$

$-877+-470\times2\rightarrow -543+(-230\times2)$

$+(-240\times2)+(-167\times2)$

$\Delta H^0 = [-1003+(-814)]-[-877+(-940)] = 0$

(b) $KCl(s)+200H_2O\rightarrow KCl(\text{in }200H_2O)\quad \Delta H^0 = ?$

$HCl(\text{in }200H_2O)+KOH(\text{in }200H_2O)\rightarrow$

$KCl(\text{in }200H_2O)+H_2O(l)\quad \Delta H^0 - 57\ \text{kJ}$

$-166+-482\rightarrow \Delta H^0 + -286$

$\Delta H^0 + (-286) — (-166)+(-482) = -57$

$\Delta H^0 = -705+286 = -419\ \text{kJ}$

Problem 2.9

Calculate ΔH^0 of the reaction

$$CH_4(g)+2O_2(g)\rightarrow CO_2(g)+2H_2O(g)\Delta H^0_{298} = -802\ \text{kJ}$$

at 1000 K using the following data.

Compound	$CH_4(g)$	$O_2(g)$	$CO_2(g)$	$H_2O(g)$
$C_P^0(\text{J K}^{-1}\ \text{mol}^{-1})$	35.6	29.4	37.1	33.6

Solution

$$\Delta C_P^0 \text{ of reaction} = C^0_{P(CO_2,g)} + 2C^0_{P(H_2O,g)} - C^0_{P(CH_4,g)} - 2C^0_{P(O_2,g)}$$

$$= 37.1+(2\times33.6)-35.6-2\times29.4 = 9.9\ \text{J K}^{-1}$$

$$\Delta H^0_{1000} = \Delta H^0_{298} + \int_{298}^{1000}\Delta C_P^0\,dT$$

$$= -802+9.9\times10^{-3}(1000-298)$$

$$= -802+6.95 = -795.1\ \text{kJ}$$

Problem 2.10

The ΔH^0_C of propane gas at 298 K is –2044 kJ. Calculate the bond energy of the O=O given the bond energies of C–H, C=O, O–H, and C–C bonds as 413, 738, 463 and 348, all in kJ mol^{-1} respectively.

Solution

Reaction

$C_3H_8(g)+5O_2(g)\rightarrow$

$3CO_2(g)+4H_2O(g)\quad \Delta H^0_{298} = -2044\ \text{kJ}$

Reactants		Products	
Bonds broken	Energy required	Bonds broken	Energy required
C–H = 8	$8 \times 413 = 3304$ kJ	C = O = 6	$6 \times 738 = 4428$ kJ
C–C = 2	$2 \times 348 = 796$ kJ	O–H = 8	$8 \times 463 = 3704$ kJ
O=O = 5	$5 \times \Delta H^0_{O=O} = ?$		

$$\Delta H^0 = -2044$$

$$= \sum \text{energy required for breaking bonds in reactants}$$

$$-\sum \text{energy required for breaking bonds in products}$$

$$= 3304 + 796 + 5 \times \Delta H^0_{O=O} - 4428 - 3704$$

$$5 \times \Delta H^0_{O=O} = 1988;\ \Delta H^0_{O=O} = \frac{1988}{5} = 398 \text{ kJ mol}^{-1}$$

NOTE

- Energy released in bonds formed in the products can also be considered.

Entropy of Mixing and Entropy Changes in Chemical Reactions

Problem 2.11

One mole of hydrogen, 2 moles of helium, and one mole of nitrogen are mixed at 300 K. What is the value of ΔS_{mixing} for this process? Assume the gases are ideal.

Solution

Mole fraction of H_2 = 1/4 =0.25; mole fraction of He = 2/4 = 0.5; mole fraction of N_2 = 1/4 = 0.25

$$\Delta S_{mix} = -R\sum_{i=1}^{i=x}\left(X_i \ln X_i\right) \text{ where } X_i = \text{mole fraction of } i$$

$$\Delta S_{mix} = -8.314 \times 2.303 \times \left[0.25 \times \log 0.25 + 0.5 \log 0.5 + 0.25 \log 0.25\right]$$

$$-19.16 \times \left[-0.150 - 0.150 - 0.150\right] = 8.62 \text{ J K}^{-1} \text{ mol}^{-1}$$

Problem 2.12

Calculate ΔS^0 for the reaction

$$N_2(g) + 3H_2(g) \rightarrow 2NH_3(g)$$

at 298 K from the following data. Also calculate ΔS^0 for the reaction at 1000 K assuming $C^0{}_P$ is independent of temperature in this range

	$N_2(g)$	$H_2(g)$	$NH_3(g)$
$S^0_{298}\left(\text{J K}^{-1}\text{ mol}^{-1}\right)$	191.6	130.7	192.7
$C^0_{P(298)}$ (J K^{-1} mol^{-1})	29.1	28.8	35.6

Solution

$$\Delta S^0 \text{ for the reaction} = (2 \times 192.7) - \left[191.6 + (3 \times 130.7)\right]$$

$$= -198.3 \text{ J K}^{-1}$$

ΔS^0 for the reaction at 1000 K

$$\partial\left(\frac{\partial S^0}{\partial T}\right)_P = \int_{T_1}^{T_2} \partial\left(\frac{\Delta C^0_P}{T}\right) dT = \Delta C^0_P \ln \frac{T_2}{T_1}$$

$$\Delta C^0_P = (2 \times 35.6) - \left(29.1 + (3 \times 28.8)\right)$$

$$= 71.2 - 115.5 = -44.3 \text{ J K}^{-1}$$

$$\Delta S^0_{1000\,K} = (-44.3 \times 2.303) \times \log \frac{1000}{298}$$

$$= -102 \times 0.526 = -53.8 \text{ J K}^{-1}$$

Problem 2.13

Calculate the entropy change for the process

$$2 \text{ moles of benzene}(l, 0°\text{C}) \rightarrow 2 \text{ moles of benzene}(s, 0°\text{C})$$

The freezing point of benzene is 5.42°C. The enthalpy of fusion of benzene at the freezing point is 9.9 kJ mol^{-1}. The heat capacity of liquid benzene is 173.3 J K^{-1} mol^{-1} and that of the solid is 134.8 J K^{-1} mol^{-1}. (Consider benzene as a supercooled liquid at 0°C.)

Solution

The following steps are considered for calculating ΔS.

Step 1: 2 moles of benzene (supercooled liquid, 0°C) → 2 moles of benzene (*s*, 5.42°C)

$$\Delta S_I = \int_{273}^{278.42} C_P(l)\left(\frac{dT}{T}\right) = C_P(l) \ln \frac{278.42}{273.0}$$

$$= 173.3 \times 2.303 \log \frac{278.42}{273.0} = 3.4 \text{ J K}^{-1}$$

For 2 moles, $\Delta S_I = 2 \times 3.4 = 6.8$ J K^{-1}

Step 2: 2 moles of benzene$(l, 5.42°\text{C}) \rightarrow$ 2 moles of benzene$(s, 5.42°\text{C})$

$$\Delta S_{II} = -\frac{(2 \times 9.9 \times 1000)}{(278.42)} = -71.1 \text{ J K}^{-1}$$

Step 3: 2 moles of benzene$(l, 5.42°\text{C}) \rightarrow$ 2 moles of benzene$(S, 0°\text{C})$

$$\Delta S_{III} = \int_{278.42}^{273.0} C_P(s)\left(\frac{dT}{T}\right) = C_P(s) \ln \frac{273.0}{278.42}$$

$$= -134.8 - 2.303 \times -0.00855 = -2.65 \text{ J K}^{-1}$$

For 2 moles, $\Delta S_{III} = 2 \times -2.65 = -5.30$ J K^{-1}

Total $\Delta S = 6.8 + (-71.1) - 5.30 = -69.6$ J K^{-1}

Problem 2.14

The standard molar entropies, S^0, of $H_2(g)$, $O_2(g)$, $H_2O_2(l)$ are 130.7, 205.1, 109.6, all in J K^{-1} mol^{-1}. Calculate ΔS^0 of the reaction,

$$H_2(g)+O_2(g)\rightarrow H_2O_2(l)$$

at 298 K. What is the significance of sign? Given $\Delta H^0_{f(H_2O_2,l)} =$ −187.8 kJ, calculate the net entropy change of the isolated system. Explain the results.

Solution

ΔS^0 of reaction = 109.6 − (130.7 + 205.1) = −226.2 J K^{-1}

From this result, one may say that as there is a decrease in entropy, the system is more ordered. It can't say anything about the spontaneity of the reaction. Since the reaction is exothermic, the heat is taken up by the surroundings.

$$\Delta S_{Surr} = \frac{187.8\times1000}{298} = 630.2 \text{ J K}^{-1}$$

$$\Delta S_{Sys} + \Delta S_{Surr} = -226.2+630.2 = 404 \text{ J K}^{-1}$$

Since ΔS of the isolated system is positive, one may say that the entropy of the isolated system increases in a spontaneous process.

Problem 2.15

One mole of acetic acid at its freezing point 289.6 K (as a solid) is added to 2 moles of acetic acid at its boiling point 391.2 K under adiabatic (isolated) conditions. What is ΔS_{total} for this process? $\Delta H^0_{f(\text{acetic acid})}$ = 11.7 kJ mol^{-1}, C_P of liquid acetic acid is 126.7 J K^{-1} mol^{-1}.

Solution

1st stage: ΔS for one mole of acetic acid (*s*) → acetic acid (*l*) at 289.6 K

$$\Delta S = \frac{1\times11.7\times1000}{289.6} = 40.4 \text{ J K}^{-1}$$

2nd stage: When one mole of acetic acid (*l*) at 289.6 K is added to 2 moles of boiling acetic acid at 391.2 K, let the resultant temperature be *t* K.

$$1\times126.7\times(t-289.6) = 2\times126.7\times(391.2-t)$$

$$t = \frac{1072}{3} = 357.3 \text{ K}$$

3rd stage: One mole of acetic acid (*l*, 289.6 K) → One mole of acetic acid (*l*, 357.3 K)

$$\Delta S = nC_P \ln\frac{T_2}{T_1} = 1\times126.7\times2.303\times\log\frac{357.3}{289.6} = 26.6 \text{ J K}^{-1}$$

Total ΔS for stages I and III = 40.4 + 26.6 = 67.0 J K^{-1}

4th stage: Two moles of acetic acid (*l*, 391.2 K) → Two moles of acetic acid (*l*, 357.3 K)

$$\Delta S = nC_P \ln\frac{T_2}{T_1} = 2\times126.7\times2.303\times\log\frac{357.3}{391.2} = -23.0 \text{ J K}^{-1}$$

$$\text{Total }\Delta S = 67.0+(-23.0) = 44.0 \text{ J K}^{-1}$$

Problem 2.16

2 moles of an ideal gas at 400 K and 2 bar undergo an adiabatic free expansion until its volume increases three-fold. Calculate (i) W (ii) Q (iii) ΔE (iv) ΔH and (v) ΔS. If the same process were carried out isothermally, what will be the values for the above quantities?

Solution

For an adiabatic free expansion, Q = 0 and W = 0, therefore $\Delta E = W = 0$.

For any isothermal process, ΔE = 0 and ΔH = 0. For an adiabatic processes,

$$\text{Since } \Delta E = nC_V(T_2-T_1) = 0,\ T_1 = T_2$$

$$\Delta H = nC_P(T_2-T_1) = 0,\ T_1 = T_2$$

$$\Delta S = nR\ln\frac{V_2}{V_1} = 2\times8.314\times2.303\times\log3 = 183 \text{ J K}^{-1}$$

For the isothermal process, all the results will be the same as there is no difference between isothermal and adiabatic free expansion.

Bond Energies from Thermochemical Data

Problem 2.17

The enthalpy change, ΔH^0, for the reaction,

$$2NH_3(g)+3\frac{1}{2}O_2(g)\rightarrow 2NO_2(g)+3H_2O(l)$$

is −700 kJ. The enthalpy of vaporisation of water is 44 kJ mol^{-1}. The bond energies of O=O, N–O (average) and O–H bonds are 498, 406, 464 respectively, all in kJ mol^{-1}. Using these data, calculate the N–H bond energy.

Solution

In calculating bond energies using ΔH^0 data, the reactants and products must all be in the gaseous state. Therefore the ΔH^0 for the reaction where $H_2O(l)$ is in the gaseous form must be taken. Thus,

(a) $2NH_3(g)+3\frac{1}{2}O_2(g)\rightarrow$

$$2NO_2(g)+3H_2O(l)\quad \Delta H^0 = -700 \text{ kJ}$$

(b) $3H_2O(l)\rightarrow 3H_2O(g)\quad \Delta H^0 = 3\times44 = 132 \text{ kJ}$

(a) + (b) gives

$$2NH_3(g)+3\frac{1}{2}O_2(g)\rightarrow$$

$$2NO_2(g)+3H_2O(g)\quad \Delta H = -568 \text{ kJ}$$

Bonds broken (reactants side)

$$\text{N–H} = 6\ 6\Delta H_{\text{N–H}} = ?$$

$$\text{O=O} = 7/2\ \Delta H_{\text{O=O}} = 7/2\times498 = 1743 \text{ kJ}$$

Bonds broken (products side)

$$\text{N–O} = 4\ \Delta H_{\text{N–H}} = 4\times406 = 1624 \text{ kJ}$$

$$\text{O–H} = 6\ \Delta H_{\text{O–H}} = 6\times464 = 2784 \text{ kJ}$$

$$(6\Delta H_{\text{N–H}} + 1743) - (1624 + 2784) = -568$$

$$6\Delta H_{N-H} = 4408 - 2311$$

$$\Delta H_{N-H} = 2097/6 = 350 \text{ kJ mol}^{-1}$$

Equilibrium Processes and Thermodynamic Quantities

Problem 2.18

2 moles of carbon tetrachloride(l) at its boiling point 349.8 K expand isothermally into an evacuated vessel till the final pressure falls to 0.1 bar. The enthalpy of vaporisation, ΔH_{vap}, at its boiling point is 29.8 kJ mol^{-1}. Calculate ΔG and ΔA for the process.

Solution

ΔG can be calculated in two stages.

1st stage: Two moles of CCl_4 (l, 349.8 K, 1 bar) $\rightarrow$ Two moles of CCl_4 (g, 349.8 K, 1 bar)

$\Delta G = 0$ because it is an equilibrium process.

2nd stage: Two moles of CCl_4 (g, 349.8 K, 1 bar) $\rightarrow$ Two moles of CCl_4 (g, 349.8 K, 0.1 bar)

$$\Delta G(\text{total}) = nRT \ln \frac{P_2}{P_1} = 2 \times 8.314 \times 349.8 \times 2.303 \times \log \frac{0.1}{10}$$

$$= -13395 \text{ J} = -13.4 \text{ kJ}$$

ΔA: $\Delta A = \Delta E - T\Delta S$ (Since the process is isothermal)

ΔE and ΔS have to be first calculated for the two stages.

ΔE: 1st stage as in ΔG calculation

$$\Delta E = Q + W = 2 \times 29.8 \times 1000 - P_{Vv}$$

$$= 59600 - (2 \times 8.314 \times 349.8)$$

$$= 59600 - 5816 = 53784 \text{ J}$$

ΔE in 2nd stage as under ΔG

Assuming CCl_4 vapour as ideal, $\Delta E = 0$ (since the process is isothermal)

Total $\Delta E = 53784$ J

ΔS: 1st stage is as under ΔG,

$$\Delta S = \frac{Q_{Rev}}{T} = \frac{29800}{349.8} = 85.2 \text{ J K}^{-1}$$

For 2 moles $\Delta S = 2 \times 85.2 = 170.4 \text{ J K}^{-1}$

2nd stage as under ΔG

$$\Delta S = nR \ln \frac{P_1}{P_2} = 2 \times 8.314 \times 2.303 \times \log \frac{0.1}{1.0} = -38.3 \text{ J K}^{-1}$$

$$\text{Total } \Delta S = 170.4 - 38.3 = 132.1 \text{ J K}^{-1}$$

$$\Delta A = 53784 - (349.8 \times 132.1)$$

$$= (53784 - 46209) \text{ J} = 7575 \text{ J} = 75.8 \text{ kJ}$$

(it is possible to calculate ΔA from the relation $\Delta A = \Delta G - P\Delta V$)

Problem 2.19

Calculate ΔG^0 and ΔH^0 for the reaction,

$$Fe_3O_4(\text{magnetite}) + 4CO(g) \rightarrow 3Fe(s) + 4CO_2(g)$$

from the following data.

	$Fe_3O_4(s)$	$CO(g)$	$Fe(s)$	$CO_2(g)$
ΔG_f^0 (kJ mol^{-1})	–1015.4	–137.2	0	q
S_{298}^0 (J K^{-1} mol^{-1})	146.4	197.7	27.3	213.7

Also calculate ΔA^0 of the reaction.

Solution

$$\Delta G^0_{\text{reaction}} = 3 \times \Delta G^0_{f(\text{Fe},s)} + 4\Delta G^0_{f(\text{CO}_2,g)} - \Delta G^0_{f(\text{Fe}_3\text{O}_4,S)} - 4\Delta G^0_{f(\text{CO},g)}$$

$$= 3 \times 0 + (4 \times -394.4) - [-1015.4 + (4 \times -137.2)]$$

$$= 1564.2 - 1577.6 = -13.4 \text{ kJ}$$

$$\Delta S^0_{\text{reaction}} = 3 \times 27.3 + (4 \times 213.7) - [146.4 + (4 \times 197.7)]$$

$$= 936.7 - 937.2 = -0.5 \text{ J K}^{-1}$$

$$\Delta H^0 = \Delta G^0 + T\Delta S^0 = -13.4 + \left[298 \times \left(\frac{-0.5}{1000}\right)\right] = -13.55 \text{ kJ}$$

For this reaction, $Q_P = \Delta H^0 = Q_V$

(since $\Delta n = 0$ in the relation $Q_P = Q_V + RT\Delta n$)

But $Q_V = \Delta E$,

$$\therefore \Delta A = \Delta E - T\Delta S = -13.55 - (298) \times \left(-\frac{0.5}{1000}\right) = -13.4 \text{ kJ}$$

Thus, $\Delta G^0 = \Delta A^0$

Problem 2.20

The equilibrium constant K_P for the reaction,

$$2SO_2(g) + O_2(g) \rightarrow 2SO_3(g)$$

is 3.42 at 727°C. In a particular study, it is proposed to pass SO_2 and O_2 gases at partial pressures of 20 bar and 10 bar and withdraw SO_3 gas at 30 bar. Calculate ΔG and comment on the feasibility of the reaction. What is K_C for the reaction?

Solution

$$Q_P \text{ for the reaction} = \frac{p^2_{SO_3}}{p^2_{SO_2} \times p_{O_2}} = \frac{30 \times 30}{20 \times 20 \times 10} = 0.225$$

$$\Delta G = RT \ln \frac{Q_P}{K_P} = 2.303 \times 8.314 \times 10^{-3} \times 1000 \log \frac{0.225}{3.42}$$

$$= 19.15 \times (-1.182) = -22.6 \text{ kJ}$$

Since ΔG is negative, the reaction is feasible under these condition

$$K_C = \frac{K_P}{(RT)^{\Delta n}} = \frac{3.42}{(0.0821 \times 1000)^{-1}}$$

$$= 3.42 \times 82.1 = 280.8 \text{ L mol}^{-1}$$

NOTE

- Equilibrium constants do not have any dimensions.

Problem 2.21

Calculate the equilibrium constant of the reaction

$$CO(g)+H_2(g)\rightarrow HCHO(g)$$

at 298 K, given $\Delta G^0_{f(HCHO,\ g)} = -110$ kJmol^{-1} and $\Delta G^0_{f(CO_2,\ g)} =$ -137.2 kJ

If α is the fraction of HCHO formed, what is its value at a total pressure of 1000 bar (1 bar ≈ 1 atm.)?

Solution

$$\Delta G^0_{reaction} = -110-(-137.2) = 27.2\,\text{kJ} = -RT\ln K_P$$

$$\log K_P = \frac{-27.2}{2.303\times 8.314\times 10^{-3}\times 298} = -4.77$$

$$K_P = 1.7\times 10^{-5}$$

If α is the fraction of HCHO formed at equilibrium,

Moles of CO = moles of $H_2 = (1-\alpha)$

Moles of HCHO = α

$$p_{CO} = p_{H_2} = \frac{1-\alpha}{2-\alpha}\times P;\ p_{HCHO}$$

$$= \frac{\alpha}{2-\alpha}\times P\,(\text{Total no. of moles} = 1-\alpha+1-\alpha+\alpha = 2-\alpha)$$

$$K_P = \frac{\alpha}{2-\alpha}\times P\times\frac{(2-\alpha)^2}{(1-\alpha)^2\times P^2} = \frac{(2-\alpha)\alpha}{(1-\alpha)^2 P}$$

Since $P = 1000$ bar; $K_P\times P = 1.7\times 10^{-5}\times 1000 = 1.7\times 10^{-2}$

$$\frac{2\alpha-\alpha^2}{1+\alpha^2-2\alpha} = 1.7\times 10^{-2};\ \text{Solving for } \alpha, \text{ we get } \alpha = 0.2$$

Problem 2.22

Given $\Delta H^0_{f(HCHO,\ g)} = -115$ kJ mol^{-1} and $\Delta H^0_{f(CO,\ g)} =$ -110 kJ mol^{-1}, at what temperature will α be 0.6 at the same pressure of 1000 bar?

Solution

$$\text{When } \alpha = 0.6,\ K_P = \frac{(2-0.6)\times 0.6}{(1-0.6)^2\times 1000} = \frac{0.84}{160} = 5.25\times 10^{-3}$$

$$\Delta H^0_{reaction} = -115-(-110) = -5.0\text{ kJ}$$

$$\log\frac{(K_P)_2}{(K_P)_1} = \frac{\Delta H^0}{2.303R}\left[\frac{1}{T_1}-\frac{1}{T_2}\right]$$

$$\log\frac{5.25\times 10^{-3}}{1.7\times 10^{-5}} = -\frac{5.0}{2.303\times 8.314\times 10^{-3}}\left[\frac{1}{298}-\frac{1}{T_2}\right]$$

$$\log 308.8 = -\frac{5.0}{0.0194}\left[\frac{1}{298}-\frac{1}{T_2}\right]$$

Solving for T_2, we get $T_2 = 76.9$ K

Absolute Entropy of Substances

Problem 2.23

Calculate the absolute entropy of sulfur at 392 K using the following data: transition temperature of rhombic to monoclinic sulfur is 368.5 K; $\Delta H_{(rhombic\rightarrow monoclinic)} = 402$ J K^{-1} mol^{-1}.

$$\int_0^{368.5} C_{P(Rh)}\frac{dT}{T} = 36.8\text{ J K}^{-1}\text{ mol}^{-1}$$

$$C^0_{P(\text{monoclinic sulphur})} = 21.2+\frac{3.9}{1000}T(\text{K})+22.3\left(\frac{T(\text{K})}{1000}\right)^2$$

In the temperature range 368.5 to 392 K.

$$\Delta H^0_{\text{fusion (monoclinic)}} = 1.73\text{ kJ mol}^{-1}\text{ at melting point 392 K}$$

Solution

$$\textbf{1}^{\textbf{st}}\textbf{ stage: } \int_0^{368.5} C_{P(Rh)}\frac{dT}{T} = 36.8\text{ J K}^{-1}$$

$$\textbf{2}^{\textbf{nd}}\textbf{ stage: } \Delta S_{tr} = \frac{402}{368.5} = 1.1\text{ J K}^{-1}$$

$$\textbf{3}^{\textbf{rd}}\textbf{stage: } S^0_{(\text{monoclinic})} = \int_{368.5}^{392}\left(21.2+0.0039T+22.3\times 10^{-6}T^2\right)\frac{dT}{T}$$

$$= 21.2+2.303\log\frac{392}{368.5}+0.0039(392-368.5)$$

$$+\frac{22.3\times 10^{-6}}{2}\left(392^2-368.5^2\right)$$

$$1.32+0.09+0.2 = 1.61\text{ J K}^{-1}$$

$$\textbf{4}^{\textbf{th}}\textbf{ stage: } \Delta S^0_{\text{fusion (monoclinic)}} = \frac{1.73\times 1000}{392} = 4.4\text{ J.K}^{-1}$$

$$\text{Total } S(\text{monoclinic liquid}) \text{ at 392 K} = 36.8+1.1+1.6+4.4$$

$$= 43.9\text{ J K}^{-1}$$

Problem 2.24

Show that the molar entropy of CO at 0 K is R ln2.

Solution

The absolute entropy of a substance is related to the thermodynamic probability (W). Assuming that CO is in a crystalline form at 0 K, 50% will have configuration of CO and the other 50% will have OC.

$$W = \frac{N!}{\frac{N}{2}!\frac{N}{2}!}$$

$$S = k\ln W = k\left[N\ln N - N - 2\left(\frac{N}{2}\ln\frac{N}{2}-\frac{N}{2}\right)\right]$$

$$S = kN\ln 2$$

where k denotes the Boltzmann constant.

Calculations Involving Equilibria and Equilibrium Constants

Problem 2.25

The equilibrium vapour pressure for the reaction,

$$CuSO_4\cdot 5H_2O(s)\rightarrow CuSO_4\cdot 3H_2O(s)+2H_2O(g)$$

at 298.0 K is 1.03×10^{-2} bar. Given the enthalpy change of the reaction is 120 kJ mol^{-1}, calculate (i) ΔG^0 (ii) ΔS^0 and (iii) ΔA^0 of the reaction.

Solution

(i) K_P of the reaction $= p^2_{H_2O} = \left(1.03\times10^{-2}\right)^2$

$$=1.06\times10^{-4}\ \text{bar}^{-2}$$

$$\Delta G^0 = -RT\ln K_P$$

$$= -8.314\times10^{-3}\times298\times2.303\log\left(1.06\times10^{-4}\right)$$

$$= -5.7\times-3.975 = +22.7\ \text{kJ}$$

(ii) $\Delta S^0 = \dfrac{\Delta H^0 - \Delta G^0}{T} = \dfrac{120-22.7}{298} = 0.33\ \text{kJ K}^{-1}$

(iii) $\Delta A^0 = \Delta E^0 - T\Delta S^0$

$$\Delta E^0 = \Delta H^0 - RT\Delta n = 120 - 8.314\times10^{-3}\times298\times2$$

$$= 120 - 5 = 115\ \text{kJ}$$

$$\Delta A^0 = 115 - (298\times0.33) = 115 - 98.3 = 16.7\ \text{kJ}$$

Problem 2.26

For the reaction

$$2P_2(g)\rightarrow P_4(g)$$

the equilibrium constant, K_P=2.0 bar^{-1} at 1157°C. (i) Calculate K_C and K_F at a pressure of 0.5 bar. (ii) At what total pressure of the system, will the degree of aggregation of P_2 be 0.8? (iii) If helium is introduced at 5 bar, when the total pressure is kept constant at 10 bar, will the yield of P_4 increase or decrease? Show by calculation. (iv) If helium is introduced at 5 bar, keeping the total volume constant, how will the equilibrium shift?

Solution

(i) $K_P = K_C(RT)^{\Delta n}$ or $K_C = \dfrac{K_P}{(RT)^{-1}}$

$$= \frac{2.0}{(0.0821\times1430)^{-1}} = 234.8\ \text{L mol}^{-1}$$

$$K_C = K_F(P)^{\Delta n};\ K_F = \frac{2.0}{(0.5)^{-1}} = 2\times0.5 = 1$$

(ii) The equation for K_P in terms of α, the degree of aggregation, is to be obtained.

No. of moles of $P_4 = \alpha$, no. moles of P_2 left $= 2(1-\alpha)$

Total no. of moles $= \alpha + 2 - 2\alpha = (2-\alpha)$

$$K_P = \frac{p_{P_4}}{p^2_{P_2}} = \frac{\dfrac{\alpha}{2-\alpha}P}{\left(\dfrac{2(1-\alpha)}{2-\alpha}P\right)^2}$$

$$= \frac{\alpha}{2-\alpha}\times P\times\frac{(2-\alpha)^2}{4(1-\alpha)^2P^2} = \frac{\alpha(2-\alpha)}{4(1-\alpha)^2P}$$

Substituting for α = 0.8 and simplifying, P = 3 bar.

(iii) First we calculate α at P =10 bar.

$$K_P\times P = 2\times10 = \frac{2\alpha-\alpha^2}{4+4\alpha^2-8\alpha}$$

that is, $81\alpha^2 - 162\alpha + 80 = 0$

$\alpha = 0.89$

When helium is introduced at 5 bar, the pressure due to reactants = 10 – 5 = 5 bar

$$K_P\times P = 2\times5 = 10 = \frac{2\alpha-\alpha^2}{4+4\alpha^2-8\alpha}$$

Solving for α, we get α = 0.84. Thus α decreases. We can also arrive at this result quantitatively by writing,

$$K_P = \frac{n_{P_4}}{n^2_{P_2}}\left(\frac{P}{N}\right)^{\Delta n}\quad \text{where } \Delta n = 1-2 = -1$$

When P is constant and N increases, the term P/N decreases. $\left(\dfrac{P}{N}\right)^{-1} = \dfrac{N}{P}$, increases. To keep K_P constant, n_{P_2} must increase. Thus α, the degree of aggregation, decreases. Under constant volume conditions, P/N remains unaffected. Therefore α remains the same.

Problem 2.27

For the reaction,

$$BaCO_3(s)\rightarrow BaO(s)+CO_2(g)$$

$\Delta G^0{}_f$ and $\Delta H^0{}_f$ data are listed below.

	$BaCO_3(s)$	$BaO(s)$	$CO_2(g)$
$\Delta G^0{}_f$ (kJ mol^{-1})	–1139	–528	–394
$\Delta H0_f$ (kJ mol^{-1})	–1219	–558	–393

Calculate ΔG^0 and ΔH^0 of the reaction. What is ΔS^0 at 298 K? Calculate the temperature at which carbondioxide pressure reaches 1 bar. (Assume ΔH^0 is independent of temperature.)

Solution

$$\Delta G^0_{\text{reaction}} = -528-398-(-1139) = 217\ \text{kJ}$$

$$\Delta H^0_{\text{reaction}} = -558-393-(-1219) = 268\ \text{kJ}$$

$$\Delta S^0_{\text{reaction}} = \frac{\Delta H^0-\Delta G^0}{T} = \frac{268-217}{298} = 0.171\ \text{kJ} = 171\ \text{J K}^{-1}$$

K_P of the reaction $= p_{CO_2(eq)}$

$$\Delta G^0 = -RT\ln K_P \text{ or } \log K_P$$

$$= -\frac{217}{2.303\times8.314\times10^{-3}\times298} = -38$$

$$K_P = 1\times10^{-38}\left(= p_{CO_2(eq)}\right)$$

$$\text{Using } \log\frac{\left(p_{CO_2}\right)_T}{\left(p_{CO_2}\right)_{298}} = \frac{\Delta H}{2.303R}\left[\frac{1}{298}-\frac{1}{T}\right]$$

$$\log\frac{1}{10^{-38}} = \frac{268}{2.303\times8.314\times10^{-3}}\left[\frac{T-298}{298\times T}\right]$$

$$\log 10^{38} = 13997 \times \frac{(T-298)}{T \times 298}$$

Solving for T, we obtain, $T = 1560$ K

Raoult's Law, Fugacity, Activities and Activity Coefficients in Liquid Mixtures

Problem 2.28

Consider a liquid mixture of two components A and B. Show that if component A obeys Raoult's law over the complete range of solvent compositions, B also must obey Raoult's law necessarily.

Solution

$$p_A = P_A^0 x_A \quad \text{(Raoult's law for } A)$$

Taking logarithms, $\ln p_A = \ln P_A^0 + \ln x_A$

$\partial \ln p_A = \partial \ln x_A$

$\left(\text{since } \partial \ln P_A^0 = 0 \text{ as } P_A^0 \text{ is constant at a given temperature}\right)$

From Duhem–Margules equation,

$$\frac{\partial \ln p_A}{\partial \ln x_A} = \frac{\partial \ln p_B}{\partial \ln x_B} = 1 \quad \left(\text{since } \frac{\partial \ln p_A}{\partial \ln x_A} = 1\right)$$

$$\partial \ln p_B = \partial \ln x_B$$

Integrating both sides, $\ln p_B = \ln x_B + \text{constant}$

At $x_B = 1$ (or pure liquid), $\ln p_B = \ln x_B + \ln P_B^0$

$\left(\text{since at } x_B = 1,\ p_B = P_B^0\right)$

$p_B = P_B^0 x_B$ which proves the applicability of Raoult's law for B

Problem 2.29

The van der Waals constants a and b for CO are 0.1505 N m^4 mol^{-2} and 3.98×10^{-5} m^3 mol^{-1}. Calculate the fugacity of the gas at 500 bar and 1000 bar at 298 K.

Solution

For a van der Waals gas, the fugacity is given by

$$\ln \frac{f}{P} = \left(b - \frac{a}{RT}\right)\frac{P}{RT} \quad \text{where } f = \text{fugacity}$$

At 500 bar, $\ln \frac{f}{P} =$

$$\left(3.98 \times 10^{-5} \left(\text{m}^3 \text{ mol}^{-1}\right) - \left(\frac{0.1505\left(\text{N m}^4 \text{ mol}^{-2}\right)}{8.314\left(\text{N m K}^{-1} \text{ mol}^{-1}\right) \times 298 \text{ K}}\right)\right)$$

$$\times \frac{500 \times 1.0135 \times 10^5 \times \text{N m}^{-2}}{8.314 \times 298 \text{ N m K}^{-1} \text{ K mol}^{-1}}$$

Note that $\left(b - \frac{a}{RT}\right)$ has the units m^3 mol^{-1}

and $\frac{P}{RT}$ has the units of m^{-3} mol so that the right hand side of the equation has no units.

On simplification, we obtain,

$$\ln \frac{f}{P} = \left(3.98 \times 10^{-5} - \left(\frac{0.1505}{2477.6}\right)\right)\frac{5.067 \times 10^7}{2477.6}$$

$$= -2.09 \times 10^{-5} \times 2.045 \times 10^4 = -0.4274$$

$$2.303 \log \frac{f}{P} = -0.4274; \ \log \frac{f}{P} = -0.1876$$

$$\frac{f}{500} = 0.652 \text{ and } f = 0.652 \times 500 = 326.1 \text{ bar}$$

At 1000 bar

$$\ln \frac{f}{P} = -2.09 \times 10^{-5} \times \frac{1000 \times 1.0135 \times 10^5}{8.314 \times 298}$$

$$= -2.09 \times 10^{-5} \times 4.09 \times 10^4 = -0.855$$

$$\log \frac{f}{P} = -\frac{0.855}{2.303} = -0.3712; \ \frac{f}{p} = 0.4254$$

and $\quad f = 1000 \times 0.4254 = 425.4$ bar

Problem 2.30

The following vapour pressure data are obtained for mixtures of chloroform–acetone at 298 K.

X_{CHCl_3}	Vapour pressure (kPa)
0.4	9.3
0.6	18.0
0.8	26.0
1.0	41.5

(a) Calculate the activity and activity coefficient of chloroform at $X_{CHCl_3} = 0.4$, 0.6, 0.8 on Raoult's law standard state. What do you conclude?

(b) Given the Henry's law constant K_H for chloroform as a solute in acetone as 16.6 kPa, calculate the activity and activity coefficients of chloroform at $X_{CHCl_3} = 0.4$, 0.6, 0.8 on Henry's law standard state.

Solution

(a)

X_{CHCl_3}	Activity of chloroform	Activity coefficient of chloroform (a_i/x_i)
0.4	9.3/41.5 = 0.224	0.224/0.40 = 0.56
0.6	18.0/41.5 = 0.434	0.434/0.60 = 0.72
0.8	26.0/41.5 = 0.626	0.626/0.80 = 0.78

Since $\gamma < 1$, the solutions show negative deviations from Raoult's law.

(b) In this case, the Henry's law constant, K_H, has to be used instead of vapour pressure of pure chloroform. Thus we use $p_i/K_H = a_i$ and $p_i/K_H x_i = \gamma_i$

X_{CHCl_3}	Activity of chloroform	Activity coefficient of chloroform (a_i/x_i)
0.4	9.3/16.6 = 0.56	0.56/0.40 = 1.40
0.6	18.0/16.6 = 1.08	1.08/0.60 = 1.80
0.8	26.0/16.6 = 1.57	1.57/0.80 = 1.96

As the mole fraction of chloroform tends to zero, its activity coefficient tends to unity as is to expected from Henry's law.

Calculations Involving Equilibrium Constants and Different Free Energies

Problem 2.31

The equilibrium constant, K_P, for the reaction

$$2SO_2(g)+O_2(g)=2SO_3(g)$$

at 1100 K is 0.521 bar^{-1}. The ΔH^0 of the reaction is –200 kJ.

(i) Calculate K_C of the reaction.
(ii) What is ΔE^0 of the reaction?
(iii) Calculate K_C at 1500 K.

Solution

$$K_P = K_C(RT)^{\Delta n}$$

$$\text{or } K_C = \frac{K_P}{(RT)^{\Delta n}} = \frac{0.521}{(0.0821\times1100)^{-1}} = 47.0 \text{ L mol}^{-1}$$

$$\Delta H^0 = \Delta E^0 + RT\Delta n$$

$$\Delta E^0 = -200-\left(8.314\times10^{-3}\times1100\times(-1)\right)$$

$$= -200-(-9.15) = -191 \text{ kJ}$$

We now use the relation $\left(\frac{\partial \ln K_C}{\partial T}\right)_P = \frac{\Delta E^0}{RT^2}$

$$\log\frac{(K_C)_2}{(K_C)_1} = \frac{\Delta E^0}{2.303R}\left[\frac{1}{T_1}-\frac{1}{T_2}\right]$$

$$\log\frac{(K_C)_2}{47.0} = \frac{-191.0}{2.303\times8.314\times10^{-3}}\left(\frac{1500-1100}{1500\times1100}\right)$$

$$= \frac{-191.0}{0.019}\times\frac{400}{1500\times1100} = -\frac{764}{313.5} = -2.44$$

$$\frac{(K_C)_2}{47.0} = 3.631\times10^{-3}$$

$$(K_C)_2 = (K_C)_{1500} = 47.0\times3.631\times10^{-3} = 1.71\times10^{-1}$$

Problem 2.32

Calculate ΔG^0 and ΔA^0 for the reaction,

$$2CH_3OH(l)+3O_2(g)\rightarrow 2CO_2(g)+4H_2O(g)$$

using the following data and also given that ΔH^0 of $CH_3OH(l)$ is –638 kJ mol^{-1}.

	CH_3OH (*l*)	O_2(*g*)	CO_2(*g*)	H_2O(*g*)
ΔG^0_f (kJ mol^{-1})	–166.3	0.0	–394.4	–228.6
S^0 (J K^{-1} mol^{-1})	126.8	205.1	197.7	188.8

Solution

$$\Delta G^0_{\text{reaction}} = 2\Delta G^0_{f(CO_2,g)} + 4\Delta G^0_{f(H_2O,g)} - 2\Delta G^0_{f(CH_3OH,l)} - 3\Delta G^0_{f(O_2,g)}$$

$$= (-2\times394.4)+(4\times-228.6)-(2\times-166.3)-3\times0$$

$$= -788.8-914.4+332.6 = -1370.6 \text{ kJ}$$

$$\Delta S^0_{\text{reaction}} = 2S^0_{(CO_2,g)} + 4S^0_{(H_2O,g)} - 2S^0_{(CH_3OH,l)} - 3S^0_{(O_2,g)}$$

$$= 2\times197.7+4\times188.8-2\times126.8-3\times205.1$$

$$= 395.4+755.2-253.6-615.3 = 281.7 \text{ J K}^{-1}$$

ΔH^0_C for 2 moles of $CH_3OH(l) = -1276$ kJ

$$\Delta H^0 = \Delta E^0 + RT\Delta n$$

$$\Delta E^0 = -1276-3\times8.314\times10^{-3}\times298 \;(\Delta n = 6-3=3)$$

$$= -1276-7.4 = -1283.4 \text{ kJ}$$

$$\Delta A^0 = \Delta E^0 - T\Delta S^0 = -1283.4-298\times281.7\times10^{-3}$$

$$= -1283.4-83.9 = -1367.3 \text{ kJ}$$

Exact Differentials, Maxwell Relations and Legendre Transformation

Problem 2.33

Show that dV is an exact differential.

Solution

As an illustration, consider one mole of an ideal gas.

$$V = \frac{RT}{P}$$

$$dV = -\left[\frac{RT}{P^2}\right]dP + \left[\frac{R}{P}\right]dP$$

$$\frac{\partial}{\partial T}\left[\frac{-RT}{P^2}\right] = \frac{-R}{P^2}$$

$$\frac{\partial}{\partial P}\left[\frac{R}{P}\right] = \frac{-R}{P^2}$$

Since the RHS is identical, dV is an exact differential.

Problem 2.34

Derive the expressions for dA and dG in terms of pressure, temperature and volume, when there is no change in the number of moles of the constituents.

Solution

Since $A = E - TS$,

$$dA = dE - TdS - SdT$$

but $TdS = dE + PdV$

$$\therefore \quad dA = -PdV - SdT$$

$$G = H - TS$$

$$dE = dH - TdS - SdT$$

But $dH = dE + PdV + VdP = TdS + VdP$

$$\therefore \quad dG = VdP - SdT$$

Problem 2.35

Show that $\left(\frac{\partial E}{\partial V}\right)_T = 0$ for ideal gases.

Solution

There are two methods of proving this.

(i) The first method invokes the rate of change of entropy with volume as shown below.

$$\partial q = dE + PdV$$

$$\therefore TdS = dE + PdV$$

$$\therefore T\frac{dS}{dV} = \frac{dE}{dV} + P = \frac{dE}{dV} + \frac{nRT}{V}$$

$$T\left(\frac{\partial S}{\partial V}\right)_T = \left(\frac{\partial E}{\partial V}\right)_T + \frac{nRT}{V}$$

$$\therefore \left(\frac{\partial S}{\partial V}\right)_T = \frac{1}{T}\left(\frac{\partial E}{\partial V}\right)_T + \frac{nR}{V}$$

At constant temepraturе, $\left(\frac{\partial S}{\partial V}\right)_T = \frac{nR}{V}$

$$\therefore \left(\frac{\partial E}{\partial V}\right)_T = 0$$

(ii) The second method involves the use of Euler's identity for the exact differential dE.

$$dE = \left(\frac{\partial E}{\partial T}\right)_V dT + \left(\frac{\partial E}{\partial V}\right)_T dV$$

$$TdS = dE + PdV$$

$$= dE + \frac{nR}{V}dV$$

$$\therefore dS = \frac{1}{T}\left(\frac{\partial E}{\partial T}\right)_V dT + \left[\frac{1}{T}\left(\frac{\partial E}{\partial V}\right)_T + \frac{nR}{V}\right]$$

But, $dS = \left(\frac{\partial S}{\partial T}\right)_V dT + \left(\frac{\partial S}{\partial V}\right)_T dV$

$$\therefore \left(\frac{\partial S}{\partial T}\right)_V = \frac{1}{T}\left(\frac{\partial E}{\partial T}\right)_V$$

and $\left(\frac{\partial S}{\partial V}\right)_T = \frac{1}{T}\left(\frac{\partial E}{\partial V}\right)_T + \frac{nR}{V}$

However, $\frac{\partial}{\partial V}\left(\frac{\partial S}{\partial T}\right) = \frac{\partial}{\partial T}\left(\frac{\partial S}{\partial V}\right)$

Hence $\frac{1}{T}\frac{\partial^2 E}{\partial V \partial T} = \frac{\partial}{\partial T}\left[\frac{1}{T}\left(\frac{\partial E}{\partial V}\right)_T\right] + 0$

$$= \frac{1}{T}\frac{\partial^2 E}{\partial V \partial T} + \left(\frac{\partial E}{\partial V}\right)_T\left(-\frac{1}{T^2}\right)$$

Hence $\left(\frac{\partial E}{\partial V}\right)_T = 0$

Problem 2.36

Show that the definitions of the internal energy, Gibbs free energy and enthalpy arise naturally from the combined first and second law of thermodynamics using Legendre transformations.

Solution

For any exact differential dY of n variables, there exist $2^n - 1$ possible Legendre transformations.

$$dY = C_1dX_1 + C_2dX_2 + \ldots + C_ndX_n$$

If dY is given by $dY = C_1dX_1 + C_2dX_2$

There exist only three new variables (since $n = 2$) denoted as T_1, T_2 and T_3. The physical interpretation of T_1, T_2 and T_3 is then provided.

$$\therefore \quad T_1 = Y - C_1X_1$$

$$T_2 = Y - C_2X_2$$

$$T_3 = Y - (C_1X_1 + C_2X_2)$$

If $dE = TdS - PdV$

$$T_1 = A = E - TS$$

$$T_2 = H = E + PV$$

$$T_3 = G = H - TS$$

The new variables T_1, T_2 and T_3 are then identified as Helmholtz free energy, enthalpy and Gibbs free energy respectively. Note that there are only three new variables arising from the exact differential dE.

Problem 2.37

What is the mathematical definition of an extensive property?

Solution

A property M is extensive if for n different systems, $M(S_1 + S_2 + S_3 + \ldots S_n) = M(S_1) + M(S_2) + M(S_3) + \ldots M(S_n)$, where S_n refers to the quantity present in the nth system.

Problem 2.38

Derive each Maxwell relation taking into account Euler's identity for exact differentials.

Consider S and V as independent variables. Since dE (or dU) is an exact differential, we can write,

$$dE = TdS - PdV; \frac{\partial^2 E}{\partial V \partial S} = \frac{\partial^2 E}{\partial S \partial V}$$

Hence

$$\left(\frac{\partial E}{\partial S}\right)_V = T; \left(\frac{\partial E}{\partial V}\right)_S = -P$$

$$\left(\frac{\partial}{\partial V}\right)_S\left(\frac{\partial E}{\partial S}\right)_V = \left(\frac{\partial}{\partial S}\right)_V\left(\frac{\partial E}{\partial V}\right)_S$$

$$\left(\frac{\partial T}{\partial V}\right)_S = -\left(\frac{\partial P}{\partial S}\right)_V$$

Consider S and P as independent variables. Since dH is an exact differential, we can write,

$$d(E+PV) = TdS + VdP;\ dE = TdS - PdV = TdS - dPV + VdP$$

$$PdV = dPV - VdP$$

$$dH = TdS + VdP$$

$$\left(\frac{\partial H}{\partial S}\right)_P = T;\ \left(\frac{\partial H}{\partial P}\right)_S = V$$

$$dH = \left(\frac{\partial H}{\partial S}\right)_P dS + \left(\frac{\partial H}{\partial P}\right)_S dP$$

$$\frac{\partial^2 H}{\partial P \partial S} = \frac{\partial^2 H}{\partial S \partial P}$$

$$\left(\frac{\partial T}{\partial P}\right)_S = \left(\frac{\partial V}{\partial S}\right)_P$$

Consider T and V as independent variables. Since dF is an exact differential,

$$dE = TdS - PdV = d(TS) - SdT - PdV$$

$$dE - TS = -SdT - PdV$$

$$dF = -SdT - PdV$$

$$dF = \left(\frac{dF}{dT}\right)_V dT + \left(\frac{dF}{dV}\right)_T dV$$

$$\left(\frac{dF}{dT}\right)_V = -S;\ \left(\frac{dF}{dV}\right)_T = -P$$

$$\frac{\partial^2 F}{\partial T \partial V} = \frac{\partial^2 F}{\partial V \partial T}$$

$$\left(\frac{dS}{dV}\right)_T = \left(\frac{dP}{dT}\right)_V$$

Consider T and P as independent variables. Since dG is an exact differential, we can write,

$$dE = TdS - PdV = d(TS) - SdT - PdV + VdP$$

$$dG = VdP - SdT$$

$$-TS = -SdT - PdV$$

$$dF = \left(\frac{dG}{dT}\right)_P dT + \left(\frac{dF}{dP}\right)_T dP$$

$$\left(\frac{dG}{dT}\right)_P = -S;\ \left(\frac{dG}{dP}\right)_T = V$$

$$\frac{\partial^2 G}{\partial T \partial P} = \frac{\partial^2 G}{\partial P \partial T}$$

$$-\left(\frac{dS}{dP}\right)_T = \left(\frac{dV}{dT}\right)_P$$

Thus, the following four Maxwell relations are obtained from various exact differentials.

$$\left(\frac{\partial T}{\partial V}\right)_S = -\left(\frac{\partial P}{\partial S}\right)_V$$

$$\left(\frac{\partial T}{\partial P}\right)_S = \left(\frac{\partial V}{\partial S}\right)_P$$

$$\left(\frac{dS}{dV}\right)_T = \left(\frac{dP}{dT}\right)_V$$

$$-\left(\frac{dS}{dP}\right)_T = \left(\frac{dV}{dT}\right)_P$$

EXERCISES

Practice Problems

1. 2 moles of an ideal gas at 47°C and a pressure of 10 bar (10 atm) is subjected to an isothermal expansion against an external pressure of 2 bar until the gas pressure reaches the external pressure. Calculate (i) $W_{rev} - W$ (ii) ΔE and ΔH (iii) Q_{rev} (iv) ΔS.

2. 5 moles of an ideal gas ($C_V = 21.0$ J K^{-1} mol^{-1}) at 400 K and 10 bar undergo an adiabatic expansion against an external pressure of one bar until its volume is tripled. Calculate (i) $\Delta H - \Delta E$ (ii) $T_{rev} - T_{irr}$ (iii) $|W_{rev}| - |W_{irr}|$ (iv) ΔS for the process.

3. Two moles of a gas at 400 K obeying van der Waals equation with constants $a = 0.2351$ N m^4 mol^{-2} and $b = 3.98\times10^{-5}$ mol^3 mol^{-1} expand isothermally and reversibly from an initial volume of 20 dm^3 to a final volume of 400 dm^3. Calculate (i) W (ii) Q (iii) ΔE and ΔH (iv) ΔS.

4. The ΔG_f^0 of $NH_3(g)$, $H_2O(g)$ and NO(g) are –16.5, –228.6, and 86.6, all in kJ mol^{-1} respectively. Calculate the equilibrium constants K_P and K_C for the reaction

$$2NH_3(g) + \frac{5}{2}O_2(g) \rightarrow 2NO(g) + 3H_2O(g)$$

5. Show that $C_P - C_V = \left(\frac{\partial V}{\partial T}\right)_P \left\{P + \left(\frac{\partial E}{\partial V}\right)_T\right\}$ and deduce from this equation that $C_P - C_V = R$ for an ideal gas.

6. One mole of benzene is heated from 300 K to 330 K. The coefficient of volume expansion is 1.24×10^{-3} deg^{-1}. The

specific volume of benzene in this temperature range is 1.136×10^{-6} m³ g⁻¹. The internal pressure $\left(\frac{\partial E}{\partial V}\right)_T$ = 3.55×10^{8} N m^{-2}. Calculate ΔH and ΔE for the process. C_P for liquid benzene is 173.0 J K^{-1} mol^{-1}.

7. For the dissociation of iodine vapour into atoms according to $I_2(g) \rightleftharpoons 2I(g)$, the following data were obtained.

Temp (°C)	900	1000
Total pressure (N m^{-2})	9300	11365
Degree of dissociation	0.34	0.52

Calculate (i) ΔH^0 (ii) ΔE^0 of the reaction.

8. The enthalpy change for the reaction
$2C(s)+3H_2(g)=C_2H_6(g)$
is –85 kJ. The enthalpy of sublimation of carbon (graphite) is 715 kJ mol^{-1}. The C–C and C–H bond energies are 348 and 436, both in kJ mol^{-1} respectively. What is the H–H bond energy?

9. 3 moles of an ideal monatomic gas (C_V = 12.5 J K^{-1} mol^{-1}) expands from an initial volume of 10 dm³ to a final volume of 50 dm³ and its temperature increases from 300 to 500 K. Calculate ΔS, ΔG and ΔA for the process. S^0 of the gas is 164.0 J K^{-1} mol^{-1} and is independent of temperature.

10. 2 moles of benzene solid at its melting point 5.4°C is allowed to melt to liquid and the liquid is heated to its boiling point of 80°C and converted to vapour. Calculate the entropy change associated with the process. The enthalpy of fusion of benzene ΔH_f^0 at the melting point is 9.9 kJ mol^{-1}; enthalpy of vapourisation ΔH_v^0 at the boiling point = 31.0 kJ mol^{-1} and the heat capacity C_P of the liquid is 118.4 J K^{-1} mol^{-1} and is independent of temperature.

11. One mole of super-cooled gallium at 293 K and 1 bar is suddenly frozen to solid under the same conditions. The melting point of gallium is 303 K and ΔH_f^0 5.6 kJ mol^{-1}. C_P of solid and liquid gallium are 102 and 26, both in J K^{-1} mol^{-1}. Calculate (i) ΔS_{system} (ii) ΔS_{surr} (iii) ΔG for the process. The S^0 values of solid and liquid gallium are 41 and 59 respectively (all in J K^{-1} mol^{-1}).

12. The van der Waals coefficients a and b for oxygen are 0.1378 N m⁴ mol^{-2} and 3.18×10^{-5} m³ mol^{-1} respectively. Calculate the fugacity and fugacity coefficient of the gas at 1000 bar at 298 K.

13. Determine the enthalpy change of the reaction $\frac{1}{2}N_2(g) + \frac{1}{2}O_2(g) \rightarrow NO(g)$ at 1000 K given $\Delta H_f^0(NO)$ = 90 kJ mol^{-1} at 298 K. The variation of heat capacity for various gases with temperature is given below (all the data are given in J K^{-1} mol^{-1}).

$C_P^0(N_2)=26.98+5.91\times10^{-3}T-3.38\times10^{-7}T^2$

$C_P^0(O_2)=25.5+13.61\times10^{-3}T-42.55\times10^{-7}T^2$

$C_P^0(NO)=26.98+5.91\times10^{-3}T-106.5\times10^{-7}T^2$

14. For the reaction $NH_3(g)+HCl(g)\rightarrow NH_4Cl(s)$ K_P at 700 K = 0.2 bar and ΔH^0 = –160.0 kJ mol^{-1}. (i) Calculate K_C (ii) If solid NH_4Cl is heated at this temperature in the presence of HCl of 0.5 bar, what are the partial pressures of NH_3 and HCl at equilibrium? (iii) What is ΔS of the reaction? (iv) Calculate ΔE of the reaction.

15. Ammonium carbamate decomposes according to $NH_4COONH_2(s)\rightarrow 2NH_3(g)+CO_2(g)$. At 298 K, the total pressure of the gases in equilibrium with the solid is 11.75 kPa. (i) Calculate K_P in units of bar³. (ii) If 4 g of carbamate decomposes in a 1 dm³ vessel at this temperature, how much is left behind at equilibrium? (iii) If the partial pressure of CO_2 is found to be 0.232 bar at a temperature of 323 K, what is K_P? Calculate ΔH^0 of the reaction.

16. For the reaction $2NOCl(g)\leftrightarrow 2NO(g)+Cl_2(g)$, $K_P = 1.81\times10^{-4}$ at 400 K. (ii) Derive an equation for K_P in terms of α, degree of dissociation, and total pressure. (ii) Making simplifying assumptions for α, calculate α at a total pressure of 506.6 kPa. (iii) At what total pressure, is α equal to 0.2? Given ΔH^0 of the reaction as 75.4 kJ, calculate ΔS^0 and ΔA^0 of the reaction.

17. Derive the thermodynamic equations of state

(i) $\left(\frac{\partial E}{\partial V}\right)_T = T\left(\frac{\partial P}{\partial T}\right)_V - P$ (ii) $\left(\frac{\partial H}{\partial P}\right)_T = V - T\left(\frac{\partial V}{\partial T}\right)_P$

Show, from these equations, that for an ideal gas there is no change in internal energy with volume or there is no change in enthalpy with pressure at constant temperature.

18. In the reaction $CO(g)+H_2O(g)\rightarrow CO_2(g)+H_2(g)$, when one mole of CO and two moles of H_2O are allowed to react at 775 K and 0.88 mole of CO_2 were formed at equilibrium, calculate the equilibrium constant of the reaction. In an experiment, it was proposed to pass CO at 2025 kPa, watervapour at 500 kPa and withdraw the products at 1000 kPa each. Calculate ΔG and state whether the reaction is feasible.

19. The following vapour pressure data were obtained for acetone–carbon disulfide mixtures at a certain temperature. The data are given below.

Mole fraction of acetone	0.0	0.1	0.3	0.5	0.7	1.0
Partial pressure of acetone (mm)	0.0	97.5	210.0	243.8	273.8	345.0
Partial pressure of CS_2 (mm)	510.0	483.8	442.5	405.0	330.0	0.0

Calculate the activity and activity coefficients of both components using Raoult's law standard states.

20. For a certain gas, the van der Waals constants, a and b, are 0.1505 N m^4 mol^{-2} and 3.98×10^{-5} m^3 mol^{-1} respectively. Calculate $\mu_{J\text{-}T}$ at 298 K, also calculate ΔH for the isothermal compression on one mole of the gas from 101.325 kPa to 5.07 10^4 kPa. The heat capacity C_P^0 of the gas is 29.2 J K^{-1} mol^{-1}.

21. Calculate ΔG and ΔS associated with the mixing of 2 moles of nitrogen and 3 moles of oxygen at 298 K and 101.325 kPa.

22. Show that the following are exact differentials: (i) dG (ii) dP (iii) dE.

23. Depict schematically, P vs V and T vs S diagrams for a Carnot cycle.

24. Show that $\Delta F = -W_{iso}^{rev}$, F denotes the Helmholtz free energy.

25. Depict schematically the variation of Gibbs free energy with temperature for solids, liquids, and gases.

26. Which of the following will violate the second law of thermodynamics?

(i) Intersection of two reversible adiabatics.

(ii) Intersection of a reversible isotherm and a reversible adiabatic.

(iii) Intersection of two reversible isotherms.

(iv) Intersection of two irreversible adiabatics.

27. If the temperature of an ideal gas is increased from T_1 to T_2, what will be the entropy change for an (i) isobaric process (ii) isometric process?

28. If $\left(\dfrac{\partial E}{\partial V}\right)_T = 0, \left(\dfrac{\partial E}{\partial P}\right)_T$ is

(i) infinity (ii) zero (iii) V (iv) negative.

29. Derive the expression for $C_p - C_v$ for the equation of state represented as

$$PV + \frac{a}{V} = RT.$$

30. In the following equation of state,

$$PV^n = \text{Constant}$$

what processes do the values 0, 1, ∞ and γ for n represent?

31. At what mole fraction will the molar entropy of mixing of ideal solutions be maximum for a two-component system?

32. Show that the maximum efficiency can be obtained only for a reversible cyclic process.

33. The molar entropy of an ideal gas as a function of the internal energy and volume varies as

$$S/R = c_1 \ln E + c_2 \ln V$$

where c_1 and c_2 are constants. Derive the internal energy of the gas using the above equation.

Multiple Choice Questions

1. Among the following, the correct statement is

(a) In an open system, transfer of mass alone occurs across the system.

(b) A system is said to be a closed system if transfer of mass but not of energy is permitted between the system and surroundings.

(c) For an isolated system no exchange of energy or mass with surroundings is possible.

(d) A closed system does not produce any significant changes in the surroundings.

2. The pair which is not an intensive property among the following is
(a) molar heat capacity and molar volume **(b)** internal energy and entropy **(c)** pressure and temperature **(d)** chemical potential and density

3. The thermodynamic condition applicable for an ideal gas is
(a) $(dH/dT)_P = 0$ **(b)** $\mu_{JT} \neq 0$ **(c)** $(dH/dV)_T \neq 0$ **(d)** $(dU/dV)_T = 0$

4. The equation that correctly represents the charateristic of a path function is

(a) $\int_{\text{State 1}}^{\text{State 2}} dq_{rev} = q_{rev}$ **(b)** $\int_{\text{State 1}}^{\text{State 2}} dW = W_2 - W_1$

(c) $\int_{\text{State 1}}^{\text{State 2}} dU = U_2 - U_1$ **(d)** $\int_{\text{State 1}}^{\text{State 2}} dG = G_2 - G_1$

5. Two moles of liquid SO_2 are evaporated at its boiling point (263 K) into vapour under reversible conditions. Its enthalpy of vaporisation is 25.0 kJ mol^{-1}. For this process
(a) $W = -8.74$ kJ **(b)** $\Delta S = 190.1$ J K^{-1} **(c)** $\Delta U = 23.8$ kJ **(d)** $\Delta G \neq 0$

6. One mole of an ideal monatomic gas at 350 K and 501.625 kPa expands against an external pressure of 101.325 kPa until its volume reaches twice the original value. For this process,
(a) $W = 1176.2$ J **(b)** $\Delta U = \Delta H \neq 0$ **(c)** $\Delta S = 5.76$ J K^{-1} **(d)** $\Delta S = 1.7$ J K^{-1}

7. Two moles of an ideal monatomic gas ($Cv = 12.5$ J K^{-1} mol^{-1}) at 40°C and occupying a volume of 10 dm^3 expands adiabatically and reversibly until its final volume is 50 dm^3. For this process,
(a) $\Delta H = -856$ J **(b)** $T_2 = 250$ K **(c)** $\Delta U = 6.5$ kJ **(d)** $\Delta S \neq 0$

8. Which of the following statements for the reversible work done in the isothermal reversible work of expansion of an ideal gas and that of a gas obeying van der Waals equation undergoing the same change of state, is true?
(a) The reversible work by a gas obeying van der Waals equation is greater than that done by an ideal gas **(b)** The reversible work done by an ideal gas is greater than that done by a gas obeying van der Waals equation **(c)** There is no difference between the reversible work done by an ideal gas and that done by a gas obeying van der Waals equation

(d) Intermolecular forces between the molecules of a gas obeying van der Waals equation do not have any effect on the work done by it in the reversible expansion from that observed in the case of an ideal gas.

9. The Joule–Thomson coefficient is defined under
(a) isobaric conditions **(b)** isothermal conditions **(c)** isentropic conditions **(d)** isoenthalpic conditions

10. The van der Waals constants a and b for oxygen are 1.360 atm L^2 mol^{-2} and 0.0318 L mol^{-1} respectively. The enthalpy change ΔH for the isothermal expansion of 2 moles of the gas (in kJ) at 400 K from 200 atm to 1 atm is
(a) 4.32 **(b)** 2.16 **(c)** 6.12 **(d)** 5.65

11. Consider the isothermal and adiabatic expansions of an ideal gas under reversible conditions starting from the same initial state (same pressure). Point out the correct statement from the following ones:
(a) The P–V curve under isothermal conditions is much steeper than the curve under adiabatic conditions **(b)** The final volume of the gas under adiabatic conditions is greater than under isothermal conditions **(c)** The work done under isothermal conditions is smaller than under adiabatic conditions for the same change of pressure **(d)** The steepness of the P–V curve under adiabatic conditions is much more than under isothermal conditions.

12. Two moles of an ideal monatomic gas (C_p = 20.8 J K^{-1} mol^{-1}) at 320 K and 10 bar pressure expand against a constant external pressure of one bar under adiabatic conditions until its volume increases to twice its initial value. For this process
(a) W = 400 J **(b)** T_2 = 220 K **(c)** ΔH = –1327 J **(d)** ΔU = –1600 J

13. Which of the following statements regarding the assignment of standard molar enthalpy H^0 of elements is correct?
(a) The enthalpy of rhombic sulphur at 298 K and 1 atm is $H^0 = 0$ **(b)** The enthalpy of diamond (a form of carbon) at 298 K and 1 atm is $H^0 = 0$ **(c)** The enthalpy of ozone (O_3) at 298 K and 1 atm is $H^0 = 0$ **(d)** The enthalpy of iodine vapour at 300 K and 1 atm is $H^0 = 0$.

14. The standard enthalpy of combustion (ΔHc) of propane is –2220 kJ mol^{-1}. Given $\Delta H^0(CO_2)$ = –394 kJ mol^{-1} and ΔH^0_f (H_2O) = –286 kJ mol^{-1} (both ΔH^0_f at 298 K), the $\Delta H^0_f(C_3H_8)$ is (in kJ mol^{-1})
(a) 138 **(b)** 318 **(c)** –212 **(d)** –106

15. The standard enthalpy of combustion of benzene(l) is –3268 kJ mol^{-1}. Given ΔH^0_f (H_2O) = –286 kJ mol^{-1} $\Delta H^0_f(CO_2, g)$ = –394 kJ mol^{-1}, the heat change of the reaction, Q_v, is (in kJ)
(a) 3722 **(b)** –3272 **(c)** 3520 **(d)** –3210

16. The standard ΔH^0_f of ammonia at 298 K is –46.4 kJ mol^{-1}. Given C_p of $N_2(g)$, $H_2(g)$ and $NH_3(g)$ as 27.1, 29.2, 26.0 (all in J K^{-1} mol^{-1}), ΔH^0_f at 1000 K is (in kJ) (assume heat capacity is independent of temperature)
(a) –275.0 **(b)** 180.4 **(c)** –136.8 **(d)** 140.6

17. From the statements given below, pick out the correct statement relating to maximum flame temperature in combustion reactions.
(a) It is the temperature attained by the products in a combustion reaction under isothermal conditions **(b)** It is the temperature attained by the reacting system at constant volume under adiabatic conditions **(c)** It is the temperature attained by the combustion of hydrocarbons in presence of excess of air at constant pressure under adiabatic conditions **(d)** Maximum flame temperature is synonymous with maximum explosion temperature in combustion reactions.

18. The ΔH^0_f (NaCl, aq) = –407 kJ mol^{-1} ΔH^0_f(HCl, aq) = –167 kJ mol^{-1}, ΔH^0_f(NaOH, aq) = –470 kJ mol^{-1}. From these data, $\Delta H^0_f(OH^-, aq)$ is (in kJ mol^{-1})
(a) –440 **(b)** –195 **(c)** –360 **(d)** –230

19. From the following data, calculate the bond enthalpy of the C=O bond.
(1) $3C(s) + 3H_2(g) + ½O_2(g) \rightarrow CH_3COCH_3(g)$ ΔH_0 = –248 kJ
(2) $C(s) \rightarrow C(g)$ ΔH_0 = 717 kJ
(3) $H_2(g) \rightarrow 2H(g)$ ΔH_0 = 436 kJ
(4) $½O_2(g) \rightarrow O(g)$ ΔH_0 = 249 kJ
(5) Bond enthalpy of the C–H bond ΔH_0 = 413 kJ
(6) Bond enthalpy of C–C bond ΔH_0 = 348 kJ
(a) 827 **(b)** 782 **(c)** 728 **(d)** 872

20. Given that in a Carnot refrigerator, T_2 and T_1 are the higher and lower temperature reservoirs, Q_1 is the heat taken by reservoir at T_1, the coefficient of performance of the machine is
(a) $W/Q_2 = (T_2 - T_1)/T_2$ **(b)** $W/Q_1 = (T_1 - T_2)/T_1$ **(c)** $-W/Q_2 = (T_1 - T_2)/T_2$ **(d)** $-W/Q_1 = (T_2 - T_1)/T_1$

NOTE

- Q_2 is heat given out by the reservoir at T_2.

21. The boiling points of liquid SO_2 and liquid H_2 are 263 K and 20 K respectively. The efficiency of the Carnot engine operating between these heat reservoirs at these temperatures is
(a) 0.96 **(b)** 0.85 **(c)** 0.92 **(d)** 0.78

22. 5 moles of an ideal gas (Cv = 12.5 J K^{-1} mol^{-1}) are heated from 300 K to 500 K and its pressure changes from 10 bar to 2 bar. ΔS for the process (in J K^{-1}) is
(a) 120.1 **(b)** 201.1 **(c)** 101.2 **(d)** 110.2

23. The enthalpy of vapourisation of water at 373 K and 1 atm is 41 kJ mol^{-1}. For the phase transformation, 2 moles of H_2O (l, 1 atm, 373 K) = 2 moles H_2O (v,1 atm, 373 K) which of the following results is correct?

(a) $\Delta S = 0$, $\Delta G = 109$ J (b) $\Delta G = -109$ J, $\Delta S = -219.8$ (c) $\Delta S = 219.8$
$\Delta G = 0$ (d) $\Delta G = 218.1$ J, $\Delta S = 0$

24. Two moles of an ideal gas (C_v =12.5 J K^{-1} mol^{-1}) at 27°C and 10 bar pressure is adiabatically expanded against a constant external pressure of 1 bar to thrice its initial volume. For this process, (assume 1 bar = 1 atm)
(a) ΔH = 1660 J **(b)** ΔS = 14.7 J K^{-1} **(c)** ΔU = 1000 J **(d)** Final T = 200 K

25. In the isothermal reversible expansion of an ideal gas from state1 to state 2
(a) $\Delta U \neq 0$ **(b)** $\Delta S = 0$ **(c)** $\Delta G = \Delta A = 0$ **(d)** $\Delta G = \Delta A = -$ve

26. 1 mole of an ideal gas at a pressure of 20 atm and 300 K undergoes an isothermal expansion against vacuum until its pressure reaches 5 atm. For this process,
(a) $\Delta S_{sys} + \Delta S_{surr}$ = 11.53 J K^{-1} **(b)** ΔG = 3458 J **(c)** $\Delta A \neq \Delta G$ **(d)** $Q \neq 0$

27. 1 mole of a piece of copper (C_p=26.6 J K^{-1} mol^{-1}) at 100°C is placed in a large heat reservoir maintained at 60°C under insulated conditions. For this process
(a) ΔH(copper) = 1.0 kJ **(b)** ΔS(copper) = 3.0 J K^{-1} **(c)** ΔS(reservoir) = –4.0 J K^{-1} **(d)** ΔS(copper) + ΔS(reservoir) = 0.2 J K^{-1}

28. The van der Waals constants a and b for carbon monoxide are 1.39 dm^2 atm mol^{-2} and 0.0391 dm^3 mol^{-1} respectively. In a certain experiment, 5 moles of the gas at 2 atm pressure are heated from 300 K to 600 K until the volume increases to twice its value. The entropy change for this process is (in J K^{-1}) (C_v of the gas = 20.8 J K^{-1} mol^{-1})
(a) 85 **(b)** 151 **(c)** 101 **(d)** 50

29. For a reaction to occur spontaneously
(a) $\Delta S_{sys} + \Delta S_{surr}$ less than 0 **(b)** $\Delta G_{(T,P)}$ less than 0 **(c)** $\Delta A_{(V,T)}$ greater than 0 **(d)** $\Delta S_{sys} + \Delta S_{surr} = 0$.

30. 5 moles of super-cooled water at 263 K are converted to ice at constant pressure of 1 atm and at the same temperature under adiabatic conditions. Given $\Delta H_{(fusion,ice)}$ = 6.0 kJ mol^{-1}) and the heat capacities of water and ice as 75.0 and 38.0 (both in J K^{-1} mol^{-1}), the entropy change of the system and surroundings put together is (in J K^{-1})
(a) –14.0 **(b)** 11.0 **(c)** 7.5 **(d)** 6.9

31. Glucose breaks down in the muscle tissue to form lactic acid according to
$C_6H_{12}O_6(l) \rightarrow 2\ CH_3CHOHCOOH(l)$
Using the data given below, at 320 K,

	ΔH^0_f(kJ mol^{-1})	S^0 (J K^{-1} mol^{-1})	$C_p(0)$ (J K^{-1} mol^{-1})
Glucose	–1273.0	209.0	219.0
Lactic acid	–674.0	192.0	128.0

indicate which of the following results is correct.
(a) $\Delta G^0 = \Delta A^0$ = –131 kJ **(b)** ΔS^0 = –200 J K^{-1} **(c)** ΔH^0 = 100 kJ **(d)** $\Delta S(0)_{surr}$ = –240 J K^{-1}

32. For the reaction $2SO_2(g) + O_2(g) \rightleftharpoons 2SO_3(g)$, K_p at 900 K is 43.56. In a given experiment, it is proposed that the reactants SO_2 and O_2 be introduced in the vessel at 20 atm and 10 atm respectively and withdraw the product SO_3 at a pressure of 15 atm. Under these conditions
(a) the reaction is not feasible **(b)** ΔG_{reac} = 78.02 kJ **(c)** reaction is feasible under the standard state and under the given conditions **(d)** ΔG^0 = 130 kJ

33. Which of the following statements is true regarding the variation of the equilibrium constant of a reaction with temperature?
(a) For an endothermic reaction, the equilibrium constant decreases with increase in temperature **(b)** For an exothermic reaction, the equilibrium constant increases with decrease in temperature **(c)** The variation of equilibrium constant of a reaction with temperature is not predictable **(d)** The equilibrium constant of a reaction does not depend on temperature.

34. For the following transformation
1 mole C_2H_5OH (l,352 K, (b.p.)(1atm) $\rightarrow$ 1 mole C_2H_5OH (v, 352 K, 1 atm) $\rightarrow$ 1 mole C_2H_5OH (v, 400 K, 1 atm) $\rightarrow$ 1 mole C_2H_5OH (v, 5 atm, 400 K)
(S^0 C_2H_5OH = 283 J K^{-1} mol^{-1}) and assume it to be independent of temperature. The total ΔG and total ΔA are in the order
(a) ΔG = 4.1 kJ, ΔA = 7.0 kJ **(b)** ΔG = 16.4 kJ, ΔA = 28.0 kJ **(c)** ΔG = –16.4 kJ
ΔA = –28.0 kJ **(d)** ΔG = –8.2 kJ, ΔA = –14.0 kJ

35. The van der Waals constants a and b for methane are a = 2.250 L^2 atm mol^{-2} and b = 0.0428 L mol^{-1} and assuming it to obey van der Waals equation, the change in ΔA when 2 moles of the gas is heated at 350 K when its volume increases from 2 L to 5 L (in J) is
(a) 3525 **(b)** –5352 **(c)** 2553 **(d)** –3255

36. 8 moles of nitrogen and 2 moles of oxygen are mixed at 300 K under constant pressure of 1 atm. The ΔG_{mix} and ΔS_{mix} in that order for the formation of the gas mixture are
(a) ΔG_{mix} = –12.52 kJ, ΔS_{mix} = 41.7 J **(b)** ΔG_{mix} = –6.26 kJ, ΔS_{mix} = 109 J **(c)** ΔG_{mix} = 25 kJ, ΔS_{mix} = 83 J **(d)** ΔG_{mix} = 12.52 kJ, ΔS_{mix} = 41.7 J

37. The ΔG^0_f of some metal oxides are given below:

Metal oxide	Cobalt oxide (CoO)	ZnO	MgO
ΔG^0_f(kJ mol^{-1})	–162.0	–262.0	–499.0

Using the above data, indicate the correct answer from the following:
(a) Zn can reduce cobalt oxide to Co but cannot reduce MgO
(b) Cobalt cannot reduce MgO but can reduce ZnO
(c) MgO can oxidise both Co and Zn
(d) Mg can reduce CoO but not reduce ZnO

38. The chemical potential, μ_I, of a species I is an
(a) extensive property **(b)** intensive property **(c)** constitutive property **(d)** colligative property

39. The standard free energy of
(a) Liquid bromine at 300 K and 1 atm is 0 **(b)** Diamond at 298 K and 1 atm is 0 **(c)** Ozone at 273 K and 1 atm is 0 **(d)** Hydrogen gas at 298 K and 1 atm is 0

40. For the reaction $N_2(g) = 2N(g)$, ΔG^0 at 2000 K = 715 kJ. The equilibrium partial pressure of N in the reaction mixture is (in atm)
(a) 2.31×10^{-9} **(b)** 1.16×10^{-10} **(c)** 4.62×10^{-10} **(d)** 8.24×10^{-9}

41. For the reaction $SO_2Cl_2(g) \rightleftharpoons SO_2(g) + Cl_2(g)$, ΔG^0 at 305 K = 8.9 kJ. Starting with 1 mol L^{-1} of SO_2Cl_2 at this temperature, the equilibrium concentrations of SO_2 and Cl_2 are
(a) $(SO_2) = (Cl_2) = 0.066$ **(b)** $(SO_2) = (Cl_2) = 0.033$ **(c)** $(SO_2) = (Cl_2) = 0.017$ **(d)** $(SO_2) = (Cl_2) = 0.068$

42. The ΔG^0 for the reactions $CaCO_3(s) \rightarrow CaO(s) + CO_2(g)$ and C(graphite) + $CO_2(g) \rightarrow 2\ CO(g)$ are 15.3 kJ and –3.8 kJ respectively at 1000 K. Solid carbon(graphite), CaO and $CaCO_3$ are mixed and allowed to equilibrate at 1000 K. The partial pressure of the gaseous product formed is (in atm)
(a) 0.232 **(b)** 0.223 **(c)** 0.189 **(d)** 0.322

43. The phosphoenol pyruvate (PEP) hydrolysis reaction according to PEP + $H_2O \rightleftharpoons$ Pyruvate + Phosphate, $\Delta G^0 = -62$ kJ is used in cells to drive the synthesis of ATP from ADP and phosphate according to ADP + Phosphate $\rightarrow$ ATP, ΔG^0 = 31 kJ through a coupled reaction. The equilibrium constant of the reaction ADP + Phosphate + PEP + $H_2O \rightarrow$ ATP + Pyruvate + Phosphate is
(a) 1.4×10^6 **(b)** 2.7×10^5 **(c)** 5.8×10^3 **(d)** 3.4×10^4

44. The standard ΔG^0 and ΔH^0 for the reaction $H_2(g) + Br_2(g) \rightarrow 2HBr(g)$ at 298 K are –53.3 kJ mol^{-1} and –36.5 kJ mol^{-1}. Using these data, the ΔG^0 of the reaction at 600 K is (in kJ)
(a) –141.0 **(b)** –184.0 **(c)** –172.0 **(d)** –282.0

45. Assuming that methane behaves as a van der Waals gas and has constants a and b equal to 2.25 L^2 atm mol^{-2}, 0.0428 L mol^{-1} respectively, calculate ΔS (in J K^{-1}) when 2 moles of methane expand from a volume of 1 L to 5 L at 400 K.
(a) 32.74 **(b)** 27.34 **(c)** 42.37 **(d)** 23.47

46. Which of the following statements/equations regarding the Duhem–Margules equation for a two component system consisting of two liquids A and B is correct? (Note that p is partial pressure and X is mole fraction in the equations below.)
(a) $dp_A/dX_A = dp_B/dX_B$ **(b)** $dp_A/dX_A \neq dp_B/dX_B$ **(c)** If one component of a binary liquid mixture A behaves ideally, the other component must also do so **(d)** $d \ln p_A/d \ln X_A \neq d \ln p_B/d \ln X_B$

47. The van der Waals constants a and b for argon are 2.321 L^2 atm mol^{-2} and 0.0398 L mol^{-1} respectively. The fugacity of the gas at 1000 atm and 300 K (in atm) is
(a) 420 **(b)** 110 **(c)** 340 **(d)** 280

48. The virial equation of state for hydrogen is $PV^* = RT\ (1 + 6.4\times10^{-4}\ P)$ valid up to a pressure of 1500 atm at 300 K. The activity coefficient of the gas under these conditions is
(a) 1.52 **(b)** 2.15 **(c)** 1.69 **(d)** 2.63

49. Point out the correct statement about the component A in a liquid mixture consisting of two liquids A and B.
(a) If the liquid mixture is not ideal, $a_A = X_A$ **(b)** The activity of A, a_A is less than X_A if the mixture shows positive deviations from Raoult's law **(c)** The activity coefficient γ_A is less than 1 if the mixture of A and B shows negative deviations from Raoult's law **(d)** The activity coefficient γ_A is greater than 1 if the mixture of A and B shows no deviation from Raoult's law.

50. It is required that a cooling machine operating in a room at 300 K has to make 9 kg of ice from water at 273 K. The amount of heat given out by the machine to the surroundings is (in kJ) (ΔH_f for water = 6.0 kJ mol^{-1})
(a) 3792 **(b)** 3297 **(c)** 3927 **(d)** 3972

51. The limiting partial molal volume of $MgSO_4$ in water is –1.4 cm^3 mol^{-1}. This value may be ascribed to
(a) The salt being a 2:2 electrolyte **(b)** The high ion pairing effect of the salt **(c)** The salt having no effect on the hydrogen bonds among the water molecules in water **(d)** the breaking down of H-bonded structure of water and the ions being hydrated.

52. The standard enthalpies (ΔH^0_f) and entropies S^0 of some compounds are given below:

	$CO_2(g)$	$H_2O(l)$	$C_6H_{12}O_6(s)$	$O_2(g)$
ΔH^0_f (kJmol^{-1})	–394.0	–286.0	–1273.0	0.0
S^0 (J K^{-1} mol^{-1})	214.0	70.0	209.0	205.0

The equilibrium constant of the reaction 6 $CO_2(g)$ + 6 $H_2O(l) \rightarrow C_6H_{12}O_6(s) + 6\ CO_2(g)$ is
(a) $10^{450.0}$ **(b)** $10^{-506.0}$ **(c)** $10^{-540.6}$ **(d)** $10^{524.9}$

53. The slopes of the plot of μ vs T(X) for the solid, liquid, vapour phases of a substance at constant pressure, decrease in the order (including the sign)
(a) X(s) greater than X(l) greater than X(v) **(b)** X(v) greater than X(l) greater than X(s) **(c)** X(l) greater than X(v) greater than X(s) **(d)** X(v) greater than X(s) greater than X(l).

54. Which of the following relations is equal to V of the system?
(a) $(dE/dV)_S$ **(b)** $(dA/dV)_T$ **(c)** $(dG/dT)_P$ **(d)** $(dH/dP)_S$

55. Which of the Maxwell's relations is correct?
(a) $(dS/dP)_T = (dV/dT)_P$ **(b)** $(dT/dV)_S = (dP/dS)_V$ **(c)** $(dS/dV)_T = (dP/dT)_V$ **(d)** $(dT/dP)_S = -(dV/dS)_P$

56. For the mixing of a number of ideal gases at constant T and P,
(a) ΔV_{mix} is positive **(b)** ΔG_{mix} is negative **(c)** ΔS_{mix} is negative **(d)** ΔH_{mix} is positive

57. Two moles of an ideal gas A and three moles of an ideal gas B are mixed at constant T and P. The entropy of mixing, ΔS_{mix}, for this process is (in J K^{-1})
(a) 27.9 **(b)** 18.7 **(c)** –24.5 **(d)** –30.6

58. Which of the following is true for the Joule–Thomson coefficient in the case of the van der Waals equation of state?

(a) directly proportional to C_p (b) inversely proportional to C_p (c) independent of C_p (d) directly proportional to T

59. For an ideal gas, the coefficient of thermal expansion equals
(a) $1/T$ (b) R/P (c) 0 (d) $1/C_p$

60. The integrating factor for the heat change (δQ) is
(a) dE (b) $1/T$ (c) dH (d) PdV

61. The natural canonical variables for the Gibbs free energy of a closed system are
(a) P and T (b) V and S (c) V and T (d) F and H

62. The equation $C_p - C_v = R$ for ideal gases is valid if
(a) C_p and C_v are independent of temperature (b) C_p and C_v are linear functions of temperature (c) C_p depends on temperature and C_v is independent of temperature (d) C_p is independent of temperature and C_v depends on temperature.

63. The plot of P vs V exhibits an inflection point at
(a) $T = T_c/2$ (b) $T = T_c/3$ (c) $T = 0$ (d) $T = T_c$.

64. Among the following, which will have the largest magnitude of the van der Waals constant b?
(a) He (b) Ne (c) Ar (d) SO_2

65. Consider the Carnot cycle represented by the temperature–entropy diagram. Which of the following is true?

(a) $\left(\frac{\partial T}{\partial S}\right)_v > \left(\frac{\partial T}{\partial S}\right)_p$ (b) $\left(\frac{\partial T}{\partial S}\right)_v = 0$

(c) $\left(\frac{\partial T}{\partial S}\right)_v = \left(\frac{\partial T}{\partial S}\right)_p$ (d) $\left(\frac{\partial T}{\partial S}\right)_v < \left(\frac{\partial T}{\partial S}\right)_p$

66. For an equilibrium process,

$$N_2 + 3H_2 \leftrightarrow 2NH_3$$

the stoichiometric coefficients v_i are 1, 3 and –2, with the chemical potentials μ_i (i = 1, 2, 3). Which of the following represents the condition for chemical equilibrium?
(a) $\Sigma \mu_i v_i = 0$
(b) $\Sigma \mu_i v_i > 0$
(c) $\mu(N_2) = \mu(H_2) = \mu(NH_3)$
(d) $\Sigma \mu_i v_i < 0$

67. In a Carnot cycle operating between T_2 and T_1, the working substance is 1 mol of He or 1 mol of N_2. Which of the following is true regarding the efficiency?
(a) Same for both He and N_2
(b) Higher for He
(c) Higher for N_2
(d) Becomes unity for N_2

Answer Keys

Practice Problems

1. (i) $|W_{rev}| - W = 4312$ J (ii) $\Delta E = \Delta H = 0$ (iii) $Q_{rev} = 8565$ J (iv) $\Delta S = 26.8$ J K^{-1}

2. (i) $\Delta H - \Delta E = -1318$ J (ii) $T_{rev} - T_{irr} = -110.7$ K (iii) $|W_{rev}| - |W_{irr}| = -18.27$ kJ (iv) $\Delta S = 37.0$ J K^{-1}

3. (i) $W_{rev} = -19.92$ kJ (ii) $Q = 19.93$ kJ (iii) $\Delta E = 4.47$ J (iv) $\Delta H = 6.42$ J (v) $\Delta S = 49.83$ J K^{-1}

4. (i) $K_P = 1.12\times10^{84}$; $K_C = 2.27\times10^{83}$ (mol/l)$^{1/2}$

5. Theoretical question

6. (i) $\Delta H = 5190$ J (ii) $\Delta E = 4019.4$ J

7. (i) $\Delta H^0 = 153.0$ kJ mol^{-1} (ii) $\Delta E^0 = 142.8$ kJ mol^{-1}

8. H–H bond energy = 349 kJ mol^{-1}

9. (i) $\Delta S = 43.5$ J K^{-1} (ii) $\Delta G = -44.84$ kJ (iii) $\Delta A = -49.8$ kJ

10. $\Delta S_{total} = 302.7$ J K^{-1}

11. (i) $\Delta S_{system} = -21.05$ J K^{-1} (ii) $\Delta S_{surr} = 21.72$ J K^{-1} (iii) $\Delta G = -180$ kJ

12. Fugacity of $O_2 = 377.6$ bar

13. $\Delta H^0_{1000\text{ K}} = 87.94$ kJ

14. (i) $K_C = 660.6$ dm^6 mol^{-2} (ii) $p\text{NH}_3 = 2.0$ bar, $p\text{HCl} = 2.5$ bar (iii) $\Delta G^0 = -9.3$ kJ (iv) $\Delta S^0 = -215.0$ J K^{-1}, $\Delta E^0 = -133.2$ kJ

15. (i) $K_P = 2.29\times10^{-4}$ bar^3 (ii) ammonium carbamate remaining = 3.88 g (iii) K_P at 323 K = 5×10^{-2} bar^3 (iv) $\Delta H^0 = 79.0$ kJ mol^{-1}

16. (i) $K_P = \frac{\alpha^3 P}{(2+\alpha)(1-\alpha)^2}$ (ii) $\alpha = 4.17\times10^{-2}$ (iii) P = 4.56 Pa (iv) $\Delta S^0 = 117$ J K^{-1} (v) $\Delta A^0 = 25.3$ kJ

17. Theoretical question

18. $K_P = 5.78$, $\Delta G^0 = -11.38$ kJ, reaction feasible.

19.

Mole fraction of acetone	0.0	0.1	0.3	0.5	0.7	1.0
Activity	0.0	0.283	0.609	0.707	0.794	1.00
Activity coefficient	0	2.83	2.03	1.41	1.13	1.00
Mole fraction of CS_2	0.0	0.1	0.3	0.5	0.7	1.0
Activity	0.0	–	0.647	0.794	0.868	1.00
Activity coefficient	0.0	–	2.16	1.59	1.24	1.00

20. $\mu_{J-T} = 0.2835$ K atm^{-1}, $\Delta H = -4133$ kJ

21. $\Delta G_{mixing} = -8359$ J, $\Delta S_{mixing} = 28.01$ J K^{-1}

22. (i) Refer (2.33) (ii) From the ideal gas equation of state, $VdP + PdV = RdT$ for 1 mole; since dV is an exact differential, dP has to be an exact differential. (iii) From the combined first and second laws of thermodynamics, $TdS = dE + PdV$. Since dS and dV are exact differentials, it follows that dE is an exact differential.

23. Theoretical question

24. Theoretical question

25. Theoretical question

26. (i)

27. Eq. 2.43 for isobaric and Eq. 2.41 for isometric process.

28. 0

29. $[C_p - C_v]/R = 1/\{1 - 2a/VRT\}$

30. $n = 0$, 1, ∞ and γ represent respectively, isobaric, isothermal, isochoric and adiabatic processes.

31. Mole fraction = 0.5

32. Theoretical question

33. $E = c_1 RT$

Multiple Choice Questions

1. (c)	**2.** (b)	**3.** (d)	**4.** (a)	**5.** (b)	**6.** (c)	**7.** (a)	**8.** (b)	**9.** (d)	**10.** (b)
11. (d)	**12.** (c)	**13.** (a)	**14.** (d)	**15.** (b)	**16.** (c)	**17.** (c)	**18.** (d)	**19.** (b)	**20.** (d)
21. (c)	**22.** (a)	**23.** (c)	**24.** (b)	**25.** (d)	**26.** (a)	**27.** (d)	**28.** (c)	**29.** (b)	**30.** (d)
31. (a)	**32.** (c)	**33.** (b)	**34.** (d)	**35.** (b)	**36.** (a)	**37.** (a)	**38.** (b)	**39.** (d)	**40.** (c)
41. (b)	**42.** (d)	**43.** (b)	**44.** (a)	**45.** (b)	**46.** (c)	**47.** (b)	**48.** (d)	**49.** (c)	**50.** (b)
51. (d)	**52.** (b)	**53.** (a)	**54.** (d)	**55.** (c)	**56.** (b)	**57.** (a)	**58.** (b)	**59.** (a)	**60.** (b)
61. (a)	**62.** (a)	**63.** (d)	**64.** (d)	**65.** (a)	**66.** (a)	**67.** (b)			

CHAPTER 3

ELECTROCHEMISTRY

CHAPTER OUTLINE

This subject finds many applications in the extraction of metals, energy devices such as lead storage cells, lithium ion batteries, corrosion prevention, electroanalysis, and so on.

3.1 Important Concepts and Conventions

Types of electrical conduction: They may be broadly divided into two categories: (i) electronic conduction and (ii) electrolytic conduction.

Metals are the best examples of electronic conductors. In this case, the electric conduction is due to the movement of electrons. Also, the electrical conductivity decreases with increase in temperature and physical effects such as heat are observed.

In the case of electrolytic conductors, the charge carries are the ions. Visible changes such as evolution of gas or deposition of metal occurs at the electrodes. The conductivity increases with increase in temperature. Aqueous solutions of acids or bases, salt solutions and molten salts are the best examples. Another class of conductors are known as semiconductors. They exhibit feeble electrical conductivity which increases with increase in temperature. Examples are ZnO, germanium and so on.

Laws of electrolysis: There are two laws – first and second laws of electrolysis – that have been proposed by Faraday.

The 1st law states that the amount of any substance deposited at an electrode is proportional to the quantity of electricity passed through the solution of electrolyte. It may be expressed as

$$W = Zit \tag{3.1}$$

where i = current strength in amperes and t = time in seconds. $W = Z$ when $it = 1$, that is, one coulomb. Z is known as the electrochemical equivalent of the metal and is equal to the weight of the substance deposited by one coulomb.

The 2nd law states that when the same quantity of electricity is passed through different solutions, the weights of different substances deposited are proportional to their chemical equivalents.

$$\frac{w_1}{w_2} = \frac{E_1}{E_2} \tag{3.2}$$

where E_1 and E_2 are the equivalent weights of substances 1 and 2. We may also write,

$$W = \frac{Eit}{96500} \quad (3.3)$$

The quantity of electricity required to deposit one gram equivalent of any metal is known as faraday (F) which is equal to 96,500 coulombs. From the second law, it can be shown that the electronic charge is equal to 1.602×10^{-19} coulombs.

Current efficiency: Sometimes during electrolysis, deviations from Faraday's laws occur. This is due to two or more reactions occurring at the electrode. In such cases, the current efficiency for a given electrode reaction is considered. It is defined as the ratio of the amount of substance actually formed at the electrode to the amount that should have been formed as per Faraday's laws.

$$\text{Current efficiency} = \frac{\text{Amount actually deposited}}{\text{Theoretical amount expected}}$$

Electrolytic conductance and related parameters: Ohm's law is applicable to solutions of electrolytes under conditions of low voltage and when the frequency of a.c. is also low. The resistance of any conductor is given by the relation

$$R = \frac{sl}{a} \quad (3.4)$$

where s = specific resistance, l = length of the conductor and a = area of cross-section of the conductor. In the case of electrolytic conductors, the more commonly used quantity is the *specific conductivity* or *specific conductance*, given by

$$\kappa = \frac{1}{s} = \frac{l}{aR} \quad (3.5)$$

where l/a = cell constant of the conductance cell and it has units of cm^{-1} or m^{-1} and κ has the units of $ohm^{-1}\ cm^{-1}$ or $ohm^{-1}\ m^{-1}$. The specific conductance is proportional to the concentration of the salt in solution.

Another quantity which is also commonly used in studies of electrolytic conductivity is the *equivalent conductance* (Λ) and is given by the relation,

$$\Lambda = \kappa \times v$$

where v = volume of the solution in m^3 containing 1 gram equivalent of the electrolyte.

$$\Lambda = \frac{\kappa}{C} \quad (3.6)$$

where C = concentration in gram equivalents m^{-3} and Λ has units of $ohm^{-1}\ m^2\ eq^{-1}$. If the concentration is expressed in geq L^{-1}, then

$$\Lambda = \frac{1000\,\kappa}{C}$$

with the units of Λ being in $ohm^{-1}\,cm^2\,eq^{-1}$. Sometimes the molar conductance of an electrolyte is specified and it is related to the equivalent conductivity in two ways depending upon whether the electrolyte is symmetrical or unsymmetrical. For a symmetrical electrolyte such as NaCl or $ZnSO_4$ where both the ions are univalent or divalent, μ, the molar conductance is given by

$$\mu = Z\Lambda \quad (3.7)$$

where Z = charge on the cation or anion. If the electrolyte is unsymmetrical and of the type M_lA_m,

$$\mu = l \times m \times \Lambda \quad (3.8)$$

where l = number of cations and m = number of anions in a molecule of electrolyte. The specific conductance of a solution containing different ions, say, in a mixture of electrolytes, is given by

$$\kappa = \sum_i \frac{c_i\lambda_i}{1000} \quad (3.9)$$

where λ_i represents the ionic conductance of the ith ion.

3.1.1 Variation of κ and Λ with Concentration

The specific conductance of a solution is measured by initially measuring the cell constant of the given conductance cell using standard solutions of potassium chloride of known specific conductance. The equivalent conductance of any desired solution of known concentration is determined by making use of this cell and measuring its resistance. The specific conductance increases with increasing concentration of any electrolyte and in dilute solutions $\kappa \propto C$.

The equivalent or molar conductivity of an electrolyte, however, decreases with increasing concentration. For strong electrolytes, like NaCl, KNO_3, and so on, Λ or μ varies only to a small extent but for weak electrolytes like organic acids, they decrease rapidly with increasing concentration. All electrolytes tend to reach a limiting value of Λ (or μ), known as the equivalent or molar conductivity at infinite dilution (Λ_0 or μ_0).

Determination of Λ_0 (or μ_0): For strong electrolytes, a plot of Λ against $\sqrt{C}$ is linear in dilute solutions and an extrapolation of this graph to $\sqrt{C} = 0$, gives Λ_0 as intercept. For weak electrolytes, the above method is not applicable and Kohlrausch's law is employed.

Kohlrausch Law of Independent Migration of Ions and Its Uses: According to this law, each ion of an electrolyte contributes a definite amount to the equivalent conductance at infinite dilution or in very dilute solutions, irrespective of the nature of the other ion. Thus,

$$\Lambda_0 = \lambda_{0_+} + \lambda_{0_-} \quad (3.10)$$

λ_{0_i} is known as the ionic conductance of the ith ion and is a definite constant for the ion and depends only on temperature.

(i) Λ_0 of weak electrolytes: For a weak electrolyte such as acetic acid, Λ_0 is given by

$$\Lambda_{0(CH_3COOH)} = \lambda_{0_{H^+}} + \lambda_{0_{CH_3COO^-}} \quad (3.11)$$

Or for a weak base, NH_4OH

$$\Lambda_{0_{(NH_4OH)}} = \lambda_{0_{NH_4^+}} + \lambda_{0_{OH^-}} \tag{3.12}$$

It is also possible to obtain the Λ_0 of such solutions by using Λ_0 data of suitable strong electrolytes, as for example in the case of NH_4OH, as

$$\Lambda_{0_{(NH_4OH)}} = \Lambda_{0_{NH_4Cl}} + \Lambda_{0_{NaOH}} - \Lambda_{0_{NaCl}} \tag{3.13}$$

(ii) Solubility of sparingly soluble salts: If S is the solubility of a sparingly soluble salt in geq dm^{-3} (or geq L^{-1}), it is given by

$$S = \frac{1000\kappa}{\Lambda_0} \tag{3.14}$$

where κ is the specific conductivity of the saturated solution of the salt corrected for the conductivity of water used and Λ_0 is the equivalent conductivity at infinite dilution of the salt. It is possible to refine the solubility data so obtained for the finite ionic concentration of the salt by employing the Debye–Hückel–Onsager equation.

3.2 Applications of Conductivity Measurements

3.2.1 Measurement of Conductance Ratio and Determination of Dissociation Constant of a Weak Acid or Weak Base

The degree of dissociation α of a weak acid or weak base may be determined from the ratio

$$\alpha = \frac{\Lambda_c}{\Lambda_0} \tag{3.15}$$

where Λ_c is the equivalent conductivity of the acid or base at a given concentration C and Λ_0 is the equivalent conductivity at infinite dilution. Once α is known, the dissociation constant can be calculated using the well known equation for κ. Note that Eq. 3.15 does not take into account the variation of ionic speeds with concentration.

3.2.2 Ionic Product of Water and Hydrolysis Constants of Salts of Weak Acids or Bases

If the specific conductivity of pure water (κ) is known and the ionic conductances at infinite dilution of H^+ and OH^- are also known, the concentration of H^+ = concentration of OH^- = C is given by,

$$C = \frac{\kappa_{H_2O}}{\Lambda_{0_{(H^+ + OH^-)}}} \text{ mol cm}^{-3} \tag{3.16}$$

$[\kappa = S\ cm^{-1}; \Lambda_0 = S\ cm^{-2}\ mol^{-1}\ (\text{or eq}^{-1})]$

Once C is known, K_W can be calculated. The hydrolysis equilibrium of the salt of a weak base and a strong acid BHCl, is given by

$$BH + Cl^- + H_2O \rightleftharpoons B + H_3O^+ + Cl^- \tag{3.17}$$

The degree of hydrolysis, α at a concentration C of the salt is given by,

$$\alpha = \frac{\Lambda_C - \Lambda'_C}{\Lambda_{HCl} - \Lambda'_C} \tag{3.18}$$

where Λ_C = equivalent conductivity of the salt at a concentration C, Λ'_C = equivalent conductivity of the salt at the same concentration in the presence of excess base B, Λ_{HCl} = equivalent conductivity of the acid at the concentration C. Using α, K_{hy} can be calculated in the usual way.

3.2.3 Conductometric Titrations

The addition of a strong base to the dilute solution of a strong acid results in the replacement of a faster moving ion such as the H^+ ion by the slower moving Na^+ ions and this results in a change of conductivity during the titration. These titrations can be employed to locate the equivalence point in the titration of a weak acid vs a strong base as well as in precipitation reactions.

3.2.4 Relation between Ionic Conductance and Ionic Mobility

The ionic conductance at infinite dilution of an ion i, is related to the ionic mobility according to

$$\lambda_{0_i} = F\omega_i^0$$

and at any other concentration of the ion, the equation

$$\lambda_i = \alpha F \omega_i \tag{3.19}$$

is valid where α is the degree of dissociation of the salt of which the ion is a part. The ionic mobility has the units $cm^2\ s^{-1}\ volt^{-1}$.

3.2.5 Debye-Hückel-Onsager (D-H-O) Equation for the Variation of Λ with Concentration

The decrease of the equivalent conductivity with concentration in the case of strong electrolytes, despite the fact that they are completely dissociated into ions in solution as well as in solid state, led the above workers to propose their interionic attraction theory. According to them, every ion in solution is surrounded by an ionic atmosphere of opposite sign due to electrostatic attractions operating between them. Under the influence of an applied electric field, there are two effects – the asymmetry or relaxation effect and the electrophoretic effect – which leads to a decrease in equivalent conductivity below the limiting value with an increase of concentration. The equation proposed by them for a uni–univalent electrolyte like NaCl or KCl is given by

$$\Lambda = \Lambda_0 - \left[\frac{82.4}{(DT)^{1/2}\eta} + \frac{8.20\times10^5}{(DT)^{3/2}}\Lambda_0\right]\sqrt{C} \tag{3.20}$$

where D = dilelectric constant of the solvent, T = temperature in K, η = visocity of the medium and C = concentration of salt in equivalent per litre.

Equation 3.20 may be abbreviated as,

$$\Lambda = \Lambda_0 - (A + B\Lambda_0)\sqrt{C} \tag{3.21}$$

where $A = \dfrac{82.4}{(DT)^{1/2}\eta}$ and $B = \dfrac{8.20\times10^5}{(DT)^{3/2}}$.

They depend only on the temperature and nature of the solvent. A plot of Λ vs $C^{1/2}$ is linear in very dilute solutions ($C \approx 10^{-3}$ eq/L) with a slope equal to ($A + B\Lambda_0$) and intercept Λ_0. Typically, for 1:1 electrolytes in acetone, at 298 K, A = 32.8 S cm^2 mol^{-1} M$^{-1/2}$ and B = 1.63 M$^{-1/2}$.

3.2.6 Wien Effect and Debye-Falkenhagen Effect

The central ion is surrounded by an ionic atmosphere of opposite sign and it derives support from the increase in conductivity observed under high field strengths (of the order of 10^5 volt cm^{-1}) and under high frequencies (of the order of 10^6 cycles s^{-1}). While, under high field strengths, the asymmetry effect is absent, both asymmetry effect and electrophoretic effect (the viscous drag of the medium) are absent under high frequencies which explains the increase of Λ of the electrolyte under these conditions.

3.2.7 Incomplete Dissociation

In the case of certain electrolytes, the plot of Λ vs $C^{1/2}$ is nearly linear but the slope is greater than would be expected from the D–H–O equation. Thus the experimental Λ is less than predicted by theory which implies that not all the ions are available for carrying the current. If α is defined as the fraction of the electrolyte dissociated into ions that are free to carry the current, the D–H–O equation is modified as

$$\Lambda = \alpha\left[\Lambda_0 - (A + B\Lambda_0)\sqrt{\alpha C}\right] \tag{3.22}$$

This equation is especially applicable in the case of electrolytes of higher valence type like $MgSO_4$, $ZnSO_4$, where in some of the ions are formed as ion–pairs which are neutral and not available for carrying the current. In the case of weak electrolytes, α represents the fraction of the electrolyte dissociated as free ions. It is possible to refine values of dissociation constants using Eq. 3.22.

3.2.8 Transport Numbers

The quantity of charge carried by an ionic species, i, in solution is proportional to the (i) concentration C_j of the ion, (ii) charge Z_j of the ion (iii) the mobility ω_j of the ion and may be expressed as

$$q_j = kC_jZ_j\omega_j \tag{3.23}$$

where k is the proportionality constant which includes time, dimensions of the electrodes, and so on. If there are different kinds of ions present in the solution, the total quantity of charge Q, carried by all ions is,

$$Q = k\sum_j C_jZ_j\omega_j \tag{3.24}$$

The fraction of the total charge carried by the jth ion is known as the transport number of the j ion and is given by,

$$t_j = \frac{q_j}{Q} = \frac{kC_jZ_j\omega_j}{k\sum_j C_jZ_j\omega_j} = \frac{C_jZ_j\omega_j}{\sum_j C_jZ_j\omega_j} \tag{3.25}$$

The sum of all t_j's in solution $\Sigma t_j = 1$. In the case of an electrolyte containing two ions, cations and anions, the transport numbers of cations and anions are given by

$$t_+ = \frac{C_+Z_+\omega_+}{C_+Z_+\omega_+ + C_-Z_-\omega_-} \text{ and } t_- = \frac{C_-Z_-\omega_-}{C_+Z_+\omega_+ + C_-Z_-\omega_-} \tag{3.26}$$

But $C_+Z_+ = C_-Z_-$ (because of electroneutrality) and thus

$$t_+ = \frac{\omega_+}{\omega_+ + \omega_-} \text{ and } t_- = \frac{\omega_-}{\omega_+ + \omega_-} \tag{3.27}$$

Considering an electrolyte at infinite dilution,

$$t_+^0 = \frac{\omega_+^0}{\omega_+^0 + \omega_-^0} \text{ and } t_-^0 = \frac{\omega_-^0}{\omega_+^0 + \omega_-^0} \tag{3.28}$$

At any other concentration, similar equations apply. Note that the transport number of an ion even at infinite dilution is not a constant for the ion (unlike ionic conductance) because it depends on the nature of the co-ion.

Determination of Transport Numbers

There are two methods available for the determination of the transport number of ions—Hittorf method and moving boundary method.

Hittorf method: In the Hittorf method, by analysing the cathode or anode compartment of the electrolysis cell and determining the concentration of anion or cation, the transport number is determined. The transport number of cation and anion are related according to

$$\frac{\text{Equivalents of electrolyte lost from anode compartment}}{\text{Equivalents of electrolyte lost from cathode compartment}} = \frac{t_+}{t_-} \tag{3.29}$$

when one faraday of charge is passed through the electrolysis cell. It is assumed in this case that the ions reaching the electrode do not attack it but get discharged. It is generally true in the case of cations but most often anions attack the anode and dissolve an equivalent amount of anode material which results in an increase of concentration of the electrolyte in the anode compartment. An example is, the electrolysis of silver nitrate solution between silver electrodes. The analysis of the anode compartment will lead to the determination of t_+ (Ag^+ ion) as indicated below.

Let N_e = number of faradays passed through solution
Equivalents of Ag^+ ion going into solution at anode = N_e
Equivalents of Ag^+ ion migrating out of anode = N_+
Net increase of Ag^+ concentration at anode = ($N_e - N_+$)

Let $N_+(f)$ be the number of equivalents of Ag^+ present for a certain weight of water after electrolysis and $N_+(i)$ be the number of equivalents of the same for the same weight of water before electrolysis. Then,

$$N_e - N_+ = N_+(f) - N_+(i)$$

$$t_+ = \frac{N_+}{N_e} = \frac{N_e - N_+(f) + N_+(i)}{N_e} \text{ and}$$

$$t_- = \frac{N_+(f) - N_+(i)}{N_e} \quad (\text{since } t_- = 1 - t_+) \tag{3.30}$$

Moving boundary method: This method essentially involves the formation of a boundary between two solutions of electrolytes XY and X′Y with a common ion say, anion Y.

Of the two solutions, the denser solution (say XY containing the experimental cation X) is placed at the bottom, the solution at the top is X′Y of a cylindrical tube with the electrodes being inserted at the two ends. If the boundary passes through a distance of d cm across a cross-sectional area of a cm^2 for the passage of Z coulombs of electricity and the electrolyte XY contains C' equivalents per cm^3, it follows that

$$C'da = t_+ \frac{Z}{F} \text{ or } t_+ \left(\text{of X ion}\right) = \frac{C'daF}{Z} \tag{3.31}$$

If the concentration of XY is expressed in equivalents/litre, C,

$$C' = \frac{C}{1000} \text{ or } t_+ = \frac{CdaF}{1000Z} \tag{3.32}$$

The transport number of an ion varies with concentration and it also varies with temperature. Abnormal variation of transport numbers is observed with concentration in certain electrolytes which is ascribed to complex formation, for example, as in the case of cadmium iodide solutions. Transport number data is used to determine the hydration numbers of ions in solutions.

3.2.9 Miscellaneous Aspects

(i) ***Abnormal λ_i^0s of H^+ and OH^- ions in water:*** The ionic conductances of H^+ and OH^- ions at 298 K in water are of the order of 350 and 200 ohm^{-1} cm^2 eq^{-1} respectively. These high values compared to other ions such as alkali metal ions, is ascribed to the special mechanism of transport of charge in aqueous acidic or basic solutions. The mechanism is known as proton jump mechanism for the transport of H^+ in aqueous solutions. Such a special mechanism is also operative in the case of movement of OH^- ions in aqueous solutions.

(ii) ***Walden's rule:*** The product of $\Lambda_0 \times \eta$ (η = viscosity of the solvent) for a given electrolyte is found to be a constant in a variety of solvents or

$$\Lambda_0 \times \eta = \text{Constant} \tag{3.33}$$

This is found to be especially valid in the case of electrolytes containing large ions such as tetraalkylammonium ions with suitable counter ions. Equation 3.33 is also valid for ionic conductivities.

(iii) ***Determination of dissociation constant from conductance data by graphical method:*** The determination of the conductance ratio, which is equal to the degree of dissociation in the case of weak electrolytes at various concentrations, enables the evaluation of the dissociation constant of a weak acid or base by a graphical approach. The degree of dissociation α, is given by

$$\alpha = \frac{\Lambda_C}{\Lambda_0} \quad \text{(see Eq. 3.15)}$$

The dissociation constant K, of a weak acid or base is given by

$$\frac{\alpha^2 C}{1-\alpha} = \frac{\alpha}{1-\alpha} \times \alpha C \tag{3.34}$$

Rearrangement and substitution of $\frac{\Lambda_C}{\Lambda_0}$ for α leads to

$$\frac{1}{\Lambda_C} = \frac{1}{\Lambda_0} + \frac{\Lambda_C C}{K\left(\Lambda_0\right)^2} \tag{3.35}$$

A plot of $\frac{1}{\Lambda_C}$ vs $\Lambda_C C$ is linear with an intercept of $\frac{1}{\Lambda_0}$ and a slope of $\frac{1}{K\left(\Lambda_0\right)^2}$. Thus K as well as Λ_0 can be obtained using such plots.

3.3 Galvanic (or Electrochemical) Cells

There are two types of cells—*galvanic cells* and *electrolytic cells*. In the former case, chemical energy of a chemical reaction occurring in the cell is converted into electrical work in the surroundings. In an electrolytic cell, a chemical reaction is forced to occur as a result of the application of an external EMF. Lead storage cell used as car battery is a classic example of a galvanic cell. The electrode at which oxidation takes place (electrons are given off) is known as the anode and it is negative while the electrode at which reduction takes place (electrons are consumed) is known as the cathode.

3.3.1 Reversible and Irreversible Cells

From a thermodynamic stand point, a reversible cell is one in which the cell reaction may be reversed by the application of an EMF infinitesimally greater than the cell EMF. If the cell EMF is infinitesimally smaller than the external EMF, the cell reaction takes place. A cell is expected to function reversibly only under conditions of extremely small currents being drawn through the cell.

In the case of irreversible cells, there is no control by which the reaction in the cell can be made to occur or stop. In such cells, a reaction occurs in the cell even without closing the circuit.

3.3.2 Reversible Electrodes

A reversible cell consists of two reversible electrodes. We find three types of electrodes in the study of galvanic cells.

(i) Electrodes of first kind are those reversible to cations. For example, zinc electrode placed in a solution of Zn^{2+} ions. Electrodes such as the hydrogen electrode are also included in this category. For example,

$$Cu^{+2} + 2e^- \rightarrow Cu \tag{3.36}$$

(ii) Electrodes of second kind are reversible to anions. Such electrodes are composed of a metal over which a layer of a sparingly soluble salt of the metal is coated and is dipped into a solution of anions. An example of such an electrode is the so called Ag, AgCl electrode and the electrode reaction is,

$$AgCl(s) + e^- \rightarrow Ag + Cl^- \tag{3.37}$$

(iii) An electrode of the third kind involves the use of an inert electrode such as Pt placed in a solution containing both the oxidised and reduced species of the metal, for example, in a mixture containing Fe(III) and Fe(II) ions. The reaction at such an electrode is

$$Fe^{+3} + e^{-} \rightarrow Fe^{+2} \qquad (3.38)$$

3.3.3 Nernst Equation

Nernst equation relates the EMF of a cell with the activities. Considering a general chemical reaction,

$$a\text{A} + b\text{B} + \ldots \rightleftharpoons x\text{X} + y\text{Y} + \ldots \qquad (3.39)$$

(where A, B, X, Y, are reactants and products, and a, b, x, y, are their stoichiometric coefficients) occurring for the passage of a faraday through the cell, its reversible EMF is related to the activities of reactants and products according to,

$$E = E^0 - \frac{RT}{nF}\ln\frac{a_X^x \times a_Y^y \times \ldots\ \ldots}{a_A^a \times a_B^b \times \ldots\ \ldots} \qquad (3.40)$$

where E^0 = standard EMF of the cell when all the activities of reactants and products are unity, it is a constant for the cell depending only on temperature and solvent; R = gas constant expressed in volt coulomb deg^{-1} mol^{-1}; F = Faraday constant; T is the temperature in K. In dilute solutions of electrolytes, when the interionic forces are small in solution, the activities may be replaced by the concentrations and Eq. 3.40 can be expressed as,

$$E = E^0 - \frac{RT}{nF}\ln\frac{c_X^x \times c_Y^y \times \ldots\ \ldots}{c_A^a \times c_B^b \times \ldots\ \ldots} \qquad (3.41)$$

3.3.4 Relation of Reversible EMF to the Thermodynamic Parameters of the Cell Reaction

The Gibbs free energy change, ΔG, at constant temperature and pressure is related to the electrical work done by the cell (as a result of the chemical reaction in the cell) according to

$$\Delta G_{T,P} = -nFE_{\text{rev}} \qquad (3.42)$$

where E_{rev} = reversible EMF of the cell, n = number of faradays. The entropy change ΔS and enthalpy change ΔH associated with the cell reaction are given by

$$\Delta S = nF\left(\frac{\partial E}{\partial T}\right)_P \qquad (3.43)$$

$$\Delta H = nF\left[T\left(\frac{\partial E}{\partial T}\right)_P - E\right] \qquad (3.44)$$

It is important to note that E is an intensive property while ΔG, ΔH and ΔS are extensive depending upon n.

3.3.5 Sign Conventions for EMF of the Cell, Half Cell and Standard Electrode Potentials

The EMF of a cell is given a positive sign when the reaction in the cell corresponds to the spontaneous cell reaction. This is also in accord with the fact that the ΔG associated with the reaction under these conditions will be negative. It is emphasised that the spontaneous reaction occurring in any cell is unique and can always be unambiguously determined by determining the polarity of electrodes in the cell by connecting it to a potentiometer and measuring its EMF by the Poggendorf's method. The EMF of the cell is expressed as the difference between two half cell potentials—that of the right hand electrode, E_R, to that of the left hand electrode, E_L, and is given by

$$E_{\text{cell}} = E_R - E_L \qquad (3.45)$$

Both half cell potentials are measured with respect to the standard hydrogen electrode (SHE) represented by,

$$\text{Pt}, H_2(g, 1\text{ atm})|H^+(a = 1) \qquad E^0 = 0 \qquad (3.46)$$

by convention at all temperatures. In a similar way, the standard EMF of the cell, E^0_{cell}, may be expressed as a difference of two standard electrode potentials E^0_R and E^0_L, these being the SEPs of the right hand and left hand electrodes respectively.

$$E^0_{\text{cell}} = E^0_R - E^0_L \qquad (3.47)$$

The SEPs are measured when the activities of all the terms in the electrode system are unity and gases, if any, are at 1 atm, pressure against the SHE.

Representation of half cell vs SHE, significance of sign of half cell potential under standard and other conditions

- The cell for representing the standard electrode potential is written with the standard hydrogen electrode on the left hand side and the half cell under consideration on the right hand side. The reaction in the cell is such that reduction occurs at the right hand electrode (at the half cell under consideration) and oxidation takes place at the SHE. This convention is followed because all half cell potentials under standard conditions or otherwise are regarded as reduction potentials.
- It is significant that this method of representation of the half cell at the right hand side has the advantage that the polarity of the electrode is identical with the polarity of the terminal to which it must be connected when measuring its EMF by Poggendorf's compensation method potentiometrically.

Example: The standard reduction potential of the Cd^{+1}/Cd system is –0.403 V at 298 K, that is, the EMF of the cell,

$$\text{Pt}, H_2(g, 1\text{ atm})\,|\,H^+(a = \text{or } c = 1)\,||\,Cd^{+2}(c = 1\text{ or}\ldots)\,|\,\text{Cd} \qquad (3.48)$$

is –0.403 V. Also, the –ve sign of the potential also coincides with the negative terminal to which this half cell must be connected when measuring its EMF by Poggendorf's method.

The standard electrode potential of the Cu^{+2}/Cu system is +0.304 V vs SHE at 298 K. This also tells us that this half cell must be connected to the +ve end in a potentiometric circuit when measuring its EMF in conjunction with SHE.

Half cell and complete cell reactions and use of Nernst equation: Consider the cell in Eq. 3.48. The reaction in the cell for passage of two faradays is,

$$H_2(g, 1\text{ atm}) + Cd^{+2}(a = 1\text{ or } c = 1) \rightarrow \text{Cd}(s) + 2H^+(a = 1) \qquad (3.49)$$

Since E of this cell is –0.403 V, ΔG of the reaction is +ve and hence the reaction is non-spontaneous. The Nernst equation for reaction (Eq. 3.49) is

$$E = E^0 - \frac{RT}{2F}\ln\frac{a_{H^+}^2 a_{Cd}}{a_{H_2} a_{Cd^{+2}}} \quad (3.50)$$

In dilute solutions, activities may be replaced by concentrations and Eq. 3.50 simplifies to

$$E = E^0 - \frac{RT}{2F}\ln\frac{C_{H^+}^2 \times 1}{1 \times C_{Cd^{2+}}} \quad (3.51)$$

In writing the Nernst equation, the following conventions are followed. (i) For gases at 1 atm, the activity or concentration is taken as unity. (ii) For metals, their activity or concentration is considered to be unity. (iii) Metal ions at unit concentration as under standard state conditions, are taken as $C = 1$ (although strictly their activity is not equal to unity).

Equation 3.51 simplifies to $E = E_0$ (since all species are at unit concentration).

For the cell corresponding to Cu^{+2}/Cu, the cell reaction is,

$$H_2(g, 1\text{ atm}) + Cu^{+2}(a = 1 \text{ or } c = 1) \rightarrow Cu(s) + 2H^+(a = 1) \quad (3.52)$$

The EMF being +0.34 V, the reaction is spontaneous as ΔG is –ve. Nernst equation can be written in a similar way as above. Since in both reactions 3.49 and 3.52, SHE is the left hand electrode, the equation for the electrode potential may be generalised as

$$E = E^0 - \frac{RT}{nF}\ln\frac{a_{red}}{a_{ox}} = E^0 - \frac{RT}{nF}\ln\frac{C_{red}}{C_{ox}} \quad (3.53)$$

Another example of a cell will now be considered where the concentration of the ions will be different from unity and how the EMF of such a cell can be calculated using Nernst equation. The cell under consideration is

$$Zn \,|\, ZnSO_4(aq, C = 0.02\text{ M}) \,||\, KCl(aq, C = 0.1\text{ M}) \,|\, AgCl(s), Ag \quad (3.54)$$

where $E^0{}_{Zn^{2+}, Zn} = -0.76$ V and $E^0{}_{AgCl, Ag} = 0.222$ V, both at 298 K. The cell reaction for the passage of 2 faradays is,

$$Zn + 2AgCl(s) \rightarrow Zn^{+2}(C = 0.02\text{ M}) + 2Ag(s) + 2Cl^-(C = 0.1\text{ M}) \quad (3.55)$$

[The half cell reactions are

$2AgCl(s) + 2e^- \rightarrow 2Ag(s) + 2Cl^-$ (i) and

$Zn^{+2} + 2e^- \rightarrow Zn(s)$ (ii)]

(i) – (ii) gives Eq. 3.55.

Application of the Nernst equation for the reaction (Eq. 3.55) gives,

$$E = E^0_{cell} - \frac{RT}{2F}\ln(a_{Zn^{2+}} \times a_{Cl^-}^2) \quad (3.56)$$

This may be rewritten in terms of concentration or molalities of ions as

$$E = E^0 - \frac{RT}{2F}\ln(m_{Zn^{2+}} \times m_{Cl^-}^2) \quad (3.57)$$

in dilute solution. A further simplification leads to

$$E = E^0 - \frac{RT}{2F}\ln m_{ZnCl_2} \quad (3.58)$$

A rigorous consideration of Eq. 3.56 must take into account mean activity, mean molality and mean activity coefficient of the electrolyte as,

$$E = E^0 - \frac{RT}{2F}\ln(a_\pm^3) = E^0 - \frac{RT}{2F}\ln(m_\pm^3 \times \gamma_\pm^3)$$

Assuming $\gamma_\pm = 1$ in dilute solution,

$$E = E^0 - \frac{RT}{2F}\ln(m_\pm^3) \quad (3.59)$$

However the use of equations of the type (Eq. 3.58) involving molality of the electrolyte will be used in further discussions. Equation 3.58 may be also obtained by writing the Nernst equation in terms of the two half cell potentials. For example,

$$E = \left(E_R^0 - E_L^0\right) - \frac{RT}{nF}\ln(m_{Cl^-}^2) + \frac{RT}{nF}\ln\frac{1}{m_{Zn^{2+}}}$$

$$= E^0_{cell} - \frac{RT}{2F}\ln(m_{Zn^{2+}} \times m_{Cl^-}^2)$$

$$= E^0_{cell} - \frac{RT}{2F}\ln m_{ZnCl_2}$$

It may be noted that $\frac{RT}{nF}\ln$ at 298 K may be given as $\frac{0.0591}{n}\log$.

3.3.6 Salient Features of Standard Reduction Potential Data

(i) The table of standard electrode potentials contains data listed from electrode systems having the most negative to the most positive values. SEPs are definite constants for the systems under consideration and depend only on temperature and solvent.

(ii) The system having the most negative electrode potential displaces the one with less negative (or more positive) electrode potential in the electrochemical series. For example, zinc displaces iron from a solution of ferrous salt and it can also displace cadmium from a solution of cadmium as both Fe/Fe^{2+} and Cd/Cd^{2+} systems have less negative potentials compared to zinc.

(iii) The larger the difference in the electrode potentials, the more complete the displacement reaction will be. For example, the displacement of Fe^{2+} ions from the ferrous salt is smaller than the displacement of Cd^{2+} ions from a cadmium salt solution.

(iv) Standard electrode potential data can be used to determine the equilibrium constants of reaction.

Example: Consider the reaction

$$Zn + 2Ag^+ \rightarrow Zn^{2+} + 2Ag \quad (3.60)$$

$$K=\frac{\left[Zn^{2+}\right][Ag]^2}{[Zn]\left[Ag^+\right]^2}=\frac{\left[Zn^{2+}\right]}{\left[Ag^+\right]^2} \quad (3.61)$$

The cell for the above reaction may be formulated as

$$Zn\,|\,Zn^{2+}(C=1\text{ M})\,||\,Ag^+(C=1\text{ M})\,|\,Ag$$

$$E^0_{cell}=E^0_R-E^0_L=0.800-(-0.760)=1.565$$

$$\Delta G^0=-nFE^0=-RT\ln K \quad (3.62)$$

$$\log K=\frac{nFE^0}{RT\times 2.303}=\frac{2\times 96500\times 1.565}{8.314\times 298\times 2.303}$$

$$=2\times\frac{1.56}{0.0591}=52.8 \text{ or}$$

$$K=6.3\times 10^{52} \quad (3.63)$$

(v) It is also possible to determine the solubility products of sparingly soluble salts using SEP data.
Consider for example, the solubility product of AgI according to

$$AgI(s)\rightleftharpoons Ag^+ + I^- \quad (3.64)$$

for which $K_{SP}=\left[Ag^+\right]\left[I^-\right]$. The two SEPs to be considered are,

(i) $AgI(s)+e^-\rightleftharpoons Ag+I^- \quad E^0=-0.15\text{ V}$

(ii) $Ag^+ + e^- \rightarrow Ag \quad E^0=+0.8\text{ V}$

(i) – (ii) gives the reaction (Eq. 3.64) for which,

$$E^0=-0.15\text{ V}-(0.8\text{ V})=-0.95\text{ V}$$

$$\log K_{SP}=-\frac{0.95}{0.0591}=-16.07 \text{ or } K_{SP}=8.5\times 10^{-15}$$

(vi) SEP data in conjunction with suitable formulation of a cell may be used to obtain the concentration of an ion such as Ag^+, Cu^{2+}, Zn^{2+} or even H^+ concentration. The requirement is the use of a suitable indicator electrode along with a reference electrode like a saturated (or 0.1 N KCl) calomel electrode. For example, the concentration of zinc ions in its salt can be determined by the following cell,

$$Zn\,|\,Zn^{2+}(C=?)\,||\,\text{SCE} \quad (3.65)$$

for which $E_{cell}=E_{SCE}-E_{Zn^{2+},Zn}$

$$=0.242-\left(E^0_{Zn^{2+},Zn}-\frac{0.0591}{2}\log\frac{1}{C_{Zn^{2+}}}\right)$$

$$=0.242-(-0.760)-\frac{0.0591}{2}\log C_{Zn^{2+}}$$

$$=1.002-\frac{0.0591}{2}\log C_{Zn^{2+}} \quad (3.66)$$

By measuring E_{cell} and giving it the proper sign, $C_{Zn^{2+}}$ in solution may be calculated using Eq. 3.66. In a similar way, it is possible to determine the H^+ ion concentration or pH of a solution using a hydrogen electrode, quinhydrone or glass electrode.

3.3.7 Reference Electrodes

The use of SHE as a reference electrode has some limitations in routine electrochemical work and thus it is common to use the calomel electrode containing saturated/1 N/0.1 N potassium chloride as the electrolyte in the sample cell. For example, in the case of the saturated calomel electrode, it is represented as,

$$\text{Hg, HgCl}(s)\,|\,\text{KCl(sat.)} \qquad E_{SCE}=0.2415\text{ V}$$

The electrode reaction being,

$$\frac{1}{2}HgCl(s)+e^-\Rightarrow Hg+Cl^- \quad (3.67)$$

If the electrolyte is 1 N KCl, $E_{SCE}=0.252$ V and if 0.1 N KCl is used then $E_{SCE}=0.334$ V. Another commonly used reference electrode is the silver, silver chloride electrode.

3.3.8 Redox Potentials

Like standard electrode potentials, the standard redox potentials of systems such as $Pt\,|\,MnO_4^-$, Mn^{2+}; $Pt\,|\,Cr_2O_7^{2-}$, Cr^{3+} and $Pt|Ce^{4+}$, Ce^{3+}, $Pt|Fe^{3+}$, Fe^{2+} are important to understand the relative strengths of the various oxidising and reducing agents. For example, the standard redox potential of the Ce^{4+}, Ce^{3+} system is 1.61 V. while that of permanganate manganous system is 1.51 V. Thus the Ce^{4+}, Ce^{3+} system is a more powerful oxidising agent towards ferrous ions compared to the MnO_4, Mn^{2+} system. They are widely employed in the study of redox reactions by potentiometry.

3.3.9 Potentiometric Titrations

The potentiometric titration assembly consists of an indicator electrode placed in a solution containing the ion being titrated. For example, a platinum electrode placed in a solution of ferrous salt is coupled with a reference electrode SCE. As an oxidising agent like $KMnO_4$ or cerium ions is added, the potential of the indicator electrode (or the EMF of the cell of which it is a part) changes gradually, but increases very sharply at the equivalence point. Beyond the equivalence point, the change in potential becomes gradual. By plotting E versus volume of titrant or $\Delta E/\Delta V$ vs volume of titrant, the equivalence point may be determined accurately.

Such titrations can be performed for determining equivalence points in acid–base titrations (using a glass electrode, or a quinhydrone electrode which is responsive to H^+ ions) and in precipitation titrations involving the titration of $AgNO_3$ vs NaCl, and so on.

3.3.10 Liquid Junction Potential

In the representation of galvanic cells used in routine electrochemical measurements (see Eqs 3.48, 3.54 and 3.65), a double vertical line between two different half cells indicates that the junction does not contribute to the overall EMF of the cell and that the junction potential is eliminated. In some other cells, a single vertical broken line between two half cells indicates a phase boundary across which transport of ions is possible. In the case where the double vertical line is present, a saturated potassium chloride bridge or saturated ammonium nitrate bridge is implied and the use of such bridges in eliminating the liquid junction potential will be explained shortly.

When two different solutions of electrolytes at the same or different concentrations are brought into contact, ions of the electrolytes migrate across the boundary at different speeds leading up to the setting up of a junction potential. The structure of the boundary and hence the concentration of the ions at different points in the boundary will depend on the method used for bridging the solutions together.

A general equation for LJP (E_L), when two electrolyte solutions containing different ionic species in contact are present, is given by,

$$E_L = -\frac{RT}{F}\int_{C_1}^{C_2}\sum_i \frac{t_i}{Z_i}\, d\ln a_i \tag{3.68}$$

Considering the junction formed between two solutions of a uni–univalent electrolyte like KCl according to

$$\begin{matrix} \mathrm{KCl}(a_1) & \vdots & \mathrm{KCl}(a_2) \\ (\text{in solution } 1) & \vdots & (\text{in solution } 2) \end{matrix} \tag{3.69}$$

The equation for LJP at the boundary is

$$E_L = -t_+\frac{RT}{F}\left\{\frac{\ln (a_+)_2 \times (a_-)_2}{(a_+)_1 (a_-)_1}\right\} + \frac{RT}{F}\ln\frac{(a_-)_2}{(a_-)_1} \tag{3.70}$$

where a_+ and a_- refer to the activities of the K^+ and Cl^- ions in solutions 1 and 2. Replacing individual activities by mean activities on the LHS and also assuming that single ion activities (RHS) are also equal to mean activities

$$E_L = -2t_+\frac{RT}{F}\ln\frac{(a_\pm)_2}{(a_\pm)_1} + \frac{RT}{F}\ln\frac{(a_\pm)_2}{(a_\pm)_1}$$

$$= (t_- - t_+)\frac{RT}{F}\ln\frac{(a_\pm)_2}{(a_\pm)_1} \tag{3.71}$$

E_L is thus seen to depend on the transport numbers of the cation and anion from Eq. 3.71. Since the transport numbers of the K^+ ion and Cl^- for KCl, NH_4^+ and NO_3^- ions in the case of NH_4NO_3 are close to each other, the use of these salt bridges is justified as $E_L \approx 0$ in such a case.

3.3.11 Concentration Cells

Such a cell is defined as one in which the EMF is due to the free energy change associated with the transfer of the electrolyte from one concentration to the other. These cells may be classified as (i) electrolyte type cells and (ii) electrode type cells. Among electrolyte type concentration cells, there are cells without transference and cells with transference.

Cells without transference: Consider the cell

$$\mathrm{Pt, H_2}(g, 1\text{ atm})\big|\mathrm{HCl}(a_\pm)_1, \mathrm{AgCl}(s)\big|$$
$$\mathrm{Ag}\big|\mathrm{AgCl}(s)\big|\mathrm{HCl}(a_\pm)_2\,|\,\mathrm{H_2}(g, 1\text{ atm}), \mathrm{Pt} \tag{3.72}$$

By considering the reaction at the two cells, which are connected in opposition to each other, the net reaction for the passage of one faraday in the cell is given by

$$\mathrm{H^+}\big((a_\pm)_2\big) + \mathrm{Cl^-}\big((a_\pm)_2\big) \rightarrow \mathrm{H^+}\big((a_\pm)_1\big) + \mathrm{Cl^-}\big((a_\pm)_1\big) \tag{3.73}$$

and the EMF due to the reaction may be expressed as E_x.

$$E_x = \frac{2RT}{F}\ln\frac{(a_\pm)_2}{(a_\pm)_1} \tag{3.74}$$

It may be noted that E_x is the EMF of the cell without transference and without liquid junction.

Cells with transference: By considering cell (Eq. 3.72) without the presence of the intervening Ag|AgCl|Ag electrode, a cell with transference given by

$$\mathrm{Pt, H_2}(g, 1\text{ atm})\big|\mathrm{HCl}(a_\pm)_1 \vdots \mathrm{HCl}(a_\pm)_2\big|\mathrm{H_2}(g, 1\text{ atm}), \mathrm{Pt} \tag{3.75}$$

is obtained. Taking into account the reaction at the electrodes and the transfer of cations (H^+ ions from the left to the right) and anions from (the right to the left) in accordance with their transference numbers, the net result for the passage of one faraday through the cell is the transfer of $t\_$ g ions of H^+ and Cl^- from the right hand to the left hand electrode given by

$$t_-\mathrm{H^+}(a_\pm)_2 + t_-\mathrm{Cl^-}(a_\pm)_2 \rightarrow t_-\mathrm{H^+}(a_\pm)_1 + t_-\mathrm{Cl^-}(a_\pm)_1 \tag{3.76}$$

and the EMF due to the reaction is

$$E_Y = \frac{2t_-RT}{F}\ln\frac{(a_\pm)_2}{(a_\pm)_1} \tag{3.76a}$$

Suppose that the transfer of ions across the boundary does not occur in the cell (Eq. 3.75), the EMF is due to the reaction at the electrodes, which is given by

$$\mathrm{H^+}\big((a_\pm)_2\big) \rightarrow \mathrm{H^+}\big((a_\pm)_1\big) \tag{3.77}$$

Therefore the EMF is given by

$$E_Z = \frac{RT}{F}\ln\frac{a_{\mathrm{H^+}}(\text{solution 2})}{a_{\mathrm{H^+}}(\text{solution 1})} \approx \frac{RT}{F}\ln\frac{(a_\pm)_2}{(a_\pm)_1} \tag{3.78}$$

In Eq. 3.78, the single ion activities are replaced by mean activities of HCl in the two solutions as an approximation.

From Eqs 3.76 and 3.77,

$$\frac{E_Y}{E_Z} = 2t_- \tag{3.79}$$

E_Y and E_Z are sometimes referred to as E_{wt} and E_{wot}. The difference $E_Y - E_Z$ gives the equation for E_L (Eq. 3.71). It is possible to have concentration cells with the end electrodes reversible to anions and the formulation of expressions for the EMFs of such cells is similar to that described above.

Electrode type concentration cells: An example of this type concentration cell is the amalgam concentration cell given by

$$\mathrm{Zn-Hg}(a_1)\big|\mathrm{ZnSO_4}(m)\big|\mathrm{Zn-Hg}(a_2) \tag{3.80}$$

The overall cell reaction for the passage of 2 faradays is

$$Zn-Hg(a_1)\rightarrow Zn-Hg(a_2) \tag{3.81}$$

The EMF of the cell (Eq. 3.80) is given by

$$E=\frac{RT}{2F}\ln\frac{a_1}{a_2} \tag{3.82}$$

A gas concentration cell, where the EMF of the cell is due to the difference in pressure of the gas at the ends, is

$$Pt,\ H_2(g,\ P_1\ atm)|HCl(m)|H_2(g,\ P_2\ atm),\ Pt \tag{3.83}$$

The EMF, of the cell is given by

$$E=\frac{RT}{2F}\ln\frac{P_1}{P_2} \tag{3.84}$$

3.3.12 Galvanic Cells as Practical Sources of Electrochemical Energy

Such cells may be broadly be classified into three types: (i) primary cells, (ii) secondary cells (iii) fuel cells.

Primary cells: The cell reaction in this type of cell is not reversible and once the cell reaction is complete, it cannot be charged again. A typical example of this class is the lechlanche cell (or dry cell) represented by,

$$Zn|NH_4Cl(20\%),ZnCl_2|MnO_2,C \tag{3.85}$$

It gives a voltage of about 1.5 V and the cell reaction is

$$Zn+2MnO_2+2H_2O\rightarrow Zn^{2+}+2MnO(OH)+2\ OH^- \tag{3.86}$$

MnO_2 acts as the depolariser in the cell since it prevents the formation of H_2 at cathode. It gets reduced preferentially at cathode.

Secondary cells: These are galvanic cells in which the normal cell reaction can be reversed by applying an external EMF. The cell can be discharged and recharged several times. Typical examples of such cells are the lead storage cell, nickel–cadmium cell, mercury cell, and so on. Let us consider the lead storage cell,

$$Pb,\ PbSO_4(s)|^-\ H_2SO_4(m\approx 2sp.gr=1.30)|^+\ PbO_2(s),\ Pb \tag{3.87}$$

The half cell and overall cell reactions are

$$PbO_2(s)+4H^++SO_4^{2-}+2e^-\rightarrow PbSO_4(s)+2H_2O$$

$$PbSO_4(s)+2e^-\rightarrow Pb(s)+SO_4^{2-}$$

$$Pb(s)+PbO_2(s)+2H_2SO_4(m)\rightarrow 2PbSO_4+2H_2O \tag{3.88}$$

Reaction (Eq. 3.88) takes place on discharge while the reverse reaction occurs on charging.

Another example of a secondary cell is,

$$Fe,\ FeO(s)|KOH(m)|Ni_2O_3(s),\ NiO(s)|\ Ni \tag{3.89}$$

in which the overall cell reaction for passage of 2 faradays is,

$$Ni_2O_3(s)+Fe(s)\rightarrow 2NiO(s)+FeO \tag{3.90}$$

The cell EMF is independent of KOH concentration as seen from Eq. 3.90.

Fuel cells: In a fuel cell, the chemical energy derived by the oxidation of the fuel is directly converted into electrical energy. The reactants in the fuel cell are supplied from outside and do not form a part of the cell. Two examples of fuel cells are (i) hydrogen–oxygen fuel cell and (ii) hydrocarbon–air fuel cell.

(i) Hydrogen–oxygen fuel cell: The cell may be represented as

$$\text{Inert electrode like } Pt|H_2(g)|KOH\ (\text{concentrated solution})|\ O_2(g),\ \text{Inert electrode like Pt} \tag{3.91}$$

In commercial cells, porous nickel electrodes are used and the electrolyte is kept around 200°C and 30 atm pressure. The half cell potentials under standard conditions are

$$\frac{1}{2}O_2(g)+H_2O(l)+2e^-\rightarrow 2OH^- \qquad (E_R^0=0.400\,V)$$

$$2H_2O(l)+2e^-\rightarrow H_2(g)+2OH^- \qquad (E_L^0=-0.83\,V)$$

The overall cell reaction is formation of water and E^0_{cell} = 1.23 V. The efficiency of a fuel cell is given by,

$$E=\frac{|\Delta G|}{\Delta H}=1-\frac{T\Delta S}{\Delta H} \tag{3.92}$$

Thus the efficiency is dependent on ΔH, ΔS and T.

(ii) Hydrocarbon–air fuel cell: Consider the methane–air fuel cell. The half cell and overall reactions are,

$$CH_4(g)+2H_2O(l)\rightarrow CO_2(g)+8H^++8e^-$$

$$2O_2(g)+8H^++8e^-\rightarrow 4H_2O(l)$$

$$CH_4(g)+2O_2(g)\rightarrow CO_2(g)+2H_2O(l) \qquad \Delta G^0=-817\,kJ$$

The EMF due to the oxidation of the fuel is

$$E=-\frac{\Delta G}{nF}=\frac{817\times 1000}{8\times 96500}=1.06\ V$$

3.4 Polarisation and Overpotential

Polarisation: In electrolyses, such as the decomposition of acidulated water using Pt or other electrodes or electrolysis of aqueous salt solutions by the application of an external EMF, a current flows through the solution and reactions occur at the electrodes. In general, whenever a current passes through a solution of an electrolyte, the electrochemical cell is said to be polarised. Polarisation encompasses the entire phenomenon arising out of the passage current through the cell. Polarisation may refer to processes occurring at the cathode (cathodic polarisation) or at the anode (anodic polarisation).

Decomposition potential of water: The minimum voltage required for the decomposition of water is 1.23 V as calculated from the reversible potentials of hydrogen and oxygen half cells, but in practice a voltage of nearly 1.7 V is required for the continuous evolution of hydrogen and oxygen at the electrodes. This voltage is known as the decomposition voltage or the decomposition potential for the electrolysis of water. The difference, which is about 0.47 V is known as the overvoltage associated with the decomposition process.

Overpotential: It is possible to consider the overpotential at an electrode instead of overvoltage of a cell. When a current is passing at an electrode, the potential of the electrode deviates from the equilibrium value (given by Nernst equation) and the difference between the actual potential and the equilibrium potential is referred to as the overpotential η.

$$\eta = V - V_0 \tag{3.93}$$

where V = actual potential at the electrode and V_0 = reversible potential. Since the rate of an electrochemical reaction is expressed as the number of equivalents of a species formed or consumed in the reaction, the current density, given as A cm^{-2} at a cathode or anode, at which the reaction is studied, becomes an important parameter in electrolysis studies. The current density is a function of the concentration of various species, potential at the electrode, and so on.

Exchange current density: When the overpotential at an electrode is small, the current density is proportional to η (at constant T), the proportionality constant, i_0, being known as exchange current density. The reversibility or irreversibility of a reaction at an electrode can be gauged from the magnitude of exchange current density. For example, i_0 for hydrogen evolution reaction at a platinum electrode is 10^{-2} A cm^{-2} while the same is as low as 10^{-14} A cm^{-2} at a mercury cathode. Because of the high exchange current density for the H^+ ion reduction on a platinum surface, the Pt electrode is used in a reversible hydrogen electrode and not mercury.

Components of overpotential: The overpotential at an electrode, η, is a sum total of several contributing factors such as ohmic, concentration and kinetic overpotentials given by,

$$\eta = \eta_{\text{ohm}} + \eta_{\text{conc}} + \eta_{\text{kin}} \tag{3.94}$$

The ohmic overpotential arises due to the iR drop between the experimental and reference electrodes (used in overpotential measurements) and can be reduced largely by reducing the distance between the electrodes. The concentration overpotential arises due to the concentration gradient of the electroactive species near the electrode and the bulk of the solution. It can also be effectively reduced by efficient stirring of the solution. The kinetic overpotential is due to the slowness of the electrode reaction.

Tafel equation: A quantitative relationship between the current density i, and overpotential η, was proposed by Tafel and is given by,

$$\frac{|\eta|}{2.303} = -\frac{RT}{\alpha nF}\log_{10} i_0 + \frac{RT}{\alpha nF}\log_{10} i \text{ or } |\eta| = 2.303\log\frac{i}{i_0} \tag{3.95}$$

where $\frac{RT}{\alpha nF}\log_{10} i_0$ and $\frac{RT}{\alpha nF}$ are the intercept and slope of the linear graph of η versus $\log_{10} i$. The Tafel slope, $\frac{RT}{\alpha nF}$, gives information on the mechanism of the electrode reaction. When η = 0, the value of j becomes equal to the exchange current density i_0.

Beneficial aspects of overvoltage: The existence of overvoltage in some cases offers advantages in the study electrochemical reactions. Two examples illustrate this aspect well.

(i) The overpotential for the reduction of H^+ ions to hydrogen gas on a mercury cathode is quite large. This observation is made use of in polarography in which a dropping mercury electrode is used as a cathode for the analytical determination of metal ions such as Cd^{+2}, Tl^+, and so on. Such a study is possible because there is no interference from hydrogen evolution in these studies.

(ii) Similarly, the high overvoltage for the reduction of H^+ ions on lead enables the charging of the lead storage battery without interference from hydrogen evolution. In the absence of such high overvoltage, during charging of the battery, electrolysis of water will also take place.

3.4.1 Electroanalytical Methods and their Importance

Electrochemical methods (or electroanalytical methods) constitute an important class of instrumental methods for analytical purposes and they involve the use of basic principles of electrochemistry. They can be broadly classified into the following categories. (i) Conductometry (briefly discussed earlier), (ii) potentiometry (briefly discussed earlier), (iii) voltammetry and polarograghy, and (iv) electrodeposition and controlled potential electrolysis.

Voltammetry and polarograghy: In voltammetry, we consider the effect of the potential of the electrode of an electrolysis cell on the current flowing through the cell. In these studies, the indicator electrode is the one whose potential is varied. The indicator electrodes may be Pt, Au and also liquids such as Hg. They may be of any size, may be stationary or in motion. The electrolyte used in the study may be stirred or unstirred. Polarography refers specifically to that branch of voltammetry wherein a dropping mercury electrode is used as the indicator electrode.

Principle of polarographic method of analysis: This method of analysis was discovered by J Heyrovsky in 1920. It is based on the measurement of current with increasing applied voltage to an electrolysis cell composed of a polarisable electrode (known as the dropping mercury electrode or DME) and a non-polarisable reference electrode such as the calomel electrode. Suppose we consider the electrolysis of a solution containing several cations like Cd^{2+}, Zn^{2+}, Tl^+, and so on. Each ion gets discharged at the cathode depending upon its reversible potential, which in turn depends on its E^0 and concentration. Gradual increase of E results in deposition of the metal with the highest reduction potential first, followed by the others. Increasing voltage results in an increase of current and as a result, the concentration of the ion being discharged becomes more and more depleted in the neighbourhood of the cathode, particularly if the solution is not stirred. At a particular stage, the current tends to reach a limiting value as all the ions reaching the cathode are discharged and any further increase in current depends on the rate of diffusion of the ion from the bulk of the solution to the surface of the electrode. This is governed by the concentration gradient between the bulk of the solution and the surface of the electrode. This method of analysis is thus based on the measurement of diffusion controlled limiting current, known as *diffusion current* i_d.

Parameters affecting i_d**:** The magnitude of i_d is related to the concentration of the electroactive species C and the complete equation governing this relation is known as the *Ilkovic equation*. It is given by,

$$i_d = 607nC\,D^{1/2}\,m^{2/3}\,t^{1/6} \tag{3.96}$$

For a given DME, m = rate of flow of Hg from dropping mercury electrode in mg s^{-1}, t = drop time (time taken between successive drops), m and t are constant for a given DME. D = diffusion coefficient of electroactive species in $cm^2\,s^{-1}$, C = concentration of electroactive species in mmol L^{-1}, i_d = average current in μA.

Thus $i_d \propto C$ and is used in analytical polarography. In order to reduce electrical migration of the electroactive species, a large concentration of an inert electrolyte known as supporting electrolyte is used in these experiments.

Electrodeposition: The electrodeposition of ions such as Cu^{2+} or Ag^+ from solutions of their salts takes place via electrolysis of these solutions using the same metal electrodes or inert electrodes such as Pt. The concepts of decomposition voltage, over voltage in an electrolysis cell or overpotential at a given electrode, say cathode or anode, are important concepts that must be borne in mind and which have been briefly referred earlier.

(i) **Calculation of the applied voltage required during electrolytic studies for deposition of an ion:** In order to find out the applied voltage necessary to carry out an electrolysis, the following parameters must be taken into account:
- concentrations of all species involved in electrode reactions,
- the IR drop across the cell,
- the cathodic and anodic overpotentials,
- pH of the solution, and
- the presence of any complexing agents which form complex ions with the electroactive species.

The applied voltage required to carry out an electrolysis reaction may be given

$$E_{\text{applied}} = E_{dv} + IR + E_{\text{conc. polarisation}} \tag{3.97}$$

where E_{dv} = decomposition voltage, the second term is IR drop across the cell, $E_{\text{conc. polarisation}}$ is concentration overvoltage. The estimation of this parameter is difficult because the concentrations of the electroactive species near the electrode surface are unknown and hence are omitted in these calculations. Thus,

$$E_{\text{applied}} = E_a - E_C + IR + \eta_a - \eta_C \tag{3.98}$$

where E_a = anode potential, E_C = cathode potential, η_a = anodic overpotential (which is a positive quantity), η_i = cathodic overpotential (which is a negative quantity).

(ii) **Relative magnitudes of anodic or cathodic potentials:** When two or more ions are present in solution, the anodic reaction occurs in such a way that the ion having the least positive decomposition potential will undergo the anodic reaction followed by an ion with a more positive decomposition potential. In an analogous manner, at the cathode, the reaction with the least negative potential takes place followed by an ion with a more negative cathodic decomposition potential.

(iii) **Controlled potential electrolysis:** The potential of the working electrode, that is, the EMF of the cell between the working electrode and an external reference electrode such as SCE is the most important factor in determining which of the two or more electrode reactions, will occur. In a controlled potential electrolysis, the potential of the working electrode is kept at a constant value by properly controlling the total voltage applied to the cell.

In such electrolyses, the current decreases with time and reaches a constant value, almost zero, in the case of a single reduction. The decay of current to a constant value (or zero) indicates that the reaction in completed. The current–time curve may be expressed by the equation

$$i_t = i_0 10^{-kt}$$

where i_0 = initial current, i_t = current at any time t and k is a constant depending on electrode area, volume of solution, rate of stirring and temperature.

Worked Out Problems

Laws of Electrolysis

Problem 3.1

(a) Adiponitrile, which is an important material required in the manufacture of Nylon 6–6, is synthesised electrolytically from acrylonitrile according to the reaction

$$2CH_2{=}CHCN + 2H^+ + 2e^- \rightarrow NCCH_2{-}CH_2{-}CH_2{-}CH_2{-}CN$$

How many (i) moles of electrons (ii) faradays (iii) coulombs are required for the production of 108 kg of adiponitrile?

(b) If the efficiency of the reaction is 70%, how many coulombs are required to produce the same 108 kg of adiponitrile?

Solution

(a) (i) $1 \text{ mole of electrons} = 6.023\times10^{23} e^-$

108 g of adiponitrile requires $2\times6.023\times10^{23} e^-$

108×10^3 g of adiponitrile requires

$$= \frac{108\times10^3\times2\times6.023\times10^{23} e^-}{108}$$

$$= 12\times10^{26} e^-$$

(ii) 108 g of adiponitrile are formed by the passage of 2F

108×10^3 g of adiponitrile requires

$$= \frac{108\times10^3\times2}{108} = 2000 \text{ F}$$

(iii) 108 g of adiponitrile are formed by the passage of 2×96500 coulombs

108×10^3 g of adiponitrile requires

$$= \frac{108\times10^3\times2\times96500}{108}$$

$$= 1930\times10^5 = 1.93\times10^8 \text{ coulombs}$$

(b) In this case, the amount of adiponitrile produced for

2×96500 coulombs is $\frac{108\times70}{100} = 75.6$ g

or, 75.6 g is produced by 2×96500 coulombs

108×10^3 g of adiponitrile is produced by

$$\frac{108\times10^3\times2\times96500}{75.6} = 2760\times10^5$$

$$= 2.76\times10^8 \text{ coulombs}$$

Problem 3.2

A current of 2 A was passed through a solution of a silver salt for coating a surface of 1 m^2 with a layer 0.01 mm thickness. Calculate the time (in s) and number of coulombs required for this deposition. (The density of silver is 10.5 kg dm^{-3}.)

Solution

Volume of silver required $= 100\times100\times0.001$

$$= 10^4\times10^{-3} = 10 \text{ cm}^3$$

Weight of silver $= 10\times10.5$

$$= 105 \text{ g}\left(105 \text{ kg dm}^{-3} = 105 \text{ g cm}^{-3}\right)$$

$10.5 \text{ kg dm}^{-3} = 10.5 \text{ g cm}^{-3}$

Since the reacion is $Ag^+ + e^- \rightarrow Ag$

108 g of Ag requires 96500 coulombs

105 g of Ag requires $\frac{105\times96500}{108} = 93817.3$ coulombs

$$\text{Time required} = \frac{93817.3}{2} = 46908.6 \text{ s} = 13.03 \text{ hours}$$

Specific and Equivalent Conductance of Electrolytes

Problem 3.3

In the study of the conductivity of a uni–univalent electrolyte at 0.05 eq dm^{-3} concentration, a cylindrical conductance cell of 1.0 m length and having electrodes of 0.002 m^2 was employed. The EMF across the cell was such that 0.50 amperes of current passed through it. If the equivalent conductance of the electrolyte is 116×10^{-4} S m^2 eq^{-1}, calculate (i) cell constant in m^{-1}, (ii) specific conductance of the solution, (iii) resistance of the solution in ohms, (iv) velocity of anions in m s^{-1} given the ionic conductance of the cation as 60×10^{-4} S m^2 eq^{-1}.

Solution

(i) $\text{Cell constant} = \frac{l}{a} = \frac{1.0\text{m}}{0.002 \text{ m}^2} = \frac{1000}{2} = 500 \text{ m}^{-1}$

(ii) $\Lambda = \frac{\kappa}{C} = \frac{l}{aR}\times\frac{1}{C}$

where k = specific conductance S m^{-1}, C = conc.in eq m^{-3}

$$= \frac{500}{R}\text{m}^{-1}\times\frac{1}{0.05 \text{ eq dm}^{-3}\times1000 \text{ dm}^3 \text{ m}^{-1}}$$

$$= \frac{500 \text{ m}^{-1}}{R\times50 \text{ ohm eq m}^{-3}}$$

$$R = \frac{500 \text{ m}^{-1}}{116\times10^{-4} \text{ S m}^2 \text{ eq}^{-1}\times50 \text{ ohm eq m}^{-3}}$$

$$= \frac{10}{116\times10^{-4}} = 8.62\times10^{-2}\times10^4 = 862 \text{ ohms}$$

(iii) $\kappa = \frac{l}{aR} = \frac{500}{862} = 5.8\times10^{-1} \text{ S m}^{-1}$

(iv) $\text{Mobility} = \frac{56\times10^{-4}}{96500} = 5.8\times10^{-8} \text{ m s}^{-1} \text{ V}^{-1}$

$$\text{Velocity of anion} = 5.8\times10^{-8}\times(862\times0.5)$$

$$= 2500\times10^{-8} = 2.5\times10^{-5} \text{ m s}^{-1}$$

Degree of Dissociation of Acids (or Bases) from Conductance Data and D-H-O Equation

Problem 3.4

The equivalent conductance of 0.1 eq dm^{-3} solution of formic acid is 19.0 S m^2 eq^{-1}, at 298 K. Given the equivalent conductances at infinite dilution of HCl, sodium formate and sodium chloride as 426.0, 105.0 and 126.5 all in S m^2 eq^{-1}, respectively, calculate the degree of dissociation of the given acid. Given A and B of D–H–O equation as 60.2 and 0.229 respectively, refine the value of α by two iterations. Calculate the dissociation constant of the acid using the final value of α. What is the per cent error in K when interionic forces are not taken into account?

Solution

$$\Lambda_{0(\text{HCOOH})} = 426.0 + 105.0 - 126.5 = 404.5 \text{ S m}^2 \text{ eq}^{-1}$$

$$\alpha = \frac{\Lambda}{\Lambda_0} = \frac{19}{404.5} = 4.7\times10^{-2}$$

$$K = \frac{(0.0472)^2 \times 0.1}{0.9528} = \frac{2.23\times10^{-4}}{0.9528} = 2.34\times10^{-4}$$

To refine this value of α, use the equation

$$\Lambda = \alpha\left[\Lambda_0 - (A + B\Lambda_0)\sqrt{\alpha C}\right]$$

1st iteration:

$$\Lambda = \alpha\left[404.5 - (60.2 + (0.229\times404.5))\sqrt{4.7\times10^{-2}\times0.1}\right]$$

$$\Lambda = \alpha\left[404.5 - (60.2 + 92.63)\times0.06855\right]$$

$$= \alpha[404.5 - 10.48] = \alpha\times394.02$$

$$\text{or, } \alpha = \frac{19.0}{394.02} = 0.0482$$

2nd iteration:

$$\Lambda = \alpha\left[404.5 - (60.2 + 92.63)\sqrt{4.82\times10^{-2}\times0.1}\right]$$

$$= \alpha\left[404.5 - (152.83\times0.0694)\right]$$

$$= \alpha[404.5 - 10.61] = 393.89$$

$$\text{Revised } \alpha = \frac{19.0}{393.89} = 0.0482$$

$$K = \frac{(0.0482)^2\times0.1}{0.9518}$$

$$= \frac{2.32\times10^{-4}}{0.9518} = 2.44\times10^{-4}$$

$$\%\text{ error} = \frac{2.44\times10^{-4} - 2.34\times10^{-4}}{2.44\times10^{-4}}\times100$$

$$= \frac{0.01\times10^{-4}}{2.44\times10^{-4}}\times10^2 = 0.41$$

Problem 3.5

In the determination of the dissociation constant of a weak acid, HA a plot of $\frac{1}{\Lambda_C}$ vs $\Lambda_C \times C$, was made (units of $\Lambda_C = \text{S cm}^2 \text{ eq}^{-1}$ and units of $C = \text{mol dm}^{-3}$) and the straight line graph yielded a intercept equal to 2.56×10^{-3} and a slope equal to 0.363. calculate Λ_0 and K, the dissociation constant of HA.

Solution

$$\text{Intercept} = \frac{1}{\Lambda_0} = 2.56\times10^{-3}$$

$$\Lambda_0 = 0.391\times10^3 = 391 \text{ S cm}^2 \text{ eq}^{-1}$$

$$\text{Slope} = \frac{1}{K(\Lambda_0)^2} = 0.363$$

$$K(\Lambda_0)^2 = 2.755 \text{ or}$$

$$K = \frac{2.755}{391\times391} = 1.8\times10^{-5}$$

Solubility Product from Molar Conductance Data

Problem 3.6

A sparingly soluble salt of type $M_2(X)_3$ was found to have a specific conductivity in its saturated solution as 7×10^{-3} S m^{-1}. The water used has a specific conductivity of 2×10^{-4} S m^{-1}. The equivalent ionic conductances of M^{3+} and X^{2-} ions at infinite dilute solutions are 80×10^{-4} and 70×10^{-4} respectively. (i) Calculate the molar conductance (μ_0) of the salt at infinite dilution. (ii) What is the solubility of the salt in mol dm^{-3}? (iii) If the solubility gives solubility of M^{3+} ions, calculate its solubility product.

Solution

Specific conductivity of saturated solution

$$= 7\times10^{-3} - 0.2\times10^{-3}$$

$$= 6.8\times10^{-3} \text{ S m}^{-1}$$

(i) μ_0 = Molar conductance of the salt

$$= 6\times150\times10^{-4} = 9\times10^{-2} \text{ S m}^2 \text{ mol}^{-1}$$

$$\text{(ii) Solubility of the salt} = \frac{6.8\times10^{-3} \text{ S m}^{-1}}{9\times10^{-2} \text{ S m}^2 \text{ mol}^{-1}}$$

$$= 7.55\times10^{-2} \text{ mol m}^{-3}$$

$$= 7.55\times10^{-5} \text{ mol dm}^{-3}$$

$$\text{(iii) } M_2(X)_3 \rightleftharpoons 2M^{3+} + 3X^{2-}$$

If s is the solubility of salt, $\left[M^{3+}\right] = 2s$; $\left[SO_4^{2-}\right] = 3s$

$$K_{SP} = \left[M^{3+}\right]^2\left[s^{2-}\right]^3$$

$$= (2s)^2(3s)^3$$

$$= 108\times s^5 = 108\times(7.55\times10^{-5})^5$$

$$= 108\times24532\times10^{-25}$$

$$= 2.65\times10^6\times10^{-25} = 2.65\times10^{-19}$$

Problem 3.7

The specific conductance of a saturated solution of lead flouride at 298 K is 489.3×10^{-4} S m^{-1}. Its solubility product at the same temperature is 3.0×10^{-8} mol^2 dm^{-6}. Calculate the equivalent ionic conductance of Pb^{2+} $\left(\frac{1}{2}\lambda^0_{Pb^{2+}}\right)$ given that of the fluoride ion as 53×10^{-4} S m^2 eq^{-1}.

Solution

$K_{SP} = 3.0\times10^{-8}$ mol^2 dm^{-6}. If s is the solubility of the salt,

$$\text{Conc.of}\left[Pb^{2+}\right] = s;\ \text{Conc. of}\left[F^-\right] = 2s$$

$$K_{SP} = s\left[2s\right]^2 = 4s^3$$

$$4s^3 = 3.0\times10^{-8}$$

$$s = \sqrt[3]{7.5\times10^{-9}}$$

$$= 1.957\times10^{-3}\ \text{mol dm}^{-3}$$

$$= 1.957\times10^{-3}\times1000 = 1.957\ \text{mol m}^{-3}$$

$$\mu_0\left(PbF_2\right) = \frac{k}{s} = \frac{489.3\times10^{-4}}{1.957}$$

$$= 250\times10^{-4}\ \text{S m}^2\ \text{mol}^{-1}$$

$$\mu_0\left(PbF_2\right) = \mu_0\left(Pb^{2+}\right) + 2\mu_0\left(F^-\right)$$

$$= \mu_0\left(Pb^{2+}\right) + 106\times10^{-4}$$

$$250\times10^{-4} = \mu_0\left(Pb^{2+}\right) + 106\times10^{-4}$$

$$\text{or, } \mu_0\left(Pb^{2+}\right) = 144\times10^{-4}\ \text{S m}^2\ \text{mol}^{-1}\ \text{or,}$$

$$\frac{1}{2}\lambda^0_{Pb^{2+}} = 72.0\ \text{S m}^2\ \text{eq}^{-1}$$

Problem 3.8

The specific conductance of a saturated solution of silver salicylate in water at 298 K is 3.25×10^{-2} S m^{-1}. The $\lambda_{0_{Ag^+}}$ and $\lambda_{0_{anion}}$ are 62×10^{-4} and 36×10^{-4}, both in S m^2 mol^{-1} respectively. Calculate the solubility of the salt in mol dm^{-3}. Also, refine the value of solubility by two iterations using the D–H–O equation with $A = 60.2$ and $B = 0.229$.

Solution

$$k = C\times\Lambda\ \text{or,}$$

$$C = \frac{\kappa}{\Lambda} = \frac{3.25\times10^{-2}}{98\times10^{-4}} = 3.32\times10^{-2}\times10^2 = 3.32\ \text{mol m}^{-3}$$

Solubility, s, is given by $= 3.32$ mol m$^{-3}\times10^{-3}$ m^3 dm^{-3}

$$= 3.32\times10^{-3}\ \text{mol dm}^{-3}$$

The D–H–O equation is given by $\Lambda = \Lambda_0 - \left(A + B\Lambda_0\right)\sqrt{C}$

Using the solubility value, we calcualte a revised value of Λ' as

$$\Lambda' = 98 - \left(60.2 + \left(0.229\times98\right)\right)\sqrt{3.32\times10^{-3}}$$

$$= 98 - \left(68.2 + 22.44\right)5.762\times10^{-2}$$

$$= 98 - 4.76 = 93.24\ \text{S cm}^2\ \text{eq}^{-1}$$

NOTE

- In using the D–H–O equation, Λ has been expressed in units of S cm^2 eq^{-1}.

1st revised value of S, that is, $S' = \frac{\kappa}{\Lambda'} = \frac{3.25\times10^{-2}\times0.01}{93.24}$

$$= 3.485\times10^{-6}\ \text{eq cm}^{-3}$$

$$= 3.485\times10^{-3}\ \text{eq dm}^{-3}$$

2nd revised value is obtained as follows Λ''

$$= 93.24 - \left(60.2 + 0.229\times93.24\right)\sqrt{3.485\times10^{-3}}$$

$$= 93.24 - \left(60.2 + 21.35\right)0.059$$

$$= 93.24 - 4.81 = 88.43\ \text{S cm}^2\ \text{eq}^{-1}$$

$$S'' = \frac{\kappa}{\Lambda''} = \frac{3.25\times10^{-4}}{88.43}$$

$$= 3.675\times10^{-6}\ \text{eq cm}^{-3} = 3.675\times10^{-3}\ \text{eq dm}^{-3}$$

A further revision can also be made in a similar manner until a constant value of s is obtained.

Conductometric Titrations

Problem 3.9

50 mL of 0.01 N solution of Na_2SO_4 is titrated with 0.1 N $Ba(OH)_2$ solution in a conductance cell. Given the λ_0 values of $\frac{1}{2}Ba^{2+} = 63.6\times10^{-4}$ S m^2 eq^{-1}; $Na^+ = 50.0\times10^{-4}$ S m^2 eq^{-1}; $\frac{1}{2}SO_4^{2-} = 80\times10^{-4}$ S m^2 eq^{-1}; $OH^- = 198.0\times10^{-4}$ S m^2 eq^{-1}, calculate (i) initial specific conductivity of the solution (ii) specific conductance of the solution at equivalence point (iii) specific conductance of the solution when 2 mL of $Ba(OH)_2$ is added after equivalence point. Assume that ionic conductances are independent of concentration and also that the contribution of $BaSO_4$ to the conductivity of the solution is negligible.

Solution

(i) Initial specific conductance of 0.01 N = 0.005 M solution of Na_2SO_4

$$= \sum_i C_i\lambda_i = C_{Na^+}\lambda_{Na^+} + C_{SO_4^{2-}}\lambda_{SO_4^{2-}}$$

where $= C =$ conc. in molm^{-3}, λ_{Na^+} = molar conductance of Na^+ and so on.

$= 2\times 0.005\times 1000\left(\text{mol dm}^{-3}\times 1000\text{ dm}^3\text{ m}^{-3}\right)\times$

$50\times 10^{-4} + 0.005\times 2\times 80\times 10^{-4}\times 1000$

$= 5\times 10^{-2} + 8\times 10^{-2} = 13\times 10^{-2}\text{ S m}^{-1} = 1.3\times 10^{-1}\text{ S m}^{-1}$

(ii) The reaction

$Na_2SO_4 + Ba(OH)_2 \rightarrow 2Na^+ + 2OH^- + BaSO_4 \downarrow$

At the equivalence point, only NaOH is present and k of solution at equivalence point $= C_{Na^+}\lambda_{Na^+} + C_{OH^-}\lambda_{OH^-} = 2\times$

$\left(\frac{0.005\times 50}{55}\times 1000\right)\times 50\times 10^{-4} + 2\times\left(\frac{0.005\times 50}{55}\times 1000\right)$

$\times 198\times 10^{-4}$

$= (2\times 0.00454\times 1000)\left[248\times 10^{-4}\right]$

$= 0.2252\text{ S m}^{-1} = 2.252\times 10^{-1}\text{ S m}^{-1}$

NOTE

- 5 mL is the volume of $Ba(OH)_2$ solution required for the complete reaction of Na_2SO_4 with $Ba(OH)_2$. The original concentration of the Na^+ ions 0.005×2, is now present in 55 mL rather than 50 mL. So, the dilution factor has to be taken into account.

(iii) In addition to NaOH, we have 2 mL of 0.05 M Ba^{2+} ions and 2 mL of 0.1 M OH^- ions in a total volume of 57 mL (50 + 5 + 2).

Conc. of $Ba^{2+} = 0.05\times\frac{2}{57} = 0.00175$ M

Conc. of $OH^- = 0.0035$ M

$\therefore$ Contribution due to added $Ba(OH)_2$

$= C_{Ba^{2+}}\lambda_{Ba^{2+}} + C_{OH^-}\lambda_{OH^-}$

$= 0.00175\times 1000\times 127.2\times 10^{-4} +$

$0.0035\times 198\times 10^{-4}\times 1000$

$= 2.23\times 10^{-2} + 6.93\times 10^{-2}$

$= 9.16\times 10^{-2}\text{ S m}^{-1}$

Total specific conductivity $= 2.252\times 10^{-1}\text{ S m}^{-1} +$

$0.916\times 10^{-1}\text{ S m}^{-1}$

$= 3.168\times 10^{-1}\text{ S m}^{-1}$

(The factor 1000 arises in converting mol dm^{-3} to mol m^{-3})

Determination of Transport Numbers

Problem 3.10

A Hittorf experiment involving electrolysis of a solution of silver nitrate using silver electrodes was carried out and a current of 2.5 mA was passed for 2 hours through the electrolysis cell. The anode solution weighed exactly 50 g before electrolysis and contained 0.540 g silver. After electrolysis, the same amount of anode solution contained 0.552 g of silver. The cathode solution contained 0.567 g of silver before electrolysis and 0.559 g silver after electrolysis. (i) Give the reactions at the anode and the cathode and explain how the silver ion concentration changes at the electrodes. (ii) Calculate the transport number of NO_3^- ion by the analysis of the anode solution and that of the Ag^+ ion by the analysis of the cathode solution.

Solution

(i) Reaction at the anode: $NO_3^- + Ag(\text{anode}) \rightarrow AgNO_3 + e^-$

Since silver dissolves at the anode, there is an increase of silver ion concentration at the anode.

(ii) Reaction at the cathode: $Ag^+ + e^- \rightarrow Ag$

As the silver ions get deposited, there is a loss in silver ion concentration at the cathode.

Number of coulombs passed through the cell

$= 2.5\times 10^{-3}\times 2\times 3600 = 18$

Weight of silver equivalent to coulombs passed

$= \frac{18\times 107.9}{96500} = 0.02$ g

$t_{NO_3^-} = \frac{\text{Increase in weight of Ag at anode}}{\text{Amount of silver to be deposited}}$

$= \frac{0.552 - 0.540}{0.02} = \frac{0.012}{0.02} = 0.60$

Analysis at cathode solution

$t_{Ag^+} = \frac{\text{Loss in weight of Ag at cathode}}{\text{Amount of silver}}$

$= \frac{0.567 - 0.559}{0.02} = \frac{0.008}{0.02} = \frac{0.8}{2} = 0.40$

Problem 3.11

In a moving boundary experiment using 0.02 M NaCl solution, a cylindrical tube of cross-sectional area of 0.107 cm^2 was used and it was found that the boundary moved 6.5 cm when a current of 2 mA was passed for half an hour. What is the transport number of the Na^+ ion?

Solution

No. of coulombs Z, passed $= 2\times 10^{-3}\times 1800 = 3.6$

Conc. of NaCl in eq cm^{-3} $= \frac{2\times 10^{-2}}{1000} = 2\times 10^{-5}$

$t_+ = \frac{2\times 10^{-5}\times 6.5\times 0.107\times 96500}{3.6} = \frac{1342\times 10^{-3}}{3.6} = 0.373$

Problem 3.12

The ionic mobilities of Ag^+ and NO_3^- ions at infinite dilution and at 298 K are 6.4×10^{-8} and 7.4×10^{-8} m^2 volt^{-1} s^{-1} respectively. Calculate the transport numbers of Ag^+ and NO_3^-

ions at infinite dilution. Suppose we have a mixture of KNO_3 and $AgNO_3$ at infinite dilution with the ionic mobility of K^+ ion as 7.62×10^{-8} m^2 volt^{-1} s^{-1} will the transport numbers of Ag^+ and NO_3^- ions be the same or different? If not, what are the new values for the Ag^+ and NO_3^- ions?

Solution

$$\lambda_{0_{Ag^+}} = 6.4\times10^{-8}\times96500 = 6.176\times10^{-3}\ \text{S m}^2\ \text{mol}^{-1}$$

$$\lambda_{0_{NO_3^-}} = 7.4\times10^{-8}\times96500 = 7.141\times10^{-3}\ \text{S m}^2\ \text{mol}^{-1}$$

$$t^0_{Ag^+} = \frac{6.176\times10^{-3}}{13.317\times10^{-3}} = 0.464;\ t^0_{NO_3^-} = \frac{7.141\times10^{-3}}{13.317\times10^{-3}} = 0.536$$

In a mixture of $AgNO_3$ and KNO_3, we have Ag^+, K^+ and NO_3^- ions.

$$\lambda_{0_{K^+}} = 7.62\times10^{-8}\times96500 = 7.353\times10^{-3}\ \text{S m}^2\ \text{mol}^{-1}$$

$$t^0_{Ag^+} = \frac{\lambda_{0_{Ag^+}}}{\lambda_{0_{Ag^+}}+\lambda_{0_{K^+}}+\lambda_{0_{NO_3^-}}} = \frac{6.176\times10^{-3}}{(6.176+7.141+7.353)\times10^{-3}}$$

$$= \frac{6.176}{20.67} = 0.300$$

$$t^0_{NO_3^-} = \frac{\lambda_{0_{NO_3^-}}}{\lambda_{0_{Ag^+}}+\lambda_{0_{K^+}}+\lambda_{0_{NO_3^-}}} = \frac{7.141\times10^{-3}}{20.67\times10^{-3}} = 0.345$$

Ionic Product of Water from Conductance Data

Problem 3.13

The specific conductivity of pure water is 6.2×10^{-6} S m^{-1} at 298 K. Relate the specific conductivity to the concentration of ions present in water and to the ionic product of water. Calculate the ionic product given $\lambda^0_{H^+} = 350\times10^{-4}$ S m^2 mol^{-1} and $\lambda^0_{OH^-} = 198\times10^{-4}$ S m^2 mol^{-1}

Solution

The specific conductance of water

$$= C_{H^+}\lambda_{H^+} + C_{OH^-}\lambda_{OH^-} \approx C_{H^+}\lambda^0_{H^+} + C_{OH^-}\lambda^0_{OH^-}$$

where $C_{H^+} = C_{OH^-}$ given in mol m^{-3}and λ^0 in S m^2 mol^{-1}

$$K_w = [H^+][OH^-] \text{and } [H^+] = [OH^-] = \sqrt{K_w}$$

$$\therefore k_{\text{water}} = \sqrt{K_w}\left(\lambda^0_{H^+} + \lambda^0_{OH^-}\right)$$

$$= \sqrt{K_w}\times548\times10^{-4}\ \text{S m}^2\ \text{mol}^{-1}$$

$$6.2\times10^{-6}\left(\text{S m}^{-1}\right) = \sqrt{K_w}\times548\times10^{-4}\ \text{S m}^2\ \text{mol}^{-1}$$

$$\sqrt{K_w} = \frac{6.2\times10^{-6}}{548\times10^{-4}}\times\frac{\left(\text{S m}^{-1}\right)}{\left(\text{S m}^2\ \text{mol}^{-1}\right)}$$

$$= 1.13\times10^{-4}\ \text{mol m}^{-3}$$

$$\sqrt{K_w} = 1.13\times10^{-4}\ \text{mol m}^{-3}\times0.001\ \text{m}^3\ \text{dm}^{-3}$$

$$= 1.13\times10^{-7}\ \text{mol dm}^{-3}$$

$$K_w = \left(1.13\times10^{-7}\right)^2 = 1.28\times10^{-14}\ \text{mol}^2\ \text{dm}^{-6}$$

Ionic Mobilities and Transport Numbers

Problem 3.14

The molar conductivity of sodium fluoride at 0.1 mol dm^{-3} concentration in water is 84.0×10^{-4} S m^2 mol^{-1} at 298 K. Given that the transference number of fluoride ion is 0.55, calculate the ionic velocities of Na^+ and F^- ions under a potential gradient of 10 V cm^{-1}. The salt is dissociated to the extent of 90%.

Solution

$$\Lambda = 84.0\times10^{-4} = \alpha F(\upsilon_+ + \upsilon_-)$$

$$= 0.90\times96500(\upsilon_+ + \upsilon_-)$$

$$(\upsilon_+ + \upsilon_-) = \frac{84.0\times10^{-4}}{0.90\times96500}$$

$$= 9.672\times10^{-8}\ \text{m}^2\ \text{V}^{-1}\ \text{s}^{-1}$$

$$\upsilon_+ = 0.45\times9.672\times10^{-8}$$

$$= 4.352\times10^{-8}\ \text{m}^2\ \text{V}^{-1}\ \text{s}^{-1}$$

$$\upsilon_- = 0.55\times9.672\times10^{-8}$$

$$= 5.32\times10^{-8}\ \text{m}^2\ \text{V}^{-1}\ \text{s}^{-1}$$

$$\text{Velocity of Na}^+\text{ ion} = 4.352\times10^{-8}\times10$$

$$= 4.352\times10^{-7}\ \text{m}^2\ \text{V}^{-1}\ \text{s}^{-1}$$

$$\text{Velocity of F}^-\text{ ion} = 5.32\times10^{-8}\times10$$

$$= 5.32\times10^{-7}\ \text{m}^2\ \text{V}^{-1}\ \text{s}^{-1}$$

Galvanic Cells and Applications

Problem 3.15

Formulate the galvanic cells in which the following cell reactions take place:

(a) $H_2(g) + Br_2(g) \rightarrow 2H^+ + 2Br^-$

(b) $2Fe^{3+} + 2I^- \rightarrow 2Fe^{2+} + I_2$

(c) $Cd(s) + Hg_2SO_4(aq) \rightarrow CdSO_4(aq) + 2Hg(l)$

(d) $2MnO_4^- + 6H^+ + 5H_2C_2O_4 \rightarrow 2Mn^{2+} + 8H_2O + 10CO_2$

(e) $2Ag + C_6H_{12}O_7 + 2H^+ \rightarrow 2Ag^+ + C_6H_{12}O_6 + H_2O$

Solution

(a) $\text{Pt}, H_2(g), H^+\text{(in solution)} \,||\, Br^-\text{(in solution)}|\,Br_2(g), \text{Pt}$

(b) $\text{Pt}, |\,I^-, I_2 \,||\, Fe^{3+}, Fe^{2+}\,|\,\text{Pt}$

(c) $Cd(s)|CdSO_4(aq)|Hg_2SO_4(aq)|Hg$

(d) $Pt|H_2C_2O_4, CO_2(g)\|MnO_4^-, Mn^{2+}|Pt$

(e) $Ag|Ag^+\|C_6H_{12}O_6(aq), C_6H_{12}O_7\text{(gluconic acid)}|Pt$

Problem 3.16

The EMF of the cell $Pt, H_2(g, 1\text{ atm})|MOH(m = 0.01)\|SCE$, where MOH is a weak base, is 0.868 V at 298 K. Given $E_{SCE} =$ 0.242 V, calculate the OH^- concentration in the cell and the dissociation constant of MOH.

Solution

$$E = E_{SCE} - E_{H_2,OH^-};\ E_{H_2,OH^-} = 0.242 - 0.868 = -0.626\text{ V}$$

$$E_{H_2,OH^-} = E^0 + \frac{RT}{F}\ln a_{H^+}$$
$$= 0 + 0.0591\log a_{H^+}$$

$$\log a_{H^+} = -\frac{0.626}{0.0591}$$
$$= -10.59 = \overline{11}.41$$
$$a_{H^+} = 2.57\times10^{-11}$$
$$a_{OH^-} = \frac{10^{-14}}{2.57\times10^{-11}}$$
$$= 3.89\times10^{-4}$$

$$K = \frac{a_{M^+}\times a_{OH^-}}{a_{MOH}} = \frac{(3.89\times10^{-4})^2}{0.01}$$
$$= \frac{15.13\times10^{-8}}{10^{-2}} = 1.51\times10^{-5}$$

Problem 3.17

For the cell, $SCE\|$soln of unknown pH (saturated with a 1 : 1 mixture of quinone : hydroquinone)$|H_2Q, Q, Pt$ the EMF is 0.221 V at 298 K. Given $E_{SCE} = 0.242$ V and $E^0_{Q,H_2Q} = 0.699$ V, calculate the pH of the solution.

Solution

$$E_{cell} = E_{H_2Q} - E_{SCE}$$
$$= 0.699 + 0.0591\log a_{H^+} - 0.242$$
$$= 0.457 - 0.0591\text{pH}$$
$$\text{pH} = \frac{0.457 - 0.221}{0.0591} = 4.00$$

Problem 3.18

When a rod of lead was placed in a solution of $Pb(NO_3)_2$ of conc. 1 M and its EMF measured against the SHE, it gave a value of 0.126 V. When the experiment was repeated with a zinc rod placed in a solution of $Zn(NO_3)_2$ of conc. 1 M, the EMF read was 0.77 V. In both cases, the SHE was the positive electrode in the EMF measurements by the Poggendorf method. (i) What are the correct signs of the electrode potentials? (ii) Write the cell corresponding to the sign vis-à-vis the SHE and also write the cell reaction. Are these reactions spontaneous? (iii) Of the two half cells, which one is a stronger oxidising agent and which is a stronger reducing agent? (iv) When the two half cells are connected, which electrode is positive and what is the spontaneous reaction in the cell? (v) If a complexing agent such as OH^- ions is added to the zinc half cell, in what direction will its half cell potential change, towards the more negative or towards the more positive side?

Solution

(i) Obviously, both the electrode potentials are negative.

(ii) $SHE\|Pb^{2+}|Pb,\quad Pb^{2+} + H_2 \to Pb + 2H^+$

Since E is negative, it is not a spontaneous reaction.

$SHE\|Zn^{2+}(C = 1.0\text{ M})|Zn,\ Zn^{2+} + H_2 \to Zn + 2H^+$

Since E is negative, it is not a spontaneous reaction.

(iii) Pb^{2+} is a stronger oxidising agent because it has a more positive electrode potential. Using a similar argument, Zn^{2+} is a stronger reducing agent, as it gets oxidised more easily because of its more negative electrode potential.

(iii) The cell is $Zn|Zn^{2+}\|Pb^{2+}(C = 1)|Pb$, the *EMF* is $-0.126 - (-0.77) = 0.644$ V. Since E is positive, the reaction $Zn + Pb^{2+}(C = 1) \to Zn^{2+}(C = 1) + Pb$, is spontaneous.

The Pb electrode is positive.

(iv) When OH^- ions are added, Zn^{2+} ion activity decreases, that is, its electrode potential becomes more positive.

Problem 3.19

(a) The standard potentials of $Sn^{4+} + 4e^- \to Sn$ and $Sn^{4+} + 2e^- \to Sn^{2+}$ are 0.006 and 0.150 V respectively at 298 K. what is the standard potential of $Sn^{2+} + 2e^- \to Sn$?

(b) The standard potentials of $Pt|Fe^{3+}, Fe^{2+}$ and $Fe^{3+} + 3e^- \to Fe$ are 0.76 V and –0.04 V respectively at 298 K. What is the standard potential of the $Fe|Fe^{2+}$ system?

(c) Will stannous ions be a better oxidising agent or Fe^{2+} in the stannous–tin and ferrous–iron system when they are combined to form a cell?

Solution

(a) $Sn^{4+} + 4e^- \to Sn \qquad E^0 = 0.006\text{ V}$

$$\Delta G_1^0 = -4\times F\times0.006$$
$$= -0.024\text{ V}\times F$$

$Sn^{4+} + 2e^- \to Sn^{2+} \qquad E^0 = 0.150\text{ V}$

$$\Delta G_2^0 = -2\times F\times0.150$$
$$= -0.300\text{ V}\times F$$

$\Delta G_1^0 - \Delta G_2^0 = 0.276F$ corresponding to the reaction $Sn^{2+} + 2e^- \to Sn$

$$E^0\text{of the reaction} = -\frac{0.276F}{2F} = -0.138\text{ V}$$

(b) $Fe^{3+} + 3e^- \rightarrow Fe \qquad E^0 = -0.04$

$$\Delta G_1^0 = -3 \times F \times -0.04$$
$$= +0.12F$$

$Fe^{3+} + e^- \rightarrow Fe^{2+} \qquad E^0 = 0.76$

$$\Delta G_2^0 = -1 \times F \times 0.76$$
$$= -0.76F$$

$$\Delta G_1^0 - \Delta G_2^0$$
$$= 0.86F \text{ corresponding reaction } Fe^{2+} + 2e^- \rightarrow Fe$$

$$E^0 \text{of the reaction} = -\frac{0.88F}{2F} = -0.44 \text{ V}$$

(c) The spontaneous reaction when the two half cells in (a) and (b) are connected is

$$Sn^{2+} + Fe \rightarrow Fe^{2+} + Sn \qquad E^0 = 0.302 \text{ V}$$

$\therefore Sn^{2+}$ is a better oxidising agent than Fe^{2+}

Problem 3.20

The standard electrode potentials, $Pt|Fe^{3+}$, Fe^{2+} and $Pt|Cr_2O_7^{2-}$, Cr^{3+} are +0.771 and +1.330 V respectively at 298 K. Formulate the cell in which the spontaneous reaction occurs and calculate the equilibrium constant of the reaction that takes place in the cell.

Solution

$Pt|Fe^{2+}, Fe^{3+}$ (at unit conc.of each ion) $||$ $Cr_2O_7^{2-}, Cr^{3+}$ (unit conc.of each ion)$|$Pt

$$E_{cell}^0 = E_R^0 - E_L^0$$
$$= 1.330 - 0.771 = 0.559 \text{ V}$$

$$E^0 = \frac{RT}{nF} \log K$$

The reaction in the cell is

$$Cr_2O_7^{2-} + 6Fe^{2+} + 14H^+ \rightarrow 2Cr^{3+} + 6Fe^{3+} + 7H_2O$$

$$0.559 = \frac{0.0591}{6} \log K$$

$$\log K = \frac{0.559 \times 6}{0.0591} = 56.75$$

$$K = 5.62 \times 10^{56}$$

Problem 3.21

The EMF of the cell $Cd|CdCl_2(aq)|AgCl(aq), Ag$ is 0.692 V at 298 K and 0.675 V at 323 K. Write the cell reaction for the passage of two faradays. Calculate ΔG^0, ΔH^0 and ΔS^0 for the above reaction at 323 K.

Solution

Cell reaction

$$Cd + 2AgCl(aq) \rightarrow CdCl_2(aq) + 2Ag$$

$$\left(\frac{\partial E}{\partial T}\right)_P = -\frac{0.692 - 0.675}{25}$$
$$= -\frac{0.017}{25} = -6.8 \times 10^{-4} \text{ V K}^{-1}$$

$$\Delta G_{323}^0 = \frac{-2 \times 96500 \times 0.675}{1000}$$
$$= -130.0 \text{ kJ}$$

$$\Delta S^0 = -2 \times 96500 \times 6.8 \times 10^{-4}$$
$$= -131.2 \text{ J K}^{-1}$$

$$\Delta H^0 = \Delta G^0 + T\Delta S^0 = -130.0 - \left(323 \times 131.2 \times 10^{-3}\right)$$
$$= -130.0 - 42.4 = -172.4 \text{ kJ}$$

Problem 3.22

The half cell potentials for hydrogen (E_{H_2,OH^-}) and oxygen (E_{O_2,OH^-}) electrodes in the alkaline medium are –0.83 V and 0.40 V at 298 K respectively. Calculate E^0 of the cell in which H_2 and O_2 combine to form water and also the ΔG^0 of the reaction.

Solution

(i) Reaction at the O_2 electrode:

$$O_2(g) + 2H_2O + 4e^- \rightarrow 4OH^- \qquad E^0 = 0.40 \text{ V}$$

(ii) Reaction at the H_2 electrode:

$$4H_2O + 4e^- \rightarrow 2H_2 + 4OH^- \qquad E^0 = -0.83 \text{ V}$$

(i) – (ii) gives

$$O_2(g) + 2H_2 \rightarrow 2H_2O(l) \quad E^0 = 0.40 - (-0.83) = 1.23 \text{ V}$$

$$\Delta G^0 = -nFE^0 = -\frac{4 \times 96500 \times 1.23}{1000} = -474.8 \text{ kJ}$$

Activity Coefficients Using Cell Potentials

Problem 3.23

The EMF of the cell $Pt, H_2(g, 1 \text{ atm})|H_2SO_4(m = 0.5)||Hg_2SO_4(s), Hg$ is 0.660 V at 298 K. The E^0 of the cell is 0.615 V. Calculate the mean molality, mean activity coefficient and mean activity of the acid.

Solution

Cell reaction: $H_2(g) + Hg_2SO_4(s) \rightarrow 2Hg + 2H^+ + SO_4^{2-}$.

$$E = E^0 - \frac{RT}{2F} \ln\left(m_{H^+}^2 \times m_{SO_4^{2-}}\right) - \frac{RT}{2F} \ln\left(\gamma_{H^+}^2 \times \gamma_{SO_4^{2-}}\right)$$

$$= E^0 - \frac{RT}{2F} \ln m_\pm^3 - \frac{RT}{2F} \ln \gamma_\pm^3$$

$$\left(H_2SO_4 \rightleftharpoons 2H^+ + SO_4^{2-} \quad m_\pm = \left(m_{H^+}^2 \times m_{SO_4^{2-}}\right)^{\frac{1}{3}}\right)$$

$$= \left((2m)^2 \times m\right)^{\frac{1}{3}}$$

$$m_\pm = 4m = 4\times 0.5 = 2$$

$$0.660 = 0.6152 - \frac{0.0591\times 3}{2}\log 2 - \frac{0.0591\times 3}{2}\log\gamma_\pm$$

$$= 0.6152 - 0.0266 - 0.0886\log\gamma_\pm$$

$$\log\gamma_\pm = -\frac{0.0714}{0.0886} = -0.806 = \bar{1}.194;$$

$$\gamma_\pm = 0.156$$

Mean activity $(a_\pm = m_\pm\gamma_\pm) = 2\times 0.156 = 0.312$

Problem 3.24

The EMF of the cell

$$\text{Pt, H}_2(g, 1\text{ atm})|\text{HBr}(m)|\text{AgBr, Ag}$$

is measured at various molalities of HBr at 298 K, as shown below.

m/mol kg^{-1}	0.020	0.050	0.080
E/V	0.279	0.235	0.217

(i) Given $\gamma_\pm(\text{HBr})$ at 0.02 m as 0.87, calculate E^0 of the cell.

(ii) Using E^0 in (i), calculate $\gamma_\pm(\text{HBr})$ at 0.05 and 0.08 m.

Solution

Cell reaction:

$$\frac{1}{2}\text{H}_2(g,1\text{atm}) + \text{AgBr}(s) \rightarrow \text{H}^+(m) + \text{Br}^-(m) + \text{Ag}$$

(i) $E = E^0 - 0.0591\log\left(m_{\text{H}^+}\times m_{\text{Br}^-}\times\gamma_{\text{H}^+}\times\gamma_{\text{Br}^-}\right)$

$$E = E^0 - 0.0591\log\left(m_\pm^2\times\gamma_\pm^2\right)$$

$$= E^0 - 0.0591\times 2\log(m_\pm\times\gamma_\pm)$$

for HBr $m_\pm = \left(m_{\text{H}^+}\times m_{\text{Br}^-}\right)^{\frac{1}{2}}$

$$= \left(m^2\right)^{\frac{1}{2}} = m$$

$$\therefore 0.279 = E^0 - 0.0591\times 2\log(0.02\times 0.87)$$

$$= E^0 - 0.1182\log 0.017$$

$$= E^0 - 0.1182\times(-1.759) = E^0 + 0.208$$

$$E^0 = 0.279 - 0.208 = 0.071\text{ V}$$

(ii) At 0.05 m

$$0.235 = 0.071 - 0.0591\times 2\log(0.05\gamma_\pm)$$

$$= 0.071 - 0.1182\log(0.05\gamma_\pm)$$

$$\log(0.05\gamma_\pm) = \frac{-0.235+0.071}{0.1182}$$

$$= -\frac{0.164}{0.1182}$$

$$= -1.387 = \bar{2}.613$$

$$0.05\gamma_\pm = 0.041 \text{ or } \gamma_\pm = \frac{0.041}{0.05} = 0.82$$

(iii) At 0.08 m

$$0.217 = 0.071 - 0.1182\log(0.08\gamma_\pm)$$

$$\log(0.08\gamma_\pm) = \frac{-0.217+0.071}{0.1182}$$

$$= -\frac{0.146}{0.1182} = -1.235 = \bar{2}.765$$

$$\gamma_\pm = \frac{0.0582}{0.08} = 0.727$$

Solubility Products

Problem 3.25

The solubility product of lead sulphate is 1.78×10^{-8} mol^2 dm^{-6} at 298 K. Formulate a cell which enables the determination of its K_{SP}. If the E^0 of Pb|Pb^{2+} half cell is –0.123 V, what will be the electrode potential of the other half cell?

Solution

The half cell reactions are

$$\text{PbSO}_4(s) + 2e^- \rightarrow \text{Pb} + \text{SO}_4^{2-} \qquad E_1^0 = ?$$

$$\text{Pb}^{2+} + 2e^- \rightarrow \text{Pb} \qquad E_2^0 = -0.126\text{ V}$$

Cell: $\text{Pb(s)}|\text{Pb}^{2+}\,||\,\text{SO}_4^{2-}|\text{PbSO}_4(s), \text{Pt}$

$$E_{\text{cell}}^0 = E_1^0 - E_2^0 = E_1^0 - (-0.126) = E_1^0 + 0.126$$

$$E_{\text{cell}}^0 = \frac{RT}{nF}\ln K_{SP}$$

$$= \frac{0.0591}{2}\log 1.78\times 10^{-8}$$

$$= \frac{0.0591}{2}\times(-7.77) = -0.230\text{ V}$$

$$E_1^0 + 0.126 = -0.230 \text{ or}$$

$$E_1^0 = E^0_{\text{Pb,SO}_4^{2-}} = -0.356\text{ V}$$

Electrode and Electrolyte Concentration Cells

Problem 3.26

The EMF of the concentration cell

$$\text{Cu}(s)|\text{CuSO}_4(\text{m} = 0.005)\vdots\text{CuSO}_4(\text{m} = 0.05)|\text{Cu}$$

is 0.030 V at 298 K. Given the mean activity coefficients of $CuSO_4$ in 0.005 and 0.05 m solutions as 0.78 and 0.52, respectively, calculate the transport number of Cu^{2+} ion.

What will the EMF of the cell be if there is no transport of ions across the boundary? What is the liquid junction potential?

Solution

$$E_{cell} = 2t_{-}\frac{RT}{2F}\ln\frac{(m_{\pm}\gamma_{\pm})_2}{(m_{\pm}\gamma_{\pm})_1}$$

$$= 2t_{-}\frac{0.0591}{2}\log\left(\frac{0.05\times0.52}{0.005\times0.78}\right)$$

$$= 2t_{-}\frac{0.0591}{2}\times0.824 = t_{-}\times0.049$$

$$t_{-} = \frac{0.030}{0.049} = 0.612 \text{ or } t_{Cu^{2+}} = 0.388$$

If there is no transport of ions across the boundary

$$E_{cell} = \frac{0.0591}{2}\log\left(\frac{0.05\times0.52}{0.005\times0.78}\right)$$

$$= \frac{0.0591}{2}\times0.824 = 0.024\text{ V}$$

$$E_{junction} = 0.030 - 0.024 = 0.006\text{ V} = 6\text{ mV}$$

Problem 3.27

Consider the cell

$Cd(Hg, 5\%\text{ by wt})|CdSO_4(0.05\text{ m})\,||\,CdSO_4(0.02\text{ m})|Cd$ $(Hg, 10\%\text{ by wt})$

(i) Write the half cell reactions and the overall cell reaction.

(ii) Calculate the EMF of the cell given the mean activity coefficients of $CdSO_4$ in 0.05 m and 0.02 m solutions as 0.20 and 0.30 respectively.

Solution

$$\text{Molality of Cd(Hg)5 wt\%} = \frac{5}{112.4}\times\frac{1000}{95} = 0.468$$

$$\text{Molality of Cd(Hg)10 wt\%} = \frac{10}{112.4}\times\frac{1000}{90} = 0.988$$

Reactions in the cell

$$Cd^{2+}(0.02\text{ m}) + 2e^{-} \rightarrow Cd(Hg, 0.988\text{ m})$$

$$Cd^{2+}(0.05\text{ m}) + 2e^{-} \rightarrow Cd(Hg, 0.468\text{ m})$$

Overall reaction:

$$Cd^{2+}(0.02\text{ m}) + Cd(Hg, 0.468\text{ m}) \rightarrow Cd^{2+}(0.05\text{ m}) + Cd(Hg, 0.988\text{ m})$$

$$E = E^0 - \frac{0.0591}{2}\log\frac{a_{Cd^{2+}(0.05\text{ m})}\times a_{Cd(0.988\text{ m})}}{a_{Cd^{2+}(0.02\text{ m})}\times a_{Cd(0.468\text{ m})}}$$

$$a_{Cd^{2+}(0.05\text{ m})} = m_{Cd^{2+}}\times\gamma_{\pm} = 0.05\times0.2 = 0.01\,(\text{assume } \gamma_{\pm} = \gamma_i)$$

$$a_{Cd^{2+}(0.02\text{ m})} = m_{Cd^{2+}}\times\gamma_{\pm} = 0.02\times0.3 = 0.006\,(\text{assume } \gamma_{\pm} = \gamma_i)$$

$$\text{or, } E = 0 - \frac{0.0591}{2}\log\frac{0.01\times0.988}{0.006\times0.468}$$

$$= -\frac{0.0591}{2}\log 3.516$$

$$= -\frac{0.0591}{2}\times0.546 = -0.016\text{ V}$$

Problem 3.28

(a) Calculate the EMF of the cell
$Ag, AgCl(s)|HCl(m = 0.1)|H_2(g, 1\text{ atm})|Pt|H_2(g, 1\text{ atm})$ $|HCl(m = 0.05)|AgCl(s)|Ag$
given the mean activity coefficients of HCl as 0.796 and 0.83 in 0.1 m and 0.05 m HCl respectively.

(b) If the hydrogen electrode in the middle is removed, the cell
$Ag, AgCl(s)|HCl(m = 0.1)\vdots HCl(m = 0.05)|AgCl(s)|Ag$
is formed and its EMF is 0.0278 V, calculate the transport number of the hydrogen ion.

(c) What is the relationship of EMF of cell (b) to
$Ag, AgCl(s)|HCl(m = 0.1)\,||\,HCl(m = 0.05)|AgCl(s)|Ag$
Using the data in (b) and (c), what additional information can be obtained? Calculate the value of the same.

Solution

(a) The overall reaction in cell (a) is

$$H^{+}(m = 0.1) + Cl^{-}(m = 0.1) \rightarrow H^{+}(m = 0.05) + Cl^{-}(m = 0.05)$$

$$E = E^0 - 0.0591\log\frac{m_{H^+}\times m_{Cl^-}(HCl = 0.05)\times\gamma'_{+}\times\gamma'_{-}}{m_{H^+}\times m_{Cl^-}(HCl = 0.1)\times\gamma''_{+}\times\gamma''_{-}}$$

$$= 0 - 0.0591\log\frac{(m_{\pm}^2\times\gamma_{\pm}^2)(HCl = 0.05)}{(m_{\pm}^2\times\gamma_{\pm}^2)(HCl = 0.1)}$$

$$= -0.0591\times2\log\frac{0.05\times0.83}{0.1\times0.796}$$

$$= -2\times0.0591\times(-0.283) = 0.0334\text{ V}$$

(b) The overall reaction is

$$t_{H^+}H^{+}(m = 0.1) + t_{H^+}Cl^{-}(m = 0.1) \rightarrow t_{H^+}H^{+}(m = 0.05) + t_{H^+}Cl^{-}(m = 0.05)$$

$$E = E^0 - 2t_{H^+}0.0591\log\frac{(m_{\pm}\times\gamma_{\pm})(HCl = 0.05)}{(m_{\pm}\times\gamma_{\pm})(HCl = 0.1)}$$

$$= -0.0591\times2t_{H^+}\log\frac{0.05\times0.83}{0.1\times0.796}$$

$$= -2\times t_{H^+}\times0.0591\times(-0.283) = 0.0334\,t_{H^+}\text{ V}$$

$$\therefore 0.0278 = 0.0334\,t_{H^+} \text{ or } t_{H^+} = \frac{0.0278}{0.0334} = 0.83$$

(c) The overall reaction is

$$Cl^-(m=0.1) \rightarrow Cl^-(m=0.05)$$

$$E = E^0 - 0.0591\log\frac{a_{Cl^-}(m_{HCl}=0.05)}{a_{Cl^-}(m_{HCl}=0.1)}$$

$$= +0.0591\log\frac{a_{\pm}(m_{HCl}=0.1)}{a_{\pm}(m_{HCl}=0.05)}$$

$$= 0.0591\log\frac{0.1\times 0.796}{0.05\times 0.83}$$

$$= 0.0591\times 0.283 = 0.0167\text{ V}$$

$$\frac{E_{cell}\text{ in(b)}}{E_{cell}\text{ in(c)}} = \frac{0.0278}{0.0167} = 2t_{H^+}$$

$$= 1.667 \text{ or } t_{H^+} = 0.83\text{ V}$$

$$E_{cell}\text{ in(b)} - E_{cell}\text{ in(c)} = \text{liquid junction potential}$$

$$= 0.0278 - 0.0167 = 0.0111\text{ V}$$

Cell Reactions in Nickel-Cadmium Cell

Problem 3.29

(a) The Ni–Cd battery may be represented as

$$\text{Ni, NiO}_2(s)|\text{aqueous KOH}|\text{Cd(OH)}_2|\text{Cd}$$

The EMF of the cell is –1.30 V. Write the half cell reactions. Given that the Ni, NiO_2 half cell potential is 0.49 V, what is the half cell potential of the Cd electrode? Does the cell potential depend upon the activity of KOH? Given $E^0_{Cd^{2+},Cd} = -0.401$ V, what is the solubility product of $Cd(OH)_2$?

Solution

$$Cd(OH)_2 + 2e^- \rightarrow Cd + 2OH^- \quad E^0 = ?$$

$$NiO_2(s) + 2H_2O + 2e^- \rightarrow Ni(OH)_2 + 2OH^- \quad E^0 = 0.49\text{ V}$$

$$E_{cell} = E_R - E_L = E_{Cd,OH^-} - E_{Ni,OH^-}$$

$$= E_{Cd,OH^-} - (0.49)$$

$$E_{Cd,OH^-} = -1.30 + 0.49$$

$$= -0.81\text{ V}$$

Since the cell reaction is

$$NiO_2(s) + 2H_2O + Cd \rightarrow Ni(OH)_2 + Cd(OH)_2$$

The cell potential does not depend upon the activity of KOH.

Solubility product of $Cd(OH)_2$

$$Cd(OH)_2 + 2e^- \rightarrow Cd + 2OH^- \qquad E^0 = -0.81\text{ V}$$

$$Cd^{2+} + 2e^- \rightarrow Cd \qquad E^0 = -0.401\text{ V}$$

Overall reaction: $Cd(OH)_2(s) \rightarrow Cd^{2+} + 2OH^-$

$$E^0 \text{ for the reaction} = -0.81 - (-0.401)$$

$$= -0.409\text{ V}$$

$$\log K_{SP} = \frac{-2\times 0.409}{0.0591}$$

$$= -13.84 = \bar{14}.16$$

$$K_{SP} \text{ of } Cd(OH)_2 = 1.44\times 10^{-14}$$

Fuel Cells

Problem 3.30

In a certain fuel cell involving ethanol as the fuel and air as the oxidising agent, (i) write the cell reaction and the half cell reactions. How many electrons are transferred? Given the ΔG_f^0 values of ethanol, water and carbon dioxide as –175.0, –237.0, –394.0 (all in kJ mol^{-1}), what is the EMF of the cell? If ΔH^0 for the combustion of ethanol is –1367 kJ mol^{-1}, calculate the thermodynamic efficiency of the fuel cell.

Solution

Reaction

$$C_2H_5OH(l) + 3O_2(g) \rightarrow 2CO_2(g) + 3H_2O(l)$$

The half reactions are:

$$C_2H_5OH + +3H_2O \rightarrow 2CO_2(g) + 12H^+ + 12e^-$$

$$3O_2(g) + 12H^+ + 12e^- \rightarrow 6H_2O$$

$$\Delta G^0 \text{ reaction} = 2\times(-394) + 3\times(-237) - (-175)$$

$$= -1324\text{ kJ}$$

$$E^0 = \frac{-\Delta G^0}{nF} = \frac{1324}{12\times 96500} = 1.14\text{ V}$$

$$\eta = \frac{|\Delta G|}{\Delta H} = \frac{1324}{1367} = 0.97$$

Exchange Current Densities for Electrode Reactions

Problem 3.31

The exchange current density, i_0, for hydrogen evolution reaction from an acid solution on a nickel electrode is 10^{-7} A cm^{-2}. The value of y in the Tafel relation is 0.13. What is the overpotential required to maintain a current density of 0.02 A cm^{-2} from an acid solution of 1 N strength?

Solution

$$\eta = y(2.303\log i - 2.303\log i_0)$$

$$= 0.13(2.303\log 10^{-2} - 2.303\log 10^{-7})$$

$$= 0.13(-4.606 - (-16.121))$$

$$= 0.13\times 11.515 = 1.497\text{ V}$$

Problem 3.32

(a) The exchange current density for hydrogen evolution reaction on a lead electrode is 10^{-13} A cm^{-2}. How much hydrogen gas is evolved on a 1 cm^2 lead electrode per second?

Solution

96500 C is required to liberate 11.2 dm^3 of H_2 at STP.

For a current density of 10^{-13} Acm^{-2}, the H_2 evolved will be

$$\frac{10^{-13}\times 11.2\ \text{A cm}^{-2}\ \text{dm}^3\ \text{cm}^2}{96500\ \text{As}} = \frac{10^{-15}\times 11.2}{965}$$
$$= 1.16\times 10^{-17}\ \text{dm}^3\ \text{s}^{-1}$$
$$= 1.16\times 10^{-14}\ \text{cm}^3\ \text{s}^{-1}$$

Problem 3.33

How much hydrogen gas is evolved on 1 cm^2 Pt electrode, if $i_0 = 10^{-2}$ As?

In this case we have

$$\frac{11.2\times 10^{-2}}{96500} = 1.16\times 10^{-4}\ \text{dm}^3$$
$$= 1.16\times 10^{-1}$$
$$= 0.116\ \text{cm}^3\ \text{s}^{-1}$$

Polarography and Analytical Applications

Problem 3.34

Consider the reaction

$$\text{Ox} + ne^- \rightleftharpoons \text{red}$$

occuring at a DME. Derive an equation relating the potential of the DME with the current flowing at the electrode. From this equation, arrive at an expression for the half wave potential and explain its significance.

Solution

We consider the case of reduction of a metal ion to the metal amalgam. The potential of cathode, E_R is given by

$$E_R = E_R^0 - \frac{RT}{nF}\ln\frac{C_R^0}{C_O^0} \quad (1)$$

where C_R^0 = concentration of the reduced species and C_O^0 = concentration of the oxidised species, both at the DME. The current, i, is limited by diffusion.

$$\therefore \quad i = k_O\left(C_O - C_O^0\right) \quad (2)$$

where C_O = concentration of the oxidised species in bulk, C_O^0 = concentration of the oxidised species at the DME and k_0 is the proportionality constant. For the limiting current, i_d, C_O^0 is very small and hence

$$i_d = k_O C_O \quad (3)$$

From Eq. (2),

$$i = i_d - k_O C_O^0 \quad (4)$$

Or,
$$C_O^0 = \frac{i_d - i}{k_O} = \frac{(i_d - i)C_O}{i_d} \quad (5)$$

If the experimental solution originally does not contain any of the reduced form, the current i is proportional to the concentration of the metal in the metal in the amalgam or

$$i = k_R C_R^0 \approx k_R C_R \quad (6)$$

Substituting Eqs (5) and (6) in Eq. (1),

$$E_R = E_R^0 - \frac{RT}{nF}\ln\frac{\left(i/k_R\right)}{\left(i_d - i/k_0\right)} \quad (7)$$

When the current is equal to one half of the diffusion current, the potential corresponding to this point is known as half wave potential $E_{1/2}$.

When $\quad i = \frac{i_d}{2},\ E = E_{1/2}$

$$E_{1/2} = E_R^0 - \frac{RT}{nF}\ln\frac{k_0}{k_R} \quad (8)$$

Substituting Eq. (8) in Eq. (7), we get

$$E_R = E_{1/2} - \frac{RT}{nF}\ln\frac{i}{i_d - i} \quad (9)$$

A plot of E_R vs $\ln\frac{i}{i_d - i}$ gives a straight line with a slope equal to $-\frac{\text{RT}}{\text{nF}}$ for the reversible reduction of ions.

Significance of $E_{1/2}$

- $E_{1/2}$ is independent of the concentration of the electroactive species.
- It is related to the standard potential E_R^0 through Eq. (8).

Since $\frac{k_0}{k_R} = \frac{D_0^{1/2}}{D_R^{1/2}}$

$$E_{1/2} \approx E^0\ (\text{if } D_0 = D_R)$$

- $E_{1/2}$ is characterstic of the substance.

Problem 3.35

Calculate the average diffusion current in a polarographic experiment with the DME using the following data. The ion getting reduced is $M^{2+} + 2e^- \rightarrow M(Hg)$. The mean rate of outflow of mercury from DME = 0.064×10^{-3} g s^{-1}, $t = 3.33$ s, diffusion coefficient, $D = 7\times 10^{-6}$ cm^2 s^{-1}; the concentration of electroactive ion $C = 5\times 10^{-3}$ mol L^{-1}.

Solution

$$i_d = 607 n D^{\frac{1}{2}} m^{\frac{2}{3}} t^{\frac{1}{6}} C$$

$$= 607\times 2\times (3.33)^{\frac{1}{6}}\times (0.064)^{\frac{2}{3}}\times \left(7\times 10^{-6}\right)^{\frac{1}{2}}\times 5 \quad \left(\begin{array}{c}\text{using}\\ \text{proper}\\ \text{units}\end{array}\right)$$

$$6070\times 0.16\times 2.642\times 10^{-3}\times 1.222 = 3.14\ \mu\text{A}$$

Problem 3.36

The following data were obtained in the reduction of a bivalent metal ion at 298 K at the DME.

Applied voltage (V)	–0.600	–0.620	–0.640	–0.660	–0.680	–0.720	–0.760	–0.800
Current (μA)	1.8	4.0	6.0	8.5	12.8	13.2	13.4	13.5

Determine the halfwave potential by a graphical plot and calculate the Gibbs free energy for the process. The potential of the reference electrode (SCE) is +0.244 V. Assume that the voltages are corrected for IR drop and the currents are corrected for residual current.

Solution

A plot of i vs E gives an S-shaped curve from which the half wave potential is determined as follows. i_d is found to be 13.8 μA. When

$$i = \frac{i_d}{2},\ E = E_{1/2}$$

or, $E = E_{1/2}$ at 6.9 μA, that is, $E_{1/2} = -0.64$ V vs SCE

In terms of the applied voltage $E_{1/2} = 0.64$ (without sign)

$$E \text{ of the cell vs } SHE\ E_{\text{appl}} = 0.244 - 0.640$$

$$= -0.396\ \text{V} \left(\text{since } E_{\text{appl}} = E_{an} - E_C\right)$$

$$\Delta G^0 = -nFE^0 = -nFE_{1/2}$$

$$= -2 \times 96500 \times (-0.396)\ \text{J}$$

$$= 76428\ \text{J} = 76.43\ \text{kJ}$$

Estimation of Overpotentials

Problem 3.37

A solution consists of a mixture of 5×10^{-2} M Ni^{2+} ions and 5×10^{-2} M in H^+ ions. This solution is electrolysed using (i) bright platinum electrode (ii) mercury electrode as cathodes in two separate experiments. Will hydrogen gas or nickel form at the cathode in the two cases? The cathodic overpotential of hydrogen (for H^+ ion reduction) at a pressure of 1.0 atm is –0.10 V and –1.05 V on platinum and mercury respectively. Assume that the temperature is 298 K. The $E^0_{Ni^{2+},Ni} = -0.236$ V.

Solution

(i) Bright Pt electrode, for hydrogen

$$E_{d.v}(\text{cathode}) = E_c + \eta_c = 0.00 + \frac{0.059}{2}\log\left(5\times10^{-2}\right)^2 - 0.10$$

$$= 0.00 + 0.059 \times (-1.301) - 0.10$$

$$= -0.077 - 0.10 = -0.177\ \text{V}$$

For Ni^{2+}

$$E_{d.v}(\text{cathode}) = -0.236 + \frac{0.059}{2}\log\left(5\times10^{-2}\right)^2$$

$$= -0.236 + 0.059 \times (-1.301)$$

$$= -0.236 - 0.077 = -0.313\ \text{V}$$

(It is assumed that $\eta = 0$ for the deposition of Ni^{2+} on Pt). Since hydrogen formation occurs at the less negative (or more positive) decomposition potential, only hydrogen gas forms.

(ii) At the Hg cathode, for hydrogen

$$E_{d.v}(\text{cathode}) = E_c + \eta_c$$

$$= 0.00 + \frac{0.059}{2}\log\left(5\times10^{-2}\right)^2 - 1.05$$

$$= 0.00 - 0.077 - 1.05 = -1.13\ \text{V}$$

For Ni^{2+}

$$E_{d.v}(\text{cathode}) = -0.236 + \frac{0.059}{2}\log\left(5\times10^{-2}\right)^2$$

$$= -0.236 - 0.077 = -0.313\ \text{V}$$

In this case also, it is assumed that $\eta = 0$ for deposition of Ni^{2+} deposition on Hg. On this basis, Ni^{2+} ions get deposited as Ni in preference to H^+ as H_2 gas because $E_{d.v}$ for Ni^{2+} is less negative (or more positive) than $E_{d.v}$ for H^+ deposition.

Problem 3.38

A 0.01 M solution of sulphuric acid is electrolysed between bright platinum electrodes at a current strength of 1.5 A at 298 K. Given that the cell resistance is 2.0 Ω and the electrolysis solution is well stirred, calculate the applied voltage necessary for continuous electrolysis. The cathodic overpotential for the reduction of H^+ ion is –0.45 V and anodic over potential for O_2 reduction is +0.42 V. The gas pressures of H_2 and O_2 are 1 atm and $E^0_{H_2,H^+} = 0.0$ V and $E^0_{O_2,H_2O}$ $\left(\text{for the reduction } \frac{1}{2}O_2 + 2H^+ + 2e^- \rightarrow H_2O\right)$ is 1.23 V.

Solution

$$E_{\text{appl}} = E_a - E_C + \eta_a - \eta_c + \text{IR}$$

For E_a, the reaction is,

$$\frac{1}{2}O_2 + 2H^+ + 2e^- \rightarrow H_2O$$

$$E_{rev(a)} = 1.23 + \frac{0.059}{2}\log(0.01\times2)^2 + \frac{0.059}{2}\log 1^2$$

$$= 1.23 + 0.059 \times \log 0.0004$$

$$= 1.23 + 0.059 \times (-3.398)$$

$$= 1.23 - 0.200 = 1.03\ \text{V}$$

$E_{rev(c)}$: For the reaction $2H^+ + 2e^- \rightarrow H_2$

$$E_{rev(c)} = 0.00 + \frac{0.059}{2}\log(0.02)^2$$

$$= 0.00 + 0.059 \times \log 0.0004 = -0.200\ \text{V}$$

$$E_{apll} = 1.03 - (-0.200) + 0.420 - (-0.450) + 2.0 \times 1.5$$
$$= 1.030 + 0.200 + 0.420 + 0.450 + 3.00 = 5.10 \text{ V}$$

Miscellaneous Problem

Problem 3.39

Assuming complete utilisation of the fuel, estimate the amount of current obtainable if hydrogen is supplied at a rate of 10 cm^3 per minute at 298 K and 1 atm. pressure using the ideal gas equation of state if the oxidation reaction is represented as $H_2 = 2\,H^+ + 2\,e^-$.

Solution

From the ideal gas equation, the number of moles at 1 atm pressure and 298 K is given by

$$n = PV/RT$$

Hence,

$$dn/dt = (P/RT) \times (dV/dT)$$
$$= (1 \text{ atm})(0.010 \text{ L min}^{-1})/[0.082 \text{ L atm mol}^{-1} \text{ K}^{-1} \times 298 \text{ K}]$$
$$= 4.1 \times 10^{-4} \text{ mol H}_2 \text{ min}^{-1}$$
$$\text{Current} = \text{number of electrons} \times F\,(dn/dt) = 1.314 \text{ A}$$

EXERCISES

Practice Problems

1. On electrolysis of an aqueous alkaline solution using a tungsten anode, the reaction
$W + 8OH^-(aq) \rightarrow WO_4^{2-}(aq) + 4H_2O + 6e^-$
occurs at the anode. How many coulombs of electricity is required if a gram of tungsten dissolves at the anode? How many moles of electrons and how many electrons are required for the process? (atomic mass W = 183.8)

2. It is desired to deposit 20 cm^2 of copper cathode with a layer of 0.02 cm thickness from 0.1 M solution of copper sulphate employing a current of 2 A. If the current efficiency is 70%, how much time is required for the above deposition to occur? If copper anode is used, write the reaction at this electrode. (atomic mass of Cu = 63.5, density of Cu = 8.96 kg dm^{-3})

3. A conductivity cell with a cell constant of 1.0 m^{-1} was used to measure the solubility of a complex cobalt salt, [tetraammineoxalato Co(III)]Cl. The ionic conductances of cobalt cation and chloride anion are 20×10^{-4} S m^2 mol^{-1} and 76.0×10^{-4} S m^2 mol^{-1} respectively at 298 K. Calculate the solubility of the salt given that the measured resistance of the solution is 26.0 ohms. Suppose the saturation of the salt was carried by dissolving in 5×10^{-4} M HCl and not pure water, what will be the solubility? $\lambda^0_{H^+}$ may be taken as 350×10^{-4} S m^2 mol^{-1}.

4. 200 cm^3 of 5×10^{-3} M H_2SO_4 was titrated with 1 M NaOH using a conductivity cell with a cell constant of 20 m^{-1}. Calculate: (i) The specific conductivity of the solution before titration. (ii) The specific conductivity of the solution at the equivalence point. (iii) What will the specific conductivity of the solution when one mL of NaOH is added after the equivalence point? The ionic conductances of H^+, OH^-, Na^+ and $SO_4{}^{2-}$, are 350×10^{-4}, 198×10^{-4}, 50×10^{-4}, 160×10^{-4} (all in S m^2 mol^{-1}) respectively.

5. The specific conductivity of a saturated solution of a soluble uni–univalent salt MX is 1.576×10^{-3} S m^{-1} at 298 K. Given the molar ionic conductances of the ions, M^+ and X^- as 64.0×10^{-4} and 74.0×10^{-4} S m^2 mol^{-1} respectively, calculate the solubility of the salt. Use the D–H–O equation where $A = 60.2$ and $B = 0.229$, to refine the solubility by two iterative steps.

6. The specific conductivity of a 5×10^{-2} N solution of propanoic acid is 3.2×10^{-4} S cm^{-1} at 298 K. The mobilities of H^+ and propionate ions are 3.63×10^{-7} and 3.73×10^{-8} m^2 V^{-1} s^{-1} respectively. Calculate the degree of dissociation of the acid taking the interionic attractive forces into account using the D–H–O equation. Calculate the dissociation constant. The constants A and B of Onsager equation are, $A = 60.2$ and $B = 0.229$.

7. An aqueous solution of HCl was electrolysed between platinum electrodes in a Hittorf cell. A silver coulometer connected in series had a deposit of silver on the cathode of 0.251 g. On analysis, the cathode compartment contained 4.986×10^{-3} eq of Cl^- ions before electrolysis and it had 4.591×10^{-3} eq of Cl^- ions after electrolysis. Write the cathode and anode reactions and calculate the transport numbers of H^+ and Cl^- ions.

8. A solution of potassium chloride was electrolysed in a Hittorf cell using silver, silver chloride electrodes. Before electrolysis, the cathode solution contained 4.476 g of KCl and after electrolysis it contained 5.147 g of KCl. In a copper coulometer connected in series, the copper deposited on cathode weighed 0.582 g. Write the cathode reaction and calculate t_{K^+} and t_{Cl^-} values.

9. The ΔG^0(solution) of a saturated solution of calcium fluoride is 59.9 kJ mol^{-1} at 298 K. The specific conductance of the saturated solution of the salt is 3.71×10^{-3} S m^{-1}. Calculate the molar conductance of CaF_2 at infinite dilution. What is the $\lambda^0_{\frac{1}{2}Ca^{2+}}$ in S m^2 eq^{-1} given that of fluoride ion is 47.0×10^{-4} S m^2 mol^{-1}.

10. The specific conductance of 0.01 M solution of H_2SO_4 is 0.86 S m^{-1} at 298 K. When it is saturated with strontium sulphate, its value rose to 0.8742 S m^{-1}. Given

$\lambda^0_{\frac{1}{2}SO_4^{2-}} = 80\times10^{-4}$ and $\lambda^0_{\frac{1}{2}Sr^{2+}} = 62\times10^{-4}$ S m^2 eq^{-1}, calculate the solubility and solubility product of $SrSO_4$. The $\lambda^0_{H^+}$ may be taken as 350×10^{-4} S m^2 eq^{-1}.

11. The cell Pt, $H_2(g, 1\text{ atm})|$solution of unknown pH || NCE (1 N KCl) was set up for determining the pH of a solution. The EMF of the cell was 0.440 V at 298 K. Given E_{NCE} = 0.280 V, calculate pH of the solution.

12. The following cell with two redox pairs Ti^{3+}, Ti^{4+} and Ce^{3+}, Ce^{4+} was set up.

$$Pt|Ti^{3+}(0.005\text{ m}), Ti^{4+}(0.02\text{ m})||Ce^{3+}(0.2\text{ m}), Ce^{4+}(0.01\text{m})|Pt$$

Given that $E^0_{Ti^{3+},Ti^{4+}}$ and $E^0_{Ce^{3+},Ce^{4+}}$ are 0.04 V and 1.61 V respectively, at 298 K, calculate the EMF of the cell. What is the equilibrium constant of the spontaneous reaction in the cell. Which electrode in the above cell is negative?

13. Calculate the EMF of the cell
$Pt, H_2(g, 1\text{ atm})|KOH(0.01\text{ M}), KCl(0.05\text{ M})|AgCl(s)$, Ag given $E^0 = 0.2240$ V at 298 K and $K_w = 10^{-14}$. Assume activity coefficients of the ions to be unity.

If the KOH and KCl solutions are replaced by a solution of HCl of unknown concentration and the measured EMF is 0.300 V, what is the concentration of HCl in the cell? What is the mean activity of the acid, if its mean activity coefficient is 0.80?

14. For the lead storage cell represented by
$Pb, PbSO_4(s)|H_2SO_4(m = 2.0)|PbSO_4(s), PbO_2(s)$, Pb

The E^0 values of Pb, $PbSO_4^-$ and Pb, PbO_2, $PbSO_4$, half cells are –0.356 V and 1.685 V at 298 K. Calculate the EMF of the cell given the mean activity coefficient of H_2SO_4 at 2.0 m is 0.100. Will the EMF be dependent on the activity of water in the solution? If the activity of water is 0.8, how will the EMF change?

15. The temperature coefficient of the lead storage cell is 2.64×10^{-5} V K^{-1}. Calculate the ΔG, ΔS and ΔH, for the cell reaction written in Problem 14.

16. Calculate the EMF of the concentration cell
$Pt, H_2(g, 5\text{ atm})|HCl(m = 0.1)||HCl(m = 0.05)|H_2(g, 2\text{ atm})$, Pt

given that the mean activity coefficients of HCl in 0.1 and 0.05 molal solutions are 0.80 and 0.83 respectively. If, in the above cell, the double lines between the acid solutions are removed, and ⋮ lines are introduced, how does the EMF change? What is its value given that $t_{H^+} = 0.85$?

17. The oxidation of hydrazine, used as a rocket fuel, occurs according to

$$NH_2\text{–}NH_2(l) + 3O_2(g) \rightarrow 2NO_2(g) + 2H_2O(l)$$

The Gibbs energy change of the reaction is –464.2 kJ. If the half cell reaction at the cathode is

$$O_2(g) + 4H^+ + 4e^- \rightarrow 2H_2O(l)$$

and its E^0 is 1.23 V, what is the other half cell reaction and what is its electrode potential?

18. The exchange current density, i_0, for the hydrogen evolution reaction from an acid solution is 10^{-2} A cm^{-2} on platinum and 10^{-14} A cm^{-2} on mercury electrode. Given the value of y (see text) as 0.03 and 0.12 on Pt and Hg, respectively, what are the values of the over potential in the two cases to obtain a current density of 0.05 A cm^{-2} from a 1 N acid solution?

19. Calculate the specific conductivity of a solution containing 1×10^{-3} M Na_2SO_4 and 2×10^{-2} M solution of K_2SO_4. The velocities of Na^+, K^+, and SO_4^- are 52.0×10^{-8}, 76×10^{-8}, 83×10^{-8} m^2 V^{-1} s^{-1} under a potential gradient of 10 V cm^{-1}. Assume that the salts are completely dissociated.

20. The following standard potential data are given. Use these data to answer the following questions:
(i) Dehydroascorbic acid + $2H^+ + 2e^- \rightarrow$ Ascorbic acid
$E^0(\text{at pH} = 0) = 0.474$ V
(ii) $Fe^{3+} + e^- \rightarrow Fe^{2+}$ $E^0(\text{at pH} = 0) = 0.771$ V
(iii) $CH_3COOH + 2H^+ + 2e^- \rightarrow CH_3CHO + H_2O$
$E^0(\text{at pH} = 0) = -0.056$ V
(iv) $CH_3CHO + 2H^+ + 2e^- \rightarrow CH_3CH_2OH$
$E^0(\text{at pH} = 0) = 0.254$ V

Questions:

(a) Can Fe^{2+} ions reduce ascorbic acid in the acid medium at 298 K? Show by formulation of the cell and calculation of E^0(cell).
(b) What is the Gibbs energy change for the reaction
$$CH_3CH_2OH + 2Fe^{3+} \rightarrow CH_3CHO + 2Fe^{2+} + 2H^+$$
(c) What is the equilibrium constant for the reaction
$$CH_3CHO + \text{dehydroascorbic acid} \rightleftharpoons CH_3COOH + \text{ascorbic acid}$$

21. (a) Explain in a few sentences, the essential basis of potentiometric titrations in determining the equivalent point in a titration.
(b) Considering the potentiometric titration of a strong acid by a strong base, show that the pH before equivalent point is given by

$$pH = -\log C^0_{acid} + \log(1-f) - \log\left(\frac{V_0}{V_0 + V_b}\right)$$

where C^0_{acid} = initial conc. of acid (eq/L), f = fraction of equivalent volume of base acid, V_0 = initial volume of acid in cm^3 and V_b = volume of base added. Show that

$$\frac{d(pH)}{df} = 0.434\left(\frac{1}{1-f}\right)$$ before the equivalence point.

(c) Show that

$$\left[\mathrm{OH^-}\right] = C_b\left(\frac{x}{V_{e.p}+x}\right) + \mathrm{H^+}$$
$$= C_b\left(\frac{x}{V_{e.p}+x}\right) + \frac{K_W}{\left[\mathrm{OH^-}\right]}$$

after the equivalent point, where C_b = conc. of base, x = excess volume of base added and $V_{e.p}$ = volume at the equivalence point.

22. Calculate the diffusion coefficient of a bivalent metal cation from the polarographic data given: drop time of capillary = 3.5 s, mass rate of flow of mercury = 2.5×10^{-3} g s^{-1}, concentration of the deploriser = 2 mM and average diffusion current = 5.0 μA.

23. A conductance study on a sample of brackish water containing equimolar concentrations of $MgCl_2$, NaCl and KCl, showed a resistance of 1600 Ω using a cell with cell constant 500 m^{-1}. Given that the limiting molar ionic conductances of Mg^{2+}, Na^+, K^+ and Cl^- ions are 106.2×10^{-4}, 50.1×10^{-4}, 73.4×10^{-4}, 76.5×10^{-4} (all in S m^2 mol^{-1}), calculate the concentration of Cl^- ions in ppm.

24. It is desired to electrolyse a 0.2 M H_2SO_4 using a platinum anode and copper cathode. The electrolysis cell has a resistance of 2.0 ohms and a current of 0.2 A is passed through cell. Calculate the voltage necessary to electrolyse the solution continuously. Use the following data in the calculation: the anodic overpotential of O_2 on Pt is + 0.42 V and the cathodic overpotential of H_2 on copper is –0.800 V. $E^0_{O_2,H_2O}$ in acid solution = 1.23 V. The hydrogen and oxygen gas pressures are maintained at 1 atm.

25. In an electrogravimetric experiment involving a mixture of 0.05 M nickel sulphate and 0.05 M copper(II) sulphate in aqueous H_2SO_4, copper starts depositing on the cathode before the deposition of nickel begins. Calculate the concentration of copper(II) ions remaining in solution when nickel starts precipitating out. Assume that there are no overpotential effects ($E^0_{Cu^{2+},Cu}$ = 0.340 V; $E^0_{Ni^{2+},\,Ni}$ = –0.236 V).

26. The ionic conductances of Na^+ and Cl^- ions are 96.4 and 50.8 (both in S cm^2 eq^{-1}) respectively. Given that the viscosity of the solution containing these ions is 1×10^{-2} poise, calculate the radii of the ions. (Hint: Use Walden's rule)

27. Consider the reduction of Pb^{2+} on a Pt electrode using a current of 2 mA. Assuming a current efficiency of 100%, estimate the number of Pb^{2+} ions involved in the reduction process per second.

28. In the potentiometric sensing of Cd^{2+} using an ion-selective electrode, the measured cell potential shifted from –471 mV to –450 mV upon adding 0.001 mol of Cd^{2+}. Estimate the approximate concentration of Cd^{2+} in (a) mol L^{-1} and (b) ppm.

29. When a buffer solution of pH = 4.0 was employed, the glass electrode developed a potential of 210 mV. The potential changed to –290 mV when the buffer was replaced with an alkaline solution. Estimate the approximate pH of the solution.

30. The molar conductance of KI in methanol is 114 S cm^2 mol^{-1} at infinite dilution. The viscosity of methanol is 0.00545 poise. From these values, estimate the molar conductance of NaCl at infinite dilution if the solvent is pyridine. (Viscosity = 0.00713 poise)

Multiple Choice Questions

1. Which of the following statements about the laws of electrolysis is correct?

(a) They provide information about the nature of ions in solution.
(b) They can be used to explain the nature of an electrochemical reaction quantitatively.
(c) It is not possible to evaluate the charge on the electron or the ions from these laws.
(d) They can be used to predict the products of electrolysis in an aqueous solution under all experimental conditions.

2. In a study on the manufacture of $KClO_3$ by the oxidation of KCl at an anode in KOH, a current of 1 A was passed for 20 hours. If the current efficiency is 80%, the amount of potassium chlorate formed is (in g) (molar mass of $KClO_3$ = 123.0)

(a) 24.50 **(b)** 6.10
(c) 18.3 **(d)** 12.20

3. Indicate the correct statement among the following dealing with specific and molar conductances of an electrolyte in an aqueous solution.

(a) Increasing the concentration of the electrolyte results in an increase of both specific and molar conductances.
(b) The specific conductance increases but the molar conductance decreases with increasing concentration.
(c) Both specific and molar conductances decrease with increasing concentration.
(d) Specific conductance decreases but molar conductance increases with increasing concentration.

4. The specific conductance of a saturated solution of barium chromate ($BaCrO_4$) at 298 K is 8.1×10^{-7} ohm^{-1} cm^{-1} after correction for water conductance. The $\lambda^0_{Ba^{2+}}$ and $\lambda^0_{CrO_4^{2-}}$ at this temperature are 64.0 and 83.0 (both in S cm^2 eq^{-1}) respectively. The Debye–Hückel constants A and B of

D–H–O equation are 60.2 and 0.229. The solubility of the salt corrected for the interionic effect (in eq L^{-1}) is

(a) 5.52×10^{-6} **(b)** 4.84×10^{-5}
(c) 2.76×10^{-5} **(d)** 3.69×10^{-4}

5. The very large value of $\lambda^0_{H^+}$ in aqueous solution in due to

(a) The movement of H^+ ions in solution towards the oppositely charged electrode.
(b) The movement of H_3O^+ ions towards the oppositely charged electrode due to the potential gradient.
(c) The transfer of the charge from one molecule of water to the next without the actual movement of the H_3O^+ species.
(d) The fast movement of the H^+ ion due to its small size across the solution to the opposite electrode.

6. The $\lambda^0_{La^{3+}}$ and $\lambda^0_{C_2O_4^{2-}}$ ions are 70.0 and 74.0 (both in ohm^{-1} cm^2 eq^{-1}) at 298 K. The molar conductance of $La_2(C_2O_4)_3$ is

(a) 432 **(b)** 864
(c) 288 **(d)** 648

7. To which ion (from those listed) is Walden's rule accurately applicable?

(a) NH_4^+ **(b)** OH^-
(c) CoO^- **(d)** $N(CH_3)_4^+$

8. In the determination of the dissociation constant of a weak acid, a graph of $1/\Lambda_c$ against $\Lambda\times C$ (where C = concentration of acid) yielded a straight line with an intercept equal to 2.59×10^{-3} and a slope equal to 0.479. The degree of dissociation of the acid at $C = 0.1$ eq L^{-1} is (C is in eq dm^{-3} and Λ is in S cm^2 eq^{-1}).

(a) 0.012 **(b)** 0.021
(c) 0.035 **(d)** 0.050

9. Which of the following statements is true in the measurement of the EMF of a Daniel cell by means of a potentiometer?

(a) The measured EMF has no special significance.
(b) The Gibbs energy change associated with the cell reaction is the maximum obtainable (including sign of ΔG) under these conditions.
(c) The Gibbs energy change associated with the cell reaction is the minimum obtainable from the cell.
(d) The electrical work done by the cell also includes P–V work.

10. During the charging of a lead storage cell,

(a) lead dioxide is removed from the system.
(b) the acid concentration in the cell decreases.
(c) water is formed.
(d) lead sulphate is consumed.

11. The molar ionic conductance of Zn^{+2} in water is 106×10^{-4} S m^2 mol^{-1} at 298 K. The speed of the ion under a potential gradient of 10 V cm^{-1} (in cm s^{-1}) is

(a) 8.6×10^{-5} **(b)** 5.5×10^{-5}
(c) 11.0×10^{-4} **(d)** 15.2×10^{-4}

12. Indicate the correct statement from the following which relates to the standard reduction potentials of the electrochemical series.

(a) The system with a less negative electrode potential displaces the one having the more negative electrode potential.
(b) The standard electrode potential data is not useful to determine the solubilities of sparingly soluble salts.
(c) The data is not useful to determine equilibrium constants of reactions.
(d) The displacement of Cu^{+2} ($E^0_{Cu^{2+}/Cu}$ = +0.34 V) from a solution of copper salt by zinc is more complete than the displacement of Sn from Sn^{2+} salt ($E^0_{Sn^{2+}/Sn}$ = –0.140 V) by the same metal.

13. The EMF of butane–air fuel cell is 1.10 V and its ΔH^0_C (enthalpy of combination) is –2878 kJ. The efficiency of the fuel cell is

(a) 0.96 **(b)** 0.85
(c) 0.98 **(d)** 0.82

14. The λ_i^0 values of Na^+, H^+, SO_4^{2-} ions at infinite dilution are 50.1, 350.0 and 80.0 (all in S cm^2 eq^{-1}) respectively. The t_i^0 values of these ions in a mixture containing all these ions in that order are

(a) 0.167, 0.104, 0.730 **(b)** 0.104, 0.167, 0.730
(c) 0.730, 0.104, 0.167 **(d)** 0.730, 0.167, 0.104

15. The decrease of equivalent conductance of an electrolyte with increasing concentration of the electrolyte is explained by

(a) The existence of a symmetrical ionic atmosphere around the central ion.
(b) The presence of asymmetry of the ionic atmosphere.
(c) The postulation of electrophoretic effect.
(d) The hypothesis that both asymmetry and electrophoretic effects operate.

16. Pick out the correct statement from the following which deals with the variation of transport number of an ion with concentration.

(a) The transport number of the H^+ ion increases with decreasing concentration of aqueous HCl solution.
(b) The transport number of Cd^{2+} decreases with increasing concentration of CdI_2 solution and even becomes negative.
(c) The transport number of ions increases quite markedly with increasing concentration of an electrolyte.
(d) Limiting transport number of ions cannot be determined from t_c vs $C^{1/2}$ plots.

17. The ionic product of water, $K_W = 1.3\times10^{-14}$ at 298 K. The limiting ionic conductances, $\lambda^0_{H^+}$ and $\lambda^0_{OH^-}$ of H^+ and OH^- ions are 350.0 and 198.0 S cm^2 mol^{-1} respectively. The specific conductivity of water (in S cm^{-1}) is

(a) 5.26×10^{-8} **(b)** 4.50×10^{-7}
(c) 6.25×10^{-8} **(d)** 5.8×10^{-7}

18. The standard electrode potentials for the following system are
$Fe^{2+} + 2e \rightarrow Fe$ $E^0 = -0.441$ V
$Fe^{3+} + e \rightarrow Fe^{2+}$ $E^0 = +0.771$ V
The standard E^0 for the reaction $Fe^{3+} + 3e \rightarrow Fe$ (in V)
(a) –0.154 **(b)** 0.128
(c) –0.095 **(d)** –0.037

19. The ion–ion atmosphere model in an electrolyte solution derives support from experimental studies such as
(a) dissociation field effect
(b) Wien effect
(c) Debye–Falkenhagen effect
(d) both Wien and Debye–Falkenhagen effects

20. The EMF of a Weston cadmium cell
Cd(Hg) | $CdSO_4.8/3\ H_2O$ (*sat.*) | $Hg_2SO_4(s)$,Hg at 298 K is 1.0183 V. The ΔS^0 of the reaction in the cell Cd(Hg) + $Hg_2SO_4(s) \rightarrow 2Hg(l) + CdSO_4(sat)$ is –9.7 J K^{-1}. The ΔH of the reaction and the temperature coefficient of the EMF of the cell in that order (in kJ and V K^{-1}) are
(a) $-199.4, -5.02\times10^{-5}$
(b) $-250.5, -3.1\times10^{-5}$
(c) $-100.2, 2.51\times10^{-5}$
(d) $180, -1.6\times10^{-5}$

21. The standard electrode potentials of Zn^{2+}/Zn = –0.763 V, Ag^+/Ag = +0.800 V, Cu^{2+}/Cu = +0.340 V, Na^{2+}/Na = –0.250 V. From these data, one can infer that
(a) Nickel will displace Zn^{2+} ions from a solution of zinc salt under standard state conditions.
(b) Copper displaces silver from a solution of silver nitrate (at C = 1 M).
(c) Silver displaces both nickel and zinc when it is placed in 1 M nickel nitrate or 1 M zinc nitrate solution.
(d) Zinc displaces nickel from a solution of 1 M $NiSO_4$ but not copper or silver from their salts.

22. In the cell with transference given by
Pt,H_2(g, 1 atm.) | HCl(m_1 = 0.05) | HCl(m_2 = 0.02) | H_2(g,1 atm.)
the mean transference number of H^+ ion = 0.75. Given: $\gamma_{\pm(HCl)}$ at 0.02 M = 0.90 and $\gamma_{\pm(HCl)}$ at 0.05 M = 0.80. The liquid junction potential at the boundary is (in mV)
(a) 8.0 **(b)** 12.0
(c) 15.0 **(d)** 10.0

23. The EMF of the cell SCE || Solution of unknown pH | H_2Q,Q | Pt is 0.163 V at 298 K. The pH of the solution is (H_2Q = hydroquinone, Q = quinone, E_{SCE} = 0.242 V, $E_{H2Q,Q}$ = +0.700 V)
(a) 4.2 **(b)** 5.0
(c) 3.6 **(d)** 6.5

24. The equilibrium constant of the reaction $2Ag^+ + C_6H_{12}O_6 + H_2O \rightarrow 2Ag + C_6H_{12}O_7 + 2H^+$ is 2.4×10^{25} and $E^0_{Ag^+/Ag}$ = +0.800 V (both at 298 K).The standard potential of Pt/$C_6H_{12}O_7$–$C_6H_{12}O_6$, H^+ system (in V) is
(a) 0.200 **(b)** 0.080
(c) 0.050 **(d)** 0.100

25. In the rechargeable lithium ion cell,
(a) the cathode is $LiCOO_2$
(b) the anode is platinum
(c) the electrolyte is aqueous lithium chloride
(d) the voltage is 2.0 V

26. In the alkaline version of dry battery
(a) the anode is graphite
(b) zinc is cathode
(c) the overall reaction is $Zn + 2MnO_2 \rightarrow ZnO + Mn_2O_3$
(d) the electrolyte is aqueous NH_4Cl

27. The standard potentials relating to the reactions are
$Cd^{2+} + 2e \rightarrow Cd$ $E^0 = -0.40$ V
$Cd(OH)_2 + 2e \rightarrow Cd + 2OH^-$ $E^0 = -0.80$ V
The solubility product of $Cd(OH)_2$ at 298 K is,
(a) 1.8×10^{-13} **(b)** 2.6×10^{-12}
(c) 5.4×10^{-11} **(d)** 1.3×10^{-14}

28. The standard potentials for the reactions
$Zn(NH_3)_4^{2+} + 2e \rightarrow Zn + 4NH_3$ $E^0 = -1.03$ V
$Zn^{2+} + 2e \rightarrow Zn$ $E^0 = -0.763$ V
The stability constant for the reaction $Zn(s) + 4NH_3(aq) \leftrightarrow Zn(NH_3)_4{}^{2+}(aq)$ at 298 K is
(a) 1.8×10^8 **(b)** 1.1×10^9
(c) 3.4×10^7 **(d)** 2.5×10^{10}

29. Iron on exposure to moist air forms rust. In this process
(a) Fe^{2+} ions are formed and water has no role.
(b) Fe^{3+} ions are reduced to Fe.
(c) Fe^{2+} ions initially formed get oxidised to Fe^{3+} which then get converted to $Fe(OH)_3$.
(d) Fe_2O_3 is not formed.

30. Which of the following characteristics of the DME is the most advantageous for studying the reduction of metal ions in polarography?
(a) Mercury forms an amalgam with many metals but there are exceptions like iron, aluminum, and so on.
(b) DME is not useful for studying reactions.
(c) The electrode provides a fresh surface each time for the reduction.
(d) The over potential for the reduction of H^+ ions on the Hg surface is high.

31. Find the correct statement regarding the half-wave potential of a redox system.
(a) It is equal to the standard potential of the system under certain conditions.
(b) It cannot be used for calculating the Gibbs energy changes associated with reduction processes.
(c) It does not include diffusion coefficients of oxidised and reduced species in the equation.
(d) For irreversible electrode reaction the $e_{1/2}$ is equal to the formal potential of the redox system.

32. Pick out the correct statement from the following about the Tafel relation (η vs log I).

(a) From the intercept of the plot, the standard exchange current density can be determined.

(b) The slope of the plot gives information on the mechanism of the reaction.

(c) The equation can be used to determine the standard potentials of electrode reactions.

(d) The equation can be used to find the products of electrolysis in electrolyses reactions involving a mixture of electrolytes.

33. The exchange current density for the hydrogen evolution reaction on a lead electrode is 10^{-14} A cm^{-2}. The volume of hydrogen evolved on a 5 cm^2 lead electrode per second is (in cm^3 s^{-1})

(a) 5.8×10^{-15} (b) 2.30×10^{-12}
(c) 11.6×10^{-14} (d) 3.0×10^{-13}

34. In the study of the reduction of a 4.0 mM solution of an electroactive ion at the DME, the average diffusion current was observed to be 4.0 μA. The DME had the following characteristics: rate of outflow of Hg = 1.27×10^{-1} mg s^{-1}, diffusion coefficient of the ion = 7×10^{-6} cm^2 s^{-1}, drop time = 3.5 s. The charge of the reduced ion is

(a) 3 (b) 2
(c) 1 (d) 4

35. The exchange current density for the reduction of O_2 to water on a polymer electrolyte fuel cell cathode (made of Pt dispersed carbon) in H_2SO_4 solution is 3×10^{-13} A cm^{-2}. The number of water molecules formed on a 2 cm^2 electrode per second is

(a) 4.5×10^6 (b) 3.74×10^8
(c) 9.4×10^7 (d) 1.9×10^6

36. The EMF of a lead storage cell
Pb,$PbSO_4(s)$ | $H_2SO_4(m = 4.0)$ | $PbSO_4$, $PbO_2(s)$ | Pb
is 2.053 V at 298 K. Given its E^0 = 2.04 V and the activity of water = 0.78, the mean activity coefficient of the acid in the cell is

(a) 0.172 (b) 0.354
(c) 0.265 (d) 0.312

37. The standard potentials of Fe^{2+} | Fe^{3+} and Ce^{3+} | Ce^{4+} system are +0.771 V and 1.610 V respectively at 298 K. Suppose a 0.1 M solution of Fe^{2+} ions is titrated with a dilute solution of Ce^{4+} ions, the potential at the equivalence point of the titration is (in V)

(a) 2.38 (b) 1.19
(c) 1.95 (d) 1.56

38. The standard EMF of the cell Pt,$H_2(g$, 1 atm.) | HCl(m) | AgCl(s), Ag at 298 K is 0.222 V. At a certain concentration, m, of the acid in the cell, the EMF is 0.2000 V. The activity of the acid in the cell is

(a) 1.95 (b) 1.68
(c) 2.14 (d) 2.36

39. The EMF of the concentration cell
Pt,$H_2(g$, 1 atm.) | HCl(m_1 = 0.05) | AgCl,Ag | AgCl | HCl(m_2 = 0.02) | $H_2(g$, 1 atm), Pt
at 298 K is (in V) [given $\gamma_\pm$ (HCl = 0.05) = 0.85, $\gamma_\pm$ (HCl = 0.02) = 0.90]

(a) 0.025 (b) –0.121
(c) –0.044 (d) 0.150

40. A commercially used electrode for pH determination is

(a) calomel electrode
(b) quinhydrone electrode
(c) hydrogen electrode
(d) glass electrode

41. The mean activity coefficient of a 0.002 M solution of Na_2SO_4 from the Debye–Hückel limiting equation is (given D–H constant A = 0.50)

(a) 0.864 (b) 0.837
(c) 0.912 (d) 0.738

42. If the activity coefficients of Zn^{2+} and NO^{3-} ions in a 0.001 M solution of $Zn(NO_3)_2$ are 0.85 and 0.90 respectively, the mean activity coefficient of the salt is

(a) 0.877 (b) 0.816
(c) 0.795 (d) 0.778

43. Among the following, which will have the most negative half wave potential at the dropping mercury electrode?

(a) –N=N– (b) S–S
(c) C–Cl (d) –C=C–

44. The increase in equivalent conductivity of electrolytes at electric fields of ~10^6 V cm^{-1} arises from

(a) Stokes–Einstein equation
(b) Wien effect
(c) Kohlrausch law
(d) Debye–Hückel–Onsager equation

45. In a concentration cell consisting of NaCl and KCl solutions, the largest magnitude of the liquid junction potential is exhibited by

(a) 0.001 M NaCl and 0.1 M KCl
(b) 0.5 M NaCl and 0.2 KCl
(c) 0.2 M NaCl and 0.3 M KCl
(d) 0.4 M NaCl and 0.4 M KCl

46. The standard reduction potential for $Ag^+ + e^- \rightarrow Ag$ is 0.80 V while that for $[Ag(S_2O_3)_2]^{3-} + e^- \rightarrow Ag + 2S_2O_3^{2-}$ is +0.01 V. The equilibrium constant for the complexation reaction $Ag^+ + 2S_2O_3^{2-} \rightarrow [Ag(S_2O_3)_2]^{3-}$ is

(a) 10^4 (b) 10^{11}
(c) 10^8 (d) 10^{13}

47. The standard reduction potentials for the following reactions are as follows:

$NO_3^- + 3H^+ + 2e^- \rightarrow HNO_2 + H_2O$ (0.94 V)

$NO_3^- + H_2O + 2e^- \rightarrow NO_2^- + 2OH^-$ (0.01 V)

Using the above data in conjunction with the ionic product of water, the dissociation constant (K_a) of HNO_2 can be estimated as

(a) 8×10^{-2} **(b)** 1×10^{-7}
(c) 3×10^{-5} **(d)** 5×10^{-4}

48. The product at the anode during the electrolysis of dilute H_2SO_4 solution using Pb electrodes is

(a) O_2 **(b)** Pb^{2+}
(c) H^+ **(d)** SO_4^{2-}

49. The electrolysis of concentrated NaCl solution can be carried out using the amalgam process (Hg electrodes) and diaphragm process (Ti anode and iron cathode). Which of the following is correct for both the processes?

(a) H_2O is formed at the cathode
(b) H_2 is liberated at the anode
(c) Na^+ is formed at the cathode
(d) Cl_2 is liberated at the anode

50. The theoretical gravimetric capacity of $LiCoO_2$ (Mol Wt = 97.9 g) (in mA kg^{-1}) is

(a) 274 **(b)** 148
(c) 170 **(d)** 956

51. Assuming the ionic conductivities of Na^+ and Cl^- ions (in S cm^2 mol^{-1}) as 50 and 76, the diffusion coefficient of 0.001 M NaCl (in 10^{-5} cm^2 S^{-1}) is

(a) 6.42 **(b)** 3.32
(c) 1.60 **(d)** 0.80

52. Consider NaBr and $BaCO_3$ electrolytes of the same concentration (0.0001 M). Let the Debye lengths be l_1 for NaBr and l_2 for $BaCO_3$. Which of the following is true?

(a) $l_1 = l_2$ **(b)** $l_1 = \sqrt{l_2}$
(c) $l_1 = l_2/\sqrt{3}$ **(d)** $l_1 = 2l_2$

53. The number of electrons involved in the oxidation of *two* moles of methanol in a methanol fuel cell using oxygen as the oxidant, is

(a) 6 **(b)** 4
(c) 12 **(d)** 8

Answer Keys

Practice Problems

1. Moles of e^- = 0.0326; number of electrons required = 1.963×10^{22}
2. Time required = 2.161 h; reaction at anode: $Cu + SO_4^{2-} \rightarrow CuSO_4 + 2e^-$
3. Specific conductivity, $\kappa = 3.846 \times 10^{-2}$ S m^{-1}; solubility = 4×10^{-3} mol dm^{-3}; solubility in presence of acid = 1.787×10^{-3} mol dm^{-3}
4. (i) Specific conductivity κ, before titration = 4.30×10^{-1} S m^{-1}
   (ii) Specific conductivity at equivalence point = 1.285×10^{-1} S m^{-1}
   (iii) Specific conductivity when 1 mL of NaOH is added after neutralisation = 2.502×10^{-1} S m^{-1}
5. (i) Solubility of MX = 1.142×10^{-4} mol dm^{-3}
   (ii) Solubility after 1st iteration = 1.15×10^{-4} mol dm^{-3}
   (iii) Solubility after 2nd iteration = 1.158×10^{-4} mol dm^{-3}
6. (i) Initial value of $\alpha = 0.0166$
   (ii) 1st revised $\alpha = 0.0167$
   (iii) 2nd revised $\alpha = 0.0169$; $K = 1.453 \times 10^{-5}$
7. (i) Cathode reaction: $H^+ + e^- \rightarrow \frac{1}{2}H_2(g)$;
   Anode reaction: $Cl^- \rightarrow \frac{1}{2}Cl_2(g) + e^-$
   (ii) $t_- = 0.17$, $t_+ = 0.83$
8. Cathode reaction: $K^+ + AgCl \rightarrow KCl + Ag^+$; $t_{K+} = 0.509$; $t_{Cl^-} = 0.491$
9. $\mu_0(CaF_2) = 1.86 \times 10^{-2}$ S m^2 mol^{-1}; $\Lambda_0(1/2\ Ca^{2+}) = 46 \times 10^{-4}$ S m^2 eq^{-1}
10. Solubility of $SrSO_4 = 5 \times 10^{-4}$ mol dm^{-3}; $K_{SP} = 2.5 \times 10^{-7}$
11. pH of the solution = 2.71
12. E of cell = 1.458; K of the reaction = 3.63×10^{26}; Ti^{3+}, Ti^{4+} electrode is negative
13. E = 1.01 V; m of HCl = 0.227; mean activity = $a_{\pm} = 0.182$
14. E = 2.024 V; EMF of cell depends on the activity of water. E^0 when water activity = 0.8 is 2.0297 V
15. $\Delta G = -391.8$ kJ; $\Delta H = -390.3$ kJ; $\Delta S = 5.1$ J K^{-1}
16. EMF of cell = –5.08 mV; EMF of new cell = 6.04 mV
17. E^0 of the half cell = 0.028 V
    Reaction of the half cell: $2NO_2(g) + 4H^+ + 4e^- \rightarrow NH_2NH_2 + 2O_2(g)$
18. Over potential on the Pt electrode = 0.048 V; overpotential on the Hg electrode = 3.51 V
19. k of the mixture = 4.233×10^{-1} S cm^{-1}

20. (a) Pt, Fe^{2+}, Fe^{3+}|| ascorbic acid, dehydroascorbic acid | Pt
$E^0_{cell} = 0.474–0.771 = –0.297$ V. Since E^0 is negative, Fe^{2+} cannot reduce ascorbic acid.

(b) ΔG^0 of the reaction = –99.8 kJ

(c) K of the reaction = 8.51×10^{17}

21. Theoretical equation

22. $D = 8.23 \times 10^{-7}$ cm^2 s^{-1}

23. Conc. of Cl^- = 835 mg dm^{-3}

24. $E_{appl} = 2.85$ V

25. $[Cu^{2+}] = 1.58 \times 10^{-21}$

26. $r_{Na^+} = 0.85$ Å; $r_{Cl^-} = 1.61$ Å

27. 6.24×10^{15}

28. (a) 3.45 (b) 38.78

29. 12.5

30. 87.1 S cm^2 mol^{-1}

Multiple Choice Questions

1. (b)	**2.** (d)	**3.** (b)	**4.** (a)	**5.** (c)	**6.** (b)	**7.** (d)	**8.** (a)	**9.** (c)	**10.** (d)
11. (b)	**12.** (d)	**13.** (a)	**14.** (b)	**15.** (d)	**16.** (b)	**17.** (c)	**18.** (d)	**19.** (d)	**20.** (a)
21. (b)	**22.** (b)	**23.** (b)	**24.** (c)	**25.** (a)	**26.** (c)	**27.** (d)	**28.** (b)	**29.** (c)	**30.** (d)
31. (a)	**32.** (b)	**33.** (a)	**34.** (b)	**35.** (d)	**36.** (a)	**37.** (b)	**38.** (d)	**39.** (c)	**40.** (d)
41. (b)	**42.** (a)	**43.** (d)	**44.** (b)	**45.** (a)	**46.** (d)	**47.** (d)	**48.** (a)	**49.** (d)	**50.** (a)
51. (c)	**52.** (d)	**53.** (c)							

CHAPTER

4 CHEMICAL KINETICS

CHAPTER OUTLINE

Estimation of Order and Molecularity

Theory of Reaction Rates and Thermodynamic Parameters

- Gas Phase Reactions
- Reactions in Liquid Phase
- Steady State Approximation

Methodology for Studying the Kinetics of Fast Reactions

- Kinetic Study of Conventional Reactions
- Kinetics of Fast Reactions
- Pressure Jump Technique
- Temperature Jump Method
- A Simple Derivation of the Marcus Cross-exchange Relation
- Michaelis–Menten Mechanism

4.1 Estimation of Order and Molecularity

A crucial step in comprehending the mechanism of chemical reactions consists in determining the order of a reaction. For various reaction orders (0, ½, 1, 2 …), the differential form yields the rate of change of the reactant concentration (Table 4.1). Any rate constant (k) will have the unit of s^{-1} along with the appropriate concentration terms. The order of any reaction can often be estimated by analysing the reactant or product at different intervals of time and employing graphical procedures (Table 4.2). It is customary to denote the initial concentration of a reactant A either as A_0 or a. The amount of the reactant remaining at time t is denoted as x.

Table 4.1 Rate equations and half lives for various reaction orders

Order	Rate equation		Common units of rate constant, k
	Differential form	Integrated form	
0	$\frac{dx}{dt} = k$	$kt = x$	$mol\ dm^{-3}\ s^{-1}$ or $atm\ s^{-1}$
½	$\frac{dx}{dt} = k(a_0 - x)^{1/2}$	$kt = (a_0 - x)^{1/2} - a_0^{1/2}$	$dm^{-3/2}\ mol^{1/2}\ s^{-1}$
1	$\frac{dx}{dt} = k(a_0 - x)$	$kt = \ln\left(\frac{a_0}{(a_0 - x)}\right)$	s^{-1}
2	$\frac{dx}{dt} = k(a_0 - x)^2$	$kt = \frac{x}{a_0(a_0 - x)}$	$atm^{-1}\ s^{-1}$ or $dm^3\ mol^{-1}\ s^{-1}$
2	$\frac{dx}{dt} = k(a_0 - x)(b_0 - x)$	$kt(a_0 - b_0) = \ln\left(\frac{b_0(a_0 - x)}{a_0(b_0 - x)}\right)$	$dm^3\ mol^{-1}\ s^{-1}$ or $atm^{-1}\ s^{-1}$
3	$\frac{dx}{dt} = k(a_0 - x)^3$	$2kt = \frac{1}{(a_0 - x)^2} - \frac{1}{a_0^2}$	$atm^{-2}\ s^{-1}$ or $dm^6\ mol^{-2}\ s^{-1}$

Table 4.2 Linear plots for deducing the reaction orders and half lives

Order	Half life	Linear plot
0	$\frac{a_0}{2k}$	x vs t
1	$\frac{\ln 2}{k}$	$\ln(a-x)$ vs t
2	$\frac{1}{ka_0}$	$\frac{1}{(a-x)}$ vs t
3	$\frac{3}{2ka_0^2}$	$\frac{1}{(a-x)^2}$ vs t
$n>1$	$\frac{\{2^{n-1}-1\}}{\{(n-1)ka_0^{n-1}\}}$	$\frac{1}{(a-x)^{n-1}}$ vs t

The half life provides an essential clue in deducing the order of a reaction. The molecularity, on the other hand, refers to the number of molecules involved in the elementary step of an overall reaction. The slowest step is referred to as the rate-determining step since it dictates the overall reaction rate.

4.2 Theory of Reaction Rates and Thermodynamic Parameters

4.2.1 Gas Phase Reactions

The two common theories of reaction rates are (i) transition state theory or activated complex theory and (ii) collision theory. The activated complex theory provides an interpretation of the activation energy (E_a) in terms of the enthalpy change ($\Delta H^\ddagger$) and change in the number of moles in the formation of the activated complex $\left(\Delta n^\ddagger\right)$. For unimolecular gas phase reactions, $\Delta n^\ddagger = 0$. For bimolecular gas phase reactions, $\Delta n^\ddagger = -1$. For reactions in solutions and unimolecular reactions, $\Delta H^\ddagger \approx E_a$.

Eyring has postulated the formation of the activated complex whose equilibrium constant is denoted as $K_{eq}^\ddagger$. The precise nature of the activated complex is dictated by the nature of the reaction (gaseous or ionic or heterogeneous). In the case of the collision theory, the molecules must collide before the reaction occurs and furthermore, only those collisions that are energetically favourable will finally yield the product. A major ingredient in this formalism is the estimation of the collision number (or frequency) which is then employed to calculate the rate constant. Furthermore, the pre-exponential factor is shown to depend on temperature (Table 4.3).

4.2.2 Reactions in Liquid Phase

The rate constants of solution-phase chemical reactions are influenced by the solvent medium as well as ionic strength. In the simple Debye–Huckel version, the activity coefficients are dependent upon the ionic strength I as

$$\log_{10} \gamma_\pm = -A|z_A z_B| I^{1/2}$$

For aqueous solutions at 25°C, the value of A is 0.509 $\text{mol}^{-1/2}\,\text{dm}^{3/2}$ with the corresponding unit for the ionic strength. For reactions in solution, when both the reactants are charged species, the activity coefficient needs to be estimated for the given concentration vis a vis the ionic strength of the solution. If one of the *reactants* is a neutral species, the rate constant does not depend upon the ionic strength. For two reactants with charges z_A and z_B, the rate constant (k) at 25°C as a function of the ionic strength can be shown to be

$$\log [k/k_0] \approx 2\times 0.509\, z_A z_B\, (I)^{1/2}$$

and k_0 denotes the rate constant under the limiting condition of $I \to 0$. Since the addition of salts increases the ionic strength, the influence of the ionic strength is sometimes referred to as *primary salt effect*. The product $z_A z_B$ can vary from +8 to –8 thus leading to a significant influence on the rate constant.

4.2.3 Steady State Approximation

An extremely powerful method of elucidating the reaction mechanism is the steady state approximation (SSA) according to

Table 4.3 Equations for estimating the rate constants from the transition state and collision theories

Transition state theory	
Formulae	**Remarks**
$k = \frac{k_B T}{h} K_{eq}^\ddagger$	Relation between rate constant and equilibrium constant of the activated complex
$\Delta G^\ddagger = -RT \ln K_{eq}^\ddagger$	For unimolecular gas phase reactions, $\Delta n^\ddagger = 0$
$\Delta G^\ddagger = \Delta H^\ddagger - T\Delta S^\ddagger$	
$k = \frac{k_B T}{h} e^{-\Delta H^\ddagger/RT} e^{\Delta S^\ddagger/R}$	For bimolecular reactions, $\Delta n^\ddagger = -1$
$E_a = \Delta H^\ddagger + RT - \Delta n^\ddagger RT$	For reactions in solution and unimolecular reactions, $\Delta H^\ddagger \approx E_a$ is sometimes assumed

(Continued)

Table 4.3 (Continued)

Collision theory	
Formulae	**Remarks**
$Z_{AA} = 2N^2\sigma^2\left(\frac{\pi RT}{M}\right)^{1/2}$	Collision frequency between two identical molecules
$N = 2pZe^{-E_a/RT}$	N is the number of molecules per cm^3, Z_{AA} will usually be ~10^{31} s^{-1}, p = probability factor or steric factor
$k_2 = \frac{4\sigma^2 L}{10^3}\left(\frac{\pi RT}{M}\right)^{1/2} e^{-E_a/RT}$	Units of k_2 will be L mol^{-1} s^{-1} if R is expressed in ergs
$Z_A = \frac{2Z_{AA}}{N^*}$	Z_A is the number of collisions made by a molecule with all other molecules per unit volume per unit time

which the concentration of the reactive intermediate is assumed to be time-independent. The methodology of employing this is shown below.

$$A \rightleftharpoons B$$

(forward and reverse rate constants k_1 and k_{-1} respectively)

$$B \xrightarrow{k_2} C$$

In order to apply the steady state approximation, the rate constant for the reaction of B to form A and C should be >> the rate constant for the formation of B from A. Hence the rate constants should obey the condition $(k_{-1} + k_2) \gg k_1$ (Table 4.4).

Table 4.4 Regime of validity for the steady state and rapid equilibrium approximations

Approximation	Condition	Rate constants
Steady state	$(k_{-1} + k_2) \gg k_1$	$\frac{k_1 k_2}{(k_{-1} + k_2)}$
Rapid equilibrium	$(k_1 + k_{-1}) \gg k_2$	$\frac{k_1 k_2}{(k_1 + k_{-1})}$
Steady state and equilibrium	$k_{-1} \gg k_1$ and $k_{-1} \gg k_2$	$\frac{k_1 k_2}{k_{-1}}$

4.2.4 Michaelis–Menten Mechanism

The most important equation to describe enzyme catalysis is that due to Michaelis and Menten. The equation governing this mechanism is:

$$E + S \underset{k_{-1}}{\overset{k_1}{\rightleftharpoons}} ES \xrightarrow{k_2} E + P$$

where E, S, ES and P denote respectively, the enzyme, substrate, enzyme–substrate complex and products. The rate of formation of P is given by $dP/dt = k_2$ [ES] and represents the velocity of the reaction. Employing the steady state approximation, the velocity can be derived as

$$v = k_2[E]_0[S]/(K_M + [S])$$

The Michaelis–Menten constant K_M equals $(k_{-1} + k_2)/k_1$. It can be shown that the maximum velocity v_{max} is reached when

$$v_{max} = k_2[E]_0 = k_{cat}[E]_0$$

$[E]_0$ denotes the initial concentration of the enzyme (t = 0). Further, the catalytic efficiency is given by k_{cat}/K_M. Among several methods of representing the enzyme kinetic data, the two most popular ones are,

(i) the double reciprocal Lineweaver–Burk plot, (1/v) vs 1/[S] (slope: K_M/v; x-axis intercept: $-1/K_M$), and

(ii) the Eadie–Hofstee method: Plot of v vs v/[S] (slope: K_M; y-axis intercept: v_{max})

There are two regimes that follow from the Michaelis–Menten analysis: the reaction is zero order when [S] >> K_M and first order when [S] << K_M.

When $k_2 \ll k_{-1}$, $K_M = K_d$, the dissociation constant of ES. Furthermore, K_M is equal to the substrate concentration when the velocity is equal to maximum velocity/2.

4.3 Methodology for Studying the Kinetics of Fast Reactions

4.3.1 Kinetic Study of Conventional Reactions

In the study of the kinetics of reactions in the laboratory, for example, the acid catalysed hydrolysis of methyl acetate or the alkaline hydrolysis of an ester such as ethyl acetate in aqueous solution at normal temperatures (around 298 K), the progress of the reaction is followed by titrating a sample of the reaction mixture with a suitable reagent. Alternatively, a change in the physical property of the solution (a change in the colour of the solution or a change in the conductivity of the solution with time) is used to follow the reaction. For the kinetic study of such reactions, the $t_{1/2}$ of the reaction is usually much larger than the time needed for mixing and follow up of the progress of the reaction.

4.3.2 Kinetics of Fast Reactions

A fast reaction is generally defined as one that occurs too fast for the normal methods of analysis to be adopted at temperatures around room temperature and using concentrations that are neither too low or too high (around 0.1 m). Such reactions usually have $t_{1/2}$ < 1 s. There are various methods for studying fast reactions. They are classified as follows.

Stopped flow and continuous flow methods: In the former case, the reactant solutions are mixed rapidly but the follow up of the reaction is carried out using special fast techniques. In the latter case, the follow up of the reaction is done without using the fast measuring techniques.

Relaxation methods: In this category, the pressure P, or temperature T of a reaction at equilibrium is rapidly altered by a certain magnitude (for example, $\Delta P = 50$ to 60 atm or $\Delta T = 5$ to 10°C using a method known as Joule heating method) and the reestablishment of the distributed equilibrium is followed as a function of time using a suitable technique.

Other methods: There are also some other methods such as the electrical field jump method, spectroscopic methods such as NMR and photochemical methods such as the flash photolysis technique for studying fast reactions. However, let us consider briefly the flow methods and relaxation methods as these constitute the some of the most commonly adopted methods in fast kinetic studies.

(i) **Continuous flow method:** Flow methods have been developed by Hartridge and Roughton. The general arrangement of *flow methods* is as follows. Two experimental solutions kept in two separate reaction chambers are driven into a special mixing chamber mechanically. The reaction mixture flows down an observation tube. In the continuous flow method, the composition at various points along the observation tube is determined and converted into a concentration–time plot.

(ii) **Stopped flow method:** A slight change in the method compared to that in the continuous flow method is adopted in the stopped flow technique. The flow of the reaction mixture is suddenly stopped at some point and the extent of reaction as a function of time is followed at that point in the observation tube (away from mixing chamber) by a physical technique.

(iii) **Relaxation methods:** As stated earlier, the chemical relaxation method is based on rapidly disturbing the position of a chemical equilibrium by changing a parameter such as temperature or pressure on the system. Obviously, following the perturbation, the system is not in stable equilibrium. However, when the displacement from the equilibrium is only to a small extent, the rate of rearrangement of the equilibrium is proportional to the instantaneous deviation from equilibrium. Figure 4.1 illustrates the change of external parameter (in this case, pressure) as well as the concentration of a given species as a function of time.

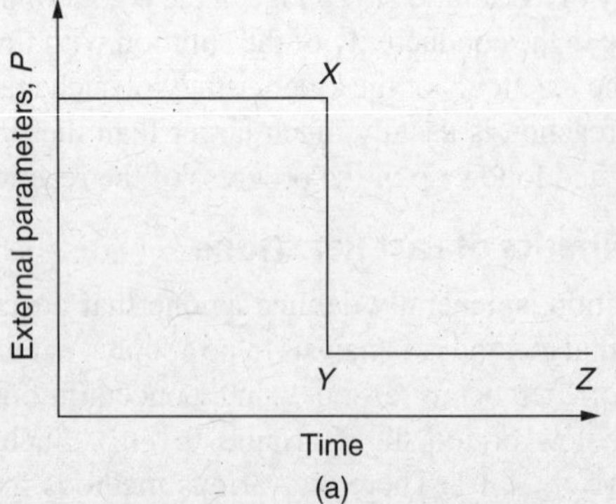

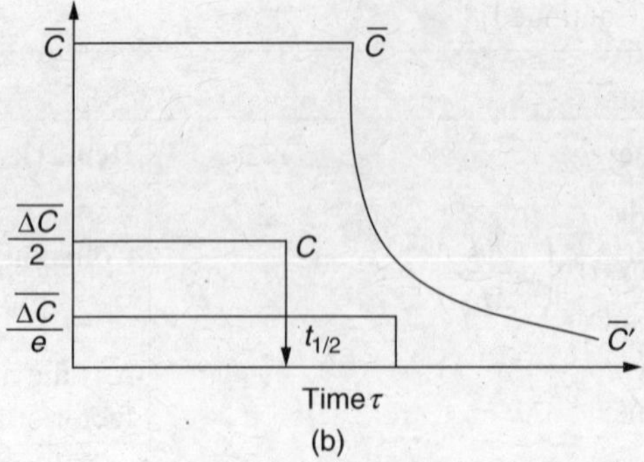

Fig. 4.1 (a) Variation of P (external parameter) with time (b) Variation of the concentration of the reactant with time

In Fig. 4.1(a), the pressure is suddenly released on the system from X to Y. In Fig. 4.1(b) the corresponding change in the concentration of the reactant species exponentially with time, is shown. $\bar{C}$ is referred to as the equilibrium concentration before disturbance or perturbation (also known as initial equilibrium concentration), C is the concentration at time t and $\bar{C}'$ is the new equilibrium concentration after perturbation (also known as final equilibrium concentration). Employing the final equilibrium concentration $\bar{C}'$ as a reference point, it is possible to arrive at an equation for the variation of concentration of the reactant, j, with time.

Writing ΔC_j = change of concentration of j from initial equilibrium to final equilibrium value, from Fig. 4.1(b) it is seen that,

$$\overline{C_j} = \overline{C_j'} + \Delta C_j \qquad (4.1)$$

Likewise, for the instantaneous concentration of j, that is, C_j

$$C_j = \overline{C_j'} + \Delta C_j \qquad (4.2)$$

The basic equation for any relaxational process is given by

$$-\frac{d\left(C-\bar{C}'\right)}{dt} = \frac{1}{\tau}\left(C-\bar{C}'\right) \qquad (4.3)$$

$$-\frac{dC}{dt} = \frac{1}{\tau}\left(C-\bar{C}'\right) \qquad \left(\text{since}\,\frac{d\bar{C}'}{dt} = 0\right) \qquad (4.4)$$

τ in Eq. 4.4 is known as *relaxation time* or time constant of the relaxation process. General integration of Eq. 4.4 gives

$$\int \frac{-dC}{\left(C-\bar{C}'\right)} = \frac{1}{\tau}\int dt$$

or,
$$-\ln\left(C-\bar{C}'\right) = \frac{t}{\tau} + \text{Constant} \qquad (4.5)$$

At $t = 0$, $C = \bar{C}$ and integration constant $= -\ln\left(\bar{C}-\bar{C}'\right)$

Using this value in Eq. 4.5, and rearranging it, the equation, $-\ln\left(C-\bar{C}'\right) = \frac{t}{\tau} - \ln\left(\bar{C}-\bar{C}'\right)$ which in turn leads to

$$-\ln\frac{\left(C-\bar{C}'\right)}{\left(\bar{C}-\bar{C}'\right)} = \frac{t}{\tau} \quad \text{or} \quad \Delta C = \Delta\bar{C}e^{\frac{-t}{\tau}} \qquad (4.6)$$

where $\Delta C = \left(C - \overline{C'}\right)$ and $\Delta\overline{C} = \left(\overline{C} - \overline{C'}\right)$. Equation 4.6 is the fundamental relation governing all relaxation processes.

Parameters affecting τ and $\Delta\bar{C}$ (known as amplitude): τ depends upon rate constant k_j, equilibrium constant k_j and also on the reaction mechanism. The amplitude $\left(\Delta\overline{C}\right)$ provides information on the equilibrium constant and other thermodynamic quantities such as ΔH and ΔV of the reaction under consideration.

(i) **Significance of τ:** From Fig. 4.1, we may define $t_{1/2}$ as $\frac{\Delta\overline{C}}{2}$, that is, the time of half reaction is equal to the time required for the initial difference $\left(C - \overline{C'}\right)$ to become half of its value. In an analogous manner, it is seen that τ is the time required for the initial difference to become $\frac{\Delta\overline{C}}{e}$. Under these conditions, from Eq. 4.6,

when $t = \tau$, $$\Delta C = \frac{\Delta\overline{C}}{e} \tag{4.7}$$

The Eq. 4.7 may be considered as the definition of τ.

(ii) **Equation for τ for a one step processes:** As mentioned, τ is a function of the rate constant, concentration of reactant species and mechanism. Therefore, the equation for relaxation time depends on the nature of reaction under consideration. For example, for a reaction $X_1 \underset{k'}{\overset{k}{\rightleftharpoons}} X_2$ the equation for $1/\tau$, (or τ) is different from that for a reaction

$$X_1 + X_2 \underset{k'}{\overset{k_2}{\rightleftharpoons}} X_3 + X_4$$

A simple reaction

$$X + Y \underset{k'}{\overset{k}{\rightleftharpoons}} Z \tag{4.8}$$

will be considered as an example for deriving the equation for $1/\tau$. The rate equation is

$$\frac{dC_X}{dt} = k'C_Z - kC_XC_Y \tag{4.9}$$

For the time dependent concentrations C_X, C_Y and C_Z, one may write

$$C_X = \overline{C'_X} + \Delta C_X;\ C_Y = \overline{C'_Y} + \Delta C_Y;\ C_Z = \overline{C'_Z} + \Delta C_Z \tag{4.10}$$

But by stoichiometry,

$$\Delta C_X = \Delta C_Y = -\Delta C_Z \tag{4.11}$$

Substituting Eq. 4.10 in Eq. 4.9,

$$\frac{d\left(\overline{C'_X} + \Delta C_X\right)}{dt} = k'\left(\overline{C'_Z} + \Delta C_Z\right) - k\left(\overline{C'_X} + \Delta C_X\right)\left(\overline{C'_Y} + \Delta C_Y\right) \tag{4.12}$$

$$\frac{d\Delta C_X}{dt} = k'\overline{C'_Z} - k\overline{C'_X}\,\overline{C'_Y} - k'\Delta C_X - k\overline{C'_X}\Delta C_Y - k\Delta C_X\overline{C'_Y} \tag{4.13}$$

Since $C_j - \overline{C'_j} \ll C_j$ in all relaxational processes, we may neglect the small quantity $k\Delta C_X\Delta C_Y$ in Eq. 4.13 also employing the stoichiometric condition (Eq. 4.11), Eq. 4.13 may be written as,

$$\frac{d\left(\Delta C_X\right)}{dt} = k'\overline{C'_Z} - k\overline{C'_X}\,\overline{C'_Y} - k'\Delta C_X - k\overline{C'_X}\Delta C_X - k\Delta C_X\overline{C'_Y}$$

By the condition of equilibrium $k'\overline{C'_Z} = k\overline{C'_X}\,\overline{C'_Y}$ or their difference is zero.

$$\frac{d\left(\Delta C_X\right)}{dt} = -\Delta C_X\left\{k' + \left(k\overline{C'_X} + k\overline{C'_Y}\right)\right\} \tag{4.14}$$

$$\frac{d\left(\Delta C_X\right)}{dt} \times \frac{1}{\Delta C_X} = -\left[k\left(\overline{C'_X} + \overline{C'_Y}\right) + k'\right] \tag{4.15}$$

$$\int \frac{d\left(\Delta C_X\right)}{\Delta C_X} = -\left[k\left(\overline{C'_X} + \overline{C'_Y}\right) + k'\right]\int dt$$

On general integration, $\ln \Delta C_X = -\frac{t}{\tau} + \text{Constant}$

At $t = 0$, Constant $= \ln \overline{\Delta C_X}$

or, $$\ln\frac{\Delta C_X}{\overline{\Delta C_X}} = -\frac{t}{\tau}$$

which may be written as

$$\Delta C_X = \overline{\Delta C_X}\, e^{-t/\tau}$$

where $$\frac{1}{\tau} = \left[k\left(\overline{C'_X} + \overline{C'_Y}\right) + k'\right] \tag{4.16}$$

From the linear plot of $\frac{1}{\tau}$ vs $\left(\overline{C'_X} + \overline{C'_Y}\right)\left(\approx \overline{C_X} + \overline{C_Y}\right)$ which is linear, the slope provides k and intercept gives k'.

4.3.3 Pressure Jump Technique

In this method, the position of equilibrium of a reaction is shifted under adiabatic condition (on changing the pressure rapidly) and this shift is expressed mathematically by the equation

$$\left(\frac{\partial \ln K}{\partial P}\right)_S = -\frac{\Delta V^0}{RT} + \frac{V\alpha}{C_p} \times \frac{\Delta H^0}{RT} \tag{4.17}$$

where ΔV^0 = change in the molar volume of the reaction, α = coefficient of thermal expansion given by $\frac{1}{V}\left(\frac{\partial V}{\partial T}\right)_P$, C_P = heat capacity, ΔH^0 = standard enthalpy change of the reaction. The contribution of the second term of Eq. 4.17 is quite small and can be neglected. Thus the above may be simplified to

$$\left(\frac{\partial \ln K}{\partial P}\right)_S = -\frac{\Delta V^0}{RT} \tag{4.18}$$

This technique is especially useful for reactions where ΔV^0 is large so that a greater shift in equilibrium can be expected for a given change of pressure. This method has been extensively employed in investigation of the mechanisms of metal complex formations such as those in beryllium sulphate, vanadyl sulphate solutions, and also in studies of hydration of carbonyl compounds, and so on. Relatively slow reactions having relaxation times of the order of 1 m s to 10 s are measured using this method.

4.3.4 Temperature Jump Method

This method is based on the displacement of the equilibrium with change of temperature according to

$$\left(\frac{\partial \ln K}{\partial T}\right)_P = \frac{\Delta H^0}{RT^2} \tag{4.19}$$

where ΔH^0 refers to the enthalpy change of the reaction. The method can be used for studying reactions whose relaxation times range from 1 μs to 1 s. The course of the reaction may be followed by using different physical parameters such as colour change through spectrophotometry, change in conductivity by using a conductance bridge, polarimetry, and so on. The T-jump method has been widely used to study a variety of reactions such as metal complex formation reactions, reactions involving micelles and a number of biochemical reactions as well.

4.3.5 Marcus Cross-exchange Relation

The rate constant k for any reaction can be written in terms of the Gibbs free energy of activation $\Delta G^{\ddagger}$ as

$$k = Ze^{-\Delta G^{\#}/RT} \tag{4.20}$$

where Z denotes the corresponding frequency factor. For a typical self-exchange electron transfer reaction such as

$$Fe^{2+} + Fe^{*3+} \rightarrow Fe^{3+} + Fe^{*2+}$$

the rate constant is k_{11}. Analogously, for another self-exchange reaction,

$$Ce^{3+} + Ce^{*4+} \rightarrow Ce^{4+} + Ce^{*3+}$$

the rate constant is k_{22}. The overall cross-exchange electron transfer reaction consisting of the above two self-exchange reactions is

$$Fe^{2+} + Ce^{4+} \rightarrow Fe^{3+} + Ce^{3+}$$

with the corresponding rate constant k_{12}. Interestingly, a simple *approximate* relation governs the theoretical prediction of k_{12}.

$$k_{12} = \left(k_{11}k_{22}K_{12}\right)^{1/2} \tag{4.21}$$

where K_{12} denotes the equilibrium constant of the cross-exchange reaction. Note that there are no adjustable parameters in Eq. 4.21. The factor *f*, consisting of work terms has been made unity while writing the above simple form. The validity of the above Marcus cross-exchange relation has been verified for a large number of electron transfer processes. The quadratic dependence of the Gibbs free energy of activation on the driving force vis a vis prediction of the 'inverted region' constitute some of the salient features of the Marcus theory.

Worked Out Problems

Estimation of Reaction Orders and Rate Constants

Problem 4.1

The half life of a first order reaction is 250 s. Estimate the (i) rate constant and (ii) time required to decrease the reactant concentration to 25% of its initial value.

Solution

(i) Half life of a first order reaction = $0.693/k$

Hence $k = 2.772\times10^{-3}$ s^{-1}

(ii) Concentration of the reactant A at any time $t = A(t)$

$$= A(t=0)\, e^{-kt}$$

$$A_0/4 = A_0\, e^{(-2.772\times10^{-3}\, t)}$$

Hence, $t = 500$ s

Problem 4.2

From the following data for a reaction between A and B, find the order of the reaction with respect to A and B and calculate the rate constant.

A (mol L^{-1})	B (mol L^{-1})	Initial Rate (mol L^{-1} s^{-1})
2.3×10^{-4}	3.1×10^{-5}	5.2×10^{-4}
4.6×10^{-4}	6.2×10^{-5}	4.2×10^{-3}
9.2×10^{-4}	6.2×10^{-5}	1.7×10^{-2}

Solution

In general, the ratio between two rates of the reaction is given by

$$\frac{R_2}{R_1} = \left(\frac{A_2}{A_1}\right)^a \left(\frac{B_2}{B_1}\right)^b$$

$$= \frac{4.2\times10^{-3}}{17\times10^{-3}} = \left(\frac{4.6\times10^{-4}}{9.2\times10^{-4}}\right)^a \left(\frac{6.2\times10^{-5}}{6.2\times10^{-5}}\right)$$

$$0.5^a = 0.247$$

$$\therefore a = 2 \text{ with respect to } [A]$$

$$\frac{4.2\times10^{-3}}{5.2\times10^{-3}} = \left(\frac{4.6\times10^{-4}}{2.3\times10^{-4}}\right)^a \left(\frac{6.2\times10^{-5}}{3.1\times10^{-5}}\right)$$

$$2^2 \times 2^b = 8.07$$

$$2^b \approx 2; b = 1 \text{ with respect to } [B]$$

$$\text{Rate} = k[A]^2[B];\ k = \frac{\text{rate}}{[A]^2[B]} = 3.17\times10^8\ (\text{mol L}^{-1})^{-2}\ \text{s}^{-1}$$

Problem 4.3

The initial concentration of a reactant is a for a first order reaction, whose half life is $t_{1/2}$. Derive the equation for $[(a/(a-x)]$ when $t = n\,t_{1/2}$ with $n > 1$.

Solution

$$k_1 = \frac{2.303}{nt_{1/2}} \log\frac{a}{a-x}$$

$$\text{Therefore } \log\left(\frac{a}{a-x}\right) = \frac{0.693n}{2.303} = 0.301\times n$$

$$\ln\left(\frac{a}{a-x}\right) = 0.693n \text{ and } \frac{a}{a-x} = \exp(0.693n)$$

Problem 4.4

For the reaction $F_2 + 2ClO_2 \rightarrow FClO_2$ at 260 K, the following data were obtained.

(F_2)	(ClO_2)	$\frac{-d[F_2]}{dt}$ (M s^{-1})
(i) 0.2	0.02	1.3×10^{-3}
(ii) 0.2	0.08	5.2×10^{-3}
(iii) 0.4	0.02	2.6×10^{-3}

Calculate the rate constant.

Solution

Using the ratio method,

From (i) and (ii), Rate $\propto$ $[ClO_2]$ and

from (i) and (iii), Rate $\propto$ $[F_2]$

Hence the reaction is first order with respect to each reactant.

$$\frac{-d[F_2]}{dt} = k_2\,[F_2]\,[ClO_2]$$

$$\text{Therefore } k_2 = \frac{1.3\times10^{-3}}{0.2\times0.02} = 0.325 \text{ M}^{-1}\text{ s}^{-1}$$

Problem 4.5

At 600 K, the rate constant of the reaction, $2NO_2 \rightarrow 2NO + O_2$ is 0.73 M^{-1} s^{-1}. How long will it take for one-tenth of a sample of $NO_2(g)$ at 402 mm of Hg and 603 K to decompose?

Solution

The number of moles of the reactant can be estimated using the ideal gas equation

$$PV = nRT$$

and

$$n = \frac{\frac{402}{760}\times1}{0.082\times603} = 0.0107$$

The amount decomposed is given by $x = a/10$

$$k_2 = \frac{1}{t}\frac{x}{a(a-x)} \quad \text{or} \quad t = \frac{1}{k_2}\frac{x}{a(a-x)}$$

or

$$t = \frac{0.1a}{k_2 a\times0.9a} = \frac{1}{9k_2 a}$$

or

$$t = \frac{1}{9\times0.73\times0.0107} = 14.22 \text{ s}$$

Problem 4.6

The reaction $A + B \rightarrow$ Products is first order with respect to each reactant with $k = 6\times10^{-3}$ L mol^{-1} s^{-1} at 25°C. Calculate $[A]$ at $t = 200$ s, if $[A]_{t=0} = 0.2$ M and $[B]_{t=0} = 6.2$ M.

Solution

Since $[B]_{t=0} > [A]_{t=0}$ the reaction may be considered as pseudo-first order, that is, the concentration of B does not change significantly as the reaction proceeds. Hence

$$\text{Rate} = k\,[A]\,[B] = 6.2\,k\,[A] = k'\,[A]$$

However $A = A_0 e^{-k't}$, where $k' = 6.2k$

$$\log[A] = \log[A_0] - \frac{6.20kt}{2.303} = -0.699 - \frac{(3.6\times10^{-2})(200)}{2.303}$$

$$A(t) \simeq 1.48\times10^{-4} \text{ M after 200 s}$$

Problem 4.7

The base-catalysed ester hydrolysis yields the following data

Time (s)	0	850	3900
Concentration of ester (M)	19.8	9.5	3.6
concentration of OH$^-$ (M)	20.9	10.5	4.7

Estimate the rate constant.

Solution

Since the ester hydrolysis under basic conditions follows a second order reaction, the constancy of a second order rate constant needs to be verified.

$$kt = \frac{1}{[B_0]-[A_0]}\ln\frac{[B]/[B_0]}{[A]/[A_0]}$$

$$850\times k = \frac{1}{1.1}\ln\frac{10.5/20.9}{9.5/19.8} = 0.0418$$

$$k = 4.921\times10^{-5} \text{ M}^{-1}\text{ s}^{-1}$$

$$3900\,k = \frac{1}{1.1}\ln\frac{4.7/20.9}{3.6/19.8} = \frac{0.2125}{1.1} = 0.193$$

$$k = 4.94\times10^{-5} \text{ M}^{-1}\text{ s}^{-1}$$

These two rate constants are nearly equal.

Problem 4.8

At 300°C, azomethane decomposes mainly according to the reaction

$$CH_3NNCH_3(g) \rightarrow C_2H_6(g) + N_2(g)$$

with a rate constant of $k = 2.4\times10^{-4}$ s^{-1}. If the activation energy for the reaction is 42 kcal mol^{-1}, estimate the temperature at which the half life of the reaction would be 139.6 seconds.

Solution

$$\text{Rate constant} = \frac{0.693}{t_{1/2}} = \frac{0.693}{139.6 \text{ s}} = 4.96\times10^{-3} \text{ s}^{-1} \text{ at the desired temperature.}$$

$$\log\frac{k_2}{k_1} = \frac{E_a}{2.303R}\left[\frac{1}{T_1} - \frac{1}{T_2}\right]$$

$$\text{That is, } \log\frac{4.96\times10^{-3}}{2.40\times10^{-4}} = \frac{42{,}000}{2.303\times1.987}\left[\frac{1}{573} - \frac{1}{T_2}\right]$$

$$\frac{1}{T_2} = \frac{1}{573} - \left[\frac{1.315\times2.303\times1.987}{42{,}000}\right]$$

Therefore $T_2 = 624.2$ K

Problem 4.9

The reaction $A \rightarrow$ Products yields the following data at 400 K.

Time (min)	0	15	75	315	1275
A (mol L^{-1})	0.24	0.12	0.06	0.03	0.015

Evaluate the rate constant.

Solution

Since $t_{1/2}$ is inversely proportional to $[A]^2$, the order of the reaction is 3.

$$k_n = \frac{1}{(n-1)t}\left[\frac{1}{A^{n-1}} - \frac{1}{A_0^{n-1}}\right]$$

When $t = t_{1/2}$, $A = \dfrac{A_0}{2}$

$$k_n = \frac{2^{n-1}-1}{(n-1)t_{1/2}A_0^{n-1}}$$

$n = 3$; $$k = \frac{2^{3-1}-1}{(3-1)15(0.24)^2} = 1.736 \text{ L}^2\text{ mol}^{-2}\text{ min}^{-1}$$

Problem 4.10

The rate constant of the decomposition of azomethane is 2.6×10^{-4} s^{-1}.

$$CH_3NNCH_3(g) \rightarrow C_2H_6(g) + N_2(g)$$

Calculate the partial pressure of azomethane, $C_2H_6(g)$ and $N_2(g)$ if the initial pressure of azomethane is 284 mm Hg and the reaction occurs for 250 s.

Solution

$$k \times t = 2.303 \log\frac{P_0}{P}$$

$$2.6\times10^{-4}\times250 = 2.303 \log\frac{284}{P}$$

Therefore $P(t) = 266.1$ mm Hg

The difference (284.0 – 266.1) corresponds to 17.9 mm Hg.

Hence the partial pressure of N_2 and C_2H_6 is 17.9 mm Hg each.

Problem 4.11

A second order reaction of the type $A + B \rightarrow$ Products, is 39% complete in 118 minutes, when the initial concentration of both A and B are 0.03 M. Calculate (i) the rate constant and (ii) the time required for the reaction to be 39% complete if the initial concentrations of both A and B are 0.15 M.

Solution

39% of 0.03 M corresponds to 0.0117; Amount reacted = x = 0.0117 M

$$k = \frac{1}{t}\frac{x}{a(a-x)} = \frac{1}{118}\times\frac{0.0117}{0.03\times0.0183} = 0.1806 \text{ min}^{-1} \quad (1)$$

For the initial concentration of 0.15 M, 39% completion corresponds to $x = 0.0585$ M. Using Eq. (1) once again,

$$0.1806 \text{ min}^{-1} = \frac{1}{t}\frac{0.0585}{0.15(0.15-0.0585)}$$

Hence $t = 23.6$ minutes.

Problem 4.12

From the following data, for a reaction between A and B, find the order of the reaction with respect to A and B and calculate the rate constant.

A (mol L^{-1})	B (mol L^{-1})	Initial rate (mol L^{-1} s^{-1})
2.2×10^{-4}	3.0×10^{-5}	5.1×10^{-4}
4.4×10^{-4}	6.0×10^{-5}	4.1×10^{-3}
8.8×10^{-4}	6.0×10^{-5}	1.6×10^{-2}

Solution

$$\frac{R_2}{R_1} = \left(\frac{A_2}{A_1}\right)^a\left(\frac{B_2}{B_1}\right)^b$$

$$\frac{4.1\times10^{-3}}{16\times10^{-3}} = \left(\frac{4.4\times10^{-4}}{8.8\times10^{-4}}\right)^a\left(\frac{6.0\times10^{-5}}{6.0\times10^{-5}}\right)^b$$

$$0.5^a = 0.256$$

Therefore $a = 2$ with respect to $[A]$

$$\frac{4.1\times10^{-3}}{0.51\times10^{-3}} = \left(\frac{4.4\times10^{-4}}{2.2\times10^{-4}}\right)^a\left(\frac{6.0\times10^{-5}}{3.0\times10^{-5}}\right)^b = 2^2\times2^b = 8.039$$

$2^b \approx 2$; $b = 1$ with respect to $[B]$.

$$\text{Rate} = k[A]^2[B]$$

$$k = \frac{\text{rate}}{[A]^2[B]} = 3.53\times10^8 \text{ M}^{-2}\text{ s}^{-1}$$

Problem 4.13

At 155°C, the gas phase decomposition of di-*t*-butyl peroxide is a first order reaction given by

$$(CH_3)_3COOC(CH_3)_3 \rightarrow 2\,CH_3COCH_3 + C_2H_6$$

The total pressure at various times in a fixed volume is as follows:

t (s)	0	180	360	540	720	900	1080
p (atm)	0.22	0.25	0.27	0.29	0.31	0.33	0.35

Calculate the first order rate constant.

Solution

At any time, the total pressure is $p = p_{DTBP} + p_A + p_E$

If y denotes the decrease in the pressure of di-*t*-butyl peroxide from the initial pressure p_0,

$$p_{DTBP} = p_0 - y$$

$$p_A = 2y$$

$$p_E = y$$

Therefore $$p = p_0 - y + 2y + y$$
$$= p_0 + 2y$$

Therefore $$y = \frac{p - p_0}{2}$$

Therefore $$p_{DTBP} = p_0 - \frac{(p-p_0)}{2}$$
$$= \frac{3p_0 - p}{2}$$

Hence a is proportional to p_0.

$(a - x)$ at any time is proportional to $\dfrac{3p_0 - p}{2}$

Hence, the first order rate constant k_1 using the data at 180 s

$$= \frac{2.303}{180}\log\frac{2\times0.22}{(3\times0.22)-0.25}$$

$$= 3.97\times10^{-4}\ \text{s}^{-1}$$

Similarly,

$$k_1 = \frac{2.303}{360}\log\frac{2\times0.22}{(3\times0.22)-0.27}$$

$$= 3.35\times10^{-4}\ \text{s}^{-1}$$

$$k_1 = \frac{2.303}{540}\log\frac{2\times0.22}{(3\times0.22)-0.29}$$

$$= 3.21\times10^{-4}\ \text{s}^{-1}$$

$$k_1 = \frac{2.303}{720}\log\frac{2\times0.22}{(3\times0.22)-0.31}$$

$$= 3.18\times10^{-4}\ \text{s}^{-1}$$

$$k_1 = \frac{2.303}{900}\log\frac{2\times0.22}{(3\times0.22)-0.33}$$

$$= 3.19\times10^{-4}\ \text{s}^{-1}$$

$$k_1 = \frac{2.303}{1080}\log\frac{2\times0.22}{(3\times0.22)-0.35}$$

$$= 3.24\times10^{-4}\ \text{s}^{-1}$$

from the given experimental data.

Problem 4.14

The time required to complete a certain fraction of a reaction depends upon the order of the reaction. If the time required to convert 75% of a reactant into products is denoted as $t_{3/4}$ and the time for 50% conversion as $t_{1/2}$, estimate the ratio, $\frac{t_{3/4}}{t_{1/2}}$ for (i) zero and (ii) first order reactions.

Solution

(i) For a zero order reaction, the rate of change of concentration (c) of the reactant is given by $-\frac{dc}{dt} = k$ where k denotes the rate constant and hence

$$c(t) = c_0 - kt \qquad (1)$$

$t_{1/2} = \frac{c_0}{2k}$ with c_0 being the initial concentration of the reactant. From Eq. (1),

$\frac{c_0}{4} = c_0 - kt_{3/4}$ for 75% completion of the reaction

Hence $t_{3/4} = \frac{3c_0}{4k}$ and $\frac{t_{3/4}}{t_{1/2}} = \frac{3}{2}$

(ii) For a first order reaction, $t_{1/2} = \frac{\ln 2}{k}$

and $t_{3/4} = \frac{\ln 4}{k}$

Hence $\frac{t_{3/4}}{t_{1/2}} = 2$

Arrhenius Equation and Temperature Dependence of Rate Constants

Problem 4.15

At 552 K, the rate constant for the decomposition of SO_2Cl_2 is 6.09×10^{-5} min^{-1}. If the energy of activation is 210 kJ mol^{-1}, calculate the frequency factor and the rate constant at 600 K.

Solution

$$k = Ae^{-E_a/RT}$$

$$\ln A = \ln k + \frac{E_a}{RT} = \ln[6.09\times10^{-5}] + \frac{210000}{8.314\times552}$$

$$= -9.7062 + 45.7583 = 36.052$$

$$A = 4.542\times10^{15}\ \text{min}^{-1} = 7.75\times10^{13}\ \text{s}^{-1}$$

$$\ln\frac{k_2}{k_1} = \frac{E_a}{R}\left[\frac{T_2 - T_1}{T_1T_2}\right]$$

$$\ln k_{600} = \ln[6.09\times10^{-5}] + \frac{210000}{8.314}\left[\frac{48}{552\times600}\right]$$

$$= 3.85\times10^{-5}\ \text{s}^{-1}$$

$$k_{600} = 0.00231\ \text{min}^{-1}$$

Problem 4.16

The rate constant for the decomposition of N_2O_5 is given by $2\times10^{13}e^{\left[-102.3\ \text{kJ mol}^{-1}\right]/RT}$ s^{-1}. Estimate the (i) rate constant at 273 K and (ii) activation energy.

Solution

(i) At 273 K, the rate constant

$$k = 2\times10^{13}e^{[-102.3\times1000]/(8.314\times273)} \cong 5.33\times10^{-7}\ \text{s}^{-1}$$

(ii) Activation energy $E_a = 102.3$ kJ mol^{-1}

Problem 4.17

A homogeneous gas phase reaction is one-third complete in 19 min at 700 K and in 39 min at 690 K. What is the activation energy of the reaction?

Solution

Whatever be the order of a reaction, the rate constants and time are inversely proportional. Hence

(i) $\ln\frac{\tau_1}{\tau_2} = \frac{E_a}{R}\left(\frac{1}{T_1} - \frac{1}{T_2}\right)$

$$\ln\frac{39}{19} = \frac{E_a}{8.314\ \text{J K}^{-1}\ \text{mol}^{-1}}\left(\frac{1}{690\ \text{K}} - \frac{1}{700\ \text{K}}\right)$$

$$E_a = 293\ \text{kJ mol}^{-1}$$

Problem 4.18

The decomposition of benezene diazonium chloride is a first order reaction with the following temperature dependence of the rate constants:

T (°C)	20	33
$k_1\times10^3$ (min^{-1})	2.62	11.62

Estimate the activation energy from this data.

Solution

$$\ln\frac{11.62\times10^{-3}}{2.62\times10^{-3}}=\frac{E_a}{2}\left(\frac{13}{293\times306}\right)$$

Therefore $E_a \sim 20.55$ kcal mol^{-1}

Problem 4.19

The activation energy of a reaction is 12 kcal mol^{-1}. If the magnitude of error in the rate constant measurement was 0.6 k, estimate the % error ($\delta k/k$) in the rate constant.

Solution

Arrhenius equation for the temperature dependence of the rate constant is

$$k = Ae^{-E_a/RT}$$

$$\ln k = \ln A - \frac{E_a}{RT}$$

On differentiation of the above equation with respect to T,

$$\frac{dk}{k}=\frac{E_a}{RT}\frac{\delta T}{T}$$

If the errors are is denoted as δk and δT,

$$\frac{\partial k}{k}=\frac{12000\text{ cal mol}^{-1}}{2\text{ cal mol}^{-1}\text{ K}^{-1}\times300\text{ K}}\times\frac{0.2}{300}=0.0399$$

Hence the magnitude of the % error is 3.99.

Problem 4.20

A proposed mechanistic scheme for the decomposition of CH_3CHO is given below:

$$CH_3CHO\xrightarrow{k_1}CH_3^{\bullet}+CHO\bullet \qquad \text{Initiation}$$

$$\left.\begin{array}{l}CH_3+CH_3CHO\xrightarrow{k_2}CH_3CO\bullet+CH_4\\ CH_3CO\bullet\xrightarrow{k_3}CH_3^{\bullet}+CO\bullet\end{array}\right\}\text{Propagation}$$

$$\left.\begin{array}{l}2CH_3^{\bullet}\xrightarrow{k_4}C_2H_6\\ CHO\bullet+\text{Wall}\xrightarrow{k_5}\text{Products}\end{array}\right\}\text{Termination}$$

The rate of the reaction for the above mechanism can be shown to be

$$\frac{-d[CH_3CHO]}{dt}=k^2\left(\frac{k_1}{2k_4}\right)^{1/2}[CH_3CHO]^{3/2}$$

Write the equation for the total activation energy for the mechanism.

Solution

For the above mechanism, the effective rate constant is,

$$k_{eff}=2k\left(\frac{k_1}{2k_4}\right)^{1/2}$$

$$k_{eff}=A_{eff}e^{-E_{aeff}/RT}$$

$$\ln k_{eff}=\ln k_2+\frac{1}{2}\ln k_1-\frac{1}{2}\ln 2-\frac{1}{2}\ln k_4$$

But $\ln k_{eff}=\ln A_{eff}-\dfrac{E_{aleff}}{RT}$

$$\therefore\quad \ln A_{eff}-\frac{E_{aleff}}{RT}=\ln A_2-\frac{E_{a2}}{RT}+\frac{1}{2}\ln A_1-\frac{1}{2}\frac{E_{al}}{RT}-\frac{1}{2}\ln 2-\frac{1}{2}\left(\ln A_4-\frac{E_{a4}}{RT}\right)$$

$$=\ln A_{eff}-\frac{1}{RT}\left(\frac{E_{al}}{2}+E_{a2}-\frac{E_{a4}}{2}\right)$$

$$E_{aleff}=E_{a2}+\frac{1}{2}(E_{al}-E_{a4})$$

Steady State and Rapid Equilibrium Approximations

Problem 4.21

Derive the expression for $\dfrac{d[\text{C}]}{dt}$ for the following mechanism

$$\text{A}\rightleftharpoons\text{B} \qquad (\text{rate constants } k_1 \text{ and } k_{-1})$$

$$\text{B}\xrightarrow{k_2}\text{C}$$

using (i) rapid equilibrium assumption and (ii) steady-state approximation. Indicate the conditions under which each assumption is valid.

Solution

For the species A, B and C, the rate of formation can be written as

$$\frac{d[\text{A}]}{dt}=-k_1[\text{A}]+k_{-1}[\text{B}]$$

$$\frac{d[\text{B}]}{dt}=k_1[\text{A}]-(k_{-1}+k_2)[\text{B}]$$

$$\frac{d[\text{C}]}{dt}=-k_2[\text{B}]$$

The *rapid* equilibrium assumption implies that the forward and reverse rate constants in the equilibrium step are much larger than the rate constant associated with the decomposition of B.

(i) Rapid equilibrium case: $(k_1+k_{-1})\gg k_2$

Since $K = [\text{B}]/[\text{A}]$, it follows that $[B] = K[A]$ and hence

$$[\text{B}]=\frac{k_1}{k_{-1}}[\text{A}]$$

Thus,

$$\frac{d[\text{C}]}{dt}\approx-\frac{k_1k_2}{k_{-1}}[\text{A}]$$

(ii) The steady state assumption implies that at a given instant of time, [B] is constant. Hence

$$\frac{d[\text{B}]}{dt}=k_1[\text{A}]-(k_{-1}+k_2)[\text{B}]\approx0$$

In this case, $(k_{-1}+k_2)\gg k_1$

Thus,

$$[\text{B}]\approx\frac{k_1[\text{A}]}{(k_{-1}+k_2)}$$

$$\frac{d[\text{C}]}{dt}\approx-\frac{k_1k_2}{k_{-1}+k_2}[\text{A}]$$

Problem 4.22

Consider $2NO \rightleftharpoons N_2O_2$ (rate constants k_1 and k_{-1}). Assuming the validity of rapid equilibrium assumption, derive the expression for $\frac{dNO_2}{dt}$.

Solution

$$N_2O_2 + O_2 \xrightarrow{k_2} 2NO_2 \text{ slow}$$

$$\frac{1}{2}\frac{dNO_2}{dt} = k_2[N_2O_2][O_2]$$

But

$$K_{eq} = \frac{k_1}{k_{-1}} = \frac{[N_2O_2]}{[NO]^2}$$

$$\therefore \quad [N_2O_2] = \frac{k_1[NO]^2}{k_{-1}}$$

$$\therefore \quad \frac{1}{2}\frac{dNO_2}{dt} = \frac{k_2k_1[NO]^2[O_2]}{k_{-1}} = k_2K_{eq}[NO]^2[O_2\cdot]$$

Problem 4.23

Consider $2NO \underset{k_{-1}}{\overset{k_1}{\rightleftharpoons}} N_2O_2$ (rate constants k_1 and k_{-1})

$$N_2O_2 + H_2 \xrightarrow{k_2} N_2O + H_2O$$

Derive the expression for $[N_2O_2]$ assuming the validity of the steady state approximation.

Solution

$$\frac{d[N_2O_2]}{dt} = k_1[NO]^2 - k_{-1}[N_2O_2] - k_2[N_2O_2][H_2] = 0$$

$$\therefore\ k_1[NO]^2 = [N_2O_2]\{k_{-1} + k_2[H_2]\}$$

$$[N_2O_2] = \frac{k_1[NO]^2}{k_{-1} + k_2[H_2]}$$

Problem 4.24

The following mechanism is suggested for the thermal decomposition of CH_3CHO.

$$CH_3CHO \xrightarrow{k_1} CH_3^{\cdot} + CHO$$

$$CH_3^{\cdot} + CH_3CHO \xrightarrow{k_2} CH_4 + CH_2CHO$$

$$CH_2CHO \xrightarrow{k_3} CH_3^{\cdot} + CO$$

$$CH_3^{\cdot} + CH_3^{\cdot} \xrightarrow{k_4} CH_3CH_3$$

Applying the steady state approximation for $CH_3^{\cdot}$ and CH_2CHO, show that $\frac{d[CH_4]}{dt} = k_{obs}[CH_3CHO]^{3/2}$.

Solution

$$CH_3CHO \rightarrow CH_4 + CO$$

Applying SSA for CH_3 and CH_2CHO, we obtain

$$\frac{d[CH_3]}{dt} = k_1[CH_3CHO] - k_2[CH_3^{\cdot}][CH_3CHO] + k_3[CH_2CHO] - 2k_4[CH_3]^2 = 0 \quad (1)$$

and

$$\frac{d[CH_2CHO]}{dt} = k_2[CH_3^{\cdot}][CH_3CHO] - k_3[CH_2CHO] = 0 \quad (2)$$

From Eqs (1) and (2),

$$[CH_3] = \left(\frac{k_1[CH_3CHO]}{2k_4}\right)^{1/2}$$

$$\frac{d[CH_4]}{dt} = k_2\left(\frac{k_1[CH_3CHO]}{2k_4}\right)^{\frac{1}{2}}[CH_3CHO] = k_{obs}[CH_3CHO]^{3/2}$$

Problem 4.25

The mechanism for the formation of NO_2 is given below:

$$2NO \underset{k_2}{\overset{k_1}{\rightleftharpoons}} (NO)_2$$

$$(NO)_2 + O_2 \xrightarrow{k_3} 2NO_2$$

Derive the equation for the rate of formation of NO_2 using the steady state approximation. Indicate the condition under which the reaction will be second order.

Solution

Applying the steady state approximation to the intermediate

$$\frac{d[(NO_2)]}{dt} = k_1[NO]^2 - k_2[(NO)_2] - k_3[(NO)_2[O_2] = 0$$

Therefore $[(NO)_2] = \frac{k_1[NO]^2}{k_2 + k_3[O_2]}$

Rate of formation of $NO_2 = 2k_3[(NO)_2][O_2]$

$$= \frac{2k_1k_3[NO]^2[O_2]}{k_2 + k_3[O_2]}$$

If $k_3[O_2] >> k_2$,

$$\text{rate} \simeq \frac{2k_1k_3[NO]^2[O_2]}{2k_3[O_2]} = 2k_1[NO]^2$$

the reaction becomes second order.

Problem 4.26

Apply the steady state approximation to Michaelis–Menten mechanism and derive (i) Michaelis–Menten constant and (ii) maximum rate of the reaction.

$$E + S \underset{k_{-1}}{\overset{k_1}{\rightleftharpoons}} P + E$$

$$k_1[E][S] = k_{-1}[ES] + k_2[ES]$$

$$[E]_t = [E] + [ES]$$

$$[E] = [E]_t - [ES]$$

$$k_1([E]_t - [ES])[S] = k_{-1}[ES] + k_2[ES]$$

$$k_1[E]_t[S] - k_1[S][ES] = k_{-1}[ES] + k_2[ES]$$

$$k_1[E]_t[S] = k_{-1}[ES] + k_2[ES] + k_1[S][ES]$$

$$[\text{ES}] = \frac{k_1[\text{E}]_t[\text{S}]}{k_{-1}+k_2+k_1[\text{S}]}$$

$$R = k_2[\text{ES}] = \frac{k_1k_2[\text{E}]_t[\text{S}]}{k_{-1}+k_2+k_1[\text{S}]}$$

$$R = \frac{k_2[\text{E}]_t[\text{S}]}{\frac{k_{-1}+k_2}{k_1}+[\text{S}]}$$

$$R = \frac{k_2[\text{E}]_t[\text{S}]}{K_M+[\text{S}]}$$

where $K_M = \frac{k_{-1}+k_2}{k_1}$

when [*S*] is large compared to $K_m: R \approx \frac{k_2[\text{E}]_t[\text{S}]}{[\text{S}]} \approx k_2[\text{E}]_t$ and the rate of the reaction under these conditions is the maximum rate,

$$R_{\max} \approx k_2[\text{E}]_t$$

$$R = \frac{R_{\max}[\text{S}]}{K_M+[\text{S}]}$$

If $[\text{S}] \approx K_M$

$$R = \frac{R_{\max}[\text{S}]}{[\text{S}]+[\text{S}]} = \frac{R_{\max}}{2}$$

when $[\text{S}] \gg K_M : R \approx \frac{k_2[\text{E}]_t[\text{S}]}{[\text{S}]} \approx k_2[\text{E}]_t$

The rate of the reaction is independent of the concentration of the substrate: the reaction follows zero order kinetics.

On the other hand, for $[\text{S}] \ll K_m$: Rate $\approx \frac{R_{\max}[\text{S}]}{K_m+[\text{S}]} \approx \frac{k_2[\text{E}]_t[\text{S}]}{K_m}$

and the reaction is first order.

Parallel and Consecutive First Order Reactions

Problem 4.27

Derive the expressions for (i) A(*t*), C(*t*) and D(*t*) and (ii) the ratio of concentrations [C]/[D].

$$\text{A} \xrightarrow{k_1} \text{C}$$

$$\text{A} \xrightarrow{k_2} \text{D}$$

$$\frac{d[\text{A}]}{dt} = -k_1[\text{A}] - k_2[\text{A}] = -(k_1+k_2)[\text{A}]$$

$$[\text{A}] = [\text{A}]_0 e^{-(k_1+k_2)t}$$

$$\frac{d[\text{C}]}{dt} = k_1[\text{A}] = \frac{k_1[\text{A}]_0}{(k_1+k_2)}\left[e^{-(k_1+k_2)t}\right]$$

$$[\text{A}]+[\text{C}]+[\text{D}] = \text{A}_0$$

$$[\text{C}] = \frac{k_1[\text{A}]_0}{(k_1+k_2)}\left[e^{-(k_1+k_2)t}\right]$$

$$[\text{D}] = \frac{k_2[\text{A}]_0}{(k_1+k_2)}\left[1-e^{-(k_1+k_2)t}\right]$$

$$\frac{[\text{C}]}{[\text{D}]} = \frac{k_1}{k_2}$$

Problem 4.28

Derive the activation energy for the scheme

$$\text{A} \xrightarrow{k_1} \text{B}$$

$$\text{A} \xrightarrow{k_2} \text{C}$$

$$-\frac{dc_\text{A}}{dt} = (k_1+k_2)c_\text{A}$$

$$E_a = RT^2\frac{1}{k}\frac{dk}{dT}$$

where *k* denotes the combined rate constant $(k_1 + k_2)$. Further,

$$E_{a1} = RT^2\frac{1}{k_1}\frac{dk_1}{dT}$$

$$E_{a2} = RT^2\frac{1}{k_2}\frac{dk_2}{dT}$$

From the above two equations,

$$\frac{dk_1}{dT} = \frac{k_1E_{a1}}{RT^2}$$

and

$$\frac{dk2}{dT} = \frac{k_2E_{a2}}{RT^2}$$

$$E_{a,obs} = RT^2\frac{1}{(k_1+k_2)}\frac{d(k_1+k_2)}{dT}$$

$$= \frac{RT^2}{(k_1+k_2)}\left[\frac{k_1E_{a_1}}{RT^2}+\frac{k_2E_{a_2}}{RT^2}\right]$$

Hence $E_{a,obs} = (k_1 E_{a1} + k_2E_{a2})/(k_1 + k_2)$

Consecutive First Order Reactions

Problem 4.29

Derive the concentrations of A, B and C as a function of time for the following consecutive first order reaction.

$$\text{A} \xrightarrow{k_1} \text{B} \xrightarrow{k_2} \text{C}$$

$$\frac{d\text{A}}{dt} = -k_1[\text{A}] \quad (1)$$

$$\frac{d\text{B}}{dt} = k_1[\text{A}] - k_2[\text{B}] \quad (2)$$

$$\frac{d\text{C}}{dt} = k_2[\text{B}] \quad (3)$$

from Eq. (1)

$$[\text{A}] = [\text{A}]_0 e^{-k_1t}$$

Substituting in Eq. (2)

$$\frac{d\text{B}}{dt} = k_1[\text{A}]_0 e^{-k_1t} - k_2[\text{B}]$$

$$\frac{dB}{dt}+k_2[B]=k_1[A]_0 e^{-k_1 t}$$

The above is a first order linear differential equation and is easy to solve for [B] as

$$[B]=\frac{k_1[A]_0}{k_2-k_1}\left(e^{-k_1 t}-e^{-k_2 t}\right)$$

Constancy of total initial concentration:

$$[A]+[B]+[C]=[A]_0$$

$$[C]=[A]_0\left[1-\frac{k_2}{k_2-k_1}e^{-k_1 t}+\frac{k_1}{k_2-k_1}e^{-k_2 t}\right]$$

Problem 4.30

Consider the consecutive reaction scheme

$$A\xrightarrow{k_1}B\xrightarrow{k_2}C$$

with $k_1 = 1.9\times10^{-3}\ s^{-1}$; $k_2 = 6.1\times10^{-5}\ s^{-1}$ at $A_0 = 0.02\ mol^{-1}$ at 0°C. Estimate the maximum concentration of B and the corresponding time.

Solution

$$t_{\max}=\frac{\ln\left(k_1/k_2\right)}{k_1-k_2}=1869.88\ s$$

$$y_{max}=\left(k_1/k_2\right)^{\frac{k_2}{k_2-k_1}}=(31.14)^{-0.033}=0.892$$

$$B_{\max}=0.892\times0.02=0.018\ mol\ L^{-1}$$

Theories of Reaction Rates

Problem 4.31

For a reaction the rate constant is $1.18\times10^{-3}\ s^{-1}$ at 300 K. If the energy of activation is 127.5 kJ mol^{-1}, estimate the following quantities: (i) A (ii) $\Delta S^\ddagger$ (iii) $K^\ddagger_{eq}$ at 300 K.

Solution

(i) $k=Ae^{-Ea/RT}$ or $A=ke^{Ea/RT}$

$$=(1.18\times10^{-3}\ s^{-1})\left(e^{\left(\frac{127.500}{8.314\times300}\right)}\right)$$

$$A=1.18\times10^{-3}\times1.589\times10^{22}=1.875\times10^{19}\ s^{-1}$$

(ii) $A=\frac{kT}{h}e^{\frac{\Delta S^\ddagger}{R}}$

$$e^{\frac{\Delta S^\ddagger}{R}}=\frac{Ah}{kT}=\frac{1.875\times10^{19}\times6.626\times10^{-34}}{1.381\times10^{-23}\times300}$$

$$=299.87\times10^4$$

$$\frac{\Delta S^\ddagger}{R}=\ln(299.87\times10^4)=14.91$$

$$\text{or } \Delta S^\ddagger=14.91\times8.314=123.99\ J\ K^{-1}$$

(iii) Rate constant $(k)=\frac{kT}{h}K^\ddagger$

$$K^\ddagger=\frac{\text{Rate constant}\times h}{kT}=\frac{1.18\times10^{-3}\times6.626\times10^{-34}}{1.381\times10^{-23}\times300}$$

$$=1.887\times10^{-16}$$

Problem 4.32

For a first order reaction, the pre-exponential factor is $4.24\times10^6\ s^{-1}$ at 100°C, At 100°C, evaluate (i) the entropy of activation and (ii) equilibrium constant if the rate constant is $2.36\times10^{-4}\ s^{-1}$.

Solution

(i) $A=\frac{k_B T}{h}e^{-\Delta S^\ddagger/R}$

$$e^{-\Delta S^\ddagger/R}=\frac{Ah}{kT}=\frac{4.240\times10^6\ s^{-1}\times6.626\times10^{-34}\ J\ s}{1.38\times10^{-23}\ J\ K^{-1}\ 373\ K}$$

$$=5.457\times10^{-7}$$

$$\Delta S^\ddagger/R=-14.42$$

(ii) $k=\frac{k_B T}{h}K^\ddagger_{eq}$

$$K^\ddagger_{eq}=\frac{2.36\times10^{-4}\times6.626\times10^{-34}}{1.38\times10^{-23}\times373}=3.037\times10^{-17}$$

Problem 4.33

The entropy of activation for a bimolecular gas phase reaction is –13.84 R. Estimate the pre-exponential factor at 298 K.

Solution

Since the reaction is bimolecular, the pre-exponential factor includes a concentration term c,

$$A=\frac{ek_B T}{ch}e^{\Delta S/R}\quad \text{where } c=1\ mol\ L^{-1}$$

$$e^{\Delta S/R}=e^{-13.84}=9.758\times10^{-17}=\frac{Ahc}{ekT}$$

$$=\frac{A\times6.626\times10^{-34}\ J\ s\times1\ mol\ L^{-1}}{2.718\times1.38\times10^{-23}\ J\ K^{-1}\times373\ K}$$

$$A=1.65\times10^7\ L\ mol^{-1}\ s^{-1}$$

Problem 4.34

Calculate the collision number Z_{AA} for the decomposition of HI at 598 K at a concentration of 1 mol L^{-1}. The hard sphere diameter of HI is 3.4 Å.

Solution

$$Z_{AA}=2N^2\sigma^2\left(\frac{\pi RT}{M}\right)^{1/2}$$

$$1\ mol\ L^{-1}=6.023\times10^{23}\ \text{molecules}\ L^{-1}$$

$$=6.023\times10^{20}\ \text{molecules cm}^{-3}$$

$$Z_{AA}=2\times(6.023\times10^{20}\ \text{molecules cm}^{-3})^2$$

$$\times(3.4\times10^{-10}\ m)^2\times\left(\frac{3.14\times8.314\ kg\ m^2\ s^{-2}\times598\ K}{128\ g}\right)^{0.5}$$

$$\simeq 10^{31}\ \text{molecules cm}^{-3}$$

Problem 4.35

The rate constants for a reaction which can proceed via two alternative mechanisms are k_a and k_c respectively. If ($\Delta S_a^\ddagger$ – $\Delta S_c^\ddagger$) = 42.84 J K^{-1} and ($\Delta H_a^\ddagger$ – ‡) = 21.92 kJ obtain k_a/k_c at 298 K.

Solution

$$k = \frac{k_B T}{h} e^{\frac{-\Delta H^\ddagger}{RT}} e^{\frac{\Delta S^\ddagger}{R}}$$

Therefore $$\frac{k_a}{k_c} = e^{\frac{-\left(\Delta H_a^\ddagger - \Delta H_c^\ddagger\right)}{RT}} e^{\frac{\left(\Delta S_a^\ddagger - \Delta S_c^\ddagger\right)}{R}}$$

$$= e^{-(21920)/(8.314 \times 298)} e^{(42.84/8.314)}$$

$$= 0.025$$

Problem 4.36

At 300°C at a concentration of 1 mol dm^{-3}, the number of molecules of HI colliding per second is 6×10^{31} cm^{-3}. The activation energy of the reaction is 200 kJ. Calculate the rate constant in molecules cm^{-3} s^{-1}.

Solution

The reaction is

$$2HI \rightarrow H_2 + I_2$$

The rate constant of a bimolecular reaction is

$$k = Z\, e^{-\frac{Ea}{RT}}$$

$$= 6 \times 10^{31} . e^{-\frac{200000}{8.314 \times 573}} = 3.51 \times 10^{13} \text{ molecules cm}^{-3} \text{ s}^{-1}$$

Problem 4.37

The entropy and enthalpy of activation ($\Delta S^\ddagger$ and $\Delta H^\ddagger$) for a reaction at 300 K are –121.47 J and 53.55 kJ mol^{-1}, respectively. What is the value of the rate constant at 300 K?

Solution

$$k_2 = \left(\frac{kT}{h}\right) e^{\frac{\Delta S^\ddagger}{R}} e^{\frac{-\Delta H^\ddagger}{RT}}$$

$$\ln k_2 = \ln \frac{kT}{h} + \frac{\Delta S^\ddagger}{R} - \frac{\Delta H^\ddagger}{RT}$$

$$= \ln \frac{1.38 \times 10^{-23} \times 300}{6.625 \times 10^{-34}} - \frac{121.47}{8.314} - \frac{53550}{8.314 \times 300}$$

or $k_2 = 1.311 \times 10^{-3}$ s^{-1}

Problem 4.38

A first order reaction has a rate constant of 0.004 s^{-1} at 400 K and an activation energy (E_a) of 98 kJ mol^{-1}. Calculate A, $\Delta H^\ddagger$, $\Delta S^\ddagger$, $\Delta G^\ddagger$ and $K^\ddagger$.

Solution

$$\ln A = \ln 0.004 + \frac{98000}{8.314 \times 400}$$

Therefore $A = 2.833 \times 10^{10}$ s^{-1}

$$E_a = \Delta H^\ddagger + RT - \Delta n^\ddagger RT$$

$$\Delta n^\ddagger = 0$$

$$E_a = \Delta H^\ddagger + RT$$

$$A = e\left(\frac{kT}{h}\right) e^{\frac{\Delta S^\ddagger}{R}}$$

or $$e^{\frac{\Delta S^\ddagger}{R}} = \frac{Ah}{ekT}$$

or $$e^{\frac{\Delta S^\ddagger}{R}} = \frac{2.833 \times 10^{10} \times 6.625 \times 10^{-34}}{2.781 \times 1.38 \times 10^{-23} \times 400}$$

Therefore $$\frac{\Delta S^\ddagger}{R} = -6.706$$

$$E_a = \Delta H^\ddagger + RT = E_a - RT$$

$$\Delta H^\ddagger = 98000 - 8.314 \times 400$$

$$\Delta H^\ddagger = 94.67 \text{ kJ mol}^{-1}$$

Since $\Delta G^\ddagger = \Delta H^\ddagger - T\Delta S^\ddagger$

$$\Delta G^\ddagger = 94670 - (400 \times -56.79) = 117.39 \text{ kJ} = -RT \ln K^\ddagger$$

$$117390 = -8.314 \times 400 \times \ln K^\ddagger$$

$$K^\ddagger = 4.67 \times 10^{-16}$$

Problem 4.39

Calculate (i) the pre-exponential factor and (ii) the rate constant for a first order gas phase reaction at 500 K, given that $\Delta G^\ddagger = 146$ kJ mol^{-1} and $\Delta H^\ddagger = 102$ kJ mol^{-1}.

Solution

$$\Delta G^\ddagger = \Delta H^\ddagger - T\Delta S^\ddagger$$

Therefore $$\Delta S^\ddagger = \frac{\Delta H^\ddagger - \Delta G^\ddagger}{T} = \frac{(102-146) \text{ kJ K}^{-1}}{500}$$

or $\Delta S^\ddagger = -88$ J K^{-1} mol^{-1}

$$A = e\left(\frac{kT}{h}\right) e^{\frac{\Delta S^\ddagger}{R}}$$

$$= 2.718\left(\frac{1.38 \times 10^{-23} \times 500}{6.626 \times 10^{-34}}\right) e^{\left(-\frac{88}{8.314}\right)}$$

or $A = 7.34 \times 10^8$ s^{-1}

$$k_2 = \left(\frac{kT}{h}\right) K^\ddagger = \left(\frac{kT}{h}\right) e^{-\frac{\Delta G^\ddagger}{RT}}$$

Therefore $$k_2 = \left(\frac{1.38 \times 10^{-23} \times 500}{6.626 \times 10^{-34}}\right) e^{\left(-\frac{146000}{8.314 \times 500}\right)}$$

$$k_2 = 5.82 \times 10^{-3} \text{ s}^{-1}$$

Since the order of magnitude of A is predominant, the term 'e' is sometimes ignored.

Problem 4.40

Write the pre-exponential factor A in terms of the partition functions for the bimolecular reaction:

$$A + B \underset{k_{-1}}{\overset{k_1}{\rightleftharpoons}} [AB]^\ddagger$$

$$[AB]^{\ddagger} \xrightarrow{k_2} \text{Products}$$

$$A = \frac{k_B T}{h} \frac{Q[AB]^{\ddagger}}{Q_A Q_B}$$

Problem 4.41

Write the equation for the rate of $\left(\frac{-dc_A}{dt}\right)$ of a bimolecular reaction A + B → Products using collision theory, assuming the steric factor to be unity.

Solution

$$\frac{-dc_A}{dt} = \sigma_{AB}\langle \upsilon_{AB}\rangle c_A c_B$$

Problem 4.42

The collision cross-section of a bimolecular reaction is 33×10^{-16} cm^2 and mean velocity is 790×10^{-2} cm s^{-1}. Estimate the pre-exponential factor in L mol^{-1} s^{-1} if the steric factor (p) is 0.5.

Solution

$$A = p \times \sigma_{AB}\langle \upsilon_{AB}\rangle N_A$$

$$= 0.5\left(33 \times 10^{-20}\ \text{m}^2\right) 790 \times 10^{-4}\ \text{m s}^{-1} \times 6.023 \times 10^{23}\ \text{mol}^{-1}$$

$$= 7.85 \times 10^{6}\ \text{L mol}^{-1}\ \text{s}^{-1}$$

where σ_{AB} denotes the collision diameter and v_{AB} is the mean velocity.

Problem 4.43

Estimate the diffusion controlled rate constant k_D (in L mol^{-1} s^{-1}) for a bimolecular reaction when the collision diameter is 10 Å and the mutual diffusion coefficient is 10^{-6} cm^2 s^{-1}.

Solution

$$k_D = 4\pi R_d D_{eff}$$

$$= 4\pi \times 5 \times 10^{-8}\ \text{cm} \times 10^{-6}\ \text{cm}^2\ \text{s}^{-1}$$

$$= 6.28 \times 10^{-13}\ \text{cm}^3\ \text{s}^{-1}$$

$$= 6.28 \times 10^{-16}\ \text{L mol}^{-1}\ \text{s}^{-1}$$

Problem 4.44

Write the equation for the pre-exponential factor A for a reaction of molecularity 3, using the transition state theory.

Solution

$$A = \frac{e^3 kT}{h(C_o)^2} e^{\Delta s^{\ddagger}/R}$$

Effect of Ionic Strength on Chemical Reactions

Problem 4.45

Consider the following substitution reaction:

$$[CoBr(NH_3)_5]^{2+} + OH^- \rightarrow [Co(OH)(NH_3)_5]^{2+} + Br^-$$

The rate constant when the ionic strength → 0, was estimated to be 20.9 L mol^{-1} s^{-1} at 25°C. Predict the rate constant when the ionic strength is 0.053 mol L^{-1}.

$$\log k = \log(20.9) + (2)(0.505)(-2)\sqrt{0.053}$$

$$= 1.32 - 0.4687 = 0.8513$$

$$k = 7.103\ \text{L mol}^{-1}\ \text{s}^{-1}$$

Marcus Cross-Exchange Relations

Problem 4.46

Consider the following self-exchange electron transfer processes:

$$[Fe(CN)_6]^{3-} + [Fe(CN)_6]^{4-} \rightarrow$$
$$[Fe(CN)_6]^{4-} + [Fe(CN)_6]^{3-}\ (k_{11} = 3 \times 10^2\ \text{M}^{-1}\ \text{s}^{-1})$$

$$[MnO_4]^- + [MnO_4]^{2-} \rightarrow$$
$$[MnO_4]^{2-} + [MnO_4]^-\ (k_{22} = 3.6 \times 10^3\ \text{M}^{-1}\ \text{s}^{-1})$$

Using the simplified Marcus relation with $f = 1$, estimate the cross-exchange electron transfer rate constant for the reaction,

$$[Fe(CN)_6]^{4-} + [MnO_4]^- \rightarrow [Fe(CN)_6]^{3-} + [MnO_4]^{2-}$$

The equilibrium constant of the above reaction has been estimated as 2400.

Solution

$$k_{12} = (k_{11} \times k_{22} \times k_{12})^{½}$$

$$k_{12} = (3 \times 10^2\ \text{M}^{-1}\text{s}^{-1} \times 3.6 \times 10^3\ \text{M}^{-1}\ \text{s}^{-1} \times 2400)^{½}$$

$$= 5.09 \times 10^4\ \text{M}^{-1}\ \text{s}^{-1}$$

Relaxation Methods

Problem 4.47

Consider the following mechanism

$$X_1 + X_2 \underset{k_1'}{\overset{k_1}{\rightleftharpoons}} X_3 \underset{k_2'}{\overset{k_2}{\rightleftharpoons}} X_4$$

When the first equilibrium is rapidly established, only one relaxation time is observed. Derive an equation for the relaxation time, τ, of the second step under these conditions.

Solution

When the first equilibrium is rapidly established, its equilibrium constant K is given by

$$K = \frac{\bar{C}_{X_3}}{\bar{C}_{X_1} \times \bar{C}_{X_2}} = \frac{k_1}{k_1'} \quad (1)$$

Taking logarithms,

$$\log K = \log \bar{C}_{X_3} - \log \bar{C}_{X_1} - \log \bar{C}_{X_2}$$

which on differentiation gives

$$\partial \log K = 0 = \partial \log \bar{C}_{X_3} - \partial \log \bar{C}_{X_1} - \partial \log \bar{C}_{X_2}$$

$$= \frac{\Delta C_{X_3}}{\bar{C}_{X_3}} - \frac{\Delta C_{X_1}}{\bar{C}_{X_1}} - \frac{\Delta C_{X_2}}{\bar{C}_{X_2}} \quad (2)$$

The rate of formation of species X_4 is given by

$$\frac{dC_{X_4}}{dt} = k_2 C_{X_3} - k_2' C_{X_4} \quad (3)$$

But

$$C_{X_3} = \bar{C}'_{X_3} + \Delta C_{X_3} \text{ and } C_{X_4} = \bar{C}'_{X_4} + \Delta C_{X_4}$$

Substituting the above in Eq. (3),

$$\frac{d\left(\bar{C}_{X_4}+\Delta C_{X_4}\right)}{dt}=\frac{d\left(\Delta C_{X_4}\right)}{dt}$$

$$=k_2\left(\bar{C}'_{X_3}+\Delta C_{X_3}\right)-k'_2\left(\bar{C}'_{X_4}+\Delta C_{X_4}\right) \quad (4)$$

Employing the condition of equilibrium

$$k_2\bar{C}'_{X_3}=k'_2\bar{C}'_{X_4} \quad (4a)$$

Equation (4) may be written as

$$\frac{d\left(\Delta C_{X_4}\right)}{dt}=k_2\frac{\Delta C_{X_3}}{\Delta C_{X_4}}\Delta C_{X_4}-k'_2\Delta C_{X_4} \quad (5)$$

From stoichiometry and balance of mass,

Total concentration of reactant $= C_{X_1}+C_{X_3}+C_{X_4}$

On differentiation of this equation

$$\Delta C_{X_1}+\Delta C_{X_3}+\Delta C_{X_4}=0$$

$$\Delta C_{X_4}=-\left(\Delta C_{X_1}+\Delta C_{X_3}\right) \quad (6)$$

Substituting Eq. (6) in Eq. (5) gives,

$$\frac{d\left(\Delta C_{X_4}\right)}{dt}=-k_2\frac{\Delta C_{X_3}}{\Delta C_{X_1}+\Delta C_{X_3}}\Delta C_{X_4}-k'_2\Delta C_{X_4} \quad (7)$$

From Eq. (2),

$$\frac{\Delta C_{X_3}}{\bar{C}_{X_3}}=\frac{\Delta C_{X_1}}{\bar{C}_{X_1}}=\frac{\Delta C_{X_2}}{\bar{C}_{X_2}} \quad (8)$$

$$=\Delta C_{X_1}\left(\frac{1}{\bar{C}_{X_1}}+\frac{1}{\bar{C}_{X_2}}\right) \quad \left(\text{since } \Delta C_{X_1}=\Delta C_{X_2}\right) \quad (9)$$

$$\text{or, } \Delta C_{X_3}=\Delta C_{X_1}\left[\frac{\left(\bar{C}_{X_1}+\bar{C}_{X_2}\right)}{\bar{C}_{X_1}\times\bar{C}_{X_2}}\right]\bar{C}_{X_3}$$

$$=\Delta C_{X_1}\left(\bar{C}_{X_1}+\bar{C}_{X_2}\right)K \quad (10)$$

$$\text{or, } \quad \frac{\Delta C_{X_1}}{\Delta C_{X_3}}=\frac{1}{K\left(\bar{C}_{X_1}+\bar{C}_{X_2}\right)} \quad (11)$$

From Eq. (7),

$$\frac{\Delta C_{X_3}}{\Delta C_{X_1}+\Delta C_{X_3}}=\frac{1}{1+\dfrac{\Delta C_{X_1}}{\Delta C_{X_3}}}=\frac{1}{1+\dfrac{1}{K\left(\bar{C}_{X_1}+\bar{C}_{X_2}\right)}}$$

$$=\frac{K\left(\bar{C}_{X_1}+\bar{C}_{X_2}\right)}{1+K\left(\bar{C}_{X_1}+\bar{C}_{X_2}\right)} \quad (12)$$

Substituting Eq. (12) into Eq. (7), we get

$$\frac{d\left(\Delta C_{X_4}\right)}{dt}=-k_2\frac{K\left(\bar{C}_{X_1}+\bar{C}_{X_2}\right)}{1+K\left(\bar{C}_{X_1}+\bar{C}_{X_2}\right)}\Delta C_{X_4}-k'_2\Delta C_{X_4} \quad (13)$$

$$=-\Delta C_{X_4}\left[k'_2+k_2\frac{K\left(\bar{C}_{X_1}+\bar{C}_{X_2}\right)}{1+K\left(\bar{C}_{X_1}+\bar{C}_{X_2}\right)}\right]$$

Put

$$\frac{1}{\tau}=\left[k'_2+k_2\frac{K\left(\bar{C}_{X_1}+\bar{C}_{X_2}\right)}{1+K\left(\bar{C}_{X_1}+\bar{C}_{X_2}\right)}\right] \quad (14)$$

Then, Eq. (13) becomes,

$$\frac{d\left(\Delta C_{X_4}\right)}{dt}=-\frac{1}{\tau}\Delta C_{X_4}$$

$$\text{or, } \quad \Delta C_{X_4}=\Delta\bar{C}_{X_4}e^{-t/\tau}$$

where τ has the significance in Eq. (14).

Problem 4.48

Relaxation studies on beryllium sulphate solution employing the P-jump technique at various concentrations of the salt gave the following relaxation times.

Conc. of $BeSO_4$ (mol L^{-1})	$10^3\times\tau$ (s)
0.001	2.86
0.002	2.70
0.0034	2.38
0.0040	2.27
0.0058	2.10
0.0078	2.04
0.0088	2.02

Assuming that the reaction proceeds via the two-step process,

$$\underset{\text{(Free ions)}}{Be^{2+}+SO_4^{2-}} \underset{k'_1}{\overset{k_1}{\rightleftharpoons}} \underset{\substack{\text{Solvent separated ion pair}\\ \text{(also known as outer sphere complex)}}}{Be^{2+}H_2OSO_4^{2-}}$$

$$\underset{k'_2}{\overset{k_2}{\rightleftharpoons}} \underset{\substack{\text{Ion pair (also known as}\\ \text{inner sphere complex)}}}{Be^{2+}SO_4^{2-}}$$

Using the equation for relaxation time obtained in the previous problem and rewriting,

$$\frac{1}{\tau}=\left[k'_2+k_2\frac{K\left(\bar{C}_{Be^{2+}}+\bar{C}_{SO_4^{2-}}\right)}{1+K\left(\bar{C}_{Be^{2+}}+\bar{C}_{SO_4^{2-}}\right)}\right]$$

with $K=\dfrac{\bar{C}_{Be^{2+}H_2OSO_4^{2-}}}{\bar{C}_{Be^{2+}}\bar{C}_{SO_4^{2-}}}=100\ M^{-1}$, determine k_2, k'_2 and K_i, the equilibrium constant of the second step of the reaction. Plot $\dfrac{1}{\tau}$ vs $\left(\bar{C}_{Be^{2+}}+\bar{C}_{SO_4^{2-}}\right)$ and employ the conditions to simplify the expression for $\dfrac{1}{\tau}$, and obtain the required results.

Solution

Rewriting the equation for $\dfrac{1}{\tau}$ as

$$\frac{1}{\tau}=k_2\left[\frac{1}{K_i}+\frac{K\left(\bar{C}_{Be^{2+}}+\bar{C}_{SO_4^{2-}}\right)}{1+K\left(\bar{C}_{Be^{2+}}+\bar{C}_{SO_4^{2-}}\right)}\right]$$

where $\quad K_i=\dfrac{k_2}{k'_2}$, we tabulate $\dfrac{1}{\tau}$ vs $\left[\bar{C}_{Be^{2+}}+\bar{C}_{SO_4^{2-}}\right]$

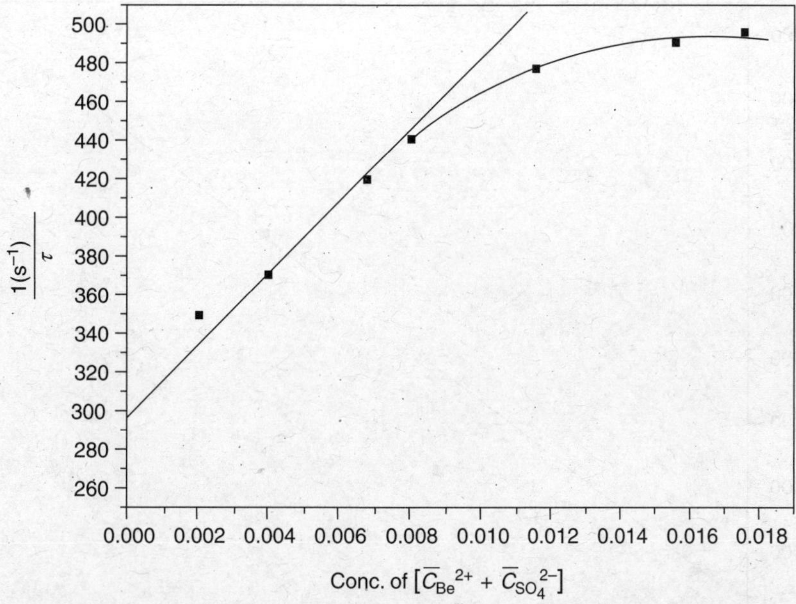

Estimation of the relaxation time using the pressure jump technique

Conc. of $\left[\bar{C}_{Be^{2+}} + \bar{C}_{SO_4^{2-}}\right]$ (mol L^{-1})	$\frac{1}{\tau}$ (s^{-1})
0.002	349.6
0.004	370.4
0.0068	420.0
0.0080	440.5
0.0116	476.2
0.0156	490.2
0.0176	495.0

A plot of the above data $\frac{1}{\tau}$ (y axis) and $\left[\bar{C}_{Be^{2+}} + \bar{C}_{SO_4^{2-}}\right]$ (x axis) gives us a graph which is linear at low concentrations (under conditions when $K\left[\bar{C}_{Be^{2+}} + \bar{C}_{SO_4^{2-}}\right]$ is << 1 in the denominator). Under these conditions,

$$\frac{1}{\tau} = k_2\left[\frac{1}{K_i} + K\left(\bar{C}_{Be^{2+}} + \bar{C}_{SO_4^{2-}}\right)\right]$$

From this linear plot, the slope is k_2K and intercept is $\frac{k_2}{K_i} = k_2'$. At high concentrations of the salt, that is, when $K\left(\bar{C}_{Be^{2+}} + \bar{C}_{SO_4^{2-}}\right) \gg 1$.

$$\frac{1}{\tau} \text{ reduces to } \frac{1}{\tau} = k_2' + k_2$$

Thus $\frac{1}{\tau}$ is independent of concentration and the plot becomes asymptotic. Indeed we find such a graph when $\frac{1}{\tau}$ is plotted $\left(\bar{C}_{Be^{2+}} + \bar{C}_{SO_4^{2-}}\right)$. From the graph, the asymptotic value, $k_2' + k_2 = 495$ and intercept $= k_2' = 295\ s^{-1}$.

$$k_2 = 495 - 295 = 200\ s^{-1}$$

$$K_i = \frac{k_2}{k_2'} = \frac{200}{295} = 0.68\ s^{-1}$$

Problem 4.49

The kinetics of the reaction of an indicator anion and hydroxyl ions was investigated by a T-jump method over a certain pH range. The following relaxation times were obtained at various hydroxyl ions concentrations keeping the indicator anion concentration at 1×10^{-4} M, the reaction being,

$$HIn^- + OH^- \underset{k_1'}{\overset{k_1}{\rightleftharpoons}} In^{2-} + H_2O$$

Conc. of [OH$^-$] (mol L^{-1})	τ (μs)
2×10^{-4}	23.0
3.2×10^{-4}	21.5
4.0×10^{-4}	18.6
5.0×10^{-4}	17.0

Derive an equation for $1/\tau$ of the reaction. Draw a suitable graph and determine k_1 and k_1' from the same.

Solution

Following the usual procedure, the equation for $1/\tau$ may be obtained as

$$\frac{1}{\tau} = k_1\left(\bar{C}_{HIn^-} + \bar{C}_{OH^-}\right) + k_1'$$

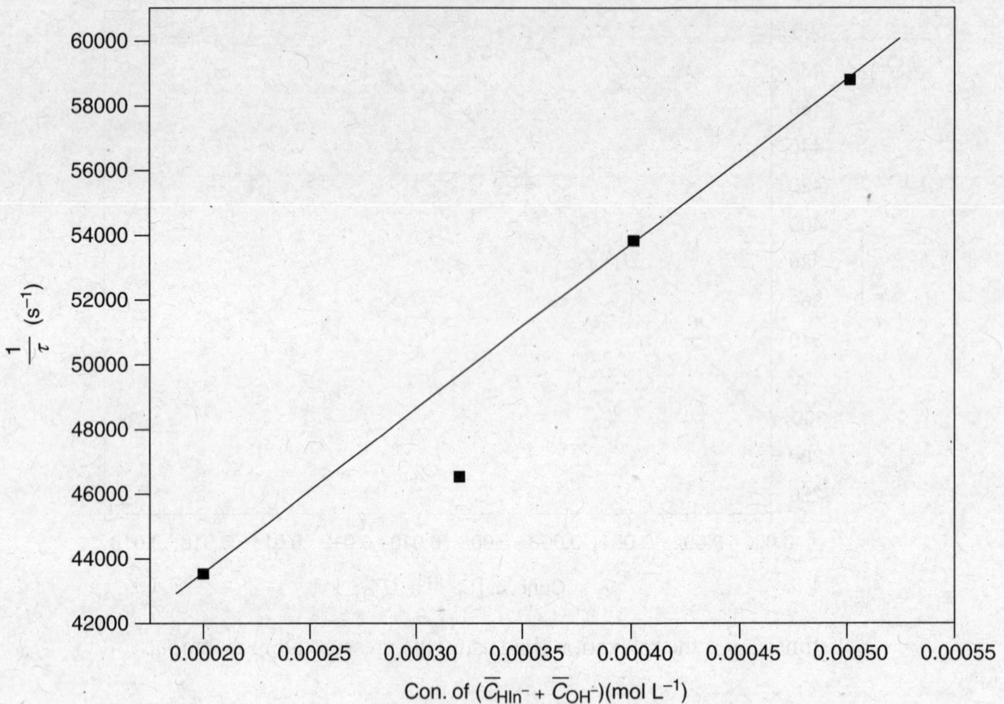

Estimation of the relaxation time using the temperature jump method

$\frac{1}{\tau}$ (s^{-1})	Conc. of $\left(\bar{C}_{HIn^-} + \bar{C}_{OH^-}\right)$ (mol L^{-1})
4.35×10^4	2×10^{-4}
4.65×10^4	3.2×10^{-4}
5.38×10^4	4.0×10^{-4}
5.88×10^4	5.0×10^{-4}

From the linear plot obtained, intercept,

$$k_1' = 2.5\times10^4\ M^{-1}\ s^{-1}$$

$$\text{Slope } k_1 = 4.8\times10^7\ M^{-1}\ s^{-1}$$

$$\text{Equilibrium constant} = \frac{4.8\times10^7}{2.5\times10^4} 1.7\times10^3$$

Problem 4.50

The decomposition of hydrogen peroxide to water and oxygen according to $H_2O_2 \rightarrow H_2O + \frac{1}{2}O_2$ was studied by a stopped flow technique at 298 K in the presence of an enzyme, catalase, as catalyst. The first order rate constant of the reaction is $2.16\times10^{-1}\ s^{-1}$. If the initial concentration of the peroxide is 4.0×10^{-4} M, calculate what percentage of peroxide has reacted at the end of 5 and 10 s.

Solution

$$k = \frac{2.303}{t}\log\frac{a}{a-x}$$

At 5 s, $$\frac{0.216\times5}{2.303} = 0.469 = \log\left(\frac{4\times10^{-4}}{4\times10^{-4}-x}\right)$$

or $$\left(\frac{4\times10^{-4}}{4\times10^{-4}-x}\right) = 2.944 \text{ from which } x = 2.64\times10^{-4}$$

$$\%\text{ of peroxide reacted} = \frac{2.64\times10^{-4}}{4\times10^{-4}}\times100 = 66$$

At 10 s, $$\frac{0.216\times10}{2.303} = 0.94 = \log\left(\frac{4\times10^{-4}}{4\times10^{-4}-x}\right)$$

or $$\left(\frac{4\times10^{-4}}{4\times10^{-4}-x}\right) = 8.71$$

$$\left(\frac{4\times10^{-4}-x}{4\times10^{-4}}\right) = 0.115;\ 1-\frac{x}{4\times10^{-4}} = 0.115$$

$$x = 3.54\times10^{-4}$$

$$\%\text{ of peroxide reacted} = \frac{3.54\times10^{-4}}{4\times10^{-4}}\times100 = 88.5$$

EXERCISES

Practice Problems

1. Derive (a) the rate constant for a reaction whose order is ½, (b) the ratio $t_{4/5}/t_{1/2}$ for a zero order reaction.

2. At 600 K, the reaction $2NO_2 \rightarrow 2NO + O_2$ was studied. 10% of the sample decomposed at 400 mmHg in 16.5 seconds. Estimate the second order rate constant.

3. In isoamyl alcohol, benzene diazonium chloride decomposed to nitrogen and other products by a first order reaction and the following kinetic data were obtained at different temperatures:

T (°C)	20	30	40	50
$k_1 \times 10^3$ (min^{-1})	2.61	11.60	47.50	192.00

(a) What is the activation energy of this reaction in kJ mol^{-1}?
(b) What is the half-life ($t_{1/2}$) of the salt in this solution at 273 K?

4. For the reaction $2NO + 2H_2 \rightarrow N_2 + 2H_2O$, the following data were obtained:

NO (M)	H_2 (M)	Rate, R (M min^{-1})
1×10^{-2}	1×10^{-3}	4.8×10^{-5}
3×10^{-2}	1×10^{-3}	43.2×10^{-5}
3×10^{-2}	2×10^{-3}	86.4×10^{-5}

Deduce the rate law for the reaction.

5. The following values were obtained for a first order reaction:

Temp (K)	0	298	318	338
k (s)	7.83×10^{-7}	3.33×10^{-5}	5.0×10^{-4}	5.0×10^{-3}

Calculate (a) the energy of activation, (b) enthalpy of activation, (c) entropy of activation, (d) the frequency factor.

6. For a first order reaction, the pre-exponential factor is 4.32×10^6 s^{-1} at 100°C. The energy of activation for the reaction is estimated to be 16.5 kcal mol^{-1}. Evaluate the (a) entropy of activation and (b) equilibrium constant at 100°C.

7. The pre-exponential factor for the tri-molecular reaction $2NO + O_2 \rightarrow 2NO_2$ is 10^9 cm^6 mol^{-2} s^{-1}. What is the value in L^2 mol^{-2} s^{-1} and cm^6 s^{-1}?

8. The bimolecular decomposition of HI is given by

$$2HI \rightarrow H_2 + I_2$$

Assuming a collision diameter of 0.35 nm for HI and an activation energy of 184 kJ mol^{-1}, calculate (a) the collision rate, (b) the reaction velocity and (c) the rate constant at 700 K and 1 atm pressure. (mol wt of HI = 127.9 g mol^{-1}, I_2 = 253.8 g mol^{-1})

9. Calculate the number of collisions per second in 1 cm^3 of nitrogen at 0°C and 1 atm pressure if the radius of N_2 molecule is 1.89 Å.

10. The kinetics of a reaction was studied at 298 and 308 K. The free energy of activation $\Delta G^\ddagger$ was 75.24 kJ mol^{-1} at 298 K and 76.08 kJ mol^{-1} at 308 K. Evaluate $\Delta H^\ddagger$ and $\Delta S^\ddagger$.

11. Consider the following reaction

$$A + OH^{-1} \underset{k_2}{\overset{k_1}{\rightleftharpoons}} B + H_2O$$

$$B + Br_2 \xrightarrow{k_3} P + Br^-$$

Derive the rate expression for the formation of *P* by applying the steady state approximation to B and deduce the limiting cases.

12. For a reaction the rate constant is 1.16×10^{-3} s^{-1} and the energy of activation is 127.5 kJ mol^{-1}. Estimate the following quantities: (a) *A* (b) $\Delta S^\ddagger$ (c) $K_{eq}^\ddagger$ at 300 K.

13. The mechanism for the formation of NO_2 is given below:

$$2NO \underset{k_2}{\overset{k_1}{\rightleftharpoons}} (NO)_2$$

$$(NO)_2 + O_2 \xrightarrow{k_3} 2NO_2$$

Derive the rate law for the formation of NO_2 using the steady state approximation.

14. Assume that the transformation of A to B occurs by both reversible first order and a reversible second order reaction involving hydrogen ions:

$$A \underset{k_2}{\overset{k_1}{\rightleftharpoons}} B$$

$$A + H^+ \underset{k_4}{\overset{k_3}{\rightleftharpoons}} B + H^+$$

What is the relationship among these four rate constants?

15. Consider the reaction scheme

$$A + B \underset{k_2}{\overset{k_1}{\rightleftharpoons}} C$$

$$C \xrightarrow{k_3} D$$

Derive the rate law using the steady state approximation for C.

16. Consider the consecutive reaction scheme $A \xrightarrow{k_1} B \xrightarrow{k_2} C$. Plot the time dependence of A, B and C when $k_1 = 2k_2$.

17. For a reversible first order reaction $A \underset{k_2}{\overset{k_1}{\rightleftharpoons}} B$, $k_1 = 10^{-2}\ s^{-1}$ and $B_{eq}/A_{eq} = 4$. If A_0 is 0.01 mol L^{-1} and $B_0 = 0$, estimate the concentration of B after 60 s.

18. The observed frequency factor for the reaction $Cl(g) + H_2(g) \rightarrow HCl(g) + H(g)$ is $1.2 \times 10^{10}\ s^{-1}$. If M = 35.453 g mol^{-1} and 2.01594 g mol^{-1} for H_2 and $\sigma = 2.00$ A and 1.500 A for Cl and H_2 respectively, what is the value of the steric factor?

19. Derive an equation for $1/\tau$ for the reaction scheme

$$X_1 + C \underset{k_1'}{\overset{k_1}{\rightleftharpoons}} X_2 + C$$

where C is the catalyst whose concentration is constant.

20. Consider the reaction scheme

$$X \underset{k_1'}{\overset{k_1}{\rightleftharpoons}} Y \underset{k_2'}{\overset{k_2}{\rightleftharpoons}} Z$$

where the first equilibrium is rapidly established. Derive the equation for $1/\tau$ under these conditions.

21. Solutions of vanadyl sulphate show a single relaxation at 298 K on subjecting them to a *P*-jump study. The relaxation behaviour has been explained due to the formation of an inner sphere complex from the free ions with the intermediate formation of an outer sphere complex in a consecutive process. The mechanism may be stated as

$$VO^{2+} + SO_4^{2-} \underset{k_1'}{\overset{k_1}{\rightleftharpoons}} \underset{\text{Outer sphere complex}}{VO^{2+}H_2OSO_4^{2-}} \underset{k_2'}{\overset{k_2}{\rightleftharpoons}} \underset{\text{Inner sphere complex}}{VOSO_4}$$

Assuming that the first equilibrium, free ions ⇌ OS complex is rapidly established,

$$\frac{1}{\tau} = k_2\left[\frac{1}{K_i} + K\left(\bar{C}_{VO^{2+}} + \bar{C}_{SO_4^{2-}}\right)\right] \text{ where } K_i = \frac{k_2}{k_2'} = 0.8$$

and the overall stability constant

$$K_\Sigma = \frac{\left[\bar{C}_{VO^{2+}H_2OSO_4^{2-}}\right] + \left[\bar{C}_{VO^{2+}SO_4^{2-}}\right]}{\left[\bar{C}_{VO^{2+}}\right]\left[\bar{C}_{SO_4^{2-}}\right]} = 200\ s^{-1}$$

What is the relation between K_Σ, K and K_i? Calculate K. Calculate τ values at concentrations of vanadyl sulphate of 0.05 M and 0.20 M. the rate constant k_2 may be taken as 1000 s^{-1}.

22. The neutralisation reaction $H^+ + OH^- \underset{k_1'}{\overset{k_1}{\rightleftharpoons}} H_2O$ has been studied by a faster technique (even compared to *P*-jump or *T*-jump method) and the relaxation time has been found to be 35×10^{-6} s at 295 K. Derive an equation for $1/\tau$ for the above scheme. Assuming $\bar{C}_{H^+} = \bar{C}_{OH^-} = 10^{-7}$ mol L^{-1} and $k_1' = 2.40 \times 10^{-5}\ s^{-1}$, calculate k_1, the rate constant for the neutralisation reaction.

23. Predict the rate constant for the reaction using the simplified Marcus relation with *f* as unity.
$[Fe(CN)_6]^{4-} + [Mo(CN)_8]^{3-} \rightarrow [Fe(CN)_6]^{3-} + [Mo(CN)_8]^{4-}$
The self-exchange rate constants $k_{11} = 3 \times 10^2\ M^{-1}\ s^{-1}$ for $[Fe(CN)_6]^{4-/3-}$ and $k_{22} = 54 \times 10^3\ M^{-1}s^{-1}$ for $[Mo(CN)_8]^{3-/4-}$. The equilibrium constant $K_{12} = 100$.

24. Consider the equilibrium between open and closed sodium ion channels across a membrane,

$$Na^+(\text{open}) \underset{k_{-1}}{\overset{k_1}{\rightleftharpoons}} Na^+(\text{closed})$$

The Gibbs free energy change at 298 K is –6.8 kJ mol^{-1}. The value of the rate constant k_1 is $4.4 \times 10^3\ s^{-1}$. Estimate the (a) rate constant k_{-1} and (b) fraction of open Na^+ channels after 100 s.

Multiple Choice Questions

1. The Michaelis–Menten mechanism for enzyme catalysis is represented by the equations

$$E + S \underset{k_{-1}}{\overset{k_1}{\rightleftharpoons}} ES$$

$$ES \xrightarrow{k_2} P$$

K_M indicates the Michaelis–Menten constant. The reaction will be zero order when
(a) $[S] >> K_M$ **(b)** $[S] << K_M$ **(c)** $[S] \rightarrow 0$ **(d)** $k_1/k_{-1} = 1$

2. For the consecutive first order reaction scheme given by

$$A \xrightarrow{k_1} B \xrightarrow{k_2} C$$

the plot of [C] vs *t* exhibits an inflection point at t_{max} corresponding to
(a) $d[B]/dt = 0$ **(b)** $d[B]/dt = \infty$
(c) $d^2[B]/dt = \infty$ **(d)** $d^2[B]/dt = 0$

3. The reaction, 2A → Products, follows second order kinetics. The initial concentration of the reactant is a_0. Which of the following is the correct equation for the half-life of the reaction?
(a) $\frac{1}{ka_0}$ **(b)** $2\times\frac{1}{ka_0}$ **(c)** $\frac{0.5}{ka_0}$ **(d)** $\frac{3}{2ka_0^2}$

4. Consider the following Michaelis–Menten mechanism given by

$$E + S \underset{K_S}{\rightleftharpoons} ES \overset{k_{cat}}{\rightleftharpoons} E + P$$

where K_S denotes the equilibrium constant and k_{cat} is the rate constant for the decomposition of the Enzyme–Substrate complex. Which of the following is the appropriate definition of the catalytic efficiency of an enzyme?
(a) K_M **(b)** k_{cat} **(c)** K_M/k_{cat} **(d)** k_{cat}/K_M

5. For obtaining accurate values of all the parameters of the Michaelis–Menten mechanism, which plot is more appropriate?
(a) Lineweaver–Burk plot
(b) Plot of velocity vs substrate concentration
(c) Plot of enzyme concentration vs substrate concentration
(d) Plot of initial velocity vs substrate concentration

6. Consider the equations $A \rightleftharpoons B$ and $B \xrightarrow{k_2} C$, with the forward and reverse rate constants being k_1 and k_{-1} respectively. If the values of k_1, k_{-1} and k_2 (in s^{-1}) are respectively 1, 0.001 and 1000, the rate determining step is,
(a) $A \rightleftharpoons B$ **(b)** $B \xrightarrow{k_2} C$
(c) Not predictable **(d)** Formation of C

7. In a closed reactor of volume of 10 L, 0.5 mol of N_2O_5 and 0.5 mol of N_2 were introduced and the decomposition of N_2O_5 occurred to completion. The ratio between the final and initial pressure (P/P_0) is given by
(a) 5/4 **(b)** 3/2 **(c)** 3/4 **(d)** 4/5

8. The relaxation time, τ, of a fast reaction is defined as the time required for $\Delta\bar{C}$ to become
(a) $\Delta\bar{C}$ //2 **(b)** $\Delta\bar{C}$ /2.303 **(c)** $\Delta\bar{C}$ /4 **(d)** $\Delta\bar{C}$ /3

9. The smallest relaxation time measurable by using a temperature jump method is (in s)
(a) 10^{-9} **(b)** 10^{-12} **(c)** 10^{-6} **(d)** 10^{-3}

10. In a bimolecular reaction

$$A + B \underset{k_{-1}}{\overset{k_1}{\rightleftharpoons}} C + D$$

the following results are obtained at constant concentration of $A = 2\times10^{-4}$ M and varying concentration of B.

Conc. of B in 10^{-4} M	$\tau\times10^6$ (s)
2.00	20.0
4.00	15.0

The values of k_1 and k_{-1} (in units of $M^{-1}\,s^{-1}$) are respectively
(a) 2.0×10^5; 8.0×10^6 **(b)** 3.2×10^4; 5.6×10^6 **(c)** 2.4×10^7; 4.5×10^5 **(d)** 8.5×10^7; 1.6×10^4

11. For the reaction

$$X \underset{k_{-1}}{\overset{k_1}{\rightleftharpoons}} Y$$

studied by the pressure jump method, the observed relaxation time is 10 m s. If 20% of X remains at equilibrium, the rate constants of k_1 and k_{-1} (in s^{-1}) are respectively
(a) 80, 20 **(b)** 60, 30 **(c)** 40, 60 **(d)** 70, 40

12. A relaxation study of beryllium sulphate solution of 4×10^{-3} M, gave a relaxation time of 2.5×10^{-3} s. If the reaction scheme is represented by

$$\underset{\text{(Free ions)}}{Be^{2+} + SO_4^{2-}} \underset{k_{-1}}{\overset{k_1}{\rightleftharpoons}} \underset{\text{Solvent separated ion pair (also known as outer sphere complex)}}{Be^{2+}H_2OSO_4^{2-}}$$

$$\underset{k_{-2}}{\overset{k_2}{\rightleftharpoons}} \underset{\text{Ion pair (also known as inner sphere complex)}}{Be^{2+}SO_4^{2-}}$$

and given that $K_1 = k_1/k_{-1} = 100\ M^{-1}$ and $K_2 = k_2/k_{-2} = 0.7$, the values of k_2 and k_{-2} are respectively
(a) 350.8, 261.3 **(b)** 250.8, 376.4 **(c)** 213.6, 305.1 **(d)** 420.2, 328.5

13. Which of the following parameters determines the extent of displacement in a temperature jump experiment?
(a) $\Delta E(0)$ **(b)** $\Delta V(0)$ **(c)** $\Delta G(0)$ **(d)** $\Delta H(0)$

14. The dependence of half-life ($t_{1/2}$) on the initial concentration of the reactant for three different reactions A, B and C is as follows: $t_{1/2}$

Reaction	Dependence on concentration
A	Independent of concentration
B	Decreases with concentration
C	Increases with concentration

The order of A, B and C is respectively
(a) 1,0 and 2 **(b)** 0,1 and 2 **(c)** 1, 2 and 3 **(d)** 0,1 and 3

15. The half-life of a reaction is found to be inversely proportional to the square root of the initial concentration of the reactant. The order of the reaction is
(a) 0 **(b)** ½ **(c)** 3/2 **(d)** 2

16. For the reaction sequence

$$A \underset{k_{-1}}{\overset{k_1}{\rightleftharpoons}} B \xrightarrow{k_2} C$$

the pre-equilibrium approximation is valid if
(a) $k_2 >> k_1$ **(b)** $k_2 = k_1$ **(c)** $k_2 >> k_{-1}$ **(d)** $k_2 << k_1$

17. For a bimolecular reaction represented as A + B → P, proceeding via an activated complex, the enthalpy change at 300 K is 15 kJ mol^{-1}. The activation energy of the reaction is, (kJ mol^{-1})
(a) 20 **(b)** 24 **(c)** 15 **(d)** 18

18. The conversion of *D*-fructose to *L*-fructose in a suitable solvent can be considered as a reversible reaction with the forward and reverse rate constants being equal $2\times10^{-6}\ s^{-1}$. The approximate time required for 90% of *D*-fructose to remain is (in min)
(a) 930 **(b)** 878 **(c)** 445 **(d)** 523

19. In the Michaelis–Menten mechanism of enzyme catalysis, the plot of velocity vs substrate concentration is
(a) sinusoidal **(b)** linear
(c) hyperbolic **(d)** bi-exponential

20. The reactions which do not have an activation barrier may be considered as
(a) diffusion-controlled **(b)** tunnelling
(c) autocatalytic **(d)** branching chain

21. Which of the following is true regarding the turnover frequency?
(a) It is directly proportional to the Michaelis–Menten constant.
(b) Indicates the catalytic activity.
(c) Total number of reactant molecules reaching the catalytic sites.
(d) Provides the magnitude of the equilibrium constant.

22. In the following catalytic reaction for R $\rightarrow$ P,

$$X_1 + R \xrightarrow{k_1} X_2$$
$$X_2 \xrightarrow{k_2} X_1 + P$$

the turnover frequency is given by
(a) k_1/k_2 (b) $k_1[R]/k_2$
(c) $k_1 + k_2[R]$ (d) $k_1k_2[R]/(k_1[R] + k_2)$

23. Which of the following statement is true for a reaction of order n with $n = 0, 1, 2...$?
(a) $t_{1/2}[A_0]^n = \text{Constant}$ (b) $t_{1/2} \propto [A_0]^{n-1}$
(c) $t_{1/2} \propto 1/[A_0]_n$ (d) $t_{1/2}\,[A_0]^{n-1} = \text{Constant}$

24. According to Lindemann mechanism, the order of the reaction will become 2 when
(a) the gas is ideal (b) pressure is higher
(c) pressure is lower (d) temperature is lower

25. The termolecular collision frequency will in general, be
(a) equal to that for bimolecular collisions
(b) greater than that for bimolecular collisions
(c) less than that for bimolecular collisions
(d) zero

26. According to Arrhenius theory for the temperature dependence of the rate constants (k) is $RT^2(d\ln k/dT) = E_a$. If the collision theory is employed, the RHS of the above equation will be
(a) the same (b) $E_a + 0.5RT$
(c) $E_a - 0.5RT$ (d) $e^{E_a/RT}$

27. Consider the following reaction at 550 K.

$$2NO_2(g) \rightarrow 2NO(g) + O_2(g)$$

The activation energy is 110 kJ mol^{-1}. The enthalpy of activation ($\Delta H^{\ddagger}$) from the transition state theory is approximately (in kJ mol^{-1}).
(a) 110.3 (b) 90.2 (c) 100.8 (d) 105.8

28. Consider the following reaction sequence:

$$X \underset{k_{-1}}{\overset{k_1}{\rightleftarrows}} Y \xrightarrow{k_2} Z$$

where $k_1 = 0.18$ min^{-1}, $k_{-1} = 0.00009$ min^{-1} and $k_2 = 0.009$ min^{-1}. Which of the following is true among the following?
(a) steady state approximation is applicable to Y
(b) the rate of formation of Y is slower than its disappearance
(c) the assumption of rapid pre-equilibrium is valid
(d) the rate-determining step cannot be inferred

29. Consider the consecutive reaction scheme given by

$$A \xrightarrow{k_1} B \xrightarrow{k_2} C$$

The reaction B $\rightarrow$ C follows second order kinetics. Which of the following is true?
(a) $C_A(t) = C_{A,0}\, e^{-k_1 t}$
(b) $C_B(t) = C_{B,0}\, e^{-k_2 t}$
(c) $C_C(t) = (C_{A,0} + C_{B,0})\, e^{-k_1 t}$
(d) $C_A(t) = C_{A,0}\, e^{-(k_1+k_2)t}$

30. Consider the following reaction sequence:

$$A \underset{k_{-1}}{\overset{k_1}{\rightleftarrows}} B \underset{k_{-2}}{\overset{k_2}{\rightleftarrows}} C$$

Starting with an initial concentration of 0.5 mole of A ($N_A = 0.5$), which of the following is valid for the concentrations of A, B and C?
(a) $N_B = 0$ (b) $N_A = 0$
(c) $N_A + N_B = N_C$ (d) $N_A + N_B + N_C = 0.5$

31. In the Michaelis–Menten mechanism of enzyme catalysis given by

$$E + S \underset{k_{-1}}{\overset{k_1}{\rightleftarrows}} ES \overset{k_2}{\longleftrightarrow} \text{Products}$$

the application of rapid pre-equilibrium approximation yields the Michaelis-Menten constant K_M as
(a) $K_M = k_1/k_{-1}$ (b) $K_M = k_{-1}+k_2/k_1$
(c) $K_M = k_2/k_{-1}$ (d) $K_M = k_1+k_2/k_{-1}$

32. Consider the following enzyme-catalysed reaction scheme

$$S + E \underset{k_{-1}}{\overset{k_1}{\rightleftarrows}} X \underset{k_{-2}}{\overset{k_2}{\rightleftarrows}} Y + E$$

Which of the following is true?
(a) $K_{eq} = k_1k_{-1}/k_2k_{-2}$ (b) $K_{eq} = k_1^2k_2^2/k_{-1}k_{-2}$
(c) $K_{eq} = k_1k_2/k_{-1}k_{-2}$ (d) $K_{eq} = k_1k_{-2}/k_{-1}k_2$

33. Consider the following proton-coupled electron transfer scheme:

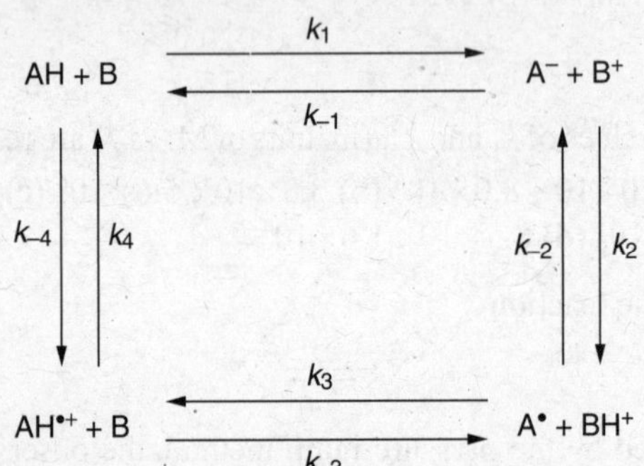

Which of the following is true according to the principle of detailed balancing?
(a) Affinities are > 0 at equilibrium
(b) All the equilibrium constants are equal
(c) All the affinities are equal
(d) All the affinities are equal to 0

Answer Keys

Practice Problems

1. (a) Theoretical question (b) $[A_0]^{1/2} - [A]^{1/2} = \frac{kt}{2}$
2. 16.48 s
3. (a) 112.9 kJ mol^{-1} (b) 8100 min
4. $k = [H_2][NO]^{-2}$
5. (a) 103 kJ mol^{-1} (b) 100.7 kJ mol^{-1} (c) 4.17×10^{13} s^{-1} (d) 16.57 J K^{-1} mol^{-1}
6. (a) 14.40 (b) 11.93×10^{-17}
7. 10^3 L^2 mol^{-2} s^{-1} and 2.76×10^{-39} cm^6 s^{-1}
8. (a) 1.02×10^{34} m^{-3} s^{-1} (b) 3.75×10^{20} mol m^{-3} s^{-1} (c) 2.05×10^{-3} dm^{-3} mol^{-1} s^{-1}
9. 10.43×10^{28} cm^{-3}s^{-1}
10. −50.32 kJ mol^{-1} K^{-1}; $\Delta S^{\ddagger} = -83.6$ J mol^{-1} K^{-1}
11. Theoretical question
12. (a) 2.582×10^{15} s^{-1} (b) 48.493 J K^{-1} (c) 1.533×10^{-16}
13. $\frac{1}{2}\frac{d[NO_2]}{dt} = k_{eff}[NO]^2[O_2]$
14. $k_1k_4 = k_2k_3$
15. Theoretical question
16. Theoretical question
17. 0.00422 M
18. 0.11
19. $k_1\bar{C}_c' + k_1'\bar{C}_c'$
20. $k_2' + k_2\frac{K}{1+K}$
21. 2.2×10^{-5} s
22. 1.43×10^{11} L mol^{-1} s^{-1}
23. 4.02×10^4 M^{-1} s^{-1}
24. (a) 2.82×10^2 s^{-1} (b) 0.94

Multiple Choice Questions

1. (a)	2. (d)	3. (c)	4. (d)	5. (b)	6. (a)	7. (a)	8. (b)	9. (c)	10. (d)
11. (a)	12. (c)	13. (d)	14. (a)	15. (c)	16. (c)	17. (a)	18. (a)	19. (c)	20. (a)
21. (b)	22. (d)	23. (d)	24. (c)	25. (c)	26. (c)	27. (c)	28. (c)	29. (a)	30. (d)
31. (a)	32. (c)	33. (d)							

CHAPTER

5 LIQUID STATE

CHAPTER OUTLINE

5.1 Important Concepts

Complexity of liquid state: The molecules of a gas move completely randomly in the available space. In the case of a crystal there is an orderly arrangement of atoms or ions and they vibrate about their fixed positions which are determined by interatomic and interionic forces.

However, in liquids, there are strong forces of cohesion which are responsible for the observed properties of the liquid state. But these forces of cohesion are unable to prevent the considerable translational motion and the resultant energy among its individual molecules. The competition between order due to these forces and the disorder due to thermal motions results in an inability to provide a theoretical model for the liquid state.

5.1.1 Internal Pressure in Liquids

There is a balance between the attractive and repulsive forces in a liquid which gives rise to *internal pressure*. It is defined by $\left(\frac{\partial E}{\partial V}\right)_T$, that is, the rate of change of internal energy with volume at a constant temperature. This can be expressed by the thermodynamic equation of state.

$$\left(\frac{\partial E}{\partial V}\right)_T = T\left(\frac{\partial P}{\partial T}\right)_V - P \tag{5.1}$$

Often, the external pressure P is quite small compared to the internal pressure and hence one may write for the internal pressure, P_{ip},

$$P_{ip} = T\left(\frac{\partial P}{\partial T}\right)_V \tag{5.2}$$

$$= T\frac{\alpha}{\beta} \tag{5.3}$$

where, α and β are the coefficients of thermal expansion and compressibility respectively. The internal pressure P_{ip} of a liquid may also be related to the intermolecular attractive term a (of van der Waals equation) by,

$$P_{ip} = \frac{a}{V^2} \tag{5.4}$$

5.1.2 Property of Liquid Surface and Surface Tension

A molecule in the bulk of a liquid is in a uniform environment and hence is attracted equally in all directions. However, a molecule at the surface has a net attraction towards the bulk of the liquid and as a consequence, the surface of a liquid always tends to contract to the smallest possible area. Thus, the drops of a liquid are always spherical. To extend the surface area of a liquid, it is necessary to do some work to bring the molecules from the bulk to the surface. This work is referred to as the *surface free energy* and the tendency of a liquid to contract is thus accompanied by a decrease in free energy.

Because of its tendency to contract, the surface of a liquid is always in a state of tension, referred to as the *surface tension*. It is symbolised by γ and has the same value at any point and in all directions along the liquid surface. It is defined as the force (in dynes) acting at right angles to a line of 1 cm length on the surface and has the unit, dyne cm^{-1}. It is numerically equal to the surface free energy given by erg cm^{-2}.

5.1.3 Interfacial Surface Energy or Interfacial Tension

When we consider two immiscible or partially immiscible liquids, the work required to enlarge the surface of separation is known as the interfacial surface energy or interfacial tension.

5.1.4 Effect of Contact Angle on Wetting

Consider a liquid (L) resting on the surface of a solid (S) and in equilibrium with an angle of contact θ (Fig. 5.1).

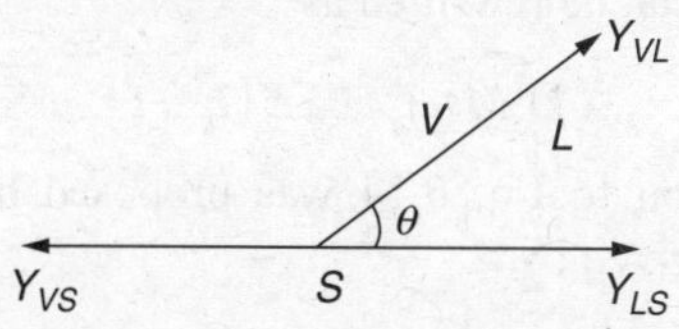

Fig. 5.1 Schematic representation of the contact angle between the solid and liquid phases

V represents the vapour of the gas (usually air). Under equilibrium, the surface tension acting in the direction of surfaces as shown are related as

$$\gamma_{VS} = \gamma_{LS} + \gamma_{VL} \cos\theta \qquad (5.5)$$

Equation 5.5 is also known as *Young's equation*. If $(\gamma_{VS} - \gamma_{LS})$ is positive, then $\cos\theta$ is positive and $\theta < 90°$ but if the above difference is negative ($\gamma_{VS} - \gamma_{LS}$ is negative), $\cos\theta$ lies between 90° and 180°. When $\theta < 90°$, wetting of the surface occurs. A classic example is water on glass. For the latter category, when θ lies between 90° and 180°, no wetting of the solid occurs as exhibited by mercury on glass or paraffin wax.

5.1.5 Dupre Equation

The work of adhesion between a solid and liquid, W_{LS}, is given by

$$W_{LS} = \gamma_{VS} + \gamma_{VL} - \gamma_{LS} \qquad (5.6)$$

5.1.6 Young's Equation in the Case of a Solid and Two Immiscible Liquids

In Fig. 5.2, the interfaces between a solid S and two immiscible liquids, X and Y are considered.

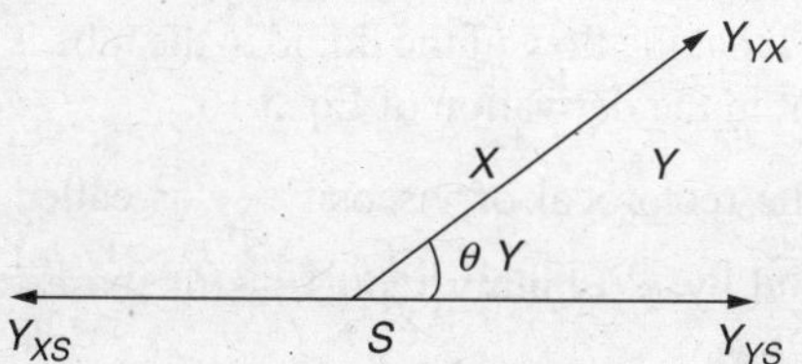

Fig. 5.2 Schematic depiction of the contact angle between a solid and two immiscible liquids

At equilibrium, the forces acting at the interfaces balance each other. An equation similar to Eq. 5.5 may be written as

$$\gamma_{XS} = \gamma_{YX} + \gamma_{YX} \cos\theta \qquad (5.7)$$

Note that $\theta_Y < 90°$ and hence the wetting of the solid becomes favourable. Since $\cos\theta$ is positive, γ_{XS} is greater than γ_{YS}.

5.1.7 Pressure Difference Inside and Outside a Spherical Bubble and its Relation to Surface Tension

As a result of surface tension, the pressure on the concave side of a spherical bubble is greater than that on the convex side. If such excess pressure on the concave side is not present, the bubble would collapse due to surface tension. If P is the excess pressure (dyne cm^{-2}) inside the sphere and γ is the surface tension acting along the circumference of the sphere whose radius is r cm, we may write,

$$P = \frac{2\gamma}{r} \left(\text{dyne cm}^{-2}\right) \qquad (5.8)$$

P is inversely proportional to r. The pressure inside a small bubble is quite large and it explains the phenomenon of *superheating* or *bumping* observed when a pure liquid is heated in a vessel with a clean and smooth surface.

5.1.8 Rise of Liquid in a Capillary and its Relation to Surface Tension

The surface of a liquid in a tube (the meniscus) is always curved. This curvature arises due to the competing forces, namely, the forces between the molecules of liquid on one hand and the forces between the molecules of liquid in contact with a solid, say glass, on the other. The nature of the meniscus depends on whether the liquid wets the surface or not. If it wets the surface of the solid, the shape of the meniscus is concave upward. In the case of a liquid that does not wet the solid, the meniscus is concave downward. Figure 5.3 illustrates the nature of contact angle θ in the two cases—(i) $\theta < 90°$ in the case of a liquid wetting the solid surface and (ii) $\theta > 90°$ in the case of a liquid not wetting the solid surface.

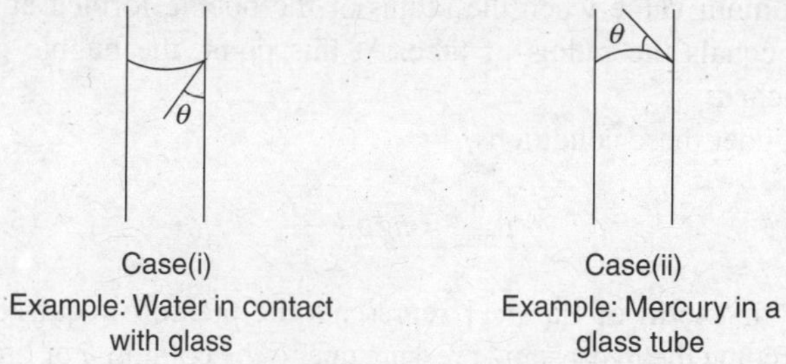

Fig. 5.3 Illustration of the contact angle (i) < 90° and (ii) > 90°

Consider a liquid (for example, water), placed in a capillary tube of small radius (say of glass) and that wets the surface. The meniscus is concave upwards. The pressure below the meniscus (convex side) is less by $\frac{2\gamma}{l}$ (when r is the radius of curvature of meniscus) than that at the same level outside the capillary tube. The pressure difference leads to the meniscus being forced up the capillary tube to a height h cm above that outside, at equilibrium. The hydrostatic pressure due to h cm of liquid is equal to the pressure difference under the meniscus and flat surface.

We may thus write

$$\frac{2\gamma}{l} = hg\rho \quad (5.9a)$$

$$\gamma = \frac{1}{2}hg\rho l \quad (5.9b)$$

In Eq. 5.9(b), ρ is the density of the liquid. In the case of a liquid such as mercury which does not wet glass, the meniscus is convex upward and here the surface of mercury in the capillary will be below that of the surrounding liquid. Since the rise or fall of a liquid in a capillary tube is associated with surface tension, the phenomenon dealing with it is known as *capillary action*.

5.2 Measurement of Surface Tension

Surface tension is measured by determining the height to which a liquid rises in a capillary tube of known radius r. Using Eq. 5.9(b), the surface tension of the liquid is calculated. However, it is more common to compare the rise of the liquid in the capillary with that of a standard liquid whose γ is known. This avoids the necessity of measuring l. If h represents the height of the liquid in a capillary tube and ρ is the density, and h_{std} and ρ_{std} are the corresponding values of the standard liquid in the same capillary, we can write,

$$\frac{\gamma}{\gamma_{std}} = \frac{h\rho}{h_{std} \times \rho_{std}} \quad (5.10)$$

Since γ_{std} and all other quantities are known, γ can be calculated using Eq. 5.10. There are several other methods of determining the surface tension.

5.2.1 Maximum Bubble Pressure Method

In this method, the pressure of air applied at the top of a fairly wide tube is increased with the tube being immersed up to a certain depth, say h cm, in the liquid. Let the radius of the tube be r cm. As the air pressure is gradually increased, it reaches a maximum value when the radius of the bubble formed at the end equals the radius of tube. At this point, the bubble gets detached.

Under these conditions,

$$p_{max} = hg\rho + \frac{2\gamma}{r} \quad (5.11)$$

The first term of Eq. 5.11 represents the pressure required to overcome the hydrostatic pressure due to the column h of liquid and the second term refers to the excess pressure on the concave side of the bubble.

5.2.2 Drop Weight Method

The basis of this method is that the weight m of a drop of a liquid falling from a vertical tube of small diameter is roughly proportional to the surface tension of the liquid. Thus if two liquids having surface tensions γ' and γ'' have average weights of drops falling from the same tube as m' and m'', the relation given below holds.

$$\frac{\gamma'}{\gamma''} = \frac{m'}{m''} \quad (5.12)$$

Variation of surface tension with temperature–Eötvos equation: The surface tension of a liquid generally decreases with increase in temperature, however, with some exceptions as in the case of liquid crystals and metals such as iron and copper. According to Eötvos, the molar surface energy of a liquid is given by $\gamma(M\upsilon)^{2/3}$ and varies linearly with temperature.

$$\gamma(M\upsilon)^{2/3} = y - Zt \quad (5.13)$$

In Eq. 5.13, γ is the surface tension of the liquid, M is the molar mass and υ is the specific volume of the liquid, the product on the LHS has the unit of energy, y and Z are constants and t is the temperature. The constant y can be evaluated by using the condition that at critical temperature, t_c, $\gamma = 0$. Therefore, from Eq. 5.13,

$$0 = y - Zt_c \text{ or } y = Zt_c$$

Equation 5.13 can be rewritten as

$$\gamma(M\upsilon)^{2/3} = Z(t_c - t) \quad (5.14)$$

A better relation to Eq. 5.14 was proposed by Ramsay and Shields and is given by,

$$\gamma(M\upsilon)^{2/3} = y - Z(t_c - 6 - t) \quad (5.15)$$

From Eq. 5.15 it is seen that when $t = t_c - 6$, $\gamma = 0$.

Z from Eqs. 5.13 and 5.15, is the slope of the linear plot of $\gamma(M\upsilon)^{2/3}$ against t. That is, it is the temperature coefficient of molar surface energy. It is found to have the value of 2.12 for non-associated liquids and less than 2.12 for associated liquids such as water or alcohols.

5.3 Viscosity of Liquids

Liquids, like gases, show resistance to flow which is characterised by viscosity. Viscosity has been defined earlier (Chapter 1 and can be considered as internal friction. The rate of flow of a liquid through a tube is related to viscosity by Poiseuille's equation given by

$$\eta = \frac{\Pi r^4 tf}{8\upsilon l} \quad (5.16)$$

where r is the radius of the tube in metre (m), t is time in s, f is the pressure driving the liquid in N m^{-2} and υ is the volume of liquid in m^3 flowing through the tube in t seconds and l is the length of the tube in metres. The unit of η may be seen from $\eta = \frac{\text{m}^4 \text{ s N m}^{-2}}{\text{m}^3 \text{ m}} = \text{N m}^{-2} \text{ s} = \text{Pa s}$. A highly viscous liquid has a large viscosity, but a highly mobile liquid has a low viscosity. It is assumed that the flow of the liquid in the tube is laminar and not turbulent in the derivation of Eq. 5.16.

Fluidity: The reciprocal of viscosity $\frac{1}{\eta}$, is called fluidity and is represented by ϕ. Fluidity represents the ease with which a liquid will flow.

Stoke's law: Viscosity exerts a drag on a body falling through a liquid. The gravitational pull exerts a downward pull and these two oppose each other. As a result, the body acquires a constant velocity while it falls through the liquid and this is known as *terminal velocity*, v. This is related to the viscosity η of liquid by the relation

$$v = \frac{2gr^2(\rho' - \rho)}{9\eta} \tag{5.17}$$

where, r = radius of the tube, ρ' = density of the spherical body, ρ = density of liquid, and g = acceleration due to gravity.

5.3.1 Determination of Viscosity

The determination of viscosity of a liquid is based on the use of Eq. 5.16. A comparative method using a viscometer is employed for this purpose. Two liquids, one a standard liquid whose viscosity (η_{std}) and density (ρ_{std}) are known and the second (unknown liquid) whose viscosity (η_{unk}) and density (ρ_{unk}) are employed for this purpose. The times of flow of the standard liquid t_{std} and that of the unknown liquid, t_{unk}, of a certain known volume of the liquids (both volumes are the same) are determined with the viscometer. The viscosities of the two liquids are related according to

$$\frac{\eta_{unk}}{\eta_{std}} = \frac{\rho_{unk} \times t_{unk}}{\rho_{std} \times t_{std}} \tag{5.18}$$

Since η_{std} and all other quantities on the RHS are known, η_{unk} can be calculated. The units for η are expressed in millipoise at a given temperature. For example, water and ethylene glycol have viscosities of 10.1 millipoise (10.1×10^{-3} poise = 10.1×10^{-4} Pa s) and 200 millipoise (0.02 Pa s).

5.3.2 Effect of Temperature on Viscosity

The effect of temperature on viscosity of a liquid is exactly opposite to that of its effect on the viscosity of a gas. While the viscosity of a gas increases with increase in temperature, the viscosity of a liquid decreases significantly with increase in temperature. It is expressed in an exponential form as

$$\eta = Xe^{E/RT} \tag{5.19}$$

or

$$\log \eta = \log X + \frac{E}{2.303RT} = Y + \frac{E}{2.303RT} \tag{5.20}$$

where X and Y are constants and E is the activation energy for the liquid flow. E is larger for associated liquids such as water or alcohols compared to non-associated liquids such as benzene. Therefore, the viscosity of an associated liquid is greater than that of a non-associated liquid.

A relationship, which is of some significance, for many non-associated liquids is given by

$$v = b + \frac{a}{\eta} = b + a\phi \tag{5.21}$$

where v is the specific volume of the liquid, a and b are constants for a given liquid. This equation is found to be valid for many non-associated liquids over a range of temperatures.

5.4 Structure of Liquids

The structure of liquids may be considered as intermediate between that of gases and solids. At the critical temperature, both the gaseous and liquid forms of a substance become indistinguishable. Similarly, there is some identity in the arrangement of molecules in the liquid and solid state near the melting point. As the temperature of the liquid is raised, its internal structure gradually changes to a disordered state near the critical temperature. In recent years, x-ray diffraction studies have yielded valuable insights into liquid structure.

5.5 Liquid Crystals

In some chemical powders, the crystalline form of the substance does not melt to the liquid phase on heating at first but goes through an intermediate state known as the mesomorphic (or paracrystalline) state. On heating further, this state reaches the liquid state at a higher temperature. The intermediate states are known as liquid crystalline states since they exhibit the properties of adjacent states. Compounds exhibiting such mesomorphic states are cholesteryl benzoate, *p*-azoxyanisole, ethyl-*p*-anisolaminocinnamate. For example, in the case of *p*-azoxyanisole, the solid melts to a liquid crystalline state at 84°C which is stable till 150°C at which temperature it undergoes a transition to an isotropic liquid. The mesomorphic state is a turbid liquid and the temperature at which the solid melts to a turbid liquid is called the *transition point* and the higher temperature at which liquid crystal melts to a pure liquid is known as the *melting point*.

From a study of several substances, it was possible to distinguish between two classes of phases—smectic and nematic.

5.5.1 Smectic Phases

In a smectic state, no normal liquid flow takes place. The movement occurs in one plane in a gliding manner. This can be observed when the drops of the liquid are touched. In a smectic phase, the structure consists of a number of planes one on top of the other. Exampes of mesomorphic materials of this type are ethyl-*p*-azoxybenzoate, *n*-octyl-*p*-azoxycinnamate.

5.5.2 Nematic Phases

Nematic liquids are similar to anisotropic liquids. (In an anisotropic substance, the velocity of light is not uniform in all directions. If a light ray enters an anisotropic crystal, it splits into two components which travel with different velocities. They show double refraction or birefringence.) Some examples of mesomorphic substances of nematic type are *p*-azoxyanisole, *p*-methoxycinnamic acid and dibenzalbenzidine.

5.5.3 Liquid Crystal Displays or Devices

The most common application of liquid crystals is in liquid crystal displays. Liquid crystal displays are used in wrist watches, pocket calculators and even in advanced computer screens. An LCD consists of two glass plates containing some liquid crystal between them. Such a device need not be bulky or heavy. Also, these devices use less power compared to the cathode ray tube (CRT).

5.6 Additive and Constitutive Properties

The physical properties of any system can be divided into three categories—additive, constitutive and colligative properties. The last category has been discussed in Chapter 9.

Additive property: This is the sum of the property under consideration of all the individual constituents of a system. A good example is mass because the *mass* of a molecule is the sum of masses of its atoms. Another property which is also largely additive, is molar volume.

Constitutive property: This property mainly depends on the arrangements of atoms in the molecule but also to a smaller extent on the nature of atoms and their number. Many physical properties are however, partly additive and partly constitutive.

5.6.1 Examples of Properties Showing Additive and Constitutive Characteristics

Molar volumes of organic liquids at their boiling points: The molar volumes of organic liquids measured at their boiling points under atmospheric pressure, $Mv = \frac{M}{d}$, (v = specific volume, d = density) are to a large extent additive. However, constitutive factors are also involved as revealed by later work. It is observed that the molar volumes of isomeric compounds of similar constitution, say, esters like methyl propionate, ethyl acetate and propyl formate, at their boiling points are found to be almost the same. An interesting observation is that in any particular homologous series, a constant difference of 22.2 cm^3 is found for each CH_2 group in the molar volume at the boiling points. Using this result, volume equivalents of various elements such as H, C, Cl, Br, O (as in –O–) and O (as in =O) have been estimated. The volume equivalents of elements may be used to calculate the molar volume of compounds at their boiling points without recourse to experiments.

Parachor, [P]: The surface tension of a liquid, γ, is related to the orthobaric densities of the liquid (ρ_l) and the saturated vapour (ρ_v) at the same temperature by an equation, known as Macleod's equation, by

$$\frac{\gamma^{1/4}}{\rho_l - \rho_v} = C \tag{5.22}$$

where, C is a constant. Equation 5.22 holds over a range of temperatures. Multiply both sides of the equation by M (the molecular weight of the substance). We get

$$\frac{M\gamma^{1/4}}{\rho_l - \rho_v} = MC = [P] = \text{Constant} \tag{5.23}$$

The constant [P] is known as *parachor*. When $\rho_v << \rho_l$, Eq. 5.23 may be rewritten as

$$\frac{M\gamma^{1/4}}{\rho_l} = [P] = V_m \times \gamma^{\frac{1}{4}} \tag{5.24}$$

where, V_m = molar volume = $\left(\frac{M}{\rho_l}\right)$. Suppose γ = 1, then $[P] = V_m$ which implies that parachor is equal to molar volume under these conditions. If V_m is expressed in cm^3 mol^{-1} and γ is expressed in dyne cm^{-1}, the units of parachor are g$^{1/4}$ s$^{-1/2}$ cm^3 mol^{-1}.

Parachor is primarily additive but also depends on structural factors within a given molecule. Similar to the case discussed under molar volumes, parachor equivalents for a number of elements have been computed. Associated liquids such as alcohol have parachors which increase with an increase in temperature.

Parachor finds application sometimes in distinguishing between alternate configurations, for example, in resolving the structure of the isocyanide radical.

Molar refraction $[R]_M$: The molar refraction of a compound is defined by the equation

$$[R]_M = \frac{n^2 - 1}{n^2 + 2} \times \frac{M}{\rho} \tag{5.25}$$

where n is the refractive index of a compound which depends on the wavelength of the light used (usually, the D-line of sodium or, α, β or γ lines of the Balmer series of hydrogen). M is the molecular weight and ρ is the density of compound. It is a characteristic property having the same value for the gaseous and liquid states and is also independent of temperature. From Eq. 5.25, molar refraction is proportional to molar volume because the refractive index term is a constant. Note that n is a dimensionless quantity. $[R]_M$ is an additive as well as a constitutive property. Just as in the case of molar volume, parachor, refraction equivalents of various elements have been derived using refractive indices measured with the wavelength of the D-line of sodium. Since molar refraction is also a constitutive property, it has been used to distinguish between alternate structures of some organic compounds.

The molar refraction $[R]_{a,b}$ of a mixture of substances a, b whose mole fractions are X_a and X_b with molar masses M_a and M_b is given by

$$[R]_{a,b} = \frac{n^2 - 1}{n^2 + 2} \times \frac{X_a M_a + X_b M_b}{\rho} \tag{5.26}$$

where, n is the refractive index and ρ is the density, both pertaining to the mixture. Equation 5.26 may also be written as

$$[R]_{a,b} = X_a [R]_a + X_b [R]_b \tag{5.27}$$

Molar polarisation, P^M: There are two effects produced when molecules of a substance are subjected to an electric field. (i) A displacement or distortion of electric charges in non-polar molecules, and (ii) a distortion of charges as well as an orientation effect in the case of polar molecules. In general, distortion of electric charges is known as *polarisation*, *P*. The molar polarisation associated with a compound is given by the Mossotti–Clausius equation as

$$P^M = \frac{D - 1}{D + 2} \times \frac{M}{\rho} \tag{5.28}$$

where, D is dielectric constant of the medium, M is the molar mass and ρ is the density of the substance. For polar substances such as HCl or acetone, P^M decreases with an increase in temperature whereas for non-polar substances such as CCl_4, P^M is independent of temperature. Since D is a dimensionless quantity and a number, P^M is also a molar volume and is additive as well as constitutive. The P^M is the sum of the induced polarisation P_{IP} and P_{OP}, orientation polarisation (under the conditions employed for measuring D),

$$P^M = P_{IP} + P_{OP} = P_{IP} + \frac{4\Pi N}{9kT} \times \mu^2 \tag{5.29}$$

The second term, P_{OP}, is shown to be equal to the term shown on the extreme right of Eq. 5.29 involving the dipole moment of the molecule. The terms N, k and T of this term are the Avogadro number, Boltzmann constant and temperature in K, respectively. If visible light of high frequency of about 10^{15} cycles s^{-1} is allowed to pass through a substance, only polarisation of electrons (P_E) in the molecule takes place and P_E is shown to be equal to the molar refraction $[R]_M$ given by Eq. 5.25. Equation 5.29 may now be expressed as

$$P^M = \frac{D-1}{D+2} \times \frac{M}{\rho} = [R]_M + \frac{4\Pi N}{9kT}\mu^2 \qquad (5.30)$$

(Note that $P_{IP} \approx P_E = [R]_M$ because, when visible light is used, only electrons undergo polarisation and not nuclei). Equation 5.30 is applicable both to non-polar and polar molecules and is used for determining the dipole moments of molecules.

Molar polarisation of mixtures of polar solvents in non-polar solvents: The total polarisation of a dilute solution of a polar substance B in a non-polar solvent A is the sum of the polarisations of the two substances. Thus, the total polarisation of mixture $P_{A,B}$ is given by

$$P_{A,B} = \frac{D-1}{D+2} \times \frac{X_A M_A + X_B M_B}{\rho} \qquad (5.31)$$

where, D = dielectric constant of the solution (measured), ρ is the density of the solution, X_A and X_B are the mole fractions of the non-polar solvent and solute respectively, and M_A and M_B are the molecular weights of solvent and solute. $P_{A,B}$ is written as

$$P_{A,B} = X_A \times P_A + X_B \times P_B \qquad (5.32)$$

In the above equation, P_A and P_B are the molar polarisations of the non-polar solvent and polar solute respectively. P_A and $P_{A,B}$ are measurable from the dielectric constant and density measurements. P_B may thus be calculated from the known mole fraction data.

Worked Out Problems

Internal Pressure of a Liquid and van der Waals Equation

Problem 5.1

(a) Explain the term internal pressure of a liquid. How it is defined mathematically? What is its physical significance?

(b) If the internal heat of vaporisation of n-hexane, L_i, is 27.5 kJ mol^{-1} and its density is 0.659 g cm^{-3} at 298 K, calculate the internal pressure of the liquid in (i) kPa (ii) atm.

Solution

(a) In the liquid state, there are strong forces of cohesion among its molecules and at the same time there is translational motion among its molecules and a certain resultant energy. The competition between these opposing forces results in internal pressure in liquids. It is defined by the equation,

$$\left(\frac{\partial E}{\partial T}\right)_V = T\left(\frac{\partial P}{\partial T}\right)_V$$

(b) $P_{IP} = \frac{L_i}{V}$ where v is the molar volume of the liquid

$$\text{or } P_{IP} = \frac{27.5 \text{ kJ mol}^{-1}}{\dfrac{86\times 10^{-3}}{659}} \times \frac{\text{kg m}^{-3}}{\text{kg mol}^{-1}}$$

$$= 21.0\times 10^5 \times 10^3 \text{ J m}^{-3}$$

$$= 21.0\times 10^8 \text{ kg m}^2 \text{ s}^{-2} \text{ m}^{-3}$$

$$= 21.0\times 10^8 \text{ Nm}^{-2} = \text{kg m}^{-1} \text{ s}^{-2}$$

$$\text{Or, } P_{IP} = 21.0\times 10^8 \text{ Pa} = 21.0\times 10^5 \text{ kPa}$$

$$= \frac{21.0\times 10^5}{101.325} = 20725 \text{ atm}$$

Problem 5.2

(a) Assuming that the van der Waals equation is applicable for a liquid (is correct under certain conditions), derive an equation for the internal pressure in terms of the van der Waals constant a.

(b) If the internal pressure of CS_2 is 3.65×10^6 kPa, calculate its internal latent heat of vaporisation L_i (kJ mol^{-1}). Its density is 1.26 g cm^{-3}.

Solution

(a) van der Waals equation: $\left(P+\frac{a}{v^2}\right)(v-b) = RT$

$$\text{Or, } \left(P+\frac{a}{v^2}\right) = \frac{RT}{(v-b)}$$

Differentiating this equation with respect to T at constant volume,

$$\left(\frac{\partial P}{\partial T}\right)_V = \frac{R}{(v-b)}$$

$$T\left(\frac{\partial P}{\partial T}\right)_V = \frac{RT}{(v-b)} = \left(P+\frac{a}{v^2}\right)$$

$$T\left(\frac{\partial P}{\partial T}\right)_V - P = \left(\frac{\partial E}{\partial V}\right)_T = \frac{a}{v^2}$$

Since the external pressure P is usually very small compared to P_{IP}

$$P_{IP} = T\left(\frac{\partial P}{\partial T}\right)_V = \frac{a}{v^2}$$

(b) We know $P_{IP} = \frac{L_i}{V}$

or $L_i = P_{i.p} \times V$

$$= 3.65\times10^6 \text{ kPa} \times \frac{0.07612 \text{ m}^3 \text{ mol}^{-1}}{1260}$$

$$= 3.65\times10^9 \text{ N m}^{-2} \text{ m}^3 \text{ mol}^{-1}$$

$$= 3.65\times10^9 \times 6.04\times10^{-5}$$

$$= 22.05\times10^4 \text{ J mol}^{-1}$$

$$= 220.5 \text{ kJ mol}^{-1}$$

Surface Tension, Contact Angle and Excess Pressure Inside a Spherical Bubble

Problem 5.3

(a) Define the term surface tension. What are its SI units? Explain why surface tension is also referred to as surface free energy.

(b) Write down the relation between the contact angle and the surface tension of the three interfaces (Young's equation).

(c) The surface tension of benzene is 28.2 dyne cm^{-1}. What is the height to which the liquid will rise in a capillary of 0.1 mm radius given that the density of benzene is 0.800 g cm^{-3}?

Solution

(a) It is the interface force (in dyne or newton) acting at right angles to a line of length one cm (or one metre) of the surface. It has units of N m^{-1} in SI units. Surface tension is numerically equal to the surface energy in erg cm^{-2} (or J m^{-2}). It is called surface free energy because the work required to extend the surface of a liquid is a form of free energy not involving *PV* work.

(b) From Fig. 5.1,

$$\gamma_{VS} = \gamma_{LS} + \gamma_{VL} Cos\theta$$

(c) $\gamma = \frac{1}{2} hg\rho l$ where the terms have their usual significance.

$\gamma = 28.2$ dyne cm^{-1}, $g = 980$ cm s^{-2}, $\rho = 0.800$ g cm^{-3}, $l = 0.01$ cm

$$h = \frac{2\gamma}{g\rho l} = \frac{2\times28.2}{980\times0.800\times0.01} \times \frac{\text{g cm s}^{-2} \text{ cm}^{-1}}{\text{cm s}^{-2} \text{g cm}^{-3} \text{ cm}} = \text{cm}$$

$$= 7.19 \text{ cm} = 0.0719 \text{ m}$$

Problem 5.4

(a) Calculate the excess pressure ΔP inside a spherical bubble of radius 0.02 cm of a liquid having a surface tension of 72.0 dyne cm^{-1}. Give your answer in N m^{-2}.

(b) The surface free energy (same as surface tension) can be related to surface enthalpy by a form of Gibbs–Helmholtz equation $\gamma = H + T\left(\frac{\partial\gamma}{\partial T}\right)_P$. For the liquid under (a), $\left(\frac{\partial\gamma}{\partial T}\right)_P = -0.15$. Calculate the enthalpy of its surface at 298 K in J m^{-2}. What is the significance of this value?

Solution

(a) $\Delta P = \frac{2\gamma}{r} = \frac{2\times72.0}{0.02} = 7200$ dyne cm^{-2}

1 N m^{-2} = 10 dyne cm^{-2}, 7200 dyne cm^{-2} = 720 N m^{-2} = 720 Pa

(b) Surface enthalpy,

$$H = \gamma - T\left(\frac{\partial y}{\partial T}\right)_P$$

$$= 72.0 - 298\times(-0.15)$$

$$= 72.0 + 44.7 = 116.7 \text{ erg cm}^{-2}$$

$$= \frac{116.7\times10^{-7}}{(0.01)^2 \text{ m}^2} = 116.7\times10^{-7}\times1\times10^4$$

$$= 116.7\times10^{-3} \text{ J m}^{-2}$$

This enthalpy may be considered as the thermal energy in J required to increase the surface area of the liquid by one square metre.

Problem 5.5

(a) The surface tension of acetone is 23.7×10^{-3} N m^{-1} at 293 K and its density is 0.780 g cm^{-3}. Calculate the maximum radius of the capillary in cm that permits the rise of the liquid to 5 cm.

(b) An excess pressure of 167.2 Pa is applied to produce a semi-spherical bubble of a liquid at the end of a capillary tube of 0.5 mm radius which is immersed in the liquid. What is the surface tension of the liquid (assume T = 293 K) in dyne cm^{-1}.

Solution

(a) $23.7\times10^{-3} = \frac{1}{2}\times5\times10^{-2}\times9.80\times780\times r$

$$r = \frac{23.7\times10^{-3}\times2}{5\times10^{-2}\times9.8\times780} = 1.24\times10^{-4}$$

$$m = 1.24\times10^{-2} \text{ cm}$$

(b) $\Delta P = \frac{2\gamma}{r}$ or $167.2(\text{Nm}^{-2}) = \frac{2\times\gamma}{0.0005 \text{ n}}$

$$\gamma = \frac{167.2\times0.0005}{2} = (\text{N m}^{-2}\times m = \text{N m}^{-1})$$

$= 0.0418$ N m^{-1} (but, 1 N m^{-1} = 10^3 dyne cm^{-1})

$$= 0.0418\times10^3 = 41.8 \text{ dyne cm}^{-1}$$

Interfacial Tension, Work of Cohesion and Adhesion, Eötvos Constant

Problem 5.6

(a) Calculate the maximum pressure in pascals required to be applied to a capillary of radius 0.03 cm and which is dipping into a liquid up to a height of 5 cm. The surface tension and the density of the liquid are 22.6×10^{-3} N m^{-1} and 791 kgm^{-3} respectively at 293 K.

(b) The interfacial tension between water and the ethylacetate is 6.80 dyne cm^{-1}. The surface tension of water and organic solvent are 72.8 and 24.0 (both in dyne cm^{-1}). Calculate (i) the work of cohesion for water and for the ester, and (ii) the work of adhesion between the two liquids (assume $T = 293$ K).

Solution

(a) $\rho_{max} = h \times g \times \rho + \frac{2\gamma}{r}$

$$= 5\times10^{-2}\times9.80\times791+\frac{2\times22.6\times10^{-3}}{0.03\times10^{-2}}$$

$$= 387.6+150.7 = 538.3 \text{ N m}^{-2} \text{ or Pa}$$

(b) (i) **Work of cohesion:** If a column of pure phase-1 is pulled apart, an area of surface 2 m^2 is formed. The work of separating the phase (as shown), is also equal to increase in the Gibb's free energy.

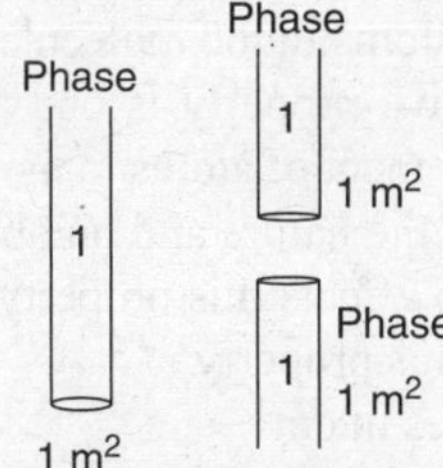

Work of cohesion for water (1) $= 2\times72.8$

$= 145.6$ dyne cm^{-1}

$= 145.6$ erg cm^{-2}

$= 145.6\times10^{-3}$ N m^{-1}

$= 145.6$ J m^{-2}

$\Delta G_1 = W_1(c) = 2\gamma^{1v}$ (when v is the vapour phase)

Work of cohesion for ethyl acetate (2):

$W_2(c) = 2\gamma = 2\times24.0$

$= 48.0\times10^{-3}$ N m^{-1}

$= 1\times48.0\times10^{-3}$ J m^{-2}

(ii) **Work of adhesion:**

$$\overline{W}_A^{12} = \gamma^{12}+\gamma^{2r}-\gamma^{12}$$

$$= (72.8+24.0-6.8)\times10^{-3}$$

$$= 90.0\times10^{-3} \text{ J m}^{-2}$$

Problem 5.7

(a) The Eötvos constant Z, valid for many liquids, is 2.27×10^{-7} J K^{-1} $mol^{2/3}$. Given that the density of CCl_4 at 40°C is 1413 kg m^{-3} and its critical temperature T_c as 283°C, calculate the surface tension of CCl_4 at 40°C. (M of $CCl_4 = 154$)

(b) The surface tension of nitrobenzene at 273 K is 46.4 dyne cm^{-1} and its density is 1.223 g cm^{-3}. Given that its critical temperature is 447°C, calculate the Eötvos constant Z in units of J K^{-1} $mol^{-2/3}$. Use this Z datum to calculate its surface tension at 293 K when its density is 1.204 g cm^{-3}.

Solution

(a) At 40°C, $d = 1413$ kg cm^{-3}

$$\text{Molar volume} = \frac{0.154}{1413} = 1.09\times10^{-4} \text{ m}^3 \text{ kg}^{-1}$$

or $\gamma(1.09\times10^{-4})^{2/3}$

$$= 2.27\times10^{-7}\times(556-313) \text{ J K}^{-1} \text{ mol}^{2/3}\times\text{K}$$

$$\gamma = \frac{2.27\times10^{-7}\times243}{\left(1.09\times10^{-4}\right)^{\frac{2}{3}}}$$

$$= \frac{2.27\times10^{-7}\times243}{2.28\times10^{-3}} = 24.19\times10^{-3} \text{ N m}^{-1}$$

$$\left(\frac{\text{N m mol}^{-2/3}}{(\text{m}^3\text{ mol}^{-1})^{2/3}} = \frac{\text{N m mol}^{-2/3}}{\text{m}^2\text{ mol}^{-2/3}} = \text{N m}^{-1}\right)$$

(b) γ of nitrobenzene at 273 K $= 46.4\times10^{-3}$ N m^{-1}, $d =$ 1223 kg m^{-3}

$$\text{Molar volume} = \frac{0.123}{1223} = 1.006\times10^{-4} \text{ m}^3 \text{ mol}^{-1}$$

$$\text{or, } 46.4\times10^{-3}\times\left(1.006\times10^{-4}\right)^{2/3} = Z(720-273)$$

$$Z = \frac{46.4\times10^{-3}\times\left(1.006\times10^{-4}\right)^{\frac{2}{3}}\times10^{-3}}{447}$$

$$= 2.25\times10^{-7} \text{ J K}^{-1} \text{ mol}^{-2/3}$$

γ at 293 K:

$$\gamma\left(\frac{0.123}{1204}\right)^{\frac{2}{3}} = 2.25\times10^{-7}(720-293)$$

$$\gamma\left(1.02\times10^{-4}\right)^{\frac{2}{3}} = 2.25\times10^{-7}\times427$$

$$\gamma = \frac{2.25\times10^{-7}\times427}{2.183\times10^{-3}}$$

$$= 440.1\times10^{-4}$$

$$= 44.01\times10^{-3} \text{ N m}^{-1}$$

$$= 44.01 \text{ dyne cm}^{-1}$$

Activation Energy of Viscous Flow, Viscosity and Fluidity of Liquids

Problem 5.8

(a) The viscosity of chloroform is 5.1×10^{-4} Pa s at 303 K and 4.65×10^{-4} Pa s at 313 K. Calculate the activation energy for viscous flow, E_{vis}, of the liquid.

(b) Calculate η at 283 K using the above result.

Solution

(a) $\log\left(\frac{4.65\times10^{-4}}{5.1\times10^{-4}}\right) = \frac{E}{2.303\times8.314}\times\frac{303-313}{303\times313}$

$$-0.0401 = \frac{E}{19.147}\times\frac{10}{94839};\ E = \frac{0.0401\times19.147\times94839}{10}$$

$$= 7281.7 \text{ J} = 7.28 \text{ kJ mol}^{-1}$$

(b) $\log\left(\frac{5.1\times10^{-4}}{\eta_1}\right) = -\frac{7.28\times10^3}{8.314\times2.303}\left(\frac{1}{303}-\frac{1}{283}\right)$

$$= \frac{-7.28\times10^3\times20}{303\times283\times2.303\times8.314}$$

$$\log\left(\frac{5.1\times10^{-4}}{\eta_1}\right) = \frac{-7.28\times10^3\times20}{19.147}\times\frac{1}{283\times2.303}$$

$$5.1\times10^{-4} = 8.153\times10^{-1}\times\eta_1$$

$$\eta_1 = \frac{5.1\times10^{-4}}{0.8153} = 6.25\times10^{-4}\ \text{Pa s}$$

Problem 5.9

(a) The densities of *n*-hexane and methanol at 293 K are 654.4 g dm^{-3} and 791.0 g dm^{-3} respectively. The viscosity of methanol at this temperature is 5.93×10^{-4} Pa s. The times required by methanol and *n*-hexane to pass between the marks of an Ostwald viscometer are 50.0 s and 32.60 s under the given conditions. What is the viscosity of *n*-hexane in Pa s?

(b) The fluidity of chloroform at 293 K is 1776 (Pa s)$^{-1}$ = Pa^{-1} s^{-1}. The constant *b* and the specific volume (at 293 K) are 4.53×10^{-4} m^3 kg^{-1} and 6.715×10^{-4} m^3 kg^{-1}. Calculate *a* of the equation relating it to fluidity. What are its units? Use these data to calculate the viscosity of chloroform at 298 K. *V* at 298 K = 6.823×10^{-4} m^3 kg^{-1}.

Solution

(a) $\frac{\eta_{\text{hexane}}}{\eta_{\text{methanol}}} = \frac{t_{\text{hexane}}\times\rho_{\text{hexane}}}{t_{\text{methanol}}\times\rho_{\text{methanol}}}$

$$\frac{\eta_{\text{hexane}}}{5.93\times10^{-4}} = \frac{32.6\times654.4}{50\times791.0}$$

$$\eta_{\text{hexane}} = \frac{5.93\times10^{-4}\times32.60\times654.4}{50\times791.0}$$

$$= 3.2\times10^{-4}\ \text{Pa s}$$

(b) $\upsilon = b + a\phi$

Using the given data at 293 K, *a* is first evaluated.

$$6.715\times10^{-4}\ (\text{m}^3\ \text{kg}^{-1})$$

$$= 4.530\times10^{-4}\ (\text{m}^3\ \text{kg}^{-1}) + a\times1776\left(\text{kg}^{-1}\ \text{m s}^{-1}\right)$$

$$a = \frac{6.715\times10^{-4}-4.530\times10^{-4}}{1776}\times\frac{\text{m}^3\ \text{kg}^{-1}}{\text{kg}^{-1}\ \text{m s}^{-1}}$$

$$= 1.23\times10^{-7}\ \text{m}^2\ \text{s}^{-1}$$

$$\phi = \frac{\upsilon - b}{a} = \frac{6.823\times10^{-4}-4.53\times10^{-4}}{1.23\times10^{-7}}$$

$$= 1.864\times10^{7}(\text{Pa s})^{-1}$$

$$= 1.864\times10^{3}\ \text{kg}^{-1}\ \text{m s}$$

$$\text{or } \eta = 5.635\times10^{-4}\ \text{kg m}^{-1}\ \text{s}^{-1}\ (\text{Pa s})$$

Molar Volume and Parachor Calculations

Problem 5.10

(a) Define the terms (i) additive property, and (ii) constitutive property with an example for each.

(b) Using the following volume equivalents of some elements and bonds, calculate the molar volumes at the boiling point of the compounds listed: (i) diethyl ether (ii) acetone (iii) ethyl acetate (iv) *n*-butyl alcohol (v) chlorobenzene. [Volume equivalents of elements: H = 4.6, C = 12.9, Cl = 22.5; –O– (ether linkage) = 7.6; O= (double bond) = 12.1; C_6H_5 ring = –15.0]

Solution

(a) (i) The sum of properties (under consideration) of all the constituents of a system is known as additive property. The mass of a molecule like, say CO_2, which is equal to 44 is the sum of the atomic masses of carbon and two oxygen atoms in the molecule.

(ii) The constitutive property is one that depends mainly on the arrangement of atoms in the molecule. However, in most cases the nature and number of atoms within a molecule also affects this property. Parachor is a good example for this property.

(b) (All molar volumes in cm^3)

(i) Diethyl ether: C_2H_5–O–C_2H_5
Total carbon = 4; total hydrogen = 10; oxygen (–O–) = 1
Molar volume of ether = 4×12.9 + 10×4.6 + 7.6 = 105.2

(ii) Acetone: CH_3–(C=O)–CH_3
Total carbon = 3; total hydrogen = 6; oxygen (=O) = 1
Molar volume of acetone = 3×12.9 + 6×4.6 + 12.1 = 78.4

(iii) Ethyl acetate: CH_3–(C=O)O–C_2H_5
Total carbon = 3; total hydrogen = 8; oxygen (–O) = 1; oxygen (–O–) = 1
Molar volume of ethyl acetate = 4×12.9 + 8×4.6 + 7.6 + 12.1 = 51.6 + 36.8 + 19.7 = 108.1

(iv) *n*-Butyl alcohol: C_4H_9OH
Molar volume = 5×12.9 + 10×4.6 + 7.6 = 118.1

(v) Chlorobenzene: C_6H_5Cl
Molar volume = 6×12.9 + 5×4.6 + 22.5 – 15.6 = 122.9 – 15.6 = 107.3

Problem 5.11

(a) The parachor equivalents of some elements and bonds are given below:
hydrogen (H) = 16.65; carbon (C) = 6.0; oxygen (O) = 20.0; nitrogen (N) =12.5; double bond (=) = 23.2; triple bond (≡) = 46.6; six-membered ring = 6.1; chlorine (Cl) = 53.8. Use the above data and calculate the parachor of nitrobenzene and ethanol.

(b) Using the parachor data, calculate the surface tensions of the above compounds. The densities of nitrobenzene and ethanol at 298 K are 1.187 g cm^{-3} and 0.785 g cm^{-3}.

Solution

(a) Nitrobenzene $C_6H_5NO_2$
There is one benzene ring, 6 carbons, 5 hydrogens, 5 double bonds, 1 nitrogen and 2 oxygens.

Parachor of nitrobenzene = 6.1 + 6×6.0 + 5×16.7 + 12.5 + 2×20.0 + 5×23.2 = 294.1

Ethanol, C_2H_5OH:

There are 6 hydrogens, 2 carbons and 1 oxygen.

Parachor of ethanol = 6×16.7 + 2×6.0 + 20.0 = 132.2

(b) Surface tension of nitrobenzene at 298 K

$$[P] = 294.1 = \frac{M\gamma^{\frac{1}{4}}}{\rho} = \frac{123}{1.187}\times\gamma^{\frac{1}{4}}$$

$$\gamma^{\frac{1}{4}} = \frac{294.1\times1.187}{123} = 2.838$$

$$\gamma = (2.838)^4 = 64.9 \text{ dyne cm}^{-1}$$

The actual value of γ of nitrobenzene at 298 K is 43.2 dyne cm^{-1}. Thus, this method does not work for the calculation of γ using parachor.

Surface tension of ethyl alcohol at 298 K

$$[P] = 132.2 = \frac{46}{0.785}\times\gamma^{\frac{1}{4}}$$

$$\gamma^{\frac{1}{4}} = \frac{132.2\times0.785}{46} = 2.256$$

$$\gamma = (2.256)^4 = 25.90 \text{ dyne cm}^{-1}$$

The actual value of γ of ethanol to nitrobenzene at 298 K = 21.8 dyne cm^{-1}

Refractive Index and Molar Refraction

Problem 5.12

(a) Define the relation between molar refraction and the refractive index of a compound and explain the terms. Does it vary with wavelength of light used? Is it an additive or a constitutive property?

(b) The refraction equivalents for the sodium D-line of a few elements, structural units and bonds are given below: hydrogen (H) = 1.1; carbon (C) = 2.4; oxygen (in –OH group) = 1.5; oxygen (in ether linkage) =1.6; oxygen (in carbonyl group) = 2.2; double bond (=) = 1.7; triple bond (≡) = 2.3; nitrogen in C≡N group = 5.4; Cl = 5.97. Calculate the molar refraction of (i) methyl ethyl ketone (also known as 2-butanone), (ii) toluene, (iii) *n*-propyl alcohol (iv) benzonitrile.

Solution

(a) The refractive index $[R]_M$ of a compound is related to its refractive index *n* by the relation

$$[R]_M = \frac{n^2-1}{n^2+2}\cdot\frac{M}{\rho}$$

Since the refractive index varies with the wavelength of light used, $[R]_M$ also varies with the wavelength of light used. $[R]_M$ is an additive as well as constitutive property.

(b) (i) Molar refraction of methyl ethyl ketone

It has 4 carbons, 8 hydrogens, 1 oxygen in C=O group and 1 double bond.

$[R]_M$ = 4×2.4 + 8×1.1 + 2.2 + 1.7 = 22.3 cm^3

(ii) Molar refraction of toluene: There are 7 carbons, 8 hydrogens and 3 double bonds.

$[R]_M$ = 7×2.4 + 8×1.1 + 3×1.7 = 30.7 cm^3

(iii) Molar refraction of *n*-propyl alcohol: It has 3 carbons, 8 hydrogens, 1 oxygen in OH– group.

$[R]_M$ = 3×2.4 + 8×1.1 + 1.5 = 17.5 cm^3

(iv) Molar refraction of benzonitrile: It has 7 carbons, 5 hydroges, 1 nitrogen and 1 triple bond.

$[R]_M$ = 7×2.4 + 5×1.1 + 5.4 + 2.3 = 30.0 cm^3

Problem 5.13

The refractive indices and densities of methyl ethyl ketone, toluene, *n*-propyl alcohol and benzonitrile are given.

	Methyl ethyl ketone	Toluene	*n*-Propyl alcohol	Benzonitrile
Refractive index (n_D)	1.3788	1.4960	1.3870	1.5289
Density (g cm^{-3})	0.805	0.870	0.803	1.01
	(At 298 K)	(At 293 K)	(At 298 K)	(At 293 K)

Calculate the $[R]_M$ values and compare them with results of previous problem.

Solution

(i) Methyl ethyl ketone

$$[R]_M = \frac{(1.3788)^2-1}{(1.3788)^2+2}\times\frac{72}{0.805}$$

$$= \frac{0.9011}{3.901}\times89.44 = 20.66 \text{ cm}^3$$

(ii) Toluene $[R]_M = \dfrac{(1.490)^2-1}{(1.490)^2+2}\times\dfrac{92}{0.870}$

$$= \frac{1.238}{4.238}\times105.75 = 30.89 \text{ cm}^3$$

(iii) *n*-propyl alcohol

$$[R]_M = \frac{(1.387)^2-1}{(1.387)^2+2}\times\frac{60}{0.803}$$

$$= \frac{0.924}{3.924}\times74.72 = 17.59 \text{ cm}^3$$

(iv) Benzonitrile $[R]_M = \frac{(1.528)^2 - 1}{(1.528)^2 + 2} \times \frac{103}{1.01}$

$$= \frac{1.335}{4.335} \times 101.98 = 31.40 \text{ cm}^3$$

While the agreement between calculated values of refraction equivalents of atoms and bonds (Problem 12) for toluene and *n*-propyl alcohol with those from the equation for $[R]_M$ are excellent, the agreement is somewhat poor in the case of methylethylketone and benzonitrile.

Problem 5.14

(a) Ethanol and benzyl alcohol are both polar in nature. Their density and refractive index data at 293 K are given.

	Ethanol (C_2H_5OH)	Benzyl alcohol ($C_6H_5CH_2OH$)
Refractive index (n_D)	1.3624	1.5396
Density (g cm^{-3})	0.789	1.044

Calculate their molar refraction values.

(b) Refractive index data may be used to calculate molecular radii of molecules assuming them to be spherical. Calculate the radii of these molecules using the given data. [Hint: Assume $[R]_M$ to be equal to $(4N\pi r^3)/3$]

Solution

(a) Ethanol

$$[R]_M = \frac{n^2 - 1}{n^2 + 2} \times \frac{M}{\rho} = \frac{(1.3624)^2 - 1}{(1.3624)^2 + 2} \times \frac{46}{0.789}$$

$$= \frac{0.856}{3.856} \times 58.30 = 12.94 \text{ cm}^3$$

Benzyl alcohol

$$[R]_M = \frac{(1.5396)^2 - 1}{(1.5396)^2 + 2} \times \frac{108}{1.044}$$

$$= \frac{1.3704}{4.3704} \times 103.45 = 32.44 \text{ cm}^3$$

(b) (i) r for ethanol

$$[R]_M = 12.94 = \frac{4}{3} \times 3.142 \times 6.023 \times 10^{23} \times r^3$$

$$r^3 = \frac{12.94}{25.23 \times 10^{23}} = 5.13 \times 10^{-24}$$

$$r = \sqrt[3]{5.13 \times 10^{-24}} = 1.73 \times 10^{-8} \text{ cm} = 1.73 \text{ Å}$$

(ii) r for benzyl alcohol

$$[R]_M = 32.44 = \frac{4}{3} \times 3.142 \times 6.023 \times 10^{23} \times r^3$$

$$= 25.23 \times 10^{23}$$

$$r^3 = \frac{32.44}{25.23 \times 10^{23}} = 1.286 \times 10^{-23}$$

$$r = \sqrt[3]{1.286 \times 10^{-23}} = 2.34 \times 10^{-8} \text{ cm} = 2.34 \text{ Å}$$

Molar and Electron Polarisation in Molecule

Problem 5.15

The density (d), refractive index (n_D) and dielectric constant (D) of two solvents at 293 K are given.

	Cyclopentane (C_5H_{10})	*n*-Heptane (C_7H_{16})
Density (g cm^{-3})	0.7454	0.6837
Refractive index (n_D)	1.4064	1.3876
Dielectric constant (D)	1.97	1.92

Calculate the molar polarisation, electron polarisation and atom polarisation of these molecules.

Solution

The molar polarisation (P^M) of cyclopentane is given by

$$P^M = \frac{D - 1}{D + 2} \times \frac{M}{\rho}$$

$$= \frac{1.97 - 1}{1.97 + 2} \times \frac{70}{0.7454}$$

$$= \frac{0.97}{3.97} \times 93.91 = 22.95 \text{ cm}^3$$

The electronic polarisation (P^E) of cyclopentane is given by

$$P^E = \frac{n^2 - 1}{n^2 + 2} \times \frac{M}{d}$$

$$= \frac{(1.4064)^2 - 1}{(1.4064)^2 + 2} \times \frac{70}{0.7457}$$

$$= \frac{0.978}{3.978} \times 93.91$$

$$= 23.09 \text{ cm}^3$$

$$P^M = P^E + P^A$$

But the difference between P^M and P^E is small and may be neglected. This is within limits of expected error. Then P^A is negligible and the molecule is obviously non-polar.

n-Heptane

$$P^M = \frac{1.92 - 1}{1.92 + 2} \times \frac{100}{0.6837}$$

$$= \frac{0.92}{3.92} \times 146.3$$

$$= 34.33 \text{ cm}^3$$

$$P^E = \frac{n^2 - 1}{n^2 + 2} \times \frac{100}{0.6837}$$

$$= \frac{(1.3876)^2 - 1}{(1.3876)^2 + 2} \times 146.3$$

$$= \frac{0.9254}{3.9254} \times 146.3$$

$$= 34.49 \text{ cm}^3$$

The same arguments as in the case cyclopentane are applicable and thus P^A is negligible. The molecule being a hydrocarbon, is non-polar.

Problem 5.16

The molar polarisation of cyclohexane at 293 K is 27.7 cm^3. A mixture of ether in cyclohexane with $X_{\text{ether}} = 0.088$ has a polarisation of 30.38 cm^3. Calculate the molar polarisation of diethyl ether.

(a) Using the refraction equivalents of various elements and bonds given earlier (see Problem 12) calculate the molar refraction of diethyl ether.

(b) What is the dipole moment in (i) debye units (ii) coulomb metre units

Solution

(a) Using $P_{A,B} = P_A X_A + P_B X_B$ (A = cyclohexane, B = diethyl ether)

$$30.38 = (1 - 0.088) \times 27.7 + 0.088 \times P_B$$

$$= 25.26 + 0.088 \times P_B$$

$$P_B = \frac{30.38 - 25.26}{0.088}$$

$$= \frac{5.12}{0.088} = 58.18 \text{ cm}^3$$

(b) Diethyl ether ($C_2H_5OC_2H_5$)

Adding refraction equivalents of C, H and O (in the ether linkage)

$$[R]_M = 4 \times 2.4 + 10 \times 1.1 + 1.6 = 22.2 \text{ cm}^3$$

$$P^0 = P^M - [R]_M = \frac{4\Pi N\mu^2}{9kT}$$

$$58.18 - 22.20 = 35.98 = \frac{4\Pi N\mu^2}{9kT}$$

$$\mu^2 = \frac{35.98 \times 9 \times 1.38 \times 10^{-16} \times 293}{4 \times 3.142 \times 6.023 \times 10^{23}}$$

$$= \frac{131006.2}{75.67} \times 10^{-39}$$

$$= 1731.3 \times 10^{-39} = 1.7313 \times 10^{-36}$$

$$\mu = \sqrt{1.731 \times 10^{-36}} = 1.351 \times 10^{-18} \text{ esu cm}$$

$$= 1.315 \text{ D}$$

$$1 \text{ D} = 10^{-30} \text{ C} \times \text{m}$$

$$\text{or } \mu = 1.315 \times 10^{-30} \text{ C} \times \text{m}$$

Problem 5.17

The permanent dipole moment of chlorobenzene is 1.55 D. Calculate its orientation polarisation (P_{OP}) at 293 K. Its dielectric constant at 293 K is 5.94 and the density at this temperature is 0.867 g cm^{-3}. What is its molar polarisation and induced polarisability?

Solution

$$P^M = \frac{D-1}{D+2} \times \frac{M}{\rho}$$

$$= \frac{5.94 - 1}{5.94 + 2} \times \frac{112.5}{0.867}$$

$$= 0.622 \times 129.76 = 807 \text{ cm}^3$$

$$P_{OP} \left(\text{or } P_D\right) = \frac{4\Pi N\mu^2}{9kT}$$

$$= \frac{4 \times 6.023 \times 10^{23} \times (1.55)^2 \times 10^{-36} \times 3.142}{9 \times 1.38 \times 10^{-16} \times 293}$$

$$= \frac{181.86 \times 10^{-36} \times 10^{23}}{3639.1 \times 10^{-36}}$$

$$= 0.0499 \times 10^3 = 49.97 \text{ cm}^3$$

$$\frac{4}{3}\Pi N\alpha = 15.9$$

$$\alpha = \frac{15.9 \times 3}{4 \times 3.141 \times 6.023 \times 10^{23}}$$

$$= 0.63 \times 10^{-23}$$

Relation of Molar Polarisation to Dipole Moment and Effect of Temperature

Problem 5.18

The density of HCl ρ is 1.64 kg m^{-3} (or g L^{-1}), its refractive index n_D at 293 K = 1.0004456 and the dielectric constant D = 1.0046. Calculate the molar polarisation P^M, molar refraction $[R]_M$ and the dipole moment μ, in debye units. (Assume d and D are also at 293 K).

Solution

$$P^M = \frac{D-1}{D+2} \times \frac{M}{\rho}$$

$$= \frac{1.0046 - 1}{1.0046 + 2} \times \frac{36.5}{0.00164}$$

$$= \frac{0.0046}{3.0046} \times 22256.1 = 34.07 \text{ cm}^3$$

$$[R]_M = \frac{n^2 - 1}{n^2 + 2} \times \frac{M}{d}$$

$$= \frac{(1.0004456)^2 - 1}{(1.0004456)^2 + 2} \times \frac{36.5}{0.00164}$$

$$= \frac{1.0009 - 1}{1.0009 + 2} \times 22256.1$$

$$= \frac{20.03}{3.000} = 6.675 \text{ cm}^3$$

$$P^M - [R]_M = 34.07\ \text{cm}^3 - 6.675\ \text{cm}^3$$

$$= 27.395\ \text{cm}^3$$

According to the Debye equation,

$$P^M - [R]_M = 27.395\ \text{cm}^3$$

$$= \frac{4\Pi N\mu^2}{9kT}$$

$$= \frac{4\times 3.142\times 6.023\times 10^{23}}{9\times 1.38\times 10^{-16}\times 293}\times\mu^2$$

$$\mu^2 = \frac{27.395\ \text{cm}^3\times 9\times 1.38\times 10^{-16}\times 293\ \text{g cm}^2\ \text{s}^{-2}\ \text{K}^{-1}}{4\times 3.142\times 6.023\times 10^{23}\ \text{molecules mol}^{-1}}$$

$$= \frac{340.246\times 10^{-16}\times 293}{75.7\times 10^{23}}$$

$$= \frac{99692.1}{75.7}\times 10^{-39}\ \text{g cm}^5\ \text{s}^{-2}$$

$$\mu = \sqrt{(1316.94\times 10^{-36}\times 10^{-3})}\ \text{g}^{\frac{1}{2}}\ \text{cm}^{\frac{5}{2}}\ \text{s}^{-1}$$

$$= 1.15\times 10^{-18}\ \text{esu cm} = 1.15\ \text{D}$$

Since esu has the units

$$\text{g}^{\frac{1}{2}}\ \text{cm}^{\frac{3}{2}}\ \text{s}^{-1} \text{(on the basis of } F = \frac{e_1e_2}{Dr^2};\ \text{dyne} = \text{g cm s}^{-2}\ \text{cm}^2 = (\text{esu})^2$$

$$\text{or esu} = \text{g}^{\frac{1}{2}}\ \text{cm}^{\frac{3}{2}}\ \text{s}^{-1}$$

$$\mu \text{ has the units of } \text{g}^{\frac{1}{2}}\ \text{cm}^{\frac{3}{2}}\ \text{s}^{-1}\times\text{cm} = \text{g}^{\frac{1}{2}}\ \text{cm}^{\frac{5}{2}}\ \text{s}^{-1}$$

Thus, it is seen that units on the LHS and RHS are the same. Note, however that this method is more appropriate for measurement of μ of polar liquids (in dilute solutions) in non-polar solvents.

Problem 5.19

The slopes of the plot of P^M vs $1/T$ for HBr and CH_3Cl are 2.727×10^3 cm^3 K^{-1} and 2.417×10^4 cm^3 K^{-1} respectively. Calculate their dipole moment in C m units.

Solution

The slope of plot of P^M vs $1/T$ is equal to $\frac{4\Pi N\mu^2}{9k}$

For HBr,

$$\mu^2 = \frac{2.727\times 10^3(\text{cm}^3\ \text{K}^1)\times 9\times 1.38\times 10^{-16}\ \text{g cm}^2\ \text{s}^{-2}\ \text{K}^1\ \text{molecule}^{-1}}{4\times 3.142\times 6.023\times 10^{23}\ \text{molecules mol}^{-1}}$$

$$\mu = \sqrt{\frac{33.87\times 10^{-13}}{75.7\times 10^{23}}}$$

$$= \sqrt{(0.447\times 10^{-36})}$$

$$= 0.67\times 10^{-18}\ \text{g}^{\frac{1}{2}}\ \text{cm}^{\frac{5}{2}}\ \text{s}^{-1}$$

$$= 0.67\times 10^{-18}\ \text{esu cm}$$

Note that esu has the units $\text{g}^{\frac{1}{2}}\ \text{cm}^{\frac{3}{2}}\ \text{s}^{-1}$.

From the equation,

$$F(\text{dyne}) = \frac{e_1e_2}{Dr^2}\frac{(\text{esu})^2}{\text{cm}^2} = \text{g}\times\text{cm}\times\text{s}^{-2}\times\text{cm}^2$$

$$= \text{g}\times\text{cm}^3\times\text{s}^{-2}$$

$$\text{esu} = \text{g}^{\frac{1}{2}}\times\text{cm}^{\frac{3}{2}}\times\text{s}^{-1}$$

$$\text{esu}\times\text{cm} = \text{g}^{\frac{1}{2}}\ \text{cm}^{\frac{5}{2}}\ \text{s}^{-1}$$

But 1 esu = u = 3.335×10^{-10} C

$$\text{esu}\times\text{cm} = \text{esu}\times 10^{-2}\ \text{m}$$

Units of μ in coulomb.m = $3.335\times 10^{-10}\ \text{C}\times 10^{-2}\ \text{m}$

$$= 3.335\times 10^{-12}\ \text{C}\times\text{m}$$

Or, $\quad 0.67\ \text{esu cm} = 0.67\times 3.335\times 10^{-12}\text{C}\times\text{m}$

$$1\ \text{D} = 10^{-18}\ \text{esu cm}$$

$$= 10^{-18}\times 0.67\times 3.335\times 10^{-12}\ \text{C}\times\text{m}$$

$$= 2.24\times 10^{-30}\ \text{C}\times\text{m}$$

For CH_3Cl, the slope of P^M vs $1/T$ = 24166.7 cm^3 K^{-1}

$$\mu^2 = \frac{24166.7\times 9\times 1.38\times 10^{-16}}{4\times 3.142\times 6.023\times 10^{23}}$$

$$= 3965\times 10^{-36}\times 10^{-3}$$

$$= 3.965\times 10^{-36}$$

$$\mu = \sqrt{3.965\times 10^{-36}}$$

$$= 1.99\times 10^{-18}\ \text{esu cm} = 1.99\ \text{D}$$

$$1\ \text{D} = 3.335\times 10^{-30}\ \text{C m}$$

$$\text{D} = 6.64\times 10^{-30}\ \text{C m (coulomb.metre)}$$

Induced Polarisability

Problem 5.20

The dielectric constant of toluene at 293 K is 2.39 and its density is 0.867 g cm^{-3}. The induced polarisability, α is 1.23×10^{-23}. Using these data calculate (i) molar polarisation (ii) molar refraction (iii) dipole moment of this compound in coulomb$\times$metre unit.

Solution

$$\text{(i)}\ P^M = \frac{D-1}{D+2}\times\frac{M}{\rho} = \frac{2.39-1}{2.39+2}\times\frac{92}{0.867}$$

$$= \frac{1.39}{4.39}\times 106.1 = 33.63$$

$$[R]_M = \frac{4}{3}\pi N\alpha$$

$$= \frac{4}{3}\times 3.141\times 6.023\times 10^{23}\times 1.23\times 10^{-23} = 31.02$$

$$P^M - [R]_M = 33.63 - 31.02 = 2.61 = \frac{4\pi N\mu^2}{9kT}$$

$$\mu^2 = \frac{2.61\times 9\times 1.38\times 10^{-16}\times 293}{4\times 3.142\times 6.023\times 10^{23}}$$

$$= \frac{9497.9}{75.67}\times 10^{-39}$$

$$= 125.52\times 10^{-39}$$

$$= 0.12552\times 10^{-36}$$

$$\mu = \sqrt{0.12552\times 10^{-36}}$$

$$= 0.354\times 10^{-18}\text{ esu cm} = 0.354\text{ D}$$

$$1\text{ D} = 3.335\times 10^{-30}\text{ C m and } \mu = 0.354\times 3.335\times 10^{-30}\text{ C m}$$

$$= 1.18\times 10^{-30}\text{ C m}$$

EXERCISES

Practice Problems

1. (a) The internal pressure of phenol is 6.64×10^5 kPa. If the density of the liquid is 1073 kg m^{-3}, calculate its internal heat of vaporisation in J mol^{-1}.

(b) The internal latent heat of vaporisation of carbon tetrachloride is 29.3 kJ mol^{-1} and its density is 1.632 g cm^{-3} at 293 K. Calculate the internal pressure of the liquid in kPa.

2. (a) The latent heat of vaporisation of liquid CO_2 is 16.0 kJ mol^{-1}. Given that the density is 1.977 kg m^{-3}, calculate the internal pressure of the liquid in kPa.

(b) Surface tension of water is 72.75×10^{-3} N m^{-1} at 293 K and the density is 1025 g dm^{-3}. To what height in cm will the liquid rise in a capillary of 0.2 mm radius?

3. (a) Excess pressure of 0.539 kPa is applied to produce a semi-spherical bubble of a liquid at the end of a capillary tube of 0.1 mm radius which is immersed in liquid. What is the surface tension of the liquid in N m^{-1}?

(b) Calculate the maximum pressure in kPa required to be applied to a capillary of radius 0.05 cm and which is dipping in a liquid up to a height of 10 cm. The surface tension and the density of the liquid are 27.0×10^{-3} N m^{-1} and 1632 kg m^{-3}.

4. (a) The interfacial tension between water and benzene is 35.0×10^{-3} N m^{-1}. The surface tensions of pure solvents, water and benzene are 72.8×10^{-3} and 28.9×10^{-3} (both in N m^{-1}) respectively. Calculate the (i) work of cohesion for water and benzene, (ii) work of adhesion between two liquids (assume $T = 293$ K).

(b) The viscosity of an organic compound is 4×10^{-4} at 273 K and 3×10^{-4} at 303 K (both in Pa s). Calculate the activation energy for viscous flow of this liquid. Using this activation energy value, calculate its viscosity at 293 K.

5. (a) The surface tension of mercury and nitromethane at 293 K and the temperature coefficients are given.

	Surface tension γ (N m^{-1})	$\frac{\partial\gamma}{\partial T}$ (N m^{-1} K^{-1})
Mercury	425.4×10^{-3}	-0.205×10^{-3}
Nitromethane	36.80×10^{-3}	-0.168×10^{-3}

Calculate their surface enthalpies and explain their significance.

(b) The surface enthalpies and temperature coefficients of surface tension of benzene and two derivatives of benzene are given.

	Surface tension γ (N m^{-1})	$\frac{\partial\gamma}{\partial T}$ (dyne cm^{-1} K^{-1})
Benzene	66.7×10^{-3}	−0.129
Chlorobenzene	68.5×10^{-3}	−0.119
Nitrobenzene	78.5×10^{-3}	−0.118

Calculate the surface tensions at 293 K in dyne cm^{-1}.

6. (a) Calculate the molar volume at the boiling points of the following compounds. (i) Benzoic acid, (ii) methyl alcohol, (iii) acetophenone, (iv) *n*-propyl acetate, (v) 2-butene.

(b) Calculate the parachor of the following compounds: (i) benzyl chloride, (ii) *n*-butanol, (iii) acetylene, (iv) ethyl acetate (use the volume equivalent data from Problem 1 and parachor equivalent data from Problem 11 of the worked out problems).

7. (a) The surface tension and density for organic compounds at 28 K are given.

	Surface tension γ (dyne cm^{-1})	d (g cm^{-3})
Ethyl acetate	23.2	0.814
n-Butyl alcohol	25.0	0.806
Dimethyl sulfoxide	42.9	1.095
Hydrazine	66.9	0.950

Calculate their parachor values.

(b) The viscosity of benzene at 293 K is 6.5×10^{-4} Pa s. The density of this compound and that of nitrobenzene at the same temperature are 879 and 1175 both in kg m^{-3}. In an experiment it was found that both benzene and nitrobenzene require 60 s and 140 s between the marks of an Ostwald viscometer. Calculate the viscosity of nitrobenzene in poise.

8. (a) Derive the units of the Eötvos constant from the relevant equation.

(b) The surface tension of ethanol and its density are 22.3×10^{-3} N m^{-1} and 789 kg m^{-3} at 293 K. If its critical temperature is 513.9 K, calculate the Eötvos constant in the relevant units. Use this data to calculate the surface tension at 300 K given its density at this temperature is 798 kg m^{-3}.

9. (a) Calculate the molar refraction values of the following compounds for which the refractive index and density data are given at 293 K.

	Refractive index (η_D)	d (g$\times$cm^{-3})
Acetic acid	1.372	1.049
Chloroform	1.447	1.490
Ether	1.354	0.714
Glycerol	1.473	1.262

(b) Calculate the molar refraction values for the same compounds as above using refraction equivalents of the various elements and bonds (for this data, see Problem 12b of worked out problems).

10. (a) Calculate the molar polarisation, electronic polarisation and atom polarisation of the following compounds using the data given at 293 K.

	Dielectric constant	Refractive index (η_D)	d (g cm^{-3})
Carbon tetrachloride	2.2	1.461	1.632
Benzene	2.3	1.501	0.879

(b) Calculate the molecular radii assuming them to be spherical.

11. The density, refractive index (n_D) and dielectric constant (D) of chlorobenzene at 293 K are 1.106 g cm^{-3}, n_D= 1.525 and D = 5.71. Calculate the molar polarisation, molar refraction and dipole moment of the compound. Assume that the temperature is 293 K.

12. The slopes of the plot of P^M vs $1/T$ for dichloroethylene and ethyl chloride are 15.6×10^3 and 25.6×10^3 (both in cm^3 K^{-1}) respectively. Calculate the dipole moments in Cm units.

13. The dielectric constant (D), refractive index (n_D) and density of ethyl acetate at 293 K are 6.08, 1.3724 and 0.901 g cm^{-3} respectively. Calculate (i) molar polarisation, (ii) molar refraction, (iii) dipole moment, and (iv) induced polarisability of the compound.

14. The surface tension of water and ethyl acetate at 293 K are 72.8×10^{-3} N m^{-1} and 23.9×10^{-3} N m^{-1} respectively. The interfacial tension between water and ethyl acetate is 6.8×10^{-3} N m^{-1}. Calculate (i) the spreading coefficient of ethyl acetate on water (ii) work of adhesion between ethyl acetate and water.

15. The density of acetone and its surface tension, both at 293 K, are 780 kg m^{-3} and 23.7×10^{-3} N m^{-1} respectively. Its vapour pressure at the same temperature is 12.36 kPa. Calculate the vapour pressure inside the droplet with radii equal to (i) 1 μm (ii) 0.01 μm.

16. (a) The surface tension of benzene is 28.9×10^{-3} N m^{-1} and its density is 879 kg m^{-3}, both at 293 K. Calculate its parachor in units of $g^{1/4}s^{-1/2}$ cm^3 mol^{-1}.

(b) Given that the Eötvos constant (also known Ramsay–Shields constant) is 2.25×10^{-7} J K^{-1} mol$^{-2/3}$, calculate the critical temperature of benzene using the modified Ramsay–Shields equation.

17. (a) Explain the term mesomorphic state, with an example. Why is a compound that exhibits this state also called *liquid crystals*?

(b) Describe the characteristics of smectic and nematic phases with examples giving their transition and melting points.

18. What are LCDs? Explain their importance by citing some examples.

19. (a) The viscosity and density of n-pentane are 2.4×10^{-4} Pa s and 626 kg m^{-3}, at 293 K. Given that the constant b in the relation connecting specific volume and fluidity is 4.53×10^{-4} m^3 kg^{-1}, calculate a for the solvent.

(b) Using the above a and b, calculate the viscosity of n-pentane in Pa s at its boiling point of 309.1 K given that its density is 611 kg m^{-3} at this temperature (assume a and b are independent of temperature).

20. Derive an equation for the ratio of the vapour pressure inside a small droplet of a liquid to the vapour pressure outside in terms of surface tension of the liquid and the radius of curvature of the droplets.

Multiple Choice Questions

1. The internal pressure of CCl_4 at 298 K is 3447×10^5 N m^{-2}. Its latent heat of vaporisation (in kJ mol^{-1}) is (density = 1.59 g cm^{-3})
   (a) 49.6 (b) 33.3
   (c) 44.5 (d) 25.8

2. The van der Waals constant a for water vapour is 0.5532 Pa m^6 mol^{-2}. Its internal pressure (in Pa) is (molar volume of liq. water $=1.8\times10^{-5}$ m^3 mol^{-1})
   (a) 1.71×10^9 (b) 7.20×10^{10}
   (c) 5.34×10^9 (d) 4.63×10^{10}

3. The van der Waals constant a for SO_2 gas is 0.680 Pa m^6 mol^{-2}. Its internal pressure (in Pa) is
   (a) 2.0×10^9 (b) 4.0×10^7
   (c) 5.6×10^{10} (d) 3.4×10^8

4. The van der Waals constant a for NH_3 is 0.4225 N m^4 mol^{-2}. Given the density of liquid ammonia at STP as 682 kg m^{-3} and the isothermal compressibility $\beta_T = 7.2\times10^{-10}$ Pa^{-1}, the isothermal coefficient of expansion of liquid ammonia at 240 K is (in K^{-1})
   (a) 3.2×10^{-2} (b) 4.4×10^{-3}
   (c) 2.03×10^{-3} (d) 3.8×10^{-1}

5. In a capillary rise experiment, water rose to a height of 3.52 cm in a capillary of radius 0.05 cm at 25°C. Given that the density of water is 1000 kg m^{-3} and $g = 9.8$ m s^{-2}, the surface tension of water (in dyne cm^{-1}) is
   (a) 75.0 (b) 86.0
   (c) 70.6 (d) 81.5

6. For benzene, the temperature coefficient of surface tension is –0.1291. Given that its surface tension at 298 K is 0.0289 N m^{-1}, the surface enthalpy of the liquid (in J m^{-2}) is
   (a) 46.95 (b) 30.84
   (c) 42.69 (d) 38.53

7. The interfacial tension between water and benzene is 35.0 dyne cm^{-1}. The surface tensions of water and benzene are 72.0 and 28.9 both in dyne cm^{-1}. The work of adhesion between the two liquids at 298 K is (in J m^{-2}) is
   (a) 6.5×10^{-2} (b) 7.9×10^{-2}
   (c) 5.8×10^{-2} (d) 8.4×10^{-2}

8. The viscosities of ethanol at 20°C and 40°C are 1.200 and 0.834 centipoise respectively. The activation energy for the viscous flow of the liquid (in kJ mol^{-1}) is
   (a) 18.8 (b) 13.9
   (c) 25.6 (d) 21.4

9. The Eötvos constant for most liquids is 2.27×10^{-7}. Given that the density and critical temperature T_c of ethanol are 800 kg m^{-3} and 516.3 K, the surface tension of the liquid (in N m^{-1}) is
   (a) 3.93×10^{-2} (b) 5.62×10^{-2}
   (c) 3.32×10^{-2} (d) 4.54×10^{-2}

10. Parachor is considered to be a/an
    (a) colligative property
    (b) additive property
    (c) constitutive property
    (d) additive as well as colligative property

11. The density of hydrazine is 1.0024 g cm^{-3} and its surface tension is 67.0 dyne cm^{-1}. The parachor of the compound is
    (a) 78.5 (b) 96.9
    (c) 85.6 (d) 91.3

12. Pick out the correct statement from the following:
    (a) The high surface tension of water is due to the strong cohesive forces rather than the adhesive forces among its molecules.
    (b) The formation of a concave meniscus in a capillary rise experiment is due to the stronger interaction among water molecules than the cohesive forces among water molecules and glass.
    (c) The higher surface tension of mercury (compared to water) is due to strong cohesive forces of intermetallic bonding among its atoms.
    (d) The stronger adhesive forces in mercury result in the formation of a concave meniscus in a capillary rise experiment using mercury.

13. The refractive index of methanol for the sodium D-line is 1.322 and its density at 298 K is 0.787 g cm^{-3}. The radius of the molecule (in angstroms) from molar refraction is
    (a) 1.754 (b) 2.369
    (c) 2.963 (d) 1.975

14. The dielectric constant of methanol is 32.0. Using the molar polarisation value P_m from this data and the (R_m) from Problem 13, the dipole moment of methanol (in debye) is
    (a) 1.56 (b) 1.19
    (c) 1.94 (d) 2.35

15. The units of the Eötvos constant Z are
    (a) kg m^2 s^{-2} mol$^{2/3}$ K (b) kg m^3 s^{-2} mol$^{2/3}$ K^{-1}
    (c) kg m^2 mol$^{-2/3}$ s^{-2} K^{-1} (d) kg m^3 s^2 mol$^{-2/3}$ K

16. The viscosities (Vis) of water (w), mercury (Hg), gasoline (ga) increase in the order
    (a) Vis(ga) $<$ Vis(w) $<$ Vis(Hg)
    (b) Vis(w) $<$ Vis(ga) $<$ Vis(Hg)
    (c) Vis(Hg) $<$ Vis(w) $<$ Vis(ga)
    (d) Vis(w) $<$ Vis(Hg) $<$ Vis(ga)

17. The surface tension of chloroform is 26.7×10^{-3} N m^{-1} and its density at 298 K is 1.49 g cm^{-3}. In a capillary rise experiment using this compound, a capillary of radius 0.05 cm was used and a rise of liquid to a height of 10 cm was reached. The maximum pressure being applied to the capillary at its base (in Pa) is
    (a) 1068.5 (b) 1860.5
    (c) 1605.8 (d) 1506.8

18. Which among the following compounds corresponds to a mesomorphic material of nematic type
(a) benzoic acid **(b)** anisole
(c) *p*-azoxy anisole **(d)** *p*-toluene sulphonic acid

19. The parachor value for toluene is 245.3. Given its surface tension as 28.5 dyne cm^{-1}, the molar volume of the liquid (in m^3 mol^{-1}) is
(a) 2.54×10^{-3} **(b)** 4.15×10^{-5}
(c) 3.46×10^{-4} **(d)** 1.06×10^{-4}

20. Which of the following is correct, regarding the temperature-derivative of γ? S denotes the interfacial entropy while G and F indicate respectively the interfacial Gibbs and Helmholtz free energies.
(a) $(\partial\gamma/\partial T)_P = -S$ **(b)** $(\partial\gamma/\partial T)_P = -(\partial G/\partial T)$
(c) $(\partial\gamma/\partial T)_P = V(\partial P/\partial T)$ **(d)** $(\partial\gamma/\partial T)_P = -(\partial F/\partial T)$

21. Among the following pairs of liquids, which system has the largest interfacial tension at 298 K?
(a) Water–carbon tetrachloride
(b) Water–nitrobenzene
(c) Water–octanol
(d) Water–heptanoic acid

Answer Keys

Practice Problems

1. (a) L_i = internal heat of vaporisation = 58.17 kJ mol^{-1}

(b) $P_{ip} = 3.105\times10^5$ kPa

2. (a) $P_{ip} = 7.19\times10^2$ kPa

(b) $h = 7.24$ cm

3. (a) $\gamma = 26.95$ dyne cm^{-1}

(b) $P_{max} = 1.71$ kPa

4. (a) Work of cohesion for water =

$$W_{w(c)} = 2\gamma^{wv} = 2\times72.8\times10^{-3}$$
$$= 145.6\times10^{-3}\ \text{J m}^{-2}$$

Work of cohesion for benzene =

$$W_{B(c)} = 2\gamma^{wv} = 2\times28.9\times10^{-3}$$
$$= 57.8\times10^{-3}\ \text{J m}^{-2}$$

Work of adhesion = $W_A^{WB} = 66.7\times10^{-3}$ J m^{-2}

(b) $E_{vis} = 6.6$ kJ mol^{-1}; $\eta_1 = 3.28\times10^{-4}$ Pa s

5. (a) Surface enthalpy of Hg = 485.5×10^{-3} J m^{-2}

Surface enthalpy of nitromethane = 86.0×10^{-3} J m^{-2}

The surface enthalpy represents the enthalpy required to extend the surface area of a liquid by l sq.m (or 1 cm^2)

(b) For benzene, γ = 29.2 dyne cm^{-2}; for chlorobenzene, γ = 33.6 dyne cm^{-1}; for nitrobenzene, γ = 43.9 dyne cm^{-1}

6. (a) (i) Molar volume of benzoic acid = 122.6 cm^3; (ii) Molar volume of CH_3OH = 38.9 cm^3; (iii) Molar volume of acetophenone = 137.1 cm^3 (iv) Molar volume of *n*-propyl acetate = 130.2 cm^3 (v) Molar volume of 2-butene = 88.4 cm^3

(b) Parachor of (i) benzylchloride = 218.5; (ii) *n*-butyl alcohol = 210.5; (iii) acetylene = 91.9; (iv) ethylacetate = 220.4

7. (a) Parachor of (i) ethylacetate = 237.3; (ii) *n*-butyl alcohol = 205.3; (iii) dimethyl sulfoxide = 178.9; (iv) hydrazine = 96.30

(b) $\eta_{\text{nitrobenzene}} = 20.2\times10^{-4}$ N m^{-2} s

8. (a) Units of Z = Nm^{-1} m^2 mol$^{-2/3}$ K^{-1}
= J mol$^{-2/3}$ K^{-1}

(b) $Z = 1.517\times10^{-7}$ J mol$^{-2/3}$K^{-1}

γ at 300 K = 21.42×10^{-3} N m^{-1}

9. (a) Molar refraction of acetic acid = 12.99 cm^3; Molar refraction of $CHCl_3$ = 21.43 cm^3; Molar refraction of ether = 22.51 cm^3; Molar refraction of ether = 22.2 cm^3; Molar refraction of glycerol = 20.45 cm^3.

(b) Molar refraction of acetic acid = 14.6 cm^3; Molar refraction of $CHCl_3$ = 21.41 cm^3; Molar refraction of ether = 22.2 cm^3; Molar refraction of glycerol = 20.5 cm^3.

10. (a) Molar polarisation P_M of CCl_4 = 26.96 cm^3; Electron polarisation P^E of CCl_4 = 25.88 cm^3; Molar polarisation P_M of benzene = 26.83 cm^3; Electron polarisation P^E of benzene = 24.64 cm^3; Atomic polarisation = 2.19 cm^3.

(b) For benzene, r = 2.138 Å; for CCl_4, r = 2.173 Å

11. (a) Molar polarisation of chlorobenzene = 62.14 cm^3

(b) Molar refraction $[R]_M$ of chlorobenzene = 31.18 cm^3

(c) Dipole moment = 1.22 D

12. (i) For dichloroethylene, $\mu = 1.6\times10^{-30}$ C m

(ii) For ethylchloride, $\mu = 2.05\times10^{-30}$ C m

13. P^M of ethylacetate = 61.4 cm^3; $[R]_M$ of ethylacetate = 22.21 cm^3; $\mu = 1.373\times10^{-30}$ C m; α = induced polarisability = 1.55×10^{-23}

14. (i) Spreading coefficient of ethylacelate on water = 42.1×10^{-3} J m^{-2}

(ii) Work of adhesion between ethylacelate and water = 89.6×10^{-3} J m^{-2}

15. (i) When $r = 1$ µm, $p = 12.38\times10^3$ Pa

(ii) When $r = 0.01$ µm, $p = 14.30\times10^3$ Pa

16. (a) Parachor of benzene = 205.7 g$^{1/4}$ s$^{-\frac{1}{2}}$ cm^3 mol^{-1}

(b) $t_c = 554.7$ K

17. Theoretical question

18. $a = 2.75\times10^{-7}$ m^2 s^{-1}

$\varnothing = 4305(\text{Pa s})^{-1}$

$\eta = 2.30\times10^{-4}$ Pa s

19. (a) $a = 3.07\times10^{-7}$ m^2 s^{-1}

(b) Viscosity $\eta = 2.32\times10^{-4}$ Pa s

20. Theoretical question

Multiple Choice Questions

1. (b)	2. (a)	3. (d)	4. (c)	5. (b)	6. (d)	7. (a)	8. (b)	9. (c)	10. (b)
11. (d)	12. (c)	13. (d)	14. (b)	15. (c)	16. (a)	17. (d)	18. (c)	19. (d)	20. (a)
21. (a)									

CHAPTER 6

SURFACE CHEMISTRY, CATALYSIS AND COLLOIDS

CHAPTER OUTLINE

Adsorption
- Adsorption of Gases by Solids
- Types of Forces in Adsorption
- BET Equation

Adsorption from Solution
- Gibbs Adsorption Equation
- Applications of Adsorption Phenomena

Electrical Phenomena at Interfaces—Electric Double Layer and its Consequences
- Electrokinetic Phenomena
- Electrocapillary Phenomena

Colloidal Systems
- Particular Aspects of Colloids
- Classification of Colloidal Systems
- Properties of Colloidal Systems
- Coagulation or Precipitation of Colloids
- Protective Action of Colloids
- Zetapotential in Relation to Colloidal Systems and their Stability
- Emulsions
- Surfactants
- Micelles
- Gels
- Isoelectric Point in Lyophilic Sols
- Applications of Colloids

6.1 Adsorption

Just as a liquid surface is in a strained state, the surface of a solid is also under strain due to a residual force field. This results in a tendency for the free energy of the surface to decrease which leads to the phenomenon of *adsorption*. This term refers to the existence of a higher concentration of a component at the surface of a liquid or solid phase than is present in the bulk of the solution. Adsorption is a spontaneous process and leads to a decrease in interfacial area and thereby to a decrease in interfacial tension. The most common types of adsorption are—(i) adsorption of gases by solids, and (ii) adsorption of solutes from solution.

6.1.1 Adsorption of Gases on Solids

Adsorption occurs on all surfaces but the effects are very significant only when the adsorbent is porous and has a very large surface area for a given mass. Examples of adsorbents include wood charcoal, coconut shell charcoal (prepared by heating them in a vacuum or in an inert atmosphere to a high temperature of 1300 K), alumina, silica gel, and so on.

Effect of temperature and pressure: The extent of adsorption of a gas increases with increase in pressure at a given temperature and it decreases with increase in temperature. Thus adsorption is an exothermic process.

Types of Adsorption Isotherms

Freundlich adsorption isotherm: It may be represented by the equation

$$\frac{a}{m} = kp^{1/n} \tag{6.1}$$

where a is the amount of gas adsorbed by m grams of adsorbent at a pressure p. k and n are constants at the given temperature and n is greater than 1. From a plot of $\log\left(\frac{a}{m}\right)$ vs $\log p$, which is linear, k and n can be determined from the intercept and slope respectively. It is generally observed that the most easily liquefiable and highly soluble gases (such as SO_2, NH_3) are the most adsorbed gases by a given adsorbent (for example, charcoal).

Langmuir adsorption isotherm: The assumptions on which this isotherm is based are:

- The adsorbent surface contains a fixed number of sites for adsorption.
- At a given temperature and pressure, only a fraction θ of the sites is occupied by adsorbed molecules leaving $(1 - \theta)$ unoccupied.
- The heat of adsorption is independent of θ and is the same at all sites.

- No interaction is possible between molecules at different sites. Employing a kinetic approach, the rate of evaporation of molecules is equal to the rate of condensation of molecules on the surface.

Thus, $k\theta = k'P(1-\theta)$

where θ is the fraction of surface covered by gas molecules, k and k' are constants for the two processes and P is the pressure of the gas.

$$\theta = \frac{k'P}{k+k'P} = \frac{ZP}{1+ZP} \quad \text{where } Z = \frac{k'}{k} \tag{6.2}$$

and is a constant referred to as the *adsorption coefficient*. Note that the Langmuir isotherm is primarily based on a continuous coverage of surface with adsorbed molecules, and the surface gets saturated when the adsorbed layer is unimolecular in thickness. θ may be considered to be proportional to V (volume of gas adsorbed at given temperature and pressure) and we can write

$$\theta = \frac{V}{V'} \tag{6.3}$$

where V' is the volume adsorbed in the unimolecular layer.

$$\frac{V}{V'} = \frac{ZP}{1+ZP} \tag{6.3a}$$

or

$$\frac{1}{V} = \frac{1}{ZPV'} + \frac{1}{V'} \tag{6.4}$$

A plot of $\frac{1}{V}$ vs $\frac{1}{P}$ is linear with the intercept equal to $1/V'$ from which V' can be evaluated.

If the cross-sectional area of the adsorbed molecule is known, the surface area of the adsorbent can be calculated. Langmuir adsorption isotherm may also be expressed in a different way in terms of the amount of a gas or solute adsorbed as

$$\frac{a}{m} = x = \frac{kP}{k'P+1} \tag{6.5}$$

where a is the amount adsorbed by m grams of the adsorbent (or x is the mass of substance adsorbed by unit mass of adsorbent), k and k' are constants, and p is the equilibrium pressure. In the case of adsorption from solution, P is replaced by C, the concentration of the solute. Two special cases of interest are:

(i) at high pressures, $k'P \gg 1, \therefore \frac{a}{m} = x = \frac{k}{k'} =$ constant, which implies that the surface is completely covered by a unimolecular layer.

(ii) At low pressures $k' \ll 1$ hence $\frac{a}{m} = x = kpi.ex \propto p$.

6.1.2 Types of Forces in Adsorption

Physisorption: If the forces of interaction between the surface of the adsorbent and the adsorbate are of van der Waals type, the adsorption is referred to as van der Waals adsorption or physical adsorption.

Chemisorption: If chemical forces are involved between the adsorbed molecules and the surface, the process is referred to as chemisorption. The following table summarises the salient features of these two types of adsorption:

	Physisorption	Chemisorption
1.	Forces are van der Waals type	Forces are chemical in nature
2.	Heats of adsorption are low and are 20 to 40 kJ mol^{-1}	Heats of adsorption are similar to those of chemical reactions, 50–400 kJ mol^{-1}
3.	Because the forces are such as those occurring in liquefaction, this adsorption does not occur at temperatures above the critical temperature of a gas	Usually a monomolecular layer is formed on the surface
4.	Multilayer adsorption occurs if the pressure of the gas is near the vapour pressure of liquid adsorbate	Highly specific in nature and depends on the nature of gas and adsorbent
5.	It is generally reversible, that is, on decreasing the pressure, the desorbed gas follows the same isotherm	It is not reversible

6.1.3 BET Equation

Applying the Langmuir approach to the concept of multilayer adsorption of gas molecules, Brunauer, Emmett and Teller (1938) derived an equation relating the volume of gas adsorbed at a pressure p and V_m, the volume of gas adsorbed in the unimolecular layer. The equation is given by

$$\frac{p}{V(p_0 - p)} = \frac{1}{V_m C} + \frac{C-1}{V_m C}\frac{p}{p_0} \tag{6.6}$$

Here, C is a constant (equal to $e^{(E'-E_{H-C})/RT}$), E' is the heat evolved when adsorption occurs forming the first layer of gas molecules, E_{H-C} is the heat of condensation of gas to liquid, and p_0 is the vapour pressure of the condensed gas at the given temperature. A plot of the LHS of Eq. 6.6 against $\frac{P}{P_0}$ is linear with an intercept of $\frac{1}{V_m C}$ and slope equal to $\frac{C-1}{V_m C}$. C and V_m may be calculated from these data. Knowing the area covered by a single molecule, the surface area of the catalyst may be obtained. The measured value of C can be used to evaluate E' from

$$E' = -\Delta H^0_{\text{vap}} + RT \ln C \tag{6.7}$$

where ΔH^0_{vap} is the heat of vaporisation of the liquefied gas.

6.2 Adsorption from Solution

If the addition of any solute to a solvent reduces the interfacial tension, there will be a greater proportion of the solute at the interface than in the bulk of solution. Such substances are referred to as *surface active*. The adsorption of phenol or acetic acid from an aqueous medium containing finely divided charcoal is a good example of positive adsorption. When the addition of a solute increases the interfacial tension, it is said to be negatively adsorbed. Electrolytes are examples of negative adsorption.

6.2.1 Gibbs Adsorption Equation

This equation may be written as

$$\Gamma_2 = -\frac{1}{RT}\left(\frac{d\gamma}{d\ln a_2}\right)_{T,P} = -\frac{a_2}{RT}\cdot\frac{d\gamma}{da_2} \tag{6.8}$$

where $\Gamma_2 = \left(n_2 - n_1\frac{n_2^0}{n_1^0}\right)/x$, $\frac{d\gamma}{da_2}$ is the rate of change of interfacial tension with activity of the solute, n_2 is the number of moles of solute in n_1 moles of solvent in the surface phase, n_2^0 is the number of moles of solute in n_1^0 moles of solvent in the bulk phase, x is the area of interface. In dilute solutions, the activity may be replaced by concentration. a_2 (or C_2) is the activity of the solute and R, T have their usual significance.

When $\frac{d\gamma}{da_2}$ is negative, Γ_2 is positive (a case of positive adsorption), and if $\frac{dr}{da_2}$ is positive, Γ_2 is negative (a case of negative adsorption).

Insoluble surface films: Certain substances like long chain fatty acids (such as stearic acid, oleic acid) spread on the surface of water when placed in water as a solution in benzene. When the benzene evaporates, a unimolecular insoluble film is formed on the surface of water with the carboxyl groups being attached to water and the hydrocarbon chain pointing upwards because they are hydrophobic. By suitable experimental methods, it is observed that the area of surface per single molecule is found to be constant for many fatty acids irrespective of their chain length, namely, about 21 Å^2, that is, 21.0×10^{-20} sq.m.

6.2.2 Applications of Adsorption Phenomena

Chromatographic adsorption and separation by ion exchange resins are two important applications that involve the use of the principle of adsorption.

Chromatographic adsorption: Separation of the constituents of a mixture in solution by an adsorbent makes use of the fact that the adsorbing capacity of the adsorbent varies for the different constituents of the mixture. If a solution of the mixture is poured down a column consisting of the powdered adsorbent, the component most readily adsorbed is retained in the upper layers while the lower layers contain those components that are less readily adsorbed. A partial separation of the components is thus achieved which can be improved by other methods. Some common adsorbents include alumina, powdered charcoal and magnesium oxide.

Ion exchange process: Certain minerals such as zeolites and synthetic resins that are organic polymers containing carboxyl (–COOH), sulfonyl (–SO_3H) or hydroxyl (–OH) groups have the capacity to exchange cations or anions with those in solution. There are cation exchange resins as well as anion exchange resins. A cation exchange resin usually consists of a number of exchangeable hydrogen ions (attached to the resin by adsorption) and when placed in a solution of a salt containing Na^+ or other cations, an exchange involving the adsorption of the cation from solution by the resin takes place releasing H^+ ions from the resin into the solution. The ion exchange process may be described by the equation

$$\begin{aligned}&H_x\text{Resin} + xNa^+\,(\text{from solution}) \rightleftharpoons\\ &Na_x\text{Resin} + xH^+\,(\text{solution})\end{aligned} \tag{6.9}$$

Other ions can be exchanged in a similar way. The adsorbing capacity of a resin for different ions depends on two factors—charge on the ion and the size of the ion in its hydrated form. The separation of rare earths into individual components is an important application of this method.

6.3 Electrical Phenomena at Interfaces–Electrical Double Layer and its Consequences

When two phases differing in chemical composition are in contact, a potential difference is developed between them. Assuming that one phase is a metal and the other is an electrolyte solution, the potential difference across the interface may be depicted as an electrical double layer. One of the phases (say the metal) is positively charged (with a deficiency of electrons) while a net negative charge accumulates on the other phase (say the solution). Helmholtz (1879) visualised this double layer in a particular way, known as the Helmholtz model (Fig. 6.1). Considering this as equivalent to a parallel plate capacitor, the potential difference across the plates (ΔE) is given by

$$\Delta E = \frac{4\pi dQ}{\varepsilon A} \tag{6.10}$$

where d is the distance separating the oppositely charged plates, ε is the dielectric constant of the medium, Q/A = surface charge density (in C cm^{-2}).

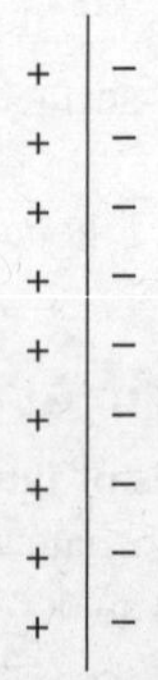

+ | –
+ | –
+ | –
+ | –
+ | –
+ | –
+ | –
+ | –
+ | –

Fig. 6.1 Helmholtz double layer

This model is too simplistic and does not take into account the thermal motion of ions in the liquid medium. Stern (1924) proposed a modification of this model and thus was born the idea of a diffuse double layer (Fig. 6.2). The potential drop across the double layer is also shown.

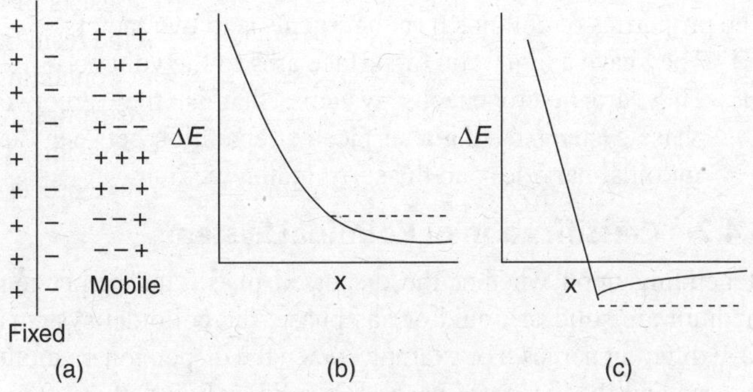

Fig. 6.2 (a) Stern's diffuse double layer
(b) Potential drop in fixed part of double layer
(c) Potential drop in mobile part of double layer

The potential drop in Fig. 6.2(b) and 6.2(c) may vary and depends on the charge distribution in fixed and mobile layers. The potential drop ζ in the diffuse layer is known as *zeta potential*. The application of a potential difference results in the movement of the diffuse layer. The zeta potential is affected by the presence of other ions, for example, by acids or bases.

6.3.1 Electrokinetic Phenomena

The existence of an electrical double layer at an interface (say solid–liquid interface, as above) leads to four effects—(i) electroosmosis, (ii) streaming potential, (iii) electrophoresis, and (iv) Dorn effect. A brief description of these effects is given below.

Electroosmosis: The movement or the motion of a liquid through membrane (or a diaphragm) under the influence of an applied electric field is known as electroosmosis. An equation similar to Eq. 6.10 may be used to express the zeta potential (ζ) as

$$\zeta = \frac{4\pi v\eta}{\varepsilon E} \tag{6.11}$$

where η = viscosity of the medium, v = uniform velocity attained by the liquid (as a result of the balancing of the electrical force causing electroosmotic flow and the frictional force between the moving liquid layer and the wall of the diaphragm), ε = dielectric constant of the medium and E = the potential gradient across the diaphragm (or membrane). The volume V of liquid transported electroosmotically per second is given by

$$V = qv \tag{6.12}$$

where q is the area of cross-section of all pores in the diaphragm and v has the same significance as described above. Substituting for v from Eq. 6.12, we have

$$\zeta = \frac{4\pi V\eta}{q\varepsilon E} \tag{6.13}$$

Expressing E as $E = I/qk$,

$$\zeta = \frac{4\pi kV\eta}{\varepsilon I} \tag{6.14}$$

where I = current strength and k = specific conductivity of the medium. From Eq. 6.14, it is seen that $V \propto I$.

Streaming potential: It is the potential difference that arises across a membrane or capillary tube when a stream of liquid is forced through it. This potential difference is due to opposite charges being built across the ends of the tube due to the diffuse double layer being swept along by the flow of the fluid. The streaming potential(s) produced may be related to the zeta potential across the diffuse double layer by the relation

$$S = \frac{\zeta P\varepsilon}{4\pi\eta k} \tag{6.15}$$

where P is the pressure driving the fluid. For a given capillary or diaphragm, it is seen that $S \propto P$.

6.3.2 Electrocapillary Phenomena

The mercury–solution interface is most advantageously employed in understanding electrified interfaces. Historically, it is known that this interface is affected by electric charge and that the interfacial tension at this interface, which is easily measurable, can be related to the potential difference applied to the interface. Further, it is possible to study the surface excess of cations or anions at this interface by simple experimental approaches.

Electrocapillary curve: The variation of the interfacial tension at such an interface (such as that at Hg–solution interface) as a function of the applied potential and the depiction of the curve of such variation with applied potential, is known as an *electrocapillary curve*. The nature of such a curve is shown in Fig. 6.3.

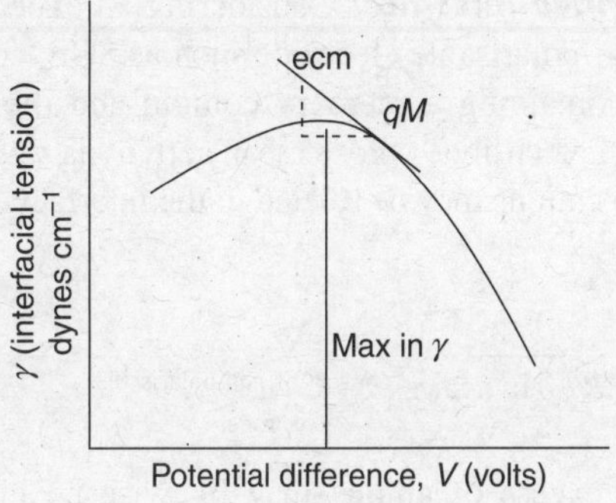

Fig. 6.3 Potential difference across the system

It is seen that the curve is parabolic in nature and that the interfacial tension reaches a maximum at a certain potential known as electrocapillary maximum (ecm). The interfacial tension also varies with the concentration of the electrolyte with the value becoming smaller with the increasing concentration of the electrolyte.

Lippmann equation: This equation is the slope of the electrocapillary curve at a given cell potential V.

$$\left(\frac{\partial r}{\partial V}\right)_{\substack{\text{constant solution}\\ \text{composition}}} = -q_M \tag{6.16}$$

By plotting the slopes of the elctrocapillary curve at various potentials, we get a linear plot representing the electrode change density q_M, as a function of V (Fig. 6.4).

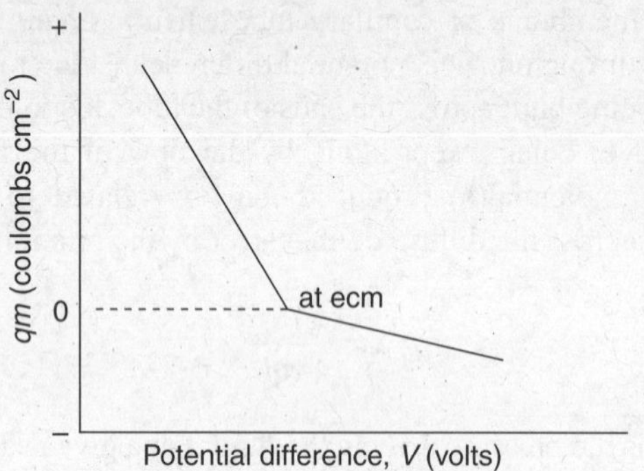

Fig. 6.4 Electrode charge density as a function of potential

It is seen from Fig. 6.4 that the excess charge on the electrode changes sign from positive to negative, passing through a value of zero charge at a certain potential. This potential difference at which $q_M = 0$ is known as potential of zero charge (PZC), when it is measured with respect to the hydrogen electrode. PZC is seen to be the potential at which ecm occurs. It can be mathematically represented as,

$$q_M = -\left(\frac{\partial r}{\partial V}\right)_{\text{constant solution composition}} = 0 \tag{6.17}$$

E_{PZC} is a very important parameter in that it is a reference potential in studies dealing with electrified interfaces.

Relation between interfacial tension and surface excess at an electrified interface: Considering an experimental set up with a non-polarisable electrode such as SCE, coupled with a capillary containing a mercury column and dipping into a solution of an electrolyte whose mean activity is $a\pm$, the surface excess of the cations may be related to the interfacial tension by the equation

$$\left(\frac{\partial r}{2RT\partial \ln a\pm}\right)_{\text{at const. } V \text{ of non-polarisable electrode, } V_-} = -\Gamma_+ \tag{6.18}$$

The surface excess of anions may likewise be given by the equation

$$\left(\frac{\partial r}{2RT\partial \ln a\pm}\right)_{\text{at const. } V \text{ of non-polarisable electrode, } V_+ \text{ like } H_2 \text{ electrode}} = -\Gamma_- \tag{6.19}$$

Note that for determining the surface excess of anions, the non-polarisable electrode should be reversible to cations like a hydrogen electrode.

6.4 Colloidal Systems

Molecular size	Colloidal particles	Suspensions
Ranges from	Ranges from	Ranges from
0.01–0.2 Å	0.5–20 Å	20–1000 Å

Definition: The colloids are an intermediate state between true solutions and coarse suspensions. The sizes of the three types of the systems are shown below and must be considered as approximate.

6.4.1 Particular Aspects of Colloids

The properties of colloids owe their genesis to two factors.

(i) They have a high ratio of surface area to the volume.

(ii) They are heterogeneous systems, that is, they are two phase systems having a surface of separation between the colloidal particles and the surrounding medium.

6.4.2 Classification of Colloidal Systems

Depending upon whether the dispersed phase and dispersion medium are solid or liquid or gas phase, the colloidal systems have different names. For example, when the dispersion medium is a gas and the dispersed phase is a solid or liquid, the system is referred as an aerosol. In the case where the dispersion medium is liquid, and dispersed phase is a liquid or gas, we have an emulsion or a foam as the colloidal system. For complete classification with examples, a standard book of physical chemistry may be consulted.

Different types of sols: (i) Lyophobic sols are those that do not have an affinity for the solvent. Metal sols or sulfur sol are good examples of this category. (ii) Lyophilic sols are those that have an affinity for the solvent. Some examples are protein sols, gums dispersed in water. The two classes of sols have characteristic differences among themselves. For example, lyophilic sols are very stable while lyophobic sols are not. Lyophobic sols require small quantities of electrolytes for coagulation while large quantities are required for coagulation in the case of lyophilic sols.

Preparation of various types of sols: Lyophilic sols such as those of starch and gelatin can be obtained by warming the substances with the dispersion medium. They are intrinsic colloids. For the preparation of lyophobic sols, the methods adopted are—dispersion methods, condensation methods in which typical chemical methods such as reduction, oxidation and double decomposition are employed. For example, a gold sol is prepared by the reduction of gold(III) chloride using formaldehyde in an aqueous medium. Sulfur sol is prepared by the oxidation of H_2S with SO_2 and silver halide sol is prepared by reacting silver nitrate solution with a suitable halide. Colloids need to be purified to prevent their coagulation and for this purpose, dialysis and ultrafiltration methods are employed.

6.4.3 Properties of Colloidal Systems

Optical properties: Colloids exhibit the *Tyndall effect.* They scatter light. When a beam of light is passed through a colloidal solution, it becomes visible when it is viewed in a direction perpendicular to the incident light.

Electrical properties: The particles of a hydrophobic sol move either towards the cathode or the anode under the application of an electric field. The direction of movement of the particles depends on the charge carried by the colloidal particles. This phenomenon is known as electrophoresis (mentioned earlier).

The electrophoretic mobility of colloidal particles is defined as the speed of the particles under an applied strength of 1 V cm^{-1}.

The charge carried by colloidal particles is due to the adsorption of ions (either positive or negative ions), usually those ions common to the ions of the colloidal system. For example, hydrous oxides of iron or aluminum sols are positively charged due to adsorption of cations while a sol like As_2S_3 is usually negatively charged due to adsorption of sulfide ions on their surface. The charge on the particles varies depending upon the method of preparation of sol.

6.4.4 Coagulation or Precipitation of Colloids

While traces of electrolytes are necessary for the stability of a sol, large amounts of electrolytes, when added, result in the precipitation of the colloidal particles. The Hardy–Schulze rule provides the necessary information governing the effect of charge on the precipitating power of an electrolyte. According to this rule, (i) an ion having a charge opposite to that of the colloidal particle is most effective in coagulation, and (ii) the coagulating power of an ion increases with increasing valency of the precipitating ion.

6.4.5 Protective Action of Colloids

When a lyophilic sol is added to a lyophobic sol, the latter becomes more stable and less sensitive to the action of electrolytes. The *gold number* of a lyophilic sol is a quantitative measure of the protective action of a lyophilic sol. It is defined as the mass in milligrams of a dry lyophilic sol which when added to 10 mL of standard gold sol prevents the precipitation of the sol on addition of 1 mL of 10% NaCl solution. Thus, the smaller the value of the gold number, the more effective is the substance in conferring protection to the lyophobic sol.

6.4.6 Zeta Potential in Relation to Colloidal Systems and their Stability

Consider a negatively charged colloidal particle (such as As_2S_3 sol). The electrical double layer surrounding the particle may be depicted as shown in Fig. 6.5.

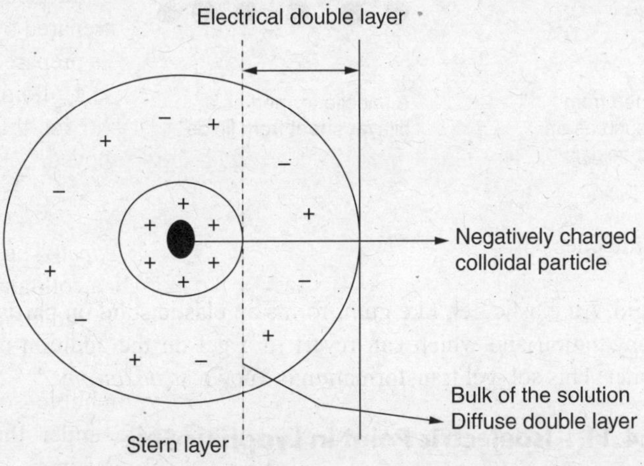

Fig. 6.5 Electrical double layer around a negatively charged colloidal particle

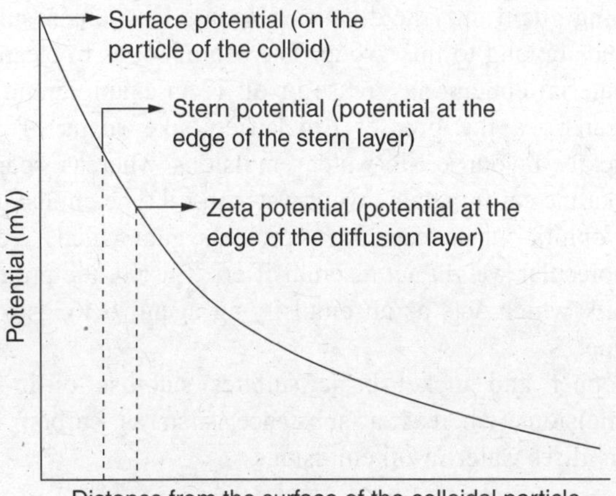

Fig. 6.6 Variation of potential with distance

If the colloidal particles in a given dispersion medium have large positive or negative zeta potentials, they will tend to repel each other and the colloid is stable. But if they have low zeta potential values, there is no force to prevent the colloidal particles coming together and the colloid flocculates. Colloidal systems which have zeta potentials more positive than +30 mV or more negative than –30 mV are stable systems. Zeta potential is affected by the pH of the medium.

The zeta potential of a colloidal system is related to the electrophoretic mobility of the colloidal particles through the relation

$$U_E = \frac{2\varepsilon\zeta f(ka)}{3\eta} \tag{6.20}$$

where U_E = electrophoretic mobility, ε = dielectric constant of medium, $f(ka)$ is a function of Debye length and has a value ranging from 1.5 to 1.0, η = viscosity of the medium. The value of 1.5 is used for $f(ka)$, for colloidal particles of size, 2×10^{-5} cm in solutions of electrolytes of concentration 10^{-3} M. For smaller particles in low dielectric constant media, using $f(K_a) = 1.0$, the zeta potential can be calculated.

6.4.7 Emulsions

An emulsion is a disperse system in which both the disperse phase and the dispersion medium are liquids. Examples of emulsions are milk, cod liver oil. In milk, globules of liquid fat are dispersed in water.

Types of emulsions: They are of two types—oil-in-water type, in which oil droplets are the disperse phase and water is the dispersion medium, and water-in-oil type, in which water droplets are the disperse phase and oil is the dispersion medium.

Emulsifying agents and stability of emulsions: Emulsions generally have poor stability and separate out on standing. They can be stabilised by adding small amounts of substances known as emulsifying agents. These emulsifying agents function by reducing the interfacial tension at the oil–water interface. Emulsifying agents are of three types.

- Long chain fatty acid salts or soaps; long chain sulfonic acids belong to this group. The use of soaps in cleaning a material containing grease or oil is an example and soap emulsifies the grease. Emulsifiers like soaps of alkali metals favour oil-in-water emulsions whereas soaps of alkaline earth metals favour water-in-oil type emulsions.
- Lyophilic substances such as proteins, gum which have high molecular weight act as emulsifiers. Casein, the protein in milk which acts as an emulsifying agent, belongs to this type.
- Copper and nickel basic sulfates stabilise oil-in-water emulsions whereas a substance such as carbon black stabilises water-in-oil emulsions.

Emulsions can be broken or de-emulsified by heating, centrifugation, and so on. They can also be broken by heating with an agent that destroys the emulsifier. For example, an acid converts a soap which is an ester, into a fatty acid which has no emulsifying power and hence the emulsion is broken.

Emulsions find many applications such as in laying roads, mining operations and formulation of drugs.

6.4.8 Surfactants

These are mostly organic compounds containing both hydrophobic groups and hydrophilic groups, that is, they are *amphiphilic* in nature. They get preferentially adsorbed at various interfaces such as oil–water, air–water, and so on. They form unimolecular layers at the interface with the hydrophilic (or polar) group pointing towards the aqueous phase and the hydrophobic hydrocarbon chain pointing towards the oil. Some examples are sodium stearate and 4-(5-dodecyl) benzene sulfonate. They can be classified into three categories.

- ***Cationic surfactants:*** These compounds give rise to cations in water. Examples include alkyl trimethyl ammonium compounds such as cetyltrimethylammonium chloride (CTAB) $C_{16}H_{33}(CH_3)_3N^+Cl^-$, octadecylammonium chloride, $C_{18}H_{37}NH_3^+Cl^-$.
- ***Anionic surfactants:*** They form anions in water. An example of this type is sodium palmitate, $C_{15}H_{31}COONa$. Salts of high molecular weight sulfonic acids, with the general formula $C_nH_{2n+1}SO_3^-M^+$ where $M^+ = Na^+$, K^+ or NH_4^+ are other examples.
- ***Non-ionic surfactants:*** Many long chain alcohols having the formula $C_nH_{2n+1}OH$ belong to this category. Other examples include polyoxyethylene glycol alkyl ethers with the formula $CH_3-(CH_2)_{10-16}-(OC_2H_4)_{1-25}-OH$.

6.4.9 Micelles

A micelle is an aggregate of surfactant molecules dispersed in a liquid medium such as water. Soaps like sodium or potassium salts of long chain fatty acids such as sodium palmitate or potassium stearate form micelles in water. Micelle formation can be detected by determing the conductivity of the solution (of soap) as a function of concentration. These compounds (soaps) behave like normal electrolytes at low concentration but as the concentration increases, micelles form and the conductivity decreases due to association. The concentration at which micelle formation begins is known as *critical micelle concentration* (CMC). When surfactants are above CMC they act as emulsifiers. Micelle formation is necessary for absorption of fat soluble vitamins and lipids in the human body.

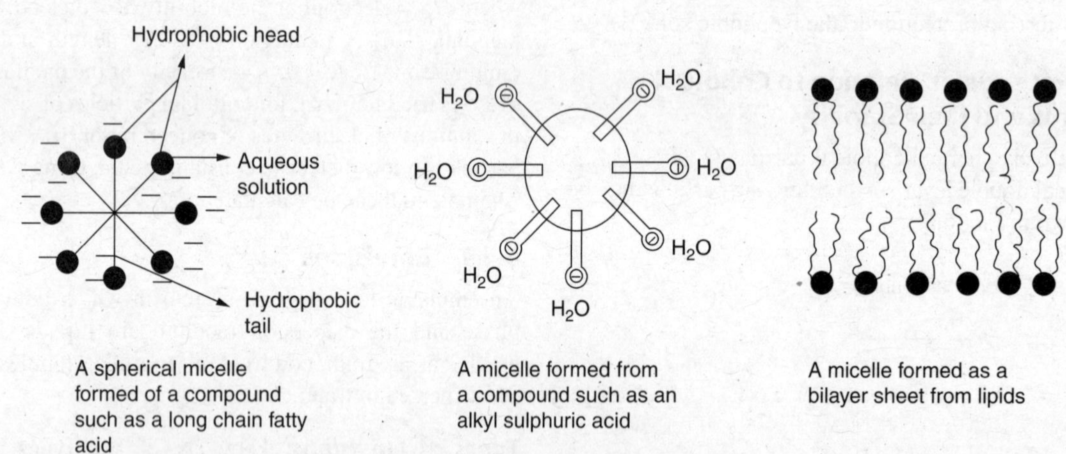

Fig. 6.7 Types of micelles

6.4.10 Gels

A liquid phase dispersed in a solid medium is referred to as a *gel*. The solid part may be powder or granulated such as natural gelatin or a hydrophilic polymer. Gels of agar agar, for example, are prepared by cooling a fairly strong dispersion of the substance in hot water. On cooling, the particles of the dispersion associate to form chains. Gels have high viscosity and swelling capacity. This swelling is known as *imbibition*. Gels can be elastic or rigid. An elastic gel, like gum, forms an elastic solid on partial dehydration and which can revert to a gel on the addition of water. This sol–gel transformation is known as *thixotropy*.

6.4.11 Isoelectric Point in Lyophilic Sols

A lyophilic sol like gelatin shows electrophoresis. The direction of movement of particles (which are charged) depends upon the pH of the medium. In an acidic medium, they acquire a positive charge and move towards the cathode in an electrophoretic experiment. In an alkaline medium, they acquire a negative charge and move towards the anode. At a particular pH, the particles do not move either towards the cathode or the anode. This pH is known as the *isoelectric point*. At this pH, there are an equal number of positive and negative charges on the molecule and hence the net charge is zero. Gelatin has an isoelectric point at pH = 4.7. Most often, proteins are unstable at the isoelectric point and tend to coagulate.

6.4.12 Applications of Colloids

They find application as thickening agents in toothpastes and lubricants. Dairy products like milk are colloidal in nature. Colloids find application in the manufacture of paints and inks. In the purification of industrial smoke, a Cottrell precipitator is used and it makes use of the charge on colloidal particles. Other applications of colloids include use in the tanning of leather and in the disposal of sewage or waste water.

Worked Out Problems

Langmuir and Freundlich Isotherms

Problem 6.1

For the adsorption of a gas on active charcoal as adsorbent, the constants k and n are found to be $10^{-3.5}$ dm^3 g^{-1}(kPa)$^{-0.8}$ and 1.25 respectively. (i) What is the volume of the gas adsorbed in cm^3 on 5 g of the adsorbent at a pressure of 53.33 kPa of Hg? (ii) What is the pressure of the gas in kPa if 40.0 mL of gas are adsorbed on 4 g of charcoal under the same conditions?

Solution

(i) $\frac{a}{x} = kp^{1/n}$; $\log\frac{a}{x} = \log k + \frac{1}{n}\log p$

$$\log\frac{a}{x} = \log 10^{-3.5} + \frac{1}{1.25}\log 53.33$$

$$= -3.5 + (0.8\times 1.727)$$

$$= -3.5 + 1.382 = -2.118 = \bar{3}.882$$

$$\frac{a}{x} = 7.621\times 10^{-3}\ \text{dm}^3$$

For 5 g of charcoal, volume adsorbed

$$= 7.621\times 10^{-3}\times 10^3\ (\text{mL})\times 5 = 38.11\ \text{mL}$$

(ii) $40\ \text{mL} = 40\times 10^{-3}\ \text{dm}^3$; $\frac{a}{x} = \frac{40\times 10^{-3}}{4} = 10^{-2}\ \text{dm}^3$

$$\log 10^{-2} = \log 10^{-3.5} + 0.8\log p = -3.5 + 0.8\log p$$

or $-2 + 3.5 = 1.5 = 0.8\log p$

$$\log p = \frac{1.5}{0.8} = 1.875$$

$$p = 75\ \text{kPa} = \frac{75}{101.325} = 0.74\ \text{atm}$$

Problem 6.2

The following data were obtained for the adsorption of nitrogen on an adsorbent at 0°C at the following pressures.

Hg (P, kPa)	4.5	7.5
Volume of gas adsorbed (V, dm^3 g^{-1})	0.007	0.0103

(i) Assuming that the volume of gas adsorbed (V) is proportional to θ, the fraction of the surface covered, calculate the values of x and z using Langmuir's adsorption isotherm.

(Hint: $V = \frac{a}{m} = k\,\theta = \frac{kZP}{1+ZP} = \frac{xP}{1+ZP}$ where $x = kZ$)

(ii) Using x and z so calculated, find the volume of gas adsorbed in cm^3 on 2 g of adsorbent at a pressure of 3.1 kPa of gas.

Solution

The two data are substituted in the relevant equation

$$0.007 = \frac{x\times 4.5}{1+(Z\times 4.5)};\ 0.0103 = \frac{x\times 7.5}{1+(Z\times 7.5)}$$

Rewriting the equations as

$$4.5\,x = 0.0315\,Z + 0.007 \quad (1)$$

$$7.5\,x = 0.0722\,Z + 0.0103 \quad (2)$$

Multiplying (1) by 7.5 and (2) by 4.5 and rearranging,

$$33.75\,x - 0.2362\,Z = 0.0525 \quad (3)$$

$$33.75\,x - 0.3474\,Z = 0.0463 \quad (4)$$

$$(3)-(4)\quad 0.1112\,Z = 0.0062 \text{ or } Z = \frac{0.0062}{0.1112} = 0.0557$$

Substituting for Z in Eq. (3), we get

$$33.75\times - 0.0132 = 0.0525;\ \text{or } x = \frac{0.0657}{33.75} = 0.00195$$

In the second part of this problem

$$V = \frac{1.95\times 10^{-3}\times 3.1}{1+(5.57\times 10^{-2}\times 3.1)} = \frac{0.006045}{1.1727}$$

$$= 5.155\times 10^{-3}\ \text{dm}^{-3} = 5.155\ \text{mL}$$

The volume of gas adsorbed on 2 g of adsorbent = 2×5.155 = 10.310 mL

Note that x has the units dm^3 g^{-1} Pa^{-1} and Z has the units Pa^{-1}.

Problem 6.3

The adsorption of methylene blue from an aqueous solution at 298 K was studied using coconut shell carbon and the following data were obtained.

MB (mg L^{-1})	MB adsorbed per unit mass of adsorbent (mg)
2.00	0.23
1.00	0.160
0.55	0.097
0.42	0.082
0.36	0.071
0.33	0.062
0.26	0.055
0.21	0.045

Calculate the values of the constants k and n of the Freundlich isotherm by a suitable graphical plot. Using these data, calculate the amount of MB adsorbed per unit mass of adsorbent when MB is present at 1.5 mg L^{-1} at equilibrium.

Solution

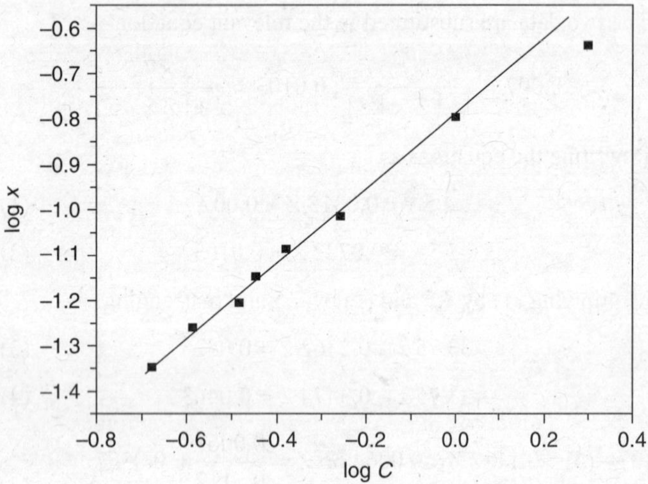

A plot of log x vs log c was made and an average value of slope equal to $\frac{1}{n}$ = 0.833 and intercept log x = –0.7 was calculated. Thus n = 1.200 and k = 0.2.

When the concentration is 1.5 mg L^{-1},

$$\log x = \log K + \frac{1}{n} \log c$$

$$\log x = \log 0.2 + \frac{1}{1.2} \log 1.5 = -0.7 + 0.833 \times 0.176$$

$$= -0.7 + 0.147 + -0.553 = \bar{1}.447$$

x = 0.278. This result is somewhat erroneous as the correct value is expected to be 0.2.

If we take n = 1.353 and K = 0.149 as reported in the literature,

$$\log x = \log 0.149 + \frac{1}{1.353} \log 1.5$$

$$= -0.827 + 0.739 \times 0.176$$

$$= -0.827 + 0.130 = -0.697$$

$$= \bar{1}.303 \text{ or, } x = 0.200$$

Problem 6.4

(a) In the study of adsorption of N_2 on silica gel, a sample of adsorbent having a surface area of 600 m^2 g^{-1} was used. If the nitrogen molecule occupies an area of 16.5 $Å^2$, calculate the volume of N_2 in the unimolecular layer at 273 K and 101.325 kPa pressure assuming that the adsorption obeys Langmuir isotherm.

(b) The volume of H_2 at 273 K and 101.325 kPa pressure required to cover a sample of adsorbent with a unimolecular layer is 0.20 dm^3 per gram of adsorbent. If H_2 occupies an area of 3.53×10^{-20} m^2, calculate the surface area of the adsorbent.

Solution

(a) Area of N_2 molecule = 16.5 $Å^2$ = 16.5×10^{-16} cm^2

$$= 16.5 \times 10^{-16} \times (10^{-2})^2$$

$$= 16.5 \times 10^{-20} \text{ m}^2$$

Number of molecules of N_2 per gram

$$= \frac{600}{16.5 \times 10^{-20}} = 36.36 \times 10^{20}$$

$$= \frac{36.36 \times 10^{20}}{6.023 \times 10^{23}} = 6.04 \times 10^{-3} \text{ mol g}^{-1}$$

One mole of gas at NTP occupies 22.414 dm^3

or 6.04×10^{-3} mol g^{-1} = $6.04 \times 10^{-3} \times 22.414 \times 1000$

$$= 135.4 \text{ cm}^3 \text{ g}^{-1}$$

(b) 0.20 dm^3 H_2 = $\frac{0.20}{22.414}$ = 8.923×10^{-3} mol

$$= 8.923 \times 10^{-3} \times 6.023 \times 10^{23}$$

$$= 53.74 \times 10^{20} \text{ molecules}$$

Each molecule occupies 3.53×10^{-20} m^2

Area occupied by gas molecules

$$= 53.76 \times 10^{20} \times 3.53 \times 10^{-20}$$

$$= 189.7 \text{ m}^2$$

Surface area of catalyst = 189.7 m^2

Problem 6.5

In the study of the adsorption of a certain gas on charcoal as adsorbent, the volumes of the gas adsorbed at various equilibrium pressures are given below.

P (kPa Hg)	13.3	40.0	66.7	93.3
Volume of gas adsorbed (dm^3)	0.01	0.025	0.037	0.046

Plot $1/V$ vs $1/P$ and determine the volume of gas adsorbed in the unimolecular layer. Determine θ at various pressures and calculate Z, the adsorption coefficient.

Solution

$1/V$ (dm^{-3})	100	40	27.2	21.7
$1/P$ (kPa^{-1})	0.075	0.025	0.015	0.011

The plot of $1/V$ vs $1/P$ is linear with an intercept $1/V' = 10$, that is, $V' = 1/10 = 0.1$ dm^3 is obtained.

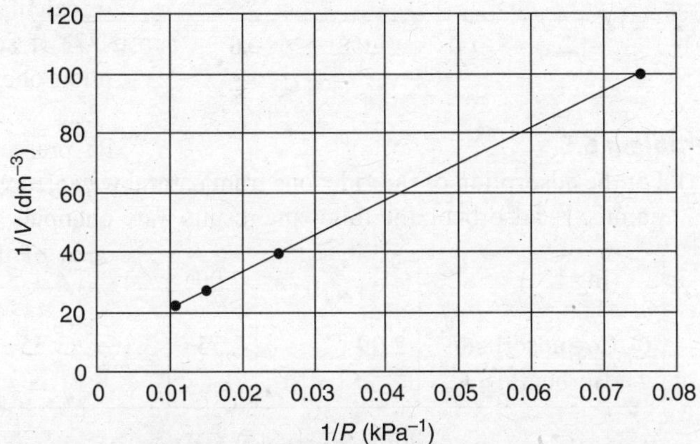

Defining $\theta = V/V'$, we can tabulate θ for various pressures.

	P	$\theta = P/P'$
1	13.3	0.1
2	40.0	0.25
3	66.7	0.37
4	93.3	0.46

Using $\theta = \dfrac{ZP}{1+ZP}$ where $Z = \dfrac{k'}{k}$

For the first data, $0.1 = \dfrac{Z\times 13.3}{1+13.3Z}$ from which $Z = 8.35\times 10^{-3}$

For the second data, $0.25 = \dfrac{Z\times 40}{1+40Z}$ from which $Z = 8.30\times 10^{-3}$

For the third data, $0.37 = \dfrac{Z\times 66.7}{1+66.7}$ from which $Z = 8.805\times 10^{-3}$

For the fourth data, $0.46 = \dfrac{Z\times 93.3}{1+(Z\times 93.3)}$ from which $Z = 9.1\times 10^{-3}$

Average Z = adsorption coefficient $\dfrac{k'}{k} = 8.64\times 10^{-3}$

Problem 6.6

Adsorption of 2-chlorophenol on active carbon generated from castor seeds resulted in the following data:

$q_e\left(\dfrac{\text{mg adsorbate}}{\text{g of adsorbent}}\right)$	16.67	25.64	40.00	41.67	47.62
C_e (equilibrium concentration of adsorbate in solution in mg L^{-1})	0.25	0.45	0.98	1.20	1.73

Assuming that Langmuir's adsorption isotherm is applicable for this case, for which the equation $\dfrac{1}{q_e} = \dfrac{1}{q_m K_L C_e} + \dfrac{1}{q_m}$ (q_m = mg adsorbate g^{-1} adsorbent in unilayer K_L = Langmuir constant, C_e = equilibrium concentration of adsorbate in solution) is valid, determine q_m and K_L using a suitable plot.

Solution

A plot of $\dfrac{1}{q_e}$ vs $\dfrac{1}{C_e}$ has been made and it is found to be linear.

The intercept $\dfrac{1}{q_m} = 0.0125$ or $q_m = 80$ mg and slope = $\dfrac{1}{q_m K_L}$ = 0.0117 or $1/K_L = 0.0117\times 80 = 0.936$, $K_L = 1.068$.

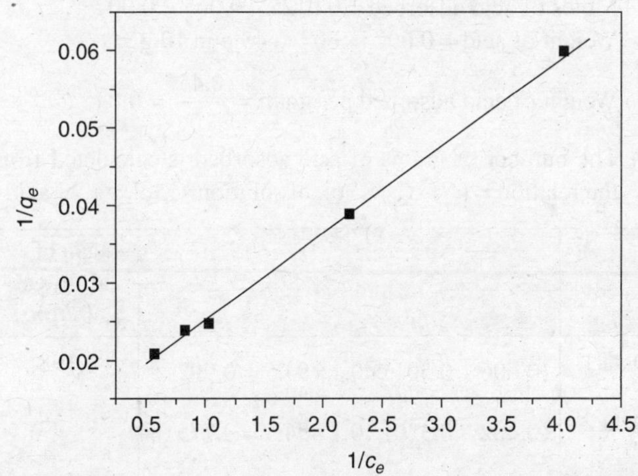

Problem 6.7

(a) In the study of the adsorption of acetic acid from an aqueous solution by activated charcoal, 100 mL of 0.5 M acetic acid solution was thoroughly shaken with 10 g of charcoal and allowed to attain equilibrium. 25 mL of the filtered solution at equilibrium required 50 mL of 0.1 N solution of sodium hydroxide for complete neutralisation. Calculate the equilibrium concentration of acetic acid and the weight of acetic acid adsorbed by unit mass of adsorbent.

(b) 100 mL of solution of monochloroacetic acid of the following concentrations were taken and thoroughly shaken with 5 g of activated charcoal obtained from rice husk at 298 K. The concentration of the acid remaining at equilibrium in this volume is also recorded.

	Initial conc. of acid (C_i)	Equilibrium conc. of acids (C_e)
1	0.606	0.501
2	0.402	0.320
3	0.305	0.252
4	0.187	0.140
5	0.111	0.079
6	0.100	0.075

Test the applicability of the Freundlich isotherm and calculate the constants k and $1/n$.

Solution

(a) At equilibrium, conc. of the acid $= \dfrac{50\times0.1}{25} = 0.2$ M

Before equilibrium, conc. of acid = 0.5 M and 25 mL contains $\dfrac{25\times0.5}{1000} = 0.0125$ mol

Moles of acid at equilibrium $= \dfrac{25\times0.2}{1000} = 0.005$

Moles of acid adsorbed = 0.0125 – 0.005 = 0.0075
Weight of acid = 0.0075×60 = 0.45 g in 10 g

Weight of acid adsorbed per gram $= \dfrac{0.45}{10} = 0.045$ g

(b) The number of grams of acid adsorbed is calculated from the relation = $(C_i - C_e)$×Vol. of solution×Mol. wt. of acid

		g/g of adsorbent (x/m)
Data 1	(0.606 – 0.501)×0.1×94.5 = 0.992	0.198
Data 2	(0.402 – 0.320)×0.1×94.5 = 0.775	0.155
Data 3	(0.305 – 0.252)×0.1×94.5 = 0.501	0.100
Data 4	(0.187 – 0.140)×0.1×94.5 = 0.444	0.089
Data 5	(0.111 – 0.079)×0.1×94.5 = 0.302	0.060
Data 6	(0.100 – 0.073)×0.1×94.5 = 0.236	0.047

	Data 1	Data 2	Data 3	Data 4	Data 5	Data 6
$\log \dfrac{x}{m}$	–0.703	–0.809	–1.00	–1.151	–1.222	–1.328
$\log C_e$	–0.300	–0.495	–0.598	–0.854	–1.102	–1.125

A plot of $\log \dfrac{x}{m}$ vs $\log C_e$ was made and was found to be linear.

Intercept = log k = –0.5 = 1.5; k = 0.316

Slope $= 1/n = \dfrac{0.85}{1.2} = 0.708$; $n = 1.41$

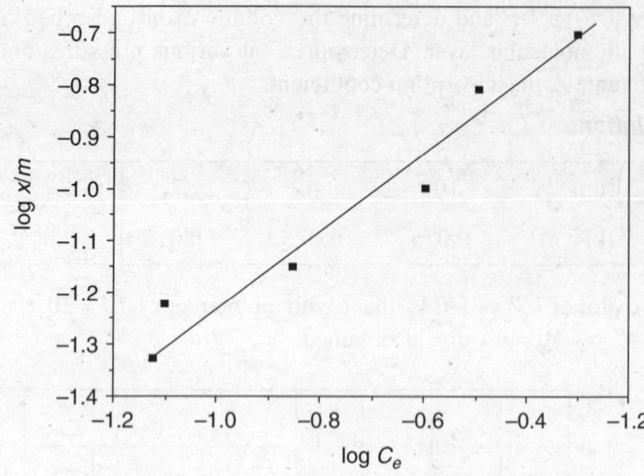

Problem 6.8

(a) For the adsorption of fluoride ions from natural water using pumice as adsorbent, the following results were obtained.

C_e (mg L^{-1})	7.94×10^{-1}	2.00	3.47	4.47
Q_e (mg adsorbed/ g adsorbent)	2.09	2.75	3.16	3.55

where C_e is the equilibrium concentration of fluoride remaining at equilibrium and Q_e is the weight of the same, adsorbed per unit mass of adsorbent. Assuming the validity of the Freundlich isotherm, make a suitable plot and determine the constants of the equation.

(b) Test the same data to check whether Langmuir's isotherm is also valid and comment on which equation better fits the results.

Solution

(a) The Freundlich isotherm is given by $q_e = KC_e^{1/n}$

Or $\log q_e = \log K + 1/n \log C_e$

$\log C_e$	–0.10	0.30	0.54	0.65
$\log q_e$	0.32	0.44	0.50	0.55

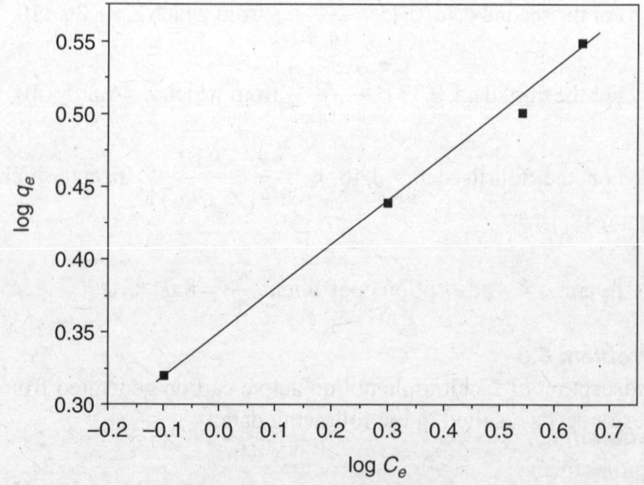

A plot of log q_e vs log C_e yielded a straight line with slope $\frac{1}{n} = 0.278$ or $n = \frac{1}{0.278} = 3.6$ and intercept, log $K = 0.32$; $K = 3.12$.

(b) The Langmuir isotherm can also be expressed in the form $q_e = \frac{q_m b C_e}{1+bC_e}$ where q_e = amount of sorbed fraction in the mass unit of adsorbent (mg/g), C_e is the equilibrium adsorbate concentration after adsorption (mg/L), q_m = adsorption capacity (amount adsorbed in unimolecular layer), b is Langmuir constant.

Rewriting the above equation,

$$\frac{1}{q_e} = \frac{1+bc_e}{q_m b C_e} \text{ or } \frac{C_e}{q_e} = \frac{1}{q_m b} + \frac{1}{q_m} C_e$$

A plot of $\frac{C_e}{q_e}$ vs Ce is linear with a slope $= \frac{1}{q_m}$ and intercept $= \frac{1}{q_m b}$.

$\frac{C_e}{q_e}$	0.38	0.727	1.098	1.259
C_e	7.94×10^{-1}	2.00	3.47	4.47

Such a plot is found to be nearly linear (see graph) but not as good as in the case of Freundlich isotherm. The intercept $\frac{1}{q_m b} = 0.13$; slope $\frac{1}{q_m} = \frac{8\times 0.1}{5.5\times 0.05} = 2.91$; $q_m =$ 0.344 mg/g.

In the unimolecular layer, there is 0.344 mg of adsorbate per g of adsorbent

$$b = \frac{1}{0.13\times 0.344} = \frac{1}{4.47\times10^{-2}} = 22.37 \text{ (units g L}^{-1}\text{)}$$

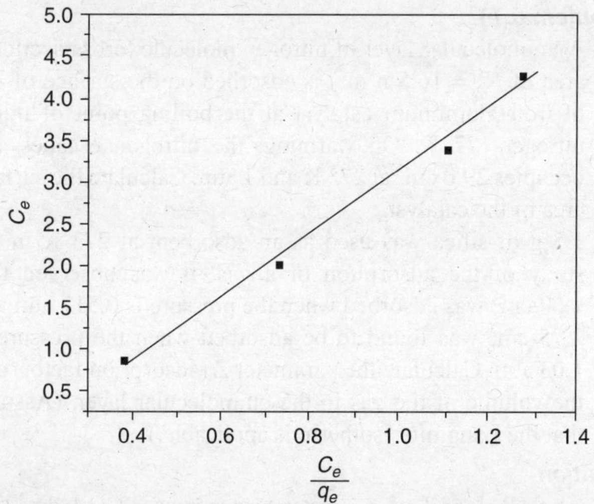

Problem 6.9

Studies on the adsorption of Rhodamine-B by Vijayakumar et al (*J. Mater. Environ. Sci.*, 3,157, 2012) from an aqueous solution on Perlite (which is a kind of siliceous rock), that has been ground and sieved to a 100 mesh size, yielded the following results at 313 K.

Equilibrium conc. C_e (mg L^{-1})	Amount adsorbed at equilibrium, q_e (mg g^{-1} of adsorbent)
3.02	8.32
7.08	16.22
12.59	23.44
19.05	29.51
27.54	37.58

Assuming that the Freundlich isotherm in the form,

$$q_e = K_{FC} \times C_e^{1/n}$$

where K_{FC} and n are constants, is applicable to the above data, determine the constants using a suitable plot of the data. What is the unit of K_{FC} (known as Freundlich constant)?

Solution

$$q_e = K_{FC} \times C_e^{1/n}$$

Taking logarithms,

$$\log q_e = \log K_{FC} + \frac{1}{n} \log C_e$$

We tabulate the given data in the logarithmic form as follows:

log C_e	log q_e
0.480	0.920
0.850	1.210
1.100	1.370
1.280	1.470
1.440	1.575

A plot of log q_e (*y*-axis) vs log C_e (*x*-axis) is found to be linear with the slope $1/n = 0.687$ or $n = 1.455$. K_{FC} has been calculated from the equation using the five data. log K_{FC} values so calculated are tabulated.

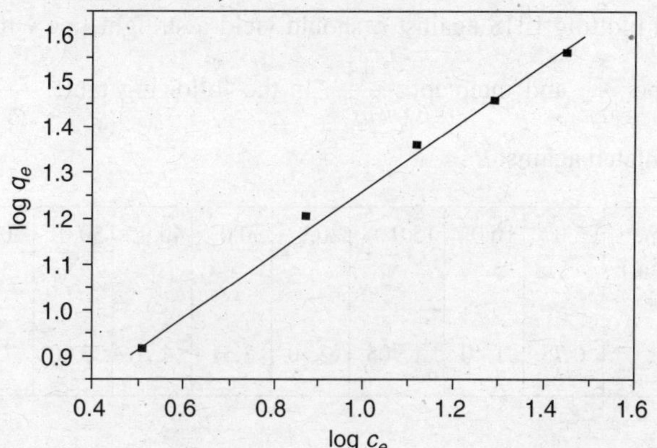

$$\log q_e = \log K_{FC} + \frac{1}{n}\log C_e$$

$\log K_{FC}$	K_{FC}
0.59	3.89
0.626	4.32
0.614	4.11
0.591	3.90
0.586	3.85

Average $K_{FC} = 4.00$. It has the units of litre$^{0.687}$ mg$^{-0.687}$ or $\left(\frac{\text{litre}}{\text{mg}}\right)^{0.687}$

Problem 6.10

The same authors (as in the previous problem) studied the kinetics of adsorption of the dye on Perlite at 303 K using an initial dye concentration of 20 mg L^{-1} and obtained the following results.

Time (min.)	5.0	10.0	15.0	20.0	30.0	40.0	50.0	60.0
q_t amount of dye adsorbed (mg g)$^{-1}$	6.667	7.143	7.895	8.000	8.571	8.421	8.330	8.276

Assuming that the data fit a second order equation given by

$$\frac{1}{q_t} = \frac{1}{k_2 q_e^2 t} + \frac{1}{q_e}$$

where, k_2 = second order rate constant, q_e = equilibrium concentration of dye adsorbed (mg/g), determine k_2 and q_e. Use a suitable form of the above equation for plotting the data and obtain the desired results.

Solution

Writing the above equation in the form

$$\frac{t}{q_t} = \frac{1}{k_2 q_e^2} + \frac{t}{q_e}$$

and plotting LHS against t, should yield a straight line with a slope $\frac{1}{q_e}$ and intercept $\frac{1}{k_2 q_e^2}$. In the following table, $\frac{t}{q_t}$ is tabulated against t.

Time (min.)	5.0	10.0	15.0	20.0	30.0	40.0	50.0	60.0
$\frac{t}{q_t}$	0.75	1.40	1.905	2.50	3.51	4.76	6.00	7.26

A plot of $\frac{t}{q_t}$ (y-axis) vs time (x-axis) yielded a straight with a slope

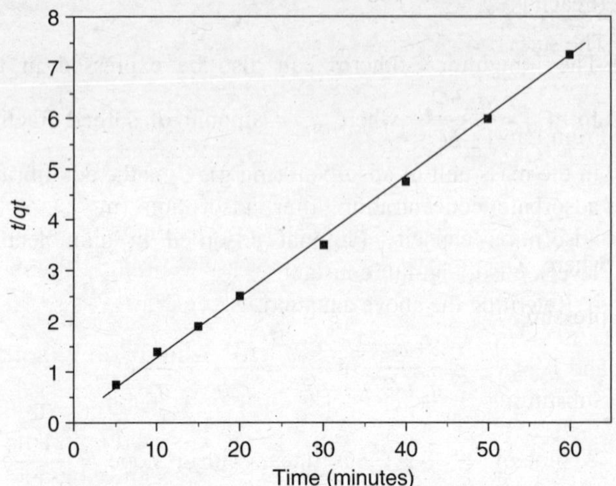

$$\frac{1}{q_e} = 0.1176; \text{ or, } q_e = \frac{1}{0.1176} = 8.503 \text{ mg g}^{-1}$$

$$\text{The intercept} = \frac{1}{k_2 q_e^2} = 0.25 \text{ or } \frac{1}{k_2 (8.503)^2} = 0.25$$

$$k_2 = \frac{1}{0.25 \times (8.503)^2} = \frac{1}{0.25 \times 72.3} = 5.53 \times 10^{-2}$$

The units of k_2:

$$0.25 \frac{\text{min}}{\text{mg g}^{-1}} = \frac{1}{k_2 \times (\text{mg g}^{-1})^2}$$

$$k_2 = \frac{1}{(\text{mg g}^{-1})^2} \times \frac{\text{mg g}^{-1}}{\text{min}} = \frac{1}{\text{mg g}^{-1}} \times \frac{1}{\text{min}}$$

k_2 has the units mg^{-1} g^{1} min^{-1}.

Problem 6.11

(a) A unimolecular layer of nitrogen molecule (cross-sectional area of N_2 = 16.5 n m^2) is adsorbed on the surface of 1 g of iron–aluminium catalyst at the boiling point of liquid nitrogen, 77 K. On warming, the nitrogen escapes and occupies 29.0 cm^3 at 273 K and 1 atm. Calculate the surface area of the catalyst.

(b) 1.5 g of silica was used as an adsorbent at 273 K in the study of the adsorption of a gas. It was observed that 1.60 cm^3 was adsorbed when the pressure is 0.517 atm and 2.75 cm^3 was found to be adsorbed when the pressure is 1.03 atm. Calculate the parameter Z (adsorption factor) and the volume of the gas in the unimolecular layer. (Assume that the Langmuir isotherm is applicable.)

Solution

(a) At STP, 1 mol of a substance occupies 22.414 dm^3 and contains 6.023×10^{23} molecules. Under these conditions 29.0×10^{-3} dm^3 contains

$$\frac{29.0\times10^{-3}\times6.023\times10^{23}}{22.414}=7.793\times10^{20} \text{ molecules}$$

If each N_2 molecule occupies $16.5 \text{ nm}^2 = 16.5\times10^{-20} \text{ m}^2$

The surface area of the catalyst

$$= 7.793\times10^{20}\times16.5\times10^{-20} = 128.6 \text{ m}^2$$

(b) From Langmuir's adsorption isotherm,

$$\theta = \frac{ZP}{1+ZP}$$

where Z = adsorption coefficient and P = equilibrium pressure. But $\theta = \frac{V}{V'}$ where, V = volume of gas adsorbed and V' = volume of gas adsorbed in an unimolecular layer. Substituting the given data in the above equation,

$$\frac{1.60}{V'}=\frac{Z\times0.517}{1+Z\times0.517} \quad (1)$$

$$\frac{2.75}{V'}=\frac{Z\times1.03}{1+Z\times1.03} \quad (2)$$

Rewriting the above equation as,

$$0.517\times ZV' = 1.60+0.827Z \quad (3)$$

$$1.03\times ZV' = 2.75+2.832Z \quad (4)$$

$$\frac{(4)}{(3)}=\frac{1.03}{0.517}=1.992=\frac{2.75+2.832Z}{1.60+0.827Z}$$

or, $3.187+1.647Z = 2.75+2.832Z$

$$2.832Z-1.647Z = 3.187-2.75$$

$$1.185Z = 0.435 \text{ or, } Z = 0.369$$

Substituting the value of Z in Eq. (2), we get,

$$V'\times0.369\times1.03 = 2.75+2.75\times0.369\times1.03$$

$$0.380\times V' = 2.75+1.045 = 3.795$$

$$V' = \frac{3.795}{0.380} = 9.987 \text{ cm}^3$$

Using Z in Eq. (1) and simplifying, we get, $V = 9.984 \text{ cm}^3$

Average of both, $V' = 9.986 \text{ cm}^3$

Problem 6.12

(a) A certain catalyst adsorbs 4.4×10^{-4} g of carbon monoxide when the pressure of the gas is 0.26 atm at a temperature of 310 K. The amount of gas adsorbed is 1.9×10^{-4} g when the pressure is 0.15 atm at the same temperature. Calculate the fraction of the surface covered at the two pressures and also the volume of gas in the unimolecular film. Assume the applicability of Langmuir isotherm.

(b) In a certain study of the adsorption of nitrogen on charcoal, it was observed that a volume of 1.0 cm³ was observed at 500 kPa and 200 K. The same extent of adsorption was observed at 250 K when the pressure was increased to 3.0×10^6 Pa. Use this data to calculate the isosteric enthalpy of adsorption for this process. (Note: Isosteric enthalpy of adsorption is the standard enthalpy of adsorption at a fixed surface coverage.)

Solution

(a) Volume of CO of at 0.26 atm. pressure

$$=\frac{nRT}{P}=\frac{4.4\times10^{-4}\times0.0821\times300}{28\times0.26}$$

$$= 1.489\times10^{-3} \text{ dm}^3$$

Volume of CO at 0.15 atm pressure

$$=\frac{nRT}{P}=\frac{1.9\times10^{-4}\times0.0821\times300}{28\times0.15}$$

$$= 1.114\times10^{-3} \text{ dm}^3$$

Applying Langmuir isotherm in the form,

$\frac{V}{V'}=\frac{ZP}{1+ZP}$ where V' = volume of gas in the unimolecular layer,

$$\frac{1.489\times10^{-3}}{V'}=\frac{Z\times0.26}{1+0.26Z} \quad (1)$$

Similarly, $$\frac{1.114\times10^{-3}}{V'}=\frac{Z\times0.15}{1+0.15Z} \quad (2)$$

Simplifying the above equation,

$$ZV'\times0.26 = 1.489\times10^{-3}+0.3871\times10^{-3}Z \quad (3)$$

$$ZV'\times0.15 = 1.114\times10^{-3}+0.1671\times10^{-3}Z. \quad (4)$$

$$\frac{(3)}{(4)}=\frac{0.26}{0.15}=1.733=\frac{1.489\times10^{-3}+0.3871\times10^{-3}Z}{1.114\times10^{-3}+0.1671\times10^{-3}Z}$$

or, $1.931\times10^{-3}+0.2896\times10^{-3}Z$

$$= 1.489\times10^{-3}+0.3871\times10^{-3}Z$$

or, $0.3871\times10^{-3}Z-0.2896\times10^{-3}Z$

$$= 1.931\times10^{-3}-1.489\times10^{-3}$$

$$10^{-3}\times0.0975Z = 0.442\times10^{-3}$$

$$Z=\frac{0.442\times10^{-3}}{0.0975\times10^{-3}} = 4.533$$

Substituting the value of Z in Eq. (1),

$$\frac{1.489\times10^{-3}}{V'}=\frac{4.533\times0.26}{1+(4.533\times0.26)}=\frac{1.178}{2.178}=0.541$$

or, $\theta = 0.541$

Using Eq. (2),

$$\frac{1.112\times10^{-3}}{V'}=\frac{4.533\times0.15}{1+(4.533\times0.15)}$$

$$=\frac{0.68}{1.68}=0.405 \text{ or } \theta = 0.405$$

$$V' \text{ in the first case} = \frac{1.489\times10^{-3}}{0.541} = 2.752\times10^{-3}$$

$$V' \text{ in the second case} = \frac{1.114\times10^{-3}}{0.405} = 2.750\times10^{-3}$$

(b) Using the Clapeyron–Clausius equation,

$$\log\frac{P_2}{P_1} = \frac{\Delta H_{ads}}{2.303R}\left[\frac{1}{T_1} - \frac{1}{T_2}\right]$$

and substituting the given data,

$$\log\frac{3.0\times10^6}{500\times10^3} = \frac{\Delta H_{ads}}{2.303\times8.314}\left[\frac{1}{200} - \frac{1}{250}\right]$$

$$\log\left(\frac{30}{5}\right) = \frac{\Delta H_{ads}}{19.147}\times\frac{50}{200\times250}$$

$$\Delta H_{ads} = \frac{0.778\times19.147\times200\times250}{50}$$

$$= 14896 \text{ J} = 14.9 \text{ kJ mol}^{-1}$$

BET Equation

Problem 6.13

The following data are obtained in the adsorption of a gas on alumina as adsorbent.

P (mm)	38.0	76.0	114.0	152	228	380
V (dm^3)	0.030	0.035	0.04	0.042	0.045	0.052

Assuming that P_0 = 760 mm Hg, plot the data using the BET equation and calculate V_m, that is, the volume of gas adsorbed in the unimolecular layer and the constant C. If the area occupied by one molecule of gas is 7.27×10^{-20} m^2, what is the surface area of the adsorbent?

Solution

A plot of $\frac{P}{V(P_0-P)}$ vs $\frac{P}{P^0}$ yielded an intercept, $\frac{1}{V_m C} = 1.0$

and slope $= \frac{C-1}{V_m C} = 40.0$.

From the intercept, $V_m C = 1.0$

and hence, $\frac{C-1}{1} = 40.0$ or, $C = 41.0$

Now $V_m = \frac{1}{41} = 0.0243$ dm^3 = 24.3 cm^3

At STP, moles of gas $= \frac{24.3\times10^{-3}}{22.414} = 1.084\times10^{-3}$

Or, number of molecules $= 1.084\times10^{-3}\times6.023\times10^{23}$

$= 6.53\times10^{20}$

Surface area of adsorbent $= 6.53\times10^{20}\times7.27\times10^{-20}$

$= 47.5$ m^2

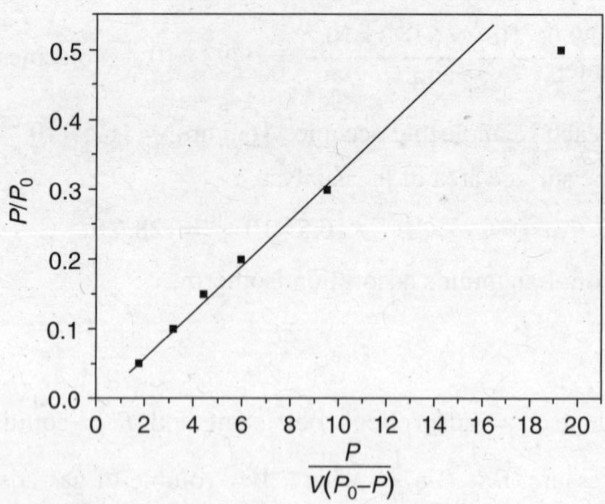

Problem 6.14

The adsorption of *n*-hexane on carbon nanofibres was studied at 303 K by gas chromatography and the following results were obtained

P/P_0	0.06	0.10	0.14	0.18	0.225
a (amount adsorbed in g)	8.51×10^{-4}	1.23×10^{-3}	1.55×10^{-3}	1.76×10^{-3}	2.07×10^{-3}

Apply the BET equation in the form

$$\frac{P/P_0}{a(1-P/P_0)} = \frac{1}{C_{BET}a_0} + \frac{(C_{BET}-1)P/P_0}{C_{BET}a_0}$$

where C_{BET} = constant of BET equation, P = equilibrium pressure, P_0 = saturation varpour pressure of adsorbed species, a = amount (g) adsorbed at relative pressure $\frac{P}{P_0}$, a_0 = amount adsorbed in unimolecular layer in g. Make a suitable plot to determine a_0 and C_{BET}.

Solution

P/P_0	0.06	0.1	0.14	0.18	0.225
$\frac{P/P_0}{a(1-P/P_0)}$	75.0	90.0	105.0	125.0	140.0

A graph of $\frac{P/P_0}{a(1-P/P_0)}$ vs P/P_0 has been made and found to be linear.

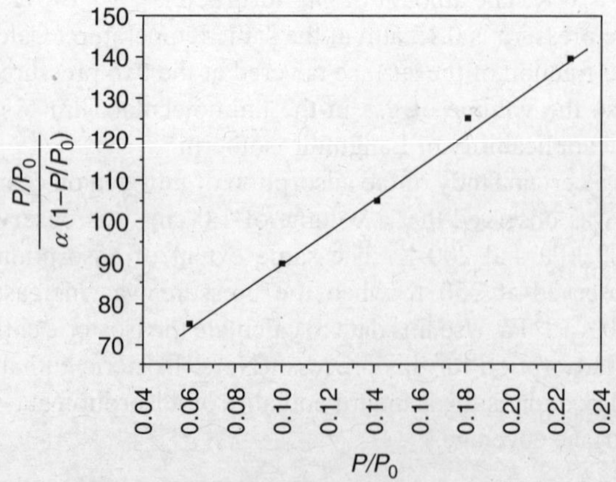

The intercept $= \dfrac{1}{C_{BET}a_0} = 52$ and slope $= \dfrac{(C_{BET}-1)}{C_{BET}a_0} = \dfrac{7.1\times 10}{9\times 0.02} = 394.4$

Or, $(C_{BET}-1)\times 52 = 394.4$

$$C_{BET} = \frac{394.4}{52} + 1 = 7.585 + 1 = 8.585$$

$$a_0 = \frac{1}{8.585\times 52} = 2.24\times 10^{-3}\text{ g}$$

$$= \frac{2.24\times 10^{-3}}{86} = 2.60\times 10^{-5}\text{ mol}$$

Surface Films

Problem 6.15

(a) 0.200 mg of palmitic acid (molecular formula = $CH_3(CH_2)_{14}COOH$, d at 25°C = 0.852 g cm^{-3}) occupies an area of 960 cm^2 on the surface of water to form a unimolecular layer. Find the cross-sectional area of the acid and its length.

(b) 0.15 mg of stearic acid (mol. formula = $CH_3(CH_2)_{16}COOH$, d at 20°C = 0.941 g cm^{-3}) whose cross-sectional area is 20 × 10^{-16} cm^2, forms a unimolecular layer on water when the surface pressure increases sharply. Find the area occupied by the film and the thickness of the film.

Solution

(a) No. of molecules in the given weight

$$= \frac{0.200\times 10^{-3}}{256}\times 6.023\times 10^{23} = 4.7\times 10^{17}$$

These molecules occupy an area A given by $4.7\times 10^{17}\times A = 960$

Area, A occupied by a molecule

$$= \frac{960}{4.7\times 10^{17}} = 20.43\times 10^{-16} = 20.43\text{ Å}^2$$

Total volume occupied by the weight given

$$= \frac{0.200\times 10^{-3}}{0.852} = 0.235\times 10^{-3}\text{ cm}^3$$

or $4.7\times 10^{17}\times 20.43\times 10^{-16}\times l = 0.235\times 10^{-3}$

or $$l = \frac{0.235\times 10^{-3}}{4.7\times 10^{17}\times 20.43\times 10^{-16}}$$

$$= \frac{0.235\times 10^{-3}}{960} = 2.45\times 10^{-4}\text{ cm}$$

(b) Volume of steric acid $= \dfrac{0.15\times 10^{-13}}{0.941} = 1.594\times 10^{-4}\text{ cm}^3$

No. of molecules in the stearic acid taken

$$= \frac{0.15\times 10^{-3}}{284}\times 6.023\times 10^{23} = 3.18\times 10^{17}$$

Area occupied by the film $= 3.18\times 10^{17}\times 20\times 10^{-16}$
$= 636\text{ cm}^2$

Volume = area × thickness = $636\times l = 1.594\times 10^{-4}$

$$l = \frac{1.594\times 10^{-4}}{636} = 2.51\times 10^{-7}\text{ cm} = 25.1\text{ Å}$$

Gibbs Adsorption Isotherm

Problem 6.16

The variation of the surface tension of aqueous surfactant solutions of the compound $CH_3(CH_2)_9(OCH_2CH_2)_5OH$ with concentration at 298 K is given.

Conc. of surfactant, C (mol m^{-2})	0.01	0.03	0.10	0.20	0.5	0.8	1.00
γ(N m^{-1})	0.0639	0.0562	0.0472	0.0416	0.034	0.030	0.0298

Determine the surface excess of the surfactant.

Solution

The Gibbs adsorption isotherm may be written as

$$\Gamma = -\frac{C}{RT}\left(\frac{\partial r}{\partial C}\right)_T = -\frac{1}{RT}\left(\frac{\partial r}{\partial \ln C}\right)_T$$

$$= -\frac{1}{2.303\,RT}\left(\frac{\partial r}{\partial \log c}\right)_T$$

A plot of γ vs log c is made and is found to be linear up to a concentration of 0.8 M with a negative slope.

$$\frac{\partial r}{\partial \log C} = -0.0156$$

Thus, $$\Gamma = -\frac{1.0}{2.303RT}\times\frac{\partial r}{\partial \log C}$$

$$= -\frac{1.0}{2.303\times 8.314\times 298}\times(-0.0156)$$

From the graph thus,

$$\Gamma = \frac{-1}{2.303\times 8.314\times 298}\times -0.0156$$
$$= 2.73\times 10^{-6}\text{ mol m}^{-2}$$

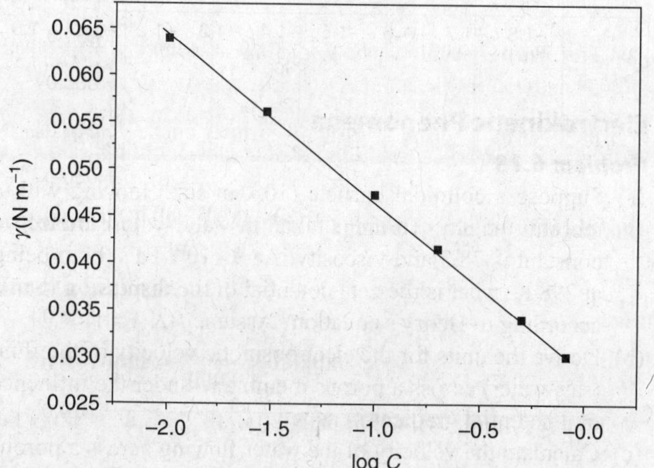

Problem 6.17

The variation of surface tension (γ) of an aqueous solution of an organic compound with a concentration C, may be expressed by the general formula γ(N m^{-1}) = 8.0×10^{-2} – 8.5×10^{-2} C which is valid in the concentration range 0.02 to 0.01 at 298 K. Calculate γ at various concentrations and make a suitable plot and calculate Γ, the average surface excess of the compound in this concentration range.

Solution

Conc. (mol dm^{-3})	0.02	0.03	0.04	0.05	0.06	0.08	0.10
γ(N m^{-1})	0.0783	0.07745	0.0766	0.07575	0.0749	0.0732	0.0715
log C	–1.700	–1.523	–1.400	–1.301	–1.222	–1.097	–1.01

A plot of γ vs log C was made which is found to be linear. The slope $\frac{\partial r}{\partial \log C}$ was found to be equal to –0.1114. Thus,

$$\Gamma = -\frac{1}{2.303\,RT}\times\frac{\partial r}{\partial \log C}$$

$$= -\frac{1}{2.303\times8.314\times298}\times -0.1114$$

$$= \frac{0.1114}{5705.85} = 1.952\times10^{-5}\ \text{mol m}^{-2}$$

(Note that Γ is independent of the concentration unit)

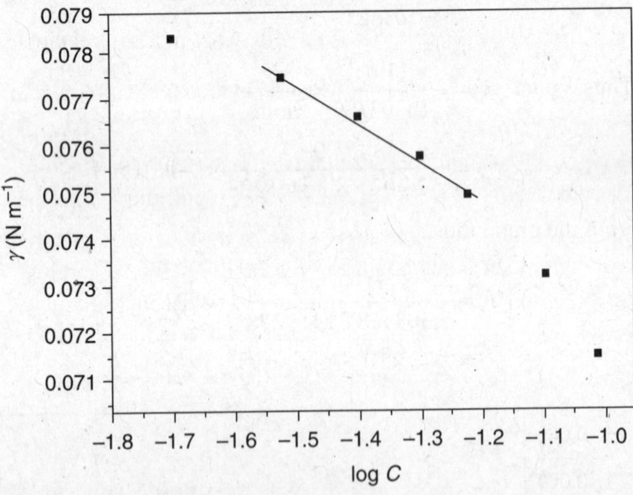

Electrokinetic Phenomena

Problem 6.18

(a) Suppose a colloidal particle (10Å in size) moving with a mobility of 3.5×10^3 m^2 s^{-1} V^{-1} in water whose dielectric constant is 78.3 and viscosity is 8.9×10^{-4} Pa s, both being at 298 K, what is the zeta potential of the dispersion in mV according to Henry's equation? Assume $f(K_a) = 1$.

(b) Derive the units for the electroosmotic velocity of the fluid (say, water) across a porous membrane under the influence of a potential gradient in m^3 s^{-1}.

(c) Calculate the velocity of the water flowing across a porous membrane under the following conditions: $\eta = 1\times10^{-1}$ Pa s, ζ = zeta potential = 50 mV, specific conductivity of the medium = 2.0 ohm^{-1}m^{-1}, ε = dielectric constant = 80.0, current strength = 1.0 A.

Solution

(a) $$\zeta = \frac{3\times\eta\omega_e}{2\times\varepsilon\times f(Ka)} = \frac{3\times3.5\times10^3\times8.9\times10^{-4}}{2\times78.3\times1.0}$$

$$= \frac{93.45\times10^{-1}}{156.6} = 0.0597\ \text{V} = 59.7\ \text{mV}$$

(b) The equation for electroosmotic flow (V) is $V = \frac{\zeta\varepsilon I}{4\pi\eta k}$

where ζ = zeta potential (volts), ε = dielectric constant, I = current strength (A), η = viscosity (Pa s or N m^{-2} s) and k = specific conductivity (ohm^{-1} m^{-1}). To derive the units of V in m^3 s^{-1}, note that voltage in volts has the SI units kg m^2 s^{-3} A^{-1}, I has the unit A, η has the unit kg m^{-1} s^{-1}, k has the unit ohm^{-1} m^{-1} and ε has the units C^2 N^{-1} m^{-2}.

(c) Substituting the given data in the equation for electroosmotic flow, it follows that electroosmotic flow =

$$\frac{\zeta\varepsilon I}{4\pi\eta k} = \frac{0.05\times80\times1.0}{4\times3.143\times0.1\times2} = 1.59\ \text{m}^3\ \text{s}^{-1}$$

Problem 6.19

(a) Calculate the streaming potential generated when an aqueous dilute salt solution having a specific conductance of 4.0 ohm^{-1} m^{-1} and dielectric constant 78.0 is forced across a membrane under a pressure difference of 15 Pa. The zeta potential of the double layer at the interface is 50 mV and the viscosity of the medium is 0.1 Pa s.

(b) What will the value of the streaming current be, if the same solution were forced under the same conditions as above, through a length of capillary tube 20 cm long and having a radius of 1 cm.

Solution

(a) $$S = \frac{\zeta P\varepsilon}{4\pi\eta K} = \frac{0.05\times15\times78}{4\times3.143\times0.1\times4.0} = 11.6\ \text{V}$$

(b) $$-I_{\text{str-curr}} = I\ \text{conduction current} = \frac{ka^2\pi S}{L}$$

where k = specific conductance of the solution in s m^{-1}; a = capillary radius in m; S = streaming potential in V and L = capillary length in m.

$$\left|I_{\text{Sti.current}}\right| = \frac{4.0\times0.01\times0.01\times3.143\times11.6}{0.2}$$

$$= 0.073\ \text{A} = 73\ \text{mA}$$

Problem 6.20

(a) The streaming current I_{str} may also be expressed as $I_{str} = |-|\frac{\varepsilon a^2\pi\Delta P\zeta}{L}$. Show that the RHS of the equation has the units of current in A.

(b) What is the value of I_{str} when ε = 78.0, a = 1 cm, ΔP = 20 Pa, η = 0.1 Pa s, L = 20 cm and ζ = 60 mV?

Solution

(a) $I_{str} = |-|\dfrac{\varepsilon a^2 \pi P \zeta}{\eta L}$

$$\frac{s^4 A^2\ kg^{-1}\ m^{-3}\ m^2\ kg\ m^{-1}\ s^{-2}\ kg\ m^2\ s^{-3}\ A^{-1}}{m\ kg\ m^{-1}\ s^{-1}}$$

$$= \frac{kg\ s^{-1} A}{kg\ s^{-1}} = A$$

(b) $I_{str} = |-|\dfrac{78.0 \times 0.01 \times 0.01 \times 3.143 \times 20 \times 0.06}{0.1 \times 0.2} = 1.47\ A$

Problem 6.21

(a) Calculate the zeta potential of colloidal particles whose size is 1×10^{-9} m and whose electrophoretic mobility is 200 $m^2\ s^{-1}\ V^{-1}$. The dielectric constant of the medium is 80.0 and the viscosity 9×10^{-4} Pa s. The debye length $1/K =$ 20 Å.

(b) Calculate the sedimentation electrical field strength ($V\ m^{-1}$) when colloidal particles having a volume fraction of 0.15 and density 3000 kg m^{-3} fall under gravitational force to form a sediment. The density of the medium is 998.0 kg m^{-3}. The specific conductivity of the dilute 1:1 electrolyte solution is 0.140 S m^{-1}. The zeta potential at the boundary of diffuse double layer is 100 mV and the viscosity of the medium is 1.0×10^{-3} Pa s ($\varepsilon_{r.s} = 1$, $\varepsilon_0 = 80 \times 8.852 \times 10^{-12}$ $s^4\ A^2\ kg^{-1}\ m^{-3}$).

Solution

$$S_p = \frac{\varepsilon_{r.s}\varepsilon_0 \Phi(\rho - \rho_0) g \zeta}{k\eta}$$

$$= \frac{1 \times 80 \times 8.854 \times 10^{-12} \times 0.15(3000 - 998.0) \times 9.81 \times 0.1}{1 \times 10^{-3} \times 0.140}$$

$= 1.49 \times 10^{-3}$ V = 1.49 mV

(a) We first calculate the charge on the colloidal particles using Stokes law, that is, $\omega_e = \dfrac{q}{6\pi R \eta}$, that is,

$$q = 200 \times 6 \times 3.143 \times 9.0 \times 10^{-4} \times 1 \times 10^{-9}$$

$$= 3.394 \times 10^{-9}\ C$$

Using this result in the equation

$$\zeta = \frac{3.394 \times 10^{-9}}{4 \times 3.143 \times 80 \times 1 \times 10^{-9}}$$

$$- \frac{3.394 \times 10^{-9}}{4 \times 3.143 \times 80 \times (1 \times 10^{-9} + 2 \times 10^{-9})}$$

$$= 3.374 \times 10^{-3} - 1.125 \times 10^{-3}$$

$$= 2.25 \times 10^{-3}\ V = 2.25\ mV$$

Problem 6.22

(a) Calculate the zeta potential at the double layer between the aqueous medium and porous membrane when a pressure difference of 20 Pa is applied across a porous cylinder of length 50 cm and radius of 0.5 cm. The streaming current observed is 1.0 A ($\varepsilon = 80$) and viscosity of the medium is 1×10^{-3} Pa s.

(b) The adsorption of dye direct yellow by coconut shell activated carbon was obtained at 283 K at various time intervals.

Time (min.)	5.0	10.0	20.0	30.0	40.0	50.0	60.0
Amount adsorbed q_t (mg/g adsorbent)	1.00	1.30	1.75	2.20	2.40	2.70	2.75

The last value at 60 minutes may be taken as the equilibrium amount adsorbed (q_e). Assuming that the relation $q_t = q_e\left[1 - e^{k_f t}\right]$ applies to the above data, determine k_f by a suitable graphical plot.

Solution

(a) $\zeta = \dfrac{I_{str} L \eta}{\varepsilon a^2 \pi \Delta P} = \dfrac{1 \times 1 \times 10^{-3} \times 0.5}{80 \times 0.005 \times 0.005 \times 3.143 \times 20}$

$= 0.00398$ V = 3.98 mV

(b) Rearranging the equation

$$\frac{q_t}{q_e} = 1 - e^{k_f t}$$

$$\frac{q_t - q_e}{q_e} = -e^{k_f t} \text{ or, } \frac{q_e - q_t}{q_e} = e^{k_f t}$$

$$\ln\left(\frac{q_e - q_t}{q_e}\right) = k_f t$$

$$\log\left(\frac{q_e - q_t}{q_e}\right) = \frac{k_f t}{2.303}$$

We record the data of $\log\left(\dfrac{q_e - q_t}{q_e}\right)$ at various time intervals.

$\left(\dfrac{q_e - q_t}{q_e}\right)$	0.636	0.527	0.364	0.200	0.127	0.018
$\log\left(\dfrac{q_e - q_t}{q_e}\right)$	−0.196	−0.278	−0.439	−0.700	−0.896	−1.745
Time, t (min)	5.0	10.0	20.0	30.0	40.0	50.0

A plot of $\log\left(\dfrac{q_e - q_t}{q_e}\right)$ (y-axis) against t (x-axis) is made and found to be linear.

$$\text{Slope} = \frac{k_f}{2.303} = 0.0208$$

or, $k_f = 2.303 \times 0.0208 = 4.8 \times 10^{-2}\ min^{-1}$

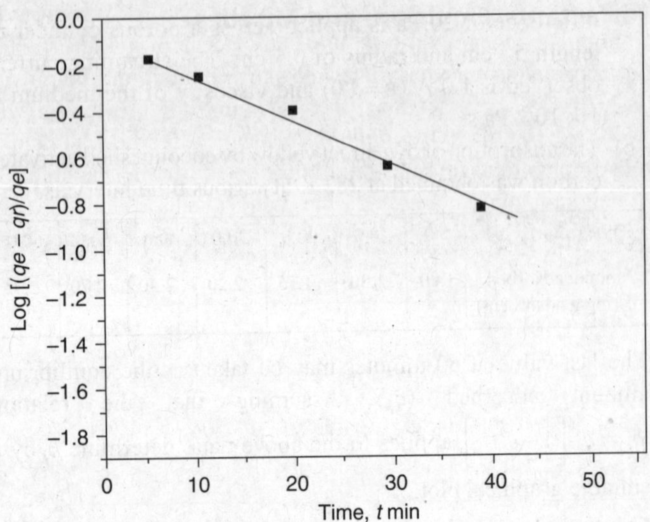

Electrocapillary Phenomenon

Problem 6.23

The electrocapillary data for γ(N m^{-1}) of Hg electrode vs SCE (V) are given below when placed in a solution of 0.1 M solution of an alkali metal chloride.

Volts/ vs SCE	–0.140	–0.340	–0.540	–0.740	–1.00	–1.40
$10^3 \times \gamma$ (N m^{-1})	3.80	4.10	4.20	4.19	3.95	3.70

Calculate q_m(C cm^{-2}) at –0.30, –0.50, –0.70, –0.90, –1.0 V by drawing suitable tangents [since $q_m = \left(\frac{\partial r}{\partial V}\right)$ constant solution composition].

Tabulate the data and calculate the potential of zero charge vs SHE by replotting q_m vs voltage.

Solution

A plot of γ vs voltage (vs SCE) is made and a parabolic curve is obtained. The slopes of the plots are taken at the various voltages mentioned and the results are given below.

–V (vs SCE)	$\left(\frac{\partial \gamma}{\partial V}\right)\left(\frac{\text{Nm}^{-1}}{\text{V}}\right)$
0.300	$+1.25\times10^{-1}$
0.500	$+0.35\times10^{-1}$
0.700	-0.20×10^{-1}
0.900	-0.70×10^{-1}
1.100	-0.80×10^{-1}

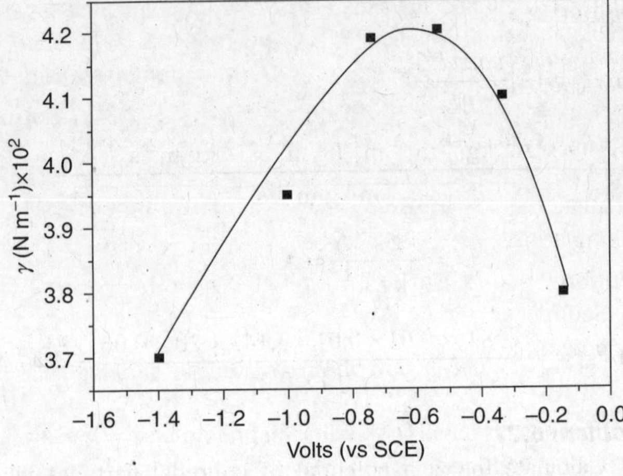

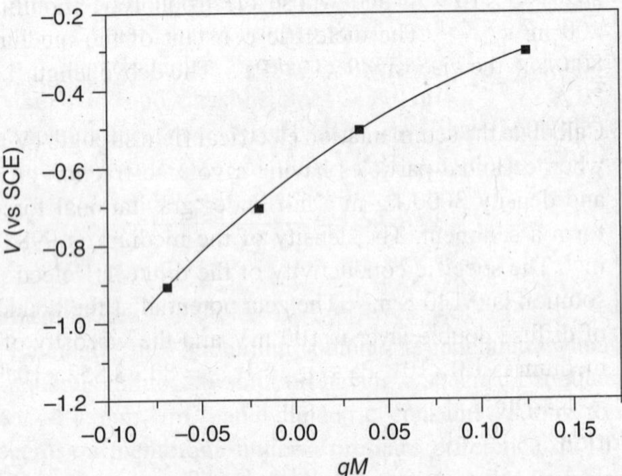

The PZC, that is, where the maximum occurs, is approximately –0.600 V vs SCE. Hence the potential of zero charge vs SHE is –0.600 V + 0.242 = –0.358 V. To convert the $\frac{\partial \gamma}{\partial V}$ of $\frac{\text{Nm}^{-1}}{\text{volts}}$ to $\frac{\partial \gamma}{\partial V}$ $\frac{\text{dynes cm}^{-1}}{\text{volts}}$ multiply it by 10^3. For example, 1.25×10^{-1} N m^{-1} V^{-1} = 1.25×10^{2} cm^{-1} V^{-1}.

Problem 6.24

(a) Calculate the sedimentation electrical field strength (sedimentation potential, V m^{-1}) when a suspension consisting of colloidal particles of volume fraction 0.1 whose density is 2800 kg m^{-3} in a medium of density of 0.998.2 kg m^{-3} falls under gravitational force and forms a sediment. The medium consists of 0.01 N KCl whose specific conductance is 0.141 S m^{-1} at 298 K. The zeta potential at the boundary of the diffuse double layer is 60 mV. The viscosity of the medium is 9×10^{-4} Pa s ($\varepsilon = 1$, $\varepsilon_0 = 78.3\times8.854\times10^{-12}$ s^4 A^2 kg^{-1} m^{-3}).

(b) The surface tensions at a mercury electrode solution interface in presence of various solutions of a 1:1 alkali metal electrolyte were measured at –1.0 V vs SCE and are given below. The mean activity of the electrolyte (rather than concentration of the electrolyte) is tabulated at 298 K.

Mean activity, $a_\pm$ of electrolyte	0.009	0.025	0.0768	0.206	0.604	1.704
γ(N m^{-1})	0.411	0.406	0.403	0.399	0.395	0.390

Calculate the surface excess of the cations at this interface by making a suitable plot of the above data.

Solution

(a) Sedimentation electrical field strength = Sedimentation potential

$$\text{Sp}\ (\text{m}^{-1}) = \frac{\varepsilon_r \varepsilon_0 (\rho - \rho_0) g\zeta}{\eta K}$$

$$= \frac{78.3 \times 8.854 \times 10^{-12} \times 0.1(2800 - 998.2) \times 9.81 \times 0.06}{0.0009 \times 0.141}$$

$$= 579384000.3 \times 10^{-12} = 5.79 \times 10^{-4}\ \text{V m}^{-1} = 0.579\ \text{mV m}^{-1}$$

(b) A plot of γ(N m^{-1}) vs log $a\pm$ was made and was found to be linear.

$$\text{Slope} = \frac{\partial \gamma}{\partial \log a \pm} = -8.65 \times 10^{-3}$$

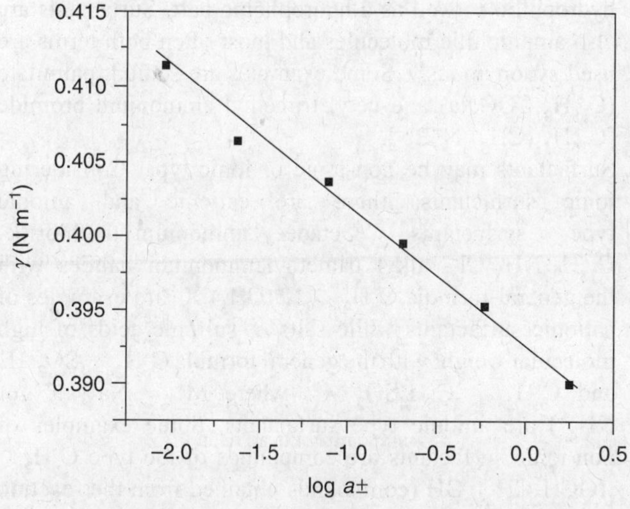

$$\text{But}\ \frac{\partial r}{2RT\partial \ln a \pm} = -\Gamma_+ = \frac{\partial r}{\partial 2.303 \log a \pm} = \frac{\partial r}{\partial \ln a \pm}$$

$$= \frac{-8.65 \times 10^{-3}}{2.303} = -3.76 \times 10^{-3}$$

$$\frac{\partial r}{2RT\partial \ln a \pm} = -\Gamma_+ = \frac{-3.76 \times 10^{-3}}{2 \times 8.314 \times 298}$$

$$= 7.59 \times 10^{-7}\ \text{mol m}^{-2}$$

or $\quad \Gamma_+ = 7.59 \times 10^{-7}\ \text{mol m}^{-2}$

[Note the units: $\dfrac{\text{N m}^{-1}}{\text{J K}^{-1}\text{mol}^{-1}} = \dfrac{\text{kg m s}^{-2} \times \text{m}^{-1}}{\text{kg m}^2\ \text{s}^{-2} \times \text{K}^{-1} \times \text{K} \times \text{mol}^{-1}}$ = mol m^{-2}]

Surfactants, Emulsions and Micelles

Problem 6.25

The interfacial tensions between water and two organic solvents along with the densities of organic solvents at 293 K are given.

Organic solvent	γ (mN m^{-1})	Density at 293 K
Benzene	35.0	0.879
Cyclohexane	50.2	0.778

Consider a 50:50 mixture of water:organic solvent (wt/wt), show that it is not possible to make an emulsion of these mixtures. Also calculate the critical surface tension that would be required to form an emulsion of each of the mixtures which have a drop size of r = 200 Å.

Solution

If the volume fraction of the disperse phase (in this case organic solvent) is smaller than the dispersion medium, then the emulsion can form. Therefore we have to calculate the volume fractions of both component mixtures.

(i) Water–benzene mixture: Assuming the water density to be 1.0,

Volume of water in the mixture = 50 cm^3

$$\text{Volume of benzene in the mixture} = \frac{50}{0.879} = 56.88\ \text{cm}^3$$

$$\text{Therefore, Volume of fraction of water} = \frac{50}{50 + 56.88} = 0.468$$

$$\text{Volume fraction of benzene} = \frac{56.88}{106.88} = 0.532$$

Since $\Phi_{\text{benzene}} > \Phi_{\text{water}}$, this emulsion will not be formed.

(ii) $\text{Volume of cyclohexane in the mixture} = \dfrac{50}{0.778} = 64.27\ \text{cm}^3$

$$\text{Volume of fraction of water} = \frac{50}{50 + 64.27} = 0.437$$

$$\text{Volume fraction of cyclohexane} = \frac{64.27}{114.27} = 0.562$$

Since $\Phi_{\text{cyclohexane}} > \Phi_{\text{water}}$, this emulsion will not be formed.

Calculation of critical surface tension required to form an emulsion in the two cases. The relevant equation is,

$$\gamma_{\text{crit}} = \frac{k_B T}{4\pi r^2}\left\{1 + \ln \Phi_{\text{org.sol}} + \left[\frac{1 - \Phi_{\text{orgsol}}}{\Phi_{\text{orgsol}}}\right] \ln\left(1 - \Phi_{\text{org.sol}}\right)\right\}$$

where k_B = Boltzmann constant, T = absolute temperature, r = radius of dispersed particles, $\Phi_{\text{org.sol}}$ = vol. fraction of organic solvent.

For the water–benzene mixture,

$$\gamma_{\text{crit}} = \frac{1.38 \times 10^{-23} \times 293}{4 \times 3.143 \times \left(2 \times 10^{-8}\right)^2}$$

$$\left\{1 + 2.303 \log 0.532 \frac{0.468}{0.532} \times 2.303 \log 0.468\right\}$$

$$= 8.04\times10^{-7}\,[1 - 0.631+(0.88\times-0.759)]$$
$$= 8.04\times10^{-7}\times-0.299$$
$$= -2.4\times10^{-7}\text{ N m}^{-1} = -2.4\times10^{-4}\text{ mN m}^{-1}$$

For the water–cyclohexane mixture,

$$\gamma_{crit} = 8.04\times10^{-7}\left[1+2.303\log0.562+\frac{0.438}{0.562}\times2.303\log0.438\right]$$
$$= 8.04\times10^{-7}\,[1 - 0.576 + (0.78\times2.303\times -0.358]$$
$$= 8.04\times10^{-7}\times-0.219 = -1.76\times10^{-7}\text{ N m}^{-1}$$
$$= -1.76\times10^{-4}\text{ mN m}^{-1}$$

Problem 6.26

(a) Define an emulsion. State the different types of emulsions. How are emulsions stabilised? Give an example.

(b) What are microemulsions? How do they differ from emulsions?

(c) Describe the disperse phase and dispersion medium in the following food colloids: (i) milk (ii) yoghurt and cheese (iii) ice cream.

Solution

(a) An emulsion is a colloidal dispersion in which both the disperse phase and dispersion medium are liquids. There are two types of emulsions—(i) water in oil type and (ii) oil in water type. The type of emulsion that results on vigorous mixing of the two liquids depends on the relative proportion of the two liquids. The one which is in excess (having a higher volume fraction) acts as the dispersion medium (or outer phase), while the one with a smaller volume fraction or smaller amount acts as disperse phase or inner medium. The nature of the emulsifying agent also has a role. For example, sodium oleate acts as a stabiliser and favours an oil-in-water emulsion while calcium oleate favours a water-in-oil emulsion.

(b) Emulsions (or macroemulsions) and microemulsions differ in the droplet size (droplets of the dispersed phase). Microemulsions have a smaller droplet size in comparison to emulsions. Macroemulsions are thermodynamically unstable while microemulsions are stable. Microemulsions consist of two types—(i) dispersed microemulsions and (ii) bicontinuous microemulsions.

In the first category, the droplets are stabilised by the presence of a surfactant while in the second category, there are continuous networks of oil and water separated by amphiphilic (a molecule containing both a solvent-loving and a solvent-repelling group) groups. Microemulsion formation often requires the presence of a co-surfactant.

(c) (i) Milk: It is an oil-in-water emulsion stabilised by protein particles. The milky appearance in milk is due to the presence of large casein micelles. The casein micelles are polydisperse associated species of some proteins bound with calcium phosphate which is colloidal in nature. These micelles are stable and can withstand high temperatures.

(ii) Yoghurt and cheese: Both are colloidal gels, formed from aggregation of casein micelles. Milk coagulated by added bacteria results in the formation of yoghurt. Cheese is prepared by adding an enzyme known as rennet that acts in the clotting of casein micelles. Casein micelles can also aggregate by variation in pH or heating to a high temperature.

(iii) Ice cream: This can be described as an emulsion as well as a foam since it is a very soft solid containing air bubbles. The fat in ice cream is dispersed in an oil–water emulsion. The content of air is usually about 50% of the total volume. There are various steps in the manufacture of ice cream such as pasteurising, homogenisation, and so on.

Problem 6.27

(a) Explain what is meant by an amphiphile. Are surfactants amphiphiles?

(b) List the different types of surfactants. Give two examples for each type of surfactant and give their formula. Mention some applications of surfactants.

(c) What are lipids? Give some examples of amphiphilic lipid molecules and outline their importance in human biology.

Solution

(a) Amphiphilic molecules are those that contain both a hydrophilic as well as a hydrophobic part. Surfactants are also amphiphilic molecules and most often both terms are used synonymously. Some examples are sodium palmitate ($C_{15}H_{31}COONa$) and cetyl trimethyl ammonium bromide [$C_{16}H_{33}(CH_3)_3N^+Br^-$].

(b) Surfactants may be non-ionic or ionic type. Considering ionic surfactants, there are cationic and anionic type surfactants. Octadecylammonium chloride, $C_{18}H_{37}NH_3^+Cl^-$, alkyl trimethylammonium halides with the general formula $C_nH_{2n+1}N^+(CH_3)_3X^-$ are examples of cationic surfactants while salts of sulfonic acids of high molecular weight with the general formula $C_nH_{2n+1}SO_3^-H^+$ and $C_nH_{2n+1}C_6H_4SO_3^-M^+$ (where $M^+ = Na^+$, K^+ or NH_4^+) are anionic type surfactants. Some examples of non-ionic surfactants are compounds of the type $C_nH_{2n+1}(OCH_2CH_2)_mOH$ (compounds obtained from the reaction of high molecular weight alcohols with many molecules of ethylene oxide) and alkyl phenol ethoxylate, having the formula $C_8H_{18}–C_6H_4–O–(OCH_2CH_2)_5–C_2H_4OH$. There are also zwitter ionic surfactants having both positive and negative groups in the molecule. Surface active agents have applications in the food industry—milk, butter milk, ice cream, and so on. They are also used in paint and varnish industries.

(c) Lipids are substances of biological origin that are soluble in organic solvents but sparingly soluble or insoluble in water. Examples of lipids include fatty acid salts with straight chain or branched chains. Glycerophospholipids are another class of lipids. These lipids are amphiphilic in nature.

Glycerophospholipids are major lipid components of cell membranes. Lecithins, which are phosphatidylcholines are also found in cells.

Problem 6.28

(a) What are micelles? Explain their structural features. Why are such structures favoured in amphiphilic molecules?

(b) What is critical micelle concentration? What experimental approaches enable the determination of CMC?

(c) The specific conductance data of the anionic surfactant, sodium dodecyl sulfate ($CH_3(CH_2)_{11}SO_4^-Na^+$) as a function of concentration is given.

Conc. (mol dm^{-3})	5.5×10^{-3}	6.25×10^{-3}	7.00×10^{-3}	7.5×10^{-3}	9.0×10^{-3}
Sp. conductivity (S m^{-1}/10^{-4})	360.0	400.0	450.0	520.0	575.0
Conc. (mol dm^{-3})	1.45×10^{-2}	2.1×10^{-2}	2.25×10^{-2}	2.55×10^{-2}	2.875×10^{-2}
Sp. conductivity (S m^{-1}/10^{-4})	725.0	850.0	900.0	960.0	1020.0

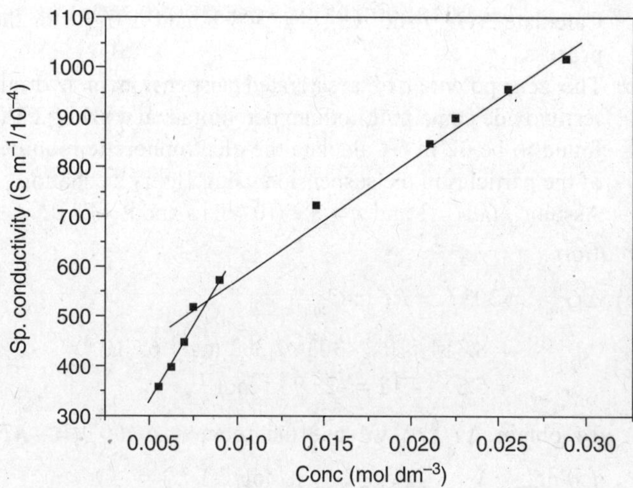

Make a suitable plot of the above data and determine the critical micelle concentration. (See below Section 6.6)

Solution

(a) Micelles are aggregates of organic molecules containing both hydrophobic and hydrophilic groups. Micelles aggregate in aqueous solution with the hydrophobic part (hydrocarbon chain) pointing inwards and the hydrophilic part (ionic part) pointing outwards towards water. The micelles are generally made up of 50 to 200 monomers at any given time and this number is known as aggregation number. The radius of a spherical micelle is about the length of a fully extended surfactant monomer. The following are some of the shapes of micelles.

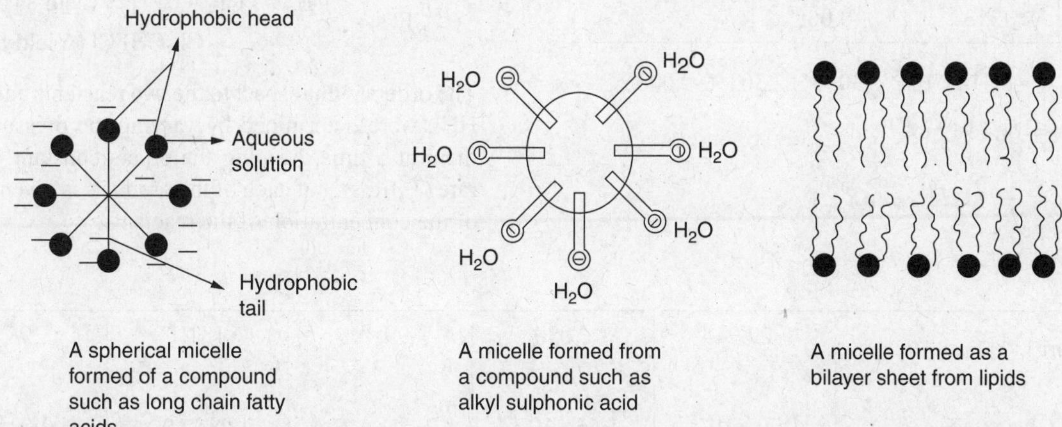

A spherical micelle formed of a compound such as long chain fatty acids

A micelle formed from a compound such as alkyl sulphonic acid

A micelle formed as a bilayer sheet from lipids

For the overall process of micelle formation, ΔG^0_{mic} is negative because $T\Delta S^0_{mic}$ is larger than the positive ΔH^0_{mic}. (Note: $\Delta G^0 = \Delta H^0 - T\Delta S^0$)

Hence micellisation is favoured due to the decrease of ΔG^0 for the overall process.

(b) For a surfactant molecule, the surface tension (or some other property like conductivity) of its aqueous solution decreases rapidly from that of the pure solvent with increasing concentration and becomes independent of concentration beyond a certain point. This concentration beyond which the property measured remains constant is known as critical micelle concentration (CMC). Surface tension, specific conductivity and osmotic pressure are some of the properties used for observing the CMC.

(c) A plot of the specific conductivity data against the concentration of surfactant was made and a break was observed in the graph at a concentration of 0.011 M. Two straight lines were observed with a difference in the slope and the intersection point of the two lines was noted at 0.011 M. Thus this concentration value is the CMC.

Problem 6.29

(a) The critical micelle concentration (CMC) of the surfactant $C_{12}H_{25}O(C_2H_4O)_8H$ varies with temperature according to the following data.

Conc. (mol dm^{-3})	Temp. (K)
8.4×10^{-5}	298
7.6×10^{-5}	303
6.8×10^{-5}	313

Calculate ΔG^0_{mic} and ΔS^0_{mic} at 303 K and ΔH^0_{mic} for the process.

(b) The zeta potential of a saturated suspension of hydrous ferric oxide containing sodium perchlorate at a pH of 2.5 is found to be 32 mV. Calculate the electrophoretic mobility of the particles in the suspension using Henry's equation. Assume $f(ka) = 1$ and $\eta = 8.9\times10^{-4}$ Pa s and $\varepsilon = 78.0$.

Solution

(a) ΔG^0_{mic} at 303 K $= RT \ln C_{cmc}$

$$= 8.314\times10^{-3}\times303\times2.303 \log(7.6\times10^{-5})$$
$$= 5.8\times-4.12 = -23.9 \text{ kJ mol}^{-1}$$

To obtain ΔH^0_{mic}, we use the relation $\Delta H^0_{mic} = -RT^2 \left(\frac{\partial \ln C_{cmc}}{\partial T}\right) = -RT^2\times2.303\left(\frac{\partial \log C}{\partial T}\right)$

Log C is tabulated at different temperatures as shown.

log C	Temp. (K)	$\frac{\partial \log C_{cmc}}{\partial T}$	
–4.080	298		
–4.120	303	–0.008	Ave = –0.0065
–4.170	313	–0.005	

$$\Delta H^0_{mic} = -8.314\times10^{-3}\times(303)^2\times2.303\times-0.0065$$
$$= 11.4 \text{ kJ mol}^{-1}$$

$$\Delta S^0_{mic} = \frac{1}{T}\,(\Delta H^0_{mic} - \Delta G^0_{mic})$$

$$= \frac{1}{303}\,[11.4-(-23.9)]\times1000$$
$$= 116.5 \text{ J K}^{-1}\text{ mol}^{-1}$$

(b) $$U_E = \frac{2\varepsilon f(Ka)\zeta}{3\eta} = \frac{2\times78.0\times1.0\times0.032}{3\times8.9\times10^{-4}}$$
$$= 1.87\times10^3 \text{ m}^2\text{ V}^{-1}\text{ s}^{-1}$$

Kinetics of Surface Catalysed Reactions

Problem 6.30

(a) Define the terms *turnover rate* and *turnover number* in the kinetic studies of surface catalysed reactions.

(b) List the parameters that affect the rate of a surface catalysed reaction.

(c) Nanchen (Ph.D. Thesis, Worcester Polytechnic Institute, USA, 2002) studied the industrially important catalytic hydrodechlorination of halosubstituted hydrocarbons in order to elucidate their mechanism. The data relating to the hydrodechlorination of the compound $CF_3–CFCl_2$ (1-dichlorotrifluoroethane) over a carbon supported palladium catalyst (with 0.5% Pd, surface area = 29 m^{2}g^{-1}) at a temperature of 403 K are shown. The reaction is

$$\underset{(\text{CFC})}{CF_3-CFCl_2} + H_2 \rightarrow \begin{matrix} CF_3-CH_3\,(\text{Yield } 4\%) \\ CF_3-CFH_2\,(\text{Yield } 84\%) + HCl \\ CF_3CHFCl\,(\text{Yield } 12\%) \end{matrix}$$

The orders with respect to the two reactants and the product, HCl, were determined by varying the pressures of one of them at a time, keeping the others constant. The turnover rate (TOR, s^{-1}) of each of the products is given as a function of the concentration of the reactants.

P_{CFC} (torr)	29.94	6.030	99.40	181.1	298.5
TOR of CF_3-CFH_2 (s^{-1})	1.00×10^{-2}	1.66×10^{-2}	2.74×10^{-2}	4.08×10^{-2}	6.08×10^{-2}
TOR of CF_3CHFCl (s^{-1})	1.66×10^{-3}	3.03×10^{-3}	5.00×10^{-3}	6.74×10^{-3}	9.10×10^{-3}
TOR of CF_3-CH_3 (s^{-1})	4.315×10^{-4}	7.482×10^{-4}	1.233×10^{-3}	1.841×10^{-3}	2.741×10^{-3}
P_{H_2} (torr)	44.67	51.89	134.2	270.1	445.3
TOR of CF_3-CFH_2 (s^{-1})	1.5×10^{-2}	2.03×10^{-2}	2.47×10^{-2}	3.69×10^{-2}	4.51×10^{-2}
TOR of CF_3CHFCl (s^{-1})	2.48×10^{-3}	2.74×10^{-3}	4.09×10^{-3}	6.74×10^{-3}	8.24×10^{-3}
TOR of CF_3-CH_3 (s^{-1})	8.26×10^{-4}	1.01×10^{-3}	1.233×10^{-3}	1.51×10^{-3}	1.84×10^{-3}
P_{HCl} (torr)	11.02	24.52	33.09	54.55	94.56

(Continued)

TOR of CF_3-CFH_2 (s^{-1})	7.43×10^{-2}	4.08×10^{-2}	3.34×10^{-2}	2.24×10^{-2}	1.50×10^{-2}
TOR of CF_3CHFCl (s^{-1})	1.36×10^{-2}	8.24×10^{-3}	5.52×10^{-3}	4.09×10^{-3}	2.74×10^{-3}
TOR of CF_3-CH_3 (s^{-1})	4.52×10^{-3}	2.24×10^{-3}	1.51×10^{-3}	9.12×10^{-4}	6.12×10^{-4}

Make plots of $\ln(p_{CFC})$ vs ln(TOR) of each of the products, $\ln p_{H2}$ vs ln(TOR) of each of the products, $\ln(p_{HCl})$ vs ln(TOR) of each of the products using the above data and find the orders of CFC, H_2 and HCl. Suggest a suitable mechanism for the reaction.

Solution

The turnover frequency (TOF) or turnover rate (TOR) are the same and their definition is,

$$\text{TOR} = \frac{\text{Number of molecules reacted}}{\text{Number of sites} \times \text{Time}}$$

Another way of defining TOR is,

$$\text{TOR} = \frac{\text{Volumetric rate of a reaction}}{\text{Number of centres/Volume}}$$

$$= \frac{\text{Moles reacted}}{\text{Volume} \times \text{Time}} \times \frac{\text{Volume}}{\text{Moles reacted}} = \text{time}^{-1}$$

The term qualifies the specific activity of a catalytic centre for a given reaction under well defined reaction conditions. Turnover number (TON) specifies the maximum use that can be made of a catalyst for a special reaction under defined conditions.

$$\text{TON} = \text{TOR}\left(\text{time}^{-1}\right) \times \text{lifetime of a catalyst}\left(\text{time}\right)$$

(b) Rates of surface catalysted reactions are influenced by the following factors: (i) surface arrangement of atoms, (ii) surface coverage, and (iii) uniformity or otherwise of the surface.

(c) The following table shows the 'ln' data of various reactants – CFC and H_2 and HCl – along with the TOR data of the products.

$\ln p_{CFC}$ (torr)	3.4	4.1	4.6	5.2	5.7
ln TOR of CF_3-CFH_2 (s^{-1})	–4.6	–4.1	–3.6	–3.2	–2.8
ln TOR of CF_3CHFCl (s^{-1})	–6.4	–5.8	–5.3	–5.0	–4.7
ln TOR of CF_3-CH_3 (s^{-1})	–7.75	–7.2	–6.7	–6.3	–5.9
$\ln[p_{H_2}]$(torr)	3.80	3.95	4.90	5.60	6.10
ln[TOR of CF_3-CFH_2] (s^{-1})	–4.20	–3.90	–3.70	–3.30	–3.10
ln[TOR of CF_3CHFCl (s^{-1})	–6.00	–5.90	–5.50	–5.00	–4.80
ln[TOR of CF_3-CH_3](s^{-1})	–7.10	–6.90	–6.70	–6.50	–6.30
$\ln[p_{HCl}]$(torr)	2.40	3.20	3.50	4.00	4.55
ln[TOR of CF_3-CFH_2](s^{-1})	–2.60	–3.20	–3.40	–3.80	–4.20
ln[TOR of CF_3CHFCl](s^{-1})	–4.30	–4.80	–5.20	–5.50	–5.90
ln[TOR of CF_3-CH_3](s^{-1})	–5.40	–6.10	–6.50	–7.00	–7.40

The $\ln(p_{CFC})$ vs $\ln(TOR)$, $\ln p_{H2}$ vs $\ln(TOR)$, $\ln(p_{HCl})$ vs $\ln(TOR)$ in all the cases were found to be linear and from the slopes, the orders were found to be 1 in the case of CFC, 0.5 in the case of H_2 and −1.0 in the case of HCl.

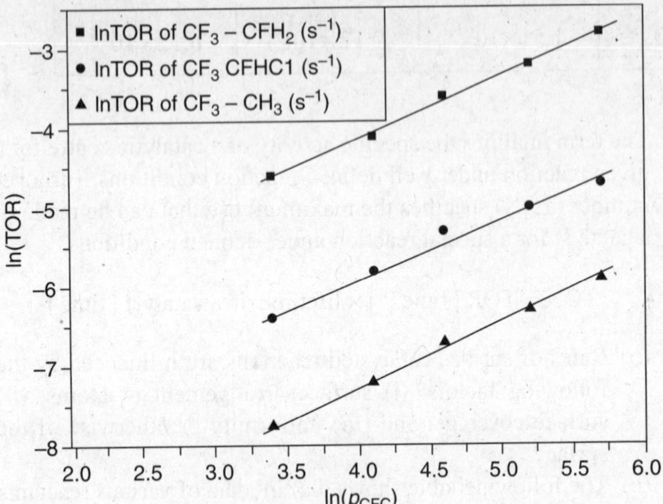

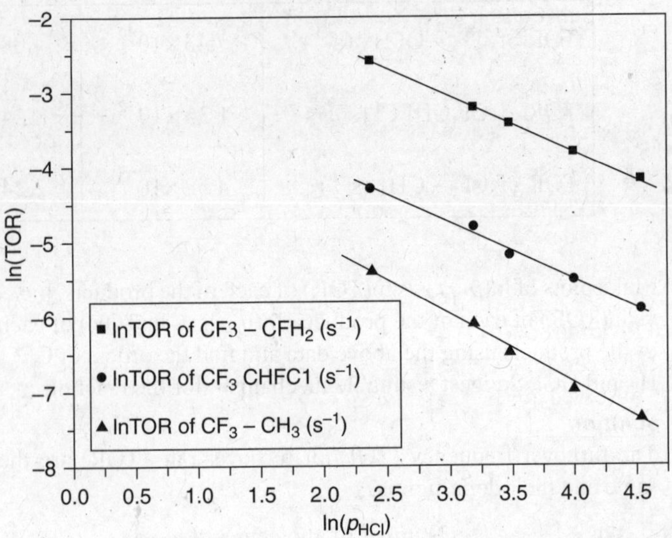

A most plausible mechanism that can be put forward as suggested by author is

$$\underset{(\text{CFC})}{\text{R}-\text{Cl}} + * \rightleftarrows \text{R}-\text{Cl}^* \quad K_{eq} \tag{1}$$

$\text{R}-\text{Cl}^* \rightarrow \ldots$ rate determining step with rate constant k_0 (2)

$$\text{H}_2 + 2* \rightleftarrows 2\text{H}^* \text{ equilibrium step} \tag{3}$$

$$\text{H}^* + \text{Cl}^* \rightleftharpoons \text{HCl} + 2* \tag{4}$$

$$2\text{HCl} + 2* \rightleftarrows \text{H}_2 + 2\text{Cl}^* \text{ equilibrium constant } K_0 \tag{5}$$

$$* + \text{Cl}^* \rightleftharpoons [\text{L}] \text{ site balance} \tag{6}$$

$\text{R}^* + \text{H}^* \rightarrow \text{RH} + 2*$ formation of hydrogenation product (7)

Note that, in the above mechanism, * represents an active site on the surface of the catalyst. $\text{R}-\text{Cl}^*$, H^* and Cl^* are adsorbed species on the surface.

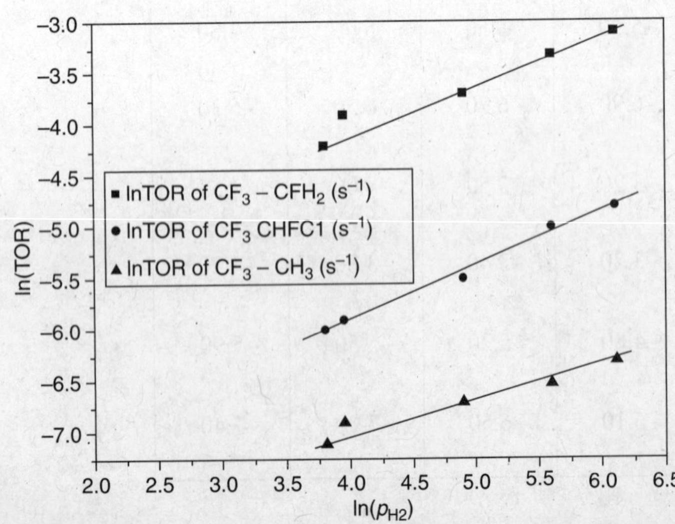

The derivation of the rate equation based on the above mechanism:

$$\text{Rate} = k_0\left[\text{R}-\text{Cl}^*\right] \tag{8}$$

But, $$\left[\text{R}-\text{Cl}^*\right] = K_{eq}[\text{R}-\text{Cl}][*] \tag{9}$$

$$\text{Rate} = k_0 K_{eq}[\text{R}-\text{Cl}][*] \tag{10}$$

From Eq. (6), $[*] = [\text{L}] - \left[\text{Cl}^*\right]$ (6a)

From Eq. (5), $K_0 = \dfrac{[\text{H}_2]\left[\text{Cl}^*\right]^2}{[\text{HCl}]^2[*]^2}$

$\left[\text{Cl}^*\right]$ from Eq. (6) is equal to,

$$\left[\text{Cl}^*\right] = \frac{[K_0]^{1/2}[\text{HCl}][*]}{[\text{H}_2]^{1/2}} \tag{11}$$

Substituting Eq. (11) in Eq. (6a), we get,

$$[*] = [\text{L}] - \frac{[K_0]^{1/2}[\text{HCl}][*]}{[\text{H}_2]^{1/2}} \tag{12}$$

$$= \frac{[\text{L}][\text{H}_2]^{1/2} - [K_0]^{1/2}[\text{HCl}][*]}{[\text{H}_2]^{1/2}} \tag{13}$$

Rewriting Eq. (13),

$$[\text{H}_2]^{1/2}[*] = [\text{L}][\text{H}_2]^{1/2} - [K_0]^{1/2}[\text{HCl}][*]$$

$$[*]\left\{[\text{H}_2]^{1/2} + [K_0]^{1/2}[\text{HCl}]\right\} = [\text{L}][\text{H}_2]^{1/2} \tag{14}$$

From Eq. (14),

$$[*] = \frac{[\text{L}][\text{H}_2]^{1/2}}{[\text{H}_2]^{1/2} + [K_0]^{1/2}[\text{HCl}]} \tag{15}$$

Substituting Eq. (15) in Eq. (10),

$$\text{Rate} = k_0 K_{eq}[\text{R}-\text{Cl}]\left[\frac{[\text{L}][\text{H}_2]^{1/2}}{[\text{H}_2]^{1/2}+[K_0]^{1/2}[\text{HCl}]}\right] \quad (16)$$

Assuming that [L] is constant, it is seen that the observed orders with respect to R–Cl, H_2 and HCl are consistent with the above mechanism, shown by Eq. (16).

Problem 6.31

In a certain enzyme-catalysed reaction, the following rates were observed at various substrate concentrations.

$[S] \times 10^3$	2.50	6.00	8.00	10.00	15.00
Rate $\times 10^6$ (mol s^{-1})	4.46	6.59	7.21	7.63	8.29

By a suitable graphical plot, determine K_M (Michaelis–Menton constant) and V_{max}.

Solution

We know $V_0 = \dfrac{V_M[S]}{K_M+[S]}$

Inverting both sides of the equation,

$$\frac{1}{V_0} = \frac{K_M+[S]}{V_M[S]} = \frac{K_M}{V_{\text{Max}}[S]} + \frac{1}{V_{\text{Max}}}$$

A plot of $\dfrac{1}{V_0}$ vs $\dfrac{1}{[S]}$ is linear according to the above equation,

with slope $= \dfrac{K_M}{V_{\text{Max}}}$ and intercept $= \dfrac{1}{V_{\text{Max}}}$.

We tabulate the data of $\dfrac{1}{V_0}$ and $\dfrac{1}{[S]}$.

1/[S]	400.0	166.67	125.0	100.0	66.67
$1/V \times 10^{-6}$	0.224	0.152	0.139	0.131	0.121

A plot of $\dfrac{1}{V}$ vs $\dfrac{1}{[S]}$ is found to be linear and the intercept is found to be

$$\frac{1}{V_{\text{Max}}} = 9.4 \times 10^4 \text{ or,}$$

$$V_{\text{Max}} = \frac{1}{9.4} \times 10^{-4} = 1.06 \times 10^{-5} \text{ mol s}^{-1}$$

$$\text{Slope} = \frac{K_M}{V_{\text{Max}}} = 325 \text{ or,}$$

$$K_M = 325 \times 1.06 \times 10^{-5} = 3.445 \times 10^{-3} \text{ M}$$

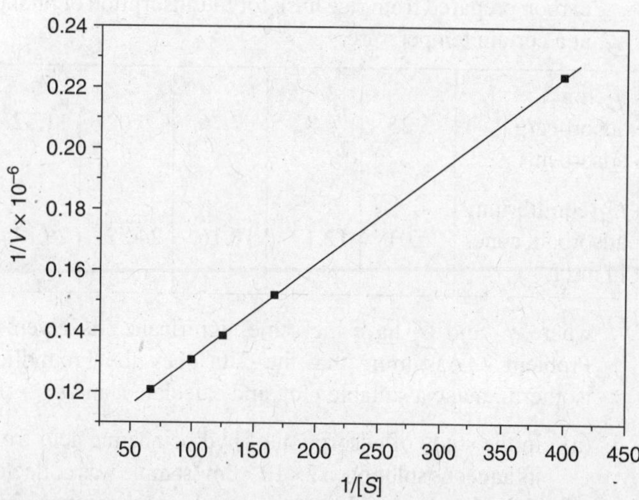

EXERCISES

Practice Problems

1. For the adsorption of nitrogen on active charcoal at –196°C, the following data were obtained at various pressures for the volume of nitrogen adsorbed.

P (kPa)	5.03	9.90	15.30	18.90	24.70
V (dm^3) (at STP)	0.023	0.028	0.033	0.036	0.042

Given the saturation pressure p_0 of liquefied gas as 97.80 kPa, use the BET equation in the form

$$\frac{Z}{1-Z} \times \frac{1}{V} = \frac{(C-1)Z}{CV_m} + \frac{1}{CV_m}$$

where $Z = p/p_0$, V_m = volume of the gas in an unimolecular layer and C = constant, calculate V_m and the surface area of the adsorbent in m^2. The cross-sectional area of the N_2 molecule = 16.2 Å^2.

2. In an experiment using activated carbon from rice husk heated at 773 K, as an adsorbent to remove phenol from waste water, the following data were obtained.

q_e (mg adsorbed/g adsorbent)	5.61	7.24	8.44	8.89	10.25
C_e (equilibrium adsorbate conc. in mg L^{-1})	16.0	21.0	27.0	32.0	41.0

Use the Langmuir adsorption isotherm in the form

$$\frac{1}{q_e} = \frac{1}{q_m K_l C_e} + \frac{1}{q_m}$$

where q_m = mg adsorbed in the unimolecular layer, K_L = Langmuir constant, determine q_m and K_L. Make a suitable plot.

3. The following data were obtained on a sample of activated carbon prepared from rice husk for the adsorption of phenol at a certain temperature.

q_e (mg adsorbed/g adsorbent)	5.25	6.82	7.76	9.02	11.22
C_e (equilibrium adsorbate conc. in mg L^{-1})	5.01	12.17	18.16	24.52	29.81

where q_m and C_e have the same significance as given in Problem 2. Assuming that the data obey the Freundlich isotherm, make a suitable plot, and calculate k and n.

4. (a) In the study of adsorption of hydrocinnamic acid from its aqueous solution, a 2×10^{-3} dm^3 sample was collected from 3.2 dm^2 of the surface of the acid solution. The concentration of the acid in the collected sample was found to be 0.075 mol m^{-3} greater than in the bulk of solution. Calculate the surface excess of the acid.
   (b) The surface tension data of aqueous solution of cetyltrimethylammonium bromide, $C_{16}(CH_3)_3NBr$, at various concentrations are given below.

C (10^{-3} mol kg^{-1})	0.100	0.200	0.300	0.400	0.500	0.600
γ(10^{-3} N m^{-1})	71.0	66.0	59.0	53.0	49.0	45.0

Plot γ vs log C and calculate the surface excess of the compound in mol m^{-2}.

5. (a) Establish the units of the electrophoretic mobility of colloidal particles from Henry's equation.
   (b) Calculate the mobility of colloidal particles whose zeta potential is 80 mV in an aqueous medium of dielectric constant 80 and viscosity 0.01 Pa s. Assume $f(ka)$ = 1.0 in the calculation and temperature as 298 K.

6. (a) Derive the units of streaming potential from the equation relating it to zeta potential, applied pressure difference, the viscosity and specific conductance of the solution.
   (b) Calculate the streaming current when an electrolytic salt solution whose specific conductivity is 2.0 ohm^{-1} m^{-1} (and having a viscosity of 0.1 Pa s) is forced through a capillary tube of 0.5 cm radius and length 10 cm. The streaming potential is 10 V.

7. (a) The sedimentation field strength (sedimentation potential, m^{-1}) is given by the equation

$$S_F = \frac{\varepsilon_r \varepsilon_0 \phi(\rho - \rho_0) g \zeta}{\eta \kappa}$$

where ε_r = relative permittivity, ρ = density of colloidal particles, ε_0 = permittivity of vacuum, ϕ = volume fraction of colloidal particles, ρ_0 = density of the medium (both densities in kg m^{-3}), g = acceleration due to gravity, ζ = zeta potential, η = viscosity of medium and κ = specific conductivity of the medium. Derive its units from the given equation.
   (b) A colloidal suspension of colloidal particles carrying a charge settles down in a 0.2 M KCl solution at 298 K in a graduated measuring cylinder. The particles have a radius of 20 μm and have a zeta potential of 50 mV and a density of 2700 kg m^{-3}. The density of the solution is 998.2 kg m^{-3} and the viscosity of the medium is 0.01 Pa s. Calculate the sedimentation potential. Assume a volume fraction of 0.05 (ε_a = $80\times8.854\times10^{-12}$, κ = 2.5 ohm^{-1} m^{-1}).

8. The interfacial tensions between water and the organic solvents carbon tetrachloride and bromobenzene are given along with their densities at 298 K.

Organic solvent	γ(mN m^{-1})	Density/298 K (g cm^{-3})
Carbon tetrachloride	45.0	1.58
Bromobenzene	38.1	1.495

State whether a 50 : 50 wt/wt mixture of these solvents with water forms an emulsion or not by suitable calculations. What is the critical interfacial tension required to form an emulsion in these cases even if the above solution shows an emulsion formation. The radius of emulsion droplets may be taken as 100 Å.

9. (a) Draw a neat sketch of an electrical double layer at a metal (say positively charged)–solution interface according to the Stern model. Show the inner and outer Helmholtz planes. Also show the region of diffuse double layer and the plane at which zeta potential exists.

(b) Calculate the potential drop at the IHP assuming it to be a parallel plate condenser given that the distance, d, across the plates is 2 Å and the surface charge density is 0.80 C m^{-2}. The dielectric constant of the medium is 80.0 and the permittivity of the vacuum is 8.852×10^{-12} F m^{-1} (or C^2 N^{-1} m^{-2}).

10. Calculate the zeta potential at the porous membrane–solution interface when a dilute solution of a salt whose specific conductance is 1.0 S m^{-1} is forced through the membrane under a pressure difference of 10 Pa. The streaming potential observed is 10 V and the relative dielectric constant is 32.0.

11. (a) For the anionic surfactant, sodium dodecylsulfate (SDS designated as S in the following equilibrium), the formation of the micelle S may be represented as

$$lS^a + (l-p)C^b \rightleftharpoons S_\alpha$$

where l = no. of monomer surfactant molecules, a and b are charges on the surfactant anion and cation, p = no. of surfactant particles p/l = degree of micellisation (or degree of dissociation of surfactant). The Gibbs free energy of micellisation (ΔG^0_{mic}) is given as

$$\Delta G^0_{\text{mic}} = RT\left(2-\frac{p}{l}\right)2.303\log C_{\text{cmc}}$$

Given the $\Delta G^0_{\text{mic}} = -20.8$ kJ mol^{-1}, $C_{\text{cmc}} = 9\times10^{-3}$ M and l = 70 (also called aggregation number of micelle), calculate the degree of dissociation of surfactant. How many surfactant molecules are involved in the micellisation?

(b) What is the entropy change ΔS^0_{mic} for the micellisation given that $\Delta H^0_{\text{mic}} = 3$ kJ mol^{-1}? The temperature is 298 K. Explain the result.

12. The variation of the surface tension with concentration of a nonionic surfactant is shown at 298 K.

Conc. (g dm^{-3})	7.9×10^{-4}	1.0×10^{-3}	1.26×10^{-3}	1.58×10^{-3}	2.0×10^{-3}	2.5×10^{-3}	3.2×10^{-3}
Surface tension (mN m^{-1})	57.5	56.0	53.8	51.3	49.5	47.6	46.0
Conc. (g dm^{-3})	5.0×10^{-3}	7.9×10^{-3}	1.26×10^{-2}	2.0×10^{-2}	3.2×10^{-2}	5.0×10^{-2}	7.9×10^{-2}
Surface tension (mN m^{-1})	42.2	38.2	36.0	33.7	32.5	31.3	31.2

(a) By a suitable plot of the data, determine the critical micelle concentration.

(b) Determine the average surface excess up to CMC in appropriate units.

13. Write a brief account of the following. (a) Zsigmondy's gold number (b) Thixotropy in gels (c) Protein sols (d) Detergency.

14. (a) The critical micelle concentration of the surfactant $C_{12}H_{25}O(C_2H_4O)_6$ is found to vary with temperature as shown.

C_{cmc} (mol dm^{-3})	8.6×10^{-5}	7.2×10^{-5}	6.8×10^{-5}	6.2×10^{-5}
Temp (K)	288	298	303	313

Determine the ΔG^0_{mic} and the equilibrium constant for micellation process at various temperatures. Determine ΔH^0_{mic} by a suitable graphical plot and also ΔS^0_{mic} at 298 K.

(b) The electrophoretic mobility of colloidal particles having a size of 100 n m was measured in a dilute electrolyte solution and found to be 5×10^5 m^2 s^{-1} V^{-1}. Calculate the zeta potential of the particles under these conditions. The dilectric constant and viscosity of the medium are 78.0 and 9×10^{-4} Pa s respectively. Assume the Debye length $(1/k)$ = 50 Å.

15. The water used in a soft drink manufacturing plant contained organic carbon as contaminant. To reduce this, activated charcoal was added to a series of flasks containg 500 mL of this water. After shaking these flasks for several hours, the amount of organic carbon remaining was analysed and the results are given below. The amount of organic carbon in the water before the addition of activated charcoal was 1.3 g dm^3.

Amount of activated carbon added (g)	4.5	10.0	14.0	20.0	28.0
Organic carbon remaining (g dm^{-3})	1.00	0.70	0.59	0.50	0.30

Plot the above data by calculating the relevant details assuming the applicability of Langmuir's isotherm and calculate the constants of the equation.

16. The table gives the electrocapillary data, that is, surface tension γ at a Hg electrode in 0.01 M NaF solution vs SCE at 298 K at various potentials.

E vs SCE (V)	–1.880	–1.750	–1.630	–1.530	–1.430	–1.380	–1.250	–1.130
γ (dyne cm^{-1})	250	270	290	310	330	341	355	368
E vs SCE (V)	–1.050	–0.950	–0.875	–0.700	–0.650	–0.625	–0.450	–0.400
γ (dyne cm^{-1})	378	388	395	402	405	407	407	405
E vs SCE (V)	–0.250	–0.125	–0.050	+0.075	+0.125	+0.250		
γ (dyne cm^{-1})	403	395	390	380	368	350		

Plot γ vs E and find out the following: (i) the potential of zero charge (PZC) vs NHE, (ii) the charge q, at –1.125 V (with the sign), (iii) the charge q, at +0.125 V (with the sign).

17. The following data are obtained for the adsorption of nitrogen on a certain adsorbant at 77 K.

Relative pressure (P/P_0)	0.085	0.133	0.154	0.210	0.235
n_a (amount adsorbed reduced to STP)	0.0168	0.0195	0.0197	0.0230	0.0243

The BET equation is given by

$$\frac{\left(P/P_0\right)}{n_a\left(1-\frac{P}{P_0}\right)} = \frac{1}{n_m C} + \frac{C-1}{n_m C} \times \frac{P}{P_0}$$

where n_m = amount adsorbed in the unimolecular layer, C is the BET constant. Draw a suitable graph of the above data and calculate C and n_m. If the cross-sectional area of the N_2 molecule is 16.2 Å^2, what is the surface area of the catalyst?

18. The interfacial tensions at a mercury electrode in the presence of various concentrations of LiCl were measured vs 1.0 N SCE at 298 K. These data as a function of the mean activity of the electrolyte (as logarithmic quantity) are given.

$\log a_\pm$ (molar scale) of electrolyte, LiCl	–2.05	–1.600	–1.108	–0.666	–0.143	0.459
$\gamma \times 10^3$ (N m^{-1})	415.0	410.0	405.0	401.0	397.0	394.3

Calculate the surface excess of cations at this interface by plotting the above data suitably.

19. The following data were obtained by the scientific group (Vijayakumar et al., *J. Mater. Environ. Sci.*, 3, 2012, 157) at 313 K for the adsorption of a basic dye Rhodamine-B from aqueous solution. The adsorbent is a powdered siliceous rock containing a high percentage of silica, alumina with small amounts of sodium and potassium oxides. The extent of adsorption was measured spectrophotometrically with the initial concentration of the adsorbent being 10 g L^{-1}.

C_e (equilibrium concentration of adsorbate mg L^{-1})	3.0	7.0	12.0	18.0	25.0
$\frac{C_e}{q_e}$	0.30	0.39	0.47	0.575	0.675

(q_e = amount adsorbed at equilibrium, mg g^{-1}). Assuming that the Langmuir isotherm in the form

$$\frac{1}{q_e} = \frac{1}{q_m} + \frac{1}{q_m b C_e}$$

where q_m = Langmuir constant, b = a parameter relating to enthalpy of adsorption and is a constant) is applicable to the data, determine q_m and b by a suitable graphical plot of the data. What are the units of q_m and b?

20. A kinetic study of the amount of adsorption with time at 303 K on the same adsorbent (see Problem 19) gave the following results with the initial concentration of the dye at 20 mg L^{-1}. The experimental value of q_e may be taken as 8.72 mg g^{-1}.

Time (min)	6.0	10.5	15.5	20.5	30.5	40.0
q_t (mg g^{-1})	6.86	7.34	7.93	8.22	8.52	8.63

Assuming that the data fit the equation

$$\frac{q_t}{q_e} = 1 - e^{-k_l t}$$

where k_l is a first order rate constant, calculate k_l from a suitable plot of the above data by rearranging the equation given above.

21. (a) A unimolecular layer of carbon monoxide (whose cross-sectional area = 0.163 n m²) is formed on the surface of a catalyst at the boiling temperature of liquid nitrogen, 77 K. On warming the catalyst, the gas escapes and occupies a volume of 45.0 cm³ at 273 K, and 1 bar (~1 atm). What is the surface area of the catalyst?

(b) The enthalpy of adsorption of ammonia on the surface of a nickel catalyst is –150 kJ mol^{-1}. Calculate the mean lifetime of an ammonia molecule on this catalyst surface at 1000 K.

22. The following data are obtained for the rate of catalytic hydrogenation on platinum supported on silica gel catalyst at various temperatures.

Temp (K)	357.1	339.0	298.6	275.9	253.2
TOR (s^{-1})	33.1	12.17	2.40	0.37	0.135

Calculate the activation energy for this chemisorption process by a suitable graphical plot.

23. For the surface catalysed reaction of the reactant X, according to the reaction scheme,

(a) $X + * \underset{k_{des}}{\overset{k_{ads}}{\rightleftharpoons}} X^*$ (reversible adsorption)

(b) $X^* \xrightarrow{k_1}$ products + * (surface reaction)

with overall reaction being $X \rightarrow$ products, derive the rate equation.

24. (a) Write down the mechanistic scheme in enzyme catalysis postulated by Michaelis and Menton.

(b) Depict the variation of the graph of V_0, the initial velocity as a function of substrate concentration. Mark the V_{max}, and indicate the significance of ½ V_{max}. How is V_0 related mathematically to the substrate concentration?

(c) Consider the reaction

$$E + S \underset{k_2}{\overset{k_1}{\rightleftharpoons}} ES \xrightarrow{k_2} P + E$$

Given that the equilibrium constant for the formation of ES is 2.5×10^3 and K_M is 6×10^{-4} s^{-1}, calculate the rate constants, k_1 and k_{-1}. Assume $k_2 = 1 \times 10^4$ s^{-1}.

Multiple Choice Questions

1. Which of the following statements regarding adsorption is correct?

(a) Adsorption is a bulk phenomenon.
(b) Adsorption leads to a decrease of the area of the interface and hence the interfacial tension.
(c) For an adsorbent to be effective, it must be porous but the surface area is of no importance.
(d) The adsorption of a gas increases with pressure but temperature changes do not influence adsorption significantly.

2. In chemisorption

(a) multimolecular layers are formed
(b) the process is reversible
(c) the process is not specific but can occur for any gas on any adsorbent
(d) the heats of adsorption are in the range 100–400 kJ mol^{-1}

3. Which of the following assumptions in the derivation of Langmuir's adsorption isotherm is incorrect?

(a) Adsorbent surface contains a limited number of sites for adsorption.
(b) Gas molecules adsorbed at different sites cannot interact.
(c) Heat of adsorption depends on the fraction of surface (θ) covered.
(d) At any given temperature T and pressure P, a fraction (θ) of the surface is covered.

4. A compound that is negatively adsorbed on finely divided charcoal is

(a) HCl **(b)** phenol
(c) oxalic acid **(d)** urea

5. The critical temperatures of Cl_2, CO_2 and NH_3 are (in K) 144, 304 and 406 respectively. The amount of adsorption of these gases under identical experimental conditions increases in the order

(a) CO_2 less than than Cl_2 less than NH_3
(b) Cl_2 less than CO_2 less than NH_3
(c) NH_3 less than CO_2 less than Cl_2
(d) CO_2 less than NH_3 less than Cl_2

6. The constants k and n for the adsorption of a gas on active charcoal are 5×10^{-4} L g^{-1} and 1.11. The volume of the gas adsorbed (in cm³) on 2 g of the adsorbent at an equilibrium pressure of 300 mm Hg is

(a) 14.96 **(b)** 19.64
(c) 31.68 **(d)** 16.94

7. In the study of adsorption of nitrogen on silica gel as adsorbent, a plot of $1/v$ (cm^{-3}) vs $1/p$ (mm^{-1}) was found to be linear whose intercept is 0.125. If the nitrogen gas molecule occupies an area of 16.5×10^{-20} m^2, the surface area of the adsorbent is (assuming a unimolecular layer is formed) (in m^2) is
(a) 42.2 (b) 35.3
(c) 53.3 (d) 33.5

8. The variation of surface tension (γ) with concentration C ($d\gamma/dC$) for butyric acid at 298 K, is –2.1 dyne cm^{-1} M^{-1}. The surface excess of this species at the interface at an equilibrium concentration of 0.1 mol dm^{-3} is in (g m^{-2})
(a) 6.58×10^{-3} (b) 5.68×10^{-4}
(c) 8.65×10^{-4} (d) 4.98×10^{-3}

9. The surface excess of a 5 M solution of NaCl at 298 K is -3.27×10^{-6} ml m^{-2}. The rate of variation of surface tension with concentration for this salt is (in dyne cm^{-1} M^{-1})
(a) 1.62×10^{-2} (b) 4.86×10^{-3}
(c) 2.61×10^{-1} (d) 1.34×10^{-1}

10. The disperse phase, dispersion medium in smoke, in that order, are
(a) solid, gas (b) gas, gas
(c) gas, solid (d) solid, solid

11. When droplets of water get dispersed in air, the colloidal system is referred to as
(a) froth (b) foam
(c) aerosol (d) solid sol

12. Freshly precipitated ferric hydroxide precipitate is converted into a sol by the process of
(a) sedimentation (b) dispersion
(c) dilution (d) peptisation

13. The coagulating power of electrolytes NaCl, K_2SO_4 and $Cs(PO_4)_3$ on an arsenious sulphide sol on which As(III) ions are adsorbed, decreases in the order
(a) $NaCl > Cs(PO_4)_3 > K_2SO_4$
(b) $Cs(PO_4)_3 > K_2SO_4 > NaCl$
(c) $K_2SO_4 > NaCl > Cs(PO_4)_3$
(d) $Cs(PO_4)_3 > NaCl > K_2SO_4$

14. The charge on colloidal particles of a given sol is determined by
(a) electrodialysis (b) electroanalysis
(c) electrophoresis (d) sedimentation

15. Phospholipids form micelles in aqueous as well as alcoholic medium. In a reverse (or inverse) micelle of this lipid in alcohol,
(a) the hydrocarbon chain points outwards
(b) the hydrophilic part has no particular preference
(c) the hydrocarbon part points inwards
(d) the phosphate groups point outwards

16. From the BET plot,
(a) it is not possible to determine the volume of gas adsorbed in the unimolecular layer
(b) it is not possible to determine the enthalpy of vaporisation of liquefied gas
(c) one can get the heat of adsorption due to formation of the unimolecular layer
(d) it is possible to determine the number of active sites of adsorption of the adsorbent

17. Pick out the correct statement about the zeta potential from the following.
(a) It is the potential drop across the fixed part of the double layer
(b) It is not affected by the presence of other ions like H+ or OH^-
(c) It decreases exponentially with distance in the mobile part of the diffuse layer
(d) It is referred to as the potential drop in the diffuse mobile layer

18. Calculate the velocity of water (in m^3 s^{-1}) flowing across a porous membrane under the following conditions: zeta potential (ζ) = 30 mv, viscosity (η) of the medium = 0.05 Pa s, dielectric constant (ε) of the medium = 78.0, current strength = 0.5 A, specific conductivity of the medium (κ) = 0.5 ohm^{-1} m^{-1}.
(a) 4.73 (b) 3.72
(c) 2.37 (d) 5.23

19. Calculate the zeta potential (in mV) at the double layer formed at the interface of water and a porous membrane given: pressure difference applied across membrane = 25 Pa, viscosity (η) = 0.02 Pa s, length of cylinder = 50 cm, radius of cylinder tube = 0.50 cm, streaming current (I) = 0.8 A, dielectric constant = 80
(a) 51 (b) 26
(c) 35 (d) 42

20. Milk is an example of
(a) lyophilić colloid (b) water in oil emulsion
(c) oil in water emulsion (d) protein sol

21. Which, among the following, is a cationic surfactant?
(a) $C_{16}H_{33}SO_3^-Na^+$
(b) $C_{18}H_{37}NH_3^+ Cl^-$
(c) $C_{15}H_{31}(OCH_2CH_3)_4OH$
(d) $C_6H_5SO_3H$

22. The effectiveness of soap, used in daily life, for cleaning is due to
(a) lather formation
(b) dissolution of the dust from the material in contact
(c) the presence of lyophilic groups which interact with the impure matter in the material
(d) its emulsifying action on oily material and the dust that is present in the materials

23. In the study of the conductivity of a solution of sodium stearate as a function of concentration of the salt,
(a) the conductivity decreases continuously
(b) the conductivity does not change significantly

(c) the conductivity decreases till CMC is reached and does not change much afterwards
(d) the conductivity rises up to a certain concentration and decreases subsequently

24. The gold numbers (in mg) of albumin (Al), dextrin (De) and gelatin (Ge) are 0.15, 13, 0.008 respectively. Their protective capacity in preventing coagulation of a lyophilic sol increases in the order
(a) Al < De < Ge (b) De < Al < Ge
(c) Ge < Al < De (d) Al < Ge < De

25. The slope of the plot of surface tension vs potential at a mercury–solution interface with 0.1 N NaCl as electrolyte, is 175 dyne cm^{-1} V^{-1}. The surface excess of the salt at the interface (in mol cm^{-2}) is
(a) 3.8×10^{-8} (b) 1.8×10^{-9}
(c) 2.5×10^{-8} (d) 2.0×10^{-7}

26. From a study of the electrocapillary curve of an electrolyte at a Hg–solution interface, one can conclude that
(a) the surface has more +ve charge as the potential is made more –ve
(b) the surface has more –ve charge as the potential is made more +ve
(c) the interface has zero charge at the potential when the surface tension reaches maximum
(d) no information on surface excess of a species can be obtained

27. Pick out the correct statement about a catalyst in a reaction.
(a) A catalyst shifts the position of equilibrium towards the products.
(b) A catalyst can initiate a reaction to occur.
(c) A positive catalyst provides a path of lower activation energy to enhance the rate of a reaction
(d) A catalyst may undergo a chemical change at the end of a reaction.

28. The reaction which is catalysed both by H_3O^+ and OH^- ions is
(a) hydrolysis of sucrose
(b) hydrolysis of urea
(c) conversion of acetone to diacetone alcohol
(d) hydrolysis of ethyl acetate

29. In a study of variation of surface tension of NaBr solutions at a certain potential of the electrode at 298 K, the following data were obtained:

Conc. of NaBr (mol dm^{-3})	4.0×10^{-4}	6.3×10^{-4}
$d\gamma/d\log a\pm$	4.46×10^{-5}	7.46×10^{-5}

(γ is in N m^{-1}).
The average surface excess of cations at the interface is
(a) -5.0×10^{-7} (b) -2.4×10^{-8}
(c) 6.0×10^{-7} (d) 4.2×10^{-8}

30. If a catalyst reduces the activation energy of a surface catalysed reaction by 10 kJ mol^{-1} as compared to an uncatalysed reaction at a temperature of 500 K, the ratio of rate constants k_{cat}/k_{uncat} is
(a) 15.7
(b) 12.4
(c) 8.5
(d) 11.1 (assume that frequency factors of both reactions to be same)

31. In the study of adsorption of a gas on active charcoal at 273 K at various pressures of the gas, a graph was drawn with $P/(x/m)$ vs P where P is the pressure of the gas and x/m refers to volume of gas (mL) adsorbed per unit mass of adsorbent. The slope and intercept were found to be 0.039 mm^{-1} g mL^{-1} and 0.028 mL^{-1} g respectively. The value of the adsorption coefficient b is
(a) 2.52 (b) 1.85
(c) 1.39 (d) 2.18
(Hint: use Langmuir's isotherm in the form $x/m = aP/(1 + bP)$ with P in mm Hg)

32. A plot of surface coverage vs bulk concentration of the adsorbate was found to be linear. This implies that the adsorption isotherm obeys
(a) BET equation
(b) Henry's law
(c) Langmuir isotherm
(d) Frumkin isotherm

33. Among the following M/NaF solution interfaces, identify the metal (M) having the least negative potential of zero charge (pzc) vs SHE?
(a) Tl (b) In
(c) Bi (d) Hg

34. The isoelectric point of a protein is 10.1. When this protein is placed in a gel of pH 7.2, which of the following is true regarding the charge associated with the protein?
(a) Zero
(b) Excess positive charge
(c) Excess negative charge
(d) Both positive and negative charges

35. The activation energy for the desorption of a physically adsorbed gas is 20 kJ mol^{-1}. If the frequency factor for the adsorption process is 10^{10} s^{-1} and assuming that the process is first order, the residence time of the gas molecules on the surface (in seconds) at 298 K is
(a) 1.2×10^{-11} (b) 2.0×10^{-8}
(c) 4.5×10^{-10} (d) 3.0×10^{-7}

36. For a gas obeying Langmuir's adsorption isotherm, 14 cm^3 of a gas (at STP) is adsorbed at a pressure of 8.0 kPa. Given that the adsorption coefficient is 0.6 kPa^{-1}, the volume of gas adsorbed in the unimolecular layer is (in cm^3)
(a) 19.6 (b) 20.5
(c) 16.9 (d) 18.4

Answer Keys

Practice Problems

1. $V_m = 27.6 \text{ cm}^3$

   Surface area of adsorbent = 120.2 Å

2. $q_m = 20 \text{ cm}^3$; $K_L = 2.44\times10^{-2}$

3. $K = 3.16$; $n = 2.83$

4. (a) surface excess = $\Gamma = 4.7\times10^{-6}$ mol m^{-2}

   (b) $\Gamma = 7.46\times10^{-6}$ mol m^{-2}

5. (a) $\text{m}\times\text{s}^{-1}$

   (b) $\text{П}_e = 426.7 \text{ m}^2 \text{ s}^{-1} \text{ V}^{-1}$

6. (a) S = volts

   (b) $-I_{\text{str}} = I_{\text{cond.current}} = 15.7$ mA

7. (a) $S_p = \text{volt}\times\text{m}^{-1}$

   (b) $S_p = 1.25\times10^{-3}$ mV

8. In both cases emulsion formation is possible since $\phi_{\text{org.solvent}} < \phi_{\text{water}}$.

   Water–CCl_4: Volume of water = 50 mL

   Volume of $CCl_4 = \dfrac{50}{1.58} = 31.65$

   Critical $\gamma_{\text{interfacial tension}}$ in the case of water – CCl_4 mixture = -56.3×10^{-4} m N m^{-1}

   Critical $\gamma_{\text{interfacial tension}}$ in the case of water – bromobenzene mixture = -54.9×10^{-4} m N m^{-1}

9. (b). $\Delta E = 2.84$ V $\left(\text{SI units kg m}^2 \text{ s}^{-3} \text{ A}^{-1}\right)$

   (c) ζ = zeta potential = 39 mV

10. (a) The degree of dissociation $\dfrac{\rho}{l} = 0.22$

    Number of molecules involved in micellisation = 15.4

    (b) $\Delta S^0_{\text{mix}} = 79.9$ J K^{-1} mol^{-1}

11. (a) CMC = 8.91×10^{-3} M

    (b) $\Gamma = 3.5\times10^{-6}$ mol m^{-2}

12. Theoretical question

13. (a) At 288 K, $\Delta G^0_{\text{mix}} = -22.44$ kJ mol^{-1}

    At 298 K, $\Delta G^0_{\text{mix}} = -23.62$ kJ mol^{-1}

    At 303 K, $\Delta G^0_{\text{mix}} = -24.19$ kJ mol^{-1}

    At 313 K, $\Delta G^0_{\text{mix}} = -25.13$ kJ mol^{-1}

    $\Delta H^0 = -7.36$ kJ mol^{-1}

    (b) $\zeta = 415$ mV

14. $\kappa = 9.25\times10^{-2}$; $k' = 1.85$

15. (i) potential of zero charge (PZC) = – 0.538 V vs SCE = –0.296 V Vs SHE

    (ii) $q = -0.1$ C m^{-2}, The slope at 0.125 V = +0.18 cm^{-2}

16. $C = 35.2$; $n_m = 2.03\times10^{-2}$ g

    Surface area of catalyst = 72.06 m^2

17. $\Gamma_+ = 7.97\times10^{-7}$ mol m^{-2}

18. $q_m = 59.52$ and it has units mg g^{-1}; $b = 0.062$;

    b has the units = l mg^{-1}

19. $k_1 = 0.089 = 8.9\times10^{-2}$ min^{-1}

20. (a) Surface area of catalyst = 197.1 m^2

    (b) 7.94×10^{-21} s

21. $E_a = 40.7$ kJ

22. Theoretical question

23. Rate = $k_{\text{ads.eq}} = \dfrac{k_{\text{ads}}}{k_{\text{des}}}$

24. (a) $E + S \rightleftharpoons ES \xrightarrow{k_2} P + E$

    (b) $V_0 = \dfrac{V_{\max}[S]}{K_m + [S]}$

    (c) $K_{\text{eq}} = \dfrac{k_1}{k_{-1}}$; $k_1 = 5\times10^7$ M^{-1} s^{-1}; $k_{-1} = 2\times10^4$ s^{-1}

Multiple Choice Questions

1. (b)	2. (d)	3. (c)	4. (a)	5. (b)	6. (d)	7. (b)	8. (c)	9. (a)	10. (a)
11. (c)	12. (d)	13. (b)	14. (c)	15. (a)	16. (c)	17. (d)	18. (b)	19. (a)	20. (c)
21. (b)	22. (d)	23. (c)	24. (a)	25. (b)	26. (c)	27. (c)	28. (d)	29. (b)	30. (d)
31. (c)	32. (b)	33. (d)	34. (b)	35. (d)	36. (c)				

CHAPTER 7

SOLID STATE

CHAPTER OUTLINE

7.1 Introduction

Solids are rigid and have a definite shape. They also occupy a certain volume and are not compressible.

Crystalline and amorphous solids: Salts like NaCl and KCl are crystalline solids. Substances like rubber and glass do not have a regular internal structure although they have a definite shape, hardness and rigidity. In crystals, the structural entities are arranged in a regular geometric configuration and this structure extends over long distances and hence they are known to have *long range order*. *Crystalline* solids are *anisotropic* while *amorphous* solids are *isotropic*. They are also of various types such as molecular, ionic, covalent and metallic solids.

Crystal lattice and unit cell: A crystal lattice may be described as an ordered three dimensional network formed by atoms, molecules or ions situated at certain points in the crystal. In a one dimensional lattice, a group of points is repeated at regular intervals along a line. When a set of points repeat in a regular manner on a plane in two dimensions, it is referred to as a two dimensional lattice. A three dimensional lattice consists of points arranged in a regular repetitive manner along three coordinate axes.

A unit cell is the smallest repeating unit in space. A repetition of several such units over and over again results in the required crystal. The unit cell of a three dimensional lattice is characterised by distances a, b and c, along its three edges and angles α, β and γ, between pairs of edges. These distances and angles are known as *lattice parameters*.

Primitive unit cells and Bravais lattices: There are seven basic unit cells—cubic, tetragonal, orthorhombic, rhombohedral, monoclinic, hexagonal and triclinic which have different combinations of edge lengths and angles between pairs of edges.

These primitive unit cells have some subdivisions. For example, a cubic lattice has primitive, body centred and face centred cubic lattice type. Table 7.1 summarises the total number of subdivisions (14) distributed among the seven primitive types. They are known as *Bravais lattices*.

Designation of lattice planes: The points in a space lattice can be arranged in a number of ways. They may be depicted in a series of parallel and equidistant planes known as *lattice planes*. A crystal lattice may be described in terms of these planes. Figure 7.1 shows the methodology of depicting the planes.

Table 7.1 Bravais lattices

S. No.	Type of system	Characteristics of axes	Angles	No. of space lattices	Examples
1	Cubic	$a = b = c$; angles equal to 90° among three axes	$\alpha = \beta = \gamma = 90°$	3	NaCl, KCl
2	Tetragonal	$a = b \neq c$; angles equal to 90° among three axes	$\alpha = \beta = \gamma = 90°$	2	TiO_2
3	Orthorhombic	$a \neq b \neq c$; angles equal to 90° among three axes	$\alpha = \beta = \gamma = 90°$	4	$BaSO_4$
4	Monoclinic	$a \neq b \neq c$; two angles equal to 90° and third one different	$\alpha = \gamma = 90°, \beta \neq 90°$	2	$Na_2SO_4 \cdot 10H_2O$
5	Rhombohedral	$a = b = c$; all three angles other than 90°	$\alpha = \beta = \gamma \neq 90°$	1	Graphite
6	Hexagonal	$a = b \neq c$; two angles equal to 90° and third one 120°	$\alpha = \beta = 90°, \gamma = 120°$	1	Mg
7	Triclinic	$a \neq b \neq c$; all three angles other than 90°	$\alpha \neq \beta \neq \gamma \neq 90°$	1	$CuSO_4 \cdot 5H_2O$

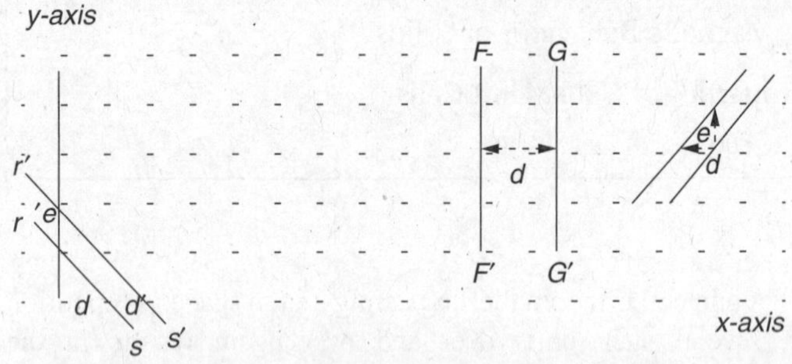

Fig. 7.1 Depicting the lattice planes

For the planes *r*, *s* (which is parallel and equidistant) and *r′s′*, the intercepts are *d* and *e*, and *d′* and *e′*. The intercepts are designated as $d{:}e{:}\infty$ and $d'{:}e'{:}\infty$ (or $2d{:}2e{:}\infty$ because $d' = 2d$ and $e' = 2e$ as seen from Fig. 7.1). Since these planes are parallel to the *Z*-axis, the last intercept is designated as ∞. For the planes FF′ and GG′, the intercept on the *X* axis is *d′* and since the intercepts on the *Y* and *Z* axes are infinity as the planes are parallel to these axes, they are designated as $d'{:}\infty{:}\infty$ (the intercept on the *X*-axis is designated as *d′* and not *d* because $d' = 2d$ as seen from Fig. 7.1). The lattice planes for the parallel lines at the right hand corner end is designated as $-d{:}e{:}\infty$ (the plane on the *X*-axis is given a negative sign because of the change of direction). Note that the intercepts of any plane on the three axes (X, Y, Z axes) are either equal to or their integral multiples, say, *kd*, *le*, *mf* where *k*, *l* and *m* are simple whole numbers. This statement is known as the *law of rational indices*. Thus, the law of rational indices refers to the intercepts of any lattice plane on the three crystallographic axes X, Y and Z. The coefficients *d*, *e* and *f* are referred to as Weiss indices.

7.2 Miller System

7.2.1 Miller System of Denoting Lattice Planes

In this notation, the reciprocals of Weiss coefficients are taken, the advantage being that we get a set of small integers. For example, in the lattice plane given in the left extreme of Fig. 7.1, for the intercepts $d : e : \infty$, the reciprocals of the Weiss indicies are $\frac{1}{1}:\frac{1}{1}:\frac{1}{\infty}$. The Miller indicies are (110). Similarly, for the set of planes FF′ and GG′, the Miller indices are (100) because the reciprocal of Weiss indices are $\frac{1}{1}:\frac{1}{\infty}:\frac{1}{\infty}$. Put in a general way, the Miller indices of a plane can be represented as (*hkl*) which refer to the reciprocals of the intercepts expressed as,

$$h = \frac{d}{\text{Intercepts of plane along the } X\text{-axis}} \tag{7.1a}$$

$$k = \frac{e}{\text{Intercepts of plane along the } Y\text{-axis}} \tag{7.1b}$$

$$l = \frac{f}{\text{Intercepts of plane along the } Z\text{-axis}} \tag{7.1c}$$

The larger the value of the Miller index, the smaller will be the value of the intercept of that plane along the given axis. The Miller indices (*hkl*) of a plane indicate the orientation of that plane in the crystal with reference to its axes.

7.2.2 Miller Indices in Primitive Cubic System

A primitive cubic lattice points towards every corner of the cube. The Miller indices of the planes that pass through these points are (100), (110) and (111) (Fig. 7.2).

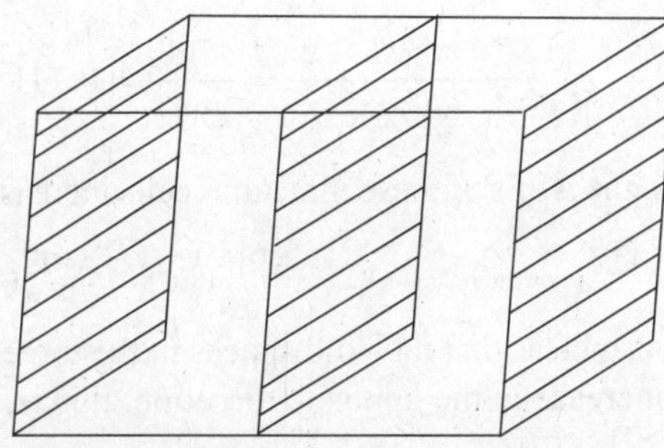

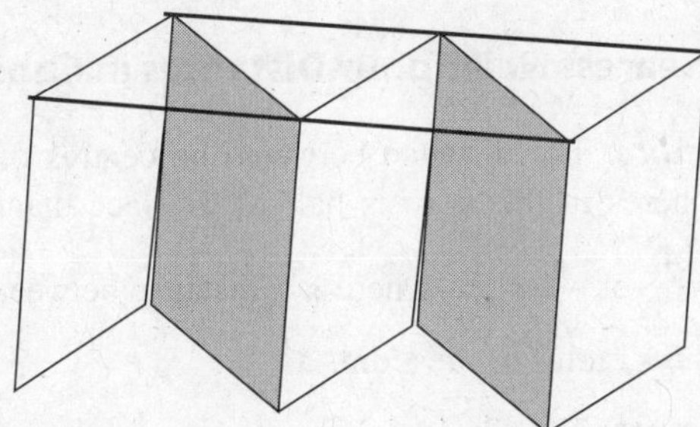

Fig. 7.2 Miller indices of the primitive cubic system

The (100) planes have intercepts on the X-axis, (110) planes have intercepts on the X and Y axes while (111) planes have intercepts on all the three axes. Similarly, planes for body centred and face centred cubic lattices can be shown.

7.2.3 Interplanar Spacing (D_{hkl}) in a Cubic Lattice

The inter planar spacing (D_{hkl}) between the hkl planes in this lattice is given by the relation

$$D_{hkl} = \frac{d}{\sqrt{h^2 + k^2 + l^2}} \tag{7.2}$$

where d is the side of the cube. Thus, for the lattice planes in Fig. 7.2, the interplanar spacings can be calculated as follows.

For $$D_{100} = \frac{d}{\sqrt{1^2 + 0 + 0}} = d$$

For $$D_{110} = \frac{d}{\sqrt{1^2 + 1^2 + 0}} = \frac{d}{\sqrt{2}}$$

For $$D_{111} = \frac{d}{\sqrt{1^2 + 1^2 + 1^2}} = \frac{d}{\sqrt{3}}$$

$$D_{100} : D_{110} : D_{111} = d : \frac{d}{\sqrt{2}} : \frac{d}{\sqrt{3}} = 1 : \frac{1}{\sqrt{2}} : \frac{1}{\sqrt{3}} = 1 : 0.707 : 0.577$$

For a face centred cubic lattice, there is an atom on each of the six faces of the cube in addition to the atoms at the corners. Thus parallel planes can be drawn in the middle between the (100) and (110) planes which have Miller indices (D_{200}) and (D_{220}). The ratio of interplanar distances,

$$D_{200} : D_{220} : D_{110} = \frac{1}{2} : \frac{1}{2\sqrt{2}} : \frac{1}{\sqrt{3}} = 1 : 0.707 : 1.155$$

For a BCC, there is an atom in the centre of the cube in addition to the atoms at the corners. Thus parallel planes can be drawn midway between (100) planes and also midway between (111) planes. They have Miller indices (200) and (222) respectively. The ratio of interplanar distances in this lattice (BCC) is,

$$D_{200} : D_{222} : D_{110} = 1 : \frac{1}{2\sqrt{3}} : \frac{1}{\sqrt{2}} = 1 : 0.577 : 1.414$$

7.3 Parameters in a Crystal

7.3.1 Number of Atoms Per Unit Cell

Table 7.2 Bravais lattices

S. no.	Type of lattice	Method of calculation	No. of atoms per unit cell
1	Primitive cubic	8 corners atoms × 1/8	1
2	Body centred cubic (BCC)	8 corners atoms × 1/8 + one centred atom	2
3	Face centred cubic (FCC)	8 corners atoms × 1/8 + 6 face atoms × ½ = 3	4
4	Edge centred cubic	8 corners atoms × 1/8 = 1 atom + 12 edges × ¼ = 3	4
5	Hexagonal closed packed lattice (HCP)	Three central atoms + six × 1/6 atoms on both top and bottom layers = 2 × 6 × 1/6 = 2 + one half atom in the centre of both top and bottom layers = 2 × ½ = 1	3 + 2 + 1 = 6

7.3.2 The Density (ρ) of a Crystal

It is given by the relation

$$\rho = \frac{nM}{Na^3}$$

where n = number of atoms per unit cell, M = molar mass, N = Avogadro's number, a = edge length of the unit cell (a^3 = volume of the unit cell). The density is usually expressed as kg m^{-3} or megagrams per m^3 (Mg m^{-3}) = g cm^{-3} (1 megagram = 10^6 g).

7.3.3 Determination of Crystal Structure by X-ray Diffraction

Bragg's equation used for this purpose is given by

$$n\lambda = 2D_{hkl} \sin\theta \tag{7.3}$$

where n = 1, 2, 3 and is called the order of diffraction, D_{hkl} is the interplanar spacing of the crystal planes with indices hkl and θ = angle of diffraction.

7.3.4 Relation Between the Lattice Constant and Atomic Radius of the Cube

Primitive cubic lattice: In this lattice, the spheres touch each other along the side of the unit cell. Thus,

Distance between the centre of the spheres = a

$$\text{Radius of the sphere} = a/2 \quad (7.4)$$

BCC lattice: In this lattice, there are two atoms per unit cell and the atoms touch each other along the body diagonal. Thus,

$$4r = \sqrt{3}a \quad (7.4a)$$

$$r = \frac{\sqrt{3}a}{4} \quad (7.4b)$$

FCC lattice: In this case, there are four atoms per unit cell and the atoms touch each other along the face diagonal. Thus,

$$4r = \sqrt{2}a \quad \text{that is, } r = \frac{\sqrt{2}a}{4} \quad (7.4c)$$

7.3.5 Closest Packed Structure

There are two types of structures—(i) hexagonally close packed (HCP) structure or the ABABAB type structure, and (b) cubic close packed (CCP) structure or the ABCABC type structure (which is the familiar FCC structure).

The number of nearest neighbours (coordinate number) in both these close packed structures is 12 whereas in the case of simple cubic and BCC structures they are six and eight respectively.

7.3.6 The Atomic Packing Factor (APF)

The atomic packing factor in a unit cell can be calculated using the relation

$$\text{APF} = \frac{\text{Volume of atoms in unit cell}}{\text{Volume of the unit cell}} \quad (7.5)$$

Primitive cubic unit cell: In this unit cell, there is only one sphere in the unit cell.

$$\text{APF}_{(\text{prim. cub. unit cell})} = \frac{\frac{4}{3}\pi\left(\frac{a}{2}\right)^3}{a^3} = \frac{4}{3}\times\frac{\frac{3.143\times a^3}{8}}{a^3} = 0.524$$

BCC unit cell: Applying Eq. 7.5 for this case, we have

$$\text{APF}_{(\text{BCC})} = \frac{2\times\frac{4}{3}\pi\left(\frac{\sqrt{3}a}{4}\right)^3}{a^3} = \frac{8.381\times 3\sqrt{3}\times a^3}{64a^3}$$

$$= \frac{8.381\times 3\times 1.732\times a^3}{64a^3} = \frac{43.55}{64} = 0.680$$

FCC unit cell: In a similar manner as shown above,

$$\text{APF}_{(\text{FCC})} = \frac{4\times\frac{4}{3}\pi\left(\frac{\sqrt{2}a}{4}\right)^3}{a^3}$$

$$= \frac{16.76\times 2\sqrt{2}\times a^3}{64a^3} = \frac{47.40}{64} = 0.741$$

Thus, the APFs of these three unit cells increase in the order

$$\text{APF}_{(\text{prim. cub. unit cell})} < \text{APF}_{(\text{BCC})} < \text{APF}_{(\text{FCC})}$$

It is also clear that the void spaces in the three types of unit cell increase in the opposite direction, that is,

$$\text{APF}_{(\text{FCC})}(0.26) < \text{APF}_{(\text{BCC})}(0.32)$$
$$< \text{APF}_{(\text{prim. cub. unit cell})}(0.477)$$

7.3.7 Nearest Neighbour Distances in Cubic Cells

FCC structure: The distance between the centres of any two nearest spheres in this case is half of the face diagonal, that is, $\frac{\sqrt{2}a}{2} = \frac{4r}{2}$ or $\frac{a}{\sqrt{2}} = 2r$ where a = distance between any two atoms and r = radius of the atom.

BCC structure: The distance between the centres of any two nearest spheres is half the body diagonal, that is, $\frac{\sqrt{3}a}{2} = \frac{4r}{2} = 2r$.

Primitive cubic structure: In this case, the centres of the atoms touch along the side of a cube (a). Table 7.3 summarises some of the results calculated above.

Table 7.3 Some important data for cubic lattice

Type of lattice	Primitive cubic	FCC	BCC
Volume of unit cell	a^3	a^3	a^3
Atomic packing factor (as fraction)	0.524	0.741	0.680
Coordination number or nearest neighbours	6	12	8
Nearest neighbour distance	a	$\frac{a}{\sqrt{2}}$	$\frac{\sqrt{3}a}{2}$
Next nearest neighbour distance	$\sqrt{2}a$	a	a

a = distance between any two atoms

7.3.8 Holes (Voids) in Cubic Closed Packed Structures

There are two types of holes present in an FCC structure—tetrahedral and octahedral holes. A tetrahedral hole is surrounded by four spheres while an octahedral hole is surrounded by six spheres. Other smaller atoms or molecules may be accommodated in these structures. It is possible to show that there are eight tetrahedral holes (per unit cell) and four octahedral holes per unit cell. The four octahedral holes arise from the fact that there are 12 edge centres ($12\times 1/4 = 3$ octahedral holes) + 1 octahedral hole in the middle of unit cell (total 4). The octahedral holes are bigger than the tetrahedral holes.

7.3.9 Packing in Ionic Compounds

Ionic solids are composed of cations and anions and they have different sizes. Because of this difference, the structure of an ionic solid is governed by geometrical considerations when the different sized ions are stacked. For example, Cs^+ and Cl^- have ionic radii of 0.174 and 0.181 (both in nm) respectively which are quite different. The radii of cations and anions have an important role in ascertaining the crystal structure of the ionic compound. Table 7.4 gives details showing the importance of radius ratios in providing a clue to the structure of the ionic compound.

Table 7.4 Effect of radius ratio on structure of ionic solids

Limiting radius ratio (r_+/r_-)	Preferred geometry	Coordination number	Type of structure	Example
Between 0.225 and 0.414	Tetrahedral	4	ZnS	CuBr, ZnS, ZnO
Between 0.414 and 0.732	Octahedral	6	NaCl	MgO, NaBr, TiO_2
0.732 and above	Cubic	8	CsCl	CsBr, NH_4Br, CaF_2

7.4 Some Typical Structures of Ionic Compounds

(i) Cesium chloride: It has a BCC structure with an 8:8 coordination. Eight Cl- ions surround the central Cs^+ ions or vice versa. The unit cell has one molecule of CsCl.

(ii) Sodium chloride: It has an FCC structure with a 6:6 coordination. The corners are occupied by Cl^- ions and they alternate with Na^+ ions in such a way that each Cl^- ion is surrounded by six Na^+ ions. This can also be alternated with Na^+ ions at the corners. There are 14 Cl^- ions and 13 Na^+ ions in the unit cell or vice versa. The unit cell contains four NaCl units. The cations being smaller, occupy the octahedral sites (when Cl^- ions occupy the corners).

(iii) Zinc sulphide: It occurs in two common types—zinc blende (also known as sphalerite) and wurtzite. In sphalerite, there is an FCC arrangement of sulphur ions at the corners (or lattice points) and zinc ions go into alternate tetrahedral voids (4). There are six sulphur ions, one at the centre of each of the six faces and 8 corner atoms contributing $6\times\frac{1}{2}+8\times\frac{1}{8}=4$ sulphur ions to the unit cell and four zinc ions in the tetrahedral holes. Thus, there are an equal number of zinc ions and sulphide ions to the unit cell to give a 4:4 coordination. The unit cell has four ZnS units. On the other hand, wurtzite is derived from an HCC array of anions and cations occupy one type of tetrahedral holes.

(iv) Calcium fluoride, CaF_2 (fluorite) structure: In this case, the ionic radius of Ca^{2+} ions is 1.26 Å and that of the fluoride ion is 1.17 Å. Thus the cation is larger and anion is smaller, the r_+/r_- = 1.17/1.26 = 0.930. It is therefore to be expected that the anions occupy all the eight tetrahedral holes while the Ca^{2+} ions occupy the corners (lattice points) and face centres. Each fluoride ion is tetrahedrally coordinated to four Ca^{2+} ions. Around each Ca^{2+} ion there are eight fluoride ions at the corners of a cube. Thus, this structure has an 8:4 (cation : anion) coordination.

(v) Antiflourite structure: The positions of cations and anions get reversed in this case when compared to the fluorite structure. The smaller cations occupy all tetrahedral voids (8) while the anions occupy the lattice points and face centres. This is exactly opposite to what happens in the fluorite structure. Examples of antifluorite structure include Na_2O, Li_2O and Rb_2S. The cation : anion coordination is 4:8.

7.5 Binding Forces in Crystals

There are five types of crystals—(i) ionic crystals, (ii) covalent crystals, (iii) molecular crystals, (iv) hydrogen bonded crystals, and (v) metallic bond containing crystals. The physical properties of the crystals depend on the type of bonding.

(i) Ionic crystals: They are held together by strong coulombic attractions of oppositely charged ions. Lattice energy determined from heat of formation and enthalpy of vaporisation measurements agree with calculations made on the basis that ions are held together by electrostatic forces. Ionic crystals are strong but brittle and have high melting points.

(ii) Covalent crystals: Diamond is an example of this type. In this case, the atoms of the crystals are held together by covalent bonds in three dimensions. They are strong and hard and also have high melting points. The crystal lattices of diamond and graphite are very different. Graphite consists of hexagonal networks. The distance between the carbon atoms in a plane is 0.142 nm but the distance between the atomic planes is 0.335 nm. While the carbon atoms are tightly held in two planes (as in diamond), the force of attraction in the third direction is weak. As a result one layer can slip over another. The planar structure explains the lubricating action of graphite. Covalently bonded crystals are insulators.

(iii) Molecular crystals: Molecular crystals are held together by van der Waals forces which are very short range forces. They arise from dipole–induced dipole and dispersion interactions. Crystals of rare gases and neutral organic molecules are examples.

(iv) Hydrogen bonded crystals: In this case, the crystals are held together by sharing of protons between electronegative atoms. Ice is a good example in which oxygen is tetrahedrally coordinated to four hydrogen atoms. These bonds are formed in proteins and polynucleotides. Hydrogen bonds are relatively weak with their strength being of the order 10 to 50 kJ mol^{-1}.

(v) Metallic crystals: Metals are good examples of this type of bond. The bonds arise due to the overlapping of wave functions of valence electrons in metals. The electrons pass through the entire volume of the crystals, sometimes referred to as electron gas. Metals are opaque, lustrous and malleable (examples are Cu and Au). They are good conductors of heat and electricity. There is a gradual transition between metallic and non-metallic character.

7.6 Imperfections (Defects) in Crystalline Solids

Few crystals are perfect or ideal in that all the unit cells have an ideal arrangement of atoms or molecules. Often, at any given temperature, a configurational disorder occurs in the crystal structure which is referred to as an *intrinsic defect*. On the other hand, it is also possible to introduce defects mechanically which are known as *extrinsic defects*. These defects profoundly influence many properties such as electrical conductivity, colour and mechanical strength. These defects may be (i) along a point (point defect), (ii) along a line (line defects), (iii) along a plane or surface (plane defect), and (d) non-stoichiometric defects.

7.6.1 Point Defects

These are broadly divided into (i) Schottky defects and (ii) Frenkel defects.

(i) Schottky defects: In a metal, it is created as a vacancy by the migration of a metal atom to the surface. In ionic crystals, both cations and anions migrate to the surface. The vacancy of a cation is accompanied by vacancy of an anion to maintain electrical neutrality. The density of crystals is reduced by the presence of this defect.

(ii) Frenkel defects: This defect is a combination of Schottky defect and an interstitial defect.

NOTE

- Interstitials are also a kind of crystallographic defect where atoms occupy a site in the crystal structure at which there is usually no atom.

Sometimes there are more number of atoms than the number of lattice sites in the crystals. Here, an ion leaves its position in the lattice and occupies an interstitial void. These defects are observed in silver halides but not in alkali halides. The cation Frenkel defect caused by Ag^+ ions may be represented by

$$Ag^+ \rightarrow V_{Ag} + Ag^+_{(\text{interstitial})}$$

where V_{Ag} is vacancy and $Ag^+_{(\text{interstitial})}$ is an interstitial Ag^+ ion. Cation Frenkel defects are common compared to anion Frenkel defects because cations are typically smaller in size compared to anions. The number of Frenkel defects present in a crystal of the MX type is given by,

$$n_f = (NN_i)^{1/2} e^{(-\Delta H_F/2RT)} \qquad (7.6)$$

where n_f = number of Frenkel defects per unit volume, N is the number of lattice sites, N_i is the number of available interstitial sites and ΔH_F is the enthalpy of formation of one mole of Frenkel defects.

7.6.2 Line Defects

Dislocations in crystals are known as line defects. They are lines through crystals along which the crystallographic symmetry is lost. Dislocations participate in the growth of crystals. They act as electrical defects in optical materials and semiconductors. There are two types of dislocations—(i) edge dislocations and (ii) screw dislocations.

(i) Edge dislocation: In this type of dislocation, the edge of an atomic plane terminates within the crystal plane instead of passing all the way through the crystal. Edge dislocation favours shear in the crystal. When shear occurs, the dislocation moves across the crystal with the top half of the crystal being displaced by one lattice distance with respect to the lower half of the crystal.

(ii) Screw dislocation: A screw dislocation forms an undulation (or a helical ramp) of a set of atomic planes above the dislocation line. Crystal growth is facilitated by this type of dislocation. Shear forces cause screw dislocations to move in a crystal. Generally, dislocations in a crystal are a mixture of both edge and screw dislocations.

7.6.3 Plane Defects

These defects include surfaces around tiny grains of crystalline metals. The boundaries between grains are under considerable strain and impurities tend to concentrate in these regions. Grain boundaries with the impurities may be found out by etching a polished metal surface. Another type of plane defect is possible in FCC packing where the normal packing sequence is ABC ABC... Suppose a B layer forms on the top of a C layer, the sequences change to ABCBACBAC... Thus, there are two different sequences of crystal planes which are mirror images and are known as twins.

7.6.4 Non-stoichiometric Defects

A number of non-stoichiometric inorganic compounds are known, which contain the constituent elements in a non-stoichiometric ratio due to defects in their crystal structure. These defects are classified as (i) metal excess defects and (ii) metal deficiency defects.

(i) Metal excess defect due to anion vacancies: In this case, negative ions are missing in the lattice sites leaving holes in which electrons are trapped to maintain electrical neutrality. An example is found among alkali halides, where F-centres are created (electrons trapped in anion vacancies) by heating an alkali metal halide in the presence of excess of metal vapours. This results in generation of different colours, for example, violet colour in KCl containing excess potassium.

(ii) Metal excess defect due to the presence of extra cations at interstitial sites: An example of this type is seen in the yellow colour developed by zinc oxide due to the loss of oxygen at high temperatures. The reaction is $ZnO \rightarrow Zn^{2+} + \frac{1}{2}O_2 + 2e^-$. There is excess zinc in the crystal due to the above reaction. The excess Zn^{2+} ions move to interstitial sites and the electrons to the neighbouring interstitial sites.

7.7 Electrical Properties of Solids

Solids may be classified as (i) conductors, (ii) semiconductors, and (iii) insulators. Conductors have a conductivity in the range of 10^7 S m^{-1}. The best examples of conductors are metals. Semiconductors have a conductivity in the range of 10^{-6} to 10^4 S m^{-1}. A good example may be cited is GaAs. Insulators have conductivity in the range 10^{-20} to 10^{-10} S m^{-1}.

7.7.1 Electrical Conduction Mechanism in Different Solids

Metals: The current is carried by electrons and hence is it called electronic conduction. The current is proportional to V, the electrical potential applied across the material. In many ionic solids, the charge carriers are ions and thus it is known as *ionic conduction*. The high conductivity of metals under a potential gradient is due to the fact that the partially filled conduction band is close to the unoccupied higher band and electrons flow easily from the former to the latter.

Semiconductors: Even in semiconductors, the gap between the conduction and valence bands is small and some electrons jump from the valence band to the conduction band. This is facilitated by an increase in temperature; thus, their conductivity increases with temperature. Semiconductors may be *intrinsic* or *extrinsic*. Si and Ge are examples of intrinsic semiconductors. In these, conduction is by electrons and holes which have opposite charges. The conductivity of intrinsic semiconductors is given by

$$S = n|e|m_e + p|e|m_h \quad (7.7)$$

where n = concentration of electrons, p = concentration of holes, m_e = mobility of electrons and m_h = mobility of holes. Since $n = p$ (because a hole is produced by the promotion of each electron to conduction band), Eq. 7.7 becomes,

$$S = 2n|e|(m_e + m_h) \quad (7.8)$$

An extrinsic semiconductor may have different concentrations of electrons and holes. They can be p-type if $p > n$ and n-type if $n > p$. They are made by doping. Doping of metals like Si or Ge with small amounts of group 15 elements such as P, As, Sb produces n-type semiconductors while doping of the same with group 13 elements such as B, Al, Ga forms p-type semiconductors. Both types of semiconductors are used in making transistors, diodes, and so on. GaAs, CdSe are fast response semiconductors.

7.7.2 Dielectric Behaviour of Solids

Application of an electric field to insulators creates dipoles due to the shift of electric charges. When the alignment of dipoles is orderly, which is observed in some substances, they are known as *piezoelectric* materials and show piezoelectric power. Substances like $BaTiO_3$ and quartz generate electricity on subjecting them to stress. Some piezoelectric materials generate electricity on heating which is known as *pyroelectricity*. When an electric field is applied across piezoelectric materials, the direction of polarisation of dipoles changes, giving rise to *ferroelectricity*. Sodium potassium tartrate is an example which shows this behaviour.

7.7.3 Magnetic Behaviour of Solids

Solids can be divided into different categories depending upon their response to a magnetic field.

(i) Diamagnetic substances: They are weakly repelled by magnetic fields. Diamagnetism arises due to the absence of unpaired electrons, for example, H^-, Li^+, F^-, O^{2-}, Mg^{2+}, and so on.

(ii) Paramagnetic materials: These materials contain permanent magnetic dipoles due to the presence of atoms, ions or molecules with unpaired electrons for example O_2, Cu^{2+}, Fe^{3+}. They are attracted by the magnetic field. However, they lose their magnetism in the absence of a magnetic field.

(iii) Ferromagnetic materials: Substances like Fe and CrO_2 show permanent magnetism when they are magnetised. This type of magnetism arises due to spontaneous alignment of magnetic moments in the same direction (↑↑ ↑↑ ↑↑).

(iv) Ferrimagnetism: It occurs when the magnetic moments are aligned in parallel and antiparallel directions in unequal number resulting in a net moment. Example Fe_3O_4 and ferrites having the formula $M^{2+}Fe_2O_4$ ($M^{2+} = Mg^{2+}$, Zn^{2+}) (↑↑ ↑↓ ↑↓).

(v) Antiferromagnetism: It occurs if the alignment of moments is in a compensatory way so as to give a net zero moment. Example MnO (↑↑ ↑↓ ↑↓...).

7.8 Properties of Cubic Unit Cells

(i) Positions of atoms in a cubic unit cell: Rectangular x, y, z axes are employed to locate atom positions (Fig. 7.3).

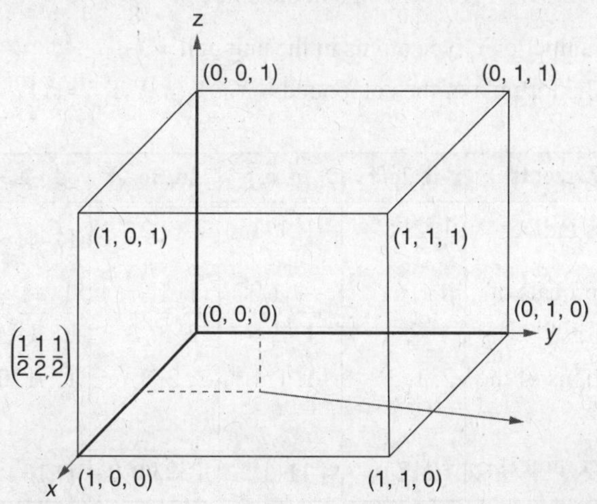

Fig. 7.3 Positions of atoms in a BCC unit cell

(ii) Directions in cubic unit cells: Sometimes it is necessary to refer to specific directions in crystal lattices. The crystallographic direction indices are vector components of the directions which are resolved along every coordinate axis. The indices are also to be indicated as smallest integers. As an example, let us consider the the direction indices of the (1,1,1) and (2,1,0) planes. The position coordinates of the cubic unit cell where the direction vector comes out of its surface is converted to integers which are the direction indices. For the (1,1,1) plane, the direction indices are [111]. Note the difference in the notation. The position coordinates of the direction vector OA are (1, ½, 0). To make them simple integers we multiply each of them by 2 to get [210] (Fig. 7.4a, 7.4b).

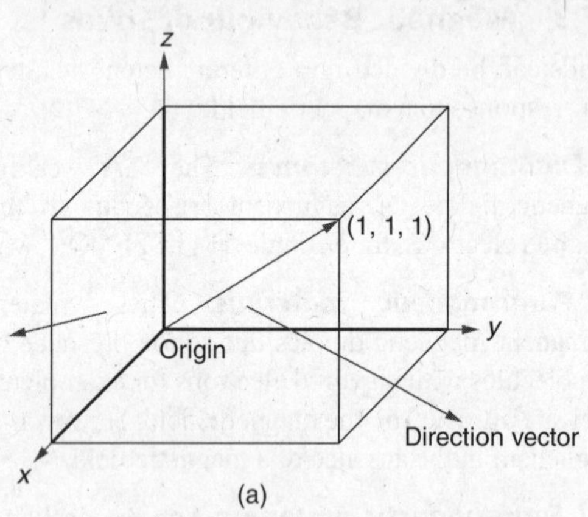

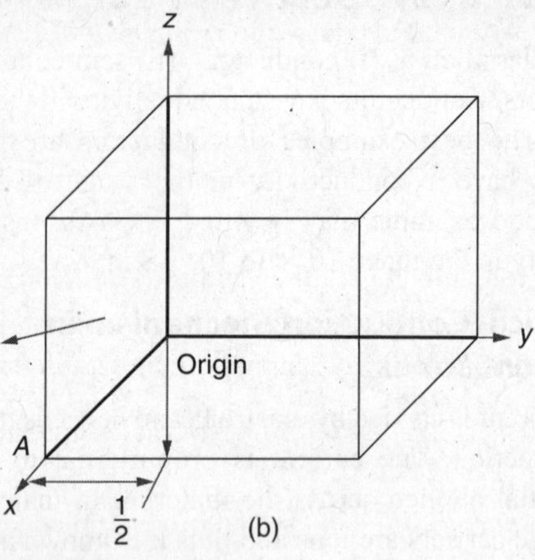

Fig. 7.4 Position coordinates of the (a) cubic cell and (b) vector OA

Worked Out Problems

Weiss, Miller Indices, FCC Lattices

Problem 7.1

(a) In an FCC lattice, X type atoms are present at the corners and Y type atoms are present at the face centres. In a given unit cell, two X type atoms are missing from the corners. What is the formula of the compound?

(b) The crystal planes in a solid intersect the crystal axes at (d, 2e, f) (–d, e, f), (2d, 3e, 2f) and (d, –e, ∞). What are the Weiss and Miller indices of these planes?

Solution

(a) Number of X type atoms in the unit cell $= \frac{6}{8} = \frac{3}{4}$

Number of Y type atoms in the unit cell = 3

The formula of the compound $= X_{3/4}Y_3 = X_3Y_{12} = XY_4$

(b)

Intercepts	d, 2e, f	–d, e, f	2d, 3e, 2f	d, –e, ∞
Weiss indices	1, 2, 1	–1, 1, 1	2, 3, 2	1, –1, ∞
Reciprocals of Weiss indices	$\frac{1}{1}, \frac{1}{2}, \frac{1}{1}$	$\frac{-1}{1}, \frac{1}{1}, \frac{1}{1}$	$\frac{1}{2}, \frac{1}{3}, \frac{1}{2}$	$\frac{1}{1}, \frac{-1}{1}, \frac{1}{\infty}$
Fractions of the above	2, 1, 2	–1, 1, 1	3, 2, 3	1, –1, 0
Miller indices	(212)	$(\bar{1}11)$	(323)	$(1\bar{1}0)$

Problem 7.2

(a) Gold crystallises as FCC at room temperature. Its atomic radius is 0.144 nm. Calculate the theoretical density of gold in kg m^{-3}. The atomic mass of gold is 197.0 and $N = 6.023\times10^{23}$.

(b) The edge length of the unit cell of vanadium is 0.304 nm and its density is 6.0×10^{-3} kg m^{-3}. Given its atomic mass as 50.9, calculate the number of atoms in the unit cell and state what type unit cell it is.

Solution

(a) $\rho = \frac{nM}{Na^3}$ For an FCC lattice, $r = \frac{\sqrt{2}a}{4}$

$$a = \frac{0.144\times4}{\sqrt{2}} = 0.407 \text{ nm}$$

$$\rho = \frac{4\times197\times10^{-3}}{6.023\times10^{23}\times\left(0.407\times10^{-9}\right)^3}$$

$$= 1.94\times10^4 \text{ kg m}^{-3}$$

(b) $n = \frac{\rho Na^3}{M} = \frac{6.023\times10^{23}\times6.0\times10^3\times\left(0.304\times10^{-9}\right)^3}{50.9\times10^{-3}}$

$$= 19.9\times10^{-1} = 1.99 \approx 2$$

Its unit cell is BCC.

Problem 7.3

Show the positions of the atoms at the lattice points and face centre points in a FCC unit cell.

Solution

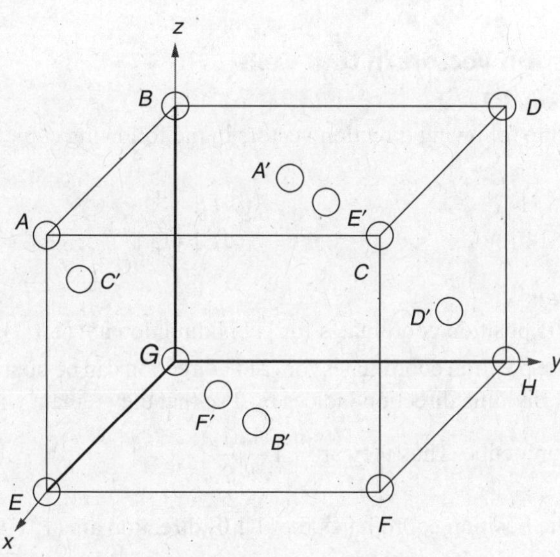

The corner atoms are ABCDEFGH. The face centre atoms are A′B′C′D′E′F′. The atom positions at the corners are: Origin G: (0,0,0); E: (1,0,0); F: (1,1,0); H: (0,1,0); A: (1,0,1); B: (0,0,1); C: (1,1,1); D: (0,1,1).

The atom positions at face centres are A′: $\left(\frac{1}{2},\frac{1}{2},\frac{1}{2}\right)$; B′: $\left(\frac{1}{2},\frac{1}{2},0\right)$; C′: $\left(\frac{1}{2},0,\frac{1}{2}\right)$; D′: $\left(\frac{1}{2},1,\frac{1}{2}\right)$; E′: $\left(0,\frac{1}{2},\frac{1}{2}\right)$; F′: $\left(1,\frac{1}{2},\frac{1}{2}\right)$.

HCP Lattices and Unit Cell Volume

Problem 7.4

Calculate the volume of the unit cell of cadmium which crystallises in HCP given its lattice constant (edge length) $a = 0.297$ nm and C (vertical height) = 0.562 nm.

Solution

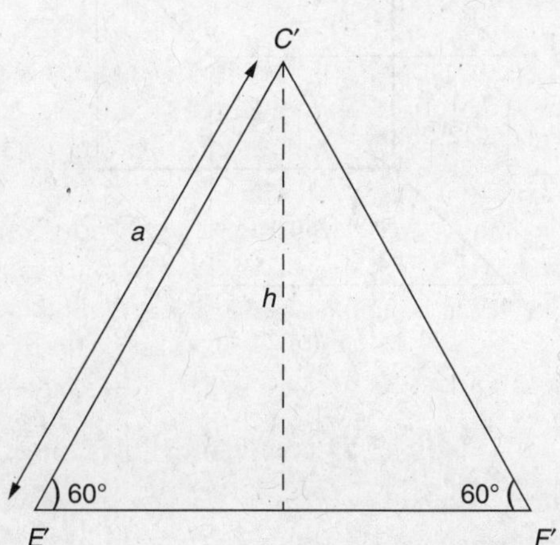

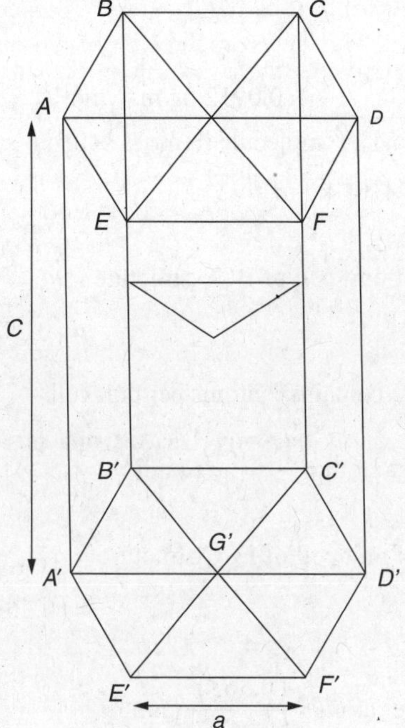

The volume of unit cell may be obtained by determining the area of the base of unit cell and multiplying by the height.

From the figure, $\angle C'E'F' = \angle C'FE' = 60°$

$$E'F' = a = 0.297 \text{ nm}$$

$$\text{Area of } \Delta C'E'F' = \frac{1}{2} \times \text{base} \times \text{height}$$

$$= \frac{1}{2} \times a \times a \sin 60 = \frac{1}{2} a^2 \sin 60$$

$$\text{Total area of base } A'B'C'E'F' = 6 \times \frac{1}{2} \times a^2 \sin 60$$

$$= 3a^2 \sin 60 = 3 \times (0.297)^2 \times 0.8660$$

$$\text{Volume of HCP unit cell} = 3 \times (0.297)^2 \times 0.8660 \times 0.562$$

$$= 0.1288\,(\text{nm})^3$$

Problem 7.5

(a) Zirconium crystallises in the HCP lattice and its lattice constant a = 0.323 nm and c (lattice vertical height) = 0.515 nm. Calculate its density in kg m^{-3}, atomic mass = 91.2.

(b) Iron undergoes a polymorphic change from BCC to FCC when heated through 1185 K. What is the change in volume when it changes from BCC to FCC? Use the following data. At 1185 K, a of BCC = 0.293 nm and a of FCC = 0.363 nm.

Solution

(a) Total area of the HCP base

$$= 6 \times \frac{1}{2} \times a^2 \sin 60 = 3a^2 \sin 60 = 3 \times a^2 \times 0.8660$$

$$\text{Volume of unit cell} = 3a^2 \sin 60 \times c$$

$$= 3 \times (0.323 \text{ nm})^2 \times 0.866 \times 0.515 \text{ nm}$$

$$= 0.136 \text{ nm}^3$$

$$\rho = \frac{nM}{Na^3}$$

$$= \frac{6 \times 0.0912 \text{ kg m}^{-3} \text{ mol}^{-1}}{6.023 \times 10^{23} \text{ molecules} \times \text{mol}^{-1} \times 0.136 \times 10^{-27} \text{ m}^3}$$

$$= 0.670 \times 10^4 \text{ kg m}^{-3}$$

(b) Volume of unit cell of BCC structure = a^3

$$= (0.293)^3 \, (\text{nm})^3$$

Since BCC contains 2 atoms per unit cell,

$$\text{volume per atom} = \frac{(0.293)^3}{2} (\text{nm})^3$$

Volume of unit cell of FCC structure = a^3

$$= (0.363)^3 \, (\text{nm})^3$$

$$\text{Volume per atom} = \frac{(0.363)^3}{4} (\text{nm})^3$$

$$\text{Change in volume} = \frac{(0.363)^3}{4} - \frac{(0.293)^3}{2}$$

$$= 0.011958 - 0.01258$$

$$= -0.000622 = -0.062\%$$

Problem 7.6

(a) Beryllium crystallises in the HCP lattice. Its lattice constants are a = 0.2286 nm and c (vertical height) = 0.3584 nm. Calculate the volume of the unit cell and its density. (Atomic mass of Be = 90)

(b) Diamond has an FCC structure. Its unit cell edge length is 0.3567 nm. Given that two atoms are present per lattice point, what is its density?

Solution

(a) Volume of unit cell

$$= 3a^2 \sin 60 \times c$$

$$= 3 \times \left(0.2286 \times 10^{-9}\right)^2 \times 0.3584 \times 10^{-9} \times 0.8660$$

$$= 0.0487 \times 10^{-27} \text{ m}^3$$

$$\rho = \frac{nM}{Na^3} = \frac{6 \times 0.09}{6.023 \times 10^{23} \times 1.195 \times 10^{-29}}$$

$$= 0.54/7.1975 \times 10^{-6} = 0.075 \times 10^{-6} = 7.5 \times 10^4 \text{ kg m}^{-3}$$

(b) For the eight lattice points, there are 4 atoms in the unit cell + there is an additional atom for each lattice point. Therefore, one must conclude that there are 8 eight atoms per unit cell.

$$\rho = \frac{nM}{Na^3} = \frac{8 \times 0.012}{6.023 \times 10^{23} \times \left(0.3567 \times 10^{-9}\right)^3}$$

$$= 0.351 \times 10^4 = 3.51 \times 10^3 \text{ kg m}^{-3}$$

Direction Vectors in Unit Cells

Problem 7.7

Show the following direction vectors in the following cubic unit cells:

(a) [111] (b) [211]

(c) [1 $\bar{1}$ 0] (d) [$\bar{1}$11]

Solution

(a) The position coordinates for [111] direction are (1,1,1).

(b) The position coordinates for [211] direction can be obtained by dividing direction indices by 2 so that they remain within a unit cube. Thus they are $\left(1, \frac{1}{2}, \frac{1}{2}\right)$.

(c) The position coordinates for [1 $\bar{1}$ 0] direction are (1, $\bar{1}$,0).

(d) The position coordinates for [$\bar{1}$ 11] direction are ($\bar{1}$,1,1).

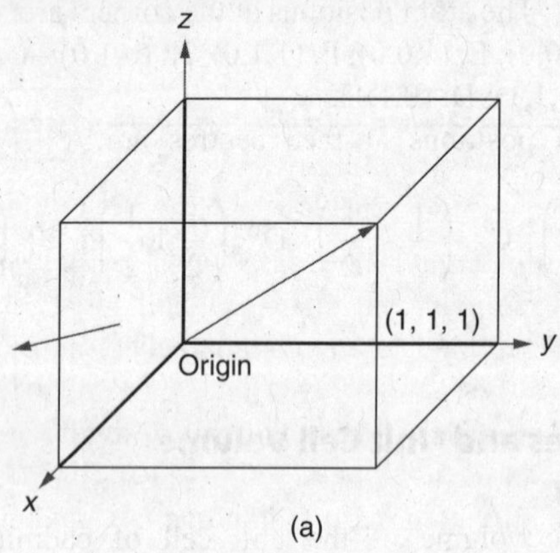

(a)

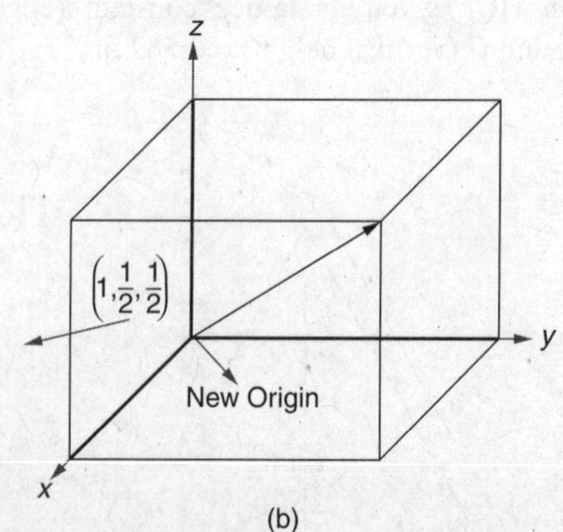

(b)

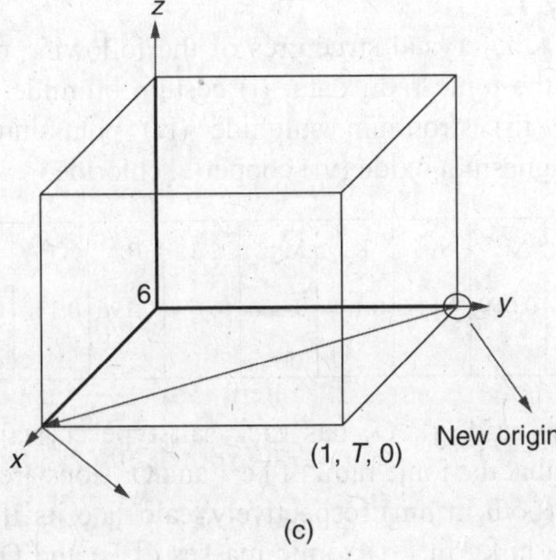

(c)

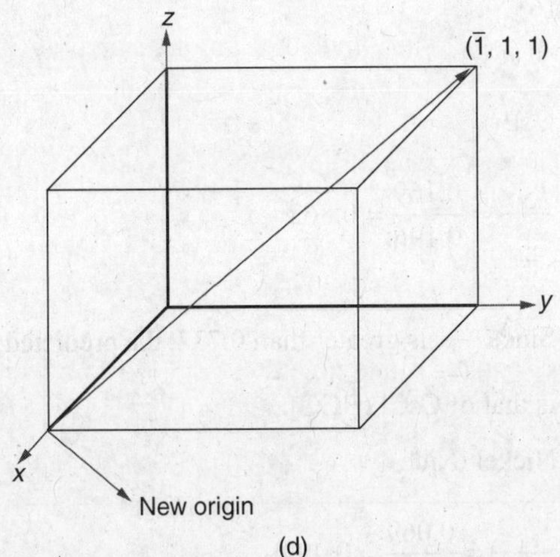

(d)

Ionic Solids, Radius Ratio and Packing Factor

Problem 7.8

Potassium chloride crystallises in FCC, where the chloride ions occupy the corners and the K^+ ions the face centres. The density of the salt is 1984 kg m^{-3}. Given the ratio $\frac{r_+}{r_-} = 0.735$, what is the radius of the anion in nm? (Atomic mass of K = 39.1)

Solution

$$\text{Density, } \rho = \frac{nM}{Na^3} = \frac{4 \times 0.0391}{6.023 \times 10^{23} \times a^3}$$

or $$a^3 = \frac{4 \times 0.0391}{6.023 \times 10^{23} \times 1984} = 1.309 \times 10^{-28}$$

$$a = \text{edge length} = \sqrt[3]{1.309 \times 10^{-28}} = 5.077 \times 10^{-10} \text{ m}$$

Also $$\{2(r_+ + r_-)\}^2 = a^2 + a^2 = 2a^2$$

$$2(r_+ + r_-) = \sqrt{2}a = 1.414 \times 5.077 \times 10^{-10}$$

$$(r_+ + r_-) = \frac{1.414 \times 5.077 \times 10^{-10}}{2} = 3.59 \times 10^{-10}$$

$$\frac{r_+}{r_-} = 0.735$$

$$r_+ + 0.735 r_- = 3.59 \times 10^{-10}$$

$$r_- = \frac{3.59 \times 10^{-10}}{1.735} = 2.07 \times 10^{-10} \text{ m} = 0.207 \text{ nm}$$

Problem 7.9

Calculate the ionic packing factor of cesium chloride given $r_{Cs+} = 0.170$ nm and $r_{Cl-} = 0.181$ nm. Assuming that the oppositely charged ions touch along the body diagonal of the unit cell, what is its theoretical density? What is its coordination and what is its nearest neighbour distance? (Molar mass of CsCl = 168.4)

Solution

IPF is defined by the equation

$$\text{IPF} = \frac{\text{No. of atoms of unit cell} \times \text{Volume of the sphere}}{\text{Volume of the unit cell}}$$

In the unit cell of CsCl, we have one Cs^+ and one Cl^- ion

Total volume of the ions

$$= \frac{\frac{4}{3} \times 3.143 \times (0.170)^3 + \frac{4}{3} \times 3.143 \times (0.181)^3}{\text{Volume of unit cell}}$$

Since in a BCC lattice the relation, $2(r_{Cs^+} + r_{Cl^-}) = \sqrt{3}a$ holds,

we have $2(0.170 + 0.181) = \sqrt{3}a$, $a = \frac{0.702}{1.732} = 0.405$ nm

or $$\text{IPF} = \frac{\frac{4}{3} \times 3.143 \times (0.004913 + 0.00593)}{(0.405)^3}$$

$$= \frac{0.0454(\text{nm})^2}{0.0664(\text{nm})^2} = 0.684$$

$$\text{Theoretical density } \rho = \frac{nM}{Na^3} = \frac{1 \times 0.1329 + 1 \times 0.0355}{6.023 \times 10^{23} \times (0.405 \times 10^{-9})^3}$$

$$= \frac{0.842 \times 10^4}{} = 8.42 \times 10^3 \text{ kg m}^{-3}$$

Coordination number in CsCl = 8 (since it is BCC)

Nearest neighbour distance

$$= \frac{1}{2} \text{ body diagonal} = \frac{2(r_{Cs^+} + r_{Cl^-})}{2} = 0.351 \text{ nm}$$

Problem 7.10

The ionic radii of Cu^+ and O^{2-} ions are 0.074 nm and 0.126 nm respectively. From the given unit cell of the binary compound of copper and oxygen, (i) determine the empirical formula of the compound, (ii) find the coordination numbers of copper and oxygen, (iii) find the edge length of the cubic unit cell and (iv) calculate the density of the compound.

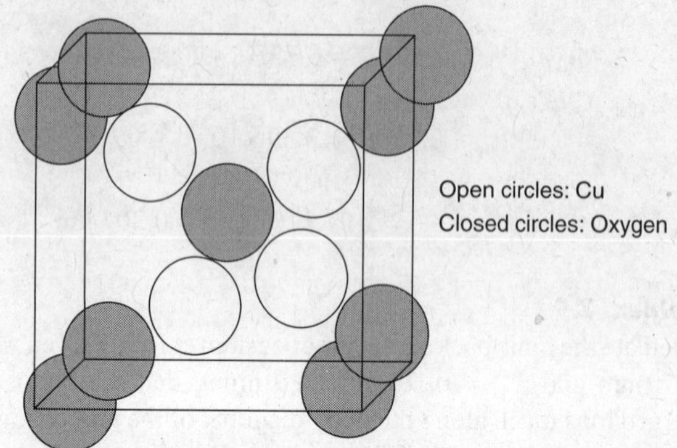

Solution

(i) There are four copper ions and one oxygen anion located in the cubic unit cell. Further there are eight oxygens at the eight corners of the unit cell as well as one oxygen ion at the centre.

Number of oxygen ions per unit cell $= 8\times\frac{1}{8}+1 = 2$

Further, there are four Cu^+ ions per unit cell. Thus, the empirical formula of the compound is $Cu_4O_2 = Cu_2O$

(ii) Coordination number of Cu^+ ion

= Anion coordination number

$\times \frac{\text{No. of anions per formula unit}}{\text{No. of cations per formula unit}}$

$= 4\times\frac{1}{2} = 2$ (see figure for the confirmation of this result)

(iii) Edge length of unit cell: The ions touch along the body diagonal of unit cell. Along the body diagonal we have two copper ions ($2r_{Cu^+} + 2r_{Cu^+}$) and one oxygen at centre ($2r_{O^{2-}}$) + one half of oxygen ions at corners ($r_{O^{2-}} + r_{O^{2-}}$)

$$\therefore \text{Body diagonal} = 2\{2r_{Cu^+}\} + \{2r_{O^{2-}}\}$$
$$= 4\{r_{Cu^+} + r_{O^{2-}}\}$$
$$= 4\times10^{-2}\{7.4+12.6\} = 0.80 \text{ nm}$$

In BCC, body diagonal

$$= \sqrt{3}a \text{ or } a = \frac{0.80}{\sqrt{3}} = \frac{0.80}{1.732} = 0.462 \text{ nm}$$

(iv) $\rho = \frac{nM}{Na^3}$

$$\rho = \frac{4\times63.55\times10^{-3} + 2\times16.00\times10^{-3}}{6.023\times10^{23}\times(0.462\times10^{-9})^3}$$
$$= \frac{0.2542+0.032}{6.023\times10^{23}\times(0.462\times10^{-9})^3}$$
$$= \frac{0.2862}{6.023\times10^{23}\times(0.462\times10^{-9})^3}$$
$$= 0.482\times10^4 = 4.82\times10^3 \text{ kg m}^{-3}$$

Problem 7.11

(a) Predict the crystal structures of the following compounds using the ionic radii data: (i) cesium bromide, (ii) nickel oxide (iii) strontium sulphide (iv) potassium chloride (v) magnesium oxide (vi) copper(I) chloride.

Ion	Cs^+	Ni^{2+}	Sr^{2+}	K^+	Mg^{2+}	Cu^+	Br^-	O^{2-}	S^{2-}	Cl^-
Radii (nm)	0.169	0.069	0.112	0.138	0.066	0.074	0.196	0.140	0.184	0.181

(b) Ferrous oxide (FeO) has rock salt type crystal structure. Given that the ionic radii of Fe^{2+} and O^{2-} ions are 0.077 and 0.140 (both in nm) respectively, calculate its IPF and the density in kg m^{-3}. (Atomic masses of Fe and O are 55.35 and 16)

Solution

(a) (i) CsBr

$$\frac{r_{Cs^+}}{r_{Br^-}} = \frac{0.169}{0.196} = 0.862$$

Since $\frac{r_+}{r_-}$ is greater than 0.732, the predicted structure is that of CsCl (BCC).

(ii) Nickel oxide

$$\frac{r_{Ni^{2+}}}{r_{O^{2-}}} = \frac{0.069}{0.140} = 0.493$$

Since $\frac{r_+}{r_-}$ lies between 0.414 and 0.732 its structure is that of NaCl (FCC).

(iii) Strontium sulphide

$$\frac{r_{Sr^{2+}}}{r_{S^{2-}}} = \frac{0.112}{0.184} = 0.609$$

Since $\frac{r_+}{r_-}$ lies between 0.414 and 0.732 the structure of SrS is FCC.

(iv) Potassium chloride

$$\frac{r_{K^+}}{r_{Cl^-}} = \frac{0.138}{0.181} = 0.762$$

Since $\frac{r_+}{r_-}$ is greater than 0.732, the predicted structure is that of CsCl (BCC).

(v) Magnesium oxide

$$\frac{r_{Mg^{2+}}}{r_{O^{2-}}} = \frac{0.066}{0.140} = 0.471$$

Since $\frac{r_+}{r_-}$ lies between 0.414 and 0.732, the structure of MgO is that of NaCl (FCC).

(vi) Copper(I) chloride

$$\frac{r_{Cu^+}}{r_{Cl^-}} = \frac{0.074}{0.181} = 0.409$$

Since $\frac{r_+}{r_-}$ is less than 0.414, the structure is that of CuCl is sphalenite (ZnS).

(b) FeO has rock salt type structure. There are four FeO units per unit cell.

$$\text{IPF} = \frac{\text{Volume of atoms in the unit cell}}{\text{Volume of the unit cell}}$$

$$\therefore \text{Volume of the unit cell} = a^3 = \left\{2\left(r_{Fe^{2+}} + r_{O^{2-}}\right)\right\}^3$$

$$= \left\{2(0.077 + 0.140)\right\}^3$$

$$= (2 \times 0.217)^3$$

$$= 0.082 \text{ nm}^3 = 0.082 \times 10^{-27} \text{ m}^3$$

$$\text{IPF} = \frac{4 \times \frac{4}{3} \times 3.143 \times (0.077)^3 + 4 \times \frac{4}{3} \times 3.143 \times (0.140)^3}{0.082}$$

$$\frac{4 \times \frac{4}{3} \times 3.143 \left\{(0.077)^3 + (0.140)^3\right\}}{0.082}$$

$$= \frac{16.763\left\{4.565 \times 10^{-4} + 2.74 \times 10^{-3}\right\}}{0.082}$$

$$= \frac{16.763 \times 3.20 \times 10^{-3}}{0.082} = \frac{0.373}{0.082} = 4.500 \text{ (absurd!!)}$$

(The value cannot be greater than 1. The equation does not seem to apply in this case!)

$$\rho = \frac{nM}{Na^3} = \frac{4 \times 0.05585 + 4 \times 0.016}{6.023 \times 10^{23} \times 0.082 \times 10^{-27}}$$

$$= \frac{0.2234 + 0.064}{6.023 \times 10^{23} \times 0.082 \times 10^{-27}}$$

$$= \frac{0.2874}{6.023 \times 10^{23} \times 0.082 \times 10^{-27}}$$

$$= 0.582 \times 10^4 = 5.82 \times 10^3 \text{ kg m}^{-3}$$

The IPF value of 4.5 possibly arises due to the fact that FeO is non-stoichiometric with the composition being closer to $Fe_{0.95}O$.

Problem 7.12

Sodium chloride crystallises in FCC and has a density of 2.2×10^3 kg m^{-3}. Calculate the cation–cation distance in the molecule (M = 58.5).

Solution

$$\text{Mass of the molecule} = \frac{58.5}{6.023 \times 10^{23}} = 9.713 \times 10^{-23}$$

$$\text{Density} = 2.2 \times 10^3 \text{ kg m}^{-3} = 2.2 \times 10^{-3} \text{ kg cm}^{-3}$$

Number of molecules in 1 cm^3

$$= \frac{\text{Density in kg cm}^{-3}}{\text{Mass of the molecule in kg molecule}^{-1}}$$

$$\frac{2.2 \times 10^{-3} \text{ kg cm}^{-3}}{9.713 \times 10^{-26} \text{ kg molecule}^{-1}} = 0.227 \times 10^{23} \text{ molecules cm}^{-3}$$

$$= 2.27 \times 10^{22} \text{ molecules cm}^{-3}$$

$$\text{The number of ions in NaCl} = 2 \times 2.27 \times 10^{22} = 4.54 \times 10^{22}$$

A unit cell of NaCl contains 4Na^+ and 4Cl^- ions

$$\therefore \text{Number of unit cells in 1 cm}^3 = \frac{4.54 \times 10^{22}}{8} = 0.568 \times 10^{22}$$

If a^3 is the volume of a unit cell, the number of unit cells in the volume

$$1 = 0.568 \times 10^{22} \times a^3$$

or
$$a = \sqrt[3]{\frac{1}{0.568 \times 10^{22}}} = \sqrt[3]{1.761 \times 10^{-22}} = 5.61 \times 10^{-8} \text{ cm}$$

The cation–cation distance d corresponds to face diagonal of the unit cell.

Thus,

$$d^2 = \left(\frac{a}{2}\right)^2 + \left(\frac{a}{2}\right)^2 = \frac{a^2}{2} \text{ or } d = \frac{a}{\sqrt{2}} = \frac{0.561}{0.414} = 0.397 \text{ nm}$$

The problem can also be solved using $\rho = \frac{nM}{Na^3}$ and calculating a. Using this a in the equation,

$$1/2 \text{ face diagonal} = \frac{\sqrt{2}a}{2} = \frac{a}{\sqrt{2}} = \frac{0.561}{1.414} = 0.397 \text{ nm}$$

Problem 7.13

(a) A crystal has unit cell dimensions of 0.1, 0.2 and 0.3 (all in nm). The plane (212) has an intercept of 0.12 nm on the *x*-axis. What are the intercepts on the *y* and *z* axes?

(b) Rubidium chloride has rock salt type structure. The ionic radii of Rb^+ and Cl^- ions are 0.166 nm and 0.181 nm respectively. From these data calculate (i) cation–anion shortest distance, (ii) edge length of the unit cell, (iii) the shortest Rb^+–Rb^+ distance, and (iv) the density of the salt. (Atomic masses: Rb = 85.5; Cl = 35.5)

Solution

(a) Let *p*, *q*, *r* be the three intercepts. Thus

$$p:q:r = \frac{0.1}{2}:\frac{0.2}{1}:\frac{0.3}{2}$$

$$\frac{p}{q} = \frac{0.1}{2} \times \frac{1}{0.2} = \frac{0.1}{0.4} = \frac{1}{4}$$

$$\frac{0.12}{q} = \frac{1}{4} \text{ or, } q = 0.48 \text{ nm}$$

$$\frac{q}{r} = \frac{0.2}{1} \times \frac{2}{0.3} = \frac{0.4}{0.3} = \frac{4}{3}$$

$$\frac{0.48}{r} = \frac{4}{3}$$

$$r = 0.48 \times \frac{3}{4} = 0.36 \text{ nm}$$

(b) (i) The cation–anion shortest distance is obtained by adding the radii

$$r_{Rb^+} + r_{Cl^-} = 0.166 + 0.181 = 0.347 \text{ nm}$$

(ii) The edge length of the unit cell is calculated from the relation

$$2\left(r_{Rb^+} + r_{Cl^-}\right) = a = 2 \times 0.347 = 0.694 \text{ nm}$$

(iii) The shortest Rb^+–Rb^+ distance

$$= \frac{1}{2} \text{ face diagonal} = \frac{1}{2}\sqrt{2}a = \frac{a}{\sqrt{2}} = \frac{0.694}{1.414} = 0.491 \text{ nm}$$

(iv) $\rho = \dfrac{nM}{Na^3}$

There are 4 Rb^+ and 4Cl^- ions in the unit cell

$$= \frac{4 \times 0.0855 + 4 \times 0.0355}{6.023 \times 10^{23} \times \left(0.699 \times 10^{-9}\right)^3} = 2.404 \times 10^3 \text{ kg m}^{-3}$$

Problem 7.14

(a) MgO has a structure similar to that of rocksalt (NaCl). Its density is 3.58×10^3 kg m^{-3}. What is its unit cell length? Calculate its atomic packing factor. (Atomic masses: Mg = 24.3; O = 16.0) ($r_{Mg2+} = 0.066$ nm; $r_{O2-} = 0.140$ nm)

(b) Show diagrammatically the (100) plane for rock salt structure, (110) plane of cesium chloride and (100) plane of the same chloride.

Solution

(a) There are 4 formula units per unit cell, that is, 4 Mg^{2+} and 4 O^{2-} ions

$$\rho = 3.58 \times 10^3 = \frac{nM}{Na^3} = \frac{4 \times 0.0243 + 4 \times 0.016}{6.023 \times 10^{23} \times a^3}$$

$$a^3 = \frac{0.1612}{6.023 \times 10^{23} \times 3.58 \times 10^3}$$

$$= 0.00748 \times 10^{-26} = 0.0748 \times 10^{-27}$$

$$a = \sqrt[3]{0.0748 \times 10^{-27}} = 0.421 \times 10^{-9} \text{ m} = 0.421 \text{ nm}$$

$$\text{APF} = \frac{\text{Volume of ions of unit cell}}{\text{Volume of unit cell}}$$

$$= \frac{4 \times \frac{4}{3} \times 3.143 \times (0.066)^3 + 4 \times \frac{4}{3} \times 3.143 \times (0.140)^3}{(0.421)^3}$$

$$= \frac{4 \times \frac{4}{3} \times 3.143 \times \left\{(0.066)^3 + (0.140)^3\right\}}{(0.421)^3}$$

$$= \frac{16.763 \times \left(2.875 \times 10^{-4} + 2.74 \times 10^{-3}\right)}{0.0746}$$

$$= \frac{16.763 \times 3.0315 \times 10^{-3}}{0.0716} = \frac{0.0508}{0.0716} = 0.710$$

(b) (i) (100) plane of rock salt structure

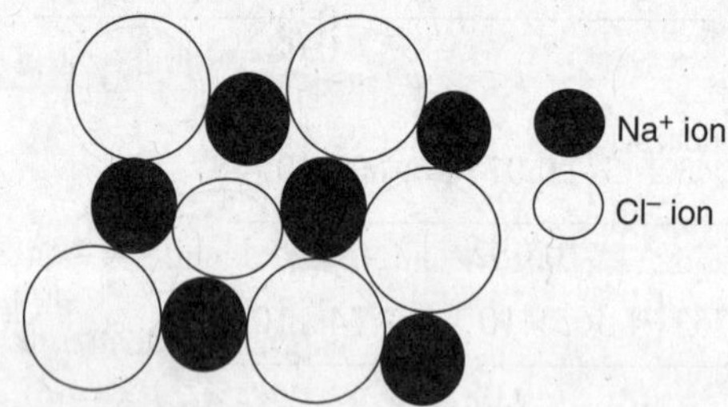

(ii) (110) plane for CsCl

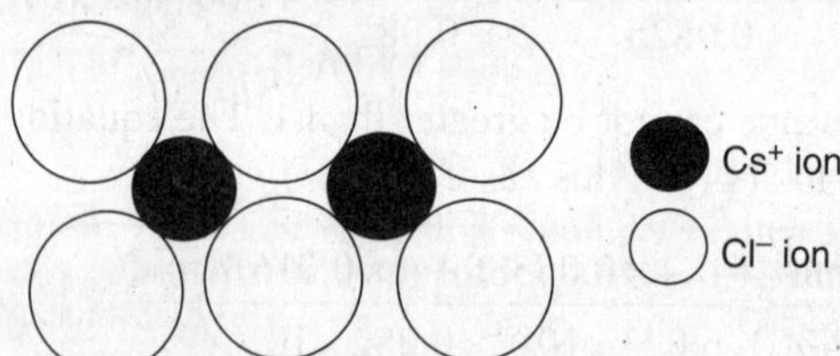

(iii) (100) plane of CsCl

BCC Lattices and Nearest Neighbour Distances

Problem 7.15

(a) Tungsten crystallises in BCC and its edge length is 0.316 nm. (i) What is the nearest neighbour distance? (ii) What is its radius?

(b) The salt sodium chloride crystallises in FCC and its edge length is 0.564 nm. The $\frac{r_c}{r_a}$ ratio is 0.536. Assuming that the Cl^- ions occupy lattice points and the Na^+ ions the octahedral voids, calculate (i) the cationic radius, (ii) the nearest neighbour distance.

Solution

(a) (i) $a = 0.316$ nm. For BCC

$$r = \frac{\sqrt{3}a}{4} = \frac{1.732 \times 0.316}{4} = 0.182 \text{ nm}$$

(ii) Nearest neighbour distance $= \frac{1}{2}$ body diagonal

$$= \frac{4r}{2} = 2r = 2 \times 0.182 = 0.364 \text{ nm}$$

(b) Edge length $= 2(r_+ + r_-) = 0.564$ nm

and $\frac{r_+}{r_-} = 0.536$ or $2(0.536r_- + r_-) = 0.564$ nm

$1.536r_- = 0.282$

Cationic radius $r_- = \frac{0.282}{1.536} = 0.184$ nm

Nearest neighbour distance $= \frac{1}{2}$ face diagonal

$$= \frac{a}{\sqrt{2}} = \frac{0.536}{\sqrt{2}} = \frac{0.536}{1.414} = 0.379 \text{ nm}$$

Problem 7.16

Tantalum crystallises in a BCC lattice and its molar mass is 181 g mol^{-1}. It density is 16.4×10^3 kg m^{-3}. (i) What is the edge length of the unit cell? (ii) Calculate the distances between the (112), (210) and (222) planes. (iii) What is the nearest neighbour distance between its atoms and what is its radius?

Solution

(i) $\rho = \frac{nM}{Na^3} = \frac{2 \times 0.181}{6.023 \times 10^{23} \times a^3}$ or $a^3 = 3.68 \times 10^{-29}$

$$a = \sqrt[3]{3.68 \times 10^{-29}} = 3.326 \times 10^{-10} \text{ m}$$

(ii) $d_{112} = \frac{a}{\sqrt{h^2 + k^2 + l^2}} = \frac{3.326 \times 10^{-10}}{\sqrt{1^2 + 1^2 + 2^2}} = 1.358 \times 10^{-10}$ m

$$d_{210} = \frac{3.326 \times 10^{-10}}{\sqrt{2^2 + 1^2 + 0}} = \frac{3.326 \times 10^{-10}}{2.236} = 1.487 \times 10^{-10} \text{ m}$$

$$d_{222} = \frac{3.326 \times 10^{-10}}{\sqrt{2^2 + 2^2 + 2^2}} = \frac{3.326 \times 10^{-10}}{3.464} = 0.96 \times 10^{-10} \text{ m}$$

(iii) Nearest neighbour distance $= \frac{1}{2}$ body diagonal

$$= \frac{\sqrt{3}a}{2} = \frac{1.732 \times 3.326 \times 10^{-10}}{2} = 2.880 \times 10^{-10} \text{ m}$$

Problem 7.17

(a) Potassium crystallises in a primitive cubic lattice and has a unit cell length of 0.335 nm. (i) What is its density in kg m^{-3}? (ii) What are the distances between the planes (111) (210) and (222)? ($M = 209$)

(b) Molybdenum forms BCC crystals at 293 K. The nearest neighbour distance between its atoms is 0.273 nm. (i) What is the radius of the atom? (ii) What is its unit length? (iii) What is its density? ($M = 95.9$)

Solution

(a) $\rho = \frac{nM}{NV}$

$$= \frac{1 \times 0.209 \text{ kg mol}^{-1}}{6 \times 10^{23} \text{ molecules} \times \text{mol}^{-1} \times (0.335 \times 10^{-9})^3 \text{ m}^3}$$

$$= 0.926 \times 10^4 \text{ kg m}^{-3}$$

$$d_{111} = \frac{a}{\sqrt{h^2 + k^2 + l^2}} = \frac{0.335}{\sqrt{1^2 + 1^2 + 1^2}}$$

$$= \frac{0.335}{\sqrt{3}} = \frac{0.335}{1.732} = 0.193 \text{ nm}$$

$$d_{210} = \frac{0.335}{\sqrt{2^2 + 1^2 + 0}} = \frac{0.335}{2.236} = 0.150 \text{ nm}$$

$$d_{222} = \frac{0.335}{\sqrt{2^2 + 2^2 + 2^2}} = \frac{0.335}{3.464} = 0.097 \text{ nm}$$

(b) Nearest neighbour distance

$$= 2r = \frac{\sqrt{3}a}{2}$$

$0.273 = 2r$; or $r = 0.137$ nm

$$a = \frac{4r}{\sqrt{3}} = \frac{4 \times 0.137}{1.732} = 0.316 \text{ nm}$$

$$\rho = \frac{nM}{Na^3} = \frac{2 \times 0.0959}{6.023 \times 10^{23} \times (0.316 \times 10^{-9})^3 \text{ m}^3}$$

$$= 1.013 \times 10^4 \text{ kg m}^{-3}$$

Problem 7.18

(a) Iron crystallises in BCC at room temperature. When X-rays of $\lambda = 0.1541$ nm were used, the diffraction from the (110) planes occurred at $\theta = 22.352°$. What is the lattice constant of iron?

(b) Calculate the linear atomic density ρ_l in the (110) direction in the crystal lattice of lead (which crystallises in FCC) and whose radius is 0.175 nm.

Solution

(a) $\lambda = 2d_{110}\sin\theta;\ d_{110} = \dfrac{\lambda}{2\sin\theta} = \dfrac{0.1541}{2\sin(22.352)}$

$= \dfrac{0.1541}{2\times 0.3803} = \dfrac{0.1541}{0.7606} = 0.2026$ nm

Since

$d_{hkl} = \dfrac{a}{\sqrt{h^2+k^2+l^2}};\ a = 0.2026\sqrt{1^2+1^2+0}$

$= 0.2026\times 1.414 = 0.2865$ nm

(b) Linear atomic density is defined as

$$\rho_l = \frac{\text{Number of atoms diameters intersected by selected length of line in the direction of interest}}{\text{Selected length of line}}$$

First let us find out the (110) direction in the given lattice.

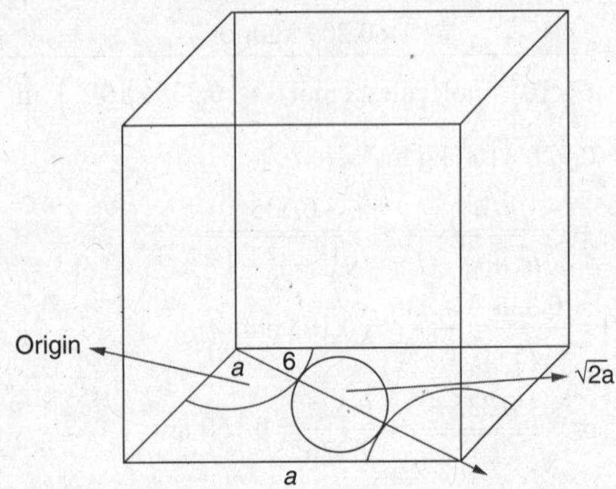

No. of atoms the direction line crosses = ½ atom + 1 full atom + ½ atom = 2 atoms

For FCC

$4r = \sqrt{2}a$ or $a = \dfrac{4r}{\sqrt{2}} = \dfrac{4\times 0.175}{1.414} = 0.495$ nm

$\rho_l = \dfrac{2 \text{ atoms}}{1.414\times 0.495(\text{nm})} = \dfrac{2}{0.70(\text{nm})} = 2.857\dfrac{\text{atoms}}{\text{nm}}$

2.857×10^9 atoms m^{-1}

Miller Indices and Bragg Equation

Problem 7.19

(a) Determine the Miller indices of the following planes. (i) A plane cuts the x-axis at 1d, the y-axis at (½)e and is parallel to the z-axis. (ii) A plane that makes intercepts on the x-axis at 2d, y-axis at 1e and z-axis at 3f of an orthorhombic crystal with $a{:}b{:}c = 4{:}3{:}2$.

(b) X-rays of wavelength of 0.154 nm impinge on the (211) plane of a sodium chloride crystal and the first order reflection was observed at an angle of 20°. What is the edge length of the crystal?

Solution

(a)

	(i)	(ii)
Intercepts	d, $\frac{1}{2}$e, ∞	2d, 1e, 3f
Weiss indices	$1, \frac{1}{2}, \infty$	$\left(\frac{2}{4}=\frac{1}{2}\right), \frac{1}{3}, \frac{3}{2}$ (Note: $a{:}b{:}c = 4{:}3{:}2$)
Reciprocals of Weiss indices	$\frac{1}{1}, \frac{2}{1}, 0$	$2, 3, \frac{2}{3}$
Fractions of the above	1, 2, 0	6, 9, 2
Miller indices	(120)	(692)

(b) $n\lambda = 2d_{211}\sin\theta;\ 1\times 0.154 = 2d_{211}\sin 20$

$d_{211} = \dfrac{1\times 0.154}{2\times 0.342} = 0.225$

$d_{211} = 0.225 = \dfrac{a}{\sqrt{h^2+k^2+l^2}} = \dfrac{a}{\sqrt{2^2+1^2+1^2}} = \dfrac{a}{\sqrt{6}}$

or $a = 0.225\times\sqrt{6} = 0.225\times 2.449 = 0.551$ nm

Problem 7.20

Draw the crystal planes which have the Miller indices (111), (221) and $(1\bar{1}0)$. Also draw a (111) plane in a BCC unit cell and list the position coordinates of atoms whose centres are intersected by this plane.

Solution

First determine the reciprocals of the MIs of the (111), (221) and $(1\bar{1}0)$ planes.

(i) For the (111) plane, the reciprocals are 1, 1, 1. The (111) plane must pass through the unit cube at intercepts on the x-axis = 1, y-axis = 1, z-axis = 1.

(ii) For the (221) plane, first determine the reciprocals of the Miller indices which are ½, ½, 1. The (221) plane must pass through a unit cube with intercepts on the x-axis = ½, y-axis = ½, z-axis = 1.

(iii) For the $(1\bar{1}0)$ plane, the reciprocal of the MIs are 1, –1, ∞. The $(1\bar{1}0)$ plane must pass through the unit cube at intercepts on the x-axis = 1, y-axis = –1, z-axis = ∞ (parallel to z-axis).

(iv) (111) Plane in a BCC unit cell: Determine the reciprocals of the Miller indices which are (111). The (111) plane must pass through a unit cube with intercepts at x-axis = 1, y-axis = 1, z-axis = 1. The atom positions which are intersected by the plane are A: (1,0,0); B: (0,0,1); C: (0,1,0); D: (½,½,½).

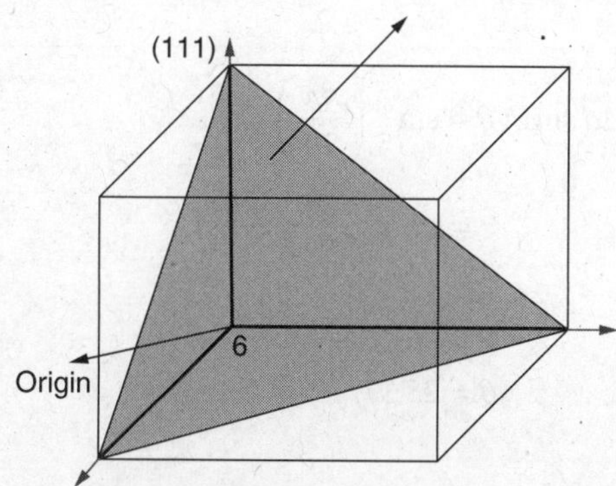

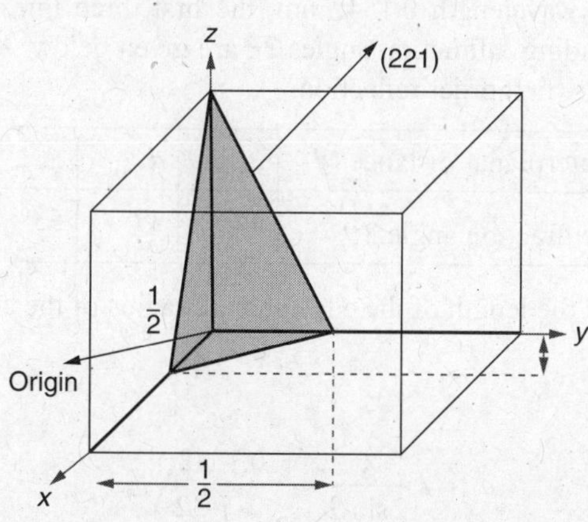

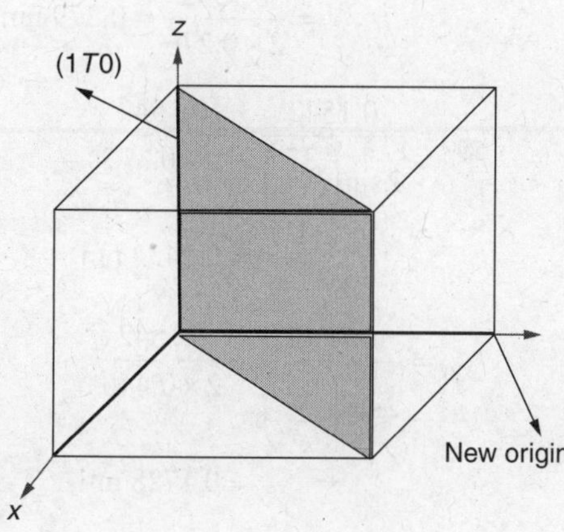

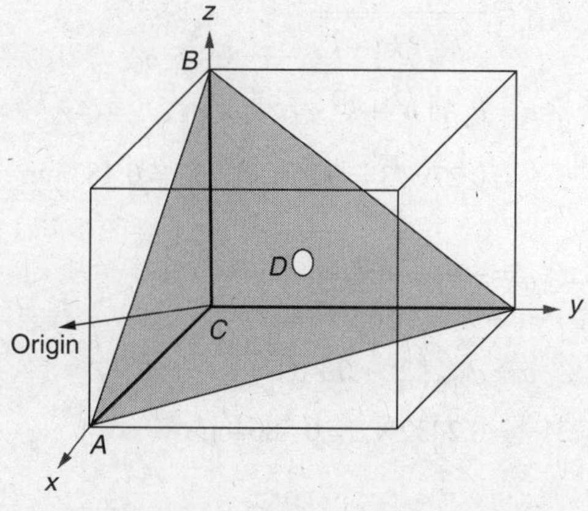

Problem 7.21

Find the Miller indices of the planes shown in the following diagrams.

(a)

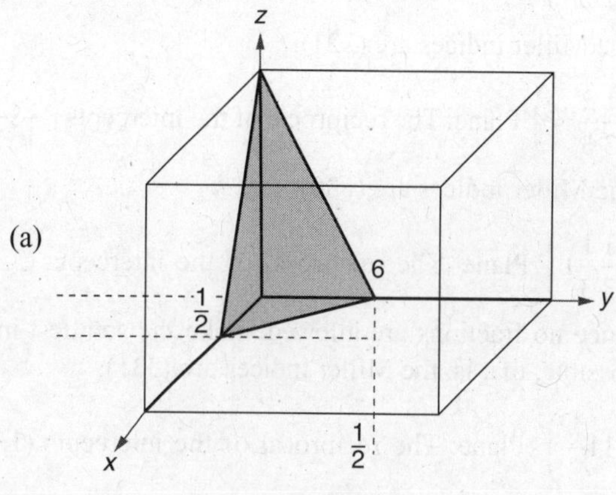

(b)

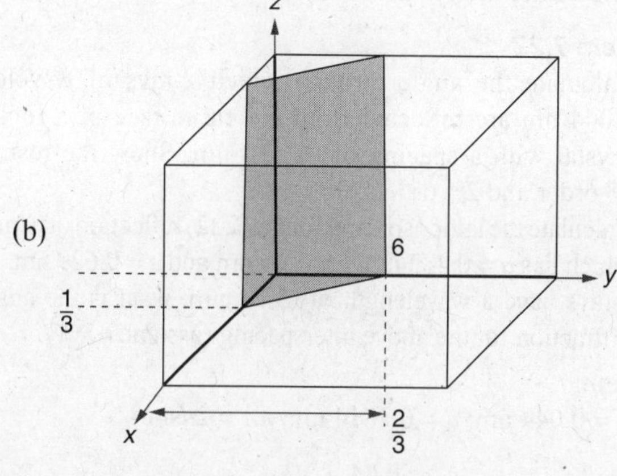

(c)

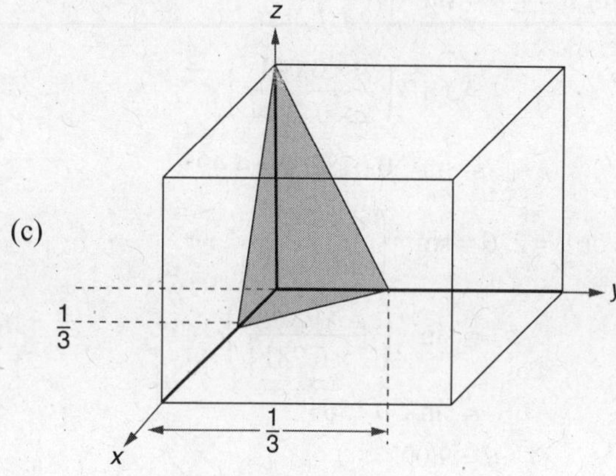

(d)

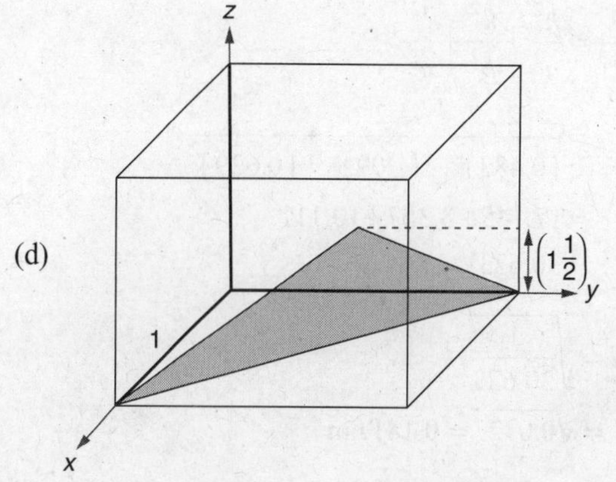

Solution

(a) $\left(\frac{1}{2}\frac{1}{2}1\right)$ Plane: The reciprocal of the intercepts (2, 2, 1).

The Miller indices are (221).

(b) $\left(\frac{1}{3}\frac{2}{3}\infty\right)$ Plane: The reciprocal of the intercepts $\left(\frac{3}{1},\frac{3}{2},0\right)$.

The Miller indices are (630).

(c) $\left(\frac{1}{3}\frac{1}{3}1\right)$ Plane: The reciprocal of the intercepts (3, 3, 1).

Since no fractions are allowed, make the smallest integers possible, that is, the Miller indices are (331).

(d) $\left(11\frac{1}{2}\right)$ Plane: The reciprocal of the intercepts (1, 1, 2).

The Miller indices are (112).

Problem 7.22

(a) Calculate the angle through which x-rays of wavelength 0.044 nm are reflected from the cubic face of a rock salt crystal with a spacing of 0.2814 nm. Show the result for 1st order and 2nd order reflections.

(b) Calculate the lattice spacing for the (222) reflection of a mineral which has $a = 0.483$ nm, $b = 1.09$ nm and $c = 0.629$ nm. If the x-rays have a wavelength of 0.154 nm, what is the angle of diffraction for the above interspacing (assume $n = 1$)?

Solution

(a) $\lambda = 0.044$ nm; $d = 0.2814$ nm; $n\lambda = 2d\sin\theta$

$$\text{For } n = 1,\ \theta = \sin^{-1}\left(\frac{n\lambda}{2d}\right)$$

$$= \sin^{-1}\left(\frac{1\times 0.044}{2\times 0.2814}\right)$$

$$= \sin^{-1} 0.0782;\ \theta = 4.49°$$

$$\text{For } n = 2,\ \theta = \sin^{-1}\left(\frac{n\lambda}{2d}\right)$$

$$= \sin^{-1}\left(\frac{2\times 0.044}{2\times 0.2814}\right)$$

$$= \sin^{-1} 0.1564$$

$$\theta = 9.000°$$

(b) $$\frac{1}{d^2} = \frac{h^2}{a^2}+\frac{k^2}{b^2}+\frac{l^2}{c^2}$$

$$= \frac{2^2}{(0.483)^2}+\frac{2^2}{(1.09)^2}+\frac{2^2}{(0.629)^2}$$

$$= 17.145 + 3.367 + 10.111$$

$$= 30.623$$

$$d = \sqrt{\frac{1}{30.623}}$$

$$= \sqrt{0.0327} = 0.181 \text{ nm}$$

$$n\lambda = 2d\sin\theta;\ \theta = \sin^{-1}\left(\frac{n\lambda}{2d}\right)$$

$$= \sin^{-1}\left(\frac{1\times 0.154}{2\times 0.181}\right)$$

$$= \sin^{-1}(0.4254)$$

$$\theta = 25.17°$$

Problem 7.23

Lead crystallises in FCC. In an x-ray study of this element using x-rays of wavelength 0.1542 nm, the first three lines and the corresponding diffraction angles 2θ are given below. Assume $n = 1$, that is, first order reflection.

Interplanar distance (d)	d_{111}	d_{200}	d_{220}
Diffraction angle 2θ	32°	37°	53°

Calculate the length of the edge and the radius of the atom.

Solution

$$d_{111} = \frac{\lambda}{2\sin\theta} = \frac{0.1542}{2\sin\left(\frac{32}{2}\right)}$$

$$= \frac{0.1542}{2\times 0.276} = 0.279 \text{ nm}$$

$$d_{200} = \frac{0.1542}{2\sin\left(\frac{37}{2}\right)} = \frac{0.1542}{2\times 0.317}$$

$$= 0.2432 \text{ nm}$$

$$d_{220} = \frac{0.1542}{2\sin\left(\frac{53}{2}\right)} = \frac{0.1542}{2\times 0.446}$$

$$= 0.1728 \text{ nm}$$

Calculation of a unit cell length,

$$d_{111} = \frac{a}{\sqrt{h^2+k^2+l^2}}$$

$$a = d_{111}\left(h^2+k^2+l^2\right)^{1/2}$$

$$= 0.279\sqrt{3} = 0.279\times 1.732 = 0.483 \text{ nm}$$

$$d_{200} = \frac{a}{\sqrt{h^2+k^2+l^2}}$$

$$a = d_{200}\left(2^2+0+0\right)^{1/2}$$

$$= 0.2432\times 2 = 0.4864 \text{ nm}$$

$$d_{220} = \frac{a}{\sqrt{h^2 + k^2 + l^2}}$$

$$a = d_{220}\left(2^2 + 2^2 + 0\right)^{\frac{1}{2}}$$

$$= 0.1728 \times 2.8284 = 0.4887 \text{ nm}$$

Average $a = 0.4860$ nm

For FCC, $\sqrt{2}a = 4r$

$$r = \frac{1.414 \times 0.4860}{4} = 0.172 \text{ nm}$$

Zinc Blende, Fluorite and Antifluorite, Calcium Fluoride Structures

Problem 7.24

Explain the unit cells of zinc blende (ZnS) and fluorite (CaF_2) and answer the following questions.

(i) What are the coordination numbers of the cation and anion in both cases?

(ii) Given the ionic radii of Ca^{2+} = 0.126 nm, F^- = 0.117 nm, Zn^{2+} = 0.074 nm S^{2-} = 0.184 nm, calculate the IPFs of CaF_2 and ZnS.

(iii) How many Ca^{2+} and F^- are present in the unit cell? How many Zn^{2+} and S^{2-} ions are present in the unit cell?

Solution

(i) Zinc blende has an FCC structure with S^{2-} ions occupying lattice points and zinc ions occupying half of the tetrahedral holes alternately. Each Zn^{2+} ion is tetrahedrally surrounded by four S^{2-} ions. Similarly, each sulphide ion is surrounded by four Zn^{2+} ions giving a 4:4 coordination. Thus, the coordination number of Zn^{2+} = 4 and the coordination number of S^{2-} = 4.

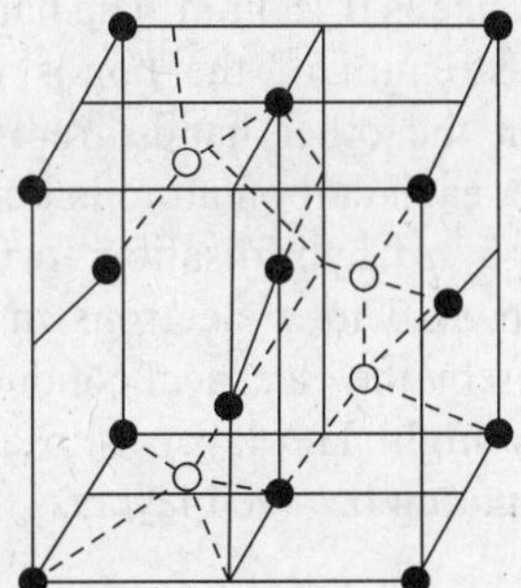

White circles denote zinc and black circles denote sulphide ions with the coordination numbers being 4.

In fluorite, Ca^{2+} ions occupy the lattice points and F^- ions occupy all tetrahedral holes. Each Ca^{2+} ion is surrounded by eight F^- ions and each F^- ion is tetrahedrally coordinated to four Ca^{2+} ions giving cation : anion = 8:4 coordination. Thus the coordination number of Ca^{2+} = 8 and F^- = 4.

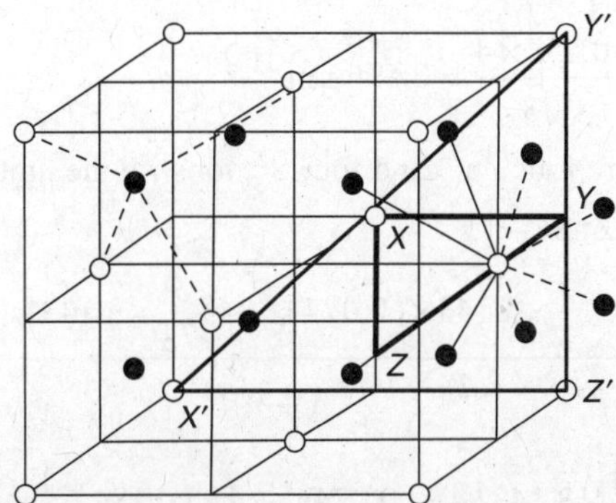

White circles denote calcium and black circles indicate fluoride ions, with the coordination numbers being 8 and 4 respectively.

The thick lines denote the right angled triangle formed by the body diagonal, face diagonal and cell edge (X′Y′Z′) the smaller triangle is (XYZ) which is one-fourth of (X′Y′Z′).

(ii) The body diagonal, face diagonal and cell edge form a right angled triangle. The smaller triangle XYZ is one fourth of the bigger triangle X′Y′Z′. The Ca^{2+} and F^- touch a short distance along the body diagonal.

$$\frac{\sqrt{3}a}{4} = \left(r_{Ca^{2+}} + r_{F^-}\right)$$

$$\therefore \quad = 0.126 + 0.117 = 0.243 \text{ nm}$$

$$a = \frac{0.243 \times 4}{\sqrt{3}} = 0.561 \text{ nm}$$

There are four Ca^{2+} and eight F^- ions in the unit cell of CaF_2.

IPF of CaF_2

$$= \frac{4 \times \frac{4}{3} \times 3.143 \times (0.126)^3 + 8 \times \frac{4}{3} \times 3.143 \times (0.117)^3}{(0.561)^3}$$

$$= \frac{\frac{4}{3} \times 3.143\left\{4 \times (0.126)^3 + 8 \times (0.117)^3\right\}}{(0.561)^3}$$

$$= \frac{\frac{4}{3} \times 3.143\{0.008 + 0.0128\}}{0.1766}$$

$$= \frac{4.19 \times 0.0280}{0.1766} = 0.493$$

IPF of ZnS: In this case also, the oppositely charged ions touch along the body diagonal and not along the face diagonal.

$$\frac{\sqrt{3}a}{4} = \left(r_{Zn^{2+}} + r_{S^{2-}}\right)$$

$$= 0.074 + 0.184 = 0.258 \text{ nm}$$

$$a = \frac{0.258 \times 4}{\sqrt{3}} = 0.596 \text{ nm}$$

There are four Zn^{2+} and four S^{2-} ions i`n the unit cell.

IPF of ZnS

$$= \frac{4 \times \frac{4}{3} \times 3.143 \times (0.074)^3 + 4 \times \frac{4}{3} \times 3.143 \times (0.184)^3}{(0.596)^3}$$

$$= \frac{\frac{4}{3} \times 3.143 \{4 \times (0.074)^3 + 4 \times (0.184)^3\}}{(0.596)^3}$$

$$= \frac{16.763\{4.052 \times 10^{-4} + 6.23 \times 10^{-3}\}}{0.212}$$

$$= \frac{0.1112}{0.212} = 0.525$$

(iii) The unit cell of CaF_2 contains four Ca^{2+} and eight F^- ions while the unit cell of ZnS contains four Zn^{2+} and four S^{2-} ions.

Problem 7.25

(a) Sodium oxide has an antifluorite structure with the edge length equal to 0.555 nm. (i) Calculate its density. (ii) Calculate the sodium–oxygen bond length. (iii) What is the oxygen–oxygen bond length?

(b) Draw a neat sketch of the unit cell of diamond. Indicate the lattice points and the occupancy of its tetrahedral holes. How is this structure different from that of graphite? Explain.

Solution

(a) In Na_2O, O^{2-} ions occupy lattice points and Na^+ ions occupy all tetrahedral holes. Each O^{2-} ion is surrounded by eight Na^+ ions and each Na^+ ion is surrounded by four O^{2-} ions thus giving a cation : anion coordination of 4:8. There are four O_2^{2-} ions and eight Na^+ ions in the unit cell. From the atomic masses of Na^+ (0.023 kg mol^{-1}) and O^{2-} (0.016 kg mol^{-1}), we have

$$\rho = \frac{8 \times 0.023 + 4 \times 0.016}{6.023 \times 10^{23} \times (0.555 \times 10^{-9})^3}$$

$$= \frac{0.184 + 0.064}{6.023 \times 10^{23} \times (0.555 \times 10^{-9})^3}$$

$$= 0.241 \times 10^4 = 2.41 \times 10^3 \text{ kg m}^{-3}$$

On making an analogy with the CaF_2 structure, the O_2^{2-} and Na^+ ions touch along the body diagonal for a certain distance.

Thus, the Na^+–O^{2-} bond distance = 0.102 + 0.140
= 0.242 nm

The shortest O^{2-}–O^{2-} distance is along the face diagonal

$$\text{The } O^{2-}\text{–}O^{2-} \text{ distance} = \frac{1}{2}\sqrt{2}a = \frac{a}{\sqrt{2}} = \frac{0.555}{1.414} = 0.393 \text{ nm}$$

(b) Diamond

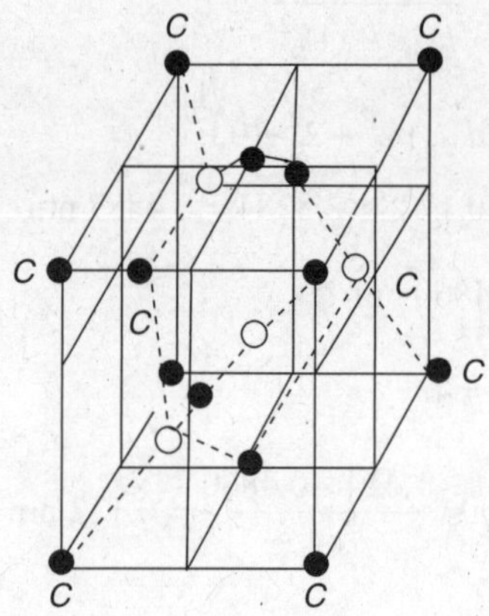

Graphite

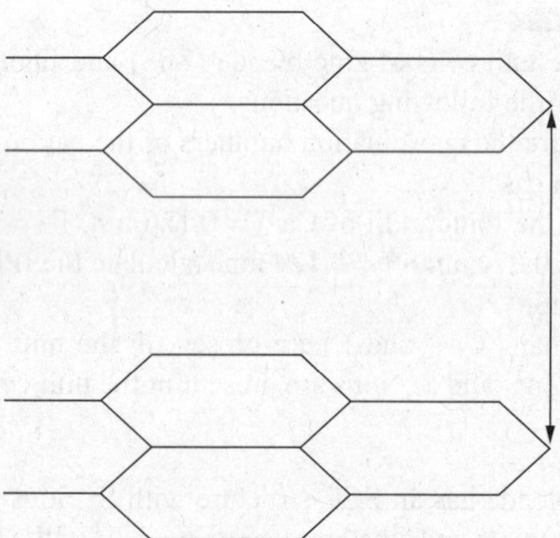

In diamond, we have a covalent bonding with the overlapping of valence orbitals between atoms. sp^3 hybridisation resulting in a tetrahedral arrangement of carbons exists in this structure. Each carbon atom is bound to four equidistant carbon atoms covalently. The C–C bond distance is 0.154 nm. The hardness of diamond is due to the strength of the bonds which is uniform throughout. On the other hand, graphite has a layered structure. Here, each carbon atom is bound to three other carbons through sp^2 hybridisation forming a hexagonal network as shown. The π electrons in the molecule are mobile and thereby the electrical conductivity of graphite is significantly high. The layers slip against each other due to weak bonding between layers.

Schottky Defects and Their Effect on Densities

Problem 7.26

(a) Chromium crystallises in BCC and has a radius of r = 0.125 nm. Its density is 7.15×10^3 kg m^{-3}. How many atoms are present in 1.00 kg of the metal?

(b) A metal (atomic mass = 39.0) crystallises in FCC and its radius is 0.231 nm. It has Schottky defects to the extent of 1%. What is its density in kg m^{-3}?

Solution

(a) Volume of 1.00 kg metal $= \dfrac{1}{7.15\times10^3} = 0.14\times10^{-3}\ \text{m}^3$

For a BCC unit cell,

$$\sqrt{3}a = 4r;\ a = \frac{4r}{\sqrt{3}} = \frac{4\times0.125}{1.732} = 0.289\ \text{nm}$$

$$\therefore\ V = a^3 = (0.289)^3 = 0.0241(\text{nm})^3 = 0.0241\times10^{-27}\ \text{m}^3$$

Number of unit cells in 1 kg metal

$$= \frac{0.14\times10^{-3}}{0.0241\times10^{-27}} = 5.81\times10^{24}$$

Number of atoms in 1 kg metal

$$= 2\times5.81\times10^{24}$$

$$= 11.62\times10^{24}\ (\text{since, for BCC } n = 2)$$

(b) The presence of Schottky defects results in a decrease in the density of the crystal. Thus, the volume of the crystal increases (under constant mass conditions). Its edge length is given by the equation

$$\sqrt{2}a = 4r$$

$$a = \frac{4\times0.231}{\sqrt{2}} = \frac{0.924}{1.414} = 0.653\ \text{nm}$$

$$\text{Theoretical (ideal) density} = \frac{nM}{Na^3}$$

$$= \frac{4\times0.039}{6.023\times10^{23}\times(0.653\times10^{-9})^3}$$

$$= 0.093\times10^4\ \text{kg m}^{-3}$$

Volume of the crystal increases in the ratio $1000:1010$

$$\frac{\text{Density of crystal containing Schottky defects}}{\text{Density of ideal crystal}} = \frac{\text{Volume of ideal crystal}}{\text{Volume of defect crystal}}$$

$$\frac{\text{Density of defect crystal}}{9.3\times10^2} = \frac{1000}{1010}$$

$$\text{Density of defect crystal} = \frac{9.3\times10^2\times1000}{1010}$$

$$= \frac{9.3\times10^5}{1010} = 0.0092\times10^5$$

$$= 9.21\times10^2\ \text{kg m}^{-3}$$

Problem 7.27

(a) The number of Schottky defects (n_s) depends on the temperature according to

$$n_s = Ne^{\left(\frac{-\Delta H_s}{2kT}\right)}$$

where ΔH_s = enthalpy of formation per defect. For CaO and LiI, the ΔH_s values are 9.8×10^{-19} J, 2.1×10^{-19} J. Calculate the ratio n_s/N at 300 K and 1000 K. Is this process endothermic or exothermic?

(b) Give reasons for the following:

(i) On heating KCl crystals in potassium vapour, the crystals turn violet.

(ii) Zinc oxide on heating to a high temperature turns yellow.

(iii) FeO is a non-stoichiometric compound with a formula $Fe_{0.95}O$.

(iv) Germanium doped with phosphorus produces an *n*-type semiconductor. If the same is doped with boron, a *p*-type semiconductor is obtained.

Solution

(a) CaO

$$\frac{n_s}{N} = e^{\left(\frac{-\Delta H_s}{2kT}\right)};\ \Delta H_s = 9.8\times10^{-19}\ \text{J}$$

$$1.602\times10^{-19}\ \text{J} = 1\ \text{eV},\ \frac{9.8\times10^{-19}}{1.602\times10^{-19}}$$

$$= 6.12\ \text{eV}$$

$$\text{Also } k = 1.38\times10^{-23}\ \frac{\text{J per molecule/ion}}{\text{ion}}$$

$$= \frac{1.38\times10^{-23}}{1.602\times10^{-19}} = 8.6\times10^{-5}\ \text{eV}$$

Now substituting eV (rather than J)

$$\frac{n_s}{N} = e^{\frac{-6.12}{2\times8.6\times10^{-5}\times300}}\ \text{at 300 K}$$

$$\frac{n_s}{N} = e^{-118.6}\ \log\frac{n_s}{N}$$

$$= \frac{-118.6}{2.303} = -51.5 = \overline{52}.5$$

$$\frac{n_s}{N} = 3.16\times10^{-52}\ (\text{at 300 K})$$

At 1000 K,

$$\frac{n_s}{N} = e^{\frac{-6.12}{2\times8.6\times10^{-5}\times1000}}$$

$$= e^{\frac{6.12}{0.192}} = e^{-31.87}$$

$$2.303 \log\frac{n_s}{N} = -31.87 \text{ or } \log\frac{n_s}{N}$$

$$= \frac{-31.87}{2.303} = -13.84 = \overline{14}.16$$

$$\frac{n_s}{N} = 1.44\times10^{-14}$$

For LiI

$$\Delta H_s = \frac{2.1\times10^{-19}}{1.602\times10^{-19}} = 1.31 \text{ eV}$$

At 300 K, $\frac{n_s}{N} = e^{\frac{-1.31}{2\times8.6\times10^{-5}\times300}}$

$$\log\frac{n_s}{N} = \frac{-25.4}{2.303}$$

$$= -11.03 = \overline{12}.97$$

$$\frac{n_s}{N} = 9.33\times10^{-12} \text{ (at 300 K)}$$

At 1000 K, $\frac{n_s}{N} = e^{\frac{-1.31}{2\times8.6\times10^{-5}\times1000}} = e^{-6.82}$

$$\log\frac{n_s}{N} = \frac{-6.82}{2.303}$$

$$= -2.96 = \overline{3}.04$$

$$\frac{n_s}{N} = 1.1\times10^{-3}$$

The process is endothermic, n_s/N is larger at the higher temperature compared to that at the lower temperature 300 K. Therefore, the process is endothermic.

(b) (i) By heating KCl in excess of alkali metal K vapour, F-centres (electrons trapped in anion vacancies) are created. Because of this, violet colour develops in this case.

(ii) When ZnO is heated to a high temperature, it decomposes resulting in the loss of oxygen. Thereby, the excess Zn atoms formed occupy the interstitials. This results in the development of yellow colour.

(iii) Although the stoichiometric formula of ironoxide is FeO, the actual stoichiometry is $Fe_{0.95}O$. The non-stoichiometry is due to ease of oxidation of Fe^{2+} to Fe^{3+} which results in the replacement of a small portion of Fe^{2+} with Fe^{3+}. Actually, for every three missing Fe^{2+} ions, the crystal contains two Fe^{3+} ions to balance the change. The formula of FeO may thus be written as $Fe_{(1-x)}O$, where x in this case is 0.05. Non-stoichiometry is prevalent in metal oxides.

(iv) Germanium is an intrinsic semiconductor. When it is doped with the group 15 element phosphorus, it has excess electrons. These electrons get transferred to Ge and results in an *n*-type semiconductor. When doped with an electron deficient element like boron (group 13 element), electrons from Ge are transferred to boron, resulting in the formation of a *p*-type semiconductor. Different combinations of *n*-type and *p*-type semiconductors are used to make transistors, diodes, and so on.

Structure of Corundum

Problem 7.28

(a) Describe briefly the structure of corundum (Al_2O_3).

(b) A compound is formed by two elements A and B. Element B forms CCP and A occupies one-third of the tetrahedral voids. What is the formula of the compound?

(c) Niobium forms BCC and its density is 8.55×10^3 kg m^{-3}. What is its atomic radius? (Atomic mass = 93)

Solution

(a) Corundum has a hexagonal close packed structure in which the Al atoms are surrounded by six O atoms. 2/3 of the octahedral voids are occupied by Al. The vacant sites are (due to Al) are arranged such that each Al shares one face with a neighbouring octahedron site.

(b) Since B forms a CCP, there are $8\times1/8$ = 1 B atom and A occupies 8/3 tetrahedral voids.

$\therefore$ Formula = $A_{\frac{8}{3}}B$ or A_8B_3

(c) $$\rho = \frac{nM}{Na^3} = 8.55\times10^3$$

$$= \frac{2\times0.093}{6.023\times10^{23}\times a^3}$$

$$a^3 = \frac{2\times0.093}{6.023\times10^{23}\times8.55\times10^3}$$

$$= 0.36\times10^{-28}$$

$$a = \sqrt[3]{0.36\times10^{-28}}$$

$$= \sqrt[3]{0.36\times10^{-27}} = 3.304\times10^{-9} \text{ m}$$

For BCC, $4r = \sqrt{3}a$

$$r = \frac{1.732\times3.304\times10^{-9}}{4}$$

$$= 1.43\times10^{-9} \text{ m} = 1.43 \text{ nm}$$

Atomic Packing Factor in Crystalline Solids

Problem 7.29

Copper(I) iodide crystallises in the zinc blende structure. The edge length of its unit cell is 0.605 nm. (i) Calculate its density. (ii) Given the radii of Cu^+ and I^- as 0.074 nm and 0.206 nm, calculate its APF. (iii) What are the shortest cation–anion and anion–anion distances? (Atomic masses: Cu = 63.5, I = 126.9)

Solution

(i) There are four formula units of CuI in the unit cell.

$$\rho = \frac{nM}{Na^3} = \frac{4\times0.0635+4\times0.1269}{6.023\times10^{23}\times\left(0.605\times10^{-9}\right)^3}$$

$$= \frac{0.7616}{6.023\times10^{23}\times0.2214\times10^{-27}}$$

$$= 0.571\times10^4 = 5.71\times10^3 \text{ kg m}^{-3}$$

$$\text{(ii) APF} = \frac{4\times\frac{4}{3}\times 3.143\times(0.074)^3 + 4\times\frac{4}{3}\times 3.143\times(0.206)^3}{(0.605\text{ nm})^3}$$

$$= \frac{16.763\times\left(4.052\times10^{-4}\text{ nm}^3 + 8.742\times10^{-3}\text{ nm}^3\right)}{0.2214\text{ nm}^3}$$

$$= \frac{0.1533}{0.2214} = 0.692$$

(iii) In an analogy with the ZnS structure, the oppositely charged ions touch along the body diagonal.

$$\frac{\sqrt{3}a}{4} = r_{Cu} + r_{I^-} = 0.280\text{ nm}$$

The cation–anion contact distance is 0.280 nm. (However this value is slightly different from that calculated from a which is 0.262 nm.)

The anion–anion touch along the face diagonal which is equal to $\sqrt{2}a$. The shortest distance is

$$\sqrt{2}\times\frac{1}{2}\times a = \frac{a}{\sqrt{2}} = \frac{0.605}{1.414} = 0.428\text{ nm}$$

Frenkel Defects in Solids and Planar Atomic Density

Problem 7.30

(a) Zirconium dioxide (ZrO_2) is good at accommodating anion Frenkel defects and has a high melting point. The enthalpy of formation of the defect is 4.10 eV and its density is 5.68×10^3 kg m^{-3}. Calculate the number of Frenkel defects in this compound at 1000 K. (Atomic masses: Zr = 91.2, O = 16.0)

(b) Write the coordinates of all the positions (the lattice points and face centres) in a FCC unit cell.

(c) Calculate the planar atomic density ρ_p on the (100) plane of the FCC lattice of copper whose lattice constant is 0.128 nm.

Solution

(a) First, we have to determine the number of lattice sites, N, per m^3 and it is given by the equation

$$N = \frac{N_{Avo}\times\rho_{ZrO_2}}{\text{Atomic weight of }ZrO_2}$$

$$= \frac{6.023\times10^{23}\times5.68\times10^3}{0.1232}$$

$$= 277.7\times10^{26} = 2.78\times10^{28}\text{ m}^{-3}$$

The number of Frenkel defects N_{Fr} is related to N according to

$$N_{Fr} = Ne^{\left(\frac{-\Delta H_f}{2kT}\right)}$$

$$N_{Fr} = 2.78\times10^{28}\times e^{-4.1/2\times8.6\times10^{-5}\times1000}$$

$$\ln N_{Fr} = \ln\left\{2.78\times10^{28}\right\} - \frac{4.1}{2\times8.6\times10^{-5}\times1000}$$

$$2.303\log N_{Fr}$$

$$= 2.303\log\left\{2.78\times10^{28}\right\} - \frac{4.1}{2\times8.6\times10^{-5}\times1000}$$

$$= 28.444 - 10.35 = 18.09$$

or $N_{Fr} = 1.23\times10^{18}$

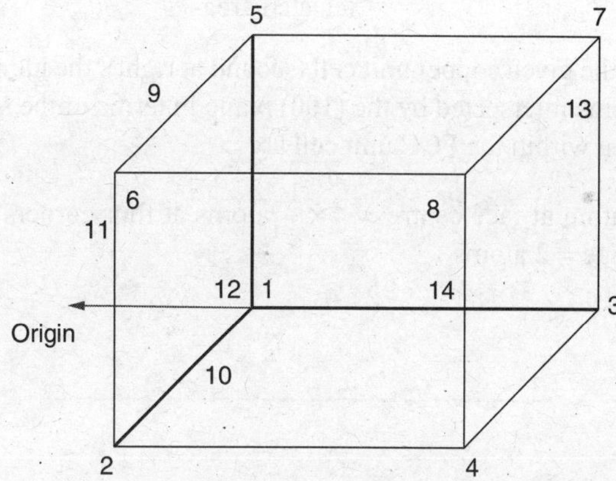

(1) (0,0,0) (2) (1,0,0) (3) (0,1,0) (4) (1,1,0) (5) (0,0,1)

(6) (1,0,1) (7) (0,1,1) (8) (1,1,1) (9) $\left(\frac{1}{2},\frac{1}{2},1\right)$

(10) $\left(\frac{1}{2},\frac{1}{2},0\right)$ (11) $\left(\frac{1}{2},0,\frac{1}{2}\right)$ (12) $\left(\frac{1}{2},1,\frac{1}{2}\right)$

(13) $\left(0,\frac{1}{2},\frac{1}{2}\right)$ (14) $\left(1,\frac{1}{2},\frac{1}{2}\right)$

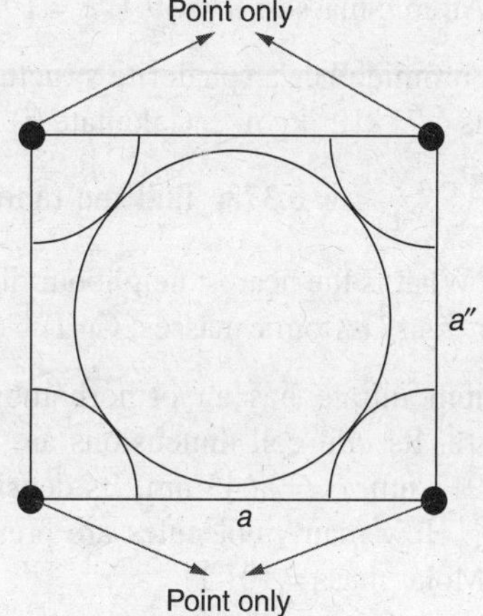

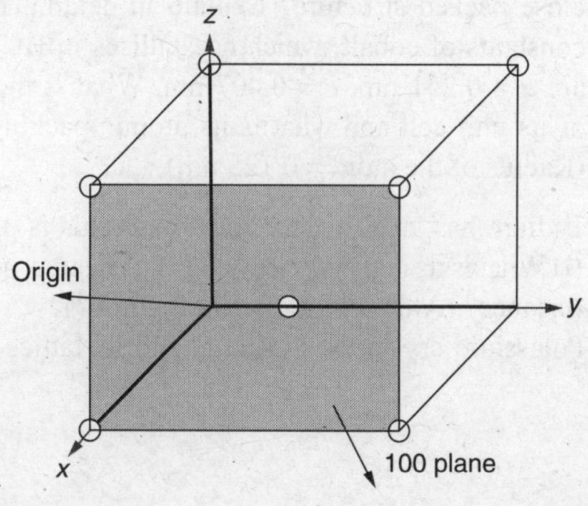

The planar atomic density ρ_p is defined as

$$\rho_p = \frac{\text{Equivalent number of atoms whose centres are intersected by a selected area}}{\text{Selected area}}$$

In the given copper unit cell (second at right), the number of atoms intersected by the (100) plane in terms of the surface area within the FCC unit cell are

1 atom at face centre $= 4 \times \frac{1}{4}$ atoms at four corners of the plane = 2 atoms

The area intersected by the plane inside the cell $= a \times a = a^2$

$$\rho_p = \frac{2 \text{ atoms}}{a^2} = \frac{2 \text{ atoms}}{\left(0.128 \times 10^{-9} \text{ m} \times 10^2 \times \frac{\text{cm}}{\text{m}}\right)^2}$$

$$= \frac{2}{1.64 \times 10^{-16}}$$

$$= 1.22 \times 10^{16} \text{ atoms cm}^{-2}$$

$$= 1.22 \times 10^{2} \text{ atoms nm}^{-2}$$

EXERCISES

Practice Problems

1. (a) A BCC lattice contains A type atoms at the corners and B type atom at the centre of the cube. If six A type corners are missing, what is the formula of the compound?
   (b) Potassium fluoride crystallises in rock salt type lattice with fluoride ions being at lattice points and potassium ions occupying cubic holes. The edge length of the unit cell is 0.534 nm and $r_{K^+}/r_{F^-} = 0.978$. What are the catonic and anionic radii? What is the density of the salt? (Atomic masses: K = 39.1, F = 19.0)

2. Copper (I) bromide has a sphalerite structure. The density of the salt is 4.71×10^3 kg m^{-3}. Calculate (i) its edge length. (ii) Given $r_{Cu^+}/r_{Br^-} = 0.378$, find the radii of cation and anion. (iii) What is the nearest neighbour distance between Cu^+ and Br^- ions? (Atomic masses: Cu(I) = 63.5; Br = 79.9)

3. (a) Potassium nitrate has an orthorhombic type crystal structure. Its unit cell dimensions are a = 0.542 nm, b = 0.917 nm, c = 0.645 nm. Its density = 2.11×10^3 kg m^{-3}. How many molecules are present in the unit cell? (Molar mass = 101.1)
   (b) How many atoms are there in the unit cell of a hexagonal close packed structure? Explain in detail. The lattice constants of cobalt, which crystallises in HCP lattice, are a = 0.251 nm, c = 0.407 nm. What is the volume of its unit cell and what is its atomic packing factor? (Radius of the atom = 0.125 nm)

4. (a) Iridium has an FCC structure. Its radius is 0.136 nm. (i) What is its density? (ii) What is its nearest neighbour distance? (Atomic mass = 192.2 gmol^{-1})
   (b) Potassium crystallises in BCC and its lattice constant is 0.533 nm. (i) Calculate its density. (ii) What is its radius? (iii) Calculate its APF. (Atomic mass = 39.1)

5. (a) Designate the positions of the points indicated in the FCC lattice shown.

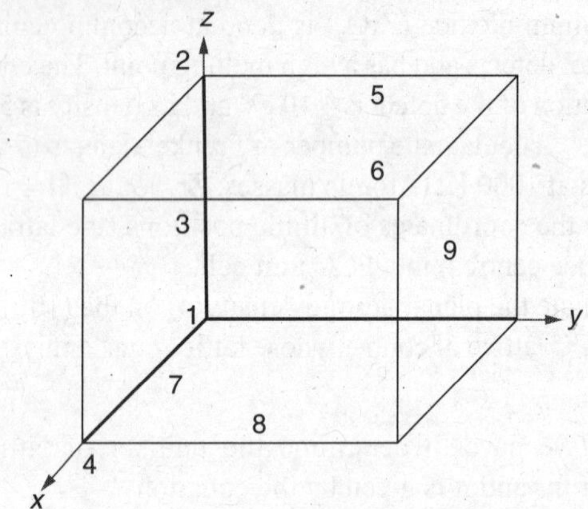

   (b) Designate the positions of points of the four atoms in the unit cell.
   (c) Designate the positions of points of the two atoms in BCC unit cell.

6. (a) The crystal planes in a crystal intersect the axes at (2d, e, f), (d, –e, f), (2d, 2e, 3f), and (d, –2e, ∞), (d, 2e, 2f). What are the Weiss and Miller indices of these planes?
   (b) Show the direction vectors in the following cubic unit cells (i) [110] (ii) [010] (iii) [112] (iv) $[01\bar{1}]$ (v) $[4\bar{1}2]$ (vi) $[1\bar{1}2]$.

7. (a) Draw the crystal planes which have the Miller indices (100), (133), $(1\overline{22})$, (101), (110).

(b) X-rays of wavelength 0.154 nm were used to study KCl crystal structure. The first order reflection from the 200 plane of the crystal occurred at 14°. Calculate (i) the edge length of the unit cell which is FCC, and (ii) the density of the crystal.

8. (a) A unit cell of sodium chloride contains 8 anions at the lattice points (corners), 6 anions at the face centres and these ions alternate with cations (13 in all) with 12 cations at the edge centres and one cation at the body centre. How many formula units of the salt are in the unit cell?

(b) Why is NaCl a poor ionic conductor?

(c) The ionic conductivity σ of an ionic conductor is given by $\sigma = nze\mu$ where σ is in $ohm^{-1}\ m^{-1}$, n = number of charge carries per unit volume, Ze is the charge and μ is the mobility in $m^2\ v^{-1}\ s^{-1}$. Given that $n = 2.23 \times 10^{28}\ m^{-3}$ and $Z = 1$, $e = 1.602 \times 10^{-19}$ coulombs and the mobility $\mu = 5.61 \times 10{-}14\ m^2\ v^{-1}\ s^{-1}$ in the case of NaCl, what is its ionic conductivity in $S\ cm^{-1}$?

(d) The variation of the ionic conductivity of a salt is given by the equation

$$\ln(\sigma T) = \ln \sigma_0 - \left(\frac{E_a}{T}\right)$$

where σ is defined as in (c) and σ_0 depends on n and Ze, and E_a is the activation energy for the ionic transport. For NaCl, the plot of log(σT) vs $1/T$ is linear and gives a slope 49,500. Assuming that its units are $J\ mol^{-1}$, express the result in eV per molecule.

9. (a) Rhodium metal has a unit cell edge length 0.381 nm and its radius is 0.134 nm. What is its unit cell type? Calculate its atomic packing actor (APF) and density. (Atomic mass = 102.9)

(b) Gold crystallises in FCC. In the X-ray study of this element with X-rays of wavelength 0.155 nm, the first three interplanar spacings and the corresponding diffraction angles 2θ are given. Assume $n = 1$. Atomic mass = 197.0.

Interplanar spacings	d_{111}	d_{200}	d_{220}
Diffraction angle 2θ	39.5°	45°	65°

Calculate (i) lattice constant, (ii) radius of the atom, (iii) the density of gold, and (iv) APF.

10. Potassium iodide has an FCC lattice. The radii of the cation and anion are 0.133 nm and 0.216 nm. (i) What is its edge length? (ii) Find the theoretical density of the crystal. (iii) What is the nearest neighbour distance? (iv) Calculate its linear atomic density in atoms/mm. (Atomic masses: $K = 39.1$; $I = 126.91$)

11. (a) Show that the minimum cation–anion ratio for a cubic structure which has a coordination number 8 is 0.732.

(b) Calculate the APF of NaBr with a cation to anion radius ratio of 0.920 and $r_+ = 0.1020$ nm.

12. (a) Cadmium sulphide has a unit cell edge length of 0.582 nm. Its density is $4.82 \times 10^3\ kg\ m^{-3}$. Find the number of Cd^{2+} and sulphide ions in the unit cell. Given the radius of Cd^{2+} and S^{2-} ions as 0.095 nm and 0.184 nm respectively, what is its APF? What are the shortest distances of Cd^{2+}–S^{2-} and Cd^{2+}–Cd^{2+} in nm? (Molar mass of CdS = 144.5)

(b) The fraction of vacancy sites in a metal is 1×10^{-10} at 773 K. what will be the fraction of vacancy sites at 1273 K?

13. (a) Cobalt has an FCC lattice with its edge length equal to 0.354 nm. What is its theoretical density? What is its HCP form density? (Atomic mass = 58.9)

(b) Deduce the formula of the Pervoskite ABX_3 where A = Ca, B = Ti, C = O.

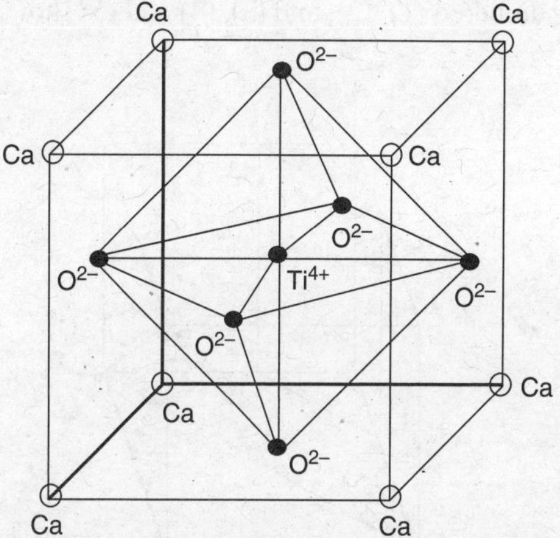

14. Uranium oxide (UO_2) has a fluorite type lattice. The radii of U^{4+} and O^{2-} ions are 0.105 nm and 0.132 nm respectively. (i) What is its APF? (ii) What is its density (iii) What is the nearest neighbour distance? (Atomic masses: U = 238.0; O = 16.0)

15. Potassium oxide (K_2O) has a typical antiflourite structure. Its unit cell length is 0.6436 nm. (i) Calculate its density. (ii) What is the potassium–oxygen length? (iii) What is the O^{2-}–O^{2-} distance? Given the radii of K^+ and O^{2-} ions as 0.133 nm and 0.140 nm, calculate its APF. (Atomic masses: K = 39.1; O = 16.0)

16. Draw the diamond cubic (dc) lattice structure and answer the following questions. (i) How many atoms are present per unit cell? (ii) What is the coordination number? (iii) How is the lattice constant related to the radius of the atom and what is its value? (iv) What is its APF? ($a = 0.3567$ nm)

17. (a) The ionic radii of Na^+ and Br^- ions are 0.095 nm and 0.196 nm respectively. Calculate the force of attraction in newtons under conditions that they touch each other?

(b) The plot of log specific conductivity ($ohm^{-1}\ cm^{-1}$) vs $1/T$ (T in K) for pure strontium fluoride in the intrinsic region is found to be linear with a slope of -96.55×10^2. Assuming that an Arrhenius type of equation (in kinetic studies) is valid, calculate the activation energy for its ion type conduction. What is the probable mechanism for its conduction?

18. Draw the tetragonal cubic cell structure of Rutile (TiO_2) considering that Ti^{4+} ions occupy lattice points and one Ti^{4+} at the centre. From the diagram, answer the following questions. (i) How many formula units of TiO_2 are present per unit cell? (ii) What are the coordination numbers of Ti^{4+} and O^{2-} ions? (iii) The radii of Ti^{4+} and O^{2-} ions are 0.061 nm and 0.141 nm respectively. Assuming that the Ti^{4+} and O^{2-} ions touch along the basal plane diagonal, what is the edge length of the unit cell? (iv) What is its density in kg m^{-3}?

19. (a) Find the Miller indices of the crystallographic planes numbered (1), (2), and (3), (4) and (5) shown.

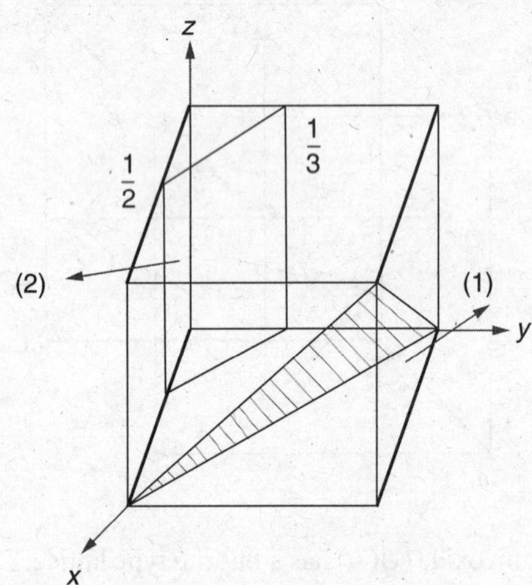

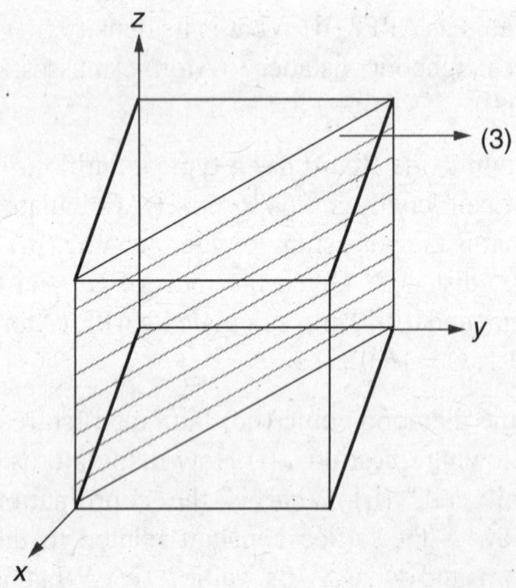

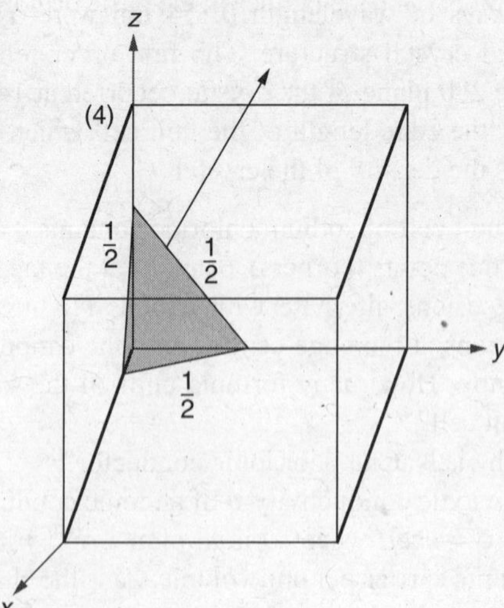

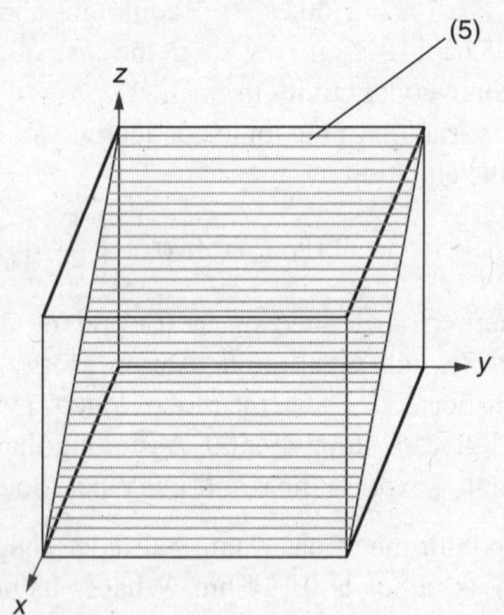

(b) Tungsten crystallises in BCC. The diffraction angles (2θ) and the corresponding interplanar spacing obtained with x-rays of 0.1541 nm are given below.

Interplanar spacings	d_{110}	d_{200}	d_{211}
Diffraction angle 2θ	40°	58°	73°

Calculate (i) lattice constant (ii) radius of the atom (iii) the density (iv) APF (atomic mass = 183.8).

20. Magnesium crystllises in HCP and its lattice constants are a = 0.321 nm and c (vertical height) = 0.521 nm. Calculate (i) its volume (ii) the density (iii) atomic packing factor (radius of the atom = 0.130 nm). (Atomic mass = 24.3)

Multiple Choice Questions

1. Find the correct statement from the following:
(a) Crystalline solids have short range order
(b) Crystals are anisotropic
(c) Rubber possesses regular internal structure
(d) Glass melts sharply at a definite temperature

2. Buckminster fullerene is a/an
(a) ionic solid
(b) covalent solid
(c) molecular solid
(d) metal

3. The number of space lattices in the basic orthorhombic unit cell is
(a) 2 (b) 1
(c) 3 (d) 4

4. Ice belongs to the unit cell type
(a) rhombohedral (b) monoclinic
(c) hexagonal (d) cubic

5. The substance which possesses a triclinic type unit cell among the following is
(a) $Na_2SO_4.10\ H_2O$ (b) $CaSO_4.2H_2O$
(c) $CuSO_4.5H_2O$ (d) K_2SO_4

6. Which among the following has a 3-fold axis of symmetry?
(a) H_2O (b) NH_3
(c) HF (d) C_6H_6

7. The planes in a crystalline compound intersect the crystal axes at 2a, 2b, 3c. The Miller indices of this plane are
(a) 223 (b) 322
(c) 232 (d) 332

8. The interplanar spacing between the (122) planes of a cubic lattice of length 400 pm is (in pm) (pm = picometre)
(a) 125.6 (b) 133.3
(c) 144.8 (d) 138.5

9. In a BCC lattice containing X and Y type atoms, X type atoms are present at the corners and Y type atoms are present at the centres. In its unit cell, suppose three atoms are missing in the corners, the formula of the compound is
(a) X_8Y_5 (b) X_3Y_5
(c) X_5Y_3 (d) X_5Y_8

10. Calcium metal has an edge length of 559 pm and its density is 1.53 g cm^{-3}. Its unit cell is of the type (molar mass = 40 g mol^{-1})
(a) BCC (b) FCC
(c) HCP (d) simple cubic

11. Pick out the correct statement regarding the sodium chloride and potassium chloride crystal structures
(a) K^+ and Cl^- have 18 electrons in their ions and have identical *x*-ray scattering powers
(b) NaCl has simple cubic structure
(c) Na^+ and Cl^- have different numbers of electrons in them but still possess identical *x*-ray scattering powers
(d) KCl has an FCC structure

12. For a simple cubic lattice, the ratios $d100 : d110 : d111$ are in the order
(a) 1.0 : 0.577 : 0.707
(b) 0.707 : 0.577 : 1.0
(c) 1.0 : 0.707 : 0.577
(d) 0.577 : 1.0 : 0.707

13. X-rays of wavelength 0.18 nm on a metal produce a diffraction angle $\theta = 24°$. Given that it is a first order reflection from (110) plane of a BCC, the edge length of the cube is (in Å)
(a) 3.24 (b) 2.13
(c) 2.85 (d) 3.12

14. The atomic packing factor (APF) in primitive cubic, BCC and FCC unit cells decreases in the order
(a) BCC > FCC > primitive cubic
(b) FCC > BCC > primitive cubic
(c) primitive cubic > BCC > FCC
(d) BCC > primitive cubic > FCC

15. The number of nearest neighbours (coordination number) in a HCP structure is
(a) 8 (b) 6
(c) 12 (d) 4

16. The ratio of radii r_{BCC}/r_{FCC} in a cubic unit cell is
(a) 1.282 (b) 1.543
(c) 1.324 (d) 1.225

17. Copper crystallises in FCC and its density is 8.94×10^3 kg m^{-3}. Given that its molar mass is 63.54×10^{-3} kg, its lattice parameter, *a* is (in nm)
(a) 0.361 (b) 0.631
(c) 0.316 (d) 0.248

18. Silver crystallises in BCC and has a density 10.7×10^3 kg m^{-3}. The distance between the nearest neighbours in the unit cell (in nm) is (molar mass: 108.0 g mol^{-1})
(a) 0.215 (b) 0.297
(c) 0.279 (d) 0.342

19. Which compound among the following has 6:3 (cation:anion) coordination?
(a) ZnS (b) TiO_2
(c) CaF_2 (d) CsCl

20. The poor electrical conductivity of NaCl is because
(a) there are Na^+ and Cl^- ions in the lattice
(b) there are no free electrons available in the molecule
(c) the oppositely charged ions are fixed in the lattice and that strong electrostatic forces exist between them
(d) the ions are not free but act as ion pairs

21. The number of atoms in the unit cell of diamond is
(a) 4 (b) 8
(c) 2 (d) 6

22. The number of tetrahedral and octahedral voids in a FCC unit cell (in that order) are
(a) 8,8 (b) 4,8
(c) 8,4 (d) 4,4

23. The percentage of the total volume occupied by the spheres in a BCC lattice is
(a) 86 (b) 74
(c) 52 (d) 68

24. The ionic radii of Cs^+ and Br^- ions are 1.69 and 1.96 Å respectively. The unit cell of cesium bromide is
(a) BCC (b) FCC
(c) HCP (d) primitive cubic

25. The lubricating property of graphite is due to
(a) the sp^2 hybridisation between the carbon atoms
(b) the mobility of the π-electrons in the sheets
(c) the weak bonding between the layers which makes the layers slip past each other
(d) the similarity of the hexagonal network as in the benzene ring

26. Lead crystallises in FCC and its lattice constant is 0.495 nm. Its linear atomic density in the (110) plane direction is
(a) 5.72×10^7 (b) 1.43×10^6
(c) 1.85×10^6 (d) 2.86×10^7

27. Which of the following is true of a Schottky defect in a crystal?
(a) It arises when an ion leaves its position in the lattice and occupies an interstitial position in the lattice
(b) It is created when an atom or ion migrates to the surface
(c) It does not affect the density of the crystal
(d) It is a defect due to an electron occupying an anion vacancy

28. Pick out the correct statement about the interstitials.
(a) Their formation is independent of the size of the atom in a metal
(b) They are not common in alloys
(c) In the zinc–copper alloy, copper atoms occupy the interstitial position by substitution
(d) It has no relation to Frenkel defects

29. The yellow colour developed by ZnO on heating to high temperatures is due to the
(a) formation of an F-centre
(b) creation of a Schottky defect
(c) development of a line defect
(d) zinc atoms occupying the interstitials

30. The ratio of vacancy density in molybdenum at 773 K and 1173 K is 2×10^{-3}. The energy for vacancy formation for this metal is (in eV)
(a) 0.075 (b) 0.094
(c) 0.057 (d) 0.088

31. A metal crystallises in FCC (molar mass = 59.0 g mol^{-1}) and has an edge length of 4×10^{-10} m. It has 0.3% Schottky defects. Its density (in kg m^{-3}) is
(a) 6206 (b) 6105
(c) 6102 (d) 6110

32. The electrical conductivity of silver metal, given its mobility, μ = 56 cm^2 V^{-1} s^{-1}, n = number of charge carriers cm^{-3} = 7.5×10^{-22}, electric charge = 1.6×10^{-19} C, is (in S cm^{-1})
(a) 7.26×10^{-40} (b) 2.76×10^{-41}
(c) 6.72×10^{-41} (d) 5.95×10^{-40}

33. The concentration of Frenkel defects in AgI at 300 K is 3.0×10^{18} cm^{-3} and its cationic mobility for the silver interstitials is 4×10^{-7} cm^2 s^{-1} V^{-1}. The ionic conductivity of the silver ion in AgI is, therefore, (in S cm^{-1})
(a) 3.15×10^{-8} (b) 1.92×10^{-7}
(c) 2.54×10^{-7} (d) 2.19×10^{-8}

34. Which of the following statements relating to semiconductors, metals and insulators is true?
(a) Semiconductors have a smaller band gap compared to insulators
(b) Insulators have empty valence bonds and filled conduction bands
(c) Metals have partially filled valence bands and hence they are highly conducting
(d) Increase of temperature decreases the conductivity of semiconductors

35. Germanium has a unit cell similar to diamond. It has a lattice parameter, a = 0.566 nm, energy of band gap = 0.67 eV, mobility of e^- = 3.9×10^3 cm^2 V^{-1} s^{-1}, mobility of holes = 1.9×10^3 cm^2 V^{-1} s^{-1}. The fraction of electrons in the valence bond excited to the conduction band is (its resistivity is 43 ohm cm)
(a) 1.84×10^{-9} (b) 2.50×10^{-9}
(c) 2.10×10^{-8} (d) 1.41×10^{-10}

36. To get a *p*-type semiconductor, silicon is doped with
(a) arsenic (b) phosphorus
(c) boron (d) magnesium

37. Iron is
(a) paramagnetic (b) ferromagnetic
(c) diamagnetic (d) antiferromagnetic

38. Sodium chloride has an FCC lattice with sodium ions at the corners and chloride ions at the face centres. Its density is 2160 kg m^3. Given its radius ratio, r_{Na^+}/r_{Cl^-} = 0.690, the radius of the anion is (in m) (molar mass: 58.5 g mol^{-1})
(a) 2.37×10^{-10} (b) 3.27×10^{-9}
(c) 2.73×10^{-9} (d) 3.15×10^{-10}

39. Magnesium crystallises in HCP and has a c/a ratio = 1.624. Its density is 1.74 g cm^{-3}. Its atomic radius is (in nm)
(a) 0.452 (b) 0.160
(c) 0.254 (d) 0.231

40. Rhenium has a HCP crystal structure with an atomic radius of 0.137 nm. Its c/a ratio is 1.615. The volume of the unit cell is (in nm^3)
(a) 4.32×10^{-1} (b) 6.83×10^{-1}
(c) 8.63×10^{-2} (d) 9.75×10^{-2}

Answer Keys

Practice Problems

1. (a) A type atoms at corners = 1/4, B present at centre = 1, formula = AB_4
(b) r_{K^+} = 0.132 nm, r_{F^-} = 0.135 nm; e = 2.54×10^3 kg m^{-3}

2. (i) a = edge length = 0.589 nm
(ii) Cu^+ = 0.080 nm, r_{Br^-} = 0.213 nm
(iii) nearest neighbour distance = 0.50 nm

3. (a) 4
(b) 6
(c) Volume of unit cell = 6.61×10^{-29} m^3
(d) APF = 0.495

4. (a) Nearest neighbour distance = 0.272 nm, ρ = 22.37×10^3 kg m^{-3}
(b) (i) e = 0.857×103 kg m^{-3},
(ii) r = 0.231 nm (iii) APF = 0.681

5. (a) Position (1): (0,0,0); Position (2): (0,0,1); Position (3): (1/2,0,1/2); Position (4): (1,0,0); Position (5): (1/2,1/2,1); Position (6): (0,1/2,1/2); Position (7): (1,1/2,1/2); Position (8): (1/2,1/2, 0); Position (9): (1/2,1,1/2)
   (b) Position of formation in FCC unit cell: (0,0,0); (1/2,1/2,0); (1/2,0,1/2); (0,1/2,1/2)
   (c) Position of two atoms in BCC unit cell: (0,0,0); (1/2,1/2,1/2)
6. (a) Weiss indices : 2,1,1; 1,–1,1; 2,2,3; 1,–2,0; 1,2,2
   Miller indices: (1,2,2); (1,–1,1) (3,3,2); ($1\bar{1}0$) (211) (c)

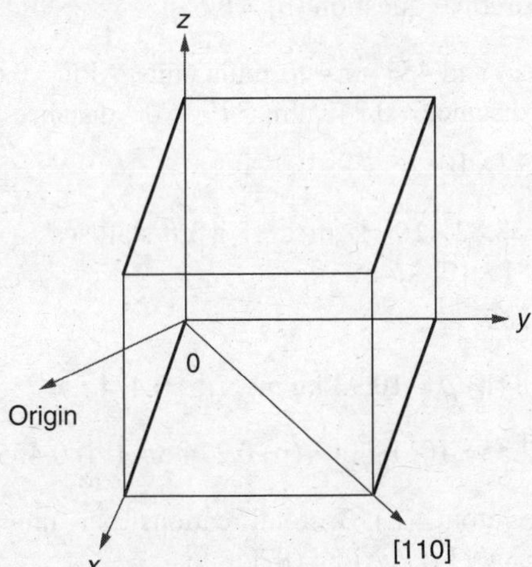

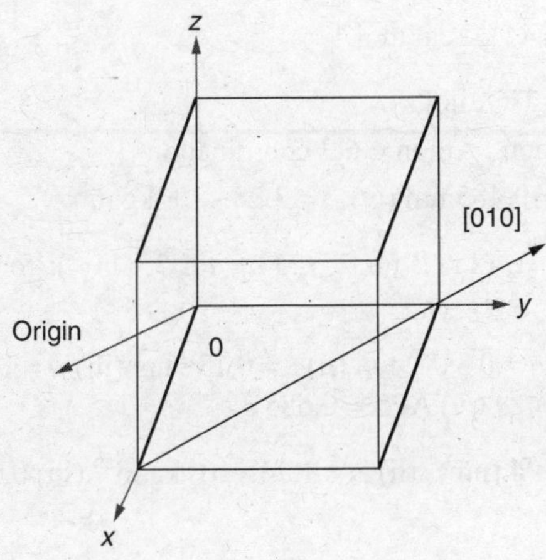

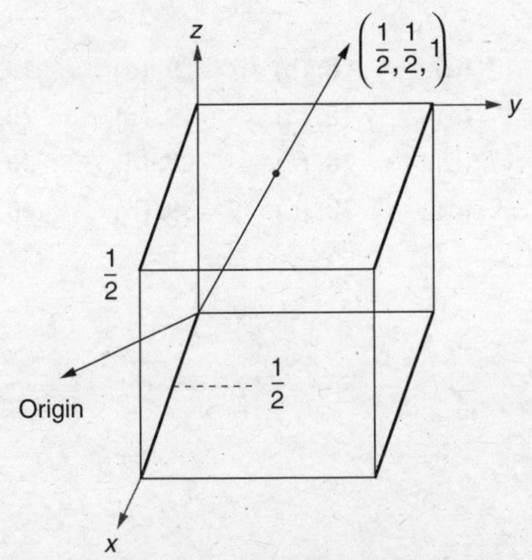

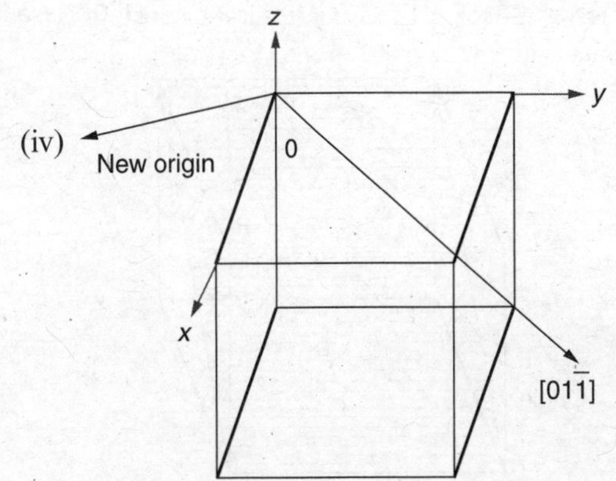

7. (a) (i)

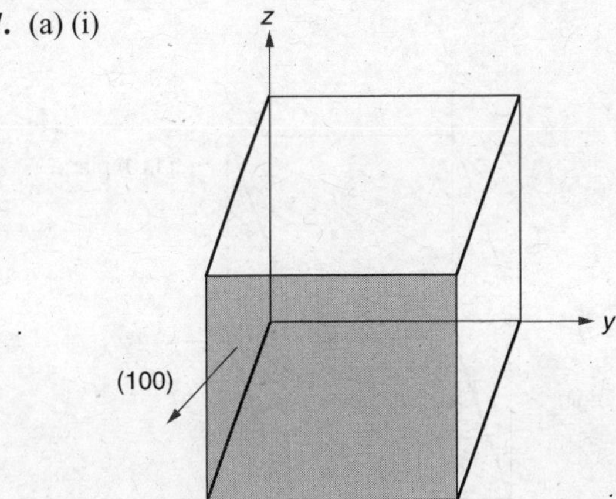

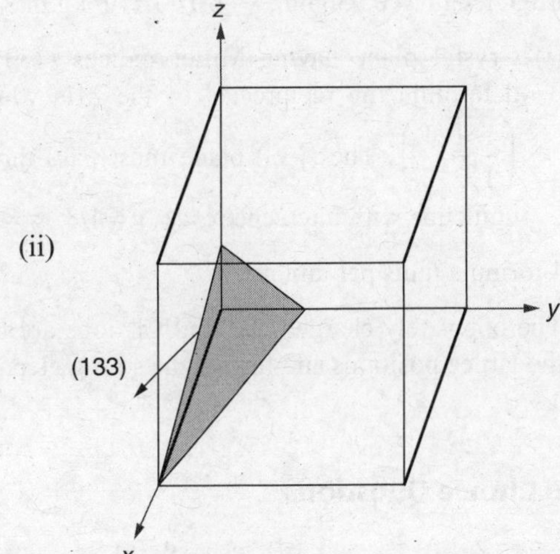

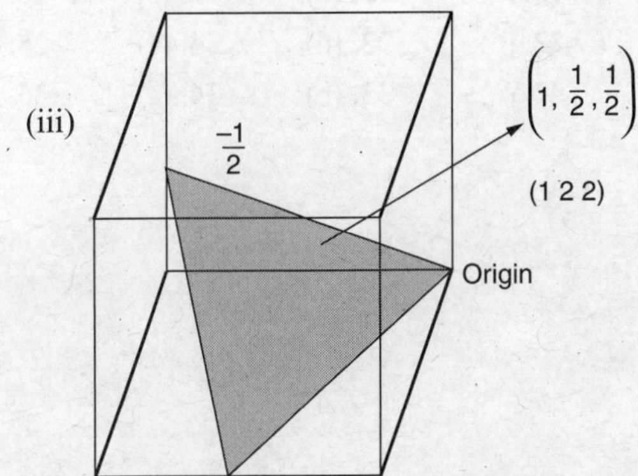

(iv)

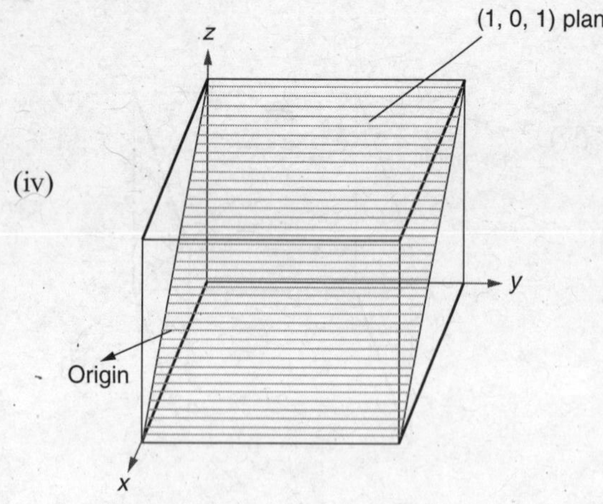

(v)

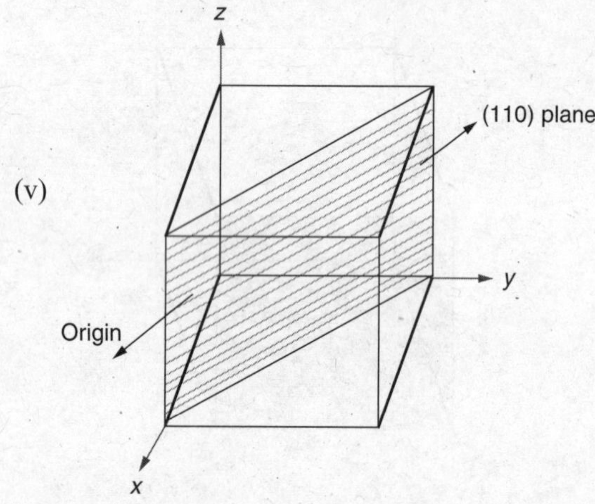

(b) Edge length = 0.636 nm; $l = 0.190 \times 10^4$ kg m^{-3}

(i) Crystal plane having Miller indices (133). First determine the reciprocals of the MIs which are $\left(\frac{1}{1}, \frac{1}{3}, \frac{1}{3}\right)$. The (133) plane must pass through a unit cube with intercepts $x = 1$, $y = 1/3$, $z = 1/3$.

8. (a) 4 formula units per unit cell

(b) The oppositely charged Na^+ and Cl^- ions are fixed in the lattice positions and further the strong electrostatic forces operating between them contribute to the poor conductivity of the salt.

(c) 2.004×10^{-6} ohm^{-1} cm^{-1}

(d) $E_a = 1.19$ eV molecule^{-1}

9. (a) (i) FCC (ii) density $\rho = 12.36 \times 10^3$ kg m^{-3}

(b) (i) 0.3971 nm (ii) $\rho = 1.993 \times 10^4$ kg m^{-3} (iii) APF = 0.741, $r = 0.1426$ nm

10. (i) a = 0.698 nm (ii) e = 3.24×10^3 kg m^{-3} (iii) nearest neighbour distance = 0.494 nm (iv) $el = 2.03 \times 10^6$ atoms mm^{-1}

11. (a) Theoretical question (b) = 0.680

12. (a) $4Cd^{2+}$ and $4S^{2-}$ or 4 formula units; APF = 0.684; Cd^{2+}–S^{2-} distance = 0.242 nm; Cd^{2+}–Cd^{2+} distance = 0.398 nm

(b) 8.472×10^{-7}

13. (a) e = 8.82×10^3 kgm^{-3}. If it crystallises in HCP, ρ = 0.881×10^4 kg m^{-3}

(b) $CaTiO_3$

14. (i) 0.588 (ii) $\rho = 10.93$ kg m^{-3} (iii) 0.474 nm

15. (i) $\rho = 2.35 \times 10^3$ kg m^{-3} (ii) 0.274 nm (iii) 0.455 nm

16. (i) One atom (ii) 3, total cations per unit cell = 8 (iii) $a = 8r/\sqrt{3}$ (iv) APF = 0.340

17. (a) $F = 2.72 \times 10^{-9}$ N

(b) $E = 44.2$ kcal mol^{-1}

18. (i) Two TiO_2 units

(ii) Cation : Anion = 6:3 coordination

(iii) $a = 0.485$ nm (iv) $\rho = 3.98 \times 10^3$ kg m^{-3}

19. (a) For 1, (111); for 2, (23∞); for 3, (11∞); for 4, (222); for 5, (1∞1)

(b) (i) $a = 0.3172$ nm (ii) $r = 0.138$ nm (iii) $\rho = 18.98 \times 10^3$ kg m^{-3} (iv) APF = 0.685

20. (i) 0.1394 (nm)^3 (ii) $\rho = 1.74 \times 10^3$ kg m^{-3} (iii) 0.527

Multiple Choice Questions

1. (b)	**2.** (c)	**3.** (d)	**4.** (a)	**5.** (c)	**6.** (b)	**7.** (d)	**8.** (b)	**9.** (d)	**10.** (b)
11. (d)	**12.** (c)	**13.** (d)	**14.** (b)	**15.** (d)	**16.** (d)	**17.** (a)	**18.** (c)	**19.** (b)	**20.** (c)
21. (b)	**22.** (c)	**23.** (d)	**24.** (a)	**25.** (c)	**26.** (a)	**27.** (b)	**28.** (c)	**29.** (d)	**30.** (a)
31. (c)	**32.** (c)	**33.** (b)	**34.** (a)	**35.** (d)	**36.** (c)	**37.** (c)	**38.** (a)	**39.** (b)	**40.** (c)

CHAPTER

8 PHOTOCHEMISTRY

CHAPTER OUTLINE

Basic Processes

Quantum Yield
- Reasons for High Quantum Yield
- Reasons for Low Quantum Yield

Chemical Actinometers
- Ferrioxalate Actinometer
- Uranyloxalate Actinometer
- Reinecke's Salt Actinometer

Photochemical Processes
- Primary Processes in Photochemical Reactions
- Secondary Photochemical Processes
- Singlet and Triplet States
- Jablonski Diagram
- Chemiluminescence
- Thermoluminescence
- Quenching
- Photosensitisation

Examples of Some Photochemical Reactions
- Decomposition of Hydrogen Iodide
- Photochemical Hydrogen–Bromine Reaction
- Reaction between H_2 and Cl_2
- Photochemical Reaction of Acetone

Flash Photolysis Technique

Photoelectrochemical Cells (PECs)
- Photogalvanic Cells

Radiation Chemistry
- Radiochemical Reactions
- Yield of Products in Radiochemical Reactions
- Fricke Dosimeter
- Reactions in the Radiolysis of Water

8.1 Basic Processes

Thermal and photochemical reactions: Ordinary chemical reactions proceed by activation through random and successive collisions between the reactant molecules. Such reactions are also referred to as *thermal reactions*. It is also possible to provide the required activation by absorption of photons of light by molecules. These photons are provided by ultraviolet, visible or infrared light. Reactions occurring under this category are known as *photochemical reactions*. Table 8.1 gives the wavelength region and the energies of the radiation.

Table 8.1 Energies of various radiations

Nature of radiation	Wavelength region (nm)	Energy (kJ mol^{-1})
Ultraviolet	1 to 400	290 to 965×10^2
Visible	400 to 700	190 to 290
Infrared	700 to 10^5	1.0 to 190

X-rays (wavelength range 0.01 nm to 1.0 nm) and γ-rays (wavelength range < 0.01 nm) are also used to activate molecules and study the reactions of the activated molecules.

Radiation chemical reactions or radiochemical reactions: Apart from the types of photons mentioned above, high energy particles such as α-rays (helium nuclei), β-rays (electrons) and deuterons are used for the activation of molecules. Because of the high energy associated with these particles, ions are often produced in the reacting systems. This class of reactions is known as *radiochemical reactions*.

Grotthuss–Draper law: This law states that only radiations that are absorbed by a reacting system can produce a photochemical change.

Beer–Lambert law: When monochromatic light passes through a medium, there is a decrease in the intensity of light which depends on the thickness of the medium as well as the concentration (or the number of absorbing molecules in the given medium). If I_0 is the intensity of the incident light and I is the intensity of the transmitted light, the quantity $\log_{10}\frac{I_0}{I}$, referred to as the absorbance A (optical density) is related to the length (or thickness) l of the solution (through which light passes) and also the concentration of the solute C in the medium. The relation between the absorbance A, l (thickness of the medium) and the concentration C of the solution is known as the Beer–Lambert law and is given by

$$\log_{10}\frac{I_0}{I} = A = \epsilon l C \tag{8.1}$$

where, ε is known as molar absorptivity (or molar extinction coefficient). Equation 8.1 is the basis for the quantitative determination of the concentration of the solute by colorimetry or spectrophotometry. Absorbance is determined directly (for example, in a spectrophotometer) for various concentrations of solute and $\log_{10} \frac{I_0}{I}$ against C is plotted. The plot is linear in dilute solutions and from the slope, εl can be determined. Since l is usually kept constant (at 1 cm or any other value depending upon the absorbing species), ε can be determined. ε has the units M^{-1} cm^{-1}. Note that absorbance is directly proportional to the concentration of solute as well as to the length of the light path.

Stark-Einstein law of photochemical equivalence: This law states that a molecule absorbs a single quantum of light which causes the reaction. The energy of the quantum e is defined by

$$e = h\nu \tag{8.2}$$

where h = Planck's constant and ν is the frequency of the radiation. The energy absorbed per mole, known as Einstein, E is given by

$$E = Nh\nu = \frac{NhC}{\lambda} \tag{8.3}$$

where N = Avogadro's number, C is the velocity of light and λ is the wavelength of energy of absorbed light. For example, for a wavelength of 400 nm, the value of E may be calculated as

$$E = \frac{6.023\times10^{23}\ \text{mol}^{-1}\times6.625\times10^{-34}\ \text{J}\times\text{s}\times2.998\times10^{8}\times \text{m}\times\text{s}^{-1}}{400\times10^{-9}\,\text{m}}$$

$$= 0.299\times10^{6}\ \text{J mol}^{-1} = 299\ \text{kJ mol}^{-1}$$

From Eq. 8.3, it is seen that E is greater, the shorter the wavelength of light.

8.2 Quantum Yield

The absorption of light by a molecule can lead to one of many processes such as chemical reaction, fluorescence, phosphorescence and energy transfer to other molecules. The yield of such processes is expressed in terms of quantum yield ϕ. It is defined by,

$$\phi = \frac{\text{Number of molecules undergoing the reaction in a given time}}{\text{Number of quanta absorbed in the same time}} \tag{8.4}$$

Another method of expressing the quantum yield is,

$$\phi = \frac{\text{Rate of the given process}}{\text{Rate of absorption of the radiation}} \tag{8.5}$$

$$= \frac{\text{Rate of given process}}{I_a} \tag{8.6}$$

where I_a = number of quanta absorbed per second. ϕ for a chemical reaction may range from a fractional value to a large number. ϕ is a large number in case of reactions involving the formation of radicals which results in a chain mechanism. The law of photochemical equivalence always applies to primary photochemical processes even though ϕ may be large due to other reactions.

8.2.1 Reasons for High Quantum Yield

When one photon decomposes or forms more than one molecule, $\phi > 1$. There are several reasons for $\phi > 1$.

(i) The photon absorbed in the primary reaction causes dissociation of one molecule of the reactant. However, the excited atoms that are formed may start a secondary reaction in which another molecule is decomposed. For example, consider the reaction of a molecule XY with a photon.

$$\left.\begin{aligned} &\text{XY} + h\nu \rightarrow \text{X} \quad \text{primary reaction} \\ &\text{XY} + \text{X} \rightarrow \text{X}_2 + \text{Y secondary reaction} \end{aligned}\right\} \tag{8.7}$$

(ii) A chain reaction occurs with the formation of many molecules per photon. Consider, for example, the reaction involving the reactants X_2 and Y_2 in which X_2 absorbs the photon.

$$\left\{\begin{aligned} &\text{X}_2 + h\nu \rightarrow 2\text{X} + \text{Y}\ (1)\ \text{Primary reaction} \\ &\text{X} + \text{Y}_2 \rightarrow \text{XY} + \text{Y}\ (2)\ \text{Secondary reaction} \\ &\text{Y} + \text{X}_2 \rightarrow \text{XY} + \text{X}\ (3)\ \text{Chain reaction} \end{aligned}\right. \tag{8.8}$$

The decomposition of HI on exposure to radiation less than 400 nm to form H_2 and I_2 and the reaction between H_2 and Cl_2 to form HCl are classical examples of this category.

8.2.2 Reasons for Low Quantum Yield

Three reasons may be cited for low quantum yields in some reactions.

(i) **Reacting molecules get deactivated:** The excited molecules formed in the primary process get deactivated before they react further. For example a molecule X on absorption of a photon in the primary process gets activated according to,

$$\text{X} + h\nu \rightarrow \text{X}^*$$

Activation occurs in the primary reaction but it gets deactivated, say, by fluorescence as follows.

$$\text{X}^* \rightarrow \text{X} + h\nu' \quad \text{Fluorescence} \tag{8.9}$$

(ii) **Possibility of reversal of a primary reaction:** In this case, the primary reaction yields a polymer but the product forms the reactants back through a thermal reaction.

$$2\text{X} \underset{\text{thermal}}{\overset{h\nu}{\rightleftharpoons}} \text{X}_2 \tag{8.10}$$

The reaction occurs until equilibrium is established.

(iii) **Dissociated species recombine:** During the primary process, the reactant molecules dissociate to form smaller fragments. They again combine to give back the reactant. For example,

$$\left.\begin{aligned} \text{XY} + h\nu &\rightarrow \text{X} + \text{Y} \\ \text{X} + \text{Y} &\rightarrow \text{XY} \end{aligned}\right\} \tag{8.11}$$

In this case, step 2 involving the fragments does not result in the formation of the product, thereby lowering the quantum yield to a large extent.

Examples of such reactions include dimerisation of anthracene and the combination of H_2 and Br_2.

8.3 Chemical Actinometers

Actinometry is a method which permits the determination of the flux of the photons for a system of specific geometry using monochromatic incident light. Actinometers are devices used for this purpose mostly in the UV–visible range. In chemical actinometers, a reference substance which undergoes a photochemical reaction and where the quantum yield is known, is used. This is usually calibrated against a standard physical device or other well studied actinometers. Several chemical actinometers are used but only some well known ones are discussed.

8.3.1 Ferrioxalate Actinometer

This actinometer is reliable for UV and visible light up to 500 nm. The potassium ferrioxalate used in this actinometer, excited on exposure to light, decomposes according to the equation,

$$Fe(C_2O_4)_3^{3-} \xrightarrow{h\nu} Fe^{2+} + C_2O_4^{-\bullet} + 2C_2O_4^{2-} \quad (8.12)$$

$$Fe(C_2O_4)_3^{3-} + C_2O_4^{-\bullet} \xrightarrow{\Delta} Fe^{2+} + 2CO_2 + 3C_2O_4^{2-} \quad (8.13)$$

The ferrous ions formed during irradiation are converted to the coloured tris-phenanthroline complex whose molar absorptivity ε =11,100 L mol^{-1} cm^{-1} at its λ_{max} = 510 nm. The number of moles of Fe^{2+} ions in the volume taken for irradiation is given by

$$\text{Moles of } Fe^{2+} = \frac{V_1 \times V_2 \times \Delta A}{1000 \times V_3 \times l \times \epsilon} \quad (8.14)$$

where V_1 = mL of solution taken for irradiation, V_2 = mL of final volume of solution taken after complexation with phenanthroline, V_3 = aliquot (mL) of irradiated solution taken for determining the Fe^{2+} ions. ΔA is the difference in absorbance between the irradiated solution and the blank (or dark) at 510 nm, l is the optical path length and ε is the molar absorptivity of the iron–phenanthroline complex at 510 nm. The moles of photons absorbed by the irradiated solution per unit time, $Nh\nu/t$ can be obtained as

$$Nh\nu/t = \frac{\text{Moles of } Fe^{2+}}{\phi_\lambda \times t \times F} \quad (8.15)$$

where ϕ_λ = quantum yield for Fe^{2+} in production at the irradiation wavelength λ, t = time of irradiation and F = fraction of light absorbed by the ferrioxalate solution (when carried in visible light irradiation).

8.3.2 Uranlyoxalate Actinometer

In this case, the photochemical reaction is

$$H_2C_2O_4 \xrightarrow{h\nu,\, UO_2^{2+}} CO_2 + CO + H_2O \quad (8.16)$$

where the uranyl ion acts as the photosensitiser. This actinometer is useful in the range λ = 208 to 426 nm and ϕ_{CO} is practically constant at 0.05 to 0.6 (in this wavelength range). The quantum yield is independent of pH in range 1 to 5.

8.3.3 Reinecke's Salt Actinometer

The function of this actinometer is based on photosubstitution reaction of Reinke's salt, $\left[Cr(NH_3)_2(SCN)_4\right]^-$ on irradiation with UV and visible light according to

$$\left[Cr(NH_3)_2(SCN)_4\right]^- + H_2O \xrightarrow{h\nu} \begin{bmatrix} Cr(NH_3)_2 \\ (SCN)_3 \\ (H_2O) \end{bmatrix} + SCN^- \quad (8.17)$$

The substitution of the ligand SCN^- by a water molecule takes place in aqueous solution. The released SCN^- is complexed with ferric nitrate and the absorbance of the blood red complex is measured at λ_{max} = 450 nm and its extinction coefficient is ε = 4300 L mol^{-1} cm^{-1}. The number of photons is thus found by the number of SCN^- ions released. The quantum yield of this reaction shows little dependence on pH in the range of λ = 316 – 750 nm.

8.4 Photochemical Processes

8.4.1 Primary Processes in Photochemical Reactions

In a photochemical process, the first step is the absorption of a quantum of radiation. This is the *primary process*. The energy absorbed in the primary process from the photon (in the UV or visible region) raises the molecule to an excited state. There are several ways by which the excited molecule may undergo further reaction. (i) There may not be any direct dissociation of the excited molecule. (ii) The absorbed energy may take the molecule to a higher energy state which is unstable and hence it breaks up into atoms or radicals. (iii) Sometimes the excited state undergoes dissociation. (iv) Sometimes, it is possible that the initially excited state may undergo transition to another higher transition state which then dissociates to produce atoms or radicals.

8.4.2 Secondary Photochemical Processes

An excited molecule formed in the primary process may sometimes emit radiation of either the same or different frequency. This emission is referred to as *fluorescence* or *phosphorescence*. These will be explained subsequently.

8.4.3 Singlet and Triplet States

Excited states of molecules may be classified as *singlet* or *triplet*, based on their electron spin angular momentum. Electrons in most organic compounds are paired and have opposite spins in bonding and non-bonding orbitals. This results in a net zero spin diamagnetic molecule for the ground state. In an applied magnetic field, they have a single energy state known as the singlet state. Electronic states having two electrons with identical spin occupy different orbitals and have a net spin. They are *paramagnetic*. In a magnetic field, such states have three energy levels which are called triplets. When a photon-induced excitation takes place, it always leads to a state of the same multiplicity, that is, a singlet to a singlet, or a triplet to a triplet. There are other ways whereby an excited state goes over to a lower state of the same multiplicity (described below).

Internal conversion: An excited state goes to a lower energy state of the same multiplicity while rapidly losing heat energy. This is also known as *relaxation*.

Radiative decay: An excited state returns to the ground state by emitting a photon.

Fluorescence: It is a spontaneous emission of a photon by an excited singlet state while reverting to the singlet ground state. The emitted radiation ceases immediately when the exciting force is removed. The intensity of fluorescent radiation is reduced when an excited atom collides with another atom. This is known as quenching.

Phosphorescence: It arises from radiative transitions from the lowest vibrational levels of the singlet ground state. This is a much slower process compared to fluorescence. Zinc sulphide and other alkaline earth sulphides exhibit phosphorescence.

Intersystem crossing: A singlet state may go over to a lower energy triplet state, this is called intersystem crossing. This process is slower than internal conversion.

8.4.4 Jablonski Diagram

A Jablonski diagram, proposed by Alexander Jablonski a Polish scientist, is primarily an energy diagram with energy levels shown schematically. Figure 8.1 shows various energy levels and the transitions occurring from the higher energy levels.

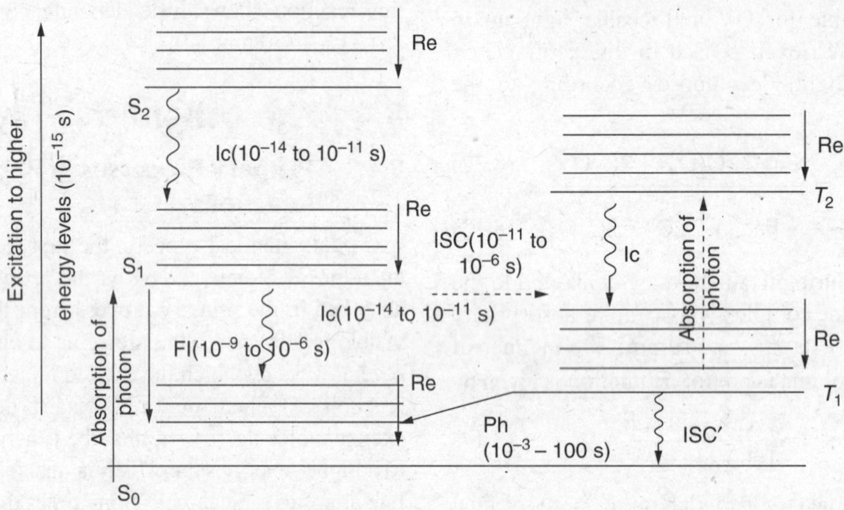

Fig. 8.1 Jablonski diagram

The time ranges of the various processes are given in brackets against each process. The non-radiative decay processes occur in a time range of 10^{-7} to 10^{-5} s (for $S_1 \rightarrow S_0$) and 10^{-3} to 10^{2} s (for $T_1 \rightarrow S_0$).

8.4.5 Chemiluminescence

When a chemical reaction is accompanied by the emission of light, this phenomenon is known as *chemiluminescence*. Consider, for example, a reactant X in its ground state is converted to an excited state product Y* via a chemical reaction. The electronically excited state Y* emits light. The reaction scheme in this process may be represented as,

$$X \rightarrow Y^* \rightarrow Y + h\nu \tag{8.18}$$

It is actually the reverse of a photochemical reaction. Since the radiation is emitted at ordinary temperatures, it is called *cold light*. Glow of yellow phosphorus in air at ordinary temperatures and glow of luciferin in fireflies apart from some chemical reactions are examples of this phenomenon.

8.4.6 Thermoluminescence

When certain crystals, such as the alkali halide crystals, are exposed to X-rays, they give rise to characteristic colours. The colouration is due to the absorption of light by the released electrons which get trapped in negative ion vacancies in the crystal lattice. On heating the crystal, the trapped electrons are released and go to a lower energy level giving off light. In the case where the crystal is heated slowly, a series of light emissions occur at definite temperatures. Apatite, calcite and some feldspars exhibit *thermoluminescence*.

8.4.7 Quenching

This refers to the process of deactivation of excited state molecules by other molecules in a solution or gas phase. The substances causing quenching are called *quenchers* and are represented as Q. The quenching of phosphorescence by a quencher enables the determination of the rate constant of the reaction of the triplet state with the quencher. If I_P^0 is the intensity in the absence of the quencher and I_P is the intensity in the presence of the quencher, the ratio I_P^0/I_P is related to the rate constant k_q of the quenching reaction by

$$\frac{I_P^0}{I_P} = 1 + k_q \times \tau_{T_1}(\mathrm{Q}) \tag{8.19}$$

τ_{T_1} is the lifetime of phosphorescence of the triplet state T_1. It may be mentioned that τ_{T_1} is related to the rate constants of the *Ph* step (k_{Ph}) and the step of ISC′ ($k_{ISC'}$) (see Jablonski diagram) via the equation,

$$\tau_{T_1} = \frac{1}{k_{Ph} + k_{ISC'}} \tag{8.20}$$

Equation 8.20 is known as the *Stern–Volmer equation*.

It is also possible that the chemical reaction of an excited molecule with another molecule results in the deactivation of the excited molecule. The excited state is then said to be quenched. This quenching process is often termed as *electronic energy transfer*. These electronic energy transfer processes are of three types—(i) radiative transfer, (ii) short range transfer, and (iii) long range resonance transfer. Considering radiative transfer, it is possible to have the excited state of a donor molecule D* emit radiation which is absorbed by an acceptor molecule A according to

$$\left.\begin{aligned} D^* &\rightarrow D + h\nu \\ A + h\nu &\rightarrow A^* \end{aligned}\right\} \tag{8.21}$$

Short range energy transfer between a donor and acceptor can occur when the distance of approach between them is of the order of a collision diameter. Actual collision between them is not required for energy transfer. Two types of reactions involving short range energy transfer among S_0, S_1, T_1 states are given below:

$$D^*(S_1) + A(S_0) \rightarrow D(S_0) + A^*(S_1) \tag{8.22}$$

$$D^*(T_1) + A(S_0) \rightarrow D(S_0) + A^*(T_1) \tag{8.23}$$

Photoperoxidation of 9,10-dimethylanthracene by singlet molecular oxygen is an example of the reaction type 8.22. In the case of long range energy transfer, a much greater distance than the collision diameter is seen between the donor and acceptor molecules. The overlap of the emission spectrum of the acceptor governs the efficiency of energy transfer.

The long range energy transfer processes are similar to Eqs. 8.22 and 8.23 with rate constants in the range 10^{10} to 10^{11} L mol^{-1} s^{-1}. The rate constant k_{ET} for the energy transfer (in reactions of the type 8.22 or 8.23 is given by

$$\frac{1}{\tau_{ET}} = \frac{1}{\tau} + k_{ET}[A] \tag{8.24}$$

where τ_{ET} = lifetime of the triplet in the presence of an acceptor A and τ have the same significance as in Eq. 8.20 (applicable in the absence of an acceptor).

8.4.8 Photosensitisation

Many molecules do not absorb radiation at a convenient wavelength for them to participate in a photochemical reaction. In such cases, another molecule absorbs the light, gets excited and transfers the energy to the original molecule which then undergoes the reaction. For example, the hydrogen molecule requires energy corresponding to a wavelength of 277.6 nm for dissociation, but it does not absorb at this wavelength. If we introduce mercury vapour into the system and expose it to radiation of 253.7 nm, mercury atoms get excited and transfer their energy to the hydrogen molecules which get dissociated into atoms. The mercury vapour is said to be the photosensitiser. Processes of this type in which the radiation is absorbed by one species and the energy is transferred to another are known as *photosensitised reactions*. Another example is

$$Hg^* + C_2H_4 \rightarrow Hg + C_2H_4^* \tag{8.25}$$

where C_2H_4 is excited to a higher state by reaction with Hg*.

Photosynthesis: A classical example of a photosensitised reaction is the action of chlorophyll as a photosensitiser in the reaction between carbondioxide and water to form carbohydrates. This reaction takes place in presence of sunlight. The reaction is

$$xCO_2 + xH_2O \xrightarrow{n\,h\nu} (CH_2O)_n + xO_2 \tag{8.26}$$

8.5 Examples of Some Photochemical Reactions

Several photochemical reactions have been studied and reported in literature. Among them may be cited—(i) the reaction between hydrogen and the halogens Cl_2, Br_2 and I_2, (ii) dimerisation of anthracene, (iii) decomposition of acetaldehyde, (iv) decomposition of acetone, (v) decomposition of NH_3, and so on. A few reactions are considered in some detail.

8.5.1 Decomposition of Hydrogen Iodide

This reaction occurs when hydrogen iodide is exposed to radiation of 280–300 nm in the gas phase or in the solution phase with hexane as solvent. The quantum yield of the reaction is 2 and the following reaction mechanism is proposed.

$$\left\{\begin{aligned} &HI \xrightarrow{h\nu} H + I \text{ rate} = I_{abs} \\ &H + HI \xrightarrow{k_2} H_2 + I \\ &\text{and } I + I \xrightarrow{k_3} I_2 \end{aligned}\right. \tag{8.27}$$

The overall reaction is

$$2HI \xrightarrow{h\nu} H_2 + I_2 \tag{8.27a}$$

For the rate of disappearance of HI, we can write

$$-\frac{d[HI]}{dt} = I_{abs} + k_2[H][HI] \tag{8.28}$$

Applying the steady state assumption for H atom concentration,

$$\frac{d[H]}{dt} = 0 = I_{abs} - k_2[H][HI]$$

$$[H] = \frac{I_{abs}}{k_2[HI]} \tag{8.29}$$

Substituting Eq. 8.29 in Eq. 8.28, we get

$$-\frac{d[HI]}{dt} = I_{abs} + I_{abs} = 2I_{abs} \tag{8.30}$$

From Eq. 8.30, it is seen that the absorption of one photon leads to the decomposition of 2 molecules of HI. The quantum efficiency of the reaction is 2.

8.5.2 Photochemical Hydrogen-Bromine Reaction

When light of short wavelength ($\lambda < 510$ nm) is passed through a mixture of H_2 and Br_2 at room temperature, a reaction takes place with the formation of HBr for which the following mechanism is suggested.

$$(1)\ Br_2 + h\nu \rightarrow 2Br \quad \text{Initiation}$$

$$\left.\begin{aligned}(2)\ Br + H_2 &\xrightarrow{k_2} HBr + H\\(3)\ H + Br_2 &\xrightarrow{k_3} HBr + Br\end{aligned}\right\}\text{Chain propagation} \quad (8.31)$$

$$(4)\ H + HBr \xrightarrow{k_4} H_2 + Br$$

$$(5)\ Br + Br \xrightarrow{k_5} Br_2 \quad \text{Termination}$$

If I_d is the intensity of light absorbed, given as einstein dm^{-3} s^{-1}, the rate of change of concentration of Br atoms (Eq. 8.31), is $2I_{ads}$. Applying the steady state approximation for Br atoms and H atoms, the rate equation may be obtained as

$$\frac{d[HBr]}{dt} = \frac{2k_2\left(\frac{2I_{abs}}{k_5}\right)^{1/2}[H_2]}{1 + \frac{k_4}{k_3}\frac{[HBr]}{[Br_2]}} \quad (8.32)$$

From Eq. 8.32, it is seen that HBr inhibits the reaction and the inhibition increases as ratio of $\frac{[HBr]}{[Br_2]}$ increases.

8.5.3 Reaction between H_2 and Cl_2

When a mixture of H_2 and Cl_2 is exposed to radiation at wavelengths less than 480 nm, Cl_2 gets decomposed in the first step according to,

$$(1)\ Cl_2 + h\nu \xrightarrow{I_{abs}} 2Cl$$

which is followed by

$$\left.\begin{aligned}(2)\ Cl + H_2 &\xrightarrow{k_2} HCl + H\\(3)\ H + Cl_2 &\xrightarrow{k_3} HCl + Cl\end{aligned}\right\} \quad (8.33)$$

$$(4)\ Cl + Cl \xrightarrow{k_4} Cl_2 \quad (\text{wall or a third body M})$$

$$(5)\ H^{\bullet} + Cl^{\bullet} \longrightarrow HCl \text{ with rate constant } k_5$$

The quantum yield of the reaction is of the order of 10^5.

Setting up steady state expressions for Cl and H atom concentrations, the rate equation may be obtained as,

$$\frac{d[HCl]}{dt} = k_2[H_2]\left[\frac{\phi I_{abs}}{k_5[M]}\right]^{\frac{1}{2}} \quad (8.34)$$

8.5.4 Photochemical Reaction of Acetone

When acetone vapour is exposed to UV light less than 330 nm at room temperature, the absorption of the energy leads to the cleavage of the –C– bond and not C=O. The reaction is

$$\begin{matrix}H_3C\\H_3C\end{matrix}C=O + h\nu \rightarrow \dot{C}H_3 + CH_3\dot{C}O \quad (8.35)$$

The acetyl radical then can decarboxylate to give CO and $\dot{C}H_3$. The methyl radicals may couple to give ethane.

8.6 Flash Photolysis Technique

This technique is a general method of providing higher concentrations of unstable intermediates than possible in conventional photochemical reactions. A powerful flash of light in the UV or visible range is used in a short duration of time and high concentrations of excited molecules may dissociate into free radicals or atoms.

This method is used for studying the kinetics of decay of excited states. Further, the absorption spectra of NH_2ClO and $C_6H_5CH_2$ radicals have been studied by this method.

The reaction of atomic selenium with ethylene to form ethylene selenide has been studied by flashing a mixture of carbon selenide and ethylene. Flash photolysis method is a means for the formation of solvated electrons e^-_{aq} in the photolysis of dissolved ions such as Cl^-, Br^-, OH^-.

8.7 Photoelectrochemical Cells (PECs)

These are essentially electrochemical cells in which light energy is used to produce an electrochemical change. In a common form of a PEC, there is a photoanode and a metal counter electrode immersed in an aqueous acidic or alkaline solution in the particular case of electrolysis of water. The photoanode is a semiconductor and the cathode may be a metal such as Pt or Cu. The semiconductor anode converts the incident photons to electrons—hole pairs which are spatially separated by an electric field inside the semiconductor. The light generated electrons travel towards the metal cathode via the external wire. A simplified diagram of a PEC is shown in Fig. 8.2.

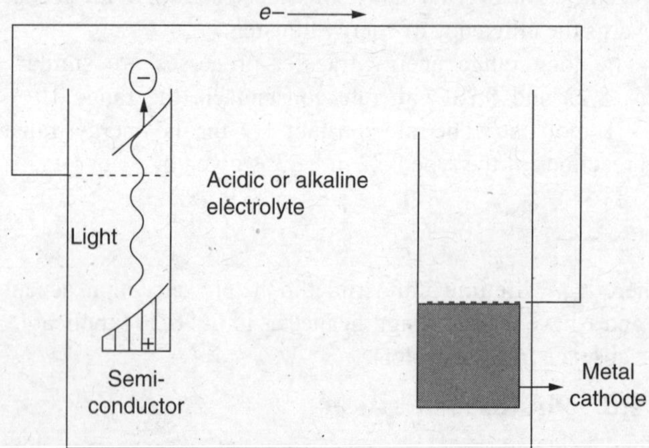

Fig. 8.2 Schematic depiction of a photoelectrochemical cell

In an acidic solution, the reactions are

$$\left.\begin{aligned}4H^+ + 4e^- &\rightarrow 2H_2 & E^0_{red} &= 0.000 \text{ V vs SHE}\\2H_2O + 4h^+ &\rightarrow 4H^+ + O_2 & E^0_{ox} &= -1.229 \text{ V vs SHE}\end{aligned}\right\} \quad (8.36)$$

In an alkaline solution, the reduction and oxidation reactions are

$$\left.\begin{array}{ll} 4H_2O+4e^- \rightarrow 2H_2+4OH^- & E^0=-0.828 \text{ V vs SHE} \\ 4OH^- +4h^+ \rightarrow 2H_2O+O_2 & E^0=-0.401 \text{ V vs SHE} \end{array}\right\} \quad (8.37)$$

The important requirements for a semiconductor photoelectrode are (i) efficient absorption of visible light and (ii) good charge transport.

8.7.1 Photogalvanic Cells

In these cells, irradiation induces a homogeneous photochemical reaction which gives rise to a current. The production of charge (or electricity) by the homogeneous chemical system follows principles of normal galvanic cells except that it is brought about photochemically. A good example of such a photogalvanic system is the iron–thionine cell first studied by E Rabinowitch. The reaction in the cell is,

$$\underset{(\text{purple})}{\frac{1}{2}\text{Th}} + Fe^{2+} \underset{\text{dark}}{\stackrel{h\nu}{\rightleftharpoons}} \underset{(\text{colourless})}{\frac{1}{2}\text{leu}-\text{Th}} + Fe^{3+} \left(\text{at pH}=2\right) \quad (8.38)$$

The equilibrium lies very much to the left in the dark. Irradiation with light of $\lambda = 478$ nm shifts the equilibrium to the right and the solution becomes colourless. Both redox couples of reaction 8.38 are electrochemically reversible and some electrical power can be drawn from the cell without separating of reaction products by photolysing the solution near one of the electrodes. The illuminated electrode becomes negative with respect to the dark one with potentials of about 0.2 V. The current can be drawn continuously on closing the circuit. The cell arrangement is shown in Fig. 8.3.

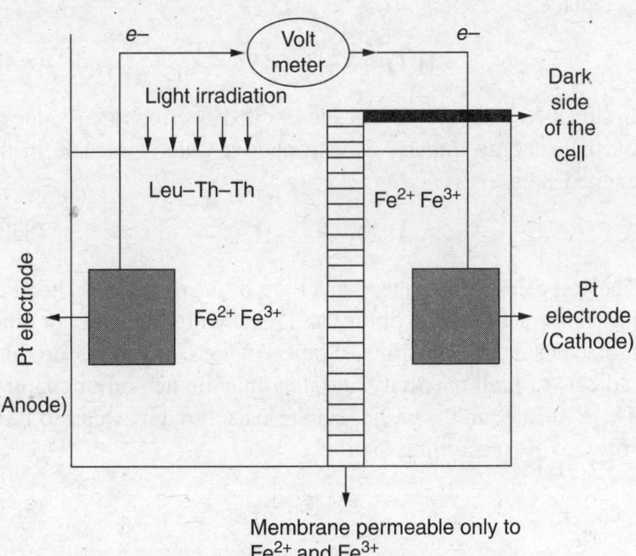

Fig. 8.3 The iron–thionine cell

As shown, the membrane is permeable only to Fe^{2+} and Fe^{3+} ions. The overall reaction is the reverse of reaction 8.38, that is, the reduction of Fe^{3+} to Fe^{2+} ions under the illumination of light.

Distinguishing features between photochemical and thermal reactions are given below.

Photochemical reactions	Thermal reactions
Occur with absorption of light radiation	Occur with absorption or evolution of heat
Presence of light is basic requirement for reactions to occur	Can take place in the dark as well as in the presence of light
Temperature has little effect on the rates of these reactions	Temperature profoundly affects the rates of these reactions
A spontaneous reaction occurring photochemically may have a positive or negative ΔG	ΔG is always negative for these reactions
Activation of these reactions is very selective. Reaction occurs when the absorbed photon excites a particular atom or a group	Activation of these reactions is not selective

8.8 Radiation Chemistry

8.8.1 Radiochemical Reactions

Reactions brought about by particles such as α-particles, β-particles, protons, deuterons, and so on, are referred to as *radiation chemical* or *radiochemical reactions.* Radiations of high energy of the order of one million electron volts (9.15×10^7 kJ mol^{-1}) are used in the study of such reactions. The primary effect of these radiations is the formation of ions, atoms or free radicals in the system.

When the dissociation of a molecule occurs in a radiation chemical reaction, it is referred to as *radiolysis*. For example, a molecule such as X, on dissociation, gives rise to

$$X \rightarrow X^+ + e^- \quad (8.39)$$

Secondary processes occur because the slow moving heavier particles act as more powerful ionising agents. As a result of the secondary processes, the number of ionised molecules is quite large.

8.8.2 Yield of Products in Radiochemical Reactions

A quantitative measure of the yield of products in radiochemical reactions is the *G value,* defined as,

$$G = \frac{\text{Number of product molecules formed}}{\text{Input of 100 eV of energy}} \quad (8.40)$$

A recent SI definition of the *G* value is, the number of molecules decomposed or formed per joule of energy absorbed. That is,

$$G = \frac{\text{Number of molecules formed}}{\text{Input of one joule of energy}} \quad (8.41)$$

If the theoretical value of G is calculated for the number of molecules formed in one mole,

$$G = \frac{6.023\times10^{23} \text{ molecules}}{\text{joule}} = \frac{6.023\times10^{23}}{(1/1.602\times10^{-19})\text{eV J}^{-1}\times\text{J}}$$

$$G = 6.023\times10^{23}\times1.602\times10^{-19}$$

$$= 9.65\times10^{4} \text{ molecules eV}^{-1}$$

$$= 9.65\times10^{6} \text{ molecules}(100 \text{ eV})^{-1}$$

G values depend on the nature of radiation, medium being irradiated and the physical state of the medium. For example, the yield of G in the irradiation of neutral water when irradiated by different kinds of radiation is given in Table 8.2.

Table 8.2 G values as μmol J^{-1} for the species

Nature of radiation	H_2O	H_2	H_2O_2	Solvated electron e_{aq}	$H^{\bullet}$	$OH^{\bullet}$
Fast electrons and also γ radiation	0.43	4.7×10^{-2}	7.3×10^{-2}	6.2×10^{-2}	0.28	2.7×10^{-3}
α-particles of 12 MeV energy	0.29	0.12	0.11	2.8×10^{-2}	5.6×10^{-3}	7×10^{-3}

The data give an idea of the yield of different products.

8.8.3 Fricke Dosimeter

It is a device for measuring the dosage of high energy radiations impinging on the system under investigation. It is a useful method to directly measure the number of reactive species in solution. The main constituent of the Fricke dosimeter is a standard solution of 1×10^{-3} M solution of ferrous ammonium sulphate in 0.8 M H_2SO_4. On irradiation of the solution, Fe^{2+} gets oxidised to Fe^{3+}. The Fe^{3+} ion develops a blue colour which can be measured spectrophotometrically. The colorimetric data is linear up to a radiation dosage of 40000 rad or 400 Gy (Gy = gray). On irradiation with high energy protons (10 MeV) or α-particles (20–40 MeV), the following reactions take place in the Fricke dosimeter.

$$H^{\bullet} + O_2 \rightarrow HO_2^{\bullet} \quad (8.42)$$

$$HO_2^{\bullet} + Fe^{2+} \rightarrow HO_2^{-} + Fe^{3+} \quad (8.43)$$

$$HO_2^{-} + H^{+} \rightarrow H_2O_2 \quad (8.44)$$

$$HO^{\bullet} + Fe^{2+} \rightarrow HO^{-} + Fe^{3+} \quad (8.45)$$

$$H_2O_2 + Fe^{2+} \rightarrow HO^{-} + Fe^{3+} + HO^{\bullet} \quad (8.46)$$

$$H^{\bullet} + H_2O \rightarrow HO^{\bullet} + H_2 \left(\text{in the absence of } O_2\right) \quad (8.47)$$

Considering all the reactions above, it is seen that (i) each $H^{\bullet}$ produces three Fe^{3+} ions, each radiolytic H_2O_2 produces two Fe^{3+} ions and each radiolytic $HO^{\bullet}$ produces one Fe^{3+} ion. Thus in the presence of O_2,

$$G(Fe^{3+}) = 2G(H_2O_2) + 3G(H^{\bullet}) + 1G(HO^{\bullet})$$

The Fricke dosimeter helps in testing other scavenger molecules and also the reactivity of these molecules.

8.8.4 Reactions in the Radiolysis of Water

Ionising radiations produce large number of secondary electrons which rapidly slow down in a thermal step to energies of almost 7.4 eV, which is the threshold energy to produce electronic transitions in water. The slowing down of both primary charged particles and secondary electrons proceeds by a transfer of energy to the medium in a series of steps. Three possible reactions may occur depending on the amount of energy transferred to the electron.

(i) Ionisation (threshold energy in water ≈13 eV).
(ii) Excitation (threshold energy in water 7.4 eV).
(iii) Transfer of thermal energy: When water is irradiated, say, with electrons of 20 keV, the following reaction takes place.

$$H_2O \xrightarrow{\text{rad}} H_2O^{\bullet+} + e^{-} \quad (8.48)$$

In step 8.46, radical ions and sub-excited electrons are produced. Further, energy transfer can produce a water molecule in the excited state.

$$H_2O \xrightarrow{\text{rad}} H_2O^{*} \quad (8.49)$$

These species are produced in a time frame of 10^{-16} s. The three initial species begin to diffuse and react with each other or other molecules in the medium. Some of these reactions produce radicals of high reactivity and they may be neutral or charged. The e^{-} formed in the earlier stages is captured by water to form solvated e_{aq}^{-} according to

$$e^{-} + H_2O \rightarrow e_{aq}^{-} \quad (8.50)$$

and e_{aq}^{-} is surrounded by a cage of water.

e^{-} can also react with H^{+} to form a radical $H^{\bullet}$.

$$e^{-} + H^{+} \rightarrow H^{\bullet} \quad (8.51)$$

The radical ion of water can dissociate to produce a H^+ ion and a hydroxyl radical $OH^\bullet$ given by

$$H_2O^{\bullet+} \rightarrow H^+ + OH^\bullet \qquad (8.52)$$

The excited H_2O^* can dissipate excess energy by bond breakage to produce H and OH radicals, as shown by the equation,

$$H_2O^* \rightarrow OH^\bullet + H^\bullet \qquad (8.53)$$

(OH bond needs 5 eV for bond breaking)

The actual concentrations of the radicals are very small compared to concentrations of ions formed in water. Thus, the three initial species H_2O^*, $H_2O^{\bullet+}$ and e^- react further to give the chemically reactive species $H^\bullet$, $OH^\bullet$ and e^-_{aq}. Both the radicals, $H^\bullet$ and $OH^\bullet$, are highly reactive and further, the $OH^\bullet$ radical is a powerful oxidising agent.

Actually after about 10^{-12} s, the chemically reactive species are still located in the vicinity of $H_2O^{\bullet+}$, e^- and H_2O^*.These and the other $H^\bullet$, $OH^\bullet$ and e^-_{aq} begin to migrate randomly from their initial positions. As this diffusion proceeds, individual pairs may come close together to react with each other. A variety of reactions are thus possible:

$$OH^\bullet + OH^\bullet \rightarrow H_2O_2 \qquad (8.54)$$

$$OH^\bullet + e^-_{aq} \rightarrow OH^- \qquad (8.55)$$

$$OH^\bullet + H^\bullet \rightarrow H_2O \qquad (8.56)$$

$$H^+ + e^-_{aq} \rightarrow H^\bullet \qquad (8.57)$$

$$e^-_{aq} + e^-_{aq} + 2H_2O \rightarrow H_2 + 2OH^- \qquad (8.58)$$

$$e^-_{aq} + H^\bullet + H_2O \rightarrow H_2 + OH^- \qquad (8.59)$$

$$H^\bullet + H^\bullet \rightarrow H_2 \qquad (8.60)$$

Most of these reactions remove chemically reactive species from the system. By ~10^{-6} s, all the reactive species are sufficiently far away so that further reactions do not occur.

Worked Out Problems

Einstein's Relation

Problem 8.1

(a) Calculate the energy of a photon in J for a radiation of 300 nm.

(b) What is the energy in einsteins for a radiation of 400 nm in kJ mol^{-1}? ($h = 6.625\times10^{-34}$ J s, $C = 3\times10^8$ m s^{-1})

Solution

(a) Energy of photon for 300 nm radiation

$$= \frac{hc}{\lambda} = \frac{6.625\times10^{-34}\times3\times10^8}{300\times10^{-9}}\ \frac{\text{J}\times\text{s}\times\text{m}\times\text{s}^*}{\text{m}}$$

$$= 6.625\times10^{-19}\ \text{J}$$

(b) Einsteins for 400 nm radiation

$$= \frac{Nhc}{\lambda} = \frac{6.023\times10^{23}\times6.625\times10^{-34}\times3\times10^8}{400\times10^{-9}}$$

$$= 29.93\times10^4\ \text{J mol}^{-1} = 299.3\ \text{kJ mol}^{-1}$$

Problem 8.2

Assuming that a photochemical reaction requires an activation energy of 85 kJ mol^{-1}, what is its (i) corresponding wavelength, (ii) wave number, and (iii) frequency?

Solution

(i) $E = 85\times1000\ \text{J} = \dfrac{Nhc}{\lambda}$

$$= \frac{6.023\times10^{23}\times6.625\times10^{-34}\times3\times10^8}{\lambda}$$

$$\lambda = \frac{6.023\times10^{23}\times6.625\times10^{-34}\times3\times10^8}{85\times10^3}\ \frac{\text{mol}^{-1}\times\text{J s}\times\text{m s}^{-1}}{\text{J mol}^{-1}}$$

$$= \frac{117\times10^{-3}}{85\times10^3} = 1.408\times10^{-6}\ \text{m} = 1408\ \text{nm}$$

(ii) Wave number $= \bar{\nu} = \dfrac{1}{\lambda} = \dfrac{1}{1.408\times10^{-6}}$

$$= 0.710\times10^6 = 7.1\times10^5\ \text{m}^{-1}$$

(iii) Frequency $= \dfrac{C}{\lambda} = \dfrac{3\times10^8}{1.408\times10^{-6}}\ \dfrac{\text{ms}^{-1}}{\text{m}}$

$$= 2.131\times10^{14}\ \text{s}^{-1}$$

Molar Absorptivity and Absorbance-Concentration Dependence

Problem 8.3

Aqueous $KMnO_4$ solution has a λ_{max} = 525 nm. When light at this wavelength is passed through a 1.6×10^{-4} M solution of $KMnO_4$ using a 1×10^{-2} m cuvette, the original intensity was reduced by 56.75% of the initial value. Calculate (i) the absorbance of the solution, (ii) the molar absorptivity of aqueous $KMnO_4$, (iii) using a cell of 5×10^{-3} m length, calculate the absorbance and percentage transmission of a 2×10^{-4} M $KMnO_4$ at the same wavelength.

Solution

(i) $\frac{I}{I_0} = \frac{56.75}{100} = 0.5675$ or, $\frac{I_0}{I} = 1.762$; $A = \log\frac{I_0}{I} = 0.246$

$$\varepsilon = \frac{A}{cl} = \frac{0.246}{0.01 \times 1.6 \times 10^{-4}}$$
$$= 0.154 \times 10^6 \text{ dm}^3 \text{ mol}^{-1} \text{ m}^{-1}$$
$$= \frac{0.154 \times 10^6 \text{ dm}^3 \text{ mol}^{-1}}{\text{m}}$$
$$= \frac{0.154 \times 10^6}{10^2 \text{ cm}}$$
$$= 1540 \text{ dm}^3 \text{ mol}^{-1} \text{ cm}^{-1}$$

(ii) $A = 1540 \times 0.5 \times 2 \times 10^{-4} = 0.154$; $\log\frac{I_0}{I} = 0.154$

$\frac{I_0}{I} = 1.425$; $\frac{I}{I_0} \times 100 = 70.2\%$

Problem 8.4

(a) Monochromatic light of a certain wavelength was passed through a solution of concentration x using a cell of 0.5 m and 25% of the incident light was absorbed. What is its absorbance? If the concentration of the solution is doubled, find the length of the cell which absorbs 60% of the incident light.

(b) The radiation given out by a monochromatic beam of 500 nm wavelength was completely absorbed by a reaction mixture. If the intensity of the radiation is 100 W, how many moles of photons are absorbed by the medium in 30 minutes? How many photons are absorbed?

Solution

(a) $\frac{I}{I_0} = 0.25$; $\frac{I_0}{I} = \frac{1}{0.25} = 4.0$; $A = \log\frac{I_0}{I} = 0.602$

In the second case,

$$\frac{I}{I_0} = \frac{60}{100} = 0.6$$

$$A = \log\frac{I_0}{I} = \log\left(\frac{1}{0.6}\right) = 0.222$$

Now using the relation, $\frac{A_1}{A_2} = \frac{\varepsilon \times x \times 0.50}{\varepsilon \times 2x \times l}$ and substituting

$$\frac{0.602}{0.222} = \frac{0.5 \times 0.5}{l}$$
$$l = \frac{0.5 \times 0.5 \times 0.222}{0.602}$$
$$= 0.0922 \text{ m} = 9.22 \text{ cm}$$

(b) $E = \frac{Nhc}{\lambda} = \frac{6.023 \times 10^{23} \times 6.625 \times 10^{-34} \times 3 \times 10^8}{500 \times 10^{-9}}$

$$= 239.4 \text{ kJ mol}^{-1}$$

Energy absorbed $= 100 \text{ Js}^{-1} \times 30 \times 60 \text{ s} = 180$ kJ

One mole of photons of $\lambda = 500$ nm has an energy of 239.4 kJ

180 kJ of energy are possessed by $\frac{180}{239.4} = 0.752$ moles of photons

Number of photons $= 0.75 \times 6.023 \times 10^{23} = 4.5 \times 10^{23}$

Quantum Yield and Extent of Reaction

Problem 8.5

(a) Photoirradiation of acetaldehyde at $\lambda = 310$ nm results in the reaction

$$CH_3CHO \xrightarrow{h\nu} CH_4 + CO$$

In a certain experiment, 200 J of light energy was absorbed in a certain interval of time. Given that the quantum yield of the reaction is 0.5, how many moles of CO are formed?

(b) 5×10^{-5} moles of a substance reacted on its exposure for 30 minutes to monochromatic light of a certain wavelength. In this length of time, the substance absorbed 10,000 photons of the light per second. What is the quantum yield of the reaction?

Solution

(a) $E = \frac{Nhc}{\lambda} = \frac{6.023 \times 10^{23} \times 6.625 \times 10^{-34} \times 3 \times 10^8}{310 \times 10^{-9}}$

$$= \frac{119.7 \times 10^{-3}}{3.1 \times 10^{-7}} = 386.1 \text{ kJ mol}^{-1}$$

$$\phi = \frac{\text{Number of moles formed}}{\text{Number of einsteins absorbed}}$$

$$0.5 = \frac{\text{Number of moles formed}}{(0.2/386.1)}$$

Number of moles of CO formed $= \frac{0.5 \times 0.2}{386.1} = 2.59 \times 10^{-4}$

(b) 5×10^{-5} moles $= 6.023 \times 10^{23} \times 5 \times 10^{-5}$

$$= 30.115 \times 10^{18} \text{ molecules}$$

Number of photons absorbed in 30 minutes $= 10000 \times 30 \times 60$

$$= 18 \times 10^6$$

$$\phi = \frac{30.115 \times 10^{18}}{18 \times 10^6} = 1.673 \times 10^{12}$$

Problem 8.6

(a) Cinnamic acid can be brominated photochemically by irradiation with light of 436 nm at room temperature. In a certain experiment lasting 18.4 min, the concentration of bromine decreased by 7.5×10^{-5} mol. Given that the intensity of light is 1.4×10^{-3} W and only 80% of the light passing through is absorbed, calculate the quantum yield of the reaction.

(b) How many moles of dibromocinnamic acid are formed if 1000 J of light energy are completely absorbed at the same wavelength?

Solution

(a) $E = \dfrac{Nhc}{\lambda} = \dfrac{6.023\times10^{23}\times6.625\times10^{-34}\times3\times10^{8}}{436\times10^{-9}}$

$= 2.75\times10^{5}\ \text{J mol}^{-1}$

Number of J absorbed in the given time $= 1.4\times10^{-3}\ \text{Js}^{-1}\times 0.8\times1104\ \text{s} = 1.236$

$$\text{Einsteins absorbed} = \frac{1.236}{2.75\times10^{5}} = 4.5\times10^{-6}$$

$$\phi = \frac{7.5\times10^{-5}}{4.5\times10^{-6}} = 16.7$$

(b) $\text{Number of einsteins absorbed} = \dfrac{1000}{2.75\times10^{5}} = 363.6\times10^{-5}$

$$\phi = 16.7 = \frac{\text{No. of moles of product formed}}{363.6\times10^{-5}}$$

No. of moles of product formed $= 16.7\times363.6\times10^{-5}$

$= 0.06$

Jablonski Diagram and Stern–Volmer Equation

Problem 8.7

(a) Using the Jablonski diagram, state the various ways by which an excited singlet molecule returns to the ground state S_0 or to the triplet state T_1. What is the order of rate constants of the various processes?

(b) What is the radiative lifetime τ_0 of a fluorescent molecule? Why is the observed lifetime τ_s less than τ_0?

(c) Define the quantum yield ϕ_F of fluorescence.

(d) The following table gives the observed lifetimes τ_s of fluorescence of a few systems along with their ϕ_F. Calculate their radiative lifetimes.

S. No	Compound	Solvent	τ_s (ns)	ϕ_F
1	Napthalene	Hexane	106	0.380
2	Anthracene	Benzene	4.0	0.240
3	Eosin	Water	4.7	0.150

Solution

(a) The various ways are (i) internal conversion through a radiationless transition whose rate constant k_{IC} ranges from 10^4 to 10^{12} s^{-1}. (ii) Fluorescence whose rate constant k_F ranges from 10^6 to 10^9 s^{-1}. (iii) Intersystem crossing from S_1 to T_1 whose rate constant k_{ISC} ranges from 10^4 to 10^{12} s^{-1}.

(b) When an excited state (a fluorescent molecule) undergoes a spontaneous emission which is not affected by any other mode of deactivation, the lifetime of the molecule under these conditions is known as *radiative lifetime* τ_0. The observed life time τ_S is always less than τ_0 due to other deactivation processes coming into play.

(c) The quantum yield for fluorescence ϕ_F is defined as the ratio of the rate of fluorescence to the total rate of deactivation of the singlet state S_1.

$$\phi_F = \frac{k_F[S_1]}{(k_{IC}+k_F+k_{ISC})[S_1]} = \frac{k_F}{(k_{IC}+k_F+k_{ISC})}$$

(d) τ_S is related to τ_0 according to $\tau_S = \tau_0\phi_F$

$$\tau_0 \text{ for naphthalene} = \frac{\tau_s}{\phi_F} = \frac{106}{0.380} = 278.9 \text{ ns}$$

$$\tau_0 \text{ for anthracene} = \frac{\tau_s}{\phi_F} = \frac{4.0}{0.24} = 16.7 \text{ ns}$$

$$\tau_0 \text{ for eosin} = \frac{\tau_s}{\phi_F} = \frac{4.7}{0.15} = 31.3 \text{ ns}$$

Problem 8.8

(a) Explain the term phosphorescence.

(b) Using the Jablonski diagram, define the phosphorescence lifetime τ_{T_1}.

(c) Write down the Stern–Volmer equation and explain the terms.

(d) The phosphorescence of an organic ketone in acetonitrile is quenched by a conjugated diene (X). If the quenching reaction is diffusion controlled and has a rate constant of 1×10^{10} dm^3 mol^{-1}s^{-1} and the lifetime of the triplet state is 6.5×10^{-8} s, calculate the ratio ϕ/ϕ_0 at concentrations of 1×10^{-3} M and 2×10^{-3} M of the quencher (ϕ = quantum yield in the presence of the quencher, ϕ_0 = quantum yield in the absence of the quencher).

Solution

(a) Phosphorescence arises from radiative transitions from the lowest vibrational level of the triplet state T_1 to the various vibrational levels of S_0. This is a much slower process compared to fluorescence. The phosphorescence spectrum is observed at longer wavelengths than the fluorescence spectrum.

(b) Phosphorescence lifetimes τ_{T_1}: This term refers to the times associated with the processes by which the triplet state T_1 goes over to the lowest singlet S_0 via radiative transition. It is defined as

$$\tau_{T_1} = \frac{1}{k_P + k'_{ISC}}$$

where k_P and k'_{ISC} are the rate constants for phosphorescence and intersystem crossing respectively.

(c) Stern–Volmer equation:

$$\frac{\phi_P^0}{\phi_P} = \frac{I_P^0}{I_P} = 1 + k_q \tau_{T_1}(q)$$

where ϕ_P^0 and ϕ_p refer to quantum yields for the phosphorescence process in the absence and in the presence of a quencher. I_P^0 and I_P are the corresponding intensities. k_q is the rate constant for the quenching process, τ_{T1} is the lifetime of phosphorescence, (q) is the concentration of the quencher.

(d) $$\frac{\phi}{\phi_0} = \frac{1}{1 + k_q \tau_{T_1}(q)}$$

For $[q] = 1\times10^{-3}$ M,

$$\frac{\phi}{\phi_0} = \frac{1}{1 + 1\times10^{10}\times6.5\times10^{-8}\times1\times10^{-3}}$$

$$= \frac{1}{6.5\times10^{-1}+1} = 0.61$$

For $[q] = 2\times10^{-3}$ M,

$$\frac{\phi}{\phi_0} = \frac{1}{1 + 1\times10^{10}\times6.5\times10^{-8}\times2\times10^{-3}}$$

$$= \frac{1}{13.0\times10^{-1}+1} = 0.435$$

Problem 8.9

(a) Poly(*p*-phenylenethylnylene)s, or PPE show extraordinary fluorescence in solution and their fluorescence is quenched by terbium chloride ($TbCl_3$) and the quenching of their fluorescence follows Stern–Volmer equation in the form

$$\frac{F_0}{F_q} = 1 + K_{SV}[q]$$

where F_0 = fluorescence in absence of a quencher, F_q = fluorescence presence of a quencher, $[q]$ is the concentration of the quencher and K_{SV} is the Stern–Volmer constant. At a concentration of 5×10^{-5} M concentration of PPE, the following data are obtained at various concentrations of $TbCl_3$.

Conc. of quencher (M)	$\frac{F_0}{F_q}$
0.00	1.0
2×10^{-6}	1.20
4×10^{-6}	1.40
6×10^{-6}	1.85

Calculate the Stern–Volmer constant by a suitable graphical plot.

(b) What is the principle of the flash photolysis technique? Give three examples of reactions studied by this method and their mechanism.

Solution

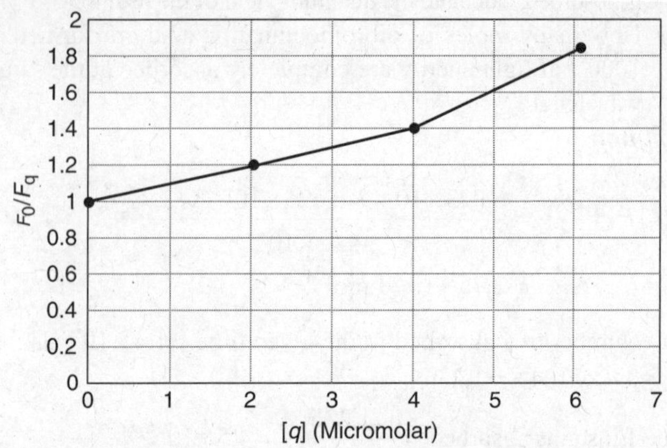

(a) From the linear plot of $\frac{F_0}{F_q}$ vs $[q]$, the Stern–Volmer constant is 1.03×10^{-5} dm^3 mol^{-1}.

(b) Flash photolysis technique involves the production of high intensity of light pulses of short durations of about 10^{-5} s. The light is absorbed by the reacting species kept in a reaction vessel and a second flash is sent through the vessel. The spectrum of the transient species is recorded.

Some reactions studied by this technique:

- The reaction of oxygen atoms and CH_2 radicals with CS_2 has been studied. The corresponding reactions are

  $O + CS_2 \rightarrow CS + OS$

  $CH_2 + CS_2 \rightarrow CS + CH_2S$

  The study of these reactions is facilitated by the stability of CS radicals which have a strong absorption band in the UV region.
- Flash photolysis of air-free solutions containing halide ions OH^- or CNS^- ions resulted in the formation of e_{aq} (solvated electrons). The existence of e_{aq} was proved for the first time by this method.
- Flash photolysis of $Hg(CH_3)_2$, azomethane, CH_3Br showed the formation CH_3 radicals in all cases because the same absorption band corresponding to the CH_3 radical was observed at 215.5 nm.

Calculations Involving Some Photochemical Reactions

Problem 8.10

(a) A mixture of hydrogen and chlorine was irradiated with radiation of wavelength 4000 Å. In a given interval of time, 7×10^{-5} J were absorbed by the reactant chlorine. If ϕ_{HCl} =

2.6×10^6, how many moles of HCl are formed per joule of energy absorbed?

(b) The photo-decomposition of acetone at a wavelength of 3000 Å occurs according to

$(CH_3)_2CO \rightarrow C_2H_6 + CO$

In a particular experiment, the intensity of radiation absorbed is found to be 10 J s^{-1} by a thermopile. 1.5×10^{-3} mol of carbon monoxide was formed at the end of 5 min. Calculate the quantum yield.

Solution

(a) $$E \text{ for } 4000 \text{ Å} = \frac{6.023 \times 10^{23} \times 6.625 \times 10^{-34} \times 3 \times 10^8}{400 \times 10^{-9}}$$

$$= 299.3 \text{ kJ mol}^{-1}$$

$$\text{No. of einsteins absorbed} = \frac{7 \times 10^{-5} \times 10^{-3}}{299.3} = 2.34 \times 10^{-10}$$

$$\phi_{HCl} = 2.6 \times 10^6 = \frac{\text{No. of moles of HCl formed}}{2.34 \times 10^{-10}}$$

$$\text{No. of moles of HCl formed} = 2.34 \times 10^{-10} \times 2.6 \times 10^6$$

$$= 6.084 \times 10^{-4}$$

No. of moles of HCl formed per J of energy

$$= \frac{6.084 \times 10^{-4}}{7 \times 10^{-5}} = 8.7$$

(b) $$E \text{ for } 3000 \text{ Å} = \frac{6.023 \times 10^{23} \times 6.625 \times 10^{-34} \times 3 \times 10^8}{300 \times 10^{-9}}$$

$$= 399 \text{ kJ mol}^{-1}$$

Energy absorbed in 5 min = $10 \times 300 = 3000$ J = 3 kJ

$$\text{No. of einsteins absorbed} = \frac{3}{399} = 7.52 \times 10^{-3}$$

$$\phi = \frac{\text{No. of moles of CO formed}}{\text{No. of einsteins absorbed}} = \frac{1.5 \times 10^{-3}}{7.52 \times 10^{-3}} = 0.2$$

Mechanisms of Photochemical Reactions

Problem 8.11

Acetaldehyde decomposes on absorption of light at wavelengths less than 3000 Å. The following steps are proposed for the mechanism of the reaction.

$$CH_3CHO + h\nu \xrightarrow{I_{abs}} CH_3 + CHO \quad (1)$$

$$CH_3CHO + CH_3 \xrightarrow{k_1} CH_3CO + CH_4 \quad (2)$$

$$CH_3CO \xrightarrow{k_2} CH_3 + CO \quad (3)$$

$$CH_3 + CH_3 \xrightarrow{k_3} C_2H_6 \quad (4)$$

Applying the steady state approximation for the CH_3 and CH_3CO radicals, show that the rate of formation of CO is first order in the concentration of acetaldehyde.

Solution

For the net rate of formation of CH_3CO, we may write

$$\frac{d[CH_3CO]}{dt} = k_1[CH_3][CH_3CHO] - k_2[CH_3CO] = 0 \quad (5)$$

$$[CH_3CO] = \frac{k_1}{k_2}[CH_3][CH_3CHO] \quad (6)$$

For the rate of formation of CH_3

$$\frac{d[CH_3]}{dt} = I_{abs} - k_1[CH_3][CH_3CHO] + k_2[CH_3CO]$$
$$- k_3[CH_3]^2 = 0 \quad (7)$$

From Eq. (6)

$$k_2[CH_3CO] = k_1[CH_3][CH_3CHO] \quad (8)$$

$$[CH_3] = \left(\frac{I_{abs}}{k_3}\right)^{1/2} \quad (9)$$

For the rate of formation of CO we write

$$\frac{d[CO]}{dt} = k_2[CH_3CO] \quad (10)$$

$$= \frac{k_2 k_1}{k_2}[CH_3][CH_3CHO] = k_1\left(\frac{I_{abs}}{k_3}\right)^{1/2}[CH_3CHO] \quad (11)$$

Problem 8.12

Hydrogen bromide decomposes on being irradiated with light of 2530 Å. The following mechanism has been put forward for the reaction.

$$HBr + h\nu \rightarrow H + Br \quad (1)$$

$$H + HBr \rightarrow H_2 + Br \quad (2)$$

$$Br + Br + M \rightarrow Br_2 + M \quad (3)$$

$$H + Br_2 \rightarrow HBr + Br \quad (4)$$

Derive the rate equation for the formation of the products.

Solution

$$\text{Rate of first step} = I_{abs}$$

$$\text{Rate of second step} = k_2[H][HBr]$$

$$\text{Rate of third step} = k_3[Br]^2[M]$$

$$\text{Rate of fourth step} = k_4[H][Br_2]$$

Let us formulate equations for steady state concentrations H and Br. For the net rate of formation of Br and applying the steady state,

$$\frac{d[Br]}{dt} = 0 = I_{abs} + k_4[H][Br_2] + k_2[H][HBr]$$
$$- k_3[Br]^2[M] \quad (5)$$

$$k_3[\mathrm{Br}]^2[\mathrm{M}] = I_{abs} + k_4[\mathrm{H}][\mathrm{Br_2}] + k_2[\mathrm{H}][\mathrm{HBr}]$$

$$[\mathrm{Br}] = \left\{\frac{I_{abs} + k_4[\mathrm{H}][\mathrm{Br_2}] + k_2[\mathrm{H}][\mathrm{HBr}]}{k_3[\mathrm{M}]}\right\}^{\frac{1}{2}} \quad (6)$$

Applying the steady state condition for the concentration of H

$$\frac{d[\mathrm{H}]}{dt} = I_{abs} - k_2[\mathrm{H}][\mathrm{HBr}] - k_4[\mathrm{H}][\mathrm{Br_2}] = 0 \quad (7)$$

$$[\mathrm{H}] = \frac{I_{abs}}{k_2[\mathrm{HBr}] + k_4[\mathrm{Br_2}]} \quad (8)$$

For the formation of Br_2,

$$\frac{d[\mathrm{Br_2}]}{dt} = +k_3[\mathrm{Br}]^2[\mathrm{M}] - k_4[\mathrm{H}][\mathrm{Br_2}]$$

Substituting for H and Br concentration from (6) and (8)

$$\frac{d[\mathrm{Br_2}]}{dt} = +k_3[\mathrm{M}]\left\{\frac{I_{abs} + k_4[\mathrm{H}](k_4[\mathrm{Br_2}] + k_2[\mathrm{HBr}])}{k_3[\mathrm{M}]}\right\}$$

$$-k_4[\mathrm{Br_2}]\left\{\frac{I_{abs}}{k_4[\mathrm{Br_2}] + k_2[\mathrm{HBr}]}\right\}$$

$$= I_{abs} + k_4\frac{I_{abs}}{k_4[\mathrm{Br_2}] + k_2[\mathrm{HBr}]}(k_4[\mathrm{Br_2}] + k_2[\mathrm{HBr}])$$

$$-k_4[Br_2]\left\{\frac{I_{abs}}{k_4[\mathrm{Br_2}] + k_2[\mathrm{HBr}]}\right\}$$

$$= I_{abs} \times k_4\left\{\frac{1}{k_4} + 1 - \frac{[\mathrm{Br_2}]}{k_4[\mathrm{Br_2}] + k_2[\mathrm{HBr}]}\right\}$$

$$= I_{abs}\left\{k_4 + 1 - \frac{k_4[\mathrm{Br_2}]}{k_4[\mathrm{Br_2}] + k_2[\mathrm{HBr}]}\right\}$$

$$= I_{abs}\left\{k_4 + 1 - \frac{k_4[\mathrm{Br_2}]}{k_4[\mathrm{Br_2}] + k_2[\mathrm{HBr}]}\right\}$$

$$= I_{abs}\left\{1 + k_4 - \frac{1}{1 + \dfrac{k_2[\mathrm{HBr}]}{k_4[\mathrm{Br_2}]}}\right\}$$

Problem 8.13

In a study of the decomposition of hydrogen iodide by irradiation with UV light of 2070 Å, it was observed that absorption of 10 J of energy resulted in the decomposition of 3.44×10^{-5} mol of HI. Calculate the number of molecules decomposed by a photon of light of this wavelength. What is the probable mechanism which agrees with this result?

Solution

10 J energy decomposes $= 3.44 \times 10^{-5} \times 6.023 \times 10^{23}$

$= 20.7 \times 10^{18}$ molecules of HI

A photon of the given wave has an energy

$$\frac{6.625 \times 10^{-34} \times 3 \times 10^8}{207 \times 10^{-9}}\,\mathrm{J} = 0.096 \times 10^{-17}$$

$$= 9.6 \times 10^{-19}\ \mathrm{J}$$

Number of photons having 10 J energy $= \dfrac{10}{9.6 \times 10^{-19}}$

$$= 1.04 \times 10^{19}$$

1.04×10^{19} photons decompose 20.7×10^{18} molecules of HI

1 photon decomposes $\dfrac{20.7 \times 10^{18}}{1.04 \times 10^{19}} \approx 2.0$

One photon decomposes 2 HI molecules.

Mechanism: The possible mechanism,

$$\mathrm{HI} \xrightarrow{h\nu} \mathrm{H} + \mathrm{I}$$

$$\mathrm{H} + \mathrm{HI} \xrightarrow{k_2} \mathrm{H_2} + \mathrm{I}$$

$$\mathrm{I} + \mathrm{I} \xrightarrow{k_3} \mathrm{I_2}$$

or,

$$\mathrm{HI} \xrightarrow{h\nu} \mathrm{HI}^*$$

$$\mathrm{HI}^* + \mathrm{HI} \rightarrow \mathrm{H_2} + \mathrm{I_2}$$

Problem 8.14

Irradiation of hydrogen peroxide with light of wavelength of 360 nm in the presence of carbon monoxide results in the formation of CO_2 and water. The following mechanism is proposed for this reaction.

$$\mathrm{H_2O_2} + h\nu \rightarrow 2\mathrm{OH}$$

$$\mathrm{CO} + \mathrm{OH} \xrightarrow{k_1} \mathrm{COOH}$$

$$\mathrm{COOH} + \mathrm{H_2O_2} \xrightarrow{k_2} \mathrm{CO_2} + \mathrm{H_2O} + \mathrm{OH}$$

$$2\mathrm{COOH} \xrightarrow{k_3} \mathrm{Products}$$

Using the steady state approximation for OH and COOH radicals, derive the rate law for the disappearance of H_2O_2.

Solution

For the steady state concentration of OH, we have

$$\frac{d[\mathrm{OH}]}{dt} = 0 = \phi I_{abs} - k_1[\mathrm{OH}][\mathrm{CO}] + k_2[\mathrm{COOH}][\mathrm{H_2O_2}] \quad (1)$$

$$[\mathrm{OH}] = \frac{\phi I_{abs} + k_2[\mathrm{COOH}][\mathrm{H_2O_2}]}{k_1[\mathrm{CO}]} \quad (2)$$

For the steady state concentration of COOH, we have

$$\frac{d[\mathrm{COOH}]}{dt} = 0 = k_1[\mathrm{OH}][\mathrm{CO}] - k_2[\mathrm{COOH}][\mathrm{H_2O_2}] - k_3[\mathrm{COOH}]^2 \quad (3)$$

Substituting for [OH] from (2) in (3), we get

$$\frac{d[\mathrm{COOH}]}{dt} = 0 = \frac{k_1[\mathrm{CO}]\{\phi I_{abs} + k_2[\mathrm{COOH}][\mathrm{H_2O_2}]\}}{k_1[\mathrm{CO}]} - k_2[\mathrm{COOH}][\mathrm{H_2O_2}] - k_3[\mathrm{COOH}]^2 \quad (4)$$

$$= \phi I_{abs} - k_3[\mathrm{COOH}]^2$$

$$[\mathrm{COOH}] = \left(\frac{\phi I_{abs}}{k_3}\right)^{1/2} \quad (5)$$

$$-\frac{d[\mathrm{H_2O_2}]}{dt} = \phi I_{abs} + k_2[\mathrm{COOH}][\mathrm{H_2O_2}] \quad (6)$$

$$= \phi I_{abs} + k_2[\mathrm{H_2O_2}]\left(\frac{\phi I_{abs}}{k_3}\right)^{\frac{1}{2}}$$

$$= \phi I_{abs} + k_2\left(\frac{\phi I_{abs}}{k_3}\right)^{1/2}[\mathrm{H_2O_2}] \quad (7)$$

Problem 8.15

Phosgene is oxidised photochemically with Cl_2 as a photosensitiser and the reaction is

$$2\mathrm{COCl_2} + \mathrm{O_2} \rightarrow 2\mathrm{CO_2}(g) + 2\mathrm{Cl_2}$$

The mechanism suggested involves the formation of Cl, ClO and COCl radicals as shown below:

$$\mathrm{COCl_2} + h\nu \rightarrow \mathrm{COCl} + \mathrm{Cl} \quad I_{abs} \quad (1)$$

$$\mathrm{COCl} + \mathrm{O_2} \xrightarrow{k_1} \mathrm{CO_2} + \mathrm{ClO} \quad (2)$$

$$\mathrm{COCl_2} + \mathrm{ClO} \xrightarrow{k_2} \mathrm{CO_2} + \mathrm{Cl_2} + \mathrm{Cl} \quad (3)$$

$$\mathrm{COCl} + \mathrm{Cl_2} \xrightarrow{k_3} \mathrm{COCl_2} + \mathrm{Cl} \quad (4)$$

$$\mathrm{Cl} + \mathrm{Cl} \xrightarrow{k_4} \mathrm{Cl_2} \quad (5)$$

Derive the rate law making suitable assumptions regarding the free radicals formed in the reaction.

Solution

$$-\frac{d[\mathrm{COCl_2}]}{dt} = I_{abs} + k_2[\mathrm{COCl_2}][\mathrm{ClO}] \quad (6)$$

$$\frac{d[\mathrm{COCl}]}{dt} = 0 = I_{abs} - k_1[\mathrm{COCl}][\mathrm{O_2}] - k_3[\mathrm{COCl}][\mathrm{Cl_2}]$$

$$[\mathrm{COCl}] = \frac{I_{abs}}{k_1[\mathrm{O_2}] + k_3[\mathrm{Cl_2}]} \quad (7)$$

$$\frac{d[\mathrm{ClO}]}{dt} = 0 = k_1[\mathrm{COCl}][\mathrm{O_2}] - k_2[\mathrm{COCl_2}][\mathrm{ClO}]$$

$$[\mathrm{ClO}] = \frac{k_1[\mathrm{COCl}][\mathrm{O_2}]}{k_2[\mathrm{COCl_2}]}$$

$$= \frac{k_1[\mathrm{O_2}]}{k_2[\mathrm{COCl_2}]} \times \frac{I_{abs}}{k_1[\mathrm{O_2}] + k_3[\mathrm{Cl_2}]} \quad (8)$$

$$\frac{d[\mathrm{CO_2}]}{dt} = k_1[\mathrm{COCl}][\mathrm{O_2}] + k_2[\mathrm{COCl_2}][\mathrm{ClO}] \quad (9)$$

Substituting for [COCl] and [ClO] from Eqs (7) and (8),

$$\frac{d[\mathrm{CO_2}]}{dt} = k_1[\mathrm{O_2}] \times \frac{I_{abs}}{k_1[\mathrm{O_2}] + k_3[\mathrm{Cl_2}]} + k_2[\mathrm{COCl_2}]\frac{k_1[\mathrm{O_2}]}{k_2[\mathrm{COCl_2}]}\,\frac{I_{abs}}{k_1[\mathrm{O_2}] + k_3[\mathrm{Cl_2}]} \quad (10)$$

$$\frac{d[\mathrm{CO_2}]}{dt} = k_1[\mathrm{O_2}] \times \frac{I_{abs}}{k_1[\mathrm{O_2}] + k_3[\mathrm{Cl_2}]} + k_1[\mathrm{O_2}]\frac{I_{abs}}{k_1[\mathrm{O_2}] + k_3[\mathrm{Cl_2}]} \quad (11)$$

$$= \frac{2I_{abs}}{1 + \dfrac{k_3[\mathrm{Cl_2}]}{k_1[\mathrm{O_2}]}}$$

Problem 8.16

(a) Explain the phenomenon of photostationary state. Give two examples to illustrate the same.

(b) Considering that the dimerisation of anthracene (denoted as X) occurs according to

$$2\mathrm{X} \rightarrow \mathrm{X_2}$$

and, that the mechanism of the reaction is,

$$\mathrm{X} + h\nu \rightarrow \mathrm{X}^* I_{abs} \quad (1)$$

$$\mathrm{X}^* + \mathrm{X} \xrightarrow{k_1} \mathrm{X_2} \quad (2)$$

$$\mathrm{X}^* \xrightarrow{k_2} \mathrm{X} + h\nu' \text{ (flourescence reaction)} \quad (3)$$

$$\mathrm{X_2} \xrightarrow{k_3} 2\mathrm{X} \quad \text{(reverse of step 1)} \quad (4)$$

derive an equation for the stationary state concentration of the dimer.

Solution

(a) Consider a chemical system in equilibrium and exposed to light of a certain wavelength which is absorbed by one of the reactants. The absorption of light by the reactant alters the rate of the forward reaction. The rate of the reverse reaction is not affected by light. Thus the equilibrium is disturbed. As a result of an increase in the concentration of products, the rate of the reverse reaction increases. After a certain time, the rates of the forward and reverse reactions balance each other with the concentration of the products being higher than in the equilibrium state. This state is referred to as the photostationary state. It is not a true equilibrium state. As an example, consider the equilibrium associated with the reaction between NO(*g*), $NO_2(g)$ and $O_3(g)$ in the lower region of earth's atmosphere. The following reactions occur in presence of light

$$NO_2(g) + h\nu \rightarrow NO(g) + O(g)$$

$$O(g) + O_2 + M \rightarrow O_3(g) + M$$

$$O_3(g) + NO(g) \rightarrow NO_2(g) + O_2(g)$$

All the above three reactions are rapid and as NO(*g*) accumulates in the first step, the rate of the third step also increases, thus regenerating $NO_2(g)$. Another example is the dimerisation of anthracene brought about by the irradiation of its solution in a suitable solvent with ultraviolet light.

$$2\text{Anthracene (X)} + h\nu \rightarrow \text{dianthracene } (X_2)$$

(b) For the net rate of formation of the dimer X_2,

$$\frac{d[X_2]}{dt} = k_1[X^*][X] - k_3[X_2] \tag{5}$$

For the steady state concentration of X^*

$$\frac{d[X^*]}{dt} = I_{abs} - k_1[X^*][X] - k_2[X^*] \tag{6}$$

Applying the steady state condition, $\frac{d[X^*]}{dt} = 0$, we have

$$0 = I_{abs} - k_1[X^*][X] - k_2[X^*] \tag{7}$$

$$[X^*] = \frac{I_{abs}}{k_1[X] + k_2} \tag{8}$$

In the photostationary state, we also have

$$\frac{d[X_2]}{dt} = 0$$

Thus, Eq. (5), $[X_2] = \frac{k_1[X^*][X]}{k_3}$ (9)

Using Eq. (8) for $[X^*]$ in Eq. (9),

$$[X_2] = \frac{k_1[X]}{k_3} \cdot \frac{I_{abs}}{\{k_1[X] + k_2\}} \tag{10}$$

$$= \frac{I_{abs}}{k_3\left\{1 + \frac{k_2}{k_1[X]}\right\}} \tag{11}$$

Under conditions when the monomer (anthracene) concentration is high, $\frac{k_2}{k_1[X]}$ in Eq. (11) is negligible, which gives,

$$[X_2] = \frac{I_{abs}}{k_3} \tag{12}$$

From Eq. (12), note that $[X_2]$ is independent of the concentration of X unlike that observed in normal kinetic studies.

Problem 8.17

(a) Erythrocin absorbs light of wavelength around 500 nm and is excited from S_0 to excited vibrational levels of S_1. Some molecules in S_1 undergo ISC to triplet state T_1. Some molecules in T_1 undergo phosphorescence (RTP = room temperature phosphorescence) while some other molecules 'back ISC' to the S_1 state and by a process of delayed fluorescence return to the S_0 state. Draw a Jablonski diagram illustrating all these processes.

(b) By employing suitable kinetic analysis, derive an equation for the ratio of intensities of delayed fluorescence and RTP.

Solution

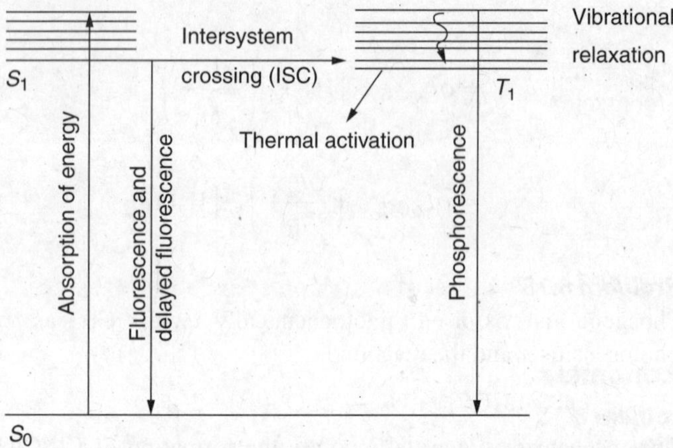

Schematic depiction of Jablonski diagram

The absorption of light energy by the singlet ground state S_0 leads to excited vibrational levels of the first excited singlet S_1. Ordinary fluorescence relates to the emission of a photon as the system returns rapidly from S_1 to S_0. However, some molecules make a transition from S_1 to the excited triplet state T_1 which is referred to as intersystem crossing (ISC). T_1 is initially formed in an excited vibrational level that rapidly relaxes. Phosphorescence occurs as T_1 returns to S_0 which is a slow process. Some molecules in the T_1 state, instead of undergoing phosphorescence, acquire enough vibrational energy and undergo back intersystem crossing to the S_1 state. They then undergo fluorescence which is known as delayed fluorescence. Delayed fluorescence occurs at the same wavelength as fluorescence.

Kinetic process: Fluorescence is a much faster process than DF and room temperature phosphorescence (RTP).

Let $[T_1]_0$ = initial concentration of molecules in the T_1 state. Triplet state molecules lose their energy by three processes—phosphorescence, triplet to singlet via ISC, and quenching. Referring to the rate constants of the three processes as k_P, k_{TS} and k_{Tq}, we can write

$$\frac{d[T_1]}{dt} = -k_P[T_1] - k_{TS}[T_1] - k_{Tq}[T_1] \quad (1)$$

Equation (1) can be solved for $[T_1]$ as

$$[T_1] = [T_1]_0 e^{-t/\tau_{RTP}} \quad (2)$$

where $$\frac{1}{\tau_{RTP}} = k_P + k_{TS} + k_{Tq}$$

τ_{RTP} is the lifetime for RTP which is directly measurable.

The net rate of formation of the S_1 state is

$$\frac{d[S_1]}{dt} = k_{TS}[T_1] - k_F[S_1] \quad (3)$$

where k_F is the rate constant for the fluorescence process. Since k_F is very large, $\frac{d[S_1]}{dt} \approx 0$ at all times. Because of that, quenching of the excited singlet is neglected

From Eq. (3),

$$[S_1] = \left(\frac{k_{TS}}{k_F}\right)[T_1] \quad (4)$$

Let i_{DF} = intensity of delayed fluorescence and i_{RTP} = intensity of room temperature phosphorescence.

$$i_{DF} = k_{TS}[S_1] = k_F[T_1] \quad (5)$$

$$i_{RTP} = k_P[T_1] \quad (6)$$

From Eqs (5) and (6), $$\frac{i_{DF}}{i_{RTP}} = \frac{k_{TS}}{k_P} \quad (7)$$

Actinometers

Problem 8.18

(a) In determining the quantum yield of Reinecke's salt ($NH_4[Cr(NH_3)_2(SCN)_4]$) actinometer, a 0.005 M solution of the salt buffered at pH 5.4 was taken and irradiated with light of 392 nm light for a certain interval of time and an energy of 4270 J was absorbed. 1 mL of the irradiated solution was diluted to 100 mL and treated with 0.1 M $Fe(NO_3)_3$ in $HClO_4$ medium. The red colour of the Fe^{3+}–SCN^- complex formed was measured at 450 nm in a 1 cm cell and the absorbance was found to be 0.194. The reaction of Reinecke's salt on irradiation is

$$\left[Cr(NH_3)_2(SCN)_4\right]^- + H_2O \xrightarrow{h\nu} \begin{bmatrix} Cr(NH_3)_2 \\ (SCN)_3 \\ (H_2O) \end{bmatrix} + SCN^-$$

Given that the molar absorptivity of the iron–thiocyanate complex is 4300 dm^3 mol^{-1} cm^{-1}, calculate the quantum yield of the actinometer.

(b) The theoretical reversible EMF for a H_2–O_2 fuel cell is 1.23 V at 298 K. What is the corresponding wavelength of light?

Solution

(a) $$E = \frac{Nhc}{\lambda} = \frac{119.7\times10^{-3}}{392\times10^{-9}} = 305 \text{ kJ mol}^{-1}$$

$$\text{Number of einsteins absorbed} = \frac{4270}{305\times10^3} = 14\times10^{-3}$$

Absorbance = 0.194 = $4300 \times C \times 1$ or, $C = 4.512\times10^{-5}$

The concentration of the original solution $= 4.512\times10^{-5}\times10^2 = 4.512\times10^{-3}$

$$\phi = \frac{4.512\times10^{-3}}{14\times10^{-3}} = 0.32$$

(b) Considering the reaction to be a two e^- process

$$2H^+ + 2e^- = H_2 \quad E^0 = 0.00 \text{ V} \quad (1)$$

$$\frac{1}{2}O_2(g) + 2H^+ + 2e^- = H_2O \quad E^0 = 1.23 \text{ V} \quad (2)$$

$$(2)–(1)\ \frac{1}{2}O_2(g) + 2H^+ = H_2O(g) \quad E^0 = 1.23 \text{ V}$$

Using the relation $E = \frac{hC}{\lambda}$

Gibbs energy for the reaction = $2\times96500\times1.23$
$= 2.374\times10^5$ J
$= 237.4$ kJ mol^{-1}

$$\lambda = \frac{6.625\times10^{-34}\times3\times10^8}{\dfrac{2.374\times10^5}{6.023\times10^{23}}}$$

$$= 8.372\times10^{-31}\times6.023\times10^{23} = 50.42\times10^{-8} \text{ m}$$

$$= 50.48\times10^{-6} \text{ cm} = 5048 \text{ Å}$$

Problem 8.19

In determining the quantum yield of the potassium ferrioxalate actinometer, the ferric ion gets reduced to Fe^{2+} ion on excitation with Hg light of wavelength 366 nm according to the reaction

$$Fe(C_2O_4)_3^{3-} + h\nu \rightarrow Fe^{2+} + C_2O_4^{-\bullet} + 2C_2O_4^{2-}$$

$$Fe(C_2O_4)_3^{3-} + C_2O_4^{-\bullet} \xrightarrow{\Delta} Fe^{2+} + 2CO_2 + 3C_2O_4^{2-}$$

The ferrous oxalate formed is complexed with 1,10-phenanthroline and the absorbance of the red coloured complex was measured at various intervals of time.

(i) Define the quantum yield of the actinometer, ϕ_{act}, if I_0 is the intensity of the radiation source and the rate of change of the concentration of the red complex with time is $n[Fe(phen)_3]^{II}/dt$.

(ii) Express the rate of variation of the complex, given above, in terms of the variation of absorbance with time.

(iii) Given the rate of increase of absorbance with time as 0.32 min^{-1} volume of solution in the experiment as 2.5×10^{-3} dm^3, molar absorptivity of the complex at 510 nm = 11,100 dm^3 mol^{-1} cm^{-1} and optical path length of the cell = 1 cm and $I_0 = 5.6\times10^{-8}$ einsteins min^{-1} cm^{-2}, calculate ϕ_{act}.

Solution

(i) $$\phi_{act} = \frac{n\left[\text{Fe}(\text{Phen})_3\right]^{II}/dt}{I_0}$$

(ii) $$\frac{n\left[\text{Fe}(\text{Phen})_3\right]^{II}}{dt} = \frac{d}{dt}(\text{abs}_{510})\times\frac{V}{\varepsilon_{510}\times l}$$

where V = volume of solution irradiated in dm^3, ε_{510} = molar absorptivity at 510 nm of [Fe(phen)$_3$]^{II} in M^{-1} cm^{-1} and l = path length of cell in cm.

(iii) $$\frac{d}{dt}(\text{abs}_{510}) = 0.32\ \text{min}^{-1}$$

$$V = 2.5\times10^{-3}\ \text{dm}^3$$

$$\varepsilon_{510} = 11{,}100\ \text{M}^{-1}\ \text{cm}^{-1}$$

$$l = 1\ \text{cm}$$

Substituting the data,

$$\frac{n\left[\text{Fe}(\text{Phen})_3\right]^{II}}{dt} = \frac{0.32\ \text{min}^{-1}\times0.0025\ \text{dm}^3}{11{,}100\ \text{dm}^3\ \text{mol}^{-1}\times1\ \text{cm}}$$

$$= 2.883\times10^{-5}\times2.5\times10^{-3}$$

$$= 7.2\times10^{-8}\ \text{mol min}^{-1}$$

$$\phi = \frac{7.2\times10^{-8}\ \text{mol min}^{-1}}{5.6\times10^{-8}\ \text{einstein min}^{-1}\ \text{cm}^2}$$

$$= 1.286\ \text{mol einstein}^{-1}\ \text{cm}^{-2}$$

G Values in Radiolysis Reactions and Fricke Dosimeter

Problem 8.20

(a) Ethylene absorbs radiation of 254 nm wavelength and decomposes to give

$$C_2H_4 \xrightarrow{h\nu} C_2H_2 + H_2$$

If the quantum yield of the reaction is equal to unity and assuming that all the light is absorbed, calculate the amount of acetylene produced in 2 hours for an absorption of 1 kJ of energy in this time.

(b) Give the SI definition of G value in radiolysis reactions. Obtain its value in molecules (eV)$^{-1}$.

(c) State the constituents of the Fricke dosimeter. What is the sequence of reactions occurring on exposure of the solution in this dosimeter to high energy α particles of 30 MeV.

Solution

(a) E for 254 nm wavelength $$= \frac{NhC}{\lambda} = \frac{119.7\times10^{-3}}{254\times10^{-9}}$$

$$= 471.2\ \text{kJ mol}^{-1}$$

$$\text{No. of einsteins absorbed} = \frac{1}{471.2} = 2.12\times10^{-3}$$

$$\phi = 1 = \frac{\text{No. of moles of } C_2H_2 \text{ produced}}{\text{No. of einstein absorbed}}$$

$$\text{No. of moles of } C_2H_2 \text{ produced} = 1\times2.122\times10^{-3}$$

$$= 2.122\times10^{-3}\ \text{mol}$$

(b) SI definition of G $$= \frac{\text{No. of molecules of product produced}}{\text{Input of 1 J of energy}}$$

$$= \frac{6.023\times10^{23}}{\left[1/1.609\times10^{-19}\right]\text{eV J}^{-1}\ \text{J}}$$

$$= 9.65\times10^4\ \text{molecule}(\text{eV})^{-1}$$

$$= 9.65\times10^6\ \text{molecule}(100\ \text{eV})^{-1}$$

(c) The Fricke dosimeter is used for measuring the dosage of high energy radiation. A solution of 0.001 M solution of ferrous ammonium sulphate in 0.8 M H_2SO_4 is used in this dosimeter. Reactions occurring in the dosimeter on subjecting to 30 MeV energy α-particles:

$$H^\bullet + O_2 \rightarrow HO_2^\bullet \quad (1)$$

$$HO_2^\bullet + Fe^{2+} \rightarrow HO_2^- + Fe^{3+} \quad (2)$$

$$HO_2^- + H^+ \rightarrow H_2O_2 \quad (3)$$

$$HO^\bullet + Fe^{2+} \rightarrow HO^- + Fe^{3+} \quad (4)$$

$$H_2O_2 + Fe^{2+} \rightarrow HO^- + Fe^{3+} + HO^\bullet \quad (5)$$

$$H^\bullet + H_2O \rightarrow HO^\bullet + H_2 \quad (\text{in the absence of } O_2)$$

Problem 8.21

(a) Define G-value in radiochemical reactions and state its significance. What are the various ways in which it is expressed?

(b) Give the equation which is used for calculating the same in a given radiochemical reaction and explain the terms.

(c) Calculate G values of the following species from the given data.

Species	λ_{max} (nm)	Molar absorptivity (M^{-1} cm^{-1})	Optical density using 0.5 cm cell	Dosage used in the pulse (Gy)
Solvated e_{aq}^-	730	18,960	0.125	35.8
OH·	263	460	0.002	18.9

Solution

(a) G-value in radiochemical reactions is defined as

$$G = \frac{\text{No. of molecules of product formed}}{\text{Input of 100 eV of energy}}$$

A more recent definition by SI is given as

$$G = \frac{\text{No. of molecules formed}}{\text{Input of joules of energy}}$$

G signifies a quantitative measure of the yield of products in radiochemical reactions.

(b) The absorbance of the solution A containing various species on pulse radiolysis is related to the G value through the relation,

$A = \varepsilon lC = \varepsilon l\rho DG$

where ε = molar absorptivity of the solution in $M^{-1}\ cm^{-1}$, l = path length of the cell in cm, D = dosage of the pulse absorbed by solution in Gy (gray) which has the units $J\ kg^{-1}$, ρ = density of solution in the case of radiolysis of water in $kg\ L^{-1}$ and the G value is in the units of $mol\ J^{-1}$.

(c) $G = \dfrac{A}{\varepsilon l\rho D}$

For the solvated electron

$$e^-_{eq}, G = \frac{0.125}{18960\times 0.5\times 0.998\times 35.8}$$

$$\left(M^{-1}\ cm^{-1}\times cm\times kg\ L^{-1}\times J\ kg^{-1}\right)$$

$$G_{e^-_{eq}} = \frac{0.125}{338705.2} = 3.69\times 10^{-7}\ mol\ J^{-1}$$

$$\left(M^{-1}\times L^{-1}\times J\times L\right)$$

Since $1.036\times 10^{-7}\ mol\ J^{-1} = 1\ molecule\,(100\ eV)^{-1}$

$$3.69\times 10^{-7}\ mol\ J^{-1} \equiv \frac{3.69\times 10^{-7}}{1.036\times 10^{-7}}$$

$$= 3.56\ molecules\,(100\ eV)^{-1}$$

[To show $1.036\times 10^{-7}\ mol\ J^{-1} \equiv 1\ molecule\,(100\ eV)^{-1}$

$1.036\times 10^{-7}\times 6.023\times 10^{23} = 6.24\times 10^{16}\ molecules\ J^{-1}$

$1\ eV = 1.602\times 10^{-19}\ J$ or, $1 \equiv 1.602\times 10^{-19}\ J\ eV^{-1}$

$$6.24\times 10^{16}\ molecules\ J^{-1} = 6.24\times 10^{16}\ molecules\ J^{-1}\times 1.602\times 10^{-19}\ J\ eV^{-1}$$

$$= 9.996\times 10^{-3}\ molecules\ eV^{-1}$$

$$= 9.996\times 10^{-3}\times 10^{2}\ molecules\,(100\ eV)^{-1}$$

$$= 9.996\times 10^{-1} \approx 1\ molecule\,(100\ eV)^{-1}]$$

(d) $G(OH.) = \dfrac{0.002}{460\times 0.5\times 0.998\times 18.9}$

$$= 4.61\times 10^{-7}\ mol\ J^{-1}$$

$$4.61\times 10^{-7}\ mol\ J^{-1} = \frac{4.61\times 10^{-7}}{1.036\times 10^{-7}}$$

$$= 4.45\ molecules\,(100\ eV)^{-1}$$

Photoelectrochemical and Photogalvanic Cells

Problem 8.22

(a) A solution absorbs radiation of 450 nm, the rate of absorbing being 5 W. How much does this correspond to in terms of $mol\ s^{-1}$?

(b) What is the difference between a photoelectrochemical cell and a photogalvanic cell? Explain the basic principle involved in the iron–thionine system. What is the Gibbs energy change associated with the reaction?

Solution

(a) $E = \dfrac{NhC}{\lambda} = \dfrac{119.7\times 10^{-3}}{450\times 10^{-9}} = 266\ kJ\ mol^{-1}$

Einsteins absorbed per $s = \dfrac{5}{266\times 10^{3}} = 0.0188\ s^{-1}$

Assuming $\phi = 1$, the number of moles of substance formed per $s = 1.88\times 10^{-2}$.

(b) In a photoelectrochemical cell, the photoanode is a semiconductor like TiO_2 or Fe_2O_3 while the cathode is a metal like Pt. The semiconductor electrode converts the incident photons to electron–hole pairs. The light generated electrons travel towards the cathode via the external wire.

In a photogalvanic cell, which is like a normal galvanic cell, irradiation induces a homogeneous photochemical reaction which gives rise to the current. In the iron–thionine photogalvanic cell, there are two redox couples Fe^{2+}–Fe^{3+} and thionine–leucothionine. The reaction in the cell is

$$\frac{1}{2}Th + Fe^{2+} \underset{dark}{\overset{hv}{\rightleftharpoons}} \frac{1}{2}leu - Th + Fe^{3+}$$

While Th is purple in colour, leucothionine is colourless. On irradiation with light of $\lambda = 478$ nm, the equilibrium is shifted to the right. The EMF generated is about 0.2 V.

$$\Delta G = -nFE = -1\times 96500\times 0.2$$

$$= 1.93\times 10^{4}\ J\ mol^{-1}$$

Chemiluminescence and Thermoluminescence

Problem 8.23

(a) Vitamin D is necessary to prevent rickets in children. Assuming that a child requires about $1\times 10^{-5}\ J\ day^{-1}$ to prevent this disease and UV light of 300 nm is used, how many quanta does it correspond to? Given the molar mass of vitamin D as 396 g mol^{-1}, how many milligrams of this vitamin are required per day for prevention of rickets in children (the data are purely hypothetical)?

(b) Explain the terms chemiluminescence and thermoluminescence with examples.

Solution

(a) $\varepsilon = hv = \dfrac{hC}{\lambda} = \dfrac{6.625\times 10^{-34}\times 3\times 10^{8}}{300\times 10^{-9}}$

$$= 6.625\times 10^{-19}\ J$$

No. of quanta in $1\times 10^{-5}\ J = \dfrac{1\times 10^{-5}}{6.625\times 10^{-19}}$

$$= 0.151\times 10^{14} = 1.51\times 10^{13}$$

$$\text{Value of } E \text{ for } 300\,\text{nm} = \frac{6.625\times10^{-34}\times3\times10^{8}\times6.032\times10^{23}}{300\times10^{-9}}$$

$$= \frac{119.7\times10^{-3}}{300\times10^{-9}}$$

$$= 39.9\times10^{4} = 399\ \text{kJ mol}^{-1}$$

$$\text{No. of einsteins required} = \frac{1\times10^{-5}\times10^{-3}}{399.0} = 2.51\times10^{-11}$$

If 396×10^{3} mg of vitamin D require 1 einstein

2.51×10^{-11} einstein of energy produce $2.51\times10^{-11}\times396\times10^{3}\approx994\times10^{-8}$ mg

Roughly, $1000\times10^{-8}=10^{-5}$ mg of vitamin D are necessary to prevent rickets.

(b) Chemiluminescence: It is the phenomenon of emission of light that accompanies a chemical reaction. For example, let a reactant X in its ground state be converted to an excited state product Y* and emit light in the process. The whole reaction may be represented as

$$X \rightarrow Y^{*} \rightarrow Y + h\nu$$

The first step may be accompanied by a decrease of free energy. Note that it is actually the reverse of a photochemical reaction. The glow of yellow phosphorous and the glow of luciferin in fireflies are two common examples. Magnesium *p*-bromophenylbromide dissolved in ether, on oxidation gives rise to chemiluminescence.

Thermoluminescence: Certain crystals like alkali halide crystals, on exposure to x-rays release electrons. The crystals get coloured due to absorption of light by the electrons. The electrons are trapped in negative ion vacancies of the crystals. On heating, the electrons are released and go to lower energy levels giving off light. This phenomenon is known as thermoluminescence. Some minerals like limestone and flourites exhibit this phenomenon even without exposure to radiation.

Reactions of Solvated Electrons

Problem 8.24

(a)

Time (ms)	Concentration × 10^5 (mol dm^{-3})
0.025	4.67
0.050	4.0
0.100	3.13
0.200	2.2
0.300	1.67

Solvated electrons e_{aq}^{-} formed in pulse radiolysis experiments are highly reactive species and react with several molecules. The reaction with cystiene anion is one of them and is given by

$$\text{cys}^{-} + e_{aq}^{-} \rightarrow \text{cys}^{2-}$$

The reaction is followed by studying the absorbance of e_{aq}^{-} at 800 nm spectrophotometrically and the data are obtained under pseudo first order conditions (at high concentration of the anion) at various times are given above.

Determine the pseudo first order rate constant by a suitable graphical plot.

(b) The triplet state decay of tris-(8-hydroxy quinoline) aluminium (III) designated as $^{3}\text{Alq}_3{}^{*}$ was studied by pulse radiolysis of argon saturated solutions of a mixture of $^{3}\text{Alq}_3{}^{*}$ (conc $= 2.25\times10^{-4}$ M) and biphenyl (conc $= 10^{-2}$ M). The triplet state of biphenyl formed decays rapidly but the triplet state of $^{3}\text{Alq}_3$ is formed according to

$$3\text{Biphenyl}^{*} + \text{Alq}_3 \rightarrow \text{Biphenyl} + 3\text{Al}_{q3}^{*}$$

and 3Al_{q3}^{*} decays also according to first order kinetics. The decay is followed by studying the absorption at its maximum, $\lambda_{max} = 510$ nm. The optical density data at various times are given below

Time (ns)	Optical density
0.414	0.700
0.838	0.650
1.281	0.600
2.093	0.520
3.560	0.400
4.994	0.310
6.920	0.220

Calculate the pseudo first order rate constant in the units s^{-1}. If the initial absorbance is 0.754 and its $\varepsilon = 3350$ M^{-1} cm^{-1}, what is the conc. of 3Al_{q3}^{*} after 3 ns?

Solution

(a)

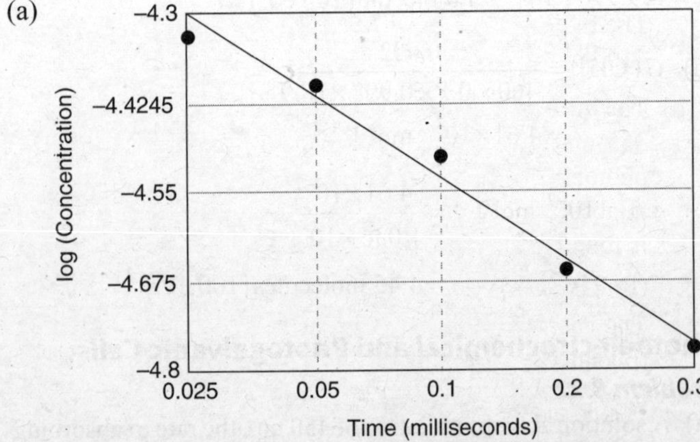

A plot of log $(a - x)$ vs time is found to be linear with an average slope equal to $-k/2.303 = -1600$ or $k = 3685$ s^{-1}.

(b) A plot of log OD vs time is made from the data shown below.

Time (ns)	0.414	0.838	1.281	2.093	3.560	4.994	6.920
Log OD	–0.155	–0.187	–0.222	–0.284	–0.398	–0.509	–0.657

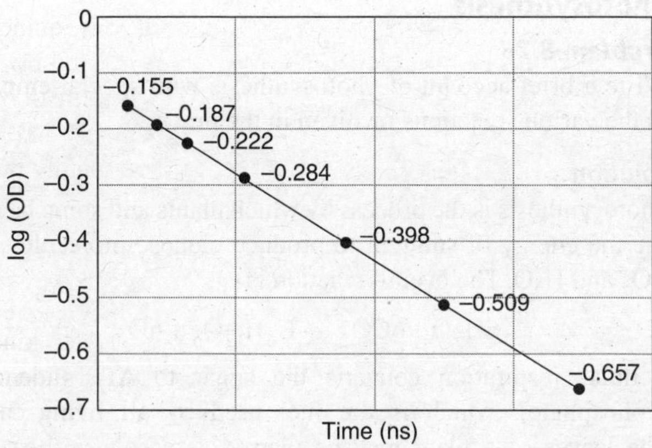

The plot is found to be linear.

$$\text{Slope} - \frac{k}{2.303} = -8.2\times10^{8} \text{ or}$$

$$k = 2.303\times8.2\times10^{8}$$

$$= 18.9\times10^{8}\ \text{s}^{-1}$$

$$\text{Initial conc} = \frac{0.754}{3350} = 2.25\times10^{-4}\ \text{M}$$

$$18.9\times10^{8} = \frac{2.303}{3\times10^{-9}}\log\frac{2.25\times10^{-4}}{(a-x)}$$

$$\log\frac{2.25\times10^{-4}}{(a-x)} = \frac{18.9\times10^{8}\times3\times10^{-9}}{2.303}$$

$$= \frac{56.7\times10^{-1}}{2.303} = 2.462$$

$$\frac{2.25\times10^{-4}}{(a-x)} = 2.897\times10^{2}$$

$$(a-x) = \frac{2.25\times10^{-4}}{2.897\times10^{2}}$$

$$= 0.777\times10^{-6} = 7.77\times10^{-6}\ \text{M}$$

Concentration of Alq_3^* at 3 ns = 7.77×10^{-6} M

Flash Photolysis

Problem 8.25

(a) The decay of the triplet state of naphthalene generated by flash photolysis was studied by following the change in its absorbance with time at 410 nm and the change in its concentration (mol dm^{-3}) recorded at various time intervals as given below.

Time (μs)	20.0	40.0	60.0	80.0	100.0	120.0	160.0	200.0
Conc. of triplet$\times10^5$	8.93	8.10	7.38	6.43	5.12	3.81	2.74	1.43

Calculate the first order rate constant by a suitable graphical plot. What is the half life of the reaction? What is the optical density of the solution at the end of two half lives? The molar absorptivity of the solution is 4200 M^{-1} cm^{-1} at 410 nm.

(b) The quenching of the triplet state of thiochromanone T_{exe} = 355 nm by phenol in acetonitrile formed by a laser flash photolysis method has been studied by following the change in absorbance at 440 nm. The observed rate constant (k_{obs}) is found to vary with the quencher concentration as shown. (Ref: A M Riveiro et al., *J. Braz. Chem. Soc.*, 21, No.8, 2010) The rate constant k_{obs} has been related to the quencher (phenol) concentration by

$$k_{obs} = k_0 + k_q[q]$$

where k_0 is rate constant in the absence of the quencher, k_q is rate constant for the decay of the triplet in the presence of the quencher and $[q]$ is the concentration of quencher. Determine k_0 and k_q by a suitable graphical method.

Concentration of phenol (M)	k_{obs} (s^{-1})$\times10^{-5}$
4×10^{-4}	4.0
6×10^{-4}	5.0
7.5×10^{-4}	5.8
1.1×10^{-3}	6.0
1.45×10^{-3}	7.0
1.8×10^{-3}	7.5
2.2×10^{-3}	9.0
2.5×10^{-3}	11.0

Solution (a)

Time (μs)	20.0	40.0	60.0	80.0	100.0	120.0	160.0	200.0
log *C*	–4.050	–4.090	–4.132	–4.190	–4.291	–4.419	–4.562	–4.845

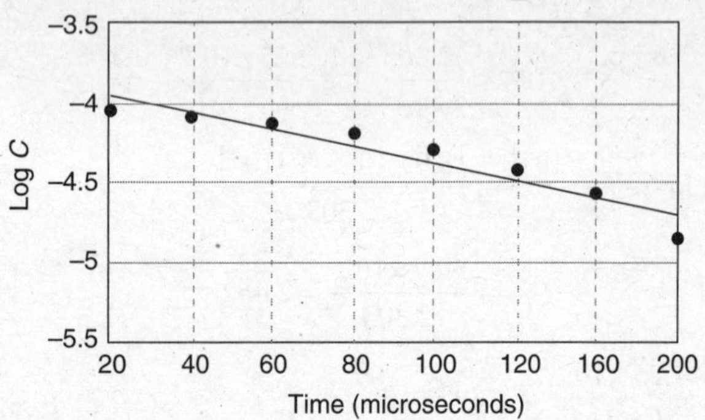

A plot of log*C* vs time yielded an approximate straight line with a slope of nearly

$$-\frac{k}{2.303} = -4368.75 \text{ or,}$$

$$k = 4368.75 \times 2.303$$

$$= 1.006 \times 10^4 \text{ s}^{-1}$$

$$t_{1/2} = \frac{0.693}{k} = \frac{0.693}{1.006 \times 10^4} = 0.69 \times 10^{-4} = 6.9 \times 10^{-5} \text{ s}$$

Time taken for two half lives $= 2 \times 6.9 \times 10^{-5}$

$$= 13.8 \times 10^{-5} \text{ s} = 138 \text{ μs}$$

log *C* at a time of 138 μs = –4.4 or, $C = 3.98 \times 10^{-5}$ M

Optical density pertaining to this conc.

$= \varepsilon Cl = 4200 \times 3.98 \times 10^{-5} \times 1 = 0.167$

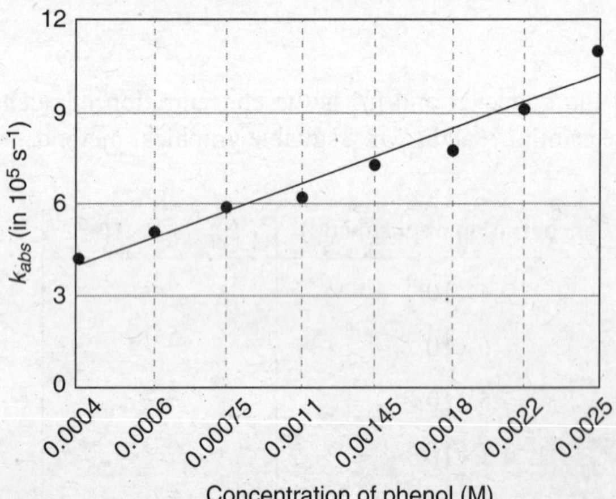

A plot of k_{obs} vs [*q*] is nearly linear with an approximate intercept

$k_0 = 3.9 \times 10^5 \text{ s}^{-1}$

and slope, $k_q = 2.72 \times 10^8 \text{ M}^{-1} \text{ s}^{-1}$

Photosynthesis

Problem 8.26

Write a brief account of photosynthesis with special emphasis of the various reactions involved in the process.

Solution

Photosynthesis is the process by which plants and some bacteria use the energy of sunlight to produce glucose molecules from CO_2 and H_2O. The overall reaction is

$$6H_2O + 6CO_2 \rightarrow C_6H_{12}O_6 + 6O_2$$

Cellular respiration converts the sugar to ATP (adenosine triphosphate), which is the fuel used by all living things. The leaves of a plant may be viewed as a solar collector of photosynthetic cells. The raw materials of photosynthesis, water and CO_2 enter the cells of the leaf and the products of photosynthesis, sugar and oxygen leave the roof.

Chlorophyll and related pigments: Chlorophyll, the green pigment common to all photosynthetic cells absorbs all wavelengths of light except green which is reflected. It is a complex molecule and has several modifications which occur among plants and some bacteria as well. Among them chlorophyll *a* and chlorophyll *b* are important. Chrophyll *a* absorbs its energy from the violet-blue and reddish orange-red part of visible light (400 to 700 nm). Chlorophyll *b* absorbs some energy in the green part of the spectrum. The energy absorbed by chlorophyll triggers a chemical reaction when it is associated with proteins embedded in a membrane (as in a chloroplast). The thylakoid in the plant leaf is the structural unit of photosynthesis. Stacks of thylakoids are known as grana and areas between grana are referred as stroma.

Stages of photosynthesis: There are two stages in photosynthesis. The first stage is the light dependent process (or light reactions) which require direct energy of light to make molecules which carry that energy and use in a second process. The second stage is a light independent process (dark reactions) which occurs when the products of the light reaction are used to form C–C covalent bonds of carbohydrates. The light reactions occur in the grana and dark reactions take place in the stroma.

- **Light reactions:** In this process, light strikes chlorophyll *a* to excite electrons to a higher energy state. The energy is converted in a series of reactions to ATP and NADPH (reduced form of nicotinamide adenine dinucleotide phosphate cation, NADP+). Water splits in the process, releasing O_2 as a byproduct.
- **Dark reactions:** Carbon dioxide from the atmosphere is captured and modified by the addition of hydrogen to form carbohydrates. This is known as carbon fixation.

The overall reaction in the light and dark reactions may be summed as

$$6CO_2 + 18ATP + 12NADPH + 12H^+ + 12H_2O \rightarrow$$
$$\text{Fructose-6-phosphate} + 18ADP + 12NADP^+ + 18H_3PO_4$$

The fructose-6-phosphate produced in the above reaction can be converted to glucose according to

$$\text{Fructose-6-phosphate} \rightarrow \text{Glucose-6-phosphate} \xrightarrow{H_2O} \text{Glucose} + H_3PO_4$$

When this reaction occurs, the overall carbon reduction cycle becomes

$$6CO_2 + 18ATP + 12NADPH + 12H^+ + 11H_2O \rightarrow \text{Glucose} + 18ADP + 12NADP^+ + 17H_3PO_4$$

If you divide the above equation by 6, we get

$$CO_2 + 3ATP + 2NADPH + 2H^+ + 2H_2O \rightarrow C(H_2O) + 3ADP + 2NADP^+ + 3H_3PO_4$$

It is seen from the above reaction that photosynthesis requires 3 moles of ATP and 2 moles of NADPH to convert one mole of CO_2 to the carbohydrate. The following diagram summarises the complete photosynthetic process in a nutshell.

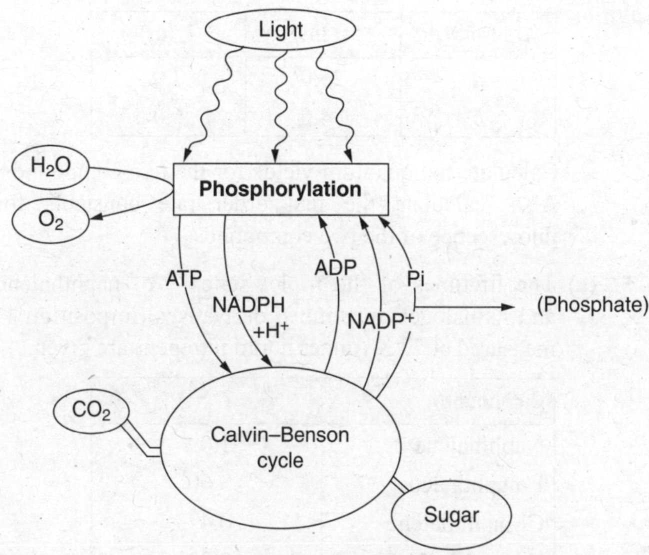

Calvin–Benson cycle

The Calvin–Benson cycle occurs in the stroma of chloroplasts. Carbon dioxide is captured by the compound ribulosephosphate which is a 5-C chemical. Six molecules of carbon dioxide enter the carbon cycle, ultimately producing one molecule of glucose.

EXERCISES

Practice Problems

1. (a) A reacting system absorbs 10^{10} ergs of radiation of wavelength 260 nm in a certain time interval. In this time, 2×10^{-4} mol of the irradiated substance reacted. Calculate the quantum efficiency of the reaction.

   (b) When a beam of monochromatic light was passed through a 0.5 m cell with a solution of concentration x, 15% of the incident light was absorbed. Suppose a cell with a concentration of x and 45% of the incident light has been absorbed, what is the length of the cell?

2. (a) Light of wavelength 250 nm was passed through a solution contained in a cell of 0.5 cm length. If the solution is of 0.001 M concentration and the intensity of light was reduced to 20% of its initial value, calculate the absorbance and molar absorptivity of the solution.

   (b) A certain photochemical reaction requires an activation energy of 100 kJ mol^{-1}. What is the wavelength corresponding to this energy in angstroms? What are the values of wavenumber and frequency of light for this wavelength?

3. Ozone is produced in the atmosphere by the reaction of oxides of nitrogen with radiation according to the scheme given below:

$$NO_2(g) + h\nu \rightarrow NO(g) + O(g)$$
$$O(g) + O_2(g) \rightarrow O_3(g) + M$$
$$NO(g) + O_3(g) \rightarrow NO_2(g) + O_2(g)$$

   (i) Given $\Delta H^0_{f(NO)}$, $\Delta H^0_{f(O)}$, $\Delta H^0_{f(NO_2)}$ as 90.3, 249.2, 33.2 (all in kJ mol^{-1}), calculate the wavelength of light which can dissociate NO_2 photochemically.

   (ii) If NO_2 has strong absorption in the region 350–450 nm, will the above reaction occur in the troposphere?

   (iii) Derive an equation for the steady state concentration of ozone using the steady state condition for NO.

4. (a) The quantum yield for the photolysis of acetone according to

$$(CH_3)_2CO \rightarrow C_2H_6(g) + CO(g)$$

   at 313 nm is 0.17. Calculate the number of molecules of CO formed per second if the intensity of radiation absorbed is 0.05 J s^{-1}.

   (b) The fluorescence lifetimes (τ_0) and the observed lifetimes (τ_s) of chlorophyll *a* and chlorophyll *b* in methanol at 298 K are given.

Chlorophyll	τ_0 (ns)	τ_s (ns)
a	25.0	7.0
b	74.0	6.0

Calculate the quantum yields for the two compounds. Also, calculate the first order rate constants for fluorescence of the two compounds.

5. (a) The lifetimes of the triplet state T_1 of naphthalene and its halogen substituted derivatives (in position 1) measured at 77 K (under liquid nitrogen) are given.

Compound	τ_{T_1} (s)
Naphthalene	1.09
F-naphthalene	3.60
Cl-naphthalene	0.473
Br-naphthalene	0.02
I-naphthalene	0.0017

If the phosphorescence is quenched by a compound *X* with a diffusion controlled rate constant of 10^{10} dm^3 mol^{-1} s^{-1}, calculate the concentration of *X* required to reduce the phosphorescence yield by 20%.

(b) The sun's radiation, effective for photosynthesis, contains light of 550 nm. Assuming that in an eight hour day the sun rays are effective and provide 20160 kJ m^{-2}, calculate the number of einsteins absorbed by any absorbing medium. Suppose $\phi_{\text{photosynthesis}} = 0.12$ and all the light energy absorbed is used by the green leaves of a plant over 1 m^2, how many moles of carbohydrate are formed in the given time?

6. Iron carbonyl, $Fe(CO)_5$, on irradiation with light of 282 nm leads to the formation of $Fe_2(CO)_9$ for which the following mechanism may be suggested.

$$Fe(CO)_5 + h\nu \rightarrow Fe(CO)_5^* \qquad \phi I_{abs}$$

$$Fe(CO)_5^* \rightarrow Fe(CO)_4 + CO \qquad k_2$$

$$Fe(CO)_4 + Fe(CO)_5^* \rightarrow Fe_2(CO)_9 \qquad k_3$$

By suitable assumptions, derive the rate law for the formation of $Fe_2(CO)_9$. Under what conditions does the rate depend only on the first step (the rate of formation of $[Fe(CO)_5]^*$)?

7. (a) Air saturated solutions of tetrachloroethylene covered with 2 mL of highly purified water were irradiated with 10 kilorad of γ-radiation from CO^{60} source for a certain time at 298 K. The main products of irradiation are found to be CCl_3COCl (trichloroacetylchloride, phosgene ($COCl_2$) and another species $COCl^\cdot$. All of them get hydrolysed by water to give trichloroacetic acid and HCl. The estimated acid content was found to be 15.0 eq. Calculate the G_{H^+} value in units of eq Gy^{-1} (eq J^{-1}). Also give the value in molecules (100 eV)$^{-1}$.

(b) Give a possible mechanism to explain the above products of irradiation.

8. It was desired to use a uranyl oxalate actinometer for the study of some photochemical reactions. The actinometer was irradiated with radiation of 400 nm for 15 minutes. The energy was absorbed at the rate of 0.2 J s^{-1}. Before irradiation, an aliquot of uranyl oxalate solution required 40.0 mL of 0.1 N $KMnO_4$ whereas after irradiation, the same aliquot required 23.5 mL of 0.1 N $KMnO_4$. Calculate the quantum yield of the actinometer.

9. Rose Bengal–oxalic acid–Tween 80 (a non-ionic surfactant) acts as a photogalvanic cell on illumination with a 200 W tungsten lamp. Rose Bengal in an ionised form (D^-) acts as a photosensitiser and the non-ionic surfactant acts as a surfactant as well as a reactant. The overall reaction in the cell is

$$D^- + S \underset{\text{dark}}{\overset{h\nu}{\rightleftharpoons}} D^{2-\cdot} + S^{+\cdot}$$

The open circuit voltage of the cell is 0.67 V.

(i) Give a sketch of the cell using a SCE as a reference electrode.

(ii) What is the Gibbs free energy change (of the process) assuming it to be a 1 e^- process.

(iii) Suggest a possible mechanism of the above reaction assuming the formation of a charge transfer complex between the excited state of D^- and surfactant S.

10. The kinetics of a generic reaction such as

$$\text{Spiro compound 1} \rightleftharpoons \text{Oxo-compound 2}$$

was studied by a flash photolysis technique by generating form 2 by flashing of compound 1. The conversion of compound 2 to compound 1 was monitored spectrophotometrically at 610 nm. The following data are obtained at 293 K.

Time (s)	10	15	20	30	40	50
log *A* (arbitarary units)	0.912	0.803	0.760	0.608	0.456	0.261

Determine the first order constant graphically. Given the activation energy of the reaction as 70 kJ mol^{-1}, what is the rate constant at 323 K? What is the frequency factor of the reaction?

11. The triplet state of benzophonone, formed by means of a Nd–YAG laser reacts with the α-hydrogen of isopropanol in the same medium according to

$$(C_6H_5)_2CO(T_1) + \left(CH_3\right)_2 CHOH \rightarrow$$

$$(C_6H_5)_2CO^\cdot H + (CH_3)_2CO^\cdot H \qquad (1)$$

At pH values > 8, the protonated benzophenone ketyl gets deprotonated through the reaction,

$$(C_6H_5)_2CO^{\bullet}H \rightarrow (C_6H_5)_2CO^{\bullet -} + H^+ \quad (2)$$

which gives benzpinacol according to the reaction

$$(C_6H_5)_2CO^{\bullet}H + (C_6H_5)_2CO^{\bullet -} \xrightarrow{k_2} (C_6H_5)_2C(OH)-(OH)C(C_6H_5)_2 \quad (3)$$

The above reaction [Eq. (3)] is followed by monitoring the decrease in concentration of the $(C_6H_5)_2CO^{-\cdot}$ species with time at 630 nm. Assuming that Eq. (3) is 2nd order in XH and X–, and using the dissociation equilibrium of Eq. (2) in the rate equation, show that

$$\frac{d[X^-]}{dt} = k_{obs}[X^-]^2 \text{ where } k_{obs} = \frac{k_2[H^+]}{K_d}.$$

where K_d denotes the dissociation constant of XH.

In a typical experiment, k_{obs} at various pH values are given.

pH	10.2	10.75	11.20	12.30	12.90
k_{obs} (M^{-1} s^{-1})	1.78×10^8	3.16×10^7	6.31×10^6	1.78×10^6	3.16×10^5

Plot the data in a suitable manner and determine k_2 at each pH given $K_d = 6\times10^{-10}$.

12. (a) In a Fricke dosimeter consisting of 1×10^{-3} M $FeSO_4$ in 0.8 M H_2SO_4 and 20 mM sodium sulphate, after subjecting the solution to cobalt-60 γ-radiation, the absorbance of Fe^{3+} ions formed was found to be 0.33 at 304 nm in a cuvette of 0.30 cm. calculate the *G*-value of the dosimeter. (ε = 2197 M^{-1} cm^{-1}, radiation dose = 300 Gy, solution density = 1.029 kg L^{-1}). What is the *G*-value in units of molecules formed per (100 eV) energy?

(b) Another radiolytic dosimeter which is also used is methyl viologen dosimeter (1,1′-dimethyl- 4,4′-bipyridinium dichloride). In this case, exposure to Co-60 γ-radiation of a 5 mM methyl viologen solution in formate buffer with a dosage of 600 Gy results in the formation of MV^+ cation and it gave an absorbance of 0.52 at 602 nm. The cuvette is 0.1 cm in length and the molar extinction coefficient of MV^+ is 13,100 M^{-1} cm^{-1} with the solution density being 1.0 kg L^{-1}. Calculate G(MV^+) in mol J^{-1} and in molecules $(100\text{ eV})^{-1}$.

(c) Assuming $G_{(H_2O_2)} = 1.2$, $G_{(HO.)} = 2.5$, Calculate $G_{(H.)}$ which is one of the products of radiolysis in the Fricke dosimeter. (Hint: consider all reaction products formed from radiolysis and relate $G(Fe^{3+})$ to the *G*s of the products.)

(d) Derive the Stern–Volmer equation by considering the formation of an excited species and its decay due to the competing processes of fluorescence and of the excited species by reaction with a quencher.

13. (a) A 2×10^{-4} M solution of a substance absorbs 40% of the incident light using a cell of 10 mm length. What will be the concentration of the same substance necessary if 95% of the light is to be absorbed using a cell of 5 mm length?

(b) A certain solution (conc. = *C*) in a cell of 20 mm length absorbs 20% of incident light. If the concentration of the solution is doubled and the cell length is reduced to 5 mm, what is the fraction of incident light that is absorbed?

14. (a) Methyl bromide vapour absorbs at 204 nm which results in dissociation to methyl radicals and bromine atoms. If the bond dissociation energy of the C–Br bond is 284.5 kJ mol^{-1}, what is the kinetic energy of the fragments if they are formed in the ground state?

(b) It is desired to produce a compound at the rate of 5×10^{-5} mol s^{-1} by exposing the desired solution to 500 nm radiation. Calculate the power required in watts.

15. (a) Explain photosynthesis briefly, along with the relevant chemical reactions.

(b) If glucose is one of the products of the reaction, how is it utilised in the plants?

(c) Explain briefly the sequence of reactions occurring in the two main stages of photosynthesis.

(d) In the study of the quenching of fluorescence of fluorescein by iodide ions, the following values for the ratio I_0/I (I_0 = fluorescence intensity in the absence of quencher, I = fluorescence intensity in the presence of quencher) were obtained at various quencher (*Q*) concentrations.

[*Q*]/M	0.005	0.010	0.015	0.020	0.025	0.030	0.040	0.050
I_0/I	1.04	1.08	1.12	1.16	1.19	1.25	1.31	1.39

Using the Stern–Volmer equation, $I_0/I = 1 + k_q\tau_0[Q]$ and making a suitable plot, determine k_q, the bimolecular quenching rate constant, given τ_0 = lifetime of excited state in the absence of quencher, as 4 ns.

16. (a) The bond dissociation energy of the O=O bond is 497 kJ mol^{-1}. Calculate the maximum wavelength of radiation at which O_2 can be dissociated.

(b) If oxygen gas does not have appreciable absorption above $\lambda = 200$ nm, will appreciable dissociation occur at the wavelength calculated above?

(c) An aqueous H_2SO_4 solution of 12 M concentration was subjected to 15 ps electron pulse radiolysis and the radical $SO_4^{-\cdot}$ was detected as one of the radiolysis products. Its absorbance at 450 nm was observed and the data are given below.

Absorbance	Extinction coefficient (M^{-1} cm^{-1})	Cell width (cm)	Dosage (Gy)	Density of acid solution (kg L^{-1})
0.01	1600	0.5	31.0	1.65

Calculate the G value in molecules $(100 \text{ eV})^{-1}$.

17. (a) The decay of the transient pair $\left(e^-_{solv} + H_3O^+\right)$ formed in 2 M H_3PO_4 on ps electron pulse radiolysis was followed by absorption measurements at 800 nm using a 0.5 cm cuvette. The absorbance data at various time intervals are given below.

Time (ns)	0.00	50.00	100.00	150.00	200.00	250.00
Absorbance	0.070	0.053	0.035	0.025	0.018	0.015

Determine the rate constant of the reaction by suitably plotting the data assuming it to be a first order process.

(b) One of the major radiolytic products of pulse radiolysis of concentrated phosphoric acid solutions is the formation of the radical $H_2PO_4^{\cdot}$. The following data for the absorption of the radical at 520 nm were obtained using 14.6 M H_3PO_4 solution.

Absorbance	Extinction coefficient (M^{-1} cm^{-1})	Cell width (cm)	Dosage (Gy)	Density of acid solution (kg L^{-1})
0.007	1850	0.5	22.0	1.68

Calculate the G value of the radical in units of mol J^{-1} and molecules $(100 \text{ eV})^{-1}$.

(c) Given that the species formed in the radiolysis of phosphoric acid solutions as $PO_4^{\cdot 2-}$, $H_2O^{\cdot +}$, $H_2PO_4^{\cdot}$, $OH^{\cdot}$, $HO_2^{\cdot}$ and $H^{\cdot}$, write down some of the possible reactions which these radicals may undergo among themselves or with other species in the reaction system. If the second order rate constants of the one of the reactions

$$OH^{\cdot} + H_2PO_4^{\cdot} \rightarrow H_3PO_5$$

is 4.0×10^9 L mol^{-1} s^{-1}, calculate the concentration of $OH^{\cdot}$ using the result obtained in (a).

18. (a) Why is doping of a semiconductor electrode in PEC cell important?

(b) Explain the phenomenon of photocorrosion in PEC cells. Give an example.

(c) (i) Represent the defect chemical reaction that occurs when ferric oxide is doped with zinc oxide.

(ii) If the above doping occurs in the presence of oxygen gas, how will the defect chemical reaction change?

19. (a) Depict schematically the photochemical processes in a photogalvanic cell which, consists of dye-dianion ($D–D^-$) system coupled with a reversible redox system (for example, Fe^{2+}–Fe^{3+}) or with an irreversible redox system (for example, EDTA–$EDTA^+$).

(b) In the photochemical reaction between hydrogen and chlorine, radiation of 440 nm was used in the irradiation study. 4.46×10^{24} molecules of HCl were formed per joule of energy of the absorbed radiation. What is the quantum yield for the process?

20. (a) Norrish et al. generated OH radicals by subjecting watervapour to flash photolysis and studied the reactions

$$OH^{\cdot} + C_2H_6 \rightarrow C_2H_5^{\cdot} + H_2O \quad (1)$$

$$OH^{\cdot} + CH_4 \rightarrow CH_3^{\cdot} + H_2O \quad (2)$$

The rate constant of the first reaction is given by

$$\log_{10} k = 11.1 - \frac{3600 \text{ J}}{2.303\, RT}$$

and that of the second reaction is given by

$$\log_{10} k = 10.7 - \frac{5000 \text{ J}}{2.303\, RT}$$

where k has the units of L mol^{-1} s^{-1}.

Calculate the rate constant of both reactions at 20°C and 40°C and also their frequency factors. What are their entropies of activation?

(b) Calculate the number of quanta of light absorbed per second in a case where 25% of energy of a 200 watt incandescent bulb is turned into visible light with an average wavelength of 500 nm.

Multiple Choice Questions

1. Which of the following statements correctly describes the characteristic of a thermal or a photochemical reaction?
(a) ΔG must be must be +ve for a thermal reaction to occur.
(b) Temperature has little effect on the rate of a photochemical reaction.
(c) The activation of a photochemical or a thermal reaction is a very selective process.
(d) Photochemical reactions do not necessarily occur in the presence of light.

2. Pick out the correct statement about the absorbance of a solution.
(a) It is proportional to the thickness of the absorbing medium.
(b) It is proportional to the concentration of the absorbing species.
(c) It is given by the equation log I_0/I (I_0 = intensity of incident light, I = intensity of transmitted light.
(d) It is given by log I/I_0.

3. The value of *einstein* for radiation of 350 nm (in kJ mol^{-1}) is
(a) 234 (b) 423
(c) 485 (d) 342

4. The molar extinction coefficient of cytosine is 6000 M^{-1} cm^{-1} at 270 nm and a certain pH. The % transmission in a 2 mm cell for a solution of concentration 5×10^{-4} M is
(a) 25.1 (b) 51.2
(c) 42.3 (d) 39.5

5. A solution of tryptophan has an absorbance of 0.50 in a cell of length 0.4 cm. If the extinction coefficient of the compound is 6400 L mol^{-1} cm^{-1}, the molar concentration of the solution is
(a) 3.54×10^{-4} (b) 1.95×10^{-4}
(c) 2.46×10^{-3} (d) 1.28×10^{-3}

6. The photochemical decomposition of HI occurs according to

$\mathrm{HI} \xrightarrow{h\nu} \mathrm{H} + \mathrm{I}$ rate = I_{abs}

$\mathrm{H} + \mathrm{HI} \xrightarrow{k_1} \mathrm{H_2} + \mathrm{I}$

$\mathrm{I} + \mathrm{J} \xrightarrow{k_2} \mathrm{I_2}$

The quantum yield of the reaction is,
(a) 1 (b) 3
(c) 2 (d) 1.5

7. Find the correct statement among the following dealing with the photochemical reaction between hydrogen and bromine.
(a) Being a chain reaction, its quantum yield is much greater than unity.
(b) The quantum yield is proportional to the concentration of bromine and inversely proportional to I_{abs}.
(c) The quantum yield is proportional to I_{abs} and inversely proportional to the concentration of hydrogen.
(d) The quantum yield is proportional to the concentration of hydrogen and inversely proportional to $I_{abs}^{1/2}$.

8. The greenish glow observed on exposure of phosphorus to air is an example of
(a) thermoluminescence
(b) a thermal reaction
(c) chemiluminescence
(d) photosensitised reaction

9. The phosphorescence exhibited by zinc sulphide or other alkaline earth sulphides may be ascribed to
(a) the transition of an excited electron from the lower vibrational levels of a triplet state to the lower vibrational levels of the singlet state
(b) the internal conversion, that is, the transition of an electron from higher vibrational levels to the lower vibrational levels in the triplet state
(c) the intersystem crossing of an excited electron from higher vibrational levels of the singlet state to those of the triplet state
(d) the relaxation of the energy of the excited electron from the higher vibrational to the lower vibrational levels in the singlet state

10. In the photochemical decomposition of dichloramine-B, Mohana et. al. (*E. J. Chem.*, 4, 502, 2007) using UV radiation of wavelength 253.7 nm, found that on absorption of 9.24×10^{-17} quanta s^{-1}, 3.1×10^{-41} moles of dichloramine-B decomposed per second (in a certain volume). The quantum yield of the reaction is
(a) 0.46 (b) 0.34
(c) 0.15 (d) 0.20

11. The C–H bond energy is 413 kJ mol^{-1}. The wavelength of light required to break this bond is (in nm)
(a) 4.2×10^4 (b) 2.9×10^5
(c) 3.6×10^4 (d) 5.6×10^5

12. An example of a photosensitised reaction is
(a) the reaction between uranyl ions and oxalic acid in an actinometer
(b) the hydrogen–chlorine reaction in the presence of light of wavelength less than 480 nm
(c) The decomposition of acetaldehyde when it is exposed to uv light of wavelength less than 300 nm
(d) The reaction between carbon dioxide and water in the presence of sunlight in green plants

13. In the simple form of a photoelectrochemical cell for splitting water,
(a) the anode is an inert platinum electrode while the cathode is a semiconductor material
(b) the anode is a semiconductor electrode while the cathode is an inert metal
(c) both the anode and the cathode are metals like platinum and copper
(d) the charge carriers are electrons in the cell

14. The uridine actinometer is based on the reaction
Uridine + Water $\xrightarrow{hv}$ Uridine hydrate
which occurs in the presence of UV radiation of 262 nm. The intensity of absorbed radiation is given by
$I = 6.0\times10^{20}/\sigma(\phi\times t)\times\log(10^{A_0}-1)(10^{A_t}-1)$
In an experiment, A_0 =1.5 $A\times t$ = 0.552 (A_0, A_t are absorbances at zero time and time t) and in a time of 3000 s, σ = molar extinction coefficient = 8000 M^{-1} cm^{-1}, ϕ = quantum yield of the reaction = 0.022. The value of I is
(a) 12.4×10^{12} (b) 21.5×10^{14}
(c) 30.6×10^{13} (d) 18.9×10^{15}

15. The delayed fluorescence exhibited by the molecules of a gas in a photochemical reaction occurs during
(a) transfer of molecules in the excited vibrational levels of the T_1 state to the S_1 state (by intersystem crossing) which then returns to the S_0 state
(b) transfer of molecules in the excited vibrational levels of the T_1 state to the ground state S_0 level
(c) transfer of molecules in the excited vibrational levels of the S_1 state to the ground state S_0 level
(d) transfer of molecules in the excited vibrational levels of the S_1 state to the excited vibrational levels of the T_1 state

16. In the presence of a quencher, disodium salt of guanosine 5-monophosphate, of concentration 2×10^{-3} M, the fluorescence lifetime of the compound, *N*-methyl acridinium iodide (τ) is found to be 27.8 ns by L K Fraiji et al., *J. Chem. Ed.*, 69, 424, 1992. Given the fluorescence lifetime in the absence of the quencher as 34.8 ns (τ_0), the quenching rate constant K_q (in units of L mol^{-1} s^{-1}) is
(a) 4.67×10^{8} (b) 2.94×10^{8}
(c) 3.62×10^{9} (d) 4.56×10^{10}

17. Solvated electrons are formed when
(a) metallic sodium is dissolved in ethanol
(b) an alkali metal is placed in a hydrocarbon solvent
(c) alkali metals react with liquid oxygen
(d) any alkali metal is dissolved in liquid ammonia

18. In the study of solvated electrons formed in the pulse radiolysis of pure hydrazine solution at a pulse rate of 10^{20} eV kg^{-1}, the spectrum showed a maximum at a wavelength of 1000 nm with an absorbance A = 0.083 using a cell of 0.08 cm. Given the density (ρ) of hydrazine as 1.004 g cm^{-3} and extinction coefficient at maximum wavelength of the solvated electron as 2.2×10^{4} M^{-1} cm^{-1}, the G value for the solvated electron (in mol J^{-1}) is
(a) 4.76×10^{-4} (b) 1.63×10^{-5}
(c) 3.65×10^{-4} (d) 2.94×10^{-6}

19. The recombination reaction $R\bullet + R\bullet \rightarrow R_2$ was studied by a 2 μs flash photolysis technique after the radical was generated. The change in absorbance was studied as a function of time as shown below:

t (μs)	0	10	15
Absorbance	0.75	0.58	0.51

The average value of the rate constant in l mol^{-1} s^{-1} from the given data is (extinction coefficient = 1.1×10^{4} L mol^{-1} cm^{-1}, l = 50 cm)
(a) 4.62×10^{10} (b) 2.23×10^{10}
(c) 5.45×10^{10} (d) 3.22×10^{10}

20. Which of the following statements is true in the application of the concept of photostationary state to the dimerisation of anthracene (A) given by

$$2A \xrightarrow{hv} A_2$$

(a) This concept is the application of stationary state to a reaction intermediate
(b) It is an equilibrium state concept
(c) It is not a true equilibrium
(d) It is a state which requires that $d(A_2)/dt = 0$ where A_2 is the dimer of anthracene

21. In the study of a certain photochemical reaction, 5×10^{20} quanta of radiation was absorbed per second. After irradiation for 30 min, 0.2 mol of the reactant reacted. The quantum yield of the reaction is
(a) 0.13 (b) 0.31
(c) 0.25 (d) 0.05

22. Which of the following is correct regarding TiO_2?
(a) photocathode (b) photoanode
(c) *p*-type semiconductor (d) band gap is 0.1 eV

23. The photoelectrochemical splitting of water by Fujishima and Honda in 1972 yielded an approximate efficiency of
(a) 10% (b) 25%
(c) 1% (d) 75%

Answer Keys

Practice Problems

1. (a) 0.091
(b) Cell length = 0.210 m

2. (a) $A = 0.699$
(b) $\bar{\nu} = 8354$ cm^{-1}; Frequency = 2.506×10^{14} s^{-1}

3. (i) $\lambda = 391$ nm
(ii) Formation of ozone is possible in the troposphere.
(iii) $[O_3] = \dfrac{k_1[NO_2]}{k_3[NO]}$

4. (a) Number of molecules of CO formed per second = 13.43×10^{15}

(b) For chlorophyll *a*, $\phi_F = 0.28;\ k_F = 4\times10^7\ s^{-1}$

For chlorophyll *b*, $\phi_F = 0.081;\ k_F = 1.35\times10^7\ s^{-1}$

5. (a) For naphthalene $[X]=3.67\times10^{-10}$ s; For F-naphthalene $[X] = 1.11\times10^{-10}$ s; For Cl-naphthalene $[X] = 8.46\times10^{-10}$ s; For Br-naphthalene $[X] = 2\times10^{-8}$ s; For I-naphthalene $[X]=2.35\times10^{-5}$ s

(b) No. of moles of carbohydrate formed per second = $0.12\times92.65\times11.1$ mol

6. $$\frac{d\left[Fe_2(CO)_9\right]}{dt}=\frac{\phi I_{abs}}{1+\dfrac{k_2}{k_3\left[Fe(CO)_4\right]}}$$

when $k_3\left[Fe(CO)_4\right]\gg k_2$,

the second term will be very small and can be neglected.

Thus $\dfrac{d\left[Fe_2(CO)_9\right]}{dt}=\phi I_{abs}$

7. (a) $G_{H^+} = 0.15$ eq J^{-1}; 0.145×10^7 molecules$(100\ eV)^{-1}$

(b) A suitable mechanism may be proposed consisting of several steps.

8. Quantum yield of the actinometer $\phi_{CO} = \phi_{CO_2} = 0.55$

9. (i) A conventional two-electrode assembly with the working and reference electrodes along with the light source as in Fig. 8.3 of the text.

(ii) $\Delta G = 64.66\ kJ\ mol^{-1}$

(iii) A suitable mechanism may be proposed

10. First order rate constant $k=3.94\times10^{-2}\ s^{-1}$

Rate constant at 323 K, $k_{323} = 7.92\ s^{-1}$

Frequency factor $A=1.41\times10^{20}\ s^{-1}$

11. Average $k_2 = 12.74\times10^8\ M^{-1}\ s^{-1}$

12. (a) $G_{Fe^{3+}} = 15.6$ molecules$(100\ eV)^{-1}$

also $= 1.62\times10^{-6}\ mol\ J^{-1}$

(b) $G_{MV^+} = 6.4$ molecules$(100\ eV)^{-1}$

(c) $G_{H^+} = 3.6$ molecules

(d) $\dfrac{I_f}{I_0}=\left[1+\dfrac{k_q}{k_f}[Q]\right]^{-1}$

13. (a) $C=\dfrac{1.30}{1110\times0.5}=2.34\times10^{-3}$

(b) Fraction of light absorbed = 0.107

14. (a) Kinetic energy of fragments is = 302.3 kJ mol^{-1}

(b) Power required = 11.97 W

15. (a), (b), (c) Theoretical questions; (d) $k_q = 1.99\times10^{-9}\ M^{-1}\ s^{-1}$

16. (a) $\lambda_{max} = 241$ nm

(b) No appreciable dissociation.

(c) $G_{(SO_4^{2-})} = 2.36$ molecules$(100\ eV)^{-1}$

17. (a) $k=6.91\times10^6\ s^{-1}$

(b) $G\left(H_2PO_4^*\right)=1.98$ molecules$(100\ eV)^{-1}$

(c) A suitable mechanism may be proposed. $\left[OH^-\right]=1.7\times10^{-3}$ M

18. Theoretical questions

19. (a) Theoretical question

(b) $\phi=\dfrac{7.4}{3.68\times10^{-6}}=2.01\times10^6$

20. (a) For the 1st reaction

k at 20°C $= 2.884\times10^{10}\ L\ mol^{-1}\ s^{-1}$

k at 40°C $= 3.16\times10^{10}\ L\ mol^{-1}\ s^{-1}$

$A=1.26\times10^{11}\ L\ mol^{-1}\ s^{-1}$

$\Delta s^{\ddagger} = -32.7$ eu $\left(J\ deg^{-1} mol^{-1}\right)$

For the 2nd reaction

k at 20°C $= 6.46\times10^9\ L\ mol^{-1}\ s^{-1}$

k at 40°C $= 7.41\times10^9\ L\ mol^{-1}\ s^{-1}$

$A=5.0\times10^{10}\ L\ mol^{-1}\ s^{-1}$

$\Delta s^{\ddagger} = -40.4$ eu $\left(J\ deg^{-1} mol^{-1}\right)$

(b) Quanta absorbed per second $= 1.26\times10^{20}$

Multiple Choice Questions

1. (b)	**2.** (c)	**3.** (d)	**4.** (a)	**5.** (b)	**6.** (c)	**7.** (d)	**8.** (c)	**9.** (a)	**10.** (d)
11. (b)	**12.** (d)	**13.** (b)	**14.** (b)	**15.** (a)	**16.** (c)	**17.** (d)	**18.** (d)	**19.** (b)	**20.** (d)
21. (a)	**22.** (b)	**23.** (c)							

CHAPTER 9

SOLUTIONS

CHAPTER OUTLINE

9.1 Important Definitions

Solution: A solution is defined as a homogeneous mixture of two or more substances whose composition may be altered over certain limits. It is an example of a homogeneous system where there is no surface of separation between the substances mixed and a uniform composition exists throughout the system. For example, a mixture of water and ethanol, sugar or sucrose dissolved in water.

Heterogeneous system: In such a system, the substances that are mixed form separate layers, and the composition is different in different phases. For example, a mixture of water and a hydrocarbon such as benzene, a mixture of two solids, water and its vapour in equilibrium. Various types of solutions are possible depending upon the physical state in which they are mixed.

Solvent and solute: The component present in a large amount is referred as the solvent and the component present in a smaller amount is known as the solute.

Saturated and unsaturated solutions: The maximum amount of a substance dissolved in a certain weight of the solvent represents its solubility. The solution is said to be saturated with respect to the solid. Any amount of solid less than the above, in the same amount of solvent represents an unsaturated solution.

9.2 Solubility

9.2.1 Ways of Expressing Solubility

Mass%: Given by the weight of solute dissolved in 100 g of solvent.

Molarity: The number of gram moles of solute dissolved in 1 dm^3 of solution. It is dependent on temperature.

Molality: The number of gram moles of solute dissolved in 1000 g of solvent. This is a preferred unit as it is independent of temperature.

Normality: Sometimes, especially in volumetric analytical methods, this unit is used. It represents the number of equivalents of the substance (may be an acid or base) dissolved in 1 dm^3 of solution.

Mole fraction: It is defined as the number of moles of the solute divided by the total number of moles of all the species (or substances) present in the solution. It is a fraction and the sum of all the fractions is unity. It is a dimensionless quantity.

9.2.2 Effect of Temperature

Solubility of a solid in a solvent is strongly affected by temperature. For substances whose enthalpy of solution is positive, the solubility increases with increase in temperature. Examples are

KCl in water, NaCl in water. For substances whose enthalpy of solution is negative, the solubility decreases with an increase in temperature. An example is the solubility of $Ce_2(SO_4)_3$ in water.

Ideal solubility: In a saturated condition of a solute in a solvent, from a thermodynamic standpoint, the chemical potential of the solute in the solid state must be equal to that of the same in the solution. Using this criterion, the ideal solubility of a solute in all solvents is given by

$$\ln X = \frac{\Delta H_{\text{fusion}}}{R}\left(\frac{1}{T_o} - \frac{1}{T}\right) \tag{9.1}$$

where X = mole fraction of solubility of the solute, ΔH_{fusion} = enthalpy of fusion of pure solute, T_0 = freezing point of pure solute and T = given temperature (or temperature under consideration). From Eq. 9.1, it is seen that the solubility of the solute is the same in all solvents with which it forms an ideal solution.

9.2.3 Raoult's Law and Ideal Solution

Raoult's law: It states that the partial vapour pressure p_2 of component 2 in a liquid mixture of 1 and 2, is proportional to the mole fraction of 2 in the liquid phase. Thus

$$p_2 = p_2^0 x_2 \tag{9.2}$$

where p_2^0 is the vapour pressure of the pure component 2, and x_2 is its mole fraction in liquid.

Ideal solution: If both the components 1 and 2 obey Raoult's law over the complete range of solvent compositions in a liquid mixture, such a solution is known as an *ideal solution*. Such mixtures are characterised by the following observations:

- the forces of cohesion between the two components are uniform,
- there is no volume change on mixing, $\Delta V_{\text{mix}} = 0$, and
- the enthalpy of mixing ΔH_{mix} is also equal to zero.

The total pressure exerted by the mixture, at any composition is given by

$$P = p_1^0 X_1 + p_2^0 X_2 = p_2^0 + X_1\left(p_1^0 - p_2^0\right) \tag{9.3}$$

Some examples of liquid mixtures which obey Raoult's law are benzene–hexane, chlorobenzene–bromobenzene, and so on.

Relation between compositions in liquid and vapour phases

if x_i and x_i' represent the mole fractions of component i in the liquid and vapour phases respectively,

$$\frac{x_i}{x_i'} = \frac{p_i/p_i^o}{p_i/P} = \frac{P}{p_i^o} \tag{9.4}$$

where $x_i' = p_i/P$ on the basis of Dalton's law and P is the total pressure of the two components in the vapour phase.

Positive and negative deviations from Raoult's law

if any component of a liquid mixture shows a partial pressure higher than that predicted by Raoult's law at any composition, the mixture is said to show positive deviations from Raoult's law. Some examples of such mixtures are benzene–ethanol, ether–acetone, acetone–CS_2, and so on.

In some cases, the partial pressure of a component shows a value less than that predicted by Raoult's law at a certain composition or over a range of compositions. Such mixtures are said to show negative deviations from Raoult's law. Some examples of this type are acetone–chloroform, pyridine–acetic acid, and so on.

Ideal dilute solution

It is not common to find many liquid mixtures where both components obey Raoult's law over the complete range of solvent compositions. However, in many cases, where the concentration of the solute is low, the solute obeys Henry's law but the solvent conforms to Raoult's law. Such solutions are referred to as ideal dilute solutions.

Henry's law

In a mixture consisting of solvent 1 and solute 2, the partial pressure of the solute 2 is proportional to its mole fraction x_2, the proportionality constant being referred to as K_H, the Henry's law constant.

$$p_2 = K_H x_2 \tag{9.5}$$

Henry's law may be expressed in another form involving the concentration of the solute.

$$c_2 = K_H' p_2 \tag{9.6}$$

where C_2 denotes the concentration of the solute in mol dm^{-3}, p_2 is the partial pressure of solute and K'_H is known as Henry's law constant (mol dm^{-3} bar^{-1} or mol dm^{-3} atm^{-1}). Henry's law is most often applied to gas solubilities in different solvents.

9.3 Colligative Properties

Definition: A colligative property is a property that depends only on the number of solute particles present in the solution and not on their nature. They involve non-volatile solutes in dilute solutions. The following are all colligative properties—(i) lowering of vapour pressure of a solution, (ii) depression of freezing point of a solution, (iii) elevation of boiling point of a solution, and (iv) osmotic pressure exerted by a solution.

All these properties are based on the fact that the chemical potential of a solvent in solution is less than that of the pure solvent as shown by the relation,

$$\mu = \mu^o + RT\ln x \tag{9.7}$$

where μ = chemical potential of the solvent in solution, μ^0 = chemical potential of pure solvent and x = mole fraction of solvent in solution.

9.3.1 Lowering of Vapour Pressure of a Solution

Raoult's law may be also expressed in another form: the relative lowering of vapour pressure caused by the addition of a non-volatile solute to a solvent is equal to the mole fraction of solute. Mathematically,

$$\frac{p_0 - p}{p_0} = \frac{n_2}{n_1 + n_2} = x_2 \tag{9.8}$$

where p_0 = vapour pressure of the solvent, p = vapour pressure of the solution and x_2 = mole fraction of the solute. Equation 9.2 can be derived from Eq. 9.8. In dilute solutions, Eq. 9.8 may be simplified as

$$\frac{P_0 - P}{P_0} \approx \frac{n_2}{n_1} = \frac{W_2 / M_2}{W_1 / M_1} \tag{9.9}$$

Equation 9.9 can be used to determine the molecular weight, M_2 of the solute. If the solute dissociates (as in the case of ionic solutes) or associates (as in the case of organic solutes in solvents like benzene), the vapour pressure is affected and the corresponding molecular weight is obtained from Eq. 9.9.

9.3.2 Freezing Point Depression

A consequence of the fact that a non-volatile solute lowers the vapour pressure of the solvent is that the freezing point of the solution is less than that of the pure solvent. Using the same thermodynamic criterion that the chemical potential of the solvent in solution is equal to the chemical potential of pure solid solvent at the freezing point, a relationship can be obtained between the freezing point depression, ΔT, and the molality of the solute m, given by

$$\Delta T_f = K_f m = K_f \frac{1000 W_2}{W_1 M_2} \tag{9.10}$$

where W_1 = weight of solvent, W_2 = weight of solute and M_2 = molar mass of the solute. M_2 can be calculated by using Eq. 9.10. K_f is known as the molal freezing point depression constant (or cryoscopic constant) and depends only on the properties of the solvent. It is given by

$$K_f = \frac{RT_0^2 M_1}{1000 \Delta H_f^0} \quad (\text{units of } K_f = \text{K kg mol}^{-1}) \tag{9.11}$$

where T_0 = freezing point of the solvent in K, M_1 = molar mass of the solvent and ΔH_f^0 = molar enthalpy of fusion of solvent. It is seen from Eq. 9.10 that ΔT_f depends on m and hence on the number of particles in solution which affects the calculated M of the solute. Thus, from the calculated values of M, it is possible to understand whether the solute dissociates or aggregates in solution.

9.3.3 Elevation of Boiling Point of a Solution

An exactly similar equation relating the elevation of the boiling point, ΔT_b, of the solution with the molality of the solute is applicable and is given by

$$\Delta T_b = K_b m = K_b \frac{1000 W_2}{W_1 M_2} \tag{9.12}$$

where K_b = molal boiling point elevation constant (or ebullioscopic constant) and m = molality of the solute. K_b depends only on the properties of the solvent and is equal to

$$K_b = \frac{RT_b^2 M_1}{1000 \Delta H_v^0} \quad (\text{units of } K_f = \text{K kg mol}^{-1}) \tag{9.13}$$

where T_b = boiling point of the solvent in K, M_1 = molar mass of the solvent and ΔH_v^0 = enthalpy of vaporisation of solvent per mole.

As in the case of freezing point depression, dissociation or association of solute in solution affects ΔT_b and hence M_2.

9.3.4 Osmotic Pressure

Osmosis: The spontaneous passage of solvent from a solution of lower concentration to a solution of higher concentration, when they are separated by a semi-permeable membrane is known as *osmosis*. This phenomenon can also be explained from the thermodynamic stand point. The chemical potential of the solvent (in a solution) is lower than that of the chemical potential of the solvent (in pure solvent). So the solvent from the pure solvent enters through the semi-permeable into the side of the solution until the chemical potentials on both sides are equal. The osmotic pressure π is related to the concentration of the solute C in exactly the same way as in the case of a gas. Thus,

$$\pi = CRT = \frac{n}{v} RT \tag{9.14}$$

where C = concentration of the solute in mol dm^{-3}, n = moles of solute and V = volume of solution in dm^3.

Osmotic pressure measurements are useful for determining the molecular weights of high polymeric substances such as proteins. In using Eq. 9.14 to calculate the molar mass of the desired substance, it is necessary to convert the height of the organic solvent (used in the experiment) in terms of the height of the mercury column using the *hdg* relation (h = height, d = density and g = acceleration due to gravity). Other colligative properties are not of much use for determining the molecular weights of large molecular weight substances.

Reverse osmosis: It is the reverse of the natural phenomenon of osmosis and is extensively used in the desalination of sea water to produce pure water. In this technique, pressure over and above that of the osmotic pressure is exerted on the salt water side of the membrane to push the water through the membrane. The pressure is usually of the order of 50 atm (or 50 bar).

9.3.5 Effect of Dissociation or Association of Solutes on Colligative Properties

(i) When an electrolytic solute, such as NaCl, dissociates in solution, the ΔT_f or ΔT_b of such a solution is greater than that observed in the case of a non-electrolyte of the same molality. The observed molecular weights are thus less than the theoretical values. A factor, known as the van't Hoff factor i, is introduced into Eqs 9.10 and 9.12 and the modified forms of these equations are

$$\Delta T'_f = iK_f m \quad \text{or} \quad \Delta T'_b = iK_b m \tag{9.15}$$

i may be equated as,

$$i = \frac{\Delta T'_f}{\Delta T_f(\text{normal})} \quad \text{or,} \quad i = \frac{\Delta T'_b}{\Delta T_b(\text{normal})} \tag{9.16}$$

where $\Delta T'_f$ and $\Delta T'_b$ are the observed values of ΔT_f and ΔT_b in the case of such solutes. i is also related to the molecular weights of the solute according to

$$i = \frac{M\,(\text{normal})\,\text{of solute}}{M'\,(\text{abnormal})\,\text{of solute}}$$

$$(M \text{ or } M' = \text{mol. wt of solute as the case may be}) \quad (9.17)$$

The van't Hoff factor, i, is also related to α (degree of dissociation) of the electrolyte and the number of ions, j, obtainable from a molecule of the electrolyte (for example, K_2SO_4, $j = 3$) according to

$$\alpha = \frac{i-1}{j-1} \quad (9.18)$$

This equation is however, not exact.

(ii) In the case where a substance forms aggregates in a solvent (for example, acetic acid as solute in benzene), the observed molecular weight is higher than the normal molecular weight and the degree of aggregation α_{agg} is given by

$$\alpha_{agg} = \frac{i-1}{\frac{1}{x}-1} \quad (9.19)$$

where x is the number of simple molecules forming the aggregate species. In the case of acetic acid in benzene, $x = 2$.

9.3.6 Osmotic Coefficient

The abnormal behaviour of an electrolytic solute can also be expressed as

$$\Delta T_f = K_f m(1-\theta) \quad (9.20)$$

Let us consider freezing point as an example. $(1 - \theta)$ is referred to as the osmotic coefficient). The mean activity coefficient of the solute can be calculated at different molalities from the equation,

$$\ln \gamma_{\pm} = -\theta - \int_0^m \frac{\theta}{m}\,dm \quad (9.21)$$

From the experimental data of ΔT_f and m, $(1 - \theta)$ is calculated and θ/m is plotted at various m values and subsequently, the integral is evaluated graphically. Thus $\gamma_{\pm}$ can be calculated from colligative property measurements.

9.4 Equilibria in Single Component Systems Such as Solid-Liquid, Liquid-Vapour Equilibria

The fundamental requirement for equilibrium between two phases (such as water–ice or water–water vapour) is that the chemical potential must be the same in the two phases at the given temperature and pressure. If the phases are represented by δ and δ',

$$\mu_\delta(T,P) = \mu_{\delta'}(T,P) \quad (9.22)$$

When the temperature and pressure are changed by a small amount, the new equilibrium, under which the chemical potentials change to new values, also requires

$$d\mu_\delta = d\mu_{\delta'} \quad (9.23)$$

where $d\mu_\delta$ and $d\mu_{\delta'}$ are the changes in μ as a result of changes in T and P. Using the relation $d\mu = VdP - SdT$, Eq. 9.23 may be obtained in the form relating change of equilibrium pressure with temperature as

$$\frac{dP}{dT} = \frac{\Delta S}{\Delta V} \quad (9.24)$$

which is known as the *Claypeyron equation*. ΔS and ΔV are the changes in entropy and volume per mole associated with the transfer of the substance from phase δ to δ'.

Since $\Delta H = T\Delta S$ for phase equilibria, Eq. 9.24 is given by

$$\frac{dP}{dT} = \frac{\Delta H}{T\Delta V} \quad (9.25)$$

For solid–liquid equilibria, Eq. 9.25 may be more explicitly given as,

$$\frac{dP}{dT} = \frac{\Delta H_{fusion}}{T\Delta V_{fusion}} \quad (9.26)$$

where ΔH_{fusion} is the enthalpy of fusion $\left(\overline{H_l} - \overline{H_s}\right)$ and ΔV_{fusion} is the change in molar volume between liquid and solid, $\left(\overline{V_l} - \overline{V_s}\right)$. In a similar way, for liquid–vapour equilibria, the above equation may be written as

$$\frac{dP}{dT} = \frac{\Delta H_{vap}}{T\Delta V_{vap}} \quad (9.27)$$

where, ΔH_{vap} and ΔV_{vap} have their usual significance. An integral form of Eq. 9.26 is sometimes useful and is given by

$$\Delta p = \frac{\Delta H_{fusion}}{\Delta V_{fusion}} \frac{\Delta T}{T_{m.p.}} \quad (9.28)$$

where ΔT is the change in melting point for a change in pressure ΔP and $T_{m.p.}$ is the melting point under the pressure at that temperature. An analogous equation (similar to Eq. 9.28) is valid in the case of liquid–vapour equilibria and may be given as

$$\Delta p = \frac{\Delta H_{vap}}{\Delta V_{vap}} \frac{\Delta T}{T_{b.p.}} \quad (9.29)$$

In the case of liquid–vapour equilibria (also for solid–vapour equilibria), Eq. 9.27 may be simplified by considering that $V_{vap} - V_{liq} \approx V_{vap}$ and assuming that the vapour behaves ideally, can be written in the form

$$\frac{d\ln p}{dT} = \frac{\Delta H_{vap}}{RT^2} \quad (9.30)$$

which, on integration gives

$$\ln \frac{p_2}{p_1} = \frac{\Delta H_{vap}}{R}\left[\frac{T_2 - T_1}{T_1 T_2}\right] \quad (9.31)$$

Equation 9.31 may be used to calculate the vapour pressure of a liquid, say at temperature T_2 given the vapour pressure at temperature T_1 and using ΔH_{vap} as a known parameter. Equation 9.31 is known as the *Clausius–Clapeyron equation*.

It is important to note that proper units must be chosen, for example, in calculating the change in melting point or boiling point (dT) with change in pressure as K Pa^{-1} or K atm^{-1} (as the case may be). For this, ΔH is usually expressed as J mol^{-1} which is equivalent to 1 kg m^2 s^{-2} mol^{-1} and ΔV in m^3 mol^{-1} so that dT/dP is calculated as K atm^{-1} (or K cm^{-1}).

9.5 Thermodynamic Aspects of Electrolyte Solutions

Electrolyte solutions, especially those of strong electrolytes, show considerable deviations from ideal behaviour (as we have seen earlier from the studies of colligative properties). This is due to the fact that the ions are charged and strong interionic forces, which operate over large distances, exist in solution. The chemical potential of an electrolyte, MA, dissociating according to

$$\mathrm{MA} \rightleftharpoons x_+\mathrm{M}^+ + y_-\mathrm{A}^-$$

where x_+ and y_- are the number of cations and anions obtained from a molecule of the electrolyte. The chemical potential is given by

$$\mu = \mu^o + RT\ln a = x_+\mu_+^o + y_-\mu_-^o + \gamma RT\ln a_\pm \qquad (9.32)$$

In Eq. 9.32, a is the activity of the electrolyte, the first two terms on the right represent the standard chemical potentials due to the cation and anion and $a_\pm$ is the mean activity of the electrolyte. Note that the activity of the electrolyte,

$$a = a_\pm^\gamma \quad \text{where } \gamma = x_+ + y_- \quad (x = \text{cations and } y = \text{anions}) \qquad (9.32a)$$

The mean activity of the electrolyte $a_\pm$ is related to the individual activities of the cation and anion by the relation

$$a_\pm = \left(a_+^{x_+} a_-^{y_-}\right)^{\frac{1}{\gamma}} \left(\gamma = x_+ + y_-\right) \qquad (9.33)$$

Expressing the activity as a product of molality and activity coefficient in Eq. 9.33, we have

$$m_\pm\gamma_\pm = \left(m_+^{x_+}\gamma_+^{x_+} m_-^{y_-}\gamma_-^{y_-}\right)^{\frac{1}{\gamma}}$$

$$= \left(m_+^{x_+} m_-^{y_-}\right)^{\frac{1}{\gamma}} \left(\gamma_+^{x_+}\gamma_-^{y_-}\right)^{\frac{1}{\gamma}} \qquad (9.34)$$

from which it can be seen that

$$m_\pm = \left(m_+^{x_+} m_-^{y_-}\right)^{\frac{1}{\gamma}} \text{ and } \gamma_\pm = \left(\gamma_+^{x_+}\gamma_-^{y_-}\right)^{\frac{1}{\gamma}} \qquad (9.35)$$

If m is the molality of the electrolyte, $m_+ = x_+ m$, $m_- = y_- m$ so that

$$m_\pm = \left(x_+^{x_+} y_-^{y_-}\right)^{\frac{1}{\gamma}} \left(m^{x_+} m^{y_-}\right)^{\frac{1}{\gamma}} = \left(x_+^{x_+} y_-^{y_-}\right)^{\frac{1}{\gamma}} m \quad (\text{since } \gamma = x_+ + y_-) \qquad (9.36)$$

9.6 Debye–Hückel Theory

9.6.1 Relation to the Activity Coefficient of an Ion in Dilute Solutions

A brief mention of the Debye–Hückel model of a central ion surrounded by the ionic atmosphere of the opposite sign has been made in Section 9.6. The deviation of electrolytic solutions from ideal behaviour arises from the strong electrostatic forces operating between the oppositely charged ions in solution. The activity coefficient of an ion (also the mean activity coefficient of an electrolyte), which is a measure of the deviation of the electrolyte from ideality, is related to the extra energy required to charge an ion in the presence of all the other ions (over and above that required to charge an isolated sphere in a medium of the given dielectric constant).

9.6.2 Debye–Hückel Limiting Law

The equation which expresses the variation of mean activity coefficient of an electrolyte as a function of the ionic strength of the medium is known as the Debye–Hückel limiting law and is given by

$$\log\gamma_\pm = -A\left|Z_+Z_-\right|\sqrt{\mu} \qquad (9.37)$$

where A is a constant depending on the temperature and the dielectric constant of the medium and Z_+, Z_- are the charges on ions. For water at 298 K, $A = 0.500$ (dielectric constant of water = 80 at this temperature). It is seen that $\log\gamma_\pm$ is a linear function of $\sqrt{\mu}$ in dilute solutions and the slope is $-A\left|Z_+Z_-\right|$. The slope is proportional to Z_+Z_-. The ionic strength of the solution is defined by

$$\mu = \frac{1}{2}\sum_i m_i Z_i^2$$

where m_i = molality of the ion i and the Z_i is its charge. Sometimes C_i instead of m_i is also used. The equation is satisfactory up to $m = 0.005$ of an electrolyte.

9.6.3 Thickness of the Ionic Atmosphere

The model envisaging a central ion surrounded by an ionic atmosphere of opposite sign (its magnitude being equal to that of the central ion) is the basis for the Debye–Hückel theory of strong electrolytes. A better understanding of the ionic atmosphere concept is possible when one attempts to calculate the charge density at a point r from a central ion using the Poisson's equation and applying the Boltzmann distribution for finding the presence of oppositely charged ions around a central ion. An expression for the thickness of the ionic atmosphere arises in calculating the charge density and it is defined as

$$x^2 = \frac{8\pi Ne^2}{1000DkT}\mu \qquad (9.38)$$

in which $1/x$ is defined as the *thickness of the ionic atmosphere* and it has the dimensions of length. It is seen that $1/x$ increases as the ionic strength of the medium decreases and in very dilute solutions, the central ion is like a point charge in comparison to the size of the ionic atmosphere.

In Eq. 9.38, N = Avogadro number, e = electronic charge, D = dielectric constant of the medium, k = Boltzmann constant, T = temperature in K and μ = ionic strength of the medium.

9.6.4 Effect of Ionic Strength on Acid-Base and Solubility Equilibria

Acid-base equilibria: Consider the dissociation equilibrium of a weak acid, HA, in aqueous solution given by

$$HA \rightleftharpoons H^+ + A^- \tag{9.39}$$

The thermodynamic dissociation constant of the acid is given by

$$K = \frac{a_{H^+} a_{A^-}}{a_{HA}} \tag{9.40}$$

where a represents the activities of the respective species. Writing Eq. 9.40 in terms of molalities and activity coefficients,

$$K = \frac{m_{H^+} m_{A^-}}{m_{HA}} \frac{\gamma_{H^+}\gamma_{A^-}}{\gamma_{HA}} = \frac{\alpha m \times \alpha m}{(1-\alpha)m} \frac{\gamma_\pm^2}{\gamma_{HA}} \tag{9.41}$$

Assuming $\alpha <<< 1$,

$$K = \alpha^2 m \frac{\gamma_\pm^2}{\gamma_{HA}} = \alpha^2 m \gamma_\pm^2$$

$$(\gamma_{HA} \approx 1 \text{ because HA is an uncharged species}) \tag{9.42}$$

From Eq. 9.42, it is seen that at a given m, since $\gamma_\pm < 1$, α must increase to keep K constant, that is, the degree of dissociation increases as the ionic strength increases. The increase in the ionic strength may be due to an increase in m, or due to the addition of an inert electrolyte. The effect of α is greater in the latter case than in the former. It is possible to calculate K by using the experimentally determined α at any given m and calculating $\gamma_\pm$ by using the D–H limiting law.

Solubility equilibria: Considering the solution equilibrium of a sparingly soluble salt such as $BaSO_4$,

$$K_{SP} = a_{Ba^{2+}} a_{SO_4^{2-}} = m_{Ba^{2+}} m_{SO_4^{2-}} \gamma_{Ba^{2+}} \gamma_{SO_4^{2-}}$$

$$K_{SP} = m^2 \gamma_\pm^2 = S^2 \gamma_\pm^2$$

$$\left(\text{since } S = \text{solubility of salt} = m_{Ba^{2+}} = m_{SO_4^{2-}}\right) \tag{9.43}$$

K_{sp} is a thermodynamic constant and depends only on temperature. In the presence of an inert salt (not having an ion common with the salt under solubility consideration), the ionic strength increases and thereby $\gamma_\pm$ decreases. Therefore to keep K_{sp} constant, the solubility S increases. Note that the increased solubility in the presence of an inert electrolyte is much smaller in comparison to the decrease in solubility observed in the presence of an ion having an ion in common with the ions of the salt involved in the study. For example, addition of K_2SO_4 on $BaSO_4$ solubility.

Worked Out Problems

Different Concentration Units and Related Examples

Problem 9.1

An aqueous solution of acetic acid at 293 K containing 82.0 g of acid per dm^3 of the solution has a density of 1.01 kg dm^{-3}. What is its concentration in (i) molarity (ii) molality (iii) mole fraction and (iv) normality?

Solution

Molar mass of acid = 60

No. of moles of acid = 82.0/60.0 = 1.367

Molarity = 1.367 mol dm^{-3}

Molality: 1010 – 82 = 928 g of water contains 1.367 moles of acid

1000 g of water contains 1000×1.367/928 = 1.473 moles

Molality = 1.473 mol kg^{-1}

Mole fraction: moles of solvent = 928/18 = 51.56; moles of acid = 1.367

Mole fraction of acid = 1.367/(51.56+1.367) = 1.367/52.927 = 0.0258

Normality: 1.367 N since the acid is monobasic.

Problem 9.2

The mole fraction of H_2SO_4 in an aqueous solution is 0.0098. Calculate the molality of the acid, molality of the H$^+$ ions assuming complete dissociation. What is the pH of the solution, if the density of the solution is 1030 kg dm^{-3} at 298 K?

Solution

Moles of acid = 0.0098; moles of water = 0.9902

Weight of water = 0.9902×18 = 17.82 g

Weight of acid = 0.0098×98 = 0.9604 g

Molality of acid = 1000×0.0098/17.82 = 0.55

Molality of H$^+$ = 0.55×2 = 1.10

Total weight of solution = 18.78 g

Volume of solution = 18.78/1.03 = 18.23 cm^3

Moles of acid in 1000 cm^3 solution = 1000×0.0098/18.23 = 0.5376

[H$^+$] = 0.5376×2 = 1.075 pH = –log[1.075] = –0.031

A pH lower than zero implies that the [H$^+$] concentration is greater than 1.

Molecular Weights Based on Lowering of Vapour Pressure, Ideal Mixtures and Raoult's Law

Problem 9.3

The vapour pressure of a solution of an organic solute in water at 300 K is 3.768 kPa, while the vapour pressure of water at the same temperature is 3.78 kPa. If 0.06 kg of solute is present in 1 kg of water, what is the molecular weight of the organic compound?

Solution

Mole fraction of the solute = $(P^0 - P)/P^0$ = (3.78 – 3.768)/3.78 = 3.175×10^{-3}

$$3.175\times10^{-3} = \frac{\frac{60}{M}}{\frac{60}{M}+\frac{1000}{18}} \approx \frac{\frac{60}{M}}{\frac{1000}{18}}$$

(assuming the solution is dilute)

$$0.1764 = \frac{60}{M} \text{ or, } M = \frac{60}{0.1764} = 340.1$$

Problem 9.4

Pentane (C_5H_{12}) and hexane (C_6H_{14}) have vapour pressures of 68.13 kPa and 20.0 kPa respectively at 298 K. 3 moles of pentane and 5 moles of hexane are mixed. For this ideal mixture, calculate the (i) partial vapour pressures and the total vapour pressure at equilibrium at this temperature, (ii) mole fractions of the two components in the vapour phase.

Solution

Mole fraction of pentane = 3/8 = 0.375
Mole fraction of hexane = 5/8 = 0.625

(i) $p_{\text{pentane}} = 68.13\times0.375 = 25.55$ kPa
$p_{\text{Hexane}} = 20.0\times0.625 = 12.5$ kPa
Total pressure = 25.55 + 12.5 = 38.05 kPa

(ii) Mole fraction of pentane in vapour pressure = 25.55/38.05 = 0.671

Mole fraction of pentane in vapour pressure = 12.5/38.05 = 0.329

Problem 9.5

Cyclopentane and hexane form ideal mixtures. Their vapour pressures in the pure state are 44.13 kPa and 20.0 kPa respectively, at 298 K. A certain mix of both shows a total vapour pressure of 29.65 kPa and the composition of the vapour is such that $X_{\text{cyclopentane}} = 0.595$. What is the composition of the original mixture?

Solution

Using Dalton's law, partial vapour pressure of cyclopentane = $0.595\times29.65 = 17.64$ kPa

Using Dalton's law, partial vapour pressure of hexane = $0.405\times29.65 = 12.00$ kPa

Using Raoult's law, $17.64 = 44.13\times X_{\text{cyclopentane}}$ (in liquid phase)

$X_{\text{cyclopentane}}$ (in liquid phase) = 17.64/44.13 = 0.400

X_{hexane} (in liquid phase) = 12.00/20.00 = 0.600

Composition of cyclopentane: hexane as mole ratio = 0.4 : 0.6 = 2 : 3

Problem 9.6

Chlorobenzene and bromobenzene mixtures form ideal solutions and their vapour pressures in the pure state at 413 K are 125.3 kPa and 66.10 kPa respectively. Calculate the composition of the liquid mixture which boils at 413 K under a pressure of 101.325 kPa. Find the composition of the vapour at this temperature.

Solution

Total vapour pressure

$$= 101.325 = p^0_{\text{Cl-B}}\times X_{\text{Cl-B}} + p^0_{\text{Br-B}}\times X_{\text{Br-B}}$$

$$= 125.2\times X_{\text{Cl-B}} + 66.10\times(1-X_{\text{Cl-B}})$$

$$= (125.2-66.10)\times X_{\text{Cl-B}} + 66.10$$

$$X_{\text{Cl-B}} \text{ in liquid phase} = \frac{35.22}{59.10} = 0.596$$

$$X_{\text{Br-B}} \text{ in liquid phase} = 0.404$$

Using Dalton's law

$$X'_{\text{Cl-B}} \text{ in vapour phase} = \frac{125.2\times0.596}{101.325} = 0.736$$

$$X'_{\text{Br-B}} \text{ in vapour phase} = \frac{66.10\times0.404}{101.325} = 0.264$$

Problem 9.7

The vapour pressure data of a few pure liquids and the total vapour pressure observed at a certain composition of the mixture are given below. State whether the mixture shows positive or negative deviations from Raoult's law and give reason for the same.

S. No.	Temperature (K)	Liquid mixtures	Vapour pressure of pure component (kPa)	Total vapour pressure (kPa) observed at $X_{0.5}$ of each (X = mole fraction)
1	305.2	Acetone + Carbon disulfide	Acetone = 44 CS_2 = 66.7	84.0
2	305.2	Acetone + $CHCl_3$	Acetone = 44 $CHCl_3$ = 37.3	32.0
3	298	Pyridine + Acetic acid	Pyridine = 2.4 Acetic acid = 2.2	1.5
4	330	Acetone + Methanol	Acetone = 105 Methanol = 74	105

Solution

(i) Acetone + CS_2: vapour pressure of the mixture at $X = 0.5$

$$P = 44\times0.5 + 66.7\times0.5 = 22 + 33.25 = 55.25 \text{ kPa}$$

Since the observed vapour pressure is greater than that predicted by Raoult's law, it shows positive deviations. These deviations are due to the fact that the cohesive forces between acetone and carbon disulfide molecules are weaker than those between acetone or CS_2 molecules among themselves.

(ii) Acetone + $CHCl_3$: vapour pressure of the mixture at $X = 0.5$

$$P = 44\times0.5 + 37.5\times0.5 = 22 + 18.7 = 40.7 \text{ kPa}$$

The observed vapour pressure is smaller than that expected if the mixture obeyed Raoult's law. Therefore, this mixture shows a negative deviation from Raoult's law. It is expected because there are strong forces of cohesion between the molecules of acetone and chloroform as compared to the forces of cohesion between the molecules of acetone or between the molecules of chloroform amongst themselves. Both acetone and chloroform molecules are polar and it is possible that H-bonding exists between the H atom of $CHCl_3$ and the O atom of acetone.

(iii) Pyridine + Acetic acid: vapour pressure of the mixture at $X = 0.5$

$$P = 2.4 \times 0.5 + 2.2 \times 0.5 = 1.2 + 1.1 = 2.3 \text{ kPa}$$

The observed vapour pressure is less than that predicted by Raoult's law; thus this mixture shows a negative deviation from Raoult's law. Pyridine is weakly basic and acetic acid is weakly acidic. Hence, it is possible that weak acid–base interactions exist among their molecules. These interactions are stronger compared to the interactions among the individual molecules of the two solvents which accounts for the observed behaviour.

(iv) Acetone + Methanol: vapour pressure of the mixture at $X = 0.5$

$$P = 105 \times 0.5 + 74 \times 0.5 = 52.5 + 37 = 89.5 \text{ kPa}$$

The observed vapour pressure is higher than that predicted by Raoult's law. Hence it shows positive deviations from Raoult's law. This is due to the fact that methanol molecules are more polar compared to acetone molecules and strong H-bonding exists between methanol molecules unlike between the acetone molecules. Thus, the dissimilar chemical nature of the solvents leads to positive deviations from Raoult's law.

Problem 9.8

The vapour pressure of water at 298 K is 3.17 kPa. A 0.2 molal solution of urea (mol. wt. = 60) is isopiestic with a solution of an organic compound containing 20.44 g in 540 g of water. What is the molecular weight of the organic compound?

Solution

Molality of urea is 0.2, that is, 0.2 mol of urea in 1000 g of water.

$$\text{Mole fraction of urea} = \frac{0.2}{0.2 + \frac{1000}{18}} = \frac{0.2}{55.75} = 3.58 \times 10^{-3}$$

$$\frac{p^0 - p}{p^0} = 1 - \frac{p}{p^0} = 3.58 \times 10^{-3} \Rightarrow \frac{p}{p^0} = 0.99642$$

Vapour pressure of the solution $= 3.17 \times 0.99642 = 3.158$ kPa

The solution of the unknown compound has the same vapour pressure, that is, 3.158 kPa, because the two solutions have the same vapour pressure.

$$\Delta p = \frac{3.17 - 3.158}{3.17} = \frac{\frac{20.44}{M}}{\frac{20.44}{M} + \frac{540}{18}} = \frac{\frac{20.44}{M}}{\frac{20.44}{M} + 30}$$

$$\frac{\frac{20.44}{M} + 30}{\frac{20.44}{M}} = 1 + \frac{30 \times M}{20.44} = \frac{3.17}{0.012} = 264.2$$

$$M = \frac{263.2 \times 20.44}{30} = 179.3$$

Problem 9.9

A mixture consisting of one mole of benzene and 4 moles of bromobenzene has a vapour pressure of 9.54 kPa at 298 K. When one mole of benzene is added to the mixture, the vapour pressure rose to 10.54 kPa.

(i) Calculate the vapour pressure of pure benzene and bromobenzene.

(ii) Calculate the mole fractions of benzene and bromobenzene in the vapour phase of the original mixture and the second mixture containing excess benzene.

(iii) What is the mole fraction ratio of benzene in the vapour phase to the liquid phase in the second mixture? Assume the mixtures are ideal.

(iv) What is the activity of benzene and what is its activity coefficient in the original mixture?

Solution

(i) Mole fraction of benzene $\frac{1}{5} = 0.20$

Mole fraction of bromobenzene = 0.8

Vapour pressure of the mixture

$$= P_B^0 \times 0.2 + P_{BrB}^0 \times 0.8 = 9.54 \text{ kPa} \qquad (1)$$

When one mole of benzene is added,

$$X_B = \frac{2}{6} = 0.33; \; X_{BrB} = \frac{4}{6} = 0.67$$

Vapour pressure of the mixture

$$= P_B^0 \times 0.33 + P_{BrB}^0 \times 0.67 = 10.54 \text{ kPa} \qquad (2)$$

Multiplying Eq. (1) by 0.33 and Eq. (2) by 0.2, we get,

$$P_B^0 \times 0.066 + P_{BrB}^0 \times 0.264 = 3.15 \qquad (3)$$

$$P_B^0 \times 0.066 + P_{BrB}^0 \times 0.134 = 2.11 \qquad (4)$$

Eq. (3) – Eq. (4)

$$P_{BrB}^0 \times 0.130 = 1.04; \; P_{BrB}^0 = \frac{1.04}{0.130} = 8.0 \text{ kPa}$$

Substituting the value in Eq. (1),

we get $P_B^0 \times 0.2 + 8 \times 0.8 = 9.54$ kPa

$$P_B^0 = \frac{3.14}{0.2} = 15.7 \text{ kPa}$$

(ii) Mole fraction of benzene in the vapour phase

$$= \frac{15.7 \times 0.2}{9.54} = \frac{3.14}{9.54} = 0.329$$

Mole fraction of bromobenzene in the vapour phase

$$= \frac{8 \times 0.8}{9.54} = \frac{6.4}{9.54} = 0.671$$

Mole fraction of benzene in the vapour phase in the second mixture

$$= \frac{15.7\times 0.33}{10.54} = \frac{5.18}{10.54} = 0.491$$

(iii) $\dfrac{\text{Ratio of } X_B \text{ in vapour phase}}{\text{Ratio of } X_B \text{ in liquid phase}} = \dfrac{0.491}{0.33} = 1.49$

(iv) $\dfrac{P}{P^0} = a;\ a = \dfrac{15.7\times 0.2}{15.7} = 0.2;\ X_B = 0.2$

$$\gamma_B = \frac{a}{X_B} = \frac{0.2}{0.2} = 1.0$$

(since these mixtures obey Raoult's law)

Ideal Solubility

Problem 9.10

(a) Naphthalene melts at 353 K. Its heat of fusion is 148.7 J g^{-1}. Calculate the ideal solubility of naphathalene at 298 K.

(b) If the ideal solubility, in mole fraction units, of diphenyl is 0.374 at 298 K and its melting point is 69°C, calculate the enthalpy of fusion of diphenyl in J g^{-1} (molecular formula of diphenyl = $C_{12}H_{10}$).

Solution

$$\ln X = \frac{\Delta H_{\text{fusion}}}{R}\left(\frac{1}{T_o} - \frac{1}{T}\right)$$

T_0 = freezing point of naphathalene, mol. wt. = 128.3

$$= \frac{148.7\times 128.3}{8.314}\left[\frac{1}{353} - \frac{1}{298}\right]$$

$$= -\left[2294.7\times\frac{55}{353\times 298}\right] = -1.199$$

$$\log X = \frac{-1.199}{2.303} = -0.502;\ X = 0.301$$

$$\ln 0.374 = \frac{\Delta H_{fusion}}{R}\left(\frac{1}{T_o} - \frac{1}{T}\right) = \frac{\Delta H_{fusion}}{8.314}\times\frac{298-342}{298\times 342}$$

$$= -\frac{\Delta H_f \times 44}{8.314\times 342\times 298\times 2.303}$$

$$-0.4271 = -\frac{\Delta H_f \times 44}{8.314\times 342\times 298\times 2.303}$$

From which $\Delta H_f = \dfrac{0.4271\times 8.314\times 342\times 298\times 2.303}{44}$

$$= 18942 \text{ J mol}^{-1}$$

$$\Delta H_f = \frac{18942}{128.2} = 147.7 \text{ J g}^{-1}$$

Henry's Law, Bunsen Coefficient

Problem 9.11

The Henry's law constant for argon in water at 303 K is 4.023×10^6 kPa. What is its Bunsen coefficient at 273 K?

Solution

Bunsen coefficient is defined as the volume of gas in cm^3 measured at 0°C and 1 atm pressure dissolved in 1 cm^3 of solvent under a partial pressure of the gas of 1 atm.

$$P = K_H X_{Ar}$$

$$101.325 = 4.023\times 10^6\times X_{Ar}$$

$$X_{Ar} = 101.325/4.023\times 10^6 = 25.19\times 10^{-6}$$

$$X_{Ar} = \frac{n_{Ar}}{n_{Ar} + n_{H_2O}} \approx \frac{n_{Ar}}{n_{H_2O}}$$

Or, $n_{Ar} = 25.19\times 10^{-6}\times 55.55 = 1.4\times 10^{-3}$ mol L^{-1}

Using the relation $C_1T_1 = C_2T_2$ (since P is same)

$$\frac{C_1}{C_2} = \frac{T_2}{T_1};\ C_2 = \frac{C_1T_1}{T_2} = \frac{1.4\times 10^{-3}\times 303}{273} = 1.554\times 10^{-3}\ \text{L}^{-1}$$

No. of moles in 1 cm^3 of solvent = $1.554\times 10^{-3}\times 10^{-3}$ = 1.554×10^{-6}

No. of litres of argon in 1 cm^3 of solvent at 273 K = $1.554\times 10^{-6}\times 22.4\times 10^3 = 34.86\times 10^{-3}$ cm^3

Therefore, Bunsen coefficient = 0.03486

Problem 9.12

The Henry's law constant for H_2S gas at 298 K is 0.179 mol dm^{-3} bar^{-1}. (i) What will be the value of this constant in units of bar? (ii) What is the solubility of the gas at a pressure of 0.25 bar? Given the first dissociation constant of H_2S as 9.1×10^{-8} (ignore the second dissociation of the gas) and considering the solubility at 0.25 bar, what is the depression in freezing point caused by H_2S? (K_f of water = 1.86 K kg mol^{-1})

Solution

(i) $K_H = 0.179$ mol dm^{-3} bar^{-1}

No. of moles of water in 1 dm^3 = $\dfrac{1000}{18} = 55.55$

Mole fraction of $H_2S = \dfrac{0.179}{0.179 + 55.55}$

$$= \frac{0.179}{55.729} = 3.21\times 10^{-3}$$

Using Henry's law in the form, $p = K'_H \times X$

When $p = 1$ bar, $K'_H = \dfrac{p}{X_{H_2S}} = \dfrac{1}{3.21\times 10^{-3}} = 312$ bar

(ii) At a pressure 0.25 bar, the concentration of H_2S in solution $= 0.179\times 0.25 = 0.045$ M

Considering the dissociation of the acid

$$H_2S(aq) \rightleftharpoons H^+ + HS^-$$

and if α is the degree of dissociation,

$$[H^+] = [HS^+] = \alpha\times 0.045;\ [H_2S] = (1-\alpha)\times 0.045$$

$$\frac{\alpha^2\times(0.045)^2}{(1-\alpha)\times 0.045} = 9.1\times 10^{-8};\ \alpha^2 \approx \frac{9.1\times 10^{-8}}{0.045} = 202.2\times 10^{-8}$$

$$\alpha = 14.22\times 10^{-4} = 1.422\times 10^{-3}$$

Total concentration of ions + Undissociated gas

$$= \left(2\times 1.422\times 10^{-3}\times 0.045\right) + 0.9986\times 0.045 = 0.04494$$

$$\Delta T_f \text{ observed} = 1.86 \times 0.04494 = 0.084°$$

The small extent of dissociation does not have much effect on ΔT. The ΔT expected is almost the same as that when the acid is undissociated.

Depression of Freezing Points and van't Hoff Factor

Problem 9.13

Calculate the weight of 1,2-ethanediol ($C_2H_6O_2$) required to be used in 5 kg of water so that the solution does not freeze above –10°C (K_f water = 1.86 K kg mol^{-1}).

Solution

$$\Delta T_f = K_f \times m; \; m = 10/1.86 = 5.376$$

5.376 moles of ethanediol are required for 1 kg water

5 kg of water require = 5×5.376 moles = 5×5.376×62 = 1667 g of ethanediol

Problem 9.14

(a) The enthalpy of fusion of acetic acid is 11.22 kJ mol^{-1} and its melting point is 289.6 K. An organic solute weighing 0.41 g was dissolved in 100 g of solvent and showed a depression of freezing point of 0.187°. What is the molar mass of the compound?

(b) Assuming that it is a 1:1 electrolyte, dissociating to the extent of 20%, calculate (i) the van't Hoff factor i, (ii) total molality of the salt, and (iii) the expected ΔT_f.

Solution

(a) $$K_f = \frac{RT_f^2 M}{1000\,\Delta H_f^o} = \frac{8.314 \times (289.6)^2 \times 60}{1000 \times 11,220}$$

$$= 3.73 \text{ K kg mol}^{-1}$$

$$\Delta T_f = K_f \times m; \; m = \frac{0.187}{3.73} = 0.05$$

1000 g of solvent contains 4.1 g of solute

Or $\frac{4.1}{\text{M}} = 0.05 \text{ M} = \frac{4.1}{0.05} = 82$

(b) If the salt, MA, dissociates as $\text{MA} \rightleftharpoons \text{M}^+ + \text{A}^-$

At equilibrium, we have $[\text{M}^+] = 0.05 \times 0.2$; $[\text{A}^-] = 0.05 \times 0.2$; $[\text{MA}] = [1 - 0.2] \times 0.05$

Or total $m = 0.01 + 0.01 + 0.04 = 0.06$

$$\propto = \frac{i-1}{\gamma - 1}; \; 0.2 = \frac{i-1}{2-1}; \; i = 1.2$$

$$\Delta T_f \,(\text{expected}) = i \times K_f \times m = 1.2 \times 3.73 \times 0.05 = 0.224°$$

Problem 9.15

5 g of sulfur was dissolved in 100 g of carbon disulfide solvent. Carbon disulfide boils at 46.3°C and its enthalpy of vaporisation is 352 J g^{-1}. Calculate the molar mass of sulfur and explain the result, given that the boiling point of the solution is 46.77°C.

Solution

Boiling point of CS_2 = 319.3 K; ΔH_f = 26.75 kJ mol^{-1}

$$K_b = \frac{RT_{bf}^2 M}{1000\,\Delta H_f^o} = \frac{8.314 \times (319.3)^2 \times 76}{1000 \times 26750} = 2.41 \text{ K kg mol}^{-1}$$

$$\Delta T_b = 0.47 = 2.41 \times m \Rightarrow m = \frac{0.47}{2.41} = 0.195$$

$$\frac{50}{M} = 0.195; \; M = 256.4$$

Since the molar mass (or at. wt.) of sulfur is 32, and the observed result is 256.4, sulfur forms aggregates in this solvent and it exists as 256.4/32 = 8, S_8 molecules.

Problem 9.16

(a) 1.5 g of benzoic acid is dissolved in 50 g of benzene and its freezing point was found to be 277.82 K. Given that K_f of benzene is 5.1 K kg mol^{-1} and its freezing point is 278.45 K, calculate the molecular weight of the solute and explain the result.

(b) If K_b of benzene is 2.53 K kg mol^{-1} and the boiling point of benzene is 80.2°C, what is the boiling point of the solution under (a)? What is the van't Hoff's factor i?

(c) If the vapour pressure of the pure benzene is 74.77 mmHg, calculate the vapour pressure of the solution given under (a). If benzoic acid behaves as normal molecules in benzene, will the vapour pressure of the solution be more or less then when it is aggregated?

Solution

(a) 1000 g of benzoic acid contains 1.5×20 = 30.0 g of benzoic acid

$$\Delta T_f = 0.63 = 5.1 \times \frac{30}{M}; \; M = 5.1 \times \frac{30}{0.63} = 242.86$$

Benzoic acid has a normal molecular weight of 122 and since the observed molecular weight is almost twice its value, it may be concluded that it exists as a dimer.

(b) Molality of the solution on the basis of observed molecular weight = 30/242.86

$$\Delta T_b = 2.53 \times \frac{30}{242.86} = 0.312°$$

Boiling point of the solution = 353.2 + 0.312 = 353.512 K

Elevation in the boiling point if it has normal molecular weight =

$$\Delta T_b = 2.53 \times \frac{30}{122} = 0.622°$$

$$\therefore \text{ van't Hoff factor } i = \frac{\Delta T_{obs}}{\Delta T_{theo}} = \frac{0.312}{0.622} = 0.502$$

(c) Number of moles of benzoic acid = 1.5/242.86 = 0.00618

Number of moles of solvent = 50/78 = 0.641

Mole fraction of the solute = 0.00618/0.641 = 0.00955

$$\frac{P^0 - P}{P^0} = 0.00955 \Rightarrow 1 - \frac{P}{P^0} = 0.00955$$

$$\Rightarrow \frac{P}{P^0} = 0.99045$$

Hence, $P = 74.77 \times 0.99045 = 74.056$ mmHg

When benzoic acid behaves as normal molecules in benzene, the vapour pressure of the solution will be less than when it is under aggregated condition.

Problem 9.17

50 g of potassium sulfate and 30 g of lithium bromide are dissolved in 1 kg of water and the resulting solution shows a freezing point of –2.569°C. If lithium bromide is completely dissociated, calculate the molar mass of K_2SO_4. What do you conclude? Calculate the value of the van't Hoff factor i and any other parameter that may be obtained from it. What is the total molal concentration of all the species from K_2SO_4? (K_f of H_2O = 1.86 K kg mol^{-1}, molar mass of K_2SO_4 = 174 and molar mass of LiBr = 86.8).

Solution

Since LiBr is completely dissociated, its molar mass calculated from freezing point measurements is = 86.8/2 = 43.4

$$\text{Total molality of the mixture} = \left(\frac{50}{M_{app}} + \frac{30}{43.4}\right)$$

$$\Delta T_f = 2.569 = \left(\frac{50}{M_{app}} + 0.691\right) \times 1.86$$

$$\Rightarrow \quad \frac{50}{M_{app}} + 0.691 = \frac{2.569}{1.86} = 1.381$$

$$M_{app} = \frac{50}{1.381 - 0.691} = \frac{50}{0.69} = 72.46$$

Since the observed molar mass of K_2SO_4 is less than its normal value, it is concluded to be dissociating in solution.

$$\text{van't Hoff factor } i = \frac{M_{normal}}{M_{app}} = \frac{174}{72.46} = 2.40$$

From this, α, the degree of dissociation of K_2SO_4 can be calculated as

$$\alpha = \frac{i-1}{v-1} = \frac{2.4-1.0}{3-1} = \frac{1.4}{2} = 0.70$$

Writing the dissociation equilibrium as

$$\begin{matrix} K_2SO_4 & \rightleftharpoons & 2K^+ & + SO_4^{2-} \\ (1-\alpha) & & 2\alpha & +\alpha \end{matrix}$$

$$\text{Total molality} = (1-0.7) + 2\times 0.7 + 1\times 0.7$$

$$= 0.3 + 1.4 + 0.7 = 2.4 \times m$$

where m = stoichoimetric molality which is 50/174 = 0.287

Therefore, total molality = 0.287×2.4 = 0.689

Problem 9.18

1,2 ethanediol ($C_6H_6O_2$) is used as an antifreeze agent in cars. In a certain cold place, it is desired that water freezes at 258 K.

(i) How much weight of ethylene glycol (or ethanediol) is needed to prevent separation of ice from 500 g of water?

(ii) Suppose the solution in (i) is cooled to 257 K, how much ice in grams separates from the solution? (K_f of water = 1.86 K kg mol^{-1})

(iii) 3.34 g of ethanediol and 0.568 g of uni-bivalent salt dissolved in 200 g of water gave a freezing point of –0.612°C. Calculate the ΔT due to the ionic compound. What is the van't Hoff factor, i, given the molar mass of compound = 142. What is the osmotic coefficient of the compound?

Solution

(i) $\Delta T_f = 1.86 \times m;\ m = \dfrac{15}{1.86} = 8.064$

$$\text{Weight of glycol needed in 500 g of water} = \frac{8.064 \times 62 \times 500}{1000} = 250 \text{ g}$$

(ii) $\text{Molality of glycol} = \dfrac{16}{1.86} = 8.602$

Wt. of glycol in 1000 g H_2O = 8.602×62 = 533.3 g

Weight of water needed to contain 250 g of glycol

$$= \frac{250 \times 1000}{533.3} = 468.8 \text{ g}$$

Weight of ice separated = 500 – 468.8 = 31.2 g

(iii) $\text{Molality of glycol} = \dfrac{3.34}{62} \times \dfrac{1000}{200} = 0.269$

ΔT due to glycol = 1.86×1.269 = 0.500°

$\therefore \Delta T$ due to ionic compound = 0.612 – 0.500 = 0.112°

$$\text{Molal conc. of ionic compound} = \frac{0.568}{142} \times \frac{1000}{200} = 0.02$$

ΔT expected from ionic compound

= 1.86×0.02 = 0.0372°

$$i = \frac{\Delta T_{obs}}{\Delta T_{theo}} = \frac{0.112}{0.0372} = 3.01$$

$$\text{Osmotic coefficient}(1-\theta) = \frac{\Delta T_f}{K_f m} = \frac{0.112}{1.86 \times 0.02}$$

$$= 3.0 \text{ (same as } i)$$

Molar Masses of Polymers from Osmotic Pressure Measurements

Problem 9.19

A 0.2% (w/v) solution of a synthetic polymer in benzene was subjected to an osmotic pressure measurement to determine its molecular weight. A rise of 0.24 cm of benzene was observed when equilibrium was attained at 298 K. Find the molecular weight of the polymer. (density of solution = 0.90 kg dm^{-3})

Solution

The height of the benzene column must be converted to the height in terms of the height of Hg column using the *hdg* relation.

$$0.24 \times 0.90 \times 980 = h_{Hg} \times 13.56 \times 980$$

$$h_{Hg} = 0.24 \times 0.90 \times \frac{980}{(13.56 \times 980)} = 0.01593 \text{ cm}$$

$$= \frac{0.01593}{76} = 2.1 \times 10^{-4} \text{ atm}$$

$$\pi = \frac{g}{M} \times \frac{1}{V} \times RT$$

where g = weight of solute of molar mass M, V = volume of solution in litres, R = gas constant and T is in K.

$$2.1\times10^{-4} = \frac{2}{M}\times\frac{1}{1V}\times0.0821\times298$$

$$\Rightarrow \quad M = 2\times0.0821\times\frac{298}{2.1\times10^{-4}} = 2.33\times10^{5}\ \text{g mol}^{-1}$$

This value is also known as the number average molecular weight.

Problem 9.20

The molecular weight of high polymeric substances is often determined using the relation $\frac{\pi}{C} = \frac{RT}{M} + AC$, that is, from a plot of π/C against C, the concentration of the polymer solution being expressed in g cm^{-3} and π expressed as cm of the solvent used. A is a constant. The plot is linear and the intercept is RT/M at $C\to0$ from which the molecular weight of the polymer is calculated. In a particular experiment at 298 K a polymer of molecular weight of 35,000 was used at a concentration of 1.35 g in 100 cm^3 of aqueous solution. What is the osmotic pressure exerted by the solution in terms of the height of water in cm? (Assume the density of solution = 1 g cm^{-3}).

Solution

$$\frac{\pi}{C_{C\to0}} = \frac{RT}{M} = \frac{0.0821\times298}{35000}$$

$$= 7\times10^{-4}\ \frac{\text{L}\times\text{atm}\times\text{K}^{-1}\times\text{mol}^{-1}\times\text{K}}{\text{g}\times\text{mol}^{-1}}$$

$$= 7\times10^{-4}\ \text{L}\times\text{atm}\times\text{g}^{-1}\times1000\ \text{cm}^3\times\text{L}^{-1}$$

$$= 7\times10^{-1}\ \text{cm}^3\times\text{atm}\times\text{g}^{-1}$$

The concentration of the polymer is 1.35 g in 100 cm^3

$= 0.0135\ \text{g}\times\text{cm}^{-3}$

$$\therefore \pi = 7\times10^{-1}\ \text{cm}^3\times\text{atm}\times\text{g}^{-1}\times0.0135\ \text{g}\times\text{cm}^{-3}$$

$$= 9.45\times10^{-3}\ \text{atm}$$

$$= 9.45\times10^{-3}\ \text{atm}\times76\ \text{Hg}\times\text{atm}^{-1} = 0.718\ \text{cm Hg}$$

Using the hdg relation $h_{Hg}\times13.56\ \text{gcm}^{-3} = h_{water}\times1.00\ \text{gcm}^{-3}$

$$h_{water} = 0.718\times13.56 = 9.74\ \text{cm water}$$

Sea Water and Reverse Osmosis

Problem 9.21

The salinity of sea water is defined as 35 g of salt per 1000 g of solution. Assuming that the major constituents of sea water are NaCl and $MgSO_4$, and that the former salt is 90 wt% of the total salt, calculate the osmotic pressure exerted by sea water at 298 K in kPa. Assume that NaCl is 80% dissociated and $MgSO_4$ is 40% dissociated. What is the pressure to be exerted on the sea water side in reverse osmosis to convert it into pure water?

Solution

Amount of total salt present in 1000 g of pure water = $35\times1000/965 = 36.27$ g
Amount of NaCl present = $36.27\times9/10 = 32.63$ g
Amount of $MgSO_4$ present = 3.64 g
Molality of NaCl = 32.63/58.5 = 0.56
Molality of $MgSO_4$ = 3.64/120.4 = 0.03

Total molal concentration of the ions and salt of NaCl = $(1-0.8)\times0.56 + 0.8\times0.56 + 0.8\times0.56$

$$= 0.112 + 0.448 + 0.448 = 1.0$$

Total molal conc. of the ions and salt of $MgSO_4 = (1-0.4)\times0.03 + 0.6\times0.03 + 0.6\times0.03$

$$= 0.018 + 0.018 + 0.018 = 0.054$$

Total conc. of the ions and salts of both = 1 + 0.054 = 1.054

$$\pi = CRT = 1.054\times0.0821\times298 = 25.79\ \text{atm}$$

$$= 25.79\times101.325 = 2613\ \text{kPa}$$

In order to convert the salt water into pure water through reverse osmosis, a pressure greater than 2613 kPa must be applied on the sea water side of the membrane to force pure water through the membrane. In calculating π, the solution is assumed to be dilute and the difference in molality and molarity is negligible.

Clausius-Clapyron Equation

Problem 9.22

Calculate the temperature at which n-butanol boils under a pressure of 8.0 kPa, given that its normal boiling point is 391 K and that its enthalpy of vapourisation is 44.4 kJ mol^{-1}. (Assume that ΔH_{vap} is independent of temperature in this temperature range.)

Solution

Using the Clapeyron–Clausius equation given by

$$\log\frac{P_2}{P_1} = \frac{\Delta H_{vap}}{2.303R}\left[\frac{1}{T_1} - \frac{1}{T_2}\right]$$

where $P_2 = 101.325$ kPa, $P_1 = 8.0$ kPa, $T_2 = 391$ K, $T_1 = ?$

$$\log\frac{101.325}{8.0} = \frac{44.4}{2.303\times8.314\times10^{-3}}\left[\frac{391-T_1}{391T_1}\right]$$

$$\frac{391-T_1}{391T_1} = \frac{1.103}{44.4}\times2.303\times8.314\times10^{-3} = 4.752\times10^{-4}$$

$$\frac{391}{391T_1} - \frac{T_1}{391T_1} = 4.752\times10^{-4}$$

$$\frac{1}{T_1} = 4.752\times10^{-4} + 25.57\times10^{-4} = 30.32\times10^{-4}$$

$$T_1 = \frac{1}{30.32\times10^{-4}} = 330\ \text{K}$$

Problem 9.23

The specific volumes of water vapour and liquid water at 373 K are 17.0×10^{-4} and 1.04×10^{-6}, both in m^3 g^{-1} respectively. The latent heat of vapourisation of water is 40.7 kJ mol^{-1}.

(i) Calculate the rate of change of vapour pressure (around boiling point) with temperature in kPa K^{-1}.
(ii) What is the boiling point of water under a pressure of 2 atm?

Solution

(i) $$\frac{dP}{dT} = \frac{\Delta H_{vap}}{T\Delta V}$$

$$= \frac{40.7\times1000\ \text{J mol}^{-1}}{373\times\left(17\times10^{-4} - 0.0104\times10^{-4}\right)\times18\ \text{K m}^3\ \text{mol}^{-1}}$$

$= 3567 \text{ J m}^{-3} \text{ K}^{-1}$

But, $0.0821 \text{ L} \times \text{atm} \times \text{K}^{-1} \text{ mol}^{-1} = 8.314 \text{ J} \times \text{K}^{-1} \text{ mol}^{-1}$

$$\frac{0.0821}{8.314} = 9.875 \times 10^{-3} \text{ L} \times \text{atm} \times \text{J}^{-1}$$

$$= 9.875 \times 10^{-3} \text{ L} \times \text{atm} \times \text{J}^{-1} \times 10^{-3} \text{ m}^3 \text{ L}^{-1}$$

$$= 9.875 \times 10^{-6} \text{ m}^3 \times \text{atm} \times \text{J}^{-1}$$

$$\frac{dP}{dT} = 3567 \text{ Jm}^{-3} \text{ K}^{-1} \times 9.875 \times 10^{-6} \text{ m}^3 \times \text{atm} \times \text{J}^{-1}$$

$$= 35224.1 \times 10^{-6} \text{ K}^{-1} \times \text{atm}$$

$$= 35224.1 \times 10^{-6} \text{ K}^{-1} \times \text{atm} \times 101.325 \text{ kPa} \times \text{atm}^{-1}$$

$$= 0.0352241 \times 101.325 = 3.57 \text{ kPa K}^{-1}$$

(ii) $\dfrac{dT}{dP} = \dfrac{1}{3.57} = 0.280 \text{ K} \times \text{kPa}^{-1}$

For an increase of pressure of 1×101.325 kPa

$dT = 0.0280 \times 1 \times 101.325 = 28.4°$

Boiling point of water under 2 atm pressure

$= 373 + 28.4 = 401.4$ K

NOTE

- Effectively, we are increasing the pressure only by 1 atm or 101.325 kPa because the given boiling point is already under 1 atm.

Problem 9.24

The density of acetic acid liquid is 1.04 g cm^{-3} and that of the solid is 1.266 g cm^{-3} at its freezing point, 16.6°C. The enthalpy of fusion of the acid is 11.7 kJ mol^{-1}. Calculate the rate of variation of its melting point with pressure in units of K kPa^{-1}. (ii) What will be its freezing point on increasing the pressure by 10 atm?

Solution

$$\frac{dT}{dP} = \frac{T\Delta V}{\Delta H} = \frac{289.6 \times \left(\frac{1}{1.04} - \frac{1}{2.66}\right) \times 60}{11700} \frac{\text{K cm}^3 \text{ mol}^{-1}}{\text{J mol}^{-1}}$$

$$= \frac{289.6 \times 0.171 \times 60}{11700} = 0.254 \text{ K cm}^3 \text{ J}^{-1}$$

But we know $\dfrac{8.314}{82.1} = 0.1013 \text{ J cm}^{-3} \text{ atm}^{-1}$

$$\frac{dT}{dP} = 0.254 \text{ K cm}^3 \text{ J}^{-1} \times 0.1013 \text{ J cm}^{-3} \text{ atm}^{-1}$$

$$= 0.0257 \text{ K atm}^{-1}$$

$$= \frac{0.0257}{101.325} \frac{\text{K} \times \text{atm}^{-1}}{\text{kPa} \times \text{atm}^{-1}} = 2.536 \times 10^{-4} \text{ K kPa}^{-1}$$

Increase in the freezing point by an increase in pressure by $10 \times 101.325 \text{ kPa} = 1013.25$ kpa

$$= 2.536 \times 10^{-4} \times 1013.25 = 0.257°$$

Therefore, melting point on increasing pressure by 10 atm = $289.6 + 0.257 = 289.857$ K

Debye–Hückel Theory of Electrolyte Solutions and Activity Ccoefficients

Problem 9.25

(a) Write down the expression for the thickness of the ionic atmosphere. Show that it has the dimensions of length.

(b) Calculate the thickness of the ionic atmosphere in (i) 0.01 M NaCl solution (ii) 0.0005 M NaCl at 298 K. What do you conclude from your results? (D of water = 80; $k = 1.38 \times 10^{-16}$ erg K^{-1})

Solution

(a) $\dfrac{1}{K}$ = Thickness of ionic atmosphere $= \left(\dfrac{1000DkT}{8\pi Ne^2\mu}\right)^{\frac{1}{2}}$

where D = dielectric constant of solvent, k = Boltzmann constant, T = temperature, N = Avogadro number, e = electric charge and μ = ionic strength.

Units of $\dfrac{1}{K}$:

$$\left(\frac{\text{no} \times \text{g} \times \text{cm} \times \text{s}^{-2} \times \text{cm} \times \text{K}^{-1} \times \text{K}}{\text{no} \times \text{no} \times \text{mol}^{-1} \times \text{g} \times \text{cm}^3 \times \text{s}^{-2} \times \text{mol} \times \text{cm}^{-3}}\right)^{\frac{1}{2}}$$

$$= \left(\text{cm}^2\right)^{1/2} = \text{cm}$$

NOTE

- The dimensions of μ have been taken as mol cm^{-3} (rather than dm^3. It does not make a difference to the dimensionality).

(b) (i) $1/K$ in 0.01 NaCl

$$\frac{1}{K} = \left[\frac{1000 \times 80 \times 1.38 \times 10^{-16} \times 298}{8 \times 3.143 \times 6.023 \times 10^{23} \times \left(4.8 \times 10^{-10}\right)^2 \times 0.01}\right]^{\frac{1}{2}}$$

$$= \left(\frac{3289.92 \times 10^{-12}}{3489.2 \times 10^3}\right)^{\frac{1}{2}} = \left(0.943 \times 10^{-13}\right)^{\frac{1}{2}}$$

$$= \left(9.43 \times 10^{-14}\right)^{\frac{1}{2}} = 3.07 \times 10^{-7} = 30.7 \times 10^{-8}$$

$$= 30.7 \text{ Å}$$

(i) $1/K$ in 0.0005 NaCl

$$= \left(\frac{3289.92 \times 10^{-12}}{3489.2 \times 10^3 \times 5 \times 10^{-4}}\right)^{\frac{1}{2}} = \left(\frac{3289.92 \times 10^{-12}}{17446.1 \times 10^{-1}}\right)^{\frac{1}{2}}$$

$$= 0.1886 \times 10^{-11} = \left(188.6 \times 10^{-14}\right)^{\frac{1}{2}}$$

$$= 13.73 \times 10^{-7} = 137.3 \text{ Å}$$

The thickness of the ionic atmosphere is so large in the second case, that the central ion may be considered a point charge in relation to the size of the ionic atmosphere.

Problem 9.26

Calculate the mean molality and the mean activity coefficient from the Debye–Hückel limiting law, mean ionic activity and the activity in 0.05 molal solutions of (i) K_2SO_4 (ii) $In_2(SO_4)_3$ solutions at 298 K. Assume complete dissociation of the salts.

Solution

(i) K_2SO_4 Mean molality, $m_\pm = \left(m_+^{x_+} m_-^{y_-}\right)^{\frac{1}{\gamma}}$

where $\gamma = x_+ + y_-$

$m_+ = 2m;\ m_- = m$

$$\therefore m_\pm = \left((2m)^2 \times m\right)^{\frac{1}{3}} = 4^{1/3} \times m = 1.567 \times 0.05 = 0.0794$$

Mean activity coefficient: $\gamma_\pm$

$$\log \gamma_\pm = -A|Z_+Z_-|\sqrt{\mu}$$

$$\mu = \frac{1}{2}\sum_i m_i Z_i^2 = \left(\frac{1}{2} \times 2m \times 1^2\right) + \left(\frac{1}{2} \times m \times 2^2\right)$$

$$= \frac{1}{2} \times 2 \times 0.05 \times 1 + \frac{1}{2} \times 0.05 \times 4 = 0.05 + 0.10 = 0.15$$

$$-\log \gamma_\pm = 0.5 \times 2 \times \sqrt{0.15} = 0.387;\ \gamma_\pm = 0.41$$

$$\text{Mean activity: } a_\pm = \left(a_+^{x_+} a_-^{y_-}\right)^{\frac{1}{\gamma}} = \left(m_+^{x_+}\gamma_+^{x_+} m_-^{y_-}\gamma_-^{y_-}\right)^{\frac{1}{3}}$$

$$= \left(m_+^{x_+} m_-^{y_-}\right)^{\frac{1}{3}} \left(\gamma_+^{x_+}\gamma_-^{y_-}\right)^{\frac{1}{3}} = m_\pm \gamma_\pm$$

$$= 0.0794 \times 0.41 = 0.0325$$

$$\text{Activity, } a = a_\pm^3 \text{ or } a = (0.0325)^3 = 3.433 \times 10^{-5}$$

(ii) $In_2(SO_4)_3$

$$In_2(SO_4)_3 \rightleftharpoons 2In^{3+} + 3SO_4^{2-}$$

Mean molality $m_\pm;\ m_+ = 2m;\ m_- = 3m$

$$m_\pm = \left(m_+^{x_+} m_-^{y_-}\right)^{\frac{1}{\gamma}} = \left[(2m)^2 (3m)^3\right]^{\frac{1}{5}} = \left(4m^2 \times 27m^3\right)^{\frac{1}{5}}$$

$$= 108^{1/5} \times m = 2.551 \times 0.05 = 0.1276$$

Mean activity coefficient: $\gamma_\pm$

$$\log \gamma_\pm = -A|Z_+Z_-|\sqrt{\mu};$$

$$\mu = \frac{1}{2}\sum_i m_i Z_i^2 = \left(\frac{1}{2} \times m_{In^{3+}} Z_{In^{3+}}^2\right) + \left(\frac{1}{2} \times m_{SO_4^{2-}} \times Z_{SO_4^{2-}}^2\right)$$

$$= \frac{1}{2} \times 2m \times 9 + \frac{1}{2} \times 3m \times 4 = 9m + 6m$$

$$= 15m = 15 \times 0.05 = 0.75$$

$$-\log \gamma_\pm = 0.5 \times 6 \times \sqrt{0.75} = 3 \times 0.866$$

$$= 2.598;\ \gamma_\pm = 0.00252$$

Mean ionic activity =

$$a_\pm = m_\pm \gamma_\pm = 0.1276 \times 0.00252 = 3.21 \times 10^{-4}$$

Activity =

$$a = a_\pm^5 = \left(3.21 \times 10^{-4}\right)^5 = 340.8 \times 10^{-20} = 3.41 \times 10^{-18}$$

Osmotic Coefficients from Freezing Point Data

Problem 9.27

The following table gives the freezing point studies on solutions of sodium chloride at two molalities and other additional data is also provided (see text).

M (mol kg^{-1})	f.p. depression ΔT_f (°)	$\int_0^m \frac{\theta}{m} dm$
5×10^{-3}	1.82×10^{-2}	1.033
1×10^{-2}	3.6×10^{-2}	1.04

Given $K_f = 1.86$ K kg mol^{-1}, calculate (i) the osmotic coefficient (ii) mean activity coefficient of NaCl at the two concentrations.

Solution

(i) Osmotic coefficient is given by $(1-\theta)$ and

$$\text{At 0.005 m, } (1-\theta) = \frac{\Delta T_f}{K_f m} = \frac{0.0182}{1.86 \times 0.005} = 1.957$$

$$\text{At 0.01 m, } (1-\theta) = \frac{0.036}{1.86 \times 0.01} = 1.936$$

(ii) Mean activity coefficient, $\gamma_\pm$
At 0.005 m,

$$2.303 \log \gamma_\pm = -\theta - \int_0^m \frac{\theta}{m} dm = 0.957 - 1.033 = -0.076$$

$$\log \gamma_\pm = -\frac{0.076}{2.303} = -0.033 = \bar{1}.967;\ \text{or } \gamma_\pm = 0.927$$

At 0.01 m,

$$2.303 \log \gamma_\pm = -\theta - \int_0^m \frac{\theta}{m} dm = 0.936 - 1.04 = -0.104$$

$$\log \gamma_\pm = -\frac{0.104}{2.303} = -0.045 = \bar{1}.955;\ \text{or } \gamma_\pm = 0.902$$

Thermodynamic Dissociation Constants of Acids and Solubility

Problem 9.28

The thermodynamic dissociation constant of monofluoro acetic acid at 298 K is 1.96×10^{-3} (that is, under conditions when the activity coefficients are unity).

(i) Calculate the degree of dissociation of the acid at 0.1 M concentration.

(ii) Use the α to determine the mean activity coefficient of the acid and calculate the classical dissociation constant at the same concentration of the acid. How has the degree of dissociation changed and by what per cent?

(iii) Suppose the acid dissociates in presence of 0.1 M concentration of a neutral salt which is completely dissociated. What is the degree of dissociation in this case and by what per cent has it changed? The salt is a 1:1 electrolyte.

Solution

(i) Thermodynamic dissociation constant $K = 1.96\times10^{-3}$. If α is the degree of dissociation,

$$K = \frac{\alpha^2 m}{1-\alpha}; \frac{\alpha^2}{1-\alpha} = \frac{1.96}{0.1}\times10^{-3} = 1.96\times10^{-2}$$

On solving for α, using

$\alpha^2 + 1.96\times10^{-2}\alpha - 1.96\times10^{-2} = 0$, we get $\alpha = 0.131$

(ii) Ionic strength,

$$\mu = \frac{1}{2}\times0.131\times0.1\times1^2 + \frac{1}{2}\times0.131\times0.1\times1^2 = 0.0131$$

$$-\log\gamma_\pm = 0.5\times1\times\sqrt{0.0131} = 0.5\times1\times0.1144 = 0.0572$$

$$\gamma_\pm = 0.877$$

Classical dissociation constant $= K' = \dfrac{K}{\gamma_\pm^2}$

(assume $\gamma_{HA} = 1$)

$$= \frac{(1.96\times10^{-3})}{(0.877)^2} = \frac{(1.96\times10^{-3})}{0.769} = 2.55\times10^{-3}$$

Calculating α' from the equation

$\alpha'^2 + 2.55\times10^{-2}\times\alpha' - 2.55\times10^{-2} = 0$

We get $\alpha' = 0.147$.

The degree of dissociation has increased by $0.147 - 0.131 = 0.016$ and the % increase $= (0.016\times100)/0.131 = 12.2$.

(iii) In the presence of 0.1 M salt and in the presence of the dissociated acid, the total ionic strength $= 0.1 + 0.0131 = 0.1131$

NOTE

- The total ionic strength may be even more due to increased dissociating acid.

$$-\log\gamma_\pm = 0.5\times1\times\sqrt{0.1131} = 0.5\times1\times0.336 = 0.168$$

$$\gamma_\pm = 0.679$$

Classical dissociation constant

$$= \frac{(1.96\times10^{-3})}{(0.679)^2} = \frac{(1.96\times10^{-3})}{0.461} = 4.25\times10^{-3}$$

Calculating α'' from

$\alpha''^2 + 4.25\times10^{-2}\times\alpha'' - 4.25\times10^{-2} = 0$

We have $\alpha'' = 0.186$

% increase in the degree of dissociation

$$= (0.186 - 0.11)\times\frac{100}{0.131} = 42$$

A considerable increase in the degree of dissociation by as much as 42% is observed in the presence of a neutral salt.

Problem 9.29

The thermodynamic solubility product, K_{SP} of silver chromate at 298 K is 7.87×10^{-12}. Calculate the solubility of the salt in the presence of 0.1 M $NaNO_3$. (Ignore the contribution of the ions of silver chromate in computing the ionic strength.) What is the per cent increase in solubility?

Solution

Solubility product, $K_{SP} = [Ag^+]^2\times[CrO_4^{2-}] = (2S)^2S = 4S^3$ where S is the solubility of the salt in g mol L^{-1}.

$$4S^3 = 7.87\times10^{-12} \Rightarrow S^3 = \frac{7.87\times10^{-12}}{4} = 1.97\times10^{-12}$$

$$S = \sqrt[3]{1.97\times10^{-12}} = 1.253\times10^{-4}\ \text{g}\frac{\text{mol}}{\text{L}}$$

Ionic strength $\mu = \frac{1}{2}\times0.1\times1^2 + \frac{1}{2}\times0.1\times1^2 = 0.1$

$$-\log\gamma_\pm = 0.5\times1\times(0.1)^{1/2} = 0.5\times1\times0.316$$

$$= -0.158 \text{ or, } \gamma_\pm = 0.695$$

Classical $K_{SP} = K'_{SP} = \dfrac{K_{SP}}{\gamma_\pm^3} = \dfrac{7.87\times10^{-12}}{(0.695)^3}$

$$= \frac{7.87\times10^{-12}}{0.336} = 23.42\times10^{-12}$$

Or, $4S^3 = 23.42\times10^{-12}$

$\Rightarrow$ $S^3 = \dfrac{23.42\times10^{-12}}{4} = 5.855\times10^{-12}$

$$S = \sqrt[3]{5.855\times10^{-12}} = 1.803\times10^{-4}\ \text{g mol L}^{-1}$$

The solubility has increased considerably compared to the value in the absence of a salt.

Activities, Activity Coefficients and Types of Standard States

Problem 9.30

The vapour pressure of pure acetone at 308 K is 45.0 kPa. The partial pressures of acetone and some acetone–carbon disulfide mixtures are given below.

S. No.	Mole fraction of acetone $X_{Acetone}$	Partial vapour pressure of acetone (kPa)
1	0.92	44.0
2	0.84	40.0
3	0.72	37.3
4	0.62	34.7
5	0.50	30.7

(a) Calculate the activity and activity coefficients of acetone at the mole fractions on Raoult's law standard state, that is, as $X_{Acetone} \to 1\ \gamma_{Acetone} \to 1$.

(b) The Henry's law standard state factor for acetone K_H(acetone) in acetone–carbon disulfide mixture at 308 K is 252 kPa. The partial vapour pressure of acetone and some acetone–carbon disulfide mixtures are given below.

S. No.	Mole fraction of acetone X_{Acetone}	Partial vapour pressure of acetone (kPa)
1	0.04	10.66
2	0.08	16.0
3	0.12	18.7
4	0.24	26.67
5	0.50	30.7

Calculate the activity and activity coefficient of acetone on Henry's law standard state, that is, as $X_{\text{Acetone}} \to 0$ $\gamma_{\text{Acetone}} \to 1$.

Solution

(a)

S. No.	Mole fraction of acetone X_{Ac}	Partial vapour pressure of acetone (*p*) (kPa)	$\frac{p}{p^0} = a_{\text{Ac}}$	$\gamma_{Ac} = \frac{a_{\text{Ac}}}{X_{\text{Ac}}}$
1	0.92	44.0	$\frac{44}{45} = 0.98$	$\frac{0.98}{0.92} = 1.06$
2	0.84	40.0	$\frac{40.0}{45} = 0.89$	$\frac{0.89}{0.84} = 1.06$
3	0.72	37.3	$\frac{37.3}{45} = 0.83$	$\frac{0.83}{0.72} = 1.15$
4	0.62	34.7	$\frac{34.7}{45} = 0.77$	$\frac{0.77}{0.62} = 1.24$
5	0.50	30.7	$\frac{30.7}{45} = 0.68$	$\frac{0.68}{0.50} = 1.306$

(b)

S. No.	Mole fraction of acetone X_{Acetone}	Partial vapour pressure of acetone (*p*) (kPa)	$\frac{p}{K_H} = a_{\text{Ac}}$	$\gamma_{Ac} = \frac{a_{\text{Ac}}}{X_{\text{Ac}}}$
1	0.04	10.7	$\frac{10.66}{252} = 0.042$	$\frac{0.042}{0.04} = 1.05$
2	0.08	16.0	$\frac{16.0}{252} = 0.063$	$\frac{0.063}{0.08} = 0.79$
3	0.12	18.7	$\frac{18.7}{252} = 0.074$	$\frac{0.074}{0.12} = 0.62$
4	0.24	26.7	$\frac{26.67}{252} = 0.106$	$\frac{0.106}{0.24} = 0.44$
5	0.50	30.7	$\frac{30.7}{252} = 0.122$	$\frac{0.122}{0.5} = 0.24$

EXERCISES

Practice Problems

1. The acid content of concentrated phosphoric acid is 85% wt/wt and its density is 1.70 kg dm^{-3} at 298 K.
   (a) What are its (i) molarity, (ii) molality, and (iii) mole fraction values (molecular weight of $H_3PO_4 = 98$)?
   (b) 2 mL of the concentrated acid is made up to 500 mL solution. Neglecting the second and third dissociation of the acid, what is the pH of the solution? Consider the solution as dilute.
   (c) According to the latest environmental protection data, the concentration of the chief green house gases – CO_2, CH_4, and N_2O – in the atmosphere are 478 ppm, 1819 ppb and 325 ppb respectively. Calculate their concentrations in molar units. (Assume the gases to be dilute in the atmosphere for calculation purposes.)

2. The vapour pressure of benzene at 300 K is 16.24 kPa. 8.21 g of naphthalene (molecular formula = $C_{10}H_8$) and 7.82 g of an unknown solute, dissolved in 50 g of benzene had a vapour pressure of 12.86 kPa.
   (i) Calculate the molecular weight of the solute.
   (ii) If the molecular formula of the compound is $C_7H_6O_2$ and if it is an organic acid, how would you account for the result in (i).
   (iii) What will be the vapour pressure of the solution if there is no change in the molecular state of the compound?

3. (a) Consider the saturation equilibrium of a solute in a solvent at definite temperature. Using the thermodynamic criterion of equilibrium under this condition, derive an equation for the solubility of the solute expressed in mole fraction units.
   (b) Calculate the ideal solubility of phenanthrene (molecular formula $C_{14}H_{10}$) at 298 K. The melting point of phenanthrene is 123.4°C and its enthalpy of fusion is 18.64 kJ mol^{-1}.

4. (a) The Henry's law constant of ammonia in water at 298 K is 27.0 mol kg^{-1} bar^{-1}.
   (i) What is its value in units of bar?
   (ii) What is its Bunsen coefficient?
   (iii) Given the dissociation constant of ammonia as 1.8×10^{-5}, what is its degree of dissociation under these conditions?
   (b) The Bunsen coefficient of xenon is 0.224 cm^3 at 273 K. Calculate the Henry's law constant in mol L^{-1} atm^{-1} and in atm.

5. (a) A solution of potassium ferricyanide $K_3Fe(CN)_6$, consisting of 0.658 g of compound dissolved in 100 g of water freezes at –0.134°C. Calculate the van't Hoff factor, *i* and the degree of dissociation of the salt. What is the total molality of the salt? (K_f of water = 1.86 K kg mol^{-1}). What do you conclude from the result?
   (b) Given the molal boiling point constant of water, K_b = 0.51 K kg mol^{-1}, what is the boiling point of the solution?
   (c) What is the osmotic pressure exerted by the solution in kPa at 298 K?

6. The following freezing point data are obtained on solutions of K_2SO_4 and $MgSO_4$ at various molalitites in aqueous medium. Other data are also provided (see text for details).

Salt	*m* (mol kg^{-1})	f.p. depression ΔT_f (°)	$\int_0^m \frac{\theta}{m} dm$	α (degree of dissociation)
K_2SO_4	1×10^{-3}	5.3×10^{-3}	1.957	0.92
	1×10^{-2}	5.0×10^{-2}	2.043	0.85
	1×10^{-1}	4.3×10^{-1}	2.163	0.66
$MgSO_4$	1×10^{-2}	2.8×10^{-2}	1.446	0.53
	1×10^{-1}	2.25×10^{-1}	1.925	0.21

The molal freezing point depression constant K_f of water is 1.86 K kg mol^{-1}.
(i) Calculate the osmotic coefficient at different *m* of the salts.
(ii) Calculate the mean activity coefficients of the salts at the given concentrations.
(iii) What are the mean ionic activities of the electrolytes at the different concentrations? Use α values if necessary.

7. (a) Calculate the thickness of the ionic atmosphere at 298 K in
   (i) 0.005 molal solution of H_2SO_4.
   (ii) 0.005 molal solution $Ca(NO_3)_2$. The mean activity coefficients are $\gamma_{\pm\, H2SO4} = 0.640$ and $\gamma_{\pm Ca(NO3)2}$ = 0.770. What are your conclusions?
   (b) Calculate the mean molality, mean ionic activity coefficient and the mean activity in 0.01 molal solution $La(NO_3)_3$.

8. (a) 10 g of human plasma fluid (molar mass = 70,000) is dissolved in 0.4 dm^3 of water. What is the osmotic pressure exerted by this fluid at 300 K in (i) atm (ii) in Pa (iii) in terms of the height of water. Does this method have an advantage over methods using other colligative properties? If so, why?
   (b) Give reasons for the following.
   (i) When a carrot is immersed in water and another carrot is immersed in salt water, it is observed that carrot in the latter case shrinks while the former is not affected.
   (ii) The proper medium for intravenous injection is 0.15 molal NaCl solution.
   (iii) Salts such as NaCl, $CaCl_2$ and so on, are used to melt ice on the streets in winter in several countries.
   (iv) Divers use helium in combination with oxygen under suitable pressurised conditions in the place of compressed air in diving operations under rivers or sea.

9. The boiling point of bromobenzene is 155°C and its molal boiling point elevation constant K_b is 6.24 K kg mol^{-1}. A mixture consisting of 0.64 g of naphthalene (molar mass = 128) and 0.60 g of pure glacial acetic acid (molar mass = 60) are dissolved in 100 g of bromobenzene and the boiling point of the resulting mixture is 155.655°C. From these data,
   (i) Calculate the apparent molar mass of acetic acid. What do you conclude? Show using appropriate chemical equations.
   (ii) Find the van't Hoff factor *i* and any other parameter to justify your conclusion.

10. The thermodynamic solubility product of calcium fluoride at 298 K is 3.2×10^{-11}. What is the solubility of salt in mol kg^{-1} in the presence of 0.05 M Na_2SO_4 (assume the salt to be completely dissociated). Consider the contribution of all ions which affect the solubility.

11. (a) The thermodynamic and classical dissociation constants of formic acid at 298 K are 1.59×10^{-4} and 1.80×10^{-4} respectively. Calculate the degree of dissociation of the acid at 0.1 M concentration of the acid from these data. What factor is responsible for the difference in α? Calculate the factor which causes the difference.
   (b) Consider the dissociation of the acid in the presence of 0.005 M potassium sulfate. What is the degree of dissociation of the acid in this case? What is the percentage change? Assume that K_2SO_4 is completely dissociated. Neglect the contribution of the ions from the dissociation of acid in the absence of salt.

12. The enthalpy of vaporisation of bromine is 15.46 kJ mol^{-1}. The specific volumes of liquid and vapour at 293 K are 0.3224×10^{-3} dm^3 g^{-1} and 0.1504 dm^3 g^{-1} respectively. Calculate (i) rate of change of vapour pressure with temperature in kPa K^{-1}, (ii) the vapour pressure of liquid

bromine at 313 K given its boiling point as 58.7°C under 1 atm., and (iii) the boiling point of bromine when the pressure is increased by 2 atm.

13. 2 mol of toluene and 3 mol of bromobenzene are mixed at 298 K and this mixture shows a vapour pressure of 6.36 kPa. The vapour pressure of the mixture rose to 6.91 kPa on adding 2.5 mol of bromobenzene. Calculate the vapour pressures of the two pure components. What are the mole fractions of the two in the vapour phase in the mixture that contains the additionally added bromobenzene?

14. Calculate the amount of glycerol (molar mass = 92), dissolved in an aqueous solution weighing 52.3 g, required to decrease the freezing of the solution by 2°C. Suppose the solution is cooled by 3°C more, find out the amount of ice separating from the solution (K_f of water = 1.86 K kg mol^{-1}).

15. 0.293 g of NaCl (molar mass = 58.5) and 0.825 g of an unknown bi-univalent electrolyte, MA_2 dissolved in 50 g of water, was found to freeze at –0.893°C. Calculate the apparent molar mass of MA_2. Given its actual molar mass as 164.1, what are the Van't Hoff factor i and the degree of dissociation of MA_2. Also calculate the total molality of the electrolyte, MA_2 from the observed results. (K_f of H_2O = 1.86 K kg mol^{-1})

16. The melting points and enthalpies of fusion of two organic compounds are given.

Name of the compound	Melting point (K)	Enthalpy of fusion (kJ mol^{-1})
Anthracene	490	28.8
Phenanthrene	396.4	18.6

Calculate the ideal solubility of the compounds at 300 K.

17. (a) The number average molecular weight of an organic polymer was obtained at 298 K by a plot of π/c against c and extrapolating it to $c \to 0$. The experiment was performed at 298 K in benzene as solvent and the extrapolated value of π/c was found to be 121.4 cm^4 g^{-1}. Calculate the molar mass of the organic polymer. (Density of Hg = 13.56 g cm^{-3}, density of benzene = 0.80 g cm^{-3})

(b) Consider an ideal dilute solution in which the solvent A obeys Raoult's law and the solute B obeys Henry's law. Derive an equation which relates the total vapour pressure, P, as a function of the mole fraction of the solute.

18. The partial vapour pressure data of a water–methanol mixture at various mole fractions at 298 K are given. The vapour pressures of water and methanol at 298 K are 3.173 kPa and 16.96 kPa respectively.

X_{H2O}	Partial pressure of H_2O (kPa)	Partial pressure of CH_3OH (kPa)
0.02	0.0973	16.622
0.100	0.4712	15.296
0.300	1.295	12.174
0.600	2.220	7.958
0.800	2.678	5.074
0.900	2.901	3.148

Calculate the activity and activity coefficient of water and methanol at the given mole fractions on Raoult's law standard state. What do you conclude and give possible reasons.

19. (a) The following table gives the vapour pressure data of a ethanol–water mixture at 323 K at various compositions of ethanol and water. The vapour pressures of the pure solvents, water and ethanol, at this temperature are 12.33 kPa and 28.13 kPa respectively.

X_{H2O}	Partial pressure of H_2O (kPa)	Partial pressure of ethanol (kPa)
0.90	11.14	10.13
0.70	10.13	14.19
0.50	8.11	16.21
0.30	6.08	20.27
0.20	4.053	22.80

Calculate the activity and activity coefficient of both the components on Raoult's law standard state.

(b) Based upon the calculations in (a), draw qualitatively, the vapour pressure–composition curves (against X_{EtOH}). Also, draw the boiling point–composition curves indicating regions of liquid and vapour phases. What conclusions can be drawn from these diagrams?

20. (a) Calculate the pressure needed to change the freezing point water by a degree (around its freezing point) using the following data. The molar volume of ice and of water at 0°C are 1.963×10^{-2} dm^3 mol^{-1} and 1.80×10^{-2} dm^3 mol^{-1} respectively and the enthalpy of fusion of ice = 6.0 kJ mol^{-1}. Also, find out the rate of change of freezing point of ice with increase in pressure.

(b) Find the pressure under which benzene boils at a temperature of 313 K given that its normal boiling point is 80°C and the latent heat of vapourisation is 30.8 kJ mol^{-1}.

Multiple Choice Questions

1. Which of the following equations between concentration (c) and molality (m) of a solution is correct? (W_x = weight of solvent, W_y = weight of solute, M_x = molecular mass of solvent, M_y = molar mass of solute, d = density of solution)

(a) $c = m[(W_x + M_y\, d)/1000W_y]$
(b) $m = (c \times 1000W_y)/(W_x \times M_y)$
(c) $c = m(W_x \times d)/(W_x + W_y)$
(d) $m = c((W_x + M_y)/1000\, W_x)$

2. The environmental protection agency of USA recommends an upper limit of 2.1×10^{-4} mol kg^{-1} of fluoride in drinking water for human consumption. Its concentration in ppb is,
(a) 6.35×10^4 **(b)** 2.86×10^6 **(c)** 5.64×10^5 **(d)** 3.99×10^3

3. From the ideal solubility relation of a solute in a solvent, one can conclude that

(a) the solubility of chemically similar solutes increases with the melting point of the solute
(b) the solubility is independent of solvent
(c) the solubility may decrease or increase with temperature
(d) the solubility is obtained in molal units

4. Ethylene chloride and ethanol have vapour pressure of 135 mm and 157.5 mm Hg respectively at 313 K. A mixture of the two liquids in which the mole fraction of ethylene chloride is 0.3 has a vapour pressure of 218 mm Hg. Hence these solvent mixtures show

(a) negative deviations from Raoult's law
(b) do not form a constant boiling mixture
(c) exhibit positive deviations from Raoult's law
(d) form ideal mixtures

5. The Henry's law constant for CO_2 in water at 298 K is 3.5×10^{-2} mol dm^{-3} atm^{-1}. The volume of the gas (in dm^3) at STP in 1 dm^3 of water at 2 atm pressure of the gas is
(a) 2.56 **(b)** 1.25 **(c)** 0.97 **(d)** 1.68

6. The Bunsen coefficient of helium in water at 298 K is $8.6 \times 10{-}3$.The solubility of the gas in water in mole fraction units is

(a) 8.50×10^{-4} **(b)** 7.55×10^{-6}
(c) 3.50×10^{-5} **(d)** 12.50×10^{-3}

7. The molar mass of the hemoglobin is 67,000 g mol^{-1}. The osmotic pressure exerted by this compound at a concentration of 10 g dm^{-3} at 300 K in terms of the height of water (in cm) is
(a) 3.79 **(b)** 2.95 **(c)** 4.84 **(d)** 4.20

8. Two solutions having the same vapour pressure are referred to as
(a) isotropic **(b)** isotonic **(c)** isentropic **(d)** isopiestic

9. A solution consisting of 18.7 g $MgCl_2$ and 31.9 g $LiClO_4$ dissolved in 1 kg of water freezes at –2.269°C. The total molal concentration of all species in the solution is (assume $LiClO_4$ is completely dissociated),
(a) 0.86 **(b)** 1.42 **(c)** 1.17 **(d)** 1.91

10. Pure water is prepared from sea water by reverse osmosis and is especially adopted in regions of scarcity of pure water. In this process

(a) the semi-permeable membrane permits water from the pure water side to the sea water side
(b) mechanically a pressure greater than the osmotic pressure is applied to the sea water side to aid the desired process.
(c) the membrane acts as a filter of the salt ion
(d) the process involves diffusion of ions from one side to the other

11. For benzene ΔH_{vap}, ΔH_{fusion} and ΔV_{vap} are all positive. But ΔV_{fusion} [$V(l) - V(s)$] is negative. Using this information, pick the correct statement from the following.

(a) Its freezing point decreases with increasing pressure.
(b) The boiling point increases with decreasing pressure.
(c) The freezing point increases with increasing pressure.
(d) Both freezing point and boiling point increase with increasing pressure.

12. 9.2 g of phenol are dissolved in 100 g of toluene (b.p. = 383.6 K) and the solution boils at 385.95 K. If the degree of aggregation of the solute is 0.72, the number of phenol molecules aggregating is (Trouton's constant of toluene = 87.3 J K^{-1} mol^{-1}, molar mass of toluene = 92.0)
(a) 3 **(b)**.1 **(c)** 4 **(d)** 2

13. At a mole faction of 0.075 of urea in an aqueous solution at 298 K, the vapour pressure of the solution is 2.814 kPa. The vapour pressure of pure water at this temperature is 3.170 kPa. Therefore, the rational activity coefficent of water in the solution is
(a) 0.89 **(b)** 1 **(c)** 0.96 **(d)** 0.78

14. Which pairs of liquid mixtures among the following show a positive deviation from Raoult's law ?

(a) pyridine–acetic acid and chloroform–diethyl ether
(b) methanol–water and ethanol–cyclohexane
(c) Benzene–$CHCl_3$ and benzene–acetone
(d) Benzene–toluene and octane–pentane

15. Glacial acetic acid freezes at 16.6°C. Its K_f = 3.6 K kg mol^{-1}. The change in its molar volume on fusion is 9.612 cm^3 mol^{-1}. From these data, the freezing point of acetic acid under a pressure of 50 atm is
(a) 15.92 **(b)** 18.75 **(c)** 16.43 **(d)** 17.81

16. The thickness of the ionic atmosphere in an electrolyte solution is

(a) proportional to the square root of the reciprocal of ionic strength at constant temperature in a given medium
(b) proportional to the ionic strength of the medium at constant temperature
(c) at constant ionic strength of the medium, it is proportional to the square root of the reciprocal of temperature
(d) it is independent of the dielectric constant of the medium at constant temperature

17. The activity of 0.01 molal solution of $La_2(SO_4)_3$ at 298 K (assuming complete dissolution of the salt) is
(a) 2.73×10^{-6} **(b)** 1.65×10^{-8}
(c) 5.45×10^{-9} **(d)** 1.34×10^{-7}

18. 12.4 g of ethylene glycol is dissolved in 112.4 g of water and the mixture is cooled to 268 K. The amount of ice (in g) separating out of the solution [K_f(H2O) =1.86 K kg mol^{-1}]
(a) 35.6 **(b)** 29.5 **(c)** 59 **(d)** 25.9

19. The thermodynamic dissociation constant of dichloroacetic acid is 3.32×10^{-2} at 298 K. At 0.1 M concentration of the acid and in the presence of 0.1 M $LiNO_3$ (which is completely dissociated), the per cent change in the degree of dissociation is
(a) 39.65 **(b)** 33.40 **(c)** 28.58 **(d)** 31.46

20. The vapour pressures of isoamyl and isobutyl alcohol at 50°C are 435 mmHg and 135 mmHg respectively. At $X=$ 0.4 of isobutyl alcohol, the mole fraction ratio of isoamyl alcohol in the liquid to vapour phase is
(a) 0.725 **(b)** 0.572 **(c)** 0.648 **(d)** 0.493

21. The freezing point data of aqueous HCl solution at two molalities of the acid are given below.

Molality of HCl	T_f(°C)	$\int_0^m \frac{\theta}{m}\,dm$
0.050	–0.183	1.152
0.104	–0.368	1.134

The mean ionic activity coefficients of HCl at 0.050 mol, 0.104 mol in that order are

NOTE

- $(1-\theta)$ is defined as the osmotic coefficient, and $K_f(H_2O)$ = 1.86 K kg mol^{-1}

(a) 0.695 and 0.628 **(b)** 0.953 and 0.861
(c) 0.832 and 0.794 **(d)** 0.794 and 0.895

Answer Keys

Practice Problems

1. (a) (i) Molality = 57.82 (ii) Molarity = 15.37 (iii) Mole fraction = 0.479

   (b) pH = 1.16

   (c) Conc. of CO_2 = 0.0109 mol dm^{-3}, Conc. of CH_4 = 1.14×10^{-4} mol dm^{-3}, Conc. of $N_2O = 7.4\times10^{-6}$ mol dm^{-3}

2. (i) Molar mass of solute = 244 (ii) Since the actual molar mass is 122, the organic acid exists as a dimer in benzene (iii) p = 13.46 kPa

3. X = ideal solubility = 0.154

4. (a) (i) 3.06 bar (ii) 0.02473 (iii) $\alpha = 8.2\times10^{-4}$

   (b) K_H = 0.01 mol L^{-1} atm^{-1}; K_H in the mole fraction unit = 5.56×10^3

5. (a) $i = 3.6$; $\alpha = 0.87$. Since $i \approx 4$, the compound gives four particles on dissociation. Total molality = 0.072

   (b) Boiling point of the solution = 373.037 K

   (c) π = 178.3 kPa

6.

Salt	Conc. of salt, m (mol kg^{-1})	Osmotic coefficient	$\gamma_\pm$	Mean activity
K_2SO_4	1×10^{-3}	2.85	0.900	1.457×10^{-3}
	1×10^{-2}	2.7	0.709	1.350×10^{-2}
	1×10^{-1}	2.32	0.430	1.047×10^{-1}

Salt	Conc. of salt, m (mol kg^{-1})	Osmotic coefficient	$\gamma_\pm$	Mean activity
$MgSO_4$	1×10^{-2}	1.505	0.390	5.30×10^{-3}
	1×10^{-1}	1.210	0.180	2.10×10^{-2}

7. (a) (i) 25.07 Å (ii) 25.07 Å
   From the equation for $1/k$, it is apparent that it is directly proportional to $T^{1/2}$ and inversely proportional to $\mu^{1/2}$. At a given temperature, $1/k$ increases with decreasing ionic strength.

   (b) $m_\pm = 0.0228$; Mean activity coefficient = 0.429; Mean activity = 9.78×10^{-3}

8. (a) π (atm) = 8.976×10^{-3}; π (Pa) = 891.25; π (height of water) = 9.065 cm

   (b) Descriptive questions

9. (i) Apparent M = 109.2. Acetic acid dimerises in benzene. $i = 0.549$; $\alpha = 0.902$.

10. Solubility on the basis of thermodynamic $K_{sp} = 2\times10^{-4}$ M

    Solubility in the presence of 0.05 M Na_2SO_4 = 5.164×10^{-4} M

11. (a) α from $K_a = 0.0392$, α from $K'_a = 0.0415$; $\gamma_\pm = 0.94$

    (b) In the presence of 0.005 M $K_2SO_4 = 0.0515$; % increase in $\alpha = 31.4$

12. (i) $\frac{dP}{dT} = 3.891$ kPa K^{-1}

    (ii) P_1 at 313 K = 543.6 mm

(iii) Boiling point of bromine = 383.78 K (when pressure is increased by 2 atm)

13. P^0 of bromobenzene = 8.015 kPa, P^0 of toluene = 3.876 kPa, $X_{toulene}$ in the vapour phase = 0.15, $X_{bromobenzene}$ in vapour phase = 0.85

14. Weight of glycerol = 4.707 g, Weight of ice separating out = 19.03 g

15. Apparent molar mass = 58.90, i = 2.786, α = 0.893, $m_{\pm}$ = 0.280

16. Ideal solubility of anthracene (mole fraction) = 1.138×10^{-2}

Ideal solubility of phenanthrene (mole fraction) = 1.632×10^{-1}

17. (a) Molar mass of organic polymer = 2.596×10^{5}

(b) $p = p_A^0 + x_B\left(K_h - p_A^0\right)$

18. H_2O–MeOH mixtures

X_{H2O}	0.020	0.100	0.300	0.600	0.800	0.900
Activity of H_2O	0.0307	0.1485	0.4081	0.700	0.8440	0.9143
Activity coefficients of H_2O	1.533	1.485	1.360	1.167	1.055	1.016
X_{CH3OH}	0.980	0.900	0.700	0.400	0.200	0.100
Activity of CH_3OH	0.980	0.902	0.718	0.469	0.299	0.186
Activity coefficient of CH_3OH	1.000	1.002	1.026	1.173	1.496	1.856

19. H_2O–EtOH mixtures

X_{H2O}	0.900	0.700	0.500	0.300	0.200
Activity of H_2O	0.903	0.822	0.658	0.493	0.329
Activity coefficients of H_2O	1.003	1.174	1.316	1.643	1.645
X_{EtOH}	0.100	0.300	0.500	0.700	0.800
Activity of EtOH	0.360	0.504	0.576	0.720	0.810
Activity coefficient of EtOH	3.60	1.68	1.15	1.03	1.013

A descriptive part of vapor pressure–composition and boiling point composition curves and discussion on the same.

20. (a) $\dfrac{dP}{dT} = -133.15 \text{ atm K}^{-1}$; $\dfrac{dT}{dP} = -7.51\times10^{-3} \text{ K atm}^{-1}$

(c) Boiling point of 313 K is achieved at a pressure of 198.8 mm Hg.

Multiple Choice Questions

1. (c) **2.** (d) **3.** (b) **4.** (c) **5.** (d) **6.** (b) **7.** (a) **8.** (d) **9.** (c) **10.** (b)
11. (a) **12.** (d) **13.** (c) **14.** (b) **15.** (d) **16.** (a) **17.** (c) **18.** (d) **19.** (b) **20.** (a)
21. (c)

CHAPTER 10

PHASE RULE AND PHASE EQUILIBRIUM

CHAPTER OUTLINE

10.1 Phase Rule and its Importance

The law of mass action has been applied to study a large number of homogeneous equilibria and also to understand some simple heterogeneous equilibria (for example, $CaCO_3$ decomposition) with the help of certain simple assumptions. However, the mass law cannot be applied to many heterogeneous equilibria, since in many cases, the number and the complexity of the different molecular aggregates in the system are not known. J W Gibbs enunciated a general theorem, known as phase rule in 1874 through which he defined the conditions of equilibrium in a heterogeneous system as a relation between the number of phases and components of the system.

10.2 Definition of Various Terms

Phase: It is defined as a certain part of a system which is homogeneous throughout and physically separated from other phases by distinct boundaries. Example: ice, water and its vapour are three phases of the same chemical substance—water. A phase must be physically homogeneous, yet it may be chemically complex.

Component: Components are defined as the smallest number of independently variable constituents which permit the composition of each phase participating in the equilibrium to be represented in the form of a chemical equation. Considering the same water system existing in the three phases – solid, liquid and vapour – the component is only one. The constituents of water, H_2 and O_2 are not to be regarded as components for two reasons: (i) they are not part of the equilibrium system and (ii) also their composition in water is not independently variable. However, if the experimental conditions are such, say high temperature, where H_2, O_2 and water are part of equilibrium, then they also have to be considered in stating the number of components.

Degrees of freedom: The number of degrees of freedom is defined as the number of variables factors, temperature, pressure and concentration of the components which must be fixed so that the condition of the system may be perfectly defined. For example, if we have a gas or vapour contained in a certain volume, its thermodynamic state is fixed if we specify two variables among the three—temperature, pressure and concentration. For a system consisting of two phases in equilibrium, say liquid water and water vapour, one degree of freedom, say temperature or pressure is enough to be specified to define the state of the system.

10.3 Gibbs Phase Rule

The phase rule is a quantitative relationship proposed by J W Gibbs and it relates the number of degrees of freedom with the number of components and phases according to

$$F = C - P + 2 \tag{10.1}$$

where, F = degrees of freedom, C = components and P = number of phases in equilibrium. The factor 2 in the above equation arises from the assumption that the state of the system is completely defined by considering the three variables temperature, pressure and composition.

Condensed phase rule: In considering systems which contain only solid and liquid phases at sufficiently high pressures to eliminate the vapour phase, the vapour phase is excluded and hence the pressure need not be specified. Under these conditions, the phase rule has the form

$$F = C - P + 1 \tag{10.2}$$

and is known as the condensed phase rule.

10.3.1 Calculation of the Number of Components, Degrees of Freedom and Phases in a System

Number of components: In calculating their number, C, the useful equation is, C = total number of components – the number of algebraic relations between them.

Number of phases: Solid, liquid and vapour phases are to be considered as separate.

(i) For the system, $2NH_3(g) \rightleftharpoons N_2(g) + 3H_2(g)$

$C = 3 - 1 = 2; P = 1, F = 2 - 1 + 2 = 3$

(ii) $CaCO_3(s) \rightleftharpoons CaO(s) + CO_2(g)$

$C = 3 - 1 = 2, P = 3, F = 2 - 3 + 2 = 1$

(iii) $2H_2(g) + O_2(g) \rightleftharpoons 2H_2O(g)$ in equilibrium at a high temperature; $C = 3 - 1 = 2, P = 1, F = 2 - 1 + 2 = 3$

(iv) $HCl(\text{in solution}) + H_2O \rightleftharpoons H_3O^+ + Cl^-$. There are four species in solution. There is one equation and in addition there is the electroneutrality condition. Therefore, $C = 4 - 2 = 2, P = 1, F = 2 - 1 + 2 = 3$. Since it is a condensed system $F = 2 - 1 + 1 = 2$.

(v) $Na_2SO_4 10H_2O(s) \rightleftharpoons Na_2SO_4(s)$

$\rightleftharpoons Na_2SO_4(\text{saturated solution})$

$C = 3 - 1 = 2, P = 3, F = 2 - 3 + 2 = 1$

Since it is also a condensed system, $F = 2 - 3 + 1 = 0$

10.3.2 Stability of a Phase With Respect To Temperature and Pressure

The thermodynamic condition for a system in equilibrium is that the chemical potential of each constituent must be the same throughout the system. The fundamental equation for the stability of a phase with respect to the temperature and pressure is based on the equation

$$d\mu = VdP - SdT \tag{10.3}$$

where S and V are the molar entropy and molar volume for a system of one component. From Eq. (3), one may write

$$\left(\frac{\partial \mu}{\partial T}\right)_P = -S \text{ and } \left(\frac{\partial \mu}{\partial P}\right)_T = V \tag{10.4}$$

Knowing that the entropy of a solid, liquid and vapour are in the order $S_{gas} >> S_{liq} >> S_{solid}$, the slopes of the plot of μ vs T for these three phases lie in the order slope$_{(gas)}$ >> slope$_{(liq)}$ >> slope$_{(solid)}$. The effect of pressure on chemical potential may be understood from Eq. 10.4. Since the molar volume is positive for any substance and also, the molar volumes of solids and liquids are small at any given temperature, the decrease in chemical potential with pressure at any given temperature for solids and liquids is quite small. For a gas, the decrease in chemical potential with decrease in pressure at given temperature is, however, quite large.

10.3.3 Phase Diagrams

A phase diagram graphically represents the existence of regions of equilibrium between various phases of a substance under different conditions of T and P. Phase diagrams are classified on the basis of the number of components in the systems. Phase diagrams are of great use for material scientists, metallurgists, mechanical and chemical engineers as well.

10.4 One-Component Systems

10.4.1 Water

For this system, the number of components $C = 1$ (which is water). For a system of one component existing only in one phase, $F = 1 - 1 + 2 = 2$, that is, the state of the system is defined by two variables. For example, the thermodynamic state of water in its vapour phase is fixed by fixing temperature and pressure (or temperature and volume). If the system consists of two phases (say liquid water and vapour, $F = 1 - 2 + 2 = 1$, that is, the system is univariant. If, for example, the temperature is arbitrarily fixed, the pressure under which water and vapour can coexist in equilibrium is also fixed. If $P = 3$, that is, when water is present in all the three phases, solid (ice), water (liquid) and vapour (gas), $F = 1 - 3 + 2 = 0$. The system is invariant. The three phases of a one component system (in this case, water) can exist in equilibrium only under one set of conditions—at one particular temperature and pressure; any variation of these factors will lead to the disappearance of one or more phases of the system.

For this system (or for any one component system), invariant systems are shown as points (point O in Fig. 10.1), univariant systems are shown by lines and bivariant systems are shown

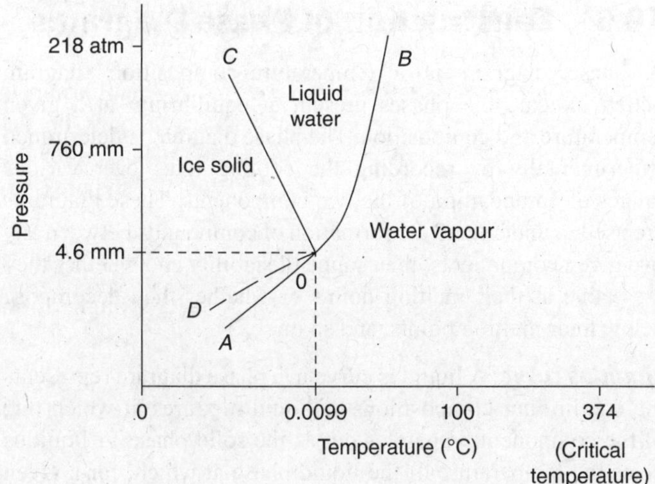

Fig. 10.1 Phase diagram of water

by areas. In Fig. 10.1, point O is known as the *triple point*, where all the three phases – solid, liquid and vapour – coexist in equilibrium. At this point $T = 0.0099°C$, $P = 4.57$ mmHg. At point O, ice and water from which all air has been expelled, is in equilibrium with water vapour which has a pressure of 4.57 mm. OD (shown by dotted lines) is the vapour pressure curve of supercooled water.

10.4.2 Sulfur

Sulfur can exist in two allotropic modifications—rhombic and monoclinic. Each form constitutes a separate phase. The two crystalline forms can coexist in equilibrium either with the liquid or vapour but not with both. The number of invariant (triple points), univariant and bivariant systems are listed (Fig. 10.2).

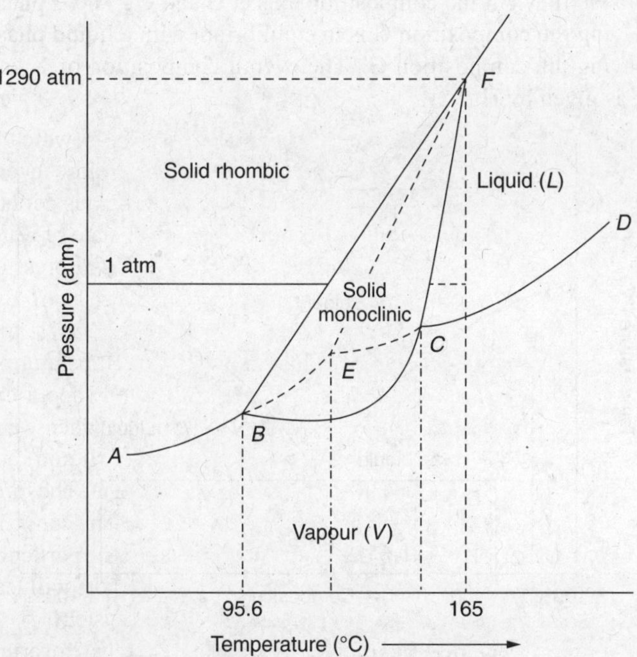

Fig. 10.2 Sulfur system

(i) Triple points (invariant systems)—4 (points B,C,F and E in diagrams). (a) Rh-sulfur–Mo-sulfur–sulfur vapour (b) Rh-sulfur–Mo-sulfur–sulfur liquid (c) Rh-sulfur–liquid sulfur–sulfur vapour (d) Mo-sulfur–liquid sulfur–sulfur vapour.

(ii) Univariant systems (one component present in two phases)—6 (lines AB, BC, AE, CF, BF, CD). (a) Rh-sulfur–vapour (b) Mo-sulfur–vapour (c) Rh-sulfur–liquid sulfur (d) Mo-sulfur–liquid sulfur (e) Rh-sulfur–Mo-sulfur (f) liquid sulfur–sulfur vapour.

(iii) Bivariant systems (one component in one phase)—4. (a) Rh-sulfur (b) Mo-sulfur (c) liquid sulfur (d) sulfur vapour.

The point of special interest in this system is the triple point relating to Rh-sulfur–Mo-sulfur–sulfur vapour. This triple point is a point of intersection of the three curves of the univariant systems Rh-sulfur–sulfur vapour, Mo-sulfur–sulfur vapour and Rh-sulfur–Mo-sulfur. The Rh-sulfur–Mo-sulfur transition is a univariant system, and the temperature at which the transition takes place depends on the pressure. This temperature, known as the *transition temperature* is fixed and constant at 1 atm pressure and it occurs at 95.6°C.

Enantiotropy and monotropy: Two interesting cases may be considered. It is possible that the transition temperature at 1 atm. lies below the melting point of the solid which is stable at normal temperatures. Alternatively, the transition point at 1 atm. may lie above the melting point of the solid which is stable at normal temperatures. In the former case, each form has a definite range of existence while in the latter case only one form is stable at all temperatures up to the melting point. The former case is *enantiotropy* and the latter *monotropy*.

Phase diagram of sulfur: AB = vapour pressure of Rh-sulfur, BC = vapour pressure of Mo-sulfur, CD = vapour pressure of liquid sulfur, BF = effect of pressure on the transition temperature, CF = effect of pressure on the melting point of Mo-sulfur. Both curves BF and CF slope to the right so that both the transition temperature and melting point of Mo-sulfur increase with increasing pressure. BE, EC, EF curves represent metastable equilibria—the region of stability of Mo-sulfur is bounded by the lines BF, CF and BC. It cannot exist at temperatures and pressure outside the triangular region BCF. Above the triple point F, occurring at 151°C and 1290 atm., solid sulfur can be obtained only in rhombic form. At F, Rh-sulfur–liquid sulfur and vapour are in equilibrium.

10.4.3 CO_2

Phase diagram of CO_2: OA = liquid–vapour curve, OB = solid–liquid curve, OC = solid–vapour curve. This is an example of a system where the solid phase undergoes sublimation (the solid directly goes into the vapour phase at ordinary temperatures and pressures). The triple point for CO_2 is 5.1 atm. and –50°C. This means that all the three phases can exist and are in equilibrium at this temperature and pressure. Liquid CO_2 cannot exist below 5.1 atm. At ordinary pressures (1 atm.), dry ice (solid CO_2) directly goes into the vapour phase (it sublimes). Note that in the case of water, ice will sublime only when the temperature and pressure are brought below the triple point.

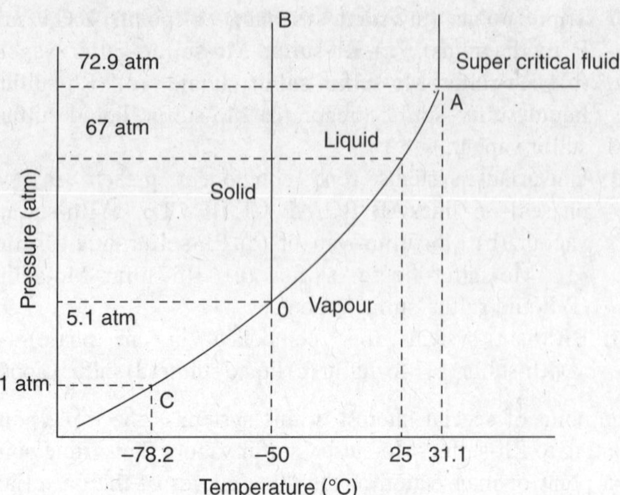

Fig. 10.3 CO_2 system

From the phase diagram of CO_2 (Fig. 10.3), it is also seen that liquid CO_2 will form at 25°C only when the pressure reaches 67 atm. Also note that the solid–liquid line (OB) slopes slightly to the right because the molar volume of liquid is greater than that of the solid.

Supercritical carbon dioxide: The critical temperature of CO_2 is only 31.1°C. CO_2 above its critical temperature possesses the solvent properties of a normal liquid and also the properties of a gas. This characteristic finds use in the removal of caffeine from coffee beans.

10.4.4 Iodine

The triple point O of iodine is at 114.1°C and 0.12 atm. The melting point of iodine is 113.8°C at 1 atm. Thus the triple point temperature and melting point differ by less than a degree. At normal temperatures, it is seen that iodine sublimes (Fig. 10.4).

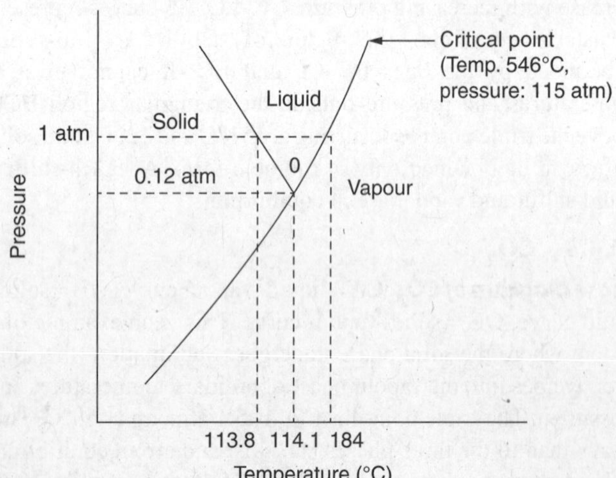

Fig. 10.4 Iodine system

10.5 Construction of Phase Diagrams

A phase diagram is a temperature–composition diagram which indicate the phases present at equilibrium at a given temperature and composition. The phase diagram is determined experimentally by recording the cooling rates over various ranges of composition of the two components. These diagrams are used to understand the formation of compounds between any two given components, their range of stability and whether they are stable at their melting points or whether they decompose below their melting points, and so on.

Liquidus curve: A liquidus curve in a phase diagram represents the equilibrium compositions of liquid mixtures at which one of the components separates out as the solid phase. A liquidus gives the temperature of the liquid phase at which, for a given overall composition, the first solid separates. A solidus gives the temperature of the solid phase at which, for a given overall composition, the first liquid forms. For example, curve XZY represents a liquidus curve in Fig. 10.5 and the line FZF′ represents a solidus curve.

Tie line: Consider the phase diagram shown in Fig. 10.5.

Curve XD represents the solubility of solid Y in solid X and this solid solution is in equilibrium with the liquid. Curve YE represents the solubility of solid X in solid Y and this solid solution is in equilibrium with the liquid. Along XZ or YZ, at any given composition and temperature, a mix of either α or β will exist as solid lumps in equilibrium with a mixture of X and Y. α represents a solid solution of Y in X. β represents a solid solution of X in Y.

A constant temperature line FF′ intersecting the two solubility curves is known as a tie line. The ends of the tie lines show the compositions of the two phases that are in equilibrium at this temperature. When the vertical lines are drawn, they cut the composition axis at G and G′. The α phase having the composition G is in equilibrium with a liquid phase having the composition G′. The overall composition of X and Y is given by H.

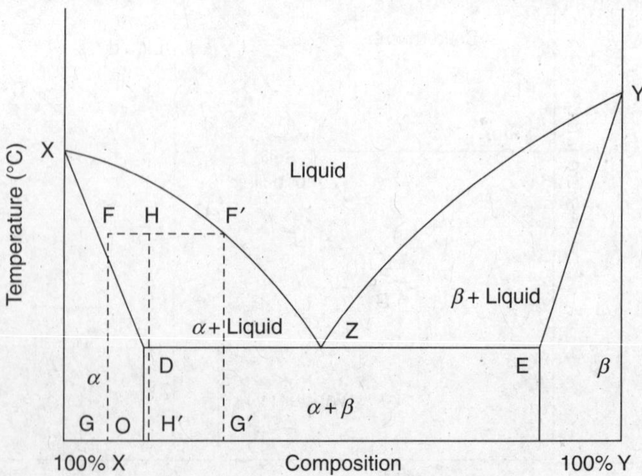

Fig. 10.5 Tie lines in a phase diagram

Lever rule: Knowing the compositions of the two phases (G and G′ in the diagram), to find how much of each phase is present, the lever rule is applied. Lever rule involves the application of the principle of conservation of mass. The proportions of the phases are given by the relative lengths of the tie lines. If FH = X and HF′ = Y, the proportions of α and liquid present are, $\frac{Y}{X+Y}$ and $\frac{X}{X+Y}$.

$$\text{Fraction of } \alpha = \frac{\text{Composition at G}' - \text{Composition at H}}{\text{Composition at G}' - \text{Composition at G}} \quad (10.5)$$

$$\text{Fraction of liquid} = 1 - (\text{fraction of } \alpha) \quad (10.5a)$$

Application of tie line and lever rule to Fig. 10.7: Suppose we cool a liquid mixture corresponding to G (shown by G′ on the composition axis) on the constant composition (isoplieth) axis GHG′. At H, solid Y separates and is in equilibrium with liquid. As the temperature is further brought down, more of Y separates and the liquid becomes richer in X and its composition moves along HZ. At any point I, JJ′ is the tie line. The amount of solid Y at I in equilibrium with the saturated solution of composition J is given by the lever rule as

$$\frac{\text{Mass of Y}}{\text{Mass of liquid}} = \frac{\text{IJ}}{\text{IJ}'} \quad (10.6)$$

10.6 Two-Component Systems

For a system of two components, the system becomes invariant when $P = 4$. It becomes univariant when $P = 3$ and the system becomes bivariant when $P = 2$. A classification of two-component systems is shown in Fig. 10.6.

The study of two-component systems consisting of solid and liquid phases only, provides information on (i) the conditions of stable existence of the participating substances, (ii) the nature and range of existence of any compounds formed. In the study of such systems, the condensed phase rule is applied because the systems are studied under constant pressure (usually atmospheric pressure) and under these conditions no vapour phase exists.

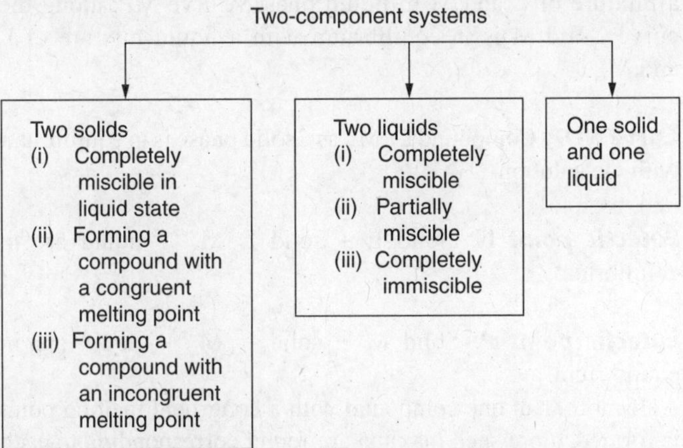

Fig. 10.6 Classification of two-component systems

10.6.1 Two Pure Solid Components Completely Miscible in the Liquid State

Only one liquid phase exists under these conditions as the two components in the liquid state are miscible in all proportions. Also, a pure component exists as a solid phase, therefore the possible phases in equilibrium are S_1–L, S_2–L, S_1–S_2–L where S_1, S_2 are the two crystalline solid phases and L is the liquid. For the two phase systems, $F = 2 - 2 + 1 = 1$. Therefore, if either temperature or composition is fixed, the other is automatically fixed. For three phase systems, S_1–S_2–L, $F = 2 - 3 + 1 = 0$. This implies that there is only one temperature at which all three phases are in equilibrium. The composition of the phases is also fixed.

Phase diagram: In Fig. 10.7, X and Y represent the melting points of pure X and Y. XZ, YZ are the freezing point curves of X and Y. Along the line XZ, solid X is in equilibrium with the liquid. Along YZ, solid Y is in equilibrium with the liquid. At point Z, the meeting point of curves XZ and YZ, both solid components are in equilibrium with a liquid of composition Z. This point Z lies at a lower temperature than the melting points of X and Y and hence is called the *eutectic point*. The region above the curves XZ and YZ represents the liquid phase (L). Below the line FF′, only the solid component mixtures of X and Y can exist as a stable solid phase. If a liquid of composition A (left of Z) is cooled, the solid component begins to crystallise out when the temperature reaches the point B and the composition of liquid varies along the curve BZ. When the composition reaches the point Z, Y also begins to crystallise out. Continuous withdrawal of heat results in the separation of X and Y together while the temperature remains constant. Since the composition also must remain constant, solid X and Y crystallise out in constant proportions. Note that the eutectic mixture of X and Y separating out at the eutectic temperature is not a compound but a mixture of the two solid phases. On cooling, a liquid mixture having the composition Z (eutectic mixture of X and Y) begins to separate out at the temperature corresponding to Z.

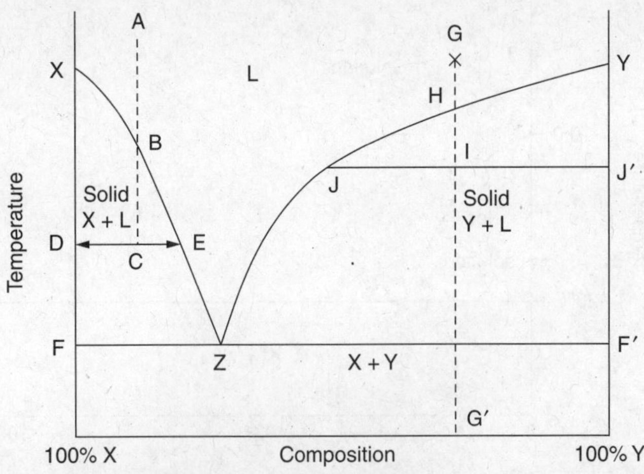

Fig. 10.7 Phase diagram of a two-component solid system

Table 10.1 Some two-component systems forming simple eutectic

Inorganic systems				
Component X	Melting point (°C)	Component Y	Melting point (°C)	Eutectic temperature (°C)
Bismuth	268	Cadmium	317	146
Silicon	1412	Aluminum	657	578
Potassium chloride	790	Silver chloride	451	306
Lead	327	Silver	961	303
Lead	327	Antimony	631	246
Organic systems				
o-Nitrophenol	44.5	*p*-Toluidine	43.3	15.6
Naphthalene	80.0	*m*-Nitrotoluene	15.6	4.7
p-Nitrotoluene	51.4	*p*-Toluidine	43.3	15.8
p-Nitrotoluene	51.4	*o*-Nitrophenol	44.5	16.9

10.6.2 Systems Involving Salt and Water (KI-H_2O System)

There is one limitation to the complete study of the salt–water system, the melting point of the salt is often so high that it is not possible to follow the system up to the melting point of the two components.

Phase diagram of KI-H_2O system: In Fig. 10.8, along AB, ice separates from the KI solution and hence it is the freezing point of the KI solution. CB is the solubility curve of KI. The steep rise of the curve CB shows that the solubility of KI increases slowly with temperature. The eutectic point B, where KI, ice and solution are in equilibrium occurs at –23°C. Ice and common salt form a eutectic at –21.1°C with the composition of the eutectic containing 23.3 wt% of the anhydrous salt.

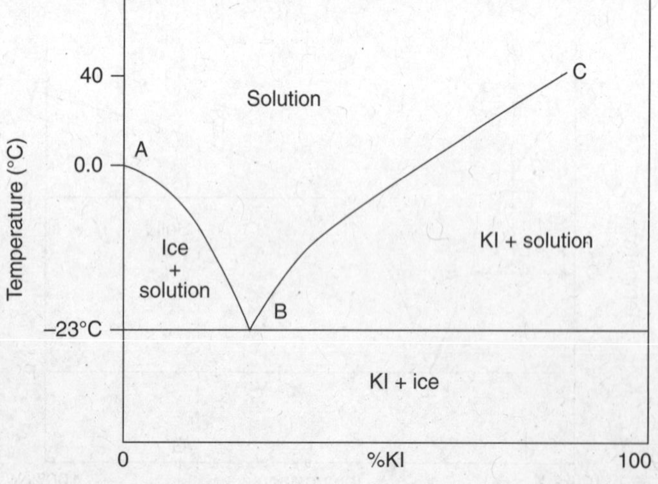

Fig. 10.8 Phase diagram of KI–H_2O system

10.6.3 Systems Involving Compound Formation with a Congruent Melting Point

When the two components can form a stable compound in equilibrium with a liquid of the same composition, that is, it possesses a *congruent melting point*, a third equilibrium curve must be added to the diagram involving simple eutectic. The compound formation shows up as a maximum in the phase diagram since addition of either pure component lowers its melting point. The maximum is called the *dystectic*.

Phase diagram: In Fig. 10.9, the points L, M, L_XM_Y (point O) are congruent melting points of solid L, M and of compound L_XM_Y.

Curve LN: Along the curve, solid L is in equilibrium with a mixture of L and M in liquid phase. Curve MP: along the curve, solid M is in equilibrium with a liquid mixture of L and M.

Curve NOP: Compound L_XM_Y as a solid phase is in equilibrium with the solution.

Eutectic point N: Solid L + solid L_XM_Y + liquid are in equilibrium.

Eutectic point P: Solid M + solid L_XM_Y + liquid are in equilibrium.

If more than one compound with a congruent melting point is formed, more such maxima are found corresponding to each compound. The point O is virtually a one component system with $F = 1 - 2 + 1 = 0$.

Table 10.2 Some two-component systems with compound formation having a congruent melting point

Component L	Melting point (°C)	Component M	Melting point (°C)	Compound	Melting point (°C)
Aluminium	657.0	Magnesium	650.0	Al_3Mg_4	463
Gold	1064.0	Tin	232.0	AuSn	425
Diphenylamine	52.8	Benzophenone	47.7	LM	40.2
Phenol	43.0	Urea	132.0	L_2M	61.0

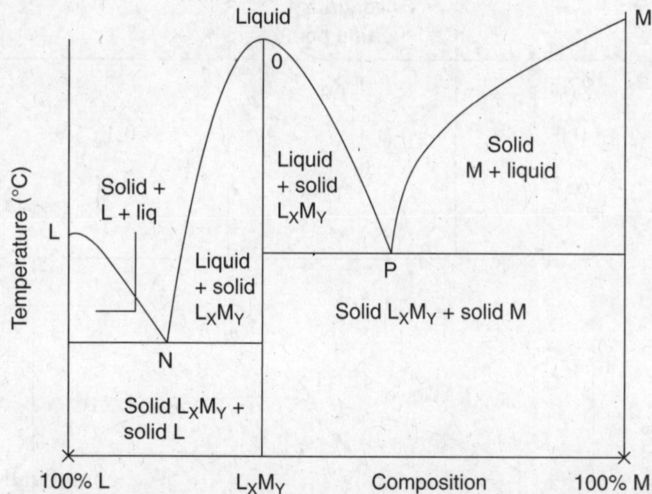

Fig. 10.9 Two-component system with compound formation

The maximum in the phase diagrams: The sharpness of the maximum in the phase diagram is a measure of the stability of the compound at its melting point. If the maximum is flat it means that the compound tends to decompose to a great extent.

10.6.4 Systems Involving Compound Formation with an Incongruent Melting Point

Sometimes the compound formed is so unstable that it decomposes completely below its melting point (Fig. 10.10). In such cases, the solid cannot be in equilibrium with a liquid of the same composition. The compound is then said to have an *incongruent melting point* or *peritectic point.*

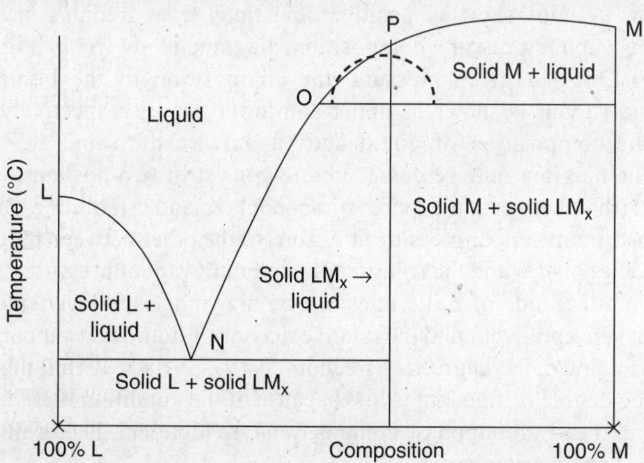

Fig. 10.10 Two-component system with formation of a compound having an incongruent melting point

Phase diagram

Along LN: Solid L and liquid are in equilibrium.

Along NO: Solid LM_X and liquid are in equilibrium.

Along MO: Solid M and liquid are in equilibrium.

O: Peritetic point. The temperature O is below the hypothetical melting point P of compound LM_X. The compound LM_X decomposes completely into solid M and a solution of composition O. O is a triple point at which solid M, solid LM_X (compound LM_X in which X = 1, 2 …) and liquid are in equilibrium, N is a triple point at which solid L, solid LM_X, liquid are in equilibrium.

10.6.5 Salt Hydrates

Salt hydrates are systems consisting of salt and water. They generally have incongruent melting points as they decompose into anhydrous salt and water or into a lower hydrate and water below their melting points.

Phase diagram of Na_2SO_4-H_2O system (Fig. 10.11)

Along curve AB: Liquid is in equilibrium with ice.

Along curve BC: Solid $Na_2SO_4.10H_2O$ is in equilibrium with liquid. It is the solubility curve of $Na_2SO_4.10H_2O$.

Point C: Incongruent melting point of $Na_2SO_4.10H_2O$. Before its melting point is reached, $Na_2SO_4.10H_2O$ decomposes to anhydrous salt and water. Here, anhydrous Na_2SO_4, $Na_2SO_4.10H_2O$ solid and liquid are in equilibrium. It is where $F = 0$.

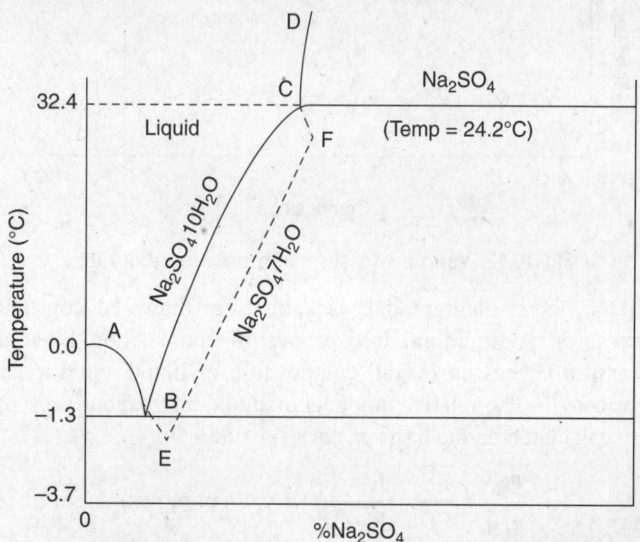

Fig. 10.11 Phase diagram of Na_2SO_4–H_2O system

Point B: here also $F = 0$ and ice, $Na_2SO_4.10H_2O$ solid and liquid are in equilibrium. The transition temperature $Na_2SO_4 10H_2O(s) \rightleftharpoons Na_2SO_4(s)$ occurs at 32.38°C. If a solution saturated with Na_2SO_4 is cooled along the curve DC, sometimes instead of $Na_2SO_4.10H_2O$ starting to deposit, the anhydrous form remains in metastable equilibrium (see dotted lines). The composition of the solution continues along F, that is, when the temperature reaches 24.2°C. At this point, a new hydrate $Na_2SO_4.7H_2O$ separates out. The metastable $Na_2SO_4.7H_2O$ is formed only if $Na_2SO_4.10H_2O$ is absent. E is a metastable eutectic at which ice, $Na_2SO_4.7H_2O$ and liquid are in metastable equilibrium. E and F are metastable triple points at which $F = 0$.

Table 10.3 Some two-component systems with compound formation with incongruent melting point

Component L	Melting point (°C)	Component M	Melting point (°C)	Compound with incongruent melting point	Melting point (°C)
Sodium	62	Potassium	97.5	L_2M	7
NaCl		Water	0.0	LM_2	0.15
Gold	1064	Antimony	631	LM_2	467
Magnesium		Nickel		LM_2	

10.6.6 Two Completely Miscible Liquids

For a two component system, $F = 2 - P + 2 = 4 - P$. Since we usually deal with liquid and vapour phases, $F = 4 - 2 = 2$. The two degrees of freedom could be vapour pressure and composition of liquid, or vapour pressure and composition of vapour. It is common to describe such systems at constant temperature as vapour pressure–composition, or temperature–composition diagrams at constant pressure (usually at 1 atm pressure).

Vapour pressure–composition diagrams can be drawn by plotting vapour pressure as a function of composition of liquid (or vapour) for solutions obeying Raoult's law (Fig. 10.12).

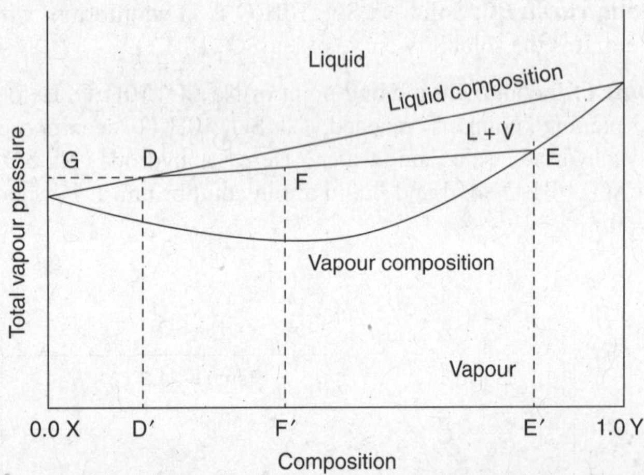

Fig. 10.12 Vapour pressure–composition diagram

In the above diagram, DE is a tie line (drawn at constant pressure). A liquid mixture of overall composition F′ at a pressure G, has an overall composition of liquid and vapour given by F. The relative amounts of liquid and vapour present are calculated on the basis of the lever rule.

$$\frac{n_{liq}}{n_{vap}} = \frac{FE}{FD} = \frac{E' - F'}{F' - D'} \; (n = \text{no. of moles}) \qquad (10.7)$$

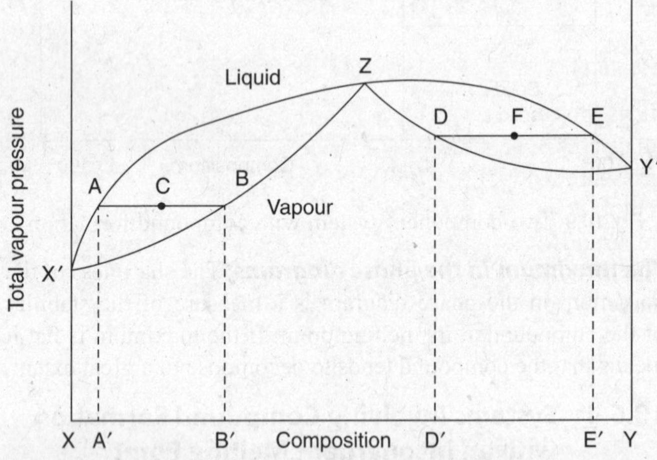

Fig. 10.13 System showing positive deviation from Raoult's law

Vapour pressure–composition diagrams for a system showing a positive deviation from Raoult's law: For systems showing positive deviations from Raoult's law, the vapour pressure–composition diagram is shown in Fig. 10.13. At a given pressure, the composition of the liquid and vapour is shown as upper and lower curves respectively. The composition of liquid and vapour are the same at Z. The diagram may be considered to consist of two portions—(i) that between the pure component X and a solution of maximum vapour pressure at Z, and (ii) the other between pure component Y and the solution of maximum vapour pressure Z. On either side of Z, tie lines can be drawn and the lever rule may be applied to find the relative masses of liquid and vapour. At point Z, the degrees of freedom $F = 2 - 2 + 1 = 1$. But, this one degree of freedom is lost because of the condition that the liquid and vapour phase compositions are identical. Therefore, Z is an invariant point.

Vapour pressure–composition diagram for a system showing a negative deviation from Raoult's law:
For systems showing negative deviations from Raoult's law (for example, H_2O–HNO_3 mixtures), the vapour pressure–composition diagram is shown in Fig. 10.14.

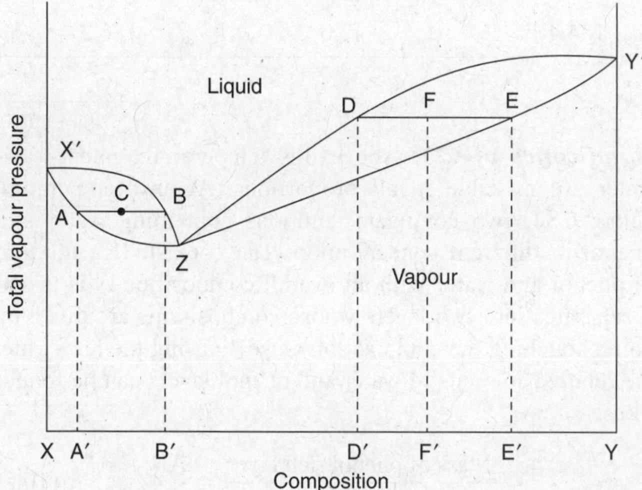

Fig. 10.14 System showing a negative deviation from Raoult's law

It is possible to construct the boiling point–composition diagrams of various mixtures of X and Y keeping the external pressure constant at 1 atm (Fig. 10.15). Typical boiling point–composition diagrams for a two-component system of liquids X and Y obeying Raoult's law and one showing a positive deviation from Raoult's law are shown in Figs 10.15 and 10.16 respectively.

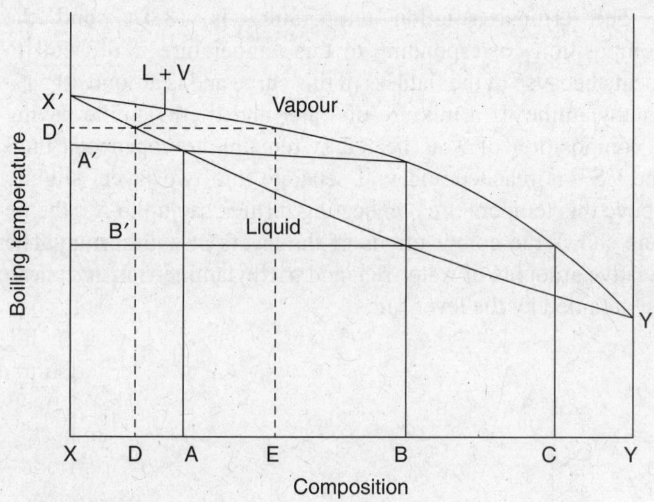

Fig. 10.15 Boiling point–composition diagram of a system obeying Raoult's law

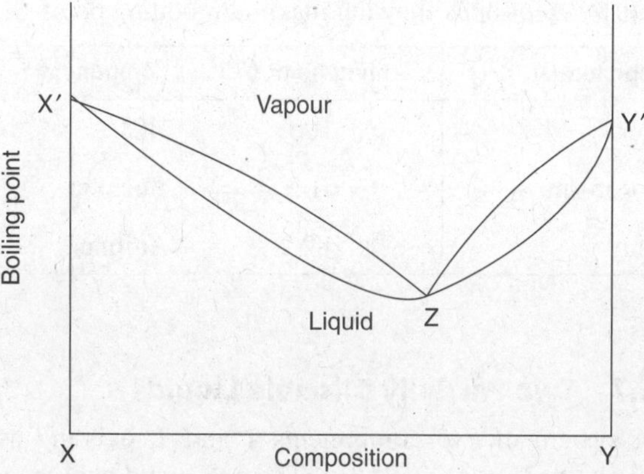

Fig. 10.16 Boiling point–composition diagram of a system showing positive deviations from Raoult's law

If a liquid of composition A is heated (Fig. 10.15, Tables 10.4 and 10.5), the vapour pressure increases until it reaches a constant value corresponding to an external pressure of 1 atm. It then begins to boil at a temperature A′. The vapours in equilibrium with the liquid have a composition B which is richer in the volatile component Y. If the vapours corresponding to B are condensed, the liquid so formed boils at B′ and the vapours formed will have a composition C which is seen to be even richer in the volatile component Y. Repetition of this process of condensation and redistillation ultimately leads to the formation of pure Y.

Considering the less volatile component X, after the first vapours are condensed (at B), let its composition be D. This residue having the composition D on boiling, shows a boiling point at D′ (the temperature at D′ is greater than the temperature at A′). The corresponding composition of vapours in equilibrium with the solution of composition D is given by E and these vapours on condensation give a liquid richer in the more volatile component Y while the residual liquid will be richer in less volatile component X. Thus, this process can lead to the separation of pure components X and Y by condensation of vapours and their redistillation.

Extending this argument to mixtures showing a positive or negative deviation from Raoult's law, it is possible to separate a liquid mixture of X and Y into one of the pure components X or Y and a mixture corresponding to composition Z. The liquid mixture corresponding to Z is known as an *azeotropic mixture* or *constant boiling mixture*. This azeotropic mixture will have a minimum in the boiling point–composition diagram for mixtures showing positive deviations from Raoult's law. For mixtures showing negative deviations from Raoult's law, a maximum is observed in the boiling point–composition diagram. An azeotropic mixture cannot be separated into the individual components by normal distillation. A few examples of minimum and maximum boiling azeotropes are shown in Tables 10.4 and 10.5.

Table 10.4 Azeotropes showing minimum boiling point

Component X	Boiling point (°C)	Component Y	Boiling point (°C)	Azeotrope (mass% X)	Boiling point (°C)
Water	100	Ethanol	78.1	4.0	78.2
Carbon disulfide	46.3	Acetone	56.2	67.0	39.3
Water	100	Pyridine	115.2	41.3	92.0

Table 10.5 Azeotropes showing maximum boiling point

Component X	Boiling point (°C)	Component Y	Boiling point (°C)	Azeotrpe (mass% X)	Boiling point (°C)
Water	100	HCl	–80.0	20.2	108.6
Chloroform	61.2	Acetone	56.1	78.5	64.4
Phenol	182.2	Aniline	184.4	42.0	186.2

10.6.7 Two Partially Miscible Liquids

For a system of two components 1 and 2, existing as two liquid phases, $F = 2 - 2 + 1 = 1$. The two liquid phases may be regarded as the solution of component 1 in component 2 and the other one as the solution of component 2 in component 1. Since $F = 1$, at each temperature, there will be a definite concentration of each of the components in the two layers and addition of excess of one merely alters their relative amounts. If the temperature is changed, the compositions of the two layers will vary. Thus two solubility curves, one for each component in the other is obtained. Examples of such systems are phenol + water, aniline + hexane, carbon disulfide + methanol.

Phenol-water system

Phenol dissolves up to 7 wt% in water at ordinary temperatures. A further addition of phenol will result in the formation of two layers, one a layer of phenol in water, and the other a layer of water in phenol. Increase of temperature results in an increase of mutual solubility and one layer is formed. A further addition of phenol is needed to separate it into two layers. By raising the temperature, it is possible to obtain a homogeneous solution. However, the composition of the two layers must be the same at a definite temperature, known as the *critical solution temperature*. Figure 10.17 shows the mutual solubility curve of the system.

At any point on the curve, $F = 2 - 2 + 1 = 1$. But at the critical solution temperature (CST), $F = 2 - 2 + 1 = 1$. However, this degree of freedom is lost because both layers have identical compositions. The composition at CST is 34.5 wt% phenol.

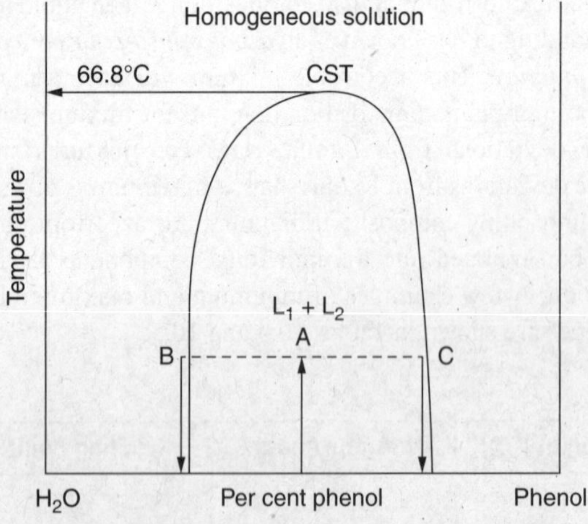

Fig. 10.17 Phase diagram of phenol–water system

Significance of CST: Above this temperature, phenol and water are miscible in all proportions. At any temperature below CST, two conjugate solutions containing water and phenol in different concentrations can coexist. If a mixture of phenol and water with an overall composition A is taken, it separates into two layers whose compositions are given by points touching the composition axis. By using the lever rule, the relative amounts of water and phenol layers can be found. Thus,

$$\frac{\text{Mass of phenol rich layer}}{\text{Mass of aqueous rich layer}} = \frac{AB}{AC} \tag{10.8}$$

10.6.8 Systems with Lower Critical Solution Temperature

Triethylamine-water system: Contrary to the phenol–water system, in some cases, the solubility decreases with increase in temperature. Triethylamine–water system is one such system where the mutual solubility decreases with increase of temperature and a *lower consolute temperature* is observed (Fig. 10.18).

The critical solution temperature is 18.5°C and the composition corresponding to this temperature is difficult to locate because of the flatness of this curve and is around 40 wt% triethylamine. If a mixture of water and triethylamine having a composition of X is heated, it remains homogeneous until the CST is reached and will separate into two layers slightly above this temperature. On heating further, say up to X″, the tie line shows the compositions of the layers in equilibrium. The relative amounts of water-rich and triethylamine-rich layers can be obtained by the lever rule.

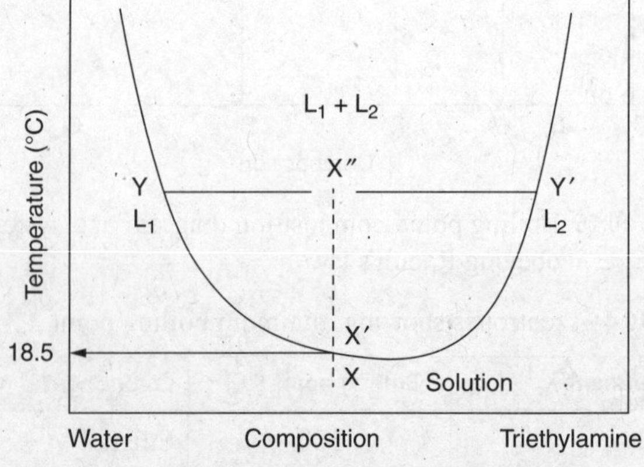

Fig. 10.18 Phase diagram of triethylamine–water system

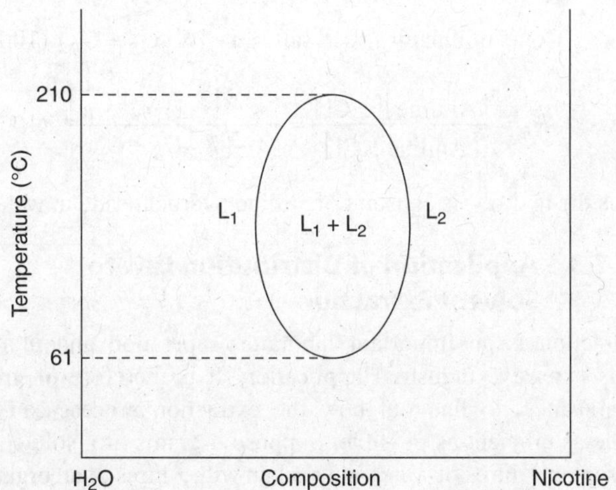

Fig. 10.19 Phase diagram of nicotine–water system

Systems are also known which show both the upper consolute and lower consolute temperatures. One such system is nicotine–water system which has an upper consolute temperature of 210°C and a lower consolute temperature of 61°C (Fig. 10.19).

A few examples of liquid mixtures showing upper consolute and lower consolute temperatures is given in Tables 10.6 and 10.7.

Table 10.6 Systems with an upper critical solution temperature

Component X	Component Y	Upper CST (°C)
Aniline	Hexane	60.0
Carbon disulfide	Methanol	40.5

Table 10.7 System with a lower critical solution temperature

Component X	Component Y	Lower CST (°C)
Water	3-Methyl piperidine	57

This system has an upper critical solution temperature too, at 235°C.

10.6.9 Two Immiscible Liquids

In a mixture of two immiscible liquids, each liquid behaves independently of the other. If the two components are A and B, the total vapour pressure P is the sum of the vapour pressures of the two pure liquids P_A^0 and P_B^0.

$$P = P_A^0 + P_B^0 \tag{10.9}$$

The mole fraction of A and B in the vapour phase, that is, X_A and X_B are given by

$$X_A = \frac{P_A^0}{P} \text{ and } X_B = \frac{P_B^0}{P} \text{ by Dalton's law} \tag{10.10}$$

The number of moles of A and B can then be expressed as

$$\frac{n_A}{n_B} = \frac{X_A}{X_B} = \frac{P_A^0}{P_B^0} \tag{10.11}$$

Equation 10.11 can be expressed in terms of the weights of A and B,

$$\frac{w_A}{w_B} = \frac{n_A M_A}{n_B M_B} = \frac{P_A^0 M_A}{P_B^0 M_B} \tag{10.12}$$

The relative weights of A and B in the vapour are thus related to their vapour pressures and molecular weights.

Equation 10.15 can be employed for steam distillation of two immiscible liquids which have either high boiling points or which decompose before their boiling point is reached. However, note that the liquid to be steam distilled should have a fairly high volatility near the boiling point of water.

10.7 Nernst Distribution Law

If, to a system of two completely immiscible liquids, some quantity of a substance soluble in both layers is added, the substance distributes itself between the two layers in a definite ratio. This is known as *Nernst's distribution law*. From a thermodynamic stand point, at equilibrium, the chemical potential of the solute must be the same in both phases. Thus

$$\mu_{\text{solute in phase 1}} = \mu_{\text{solute in phase 2}} \tag{10.13}$$

But

$$\mu_{(1)}^0 + RT \ln a_{(1)} = \mu_{(2)}^0 + RT \ln a_{(2)} \tag{10.14}$$

Since the chemical state in the standard state is constant at given temperature and pressure, we may write

$$RT \ln a_{(1)} - RT \ln a_{(2)} = \mu_{(2)}^0 - \mu_{(1)}^0 \tag{10.15}$$

$$\ln \frac{a_{(1)}}{a_{(2)}} = \frac{\mu_{(2)}^0 - \mu_{(1)}^0}{RT} \tag{10.16}$$

$$\frac{a_{(1)}}{a_{(2)}} = e^{\frac{\mu_{(2)}^0 - \mu_{(1)}^0}{RT}} = \text{Constant} \tag{10.17}$$

In dilute solutions, the activities may be replaced by molalities (m) or concentrations (C).

$$\frac{C_{(1)}}{C_{(2)}} = K \tag{10.18}$$

which proves the distribution law. It may be mentioned that the distribution law applies to the same molecular species in the two layers. If a solute undergoes dissociation or association in any one of the solvents, this has to be taken into consideration in applying the distribution law. This is used in the determination of equilibrium constants of reactions or association constants. For example, it is possible to determine the dissociation constant of the reaction

$$KI_3 \rightleftharpoons KI + I_2 \tag{10.19}$$

in water or the association constant of benzoic acid when it is distributed between benzene and water from distribution measurements.

10.7.1 Equation for Association Constant of Benzoic Acid from Distribution Studies

When benzoic acid is distributed between water and benzene, it gets associated in the benzene layer and equilibrium exists between associated and unassociated molecules in this layer. Representing the dissociation of the associated species by the reaction

$$(C_6H_5COOH)_n \rightleftharpoons nC_6H_5COOH \quad (10.20)$$

$C_{(2)}$ is the concentration of benzoic acid in the benzene layer and α is the degree of dissociation of associate molecules at equilibrium

Conc. of associated molecules at equilibrium = $C_{(2)}(1-\alpha)$ (10.21)

Conc. of umassociated molecules at equilibrium = $\alpha \times C_{(2)} \times n$ (10.22)

$$K_a = \text{Equilibrium constant of Eq.10.20} = \frac{(\alpha \times C_{(2)} \times n)^n}{C_{(2)}(1-\alpha)} \quad (10.23)$$

Conc. of unassociated molecules in benzene layer

$$= \alpha \times C_{(2)} \times n = \sqrt[n]{K_a C_{(2)}(1-\alpha)} \quad (10.24)$$

If $C_{(1)}$ is the concentration of unassociated molecules in the aqueous layer,

$$\text{Distribution coefficient} = K = \frac{C_{(1)}}{\sqrt[n]{K_a C_{(2)}(1-\alpha)}} \quad (10.25)$$

$$\text{If } \alpha \ll 1, K' = \frac{C_{(1)}}{\sqrt[n]{C_{(2)}}} \text{ where } K' = K\sqrt[n]{K_a} \quad (10.26)$$

$C_{(1)}$ and $C_{(2)}$ are experimentally determined and n can be found by trial and error.

10.7.2 Hydrolysis Constant from Distribution Studies

It is also possible to determine the hydrolysis constant of salts such as aniline hydrochloride by employing the distribution law. Aniline hydrochloride on hydrolysis gives aniline and hydrochloric acid. On shaking the salt in water with an immiscible solvent like benzene, aniline distributes itself between the aqueous and benzene layers. The acid, HCl and the salt are insoluble in the benzene layer.

Let K_X be the distribution coefficient of aniline between water and benzene. Thus

$$K_X = \frac{C_W}{C_B} \quad (10.27)$$

where C_w = concentration of aniline in water and C_B = concentration of aniline in benzene. From Eq. 10.27,

$$C_W = K_X C_B \quad (10.27a)$$

Let a be the initial concentration of aniline hydrochloride. The concentration of the acid in water, C_{HCl} is equal to the sum of the concentration of aniline in water and benzene. Thus,

$$C_{HCl} = K_X C_B + C_B \quad (10.28)$$

Conc. of unhydrolysed salt = $a-(K_X C_B + C_B)$ (10.29)

$$\text{Now,} \quad K_h = \frac{[\text{Aniline}][\text{HCl}]}{[\text{Aniline HCl}]} = \frac{C_W \times [K_X C_B + C_B]}{a-(K_X C_B + C_B)} \quad (10.30)$$

K_h is the hydrolysis constant of aniline hydrochloride in water.

10.7.3 Application of Distribution Law to Solvent Extraction

Extraction is an important laboratory operation and it also finds extensive industrial application. It is therefore of great significance, to find out how the extraction process can be made as efficient as possible. Suppose a grams of a solute are present in V litres of water and shaken with x litres of an organic component. Let W_1 g of the solute remain in the aqueous layer after the first extraction.

The amount of solute extracted into organic layer = $(a - W_1)$
From the distribution law, the partition coefficient,

$$K_p = \frac{C_{\text{water}}}{C_{\text{organic}}} = \frac{W_1/V}{(a-W_1)/x} \quad (10.31)$$

$$\frac{(a-W_1)}{W_1} = \frac{x}{K_p V}, \text{ or } \frac{a}{W_1} = \frac{K_p V + x}{K_p V} \quad (10.32)$$

$$W_1 = a\frac{K_p V}{K_p V + x} \quad (10.33)$$

Equation 10.32 gives the weight of the solute left in the aqueous layer after the first extraction. After second extraction, the weight of solute left in the aqueous layer W_2 is given by

$$W_2 = W_1\left(\frac{K_p V}{K_p V + x}\right) \quad (10.34)$$

Note that same x litres of organic solvent is used in the second extraction. Equation 10.33 may be rewritten as

$$W_2 = a\left(\frac{K_p V}{K_p V + x}\right)^2 \quad (10.35)$$

After n extractions, the weight W_n of solute left in the aqueous layer is given by

$$W_n = a\left(\frac{K_p V}{K_p V + x}\right)^n \quad (10.36)$$

If we use all the volume (nx litres) of organic solvent in one step, the amount of solute W_n' still remaining without being extracted is given by

$$W_n' = a\frac{K_p V}{K_p V + nx} \quad (10.37$$

For a given value of K_p, the greater the value of n, the smaller the W_n will be because the term in brackets of Eq. 10.35 is a fraction. Also, as the value of n becomes larger, it is preferable to keep the value of x small. Also note that,

$$\frac{K_pV}{K_pV+nx}\text{(from a single extracion)} > \left(\frac{K_pV}{K_pV+x}\right)^n$$

$$\text{(from Eq.10.37)} \qquad \text{(from Eq.10.36)}$$

$$\therefore W_n' > W_n$$

Thus, it is recommended that a larger number of extractions with smaller volumes of organic solvent be performed rather than a single extraction with a large volume of organic solvent.

10.8 Three-Component System

For a system of three components, the maximum number of degrees of freedom is 4 when they are in a single phase as seen from the equation $F = 3 - 1 + 2 = 4$. Fixing two of the variables T and P enables one to represent the conditions of equilibrium on a planar diagram using rectangular coordinates but it is convenient to use a triangular diagram involving an equilateral triangle. The variables are the three compositions X_1, X_2 and X_3 (mole fractions). Specifying two of the variables fixes the third from the relation $X_1 + X_2 + X_3 = 1$. The salient features of the triangular diagram (Fig. 10.20) are described below.

(i) The apices of the triangle represent the pure components.

(ii) The lines parallel to XY represent various percentages of Z. Any point on line XY represents 0% Z. Point A, for example, represents 30% Z.

(iii) The length perpendicular to a given side of the triangle represents the percentage of the component at the vertex opposite to that side. Thus, the length AB gives the percentage of X. Length AC represents the percentage of Z and length AD represents the percentage of Y.

(iv) The sum of the lengths of the perpendiculars AB + AC + AD = length XY (or YS or ZX). The composition of the system can be represented by a point within the triangle.

(v) Two points, say E and F representing definite compositions are mixed, the compositions of the mixture will be on the line (G) connecting the points. Similarly, if three compositions H, I, J are mixed together, the overall compositions of the mixture will be within the triangle formed by them.

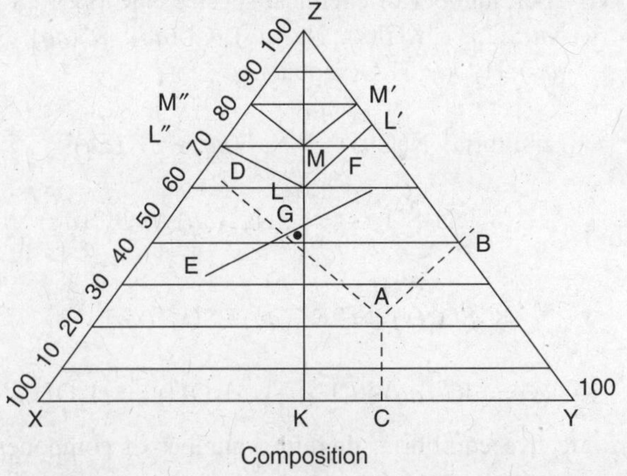

Fig. 10.20 Phase diagram of a three-component system

(vi) All systems represented by points on a line through a vertex contain the other two components in the same ratio. For example, all systems represented by points on line ZK contain X and Y components in the same ratio. The perpendiculars drawn LL′, LL″ and MM′, MM″ are related in the following way.

$$\frac{MM'}{LL'} = \frac{MM''}{LL''} \text{ also } \frac{MM'}{MM''} = \frac{LL'}{LL''}$$

10.8.1 Liquid-Liquid Equilibria

A good example of this class is the $CHCl_3$–CH_3COOH–H_2O system. In this case, $CHCl_3$ and CH_3COOH, and water and acetic acid are completely miscible while $CHCl_3$ and H_2O are not miscible. The phase diagram of this system is shown in Fig. 10.21.

Points X and Y represent the chloroform-rich and aqueous-rich layers in the absence of acetic acid. If the overall composition is represented by Z, by the application of lever rule, it is seen there is more of layer Y and less of X.

$$\frac{\text{Mass of layer X}}{\text{Mass of layer Y}} = \frac{ZY}{ZX} \tag{10.38}$$

Addition of a little acetic acid results in the composition line moving towards Z′, that is, the line connecting Z to the apex opposite. The composition of the two layers moves along X′Y′. Since the X′Z′ line increases in length, the water layer size grows indicating that acetic acid goes preferentially to the aqueous layer than to the $CHCl_3$ layer. Continued addition of acetic acid moves the composition along X″Y″, the line tilting more towards the $CHCl_3$ apex. The water rich layer grows in size while the $CHCl_3$ rich layer diminishes. At Y‴, actually only a trace of $CHCl_3$ layer is present. Above Y‴, only a homogeneous solution is present. Since the tie lines are not parallel, the conjugate solutions having the same composition lie to one side of the curve, that is, at Z′. This point is known as *plait point*.

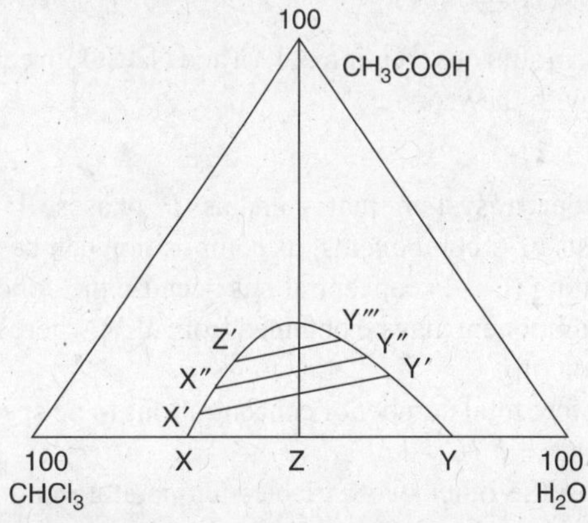

Fig. 10.21 Phase diagram of $CHCl_3$–CH_3COOH–H_2O system

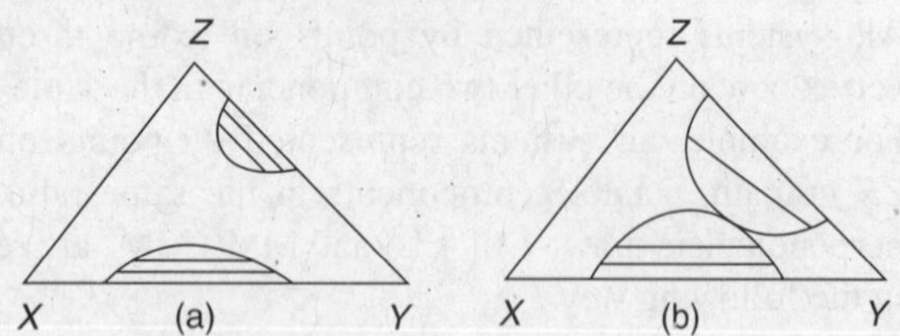

Fig. 10.22 Two pairs of partially miscible liquids

10.8.2 Two Pairs of Partially Miscible Liquids

Consider Fig. 10.20. Let pairs *XY* and *YZ* be partially miscible. In such a case, the phase diagram will have the following shape (Fig. 10.22a). In some cases, the binodal curves shown in Fig. 10.20, may overlap at lower temperatures. In such a case, the plait points may join one another and the phase diagram has the form shown in Fig. 10.22b.

10.8.3 Other Three-Component Systems of Interest

Systems containing two salts and water are of great interest. There are several possibilities in the study of such systems.

(i) Two salts having a common ion and water: An example of this kind of system is NH_4Cl–$(NH_4)_2SO_4$–H_2O where NH_4^+ is the common ion and the presence of one salt influences the solubility of the other.

(ii) Two salts form a compound: This is a double salt. In this case, the solubility of the compound may also show up as a line in the phase diagram. Examples are (a) $AgNO_3$–NH_4NO_3–H_2O in which the double salt formed is $AgNO_3.NH_4NO_3$ and (b) K_2SO_4–Na_2SO_4–H_2O in which the double salt with the composition $3K_2SO_4.Na_2SO_4$ is referred to as glaserite.

(iii) One of the salts forms a hydrate: An example of this is Na_2SO_4–NaCl–H_2O in which $N_{a2}SO_4.10H_2O$ is the hydrate formed.

(iv) A double salt hydrate and a simple hydrate are present: An example of this is $NaNO_3$–Na_2SO_4–H_2O, in which $NaNO_3.Na_2SO_4.H_2O$ and $Na_2SO_4\,10H_2O$ are formed.

The phase diagrams in cases (ii) and (iii) have different shapes and they also differ from case (i).

Worked Out Problems

Gibbs Phase Rule and its Applications

Problem 10.1

(a) Derive the Gibbs phase rule relating the number of components (*C*), phases (*P*) and the degrees of freedom (*F*).

(b) Calculate the number of components (*C*), phases (*P*) and the degrees of freedom (*F*) giving sufficient details where necessary.

(i) $N_2(g)+O_2(g)$

(ii) S (rhombic) $\rightleftharpoons S$ (monoclinic) $\rightleftharpoons S$ (liquid)

(iii) $NH_4Cl(s) \rightleftharpoons NH_3(g)+HCl(g)$

(iv) $KCl(s) \rightleftharpoons$ KCl (saturated solution) at 298 K

(v) $I_2(s) \rightleftharpoons I_2(g)$

(vi) An aqueous solution of KCl and Na_2SO_4 in equilibrium with the vapour.

Solution

(a) Consider a system that contains *P* phases. If a phase consists of *C* components, its composition can be specified by stating (*C* – 1) concentrations because the concentration of a component may be obtained from $\Sigma_i X_i$ where X_i = mole fraction of *i*.

Therefore total number of concentrations to be specified for *P* phases = *P* (*C* – 1)

Adding the other two variables temperature and pressure, the number of variables = *P* (*C* – 1) + 2 (1)

Considering the number of relationships that must be specified at equilibrium, if there are *C* components, and *P* phases, the number of independent equations = *C* (*P* – 1) (2)

(This is because if there are *P* phases, there are only (*P* – 1) equilibrium relationships for each component.)

The difference between the number of variables and the number of independent equations is the number of independent variables or degrees of freedom *F*.

$$F=[P(C-1)+2]-\{C(P-1)\}=C-P+2 \qquad (3)$$

(b) (i) $C=2,\ P=1;\ F=2-1+2=3$

(ii) $C=1, P=3; F=1-3+2=0$

(iii) $C=1, P=2; F=1-2+2=1$ (assuming $p_{NH_3}=p_{HCl}$)

(iv) $C=2,\ P=2; F=C-P+1=2-2+1=1$

(v) $C=1,\ P=2; F=1-2+2=1$

(vi) Total number of chemical species NaCl(s), $K_2SO_4(s)$, $Na_2SO_4(s)$, KCl(*s*), $Na^+(aq)$, $Cl^-(aq)$, $K^+(aq)$, $SO_4^{2-}(aq)$, $H_2O(l)$, H_2O(vapour).

$$\text{Equilibria: } NaCl(s) \overset{H_2O}{\rightleftharpoons} Na^+(aq)+Cl^-(aq)$$

$$Na_2SO_4(s) \overset{H_2O}{\rightleftharpoons} 2Na^+(aq)+SO_4^{2-}(aq)$$

$$K_2SO_4(s) \overset{H_2O}{\rightleftharpoons} 2K^+(aq)+SO_4^{2-}(aq)$$

$$KCl(s) \overset{H_2O}{\rightleftharpoons} K^+(aq)+Cl^-(aq),\ H_2O(l) \rightleftharpoons H_2O(v)$$

There are five equilibria, therefore number of components = $8-6=2$ (also considering conditions of electroneutrality). Phases, $P=4$ (two solid phases, one liquid phase, one vapour phase)

$$F=2-4+2=0$$

Phase Diagram of a One-Component System

Problem 10.2

Draw the phase diagram of a one-component system that has a triple point at 516.8 kPa and 223 K. It has a critical temperature of 304.1 K and 7397 kPa. Indicate the supercritical region on the diagram. How is the phase in the supercritical region described? Mark all regions on the diagram.

Solution

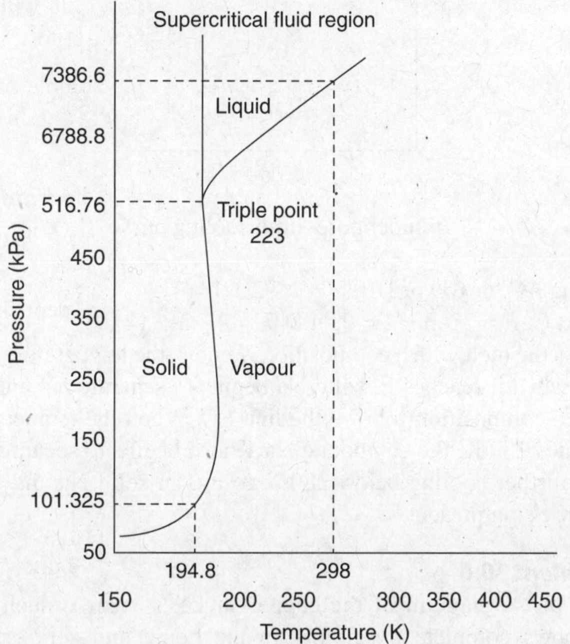

Phase diagram of an unknown one-component system

The phase in the supercritical region is referred to as a supercritical fluid. This is neither liquid nor vapour. The triple point is at 516.8 kPa and 223 K. At 101.325 kPa (1 atm), the temperature is 194.8 K, that is, –78.2°C. This substance does not exist as liquid at 1 atm.

Phase Diagram of a Two-Component System

Problem 10.3

(a) Draw the phase diagram of the lead–silver system using the following data. The melting point of Pb and Ag are 600 K and 1234 K and the eutectic temperature and composition are 576 K and 97.4 wt% Pb. Mark all the regions in the diagram. Calculate the degree of freedom in a region, along a line and at the eutectic point. How do you show that the eutectic composition is a mixture and not a compound?

(b) Consider a point A in the liquid region and explain sequentially what happens as it is cooled to a point B below the eutectic temperature.

Solution

(a) The diagram is shown below. *F* is calculated for various regions and shown in the diagram.

Differences between a eutectic composition and a compound:

- The eutectic composition varies with pressure and in the case of a compound, this is not so.
- Addition of either component to the eutectic composition at eutectic temperature raises the temperature whereas in the case of a compound, addition of another substance in small quantities lowers its freezing point.

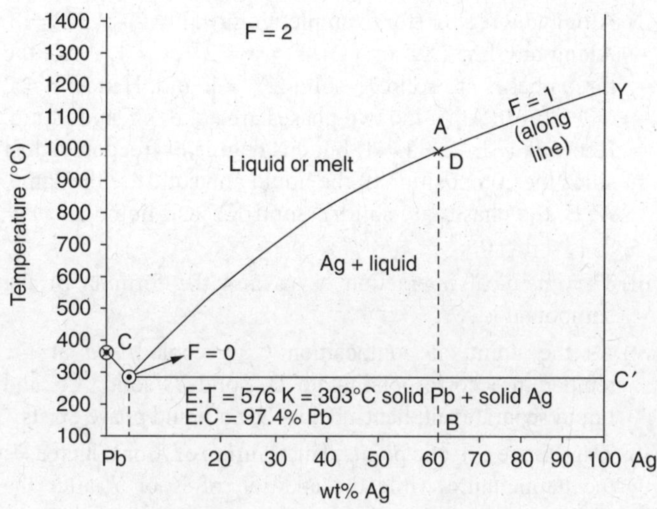

Phase diagram of lead–silver system

(b) Consider a melt having 60 wt% Ag composition shown at A in the diagram. On cooling the melt, silver starts to come out of the melt as a solid at point D, and as further cooling takes place the composition of the liquid melt follows the curve and when the temperature reaches point C solid lead also separates out. At C, solid lead, solid silver and the liquid are in equilibrium.

Problem 10.4

The phase diagram of a two-component system involving components X and Y is given below:

(i) Mark the phases in the various parts of the diagram.

(ii) Calculate the degrees of freedom in a region, along a line and at the points Z, A, B. What is the significance of the value of F at these points?

(iii) Is there a chemical interaction between X and Y? If so, what is the formula of the phase formed at A?

(iv) What will happen if a liquid melt of composition corresponding to C is cooled?

(v) How are the compositions at Z and B different from that of A? How do you differentiate them through other tests?

Solution

(i)

Phase diagram of X–Y component system

(ii) Along any region (for example, liquid), F = 2 – 1 + 2 = 2. Along any line (XZ or YB), F = 2 – 2 + 1 = 1. At Z, the three phases are solid X, solid XY + liquid. Hence, F = 2 – 3 + 1 = 0. At A, the two phases are solid XY and liquid. Hence, F = 2 – 2 + 1 = 1, but this degree of freedom is lost since the composition of the liquid and solid are the same. At B, the phases are solid Y, solid XY and liquid and F = 2 – 3 + 1 = 0.

(iii) The chemical interaction exists and the formula of the compound is XY.

(iv) As the liquid at composition C is cooled and at the temperature corresponding to B, solid Y, solid XY and liquid separate out; hence below B, no liquid phase exists.

(v) Z and B are eutectic points. The addition of X or Y increases the temperature while the addition of X or Y after the compound formation, decreases the temperature.

Problem 10.5

The phase diagram of the potassium–sodium system has the following features. The melting point of pure sodium, 370.5 K, melting point of pure potassium, 335 K. A compound Na_2K with an incongruent melting point of 280 K is formed (the compound decomposes before its melting point is reached). A eutectic point is observed at 246 K and at a mol% composition of 29% Na. Draw the phase diagram of temperature–mol% of a component using the above data. Calculate F along a line, at the eutectic point and at the point of decomposition of Na_2K. Mark all regions in the diagram. Consider a point X in the melt region to the right of incongruent melting point and explain what happens as the melt cools down. Draw the temperature–time cooling curve to illustrate the changes.

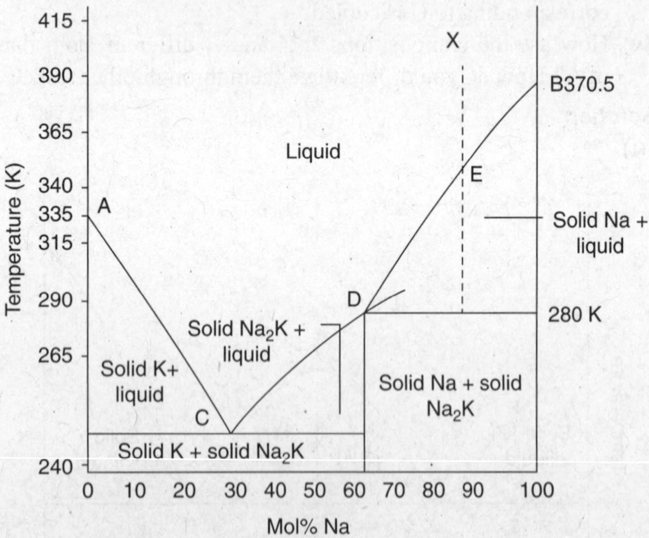

Phase diagram of potassium–sodium system

Solution

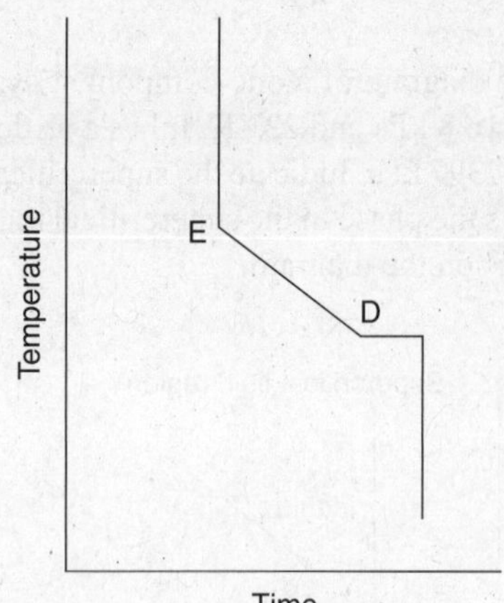

Temperature–time cooling curve

Along AC, or CD or DB, $F = 2 - 2 + 1 = 1$

At C, $F = 2 - 3 + 1 = 0$; at D, $F = 2 - 3 + 1 = 0$

As the melt with a composition X cools, the temperature falls and when it reaches E, solid Na begins to separate out and the liquid composition follows the line ED. When the temperature reaches 280 K, the compound Na_2K also begins to separate out and further cooling below 280 K results in solid Na and solid Na_2K being present.

Problem 10.6

The phase diagram of the copper–nickel system is such that it shows complete miscibility in the liquid and solid states. The diagram is shown below.

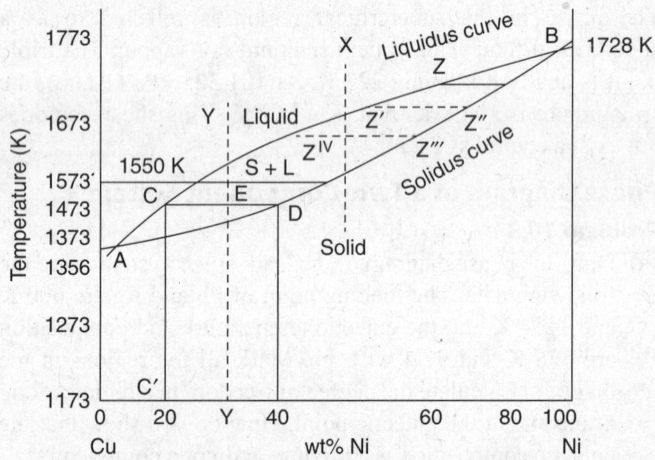

Phase diagram of the copper–nickel system

(a) Show the cooling curve, that is, temperature–time curve (i) when pure molten copper is cooled (ii) when an alloy of 40 wt% copper in molten state is cooled.

(b) Calculate F along a line and in the region between the curves.

(c) Consider the point Z corresponding to 30 wt% copper in overall composition in molten state is gradually cooled. Describe the changes that occur.

(d) Consider a copper–nickel alloy having 70 wt% copper shown at point Y. Consider 1 kg of this alloy at 1473 K, calculate (approximately) how much of solid and liquid phases are present.

Solution

(a) The cooling curves are shown below.

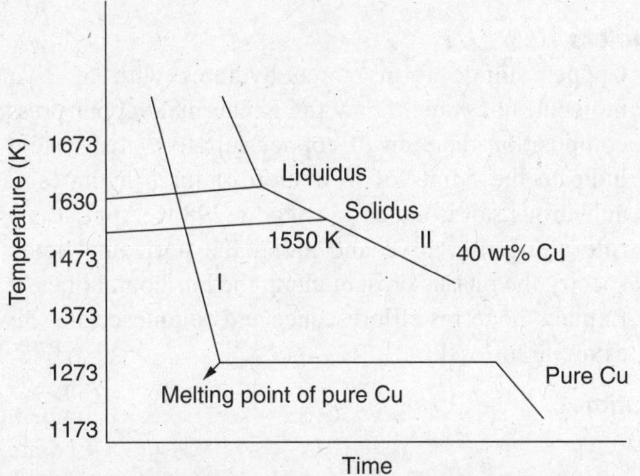

Curve I represents the cooling curve of pure copper melt. It starts to solidify at 1356 K. At this point, there is a change in the slope of the curve due to release of heat. The curve flattens out as the melt solidifies to pure Cu. The temperature then falls down further. Curve II represents the cooling curve of 40 wt% Cu alloy. Alloys generally solidify over a range of temperatures. The first solid appears at 1630 K and goes on till 1550 K, below which the pure solid 40 wt% Cu alloy is present.

(b) Along the liquidus curve, $F = 2 - 2 + 1 = 1$. Similarly along the solidus curve $F = 2 - 2 + 1 = 1$.

(c) When a melt of composition Z (see diagram) is cooled, it touches curve liquidus at Z′ and a solid of composition Z″ begins to crystallise out from the melt. As the melt is further cooled, more of solid solution separates out (whose composition will also vary) and between Z′ and Z″, the two phases (melt and solid solution) exist. Below Z‴, only the solid phase exists (solid solution of Cu and Ni). It may be noted that at Z‴, the solid solution is in equilibrium with a melt of composition Z^{IV}.

(d) From the diagram:

Composition at Y = 70 wt% Cu, composition at C′ = 81 wt% Cu, composition at D′ = 64 wt% Cu. Amounts of phases:

$$\% \text{ liquid phase} = \frac{YD'}{C'D'} \times 100 = \frac{64-70}{64-81} \times 100 = 35.3$$

$$\% \text{ solid phase} = \frac{YC'}{C'D'} \times 100 = \frac{70-81}{64-81} \times 100 = 64.7$$

In 1 kg of alloy melt at 1473 K,

$$\text{Weight of liquid phase} = \frac{1000 \times 35.3}{100} = 353 \text{ g}$$

$$\text{Weight of solid phase} = \frac{1000 \times 64.7}{100} = 647 \text{ g}$$

Solubility in Salt–Water System

Problem 10.7

The following solubility data are obtained on $(NH_4)_2SO_4$–water system at various temperatures.

Temp (K)	267.5	262.0	255.0	273.0	293	313	333	353	373
Mol% $(NH_4)_2SO_4$	2.65	5.19	7.56	8.85	9.36	9.95	10.70	11.51	12.35
Solid phase	Ice	Ice	Ice	$(NH_4)_2SO_4$	$(NH_4)_2SO_4$	$(NH_4)_2SO_4$	$(NH_4)_2SO_4$	$(NH_4)_2SO_4$	$(NH_4)_2SO_4$

Draw the phase diagram. Mark the various regions. Determine the eutectic point and eutectic composition. Considering a point corresponding to overall 15.0 mol% $(NH_4)_2SO_4$ on the composition axis at 283 K, calculate the weight per cent of ammonium sulfate solid and weight per cent of solution using the lever rule.

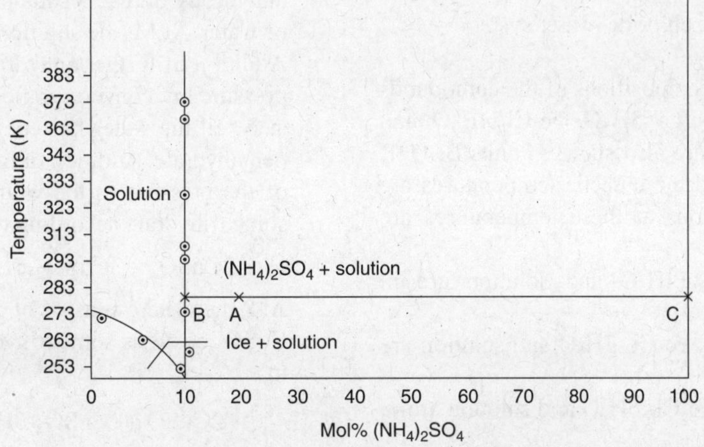

Phase diagram of the $(NH_4)SO_4$–H_2O system

Solution

The phase diagram is shown. Eutectic temperature = 258 K, eutectic composition = 9.0 mol% $(NH_4)_2SO_4$. At 15 mol% $(NH_4)_2SO_4$ and 283 K (point A in diagram).

$$\frac{AC}{AB} = \frac{\text{Parts of solution}}{\text{Parts of solid}} = \frac{85.0}{8.0}$$

$$\text{Per cent of solid} = \frac{100 \times 8}{93} = 8.6$$

$$\text{Per cent of solution} = \frac{100 \times 85.0}{93} = 91.4$$

Problem 10.8

Draw the phase diagram of ferric chloride (Fe_2Cl_6)–H_2O system with the following features.

There are four congruently melting hydrates formed with the compositions $Fe_2Cl_612H_2O$ (m.p. 37°C); $Fe_2Cl_67H_2O$ (m.p. 32.5°C); $Fe_2Cl_65H_2O$ (m.p. 56°C); $Fe_2Cl_64H_2O$ (m.p. 73.5°C). There are five eutectic points at –55°C, 27.5°C, 30.0°C, 55.0°C, 66.0°C. Mark the phases in various regions of the diagram and calculate F at the eutectic points and maxima. Explain the significance of the maxima and eutectic points in the diagram.

Solution

The phase diagram is shown below.

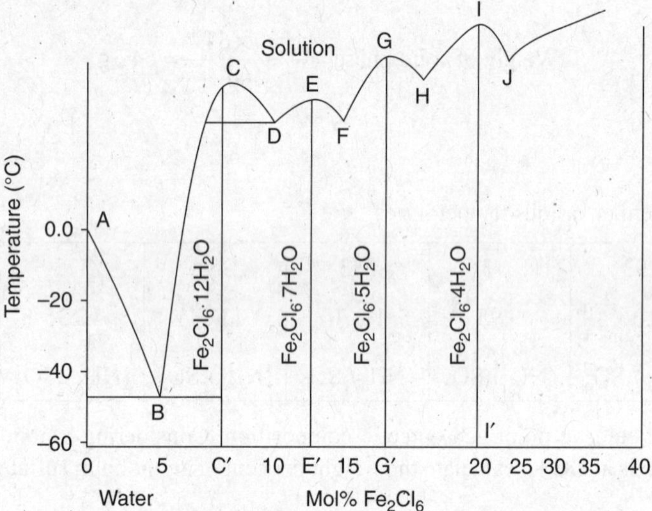

Phase diagram of ferric chloride–water system

Points C, E, G, I represent the compositions of the compounds $Fe_2Cl_6.12H_2O$, $Fe_2Cl_6.7H_2O$, $Fe_2Cl_6.5H_2O$, $Fe_2Cl_6.4H_2O$ and their congruent melting points are also shown. Points B, D, F, H, J are the eutectic points and the eutectic temperatures are shown. The eutectic compositions at these temperatures are approximate.

At B → ice, solid $Fe_2Cl_6.12H_2O$ and solution are in equilibrium $F = 0$.

At D → solid $Fe_2Cl_6.12H_2O$, $Fe_2Cl_6.7H_2O$ and solution are in equilibrium $F = 0$.

At F → solid $Fe_2Cl_6.7H_2O$, $Fe_2Cl_6.5H_2O$ and solution are in equilibrium $F = 0$.

At H → solid $Fe_2Cl_6.5H_2O$, $Fe_2Cl_6.4H_2O$ and solution are in equilibrium $F = 0$.

At I → solid $Fe_2Cl_6.4H_2O$, Fe_2Cl_6 and solution are in equilibrium $F = 0$.

At the eutectic points B, D, F, H, J, addition of any solids raises the temperature where as the compositions at maxima C, E, G, I are pure compounds and they are in equilibrium with a saturated solution of a particular composition. Addition of a foreign substance is small amounts lowers the freezing points of these compounds. At the maxima $F = 0$.

Problem 10.9

(a) Copper sulfate forms three hydrates with 5, 3 and 1 molecules of water. Draw the isothermal vapour pressure–composition diagram of copper sulfate–water system and indicate the points of formation of three hydrates as the anhydrous salt is added to water at 298 K. Apply the phase rule along the curve and along the horizontal lines and specify the phases present along the horizontal lines.

(b) Explain the terms efflorescence and deliquescence. Give an example for each.

Solution

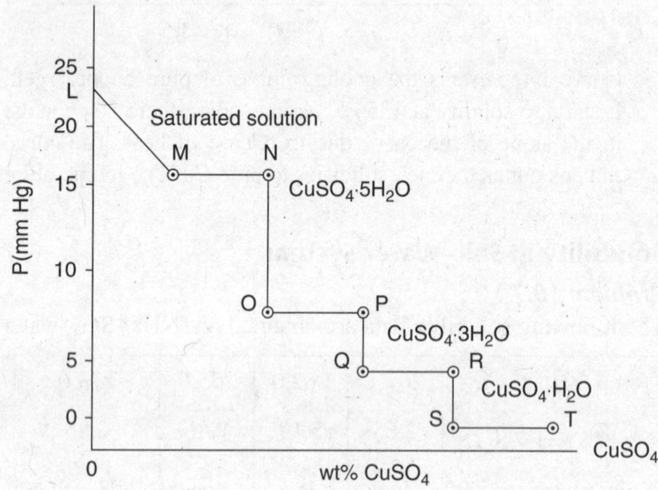

Phase diagram of $CuSO_4$–H_2O system

(a) The phase diagram is shown. In the region LM, addition of anhydrous $CuSO_4$ results in a lowering of vapour pressure of water. At M, the solution is saturated with $CuSO_45H_2O$. Addition of further anhydrous $CuSO_4$ does not change the pressure but converts some of the solution to $CuSO_45H_2O$. at N, all the water has combined with $CuSO_4$ to form the pentahydrate. Addition of more $CuSO_4$ results in a lowering of the pressure to the value at OP with the formation of some trihydrate according to the reaction

$$2CuSO_4 + 3CuSO_4.5H_2O \rightarrow 5CuSO_4.3H_2O$$

At P, the system consists of $CuSO_43H_2O$. Further addition of $CuSO_4$ converts some trihydrate to monohydrate according to

$$2CuSO_4 + CuSO_4.3H_2O \rightarrow 3CuSO_4.H_2O$$

And the pressure drops to Q. The formation of $CuSO_4H_2O$ is complete at R. The pressure drops down to S finally. Along LM: $F=2-2+1=1$; along MN: $F=2-3+1=0$ (the phases are saturated solution, $CuSO_45H_2O$ and vapour), along OP: $F=2-3+1=0$ (the phases are $CuSO_45H_2O$, $CuSO_43H_2O$ and vapour), along QR: $F = 2 - 3 + 1 = 0$ (the phases are $CuSO_43H_2O$, $CuSO_4H_2O$ and vapour), along ST: $F = 2 - 3 + 1 = 0$ (the phases are $CuSO_4$ H_2O, $CuSO_4$ and vapour). The establishment of constant pressure in a salt hydrate system requires the presence of three phases.

(b) **Efflorescence:** It refers to the phenomenon of a salt hydrate losing some water vapour if the vapour pressure of water in the air is below the dissociation pressure of the hydrate. For example, $Na_2SO_410H_2O$ has a dissociation pressure greater than the aqueous vapour pressure at room temperature. Therefore it effloresces.

Deliquescence: If the vapour pressure of water in air exceeds the vapour pressure of the saturated solution of a salt, the salt will get covered with a layer of saturated solution on exposure of the salt to air. This is known as deliquescence. For example $CaCl_2.6H_2O$ deliquesces.

Vapour Phase–Composition, Boiling Point–Composition Diagrams for Different Systems

Problem 10.10

(a) Draw the vapour pressure–composition diagram of benzene–toluene mixtures, which obey Raoult's law over the complete range of solvent compositins at 300 K. Mark the various regions. Considering a point *a* in the liquid + vapour region at overall 60 mol% toluene, calculate the relative amounts of vapour and liquid using the lever rule.

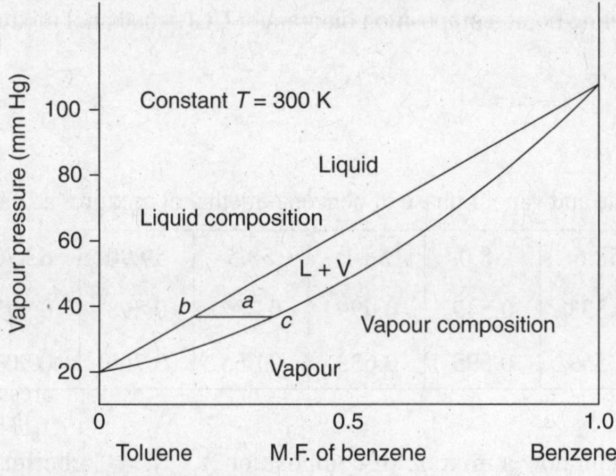

Vapour pressure–composition diagram of benzene–toluene mixtures

(b) Draw the boiling point–composition diagram of this mixture at 1 atm and mark various regions. At a boiling temperature of 368 K, calculate the mole fraction of toluene in the liquid phase of the benzene–toluene mixture. Also, calculate the mole fraction of toluene in the vapour phase. The boiling points of benzene and toluene are 353.1 K and 383.6 K respectively. At 368 K, the vapour pressure of toluene = 0.631 atm, vapour pressure of benzene = 1.56 atm.

Solution

(a) From the phase diagram we see that, in the liquid region, $F=2-1+1=2$. In the vapour region also, $F=1$. In the liquid + vapour region, $F = 2 - 2 + 1 = 1$. At *b*, the composition = 90 mol% toluene; at *a*, the composition = 60 mol% toluene; at *c*, the composition = 35 mol% toluene.

$$ab = 60 - 90 = -30;\ ac = 35 - 60 = -25;\ \frac{ab}{ac} = \frac{30}{25}$$

$$\frac{\text{Mol \% liquid phase}}{\text{Mol \% vapour phase}} = \frac{30}{25}$$

$$\text{\% liquid phase} = \frac{30}{55}\times 100 = 54.54$$

$$\text{\% vapour phase} = \frac{25}{55}\times 100 = 45.45$$

(b) Boiling point–composition diagram is shown below. Toluene is designated as component 1 and benzene as component 2.

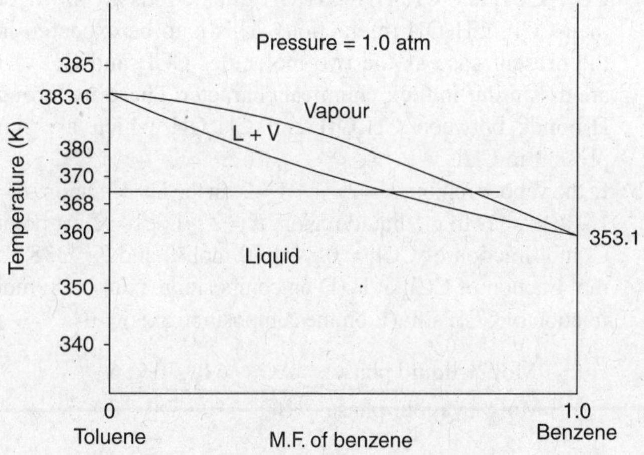

Boiling point–composition diagram of benzene–toluene mixtures

$$x_1 = x_{\substack{\text{toluene}\\ \text{(in liq)}}} = \frac{p - p_2^0}{p_1^0 - p_2^0} = \frac{1.0-1.56}{0.631-1.56} = \frac{0.56}{0.929} = 0.603$$

$$y_1 = y_{\substack{\text{toluene}\\ \text{(in liq)}}} = \frac{x_1 p_1^0}{p_2^0 + (p_1^0 - p_2^0)x_1}$$

$$= \frac{0.603\times 0.631}{1.56+(0.631-1.56)0.603}$$

$$= \frac{0.380}{1.56-(0.603\times 0.929)}$$

$$= \frac{0.380}{1.56-0.560} = 0.380$$

Problem 10.11

(a) Give two examples each of liquid mixtures (i) showing a minimum (ii) showing a maximum in the boiling point–composition diagram. Depict the boiling point–composition diagram of a mixture showing a minimum in the boiling point. Give reasons for the observation of such a minimum in the diagram. Mark the various regions in the diagram.

(b) Calculate the degrees of freedom in the various regions. Choose 70 mol% of a certain component (overall composition) and calculate the percentage of liquid and vapour for this composition.
(c) Take a point in the liquid region below the curves and explain the sequence of events for separating the mixtures into the components by fractional distillation.

Solution

(a) Examples of mixtures showing a minimum in boiling point–composition diagram: (i) water–ethanol (minimum boiling point: 351.1 K) (ii) CCl_4–CH_3OH (minimum boiling point: 328.7 K). Examples of mixtures showing maximum in boiling point–composition diagram: (i) water–HNO_3 (maximum boiling point: 393.5 K) (ii) $CHCl_3$–CH_3COCH_3 (maximum boiling point: 337.4 K). A minimum in a boiling point–composition diagram corresponds to a maximum in the vapour pressure–composition diagram. Such positive deviations from Raoult's law occur when CCl_4–CCl_4 and CH_3OH–CH_3OH interactions are stronger than CCl_4–CH_3OH interactions. This is to be expected in the present case as the two molecules CCl_4 and CH_3OH are dissimilar in their chemical character. There are strong H-bonds between CH_3OH and CH_3OH which are not present in CCl_4.

(b) In the vapour region, $F = 2 - 1 + 1 = 2$. In the L + V region, $F = 2 - 2 + 1 = 1$. In the liquid region, $F = 2 - 1 + 1 = 2$. At point C, mol fraction of CCl_4 = 0.7 (or 70 mol%) and T = 338 K; mol fraction of CCl_4 at B (D on composition axis) = 78; mol fraction of CCl_4 at E (E on the composition axis) = 64.

$$\frac{\text{Mol \% liquid phase}}{\text{Mol \% vapour phase}} = \frac{AC}{BC} = \frac{64-70}{70-78} = \frac{6}{8}$$

$$\% \text{ liquid} = \frac{6}{14} \times 100 = 42.86$$

$$\% \text{ vapour} = \frac{8}{14} \times 100 = 57.14$$

(c) Consider point E′, (about 22 mol% CCl_4). As the mixture is heated and reaches F, the boiling point of 332 K is observed and the vapour has the composition indicated by G. Condensing this mixture (see G′) and fractionating again results ultimately in the formation of azeotrope at D as the distillate and the residue in the distillation vessel is pure methanol. By this procedure, one can get only an azetrope and one of the pure components. It is not possible to obtain the pure components completely.

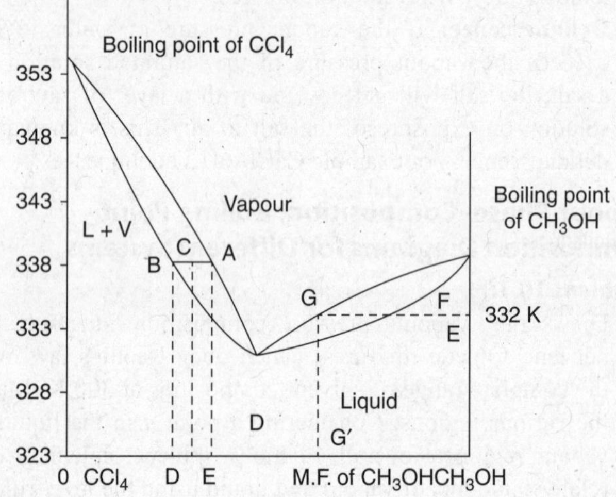

Boiling point–composition diagram of CCl_4–methanol mixtures

Problem 10.12

The following table gives the boiling point–composition data in liquid phase and vapour phase of benzene–methanol mixtures at 1 atm.

Temp (°C)	70.7	66.4	62.9	60.2	58.6	58.0	58.1	58.5	59.90	62.70
X_{MeOH} in liquid phase	0.026	0.050	0.088	0.164	0.333	0.549	0.699	0.782	0.898	0.973
X_{MeOH} in vapour phase	0.267	0.371	0.457	0.526	0.559	0.595	0.633	0.665	0.760	0.907

Draw the boiling point–composition phase diagram of these mixtures (boiling point of benzene: 80.2°C; boiling point of methanol: 64.7°C). Mark the various regions and calculate F. Calculate the percentages of liquid and vapour at point A in the diagram corresponding to X_{MeOH} = 0.2 and temperature 62°C. Explain how a mixture of composition X_{MeOH} = 0.4 boiling at 60°C can be separated by fractional distillation. What is the azeotropic temperature and what is the composition at this point?

Solution

The boiling point–composition diagram is shown below.

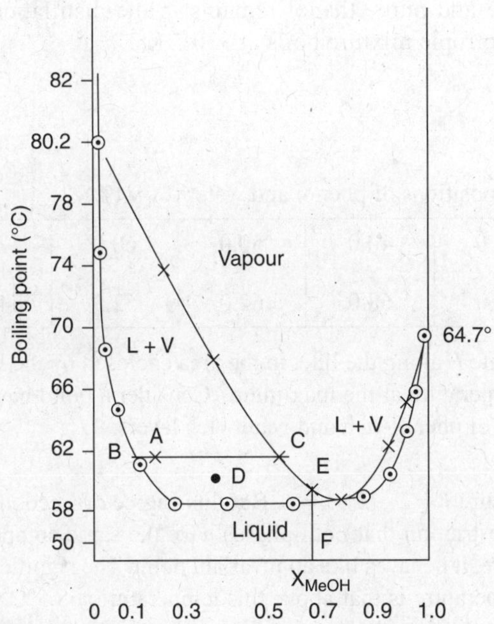

Boiling point–composition diagram of benzene–methanol mixtures

In the vapour region, $F = 2 - 1 + 1 = 2$; in the L + V region, $F = 2 - 2 + 1 = 1$; in the liquid region, $F = 2 - 1 + 1 = 2$. At 62°C and $X_{MeOH} = 0.2$, a tie line is drawn to connect the liquid and vapour lines at points B and C.

$$\frac{AC}{AB} = \frac{3 \text{ div}}{0.9 \text{ div}}$$

$$\frac{\text{Mole fraction of liquid phase}}{\text{Mole fraction of vapour phase}} = \frac{0.5 - 0.2}{0.2 - 0.11} = \frac{0.3}{0.09}$$

$$\text{Mol \% of liquid phase} = \frac{0.3}{0.39} \times 100 = 76.9$$

$$\text{Mol \% of vapour phase} = \frac{0.09}{0.39} \times 100$$

$$= 23.08 \approx 23.1$$

The point at $X_{MeOH} = 0.4$ is marked as D. The mixture boiling at D (62°C), has a vapour composition marked E which on condensation gives a liquid whose composition is $X_{MeOH} = 0.58$ and it boils at 58°C (point F). this point is very close to the azeotropic composition. Further boiling of the condensed vapour will result in the azeotropic composition and pure methanol being formed.

Azetropic temperature = 58°C; azeotropic composition = $X_{MeOH} = 0.65$

Problem 10.13

The vapour–liquid composition of ethanol–*n*-hexane mixtures and the corresponding boiling points at 1 atm pressure are given below.

Boiling point (K)	Wt% hexane in liquid	Wt% hexane in vapour
351.3	0.0	0.0
349.0	1.85	16.41
346.2	3.68	30.90
340.4	10.66	51.80
334.8	25.10	68.0
331.7	48.28	76.33
331.4	60.66	76.87
331.1	72.74	77.64
331.0	79.15	78.40
331.3	83.13	79.15
331.5	85.89	79.52
332.2	94.27	82.07
336.5	99.46	90.95
339.7	99.68	96.41
341.7	100.0	100.0

Draw the phase diagram and label the various regions. Comment on the ideal and non-ideal nature of the mixtures and give reasons. If a mixture of 25 wt% hexane and 75% ethanol is heated to boiling, what is the composition of the vapour formed? What is the ultimate composition of the distillate and residue?

Solution

The phase diagram is shown below with the various regions.

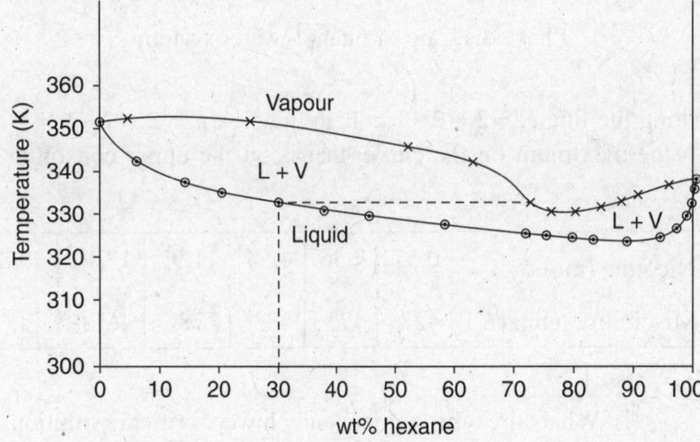

Vapour–liquid composition diagram of ethanol–*n*-hexane mixtures

The vapour region (above the curves) and the liquid region (below the curves) $F = 2 - 1 + 1 = 2$. In the L + V region, $F = 2 - 2 + 1 = 1$. These mixtures show a positive deviation from Raoult's law. This is because *n*-hexane is a hydrocarbon and ethanol is a hydroxylic solvent. They are dissimilar in their chemical nature and repulsions arise in their molecules. A 25 wt% hexane–ethanol mixture on heating boils at 334.5 K and the vapour in equilibrium with it has a composition of 67.5 wt% hexane. The condensate of that vapour on fractionation boils at 331 K and ultimately results in azeotropic compositions of 77.5 wt% hexane as distillate and pure ethanol remains in the distillation flask. The azeotropic mixture boils at 330.5 K.

Phase Diagrams of Partially Miscible Systems

Problem 10.14

The following miscibility temperatures are observed on mixing various compositions of phenol and water (by wt%)

Phenol (Wt%)	12.0	15.0	20.0	30.0	40.0	50.0	60.0	70.0
Misicibility temperature (°C)	51.0	52.0	62.0	64.0	68.0	62.0	52.5	45.0

Construct the phase diagram, mark the phases in various regions, and calculate F along the line, in the area enclosed by the line and also at the maximum point on the curve. What is the significance of the temperature at the maximum? Consider a point having the overall composition of 30 wt% phenol at 50°C and find the relative amounts of phenol-rich and water-rich layers.

Solution

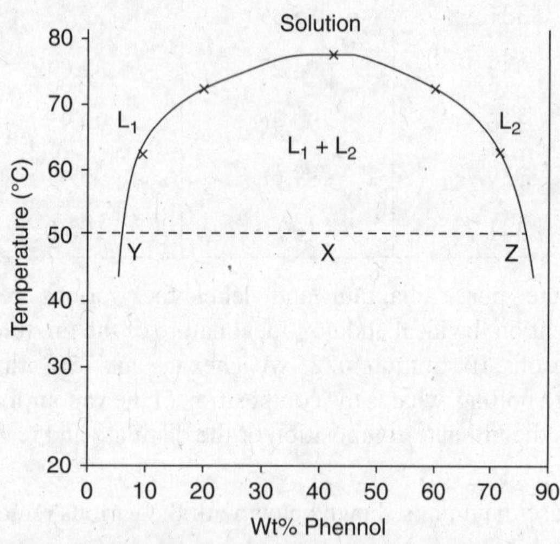

Phase diagram of phenol–water system

Along the line $F = 2 - 2 + 1 = 1$, in the area $F = 2 - 2 + 1 = 1$. At the maximum on the curve, that is, at the upper consolute temperature $F = 2 - 2 + 1 = 1$. But this degree of freedom is lost by the restriction that both phases have the same composition. Therefore, it behaves like an invariant point. The significance of this temperature is that above this temperature, 68.5°C, phenol and water are completely miscible in all proportions. At point X, (30 wt% phenol) and at 50°C, a tie line is drawn which cuts the curve at Y and Z.

At Y, composition of phenol = 12 wt%; at Z, composition of phenol = 65 wt%

Mass of phenol-rich layer = XY =30 – 12 = 18; Mass of water-rich layer = XZ = 65 – 30 = 35

% Phenol-rich layer = (18/53)×100 = 33; % Water-rich layer = (35/53)×100 = 66.04

Problem 10.15

(a) Define the lowest consolute temperature. Give an example of such a system. Draw the phase diagram of the system and show the phases present in different parts of the diagram. Calculate F in a region, along a line and at the lowest temperature, that is, lower critical solution temperature.

(b) Use the following data to construct the water–nicotine phase diagram and answer the questions given below:

Nicotine (mol%)	0.57	2.36	4.54	11.97	17.14	30.81	33.6	30.81	20.60	13.38	4.54	2.70	1.22	0.95	0.71
Miscibility temp (K)	428	473	483	478	461	448	411	369	343	335	333	338	351	355	371

(i) What are the upper and lower critical solution temperatures?

(ii) What are the compositions of the conjugate solutions at 450 K?

(iii) At 20 mol% nicotine and 375 K, what are the relative masses of nicotine-rich and water-rich layers in terms of the segments of lines drawn across the diagram?

(iv) At 15 mol% nicotine overall composition and 400 K, what are the compositions of the layers in equilibrium?

(v) At 35 mol% nicotine overall composition and 450 K, what are the compositions of layers in equilibrium?

Solution

(a) The lowest consolute temperature is the lowest temperature below which the two liquids are completely miscible in all proportions. The triethylamine–water system is an example of such a system.

Within the curve, $F = 2 - 2 + 1 = 1$; outside the curve, $F = 2 - 1 + 1 = 2$

At LCST, $F = 2 - 2 + 1 = 1$, But this degree of freedom is lost because the composition of both the layers is the same.

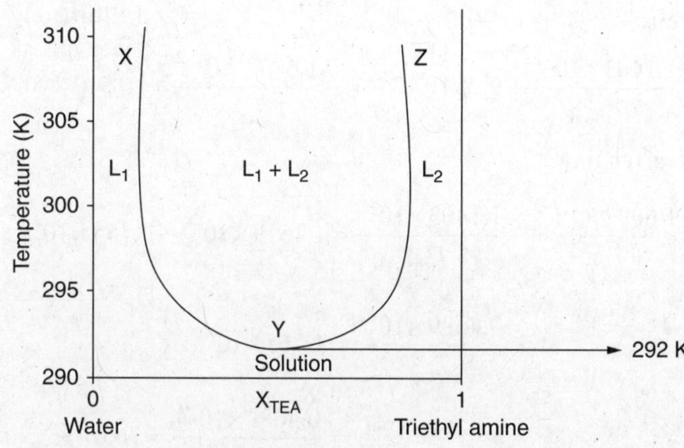

Phase diagram of triethylamine–water system

(b)

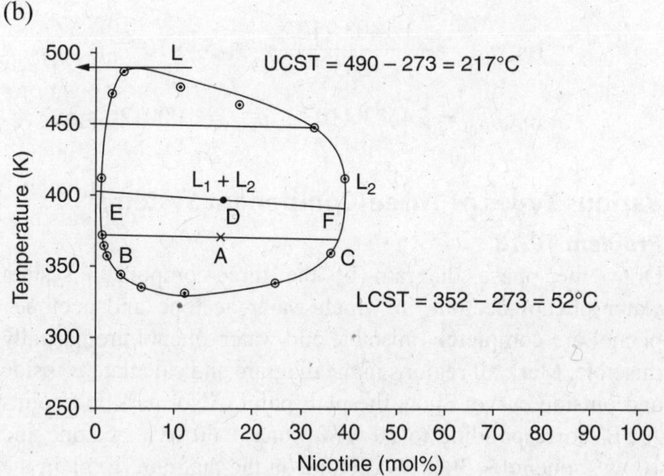

Phase diagram of nicotine–water system

Along a line, $F = 2 - 2 + 1 = 1$, and the phases are in equilibrium; in a region (within the curve) $F = 2 - 2 + 1 = 1$. But this degree of freedom is lost because the composition of the two phases is the same.

(i) At the upper critical solution temperature, $T = 217°C$; at the lower critical solution temperature, $T = 52°C$.

(ii) Compositions of conjugate solutions at 450 K—2 mol% nicotine and 32 mol% nicotine.

(iii) At 20 mol% nicotine and 375 K.

$$\frac{\text{mol\% nicotine-rich layer}}{\text{mol\% water-rich layer}} = \frac{AB}{AC} = \frac{20-2}{32-20} = \frac{18}{12}$$

$$\text{wt\% nicotine-rich layer} = \frac{18}{30} \times 100 = 60\%$$

$$\text{wt\% water-rich layer} = \frac{12}{30} \times 100 = 40\%$$

(iv) At 15 mol% nicotine and 400 K (point D in Fig.), a tie line is drawn which meets the curve at E and F. the compositions of the layers at these points are 3 mol% nicotine (water-rich layer) and 33 mol% nicotine (in the nicotine-rich layer).

(v) At 35 mol% nicotine and 450 K, there is only one layer having the same composition.

Problem 10.16

From the data relating to the partially immiscible liquid mixture of carbon disulfide–methyl acetate at 1 atm, draw the phase diagram of the system. The upper critical solution temperature $= -53°C$ at $X_{CS2} = 0.65$.

Composition of conjugate solutions given as X_{CS2}

Temperature (°C)	Methyl acetate-rich layer	Carbon disulfide-rich layer
–83	0.1	0.95
–73	0.2	0.90
–63	0.3	0.82

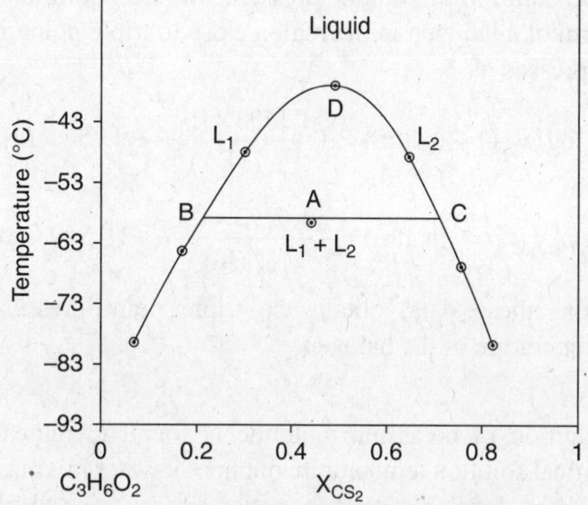

Phase diagram of methylacetate–carbon disulfide system

Mark the phases in various regions and calculate the degrees of freedom in a region, along the line and at the upper critical solution temperature. Suppose we have a mixture of the two liquids at $X_{CS2} = 0.5$ at a temperature of –63°C. Calculate the percentage amounts of CS_2-rich and methyl acetate-rich layers.

Solution

The phase diagram is shown above.

In the liquid region: $F = 2 - 1 + 1 = 2$

Along the curve: $F = 2 - 2 + 1 = 1$

At the point D, which is UCST, $F = 2 - 2 + 1 = 1$. But this one degree of freedom is lost because of the restriction that the two layers have identical composition. Consider a point at temperature of –63°C, $X_{CS2} = 0.5$(overall). It is in the two-phase region and separates into two layers at B and C.

At B, $X_{CS2} = 0.3$ (methyl acetate-rich layer)

At C, $X_{CS2} = 0.85$ (CS_2-rich layer)

$$\frac{\text{Amount of carbon disulfide-rich layer}}{\text{Amount of methyl acetate-rich layer}} = \frac{0.5-0.3}{0.85-0.50} = \frac{0.2}{0.35}$$

$$\text{\% Methyl acetate-rich layer} = \frac{0.35}{0.55} \times 100 = 63.64$$

$$\text{\% } CS_2\text{-rich layer} = \frac{0.2}{0.2+0.35} = 36.36$$

Problem 10.17

(a) Describe the effect of addition of a salt such as sodium or potassium chloride on the miscibility temperature of phenol–water mixtures. How do you explain the observed behaviour?

(b) The following results are obtained on adding various concentrations of potassium chloride on the critical solution temperature of the phenol–water system.

KCl (%)	0.0	0.5	1.0	1.5	2.0
CST (°C)	68.6	70.5	74.0	76.5	79.0

In a certain experiment, an unknown solution of KCl gave a critical solution temperature of 75°C. What is the concentration of the salt?

(c) The equilibrium vapour pressure for the liquid and solid form of a halogen in the region close to triple point may be expressed as

$$\log p_{\text{liq}}\,(p_{\text{liq}}\text{in atm}) = -\frac{1.1403\times10^{-2}}{T(\text{K})} + 9.754\times10^{-5}$$

$$\log p_{\text{solid}}\,(p_{\text{solid}}\text{in atm}) = -\frac{1.6092\times10^{-2}}{T(\text{K})} + 1.1519\times10^{-4}$$

From these data, obtain the triple point pressure and temperature of the halogen.

Solution

(a) Addition of potassium chloride is found to increase the critical solution temperature of phenol–water mixtures. The water molecules hydrate the cation and anion and hydration shells form around them preferentially. This reduces the solubilities of phenol in water and results in an increase of critical solution temperature, as is observed.

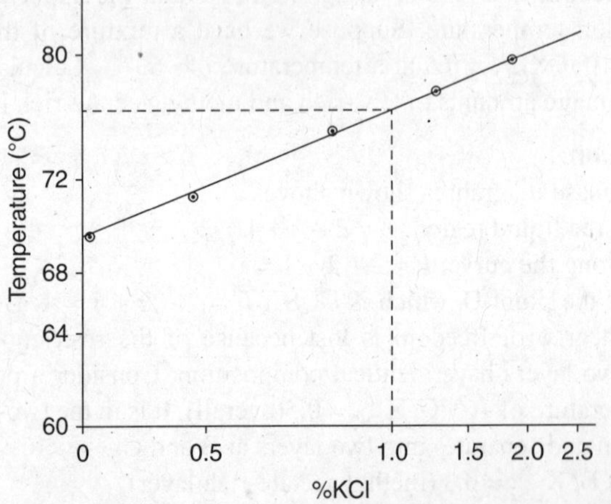

Plot of CST vs % KCl in the phenol–water system

(b) The plot of critical solution temperature vs %KCl is observed to be linear. From this graph, the per cent of KCl corresponding to CST of 75°C can be read off and found to be 1.25% KCl.

(c) The two vapour pressure relations may be equated to each other since all the three phases are in equilibrium with each other.

Thus

$$-\frac{1.1403\times10^{-2}}{T_{\text{trip}}} + 9.754\times10^{-5} = -\frac{1.6092\times10^{-2}}{T_{\text{trip}}} + 1.1519\times10^{-4}$$

Rearranging

$$\frac{1.6092\times10^{-2}}{T_{\text{trip}}} - \frac{1.1403\times10^{-2}}{T_{\text{trip}}} = 1.1519\times10^{-4} - 9.754\times10^{-5}$$

$$\frac{0.4689\times10^{-2}}{T_{\text{trip}}} = 1.765\times10^{-5} \text{ or,}$$

$$T_{\text{trip}} = \frac{0.4689\times10^{-2}}{1.765\times10^{-5}} = 265.7\text{ K}$$

Substituting T_{trip} in the equation for log (p_{lid})

$$\log p_{\text{trip}} = -\frac{1.1403\times10^{-2}}{265.7} + 9.754\times10^{-5}$$

$$\log p_{\text{trip}} = 5.4623\times10^{-5};\; p_{\text{trip}} = 1.000126\text{ atm}$$

Various Types of Three-Component Systems

Problem 10.18

Draw the phase diagram of the three-component system water–phenol–acetone in which water–acetone and acetone–phenol are completely miscible and water–phenol are partially miscible. Mark all regions in the diagram and calculate F inside and outside curve. Show the plait point. Also mark the points A′, B′ corresponding to 20 wt% water + 60 wt% acetone and 10 wt% phenol + 30 wt% acetone on the diagram. Explain the significance of the line connecting an apex to the opposite side.

Solution

The triangular phase diagram is shown below. In the two-phase region, $F = 3 - 2 + 0 = 1$ (since the temperature and pressure are constant). In the single phase region above the curve, $F = 3 - 1 + 0 = 2$. Points A′ and B′ corresponding to the two compositions are marked.

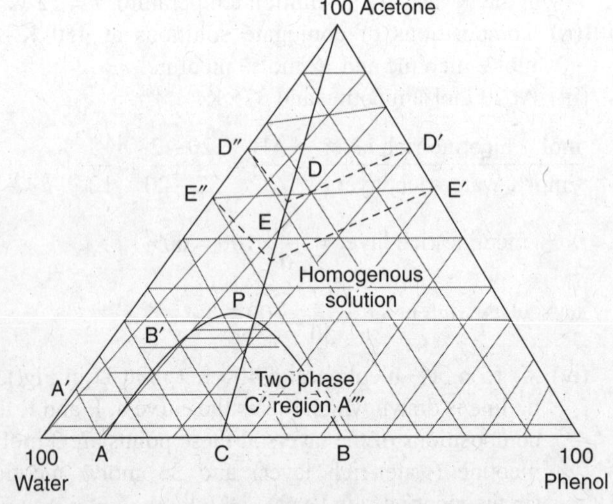

Phase diagram of water–phenol–acetone system

Points A and B represent the conjugate layers in the absence of acetone. The apex at 100% acetone is connected to point C on the diagram. The significance of this line is that any point on this line contains water and phenol in the same ratio. Taking points D and E on the line (connecting apex at acetone to point C), draw perpendiculars DD″, DD′ and EE″, EE′. The relation $\frac{DD''}{EE''} = \frac{DD'}{EE'}$ holds. Addition of acetone changes the compositions to A″ and A‴. This line is not parallel to AB. The line corresponding to aqueous layer, that is, C′A‴ is longer than the phenol layer (line C′A‴) indicating that acetone goes preferentially into the aqueous layer. As more acetone is added, the phenol layer shrinks in size and the aqueous layer grows. The plait point, P (the point at which both conjugate solutions have the same composition) is not at the top but to the side as the tie lines are not parallel.

Problem 10.19

From the following data relating to $NaNO_3$–KNO_3–H_2O system at 323 K, draw the phase diagram of the system, mark the regions and calculate F in the various regions and along the lines.

	$NaNO_3$	KNO_3	H_2O	Solid phase
Mole %	19.4	0.0	80.6	$NaNO_3$
	19.24	4.19	76.56	$NaNO_3$
	19.04	8.00	72.95	$NaNO_3$
	18.73	11.45	69.62	$NaNO_3 + KNO_3$
	13.62	11.31	75.07	KNO_3
	7.29	11.52	81.12	KNO_3
	0.00	13.26	86.74	KNO_3

Convert all the mol% data into the corresponding wt% and use wt% data in plotting the phase diagram. Consider the point G on the line DJ. If a solution of composition G is isothermally evaporated, what are sequences of changes observed?

Solution

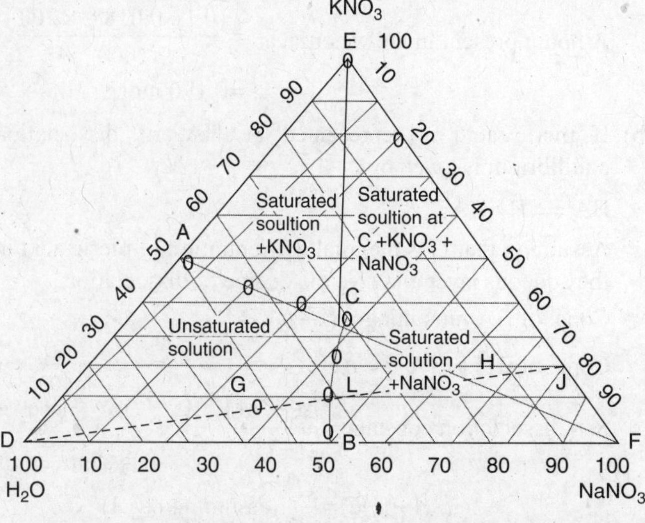

Phase diagram of $NaNO_3$–KNO_3–H_2O system

Region	Nature of phases	F (Degree of freedom)
ACBD	Unsaturated solution	$F = 3 - 1 + 0 = 2$
AEC	Saturated solution + KNO_3	$F = 3 - 2 + 0 = 1$
BCF	Saturated solution + $NaNO_3$	$F = 3 - 2 + 0 = 1$
CEF	Saturated solution + KNO_3 + $NaNO_3$	$F = 3 - 3 + 0 = 0$

When a solution of composition G is isothermally evaporated, the temperature falls and the composition of the solution changes along the line GHJ, and when it reaches H, $NaNO_3$ crystallises and the liquid composition moves along the line HC. At point I, KNO_3 also begins to crystallise. Further evaporation results in the deposition of both KNO_3 and $NaNO_3$ and when point J is reached, the solution disappears completely and only a mixture of KNO_3 and $NaNO_3$ remains.

Problem 10.20

Draw the triangular phase diagram of a three-component system consisting of two salts and water in which a double salt is formed in which the compound saturates congruently. Mark the various regions of the diagram and calculate the degrees of freedom in the various regions. Explain the sequence of changes when a solution of composition above the curve is isothermally evaporated.

Solution

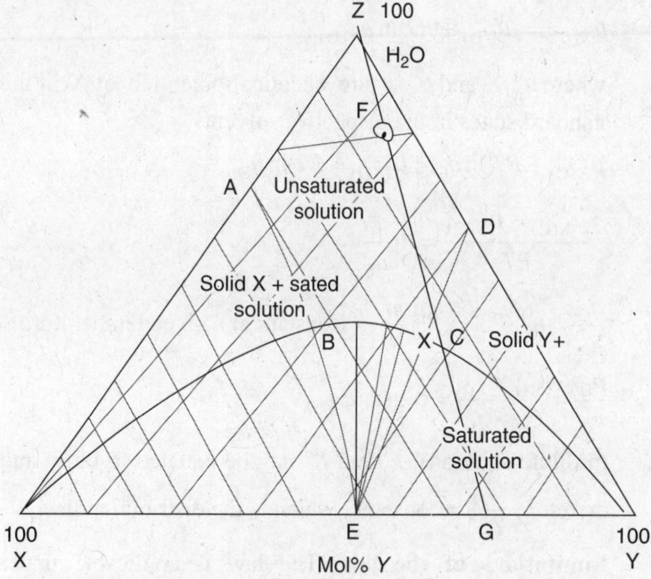

Phase diagram of a three-component system consisting of two salts (X and Y) and H_2O

Region ABCDZ: $F = 3 - 1 + 0 = 2$
Region XBE: $F = 3 - 3 + 0 = 0$
Region ABX: $F = 3 - 2 + 0 = 1$
Region BCE: $F = 3 - 2 + 0 = 1$
Region CDY: $F = 3 - 2 + 0 = 1$
Region ECY: $F = 3 - 3 + 0 = 0$

When the solution of composition F is isothermally evaporated, the composition changes along the line FCG. At C, solid XY is formed and further cooling results in the formation of Y and further evaporation results in formation of solid X and Y at G. The solution phase disappears completely at G.

Nernst Distribution Law, Immiscible Liquids and Steam Distillation

Problem 10.21

(a) Derive the Nernst distribution law using relevant thermodynamic principles. Are there any limitations to this law?

(b) The partition coefficient of ammonia between chloroform and water defined as

$$K_{\text{dist}} = \frac{[\text{Conc. in CHCl}_3]}{[\text{Conc. in H}_2\text{O}]} = 3.85\times10^{-2}$$

at 298 K. If 100 mL of 0.1 M solution of NH_3 in water is shaken with 100 mL of $CHCl_3$, calculate the amount of ammonia in chloroform layer. Why is ammonia present in a larger amount in the aqueous layer than in the chloroform layer?

Solution

(a) If a solute X is distributed between two immiscible solvents C and D, the chemical potential of X in the solvent C is equal to the chemical potential of the same in solvent D by the condition of equilibrium.

$$\mu_{\text{X(C)}} = \mu_{\text{X(D)}}$$

$$\mu_{\text{X(C)}} = \mu^0_{\text{X(C)}} + RT\ln a_{\text{C}}$$

$$\mu_{\text{X(D)}} = \mu^0_{\text{X(D)}} + RT\ln a_{\text{X(D)}}$$

where $\mu^0_{\text{X(C)}}$ and $\mu^0_{\text{X(D)}}$ are chemical potentials of X in the standard states in the respective solvents.

$$\mu^0_{\text{X(C)}} + RT\ln a_{\text{C}} = \mu^0_{\text{X(D)}} + RT\ln a_{\text{D}}$$

$$\frac{\mu^0_{\text{X(C)}} - \mu^0_{\text{X(D)}}}{RT} = \ln\frac{a_{\text{D}}}{a_{\text{C}}}$$

$$\frac{a_{\text{D}}}{a_{\text{C}}} = e^{(\mu^0_{\text{X(C)}} - \mu^0_{\text{X(D)}})/RT} = \text{Constant at a given temperature}$$

$$\frac{a_{\text{D}}}{a_{\text{C}}} = \frac{C_{\text{D}}\times f_{\text{D}}}{C_{\text{C}}\times f_{\text{C}}} = \text{constant}$$

In dilute solutions, f_{D} and f_{C} may be equated to unity and therefore $\frac{C_{\text{D}}}{C_{\text{C}}}$ = constant, which is the distribution law.

Limitations of the law: The law is applicable under conditions when the species X exists in the same molecular state in the two solvents. If the solute X dissociates or associates in any of the solvents, the law fails.

(b) $$K'_{\text{dist}} = \frac{[\text{Conc. in H}_2\text{O}]}{[\text{Conc. in CHCl}_3]} = \frac{1}{3.85\times10^{-2}} = 25.97$$

Let the concentration in the $CHCl_3$ layer be C'. Then the concentration in $CHCl_3$ = 0.1/25.97 = 0.00385

Concentration in $CHCl_3$ = 3.85×10^{-3} in 100 mL of $CHCl_3$. The amount present is $3.85\times10^{-3}\times100/1000 = 3.85\times10^{-4}$ mol

Ammonia forms hydrogen bonds with water, which results in its larger solubility in water.

Problem 10.22

(a) The distribution coefficient of succinic acid between ether and water $\left(\frac{C_{\text{ether}}}{C_{\text{water}}}\right)$ is 0.182. 200 mL of 0.1 M succinic acid is shaken with 100 mL of ether. Find the concentration of the acid in both layers. (Assume negligible ionisation of the acid in the aqueous layer.) What are the amounts present in each layer?

(b) Picric acid distributes between benzene and water and the following data gives their concentrations in both layers at 298 K.

S. No.	Aqueous layer	Benzene layer
1	2.08×10^{-3}	9.32×10^{-4}
2	3.23×10^{-3}	2.26×10^{-3}
3	6.85×10^{-3}	1.01×10^{-2}
4	9.62×10^{-3}	2.0×10^{-2}

Picric acid exists as single molecule in benzene layer. Use the data to determine the nature of picric acid in the aqueous layer.

Solution

(a) Distribution coefficient $= K_D = \left(\frac{C_{\text{water}}}{C_{\text{ether}}}\right) = \frac{1}{0.182} = 5.50$

$$K_D = \frac{0.1}{C_{\text{ether}}};\ C_{\text{ether}} = \frac{0.1}{5.50} = 1.82\times10^{-2}$$

$$\text{Amount present in the ether layer} = \frac{1.82\times10^{-2}\times100}{1000}$$

$$= 1.82\times10^{-3}\ \text{mol}$$

$$\text{Amount present in the water layer} = \frac{(0.1-0.0182)\times200}{1000}$$

$$= 0.196\ \text{mol}$$

(b) If picric acid is represented by HA, its dissociation equilibrium is given by

$$\text{HA} \rightleftharpoons \text{H}^+ + \text{A}^-$$

Assuming that C is the total concentration of picric acid in the aqueous layer and α is the degree of dissociation,

Conc. of HA remaining = $(1-\alpha)C$

Conc. of $[\text{H}^+] = [\text{A}^-] = \alpha C$

$$K = \text{dissociation constant} = \frac{\alpha C\times\alpha C}{(1-\alpha)C}$$

$$(1-\alpha)C = \frac{C^2}{K}\ (\text{assuming } \alpha \approx 1)$$

$$K_X = \frac{\text{Conc. in benzene}}{\text{conc. in water}} = \frac{C_B}{\frac{C^2}{K}}$$

$$= \frac{C_B \times K}{C^2}$$

$$\frac{C_B}{C^2} = \frac{K_X}{K} = \text{constant}$$

From the given data, we calculate $\frac{C_B}{C^2}$ as follows

S. No.	Aqueous layer	Benzene layer			C_B/C_{W^2}	C_B/C_W
1	2.08×10^{-3}	9.32×10^{-4}	$\frac{9.32\times10^{-4}}{(2.08\times10^{-3})^2}$	$\frac{9.32\times10^{-4}}{4.324\times10^{-6}}$	215.0	0.448
2	3.23×10^{-3}	2.26×10^{-3}	$\frac{2.26\times10^{-3}}{(3.23\times10^{-3})^2}$	$\frac{2.26\times10^{-3}}{1.043\times10^{-5}}$	216.7	0.700
3	6.85×10^{-3}	1.01×10^{-2}	$\frac{1.01\times10^{-1}}{(6.85\times10^{-3})^2}$	$\frac{1.01\times10^{-1}}{4.692\times10^{-5}}$	215.3	1.474
4	9.62×10^{-3}	2.0×10^{-2}	$\frac{2.0\times10^{-2}}{(9.62\times10^{-3})^2}$	$\frac{2.0\times10^{-2}}{9.254\times10^{-5}}$	216.1	2.079

It is seen that while the $\frac{C_B}{C_W}$ values show a continuous increase, $\frac{C_B}{C_W^2}$ is fairly constant indicating that picric acid dissociates in water to a large extent.

Problem 10.23

(a) The partition coefficient of ethanol between water and CCl_4, given as $K_{p.c} = \frac{\text{Conc in water}}{\text{Conc in } CCl_4} = 40.8$ at 298 K. 2 g of ethanol are shaken with a mixture of 25 mL of water and 100 mL CCl_4. Calculate the concentration of ethanol in the two layers.

(b) The following results were obtained for the distribution of benzoic acid between water and chloroform at 298 K.

Conc. of benzoic acid (mol/litre)	Aqueous layer (C_{water})	Chloroform layer (C_{CHCl_3})
1	2.11×10^{-3}	5.184×10^{-3}
2	3.54×10^{-3}	1.562×10^{-2}
3	1.25×10^{-2}	1.80×10^{-1}

Assuming that benzoic acid does not dissociate in water, what conclusions can you draw regarding the nature of benzoic acid in chloroform?

Solution

(a) If 2 g ethanol (0.0435 moles is dissolved in 25 mL of water), its concentration

$$= \frac{0.0435\times1000}{25} = 1.74 \text{ M}$$

and it is shaken with 100 mL CCl_4.

$$\text{Conc. in } CCl_4 = \frac{\text{Conc in water}}{K_{p.c}} = \frac{1.74}{40.8} = 0.0426$$

$\therefore$ Conc. remaining in water = 1.74 – 0.043 = 1.697

In CCl_4 = 0.043 M

(b) Let benzoic acid be represented by A and let n molecules of benzoic acid be associated in $CHCl_3$ to give a species $(A)_n$. If C is the stoichiometric concentration of benzoic acid in chloroform and the association is represented by the equilibrium

$$\begin{array}{cc} nA & \rightleftharpoons (A)_n \\ C(1-\alpha) & C(\alpha/n) \end{array}$$

where α represents the degree of association of the acid in $CHCl_3$.

The equilibrium constant K is given by

$$K = \frac{C\left(\frac{\alpha}{n}\right)}{\{C(1-\alpha)\}^n}$$

$$C(1-\alpha) = \left(\frac{C\alpha}{nK}\right)^{\frac{1}{n}}$$

From the distribution law, the dissociation coefficient, K_x is given by

$$K_X = \frac{[\text{Conc. of water}](C_W)}{[\text{Conc.of } CHCl_3]}$$

where the concentrations represent the unassociated species of benzoic acid in the two solvents.

$$K_X = \frac{C_W}{\left(\frac{C\alpha}{nK}\right)^{\frac{1}{n}}}$$

$$= \frac{C_W}{C^{\frac{1}{n}}} \times \left(\frac{nK}{\alpha}\right)^{\frac{1}{n}}$$

$$= \frac{C_W}{C^{\frac{1}{n}}} \times \text{constant (assuming } \alpha \text{ is large and close to 1)}$$

$$\frac{K_X}{\text{Constant}} = K_X' = \frac{C_W}{C^{\frac{1}{n}}}$$

We now calculate for which value of n, K_x' remains constant. Assuming dimerisation ($n = 2$), the calculated values of K_x' are as follows:

S. No.		$C_W/C^{1/n}$	C_W/C
1	$\frac{2.11\times10^{-3}}{(5.184\times10^{-3})^{1/2}}$	0.0293	0.407
2	$\frac{3.54\times10^{-3}}{(1.562\times10^{-2})^{1/2}}$	0.0283	0.227
3	$\frac{1.25\times10^{-2}}{(1.80\times10^{-1})^{1/2}}$	0.0295	0.069

As can be seen, the calculation involving $n = 2$ gave constant values of K_x'. Therefore, it is clear that benzoic acid exists as the dimerised species in $CHCl_3$.

Problem 10.24

(a) The equilibrium constant of the reaction $KI + I_2 \rightleftharpoons KI_3$ in water may be determined by partition measurements of I_2 between an organic solvent like CCl_4 and water followed by partition experiments of I_2 between CCl_4 and an aqueous solution of KI of definite concentration. Derive an equation for the equilibrium constant of the reaction (K_{eq}) using the following nomenclature. (Distribution coefficient of iodine between CCl_4 and water = K_{DC}; concentration of KI taken in experiment = C_i; concentration of I_2 in CCl_4 found in experiments dealing with partition between CCl_4 and KI of concentration $C_i = C_y$; concentration of iodine found in aqueous layer containing KI = C_z)

(b) Nitrobenzene and water form immiscible liquids. On steam distillation, nitrobenzene boils at 371 K. The vapour pressures of water and nitrobenzene are 94.94 kPa and 2.53 kPa respectively. To obtain 150 g of nitrobenzene in the distillate, how many grams of steam would condense?

Solution

(a) From the analysis of CCl_4 and aqueous layers in the distribution of I_2 between CCl_4 and aqueous KI.

Concentration of I_2 in the CCl_4 layer = C_y
Concentration of I_2 (total) in the aqueous layer = C_z
From the above two data and from known K_{DC},

$$\text{Concentration of free iodine in aqueous KI layer} = \frac{C_y}{K_{DC}}$$

$$\text{Conc. of bound iodine as triiodide} = \left(C_Z - \frac{C_y}{K_{DC}}\right)$$

Original conc. of KI = C_i

Conc. of KI remaining at equilibrium in the aqueous layer

$$= C_i - \left(C_Z - \frac{C_y}{K_{DC}}\right)$$

$$\text{Conc. of } KI_3 \text{ formed at equilibrium} = \left(C_Z - \frac{C_y}{K_{DC}}\right)$$

$\therefore K_{eq}$ of the reaction $KI + I_2 \rightleftharpoons KI_3$

$$\therefore K_{eq} = \frac{\left(C_Z - \frac{C_y}{K_{DC}}\right)}{\left(\frac{C_y}{K_{DC}}\right)\times\left(C_i - \left(C_Z - \frac{C_y}{K_{DC}}\right)\right)}$$

$$\text{(b)} \quad \frac{W_{\text{nitrobenzene}}}{W_{\text{water}}} = \frac{\text{Vapour pressure of nitrobenzene} \times \text{Mol. wt of nitrobenzene}}{\text{Vapour pressure of water} \times \text{Mol. wt of water}}$$

Substituting the relevant data,

$$\frac{150}{W_{\text{water}}} = \frac{2.53\times123}{94.94\times18}$$

$$W_{\text{water}} = \frac{94.94\times18\times150}{2.53\times123} = 823.7\text{ g}$$

Problem 10.25

(a) Analysis of the distillate on steam distillation of aniline was found to contain 27 wt% aniline. Calculate the vapour pressure of aniline at the distillation temperature under 101.325 kPa.

(b) In the steam distillation of an organic compound at 371 K, the wt% of the organic compound in the distillate is found to be 43.75. Given that the vapour pressure of water at the boiling point is 94.94 kPa, calculate the molar mass of organic compound.

Solution

$$\text{(a)} \quad \frac{27}{73} = \frac{p^0_{\text{aniline}}\times93}{94.94\times18}; \; p^0_{\text{aniline}} = \frac{27\times94.94\times18}{93\times73} = \frac{46140.84}{6789} = 6.8$$

$$\text{(b)} \quad \frac{43.75}{56.25} = \frac{6.385\times M}{94.94\times18}; \; M = \frac{43.75\times94.94\times18}{6.385\times56.25} = \frac{74765.25}{359.156} = 208$$

Problem 10.26

(a) A solution of iodine in carbon tetrachloride is shaken with water at 298 K. When equilibrium was established, the aqueous layer was found to contain 0.236 mmol of iodine per litre and the organic layer contained 0.0201 mol per litre of iodine. Calculate the distribution coefficient of iodine between CCl_4 and water (K_D).

(b) In another experiment 100 mL of 0.1 M solution of potassium iodide is shaken with 100 mL of a solution of carbon tetrachloride containing iodine at the same temperature as above. At equilibrium, 20 mL of the CCl_4 layer required 17.0 mL of 0.1 N sodium thiosulfate for the reaction with iodine. 25 mL of the aqueous layer required 10.3 mL of 0.1 N sodium thiosulfate for complete reaction with iodine. Calculate the equilibrium constant of the reaction $KI + I_2 \rightleftharpoons KI_3$ occuring in the aqueous layer.

Solution

(a) Distribution coefficient (K_D) of iodine between CCl_4 and water is

$$= \frac{[\text{iodine}]_{CCl_4}}{[\text{iodine}]_{H_2O}} = \frac{0.0201}{0.236 \times 10^{-3}} = 85.2$$

(b) Strength of iodine in the CCl_4 layer $= \dfrac{17.0 \times 0.1}{20} = 0.085$

Conc. of iodine in the aqueous layer $= \dfrac{0.085}{85.2} = 9.98 \times 10^{-4}$

Strength of iodine in the aqueous layer $= \dfrac{10.3 \times 0.1}{25} = 0.0412$

This concentration corresponds to the total strength of free iodine and bound iodine (as I_3^-)

Conc. of bound iodine $I_3^- = 0.0412 - 0.000998 = 0.0402$

Conc. of KI remaining in the aqueous layer $= 0.1 - 0.0402$

$= 0.0598$

$$\therefore K_{\text{eq.const}} = \frac{[I_3^-]}{[I^-][I_2]} = \frac{0.0402}{0.0598 \times 9.98 \times 10^{-4}}$$

$$= \frac{4.02 \times 10^{-2}}{5.98 \times 10^{-2} \times 9.98 \times 10^{-4}}$$

$$= \frac{4.02}{5.98 \times 9.98 \times 10^{-4}} = 673.6$$

Gas Solubility Relation

Problem 10.27

(a) Consider methane gas in equilibrium with its solution in benzene at constant temperature. What is F for this system? Derive a simple relation between the pressure and concentration of the gas in solution.

(b) Account for the following observations.

(i) Solid CO_2 sublimes at 1 atm pressure.

(ii) The triple point of iodine is 387 K at which the equilibrium pressure is 12.0 kPa. Solid iodine sublimes readily on heating in an open vessel.

(iii) The minimum boiling azeotrope of carbon disulfide and acetone having a boiling point of 312.8 K and a composition of 67 wt% carbon disulfide, behaves as an invariant point.

Solution

(a) When methane gas is in equilibrium with its solution in benzene, the equilibrium condition that

$\mu_{CH_4(g)} = \mu_{CH_4(\text{benzene})}$ applies

$$\mu^0_{CH_4(g)} + RT \ln p = \mu^0_{CH_4(\text{benzene})} + RT \ln a_{CH_4(\text{benzene})}$$

$\mu^0_{CH_4(g)}$ and $\mu^0_{CH_4(\text{benzene})}$ depend on T and P and not on composition. Defining that

$$-\mu^0_{CH_4(g)} + \mu^0_{CH_4(\text{benzene})} = RT \ln K_{CH_4}$$

where RT in K_{CH_4} is a constant at given temperature, we can write

$$RT \ln p = RT \ln K_{CH_4} + RT \ln a_{CH_4}$$

$$\ln p = \ln K_{CH_4} + \ln a_{CH_4}$$

$$\text{or } p = K_{CH_4} \times a_{CH_4} = K_{CH_4} \times m_{CH_4} \times \gamma_{CH_4}$$

In dilute solution, $\gamma_{CH_4} \approx 1$, or $p = K_{CH_4} \times m_{CH_4}$

$p \propto m_{CH_4}$ which is a form of Henry's law.

(b) (i) The triple point of CO_2 is at 5.1 atm and –50°C. Only at this temperature and pressure can all the three phases exist in equilibrium. So at 1 atm, the liquid phase is no longer present and hence solid CO_2 sublimes at 1 atm.

(ii) The triple point of iodine is at 378 K at a pressure of 12.0 kPa. On heating solid iodine in an open vessel (exposed to atmospheric pressure), it goes directly into the vapour phase as the liquid phase does not exist at this pressure.

(iii) For the mixtures of CS_2 and acetone, which has a minimum boiling point at 312.8 K and a composition of 67 wt% CS_2, application of phase rule gives $F = 2 - 2 + 1 = 1$. But this one degree of freedom is lost due to the fact that at the azeotropic composition, both liquid and vapour have the same composition. Therefore it behaves like an invariant point.

Maximum Extraction of a Solute from an Immiscible Liquid Mixture

Problem 10.28

(a) The partition coefficient of H_2S between benzene and water $\left(K_p = \dfrac{[H_2S]_{\text{benzene}}}{[H_2S]_{\text{water}}}\right)$ is 5.90 at 298 K. Using a 0.1 M solution of H_2S in water, calculate the minimum volume required to extract 95% H_2S into the benzene layer in a single step.

(b) The total volume of benzene required to extract 95% H_2S in four steps, each time using the same volume of benzene may also be computed.

(c) Using 275 mL of benzene each time, calculate the number of extractions required to extract 95% H_2S into the benzene layer.

Solution

$$K_P' = \frac{[H_2S]_{water}}{[H_2S]_{benzene}} = \frac{1}{5.90} = 0.169$$

Let us assume that we have taken one litre of an aqueous solution containing 0.1 mol of H_2S.

(a) If 95% of it is to be extracted into the benzene layer,
No. of moles left in the aqueous layer = 0.1 – 0.095 = 0.005
We use the relation,

$$W_n' = 0.005 = 0.1\left(\frac{K_pV}{K_pV+nx}\right) = 0.1\left(\frac{0.169\times 1}{0.169\times 1+nx}\right)$$

$$0.05 = \frac{0.169}{0.169+nx};\ 0.05\times 0.169 + 0.05(nx) = 0.169$$

$$0.05\,(nx) = 0.169 - 0.00845 = 0.16055$$

$$(nx) = \frac{0.16055}{0.05} = 3.21\text{ L}$$

To extract 95% H_2S into the benzene layer in one extraction, 3.21 L of benzene is required.

(b) We use the expression

$$W_n = a\left(\frac{K_pV}{K_pV+x}\right)^n$$

where $W_n = 0.005; a = 0.1; K_p = 0.169,\ V = 1.0; x = ?\ n = 4$

$$\text{or } 0.005 = 0.1\left(\frac{0.169\times 1}{0.169\times 1+x}\right)^4;\ 0.05 = \left(\frac{0.169\times 1}{0.169\times 1+x}\right)^4$$

$$\log 0.05 = 4\log\left(\frac{0.169\times 1}{0.169\times 1+x}\right)$$

$$\log\left(\frac{0.169\times 1}{0.169+x}\right) = \frac{1}{4}\log 0.05 = -\frac{1.3010}{4}$$

$$\frac{0.169}{0.169+x} = 0.4728; 0.4728x = 0.169 - 0.08 = 0.089$$

$$x = 0.188\text{ L}$$

∴ The total volume of benzene required to remove 95% H_2S into the benzene layer is $4x = 4\times 0.188 = 0.752$ L

(c) For this calculation we have

$$\text{or } 0.005 = 0.1\left(\frac{0.169\times 1}{0.169+0.275}\right)^n;\ 0.05 = \left(\frac{0.169}{0.444}\right)^n$$

$$\log 0.05 = n\log 0.3806; -1.3010 = nx - 0.4195$$

$$n = \frac{1.3010}{0.4195} = 3.1 \approx 3.0$$

Three extractions, using 0.275 litres of benzene each time, enables 95% H_2S to be extracted into the benzene layer.

Ideal Solubility Relation

Problem 10.29

The melting point of cadmium and bismuth are 593 K and 543 K respectively. The enthalpy of fusion of cadmium is 6.3 kJ mol^{-1} and the enthalpy of fusion of bismuth is 10.9 kJ mol^{-1}. The following freezing points of certain melts of Cd and Bi have been observed.

Melts of Bi containing Cd	A_1	A_2	A_3	A_4	Pure Bi
Freezing point (K)	325.6	393.5	448.0	487.1	543.0
Melts of Cd containing Bi	B_1	B_2	B_3	B_4	Pure Cd
Freezing point (K)	262.4	345.3	423.7	505.1	593.0

Calculate the mole fraction solubilities of bismuth from the first set of data and mole fraction solubilities of cadmium from the second set of data. Determine the eutectic temperature and eutectic composition from the temperature–composition plot. Assume that cadmium and bismuth melts obey the law of ideal solubility.

Solution

First set of data:

$$\text{Use the equation: } \log X_{Bi} = \frac{\Delta H_{fusion}}{2.303R}\left[\frac{1}{T_0} - \frac{1}{T}\right]$$

$$T_0 = \text{m.p. of Bi; } T = \text{observed f.p.}$$

$$\log X_{Bi} = \frac{10.9}{2.303\times 8.314\times 10^{-3}}\left[\frac{1}{543} - \frac{1}{325.6}\right]$$

$$= \frac{10.9}{0.01915}\times\frac{-217.4}{543\times 325.6} = \frac{-2369.66}{3385.73} = -0.6999$$

$$A_1 \text{ or } X_{Bi} = 0.200$$

$$A_2 = \log X_{Bi} = \frac{10.9}{2.303\times 8.314\times 10^{-3}}\left[\frac{1}{543} - \frac{1}{393.5}\right]$$

$$= \frac{-569.2\times 149.5}{543\times 393.5} = -0.3982$$

$$X_{Bi} = 0.400$$

$$A_3 = \log X_{Bi} = \frac{10.9}{2.303\times 8.314\times 10^{-3}}\left[\frac{1}{543} - \frac{1}{448}\right]$$

$$= \frac{-569.2\times 95}{543\times 448} = -0.2223$$

$$X_{Bi} = 0.600$$

$$A_4 = \log X_{Bi} = \frac{10.9}{2.303\times 8.314\times 10^{-3}}\left[\frac{1}{543} - \frac{1}{497.1}\right]$$

$$= \frac{-569.2\times 45.9}{543\times 497.1} = -0.0968$$

$$X_{Bi} = 0.800$$

Second set of data:

Use of equation:

$$\log X_{Cd} = \frac{\Delta H_{fusion}}{2.303R}\left[\frac{1}{T_0} - \frac{1}{T}\right]$$

T_0 = m.p. of Cd; T = observed f.p.

$$B_1 = \log X_{Cd} = \frac{6.3}{2.303 \times 8.314 \times 10^{-3}}\left[\frac{1}{593} - \frac{1}{262.4}\right]$$

$$= \frac{-328.98 \times 330.6}{593 \times 262.4} = -0.6990$$

B_1 or $X_{Cd} = 0.200$

$$B_2 = \log X_{Cd} = \frac{6.3}{2.303 \times 8.314 \times 10^{-3}}\left[\frac{1}{593} - \frac{1}{345.6}\right]$$

$$= \frac{-328.98 \times 247.7}{593 \times 345.6} = -0.3973$$

B_2 or $X_{Cd} = 0.400$

$$B_3 = \log X_{Cd} = \frac{6.3}{2.303 \times 8.314 \times 10^{-3}}\left[\frac{1}{593} - \frac{1}{423.7}\right]$$

$$= \frac{-328.98 \times 169.3}{593 \times 423.7} = -0.2217$$

B_3 or $X_{Cd} = 0.600$

$$B_4 = \log X_{Cd} = \frac{6.3}{2.303 \times 8.314 \times 10^{-3}}\left[\frac{1}{593} - \frac{1}{505.1}\right]$$

$$= \frac{-328.98 \times 87.9}{593 \times 505.1} = -0.0965$$

B_4 or $X_{Cd} = 0.800$

A plot of X_{Bi} vs T and X_{Cd} vs T is made and the meeting point of the two curves gives the eutectic temperature = 410 K and eutectic composition = 55 mol% Cd. (See phase diagram below)

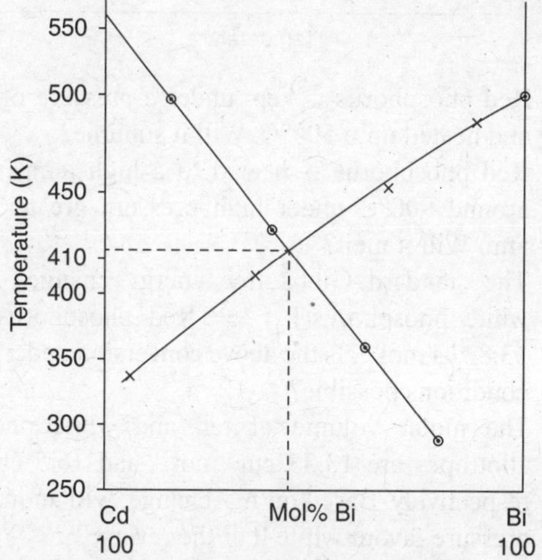

Phase diagram of Cd–Bi system

EXERCISES

Practice Problems

1. Calculate the number of components (C), number of phases (P) and the number of degrees of freedom (F) in the following cases.

(i) $2NH_3(g) \rightleftharpoons N_2(g) + 3H_2(g)$ in which equilibrium is established.

(ii) $H_2(g) + ½O_2(g) \rightleftharpoons H_2O(g)$ at ordinary temperatures.

(iii) Zinc and cadmium metals forming various liquid mixtures or melts at 1 atm.

(iv) $CH_3COOH + H_2O \rightleftharpoons CH_3COO^- + H^+$ in equilibrium at 1 atm. and 298 K.

(v) An aqueous solution of sodium nitrate and potassium nitrate in equilibrium with the vapour phase.

(vi) $Fe_3O_4(s) + H_2(g) \rightleftharpoons 3FeO(s) + H_2O(g)$ at equilibrium at a high temperature of 1000 K.

(vii) Solid copper sulfate dissolved in aqueous ammonium hydroxide and benzene at 1 atm. and at 30°C.

(viii) White phosphorus $\rightleftharpoons$ Red phosphorus

2. (a) Pure lead is melted and heated above its melting point of 600 K and gradually cooled until it solidifies. Draw the temperature–time cooling curve and explain the features of the curve.

(b) Suppose the same temperature–time cooling curve for a mixture of two components (say Pb and Sb), where only one pure component separates out of this system, is plotted, draw the nature of this curve and explain the features.

3. (a) The phase diagram of a pure compound has a triple point at 197 K and 10.13 kPa. It has a normal melting point of 200 K and a normal boiling point of 263 K. Draw the phase diagram of the compound and label all the regions. Calculate F in a region, along a line and at the triple point. If the solid compound at a pressure of 91 kPa and 240 K is taken, will it sublime?

(b) The phase diagram of phosphorus (red form) is qualitatively shown in the following diagram. Answer the questions based on the diagram.

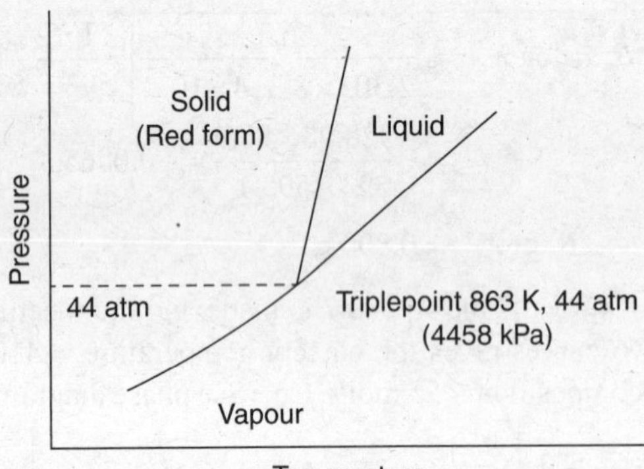

(i) Red phosphorus is kept under a pressure of 10 atm. and heated upto 500°C. Will it sublime?

(ii) Red phosphorus is heated to a high temperature of around 900°C under high pressure greater than 50 atm. Will it melt?

(iii) The standard Gibbs free energy change, ΔG^0 for white phosphorus(P_4) $\rightleftharpoons$ Red phosphorus (P_4) is 13.82 kJ mol^{-1}. Is the above conversion under standard conditions possible?

(iv) The molar volume of red and white phosphorus allotropes are 13.33 cm^3 mol^{-1} and 16.5 cm^3 mol^{-1} respectively. Based on ΔG change, will an increase of pressure favour white P or the reverse?

(v) The enthalpy of vaporisation of red phosphorus is 133.5 kJ mol^{-1}. What will be the necessary temperature in order that red phosphorus may sublime under a pressure of 20 atm? The vapor pressure of red phosphorus at 450°C is 1.26 atm.

4. The following freezing point data has been obtained on a series of napthalene and biphenyl mixtures as given below.

Naphthalene (mol%)	100	74.2	59.4
Freezing point (K)	353.5	337.1	324.6
Biphenyl (mol%)	100	71.4	55.5
Freezing point (K)	342.1	318.1	311.1

Draw the phase diagram and label all the regions. Calculate F in a region along the line and at the lowest temperature point. What are the values of eutectic temperature and eutectic composition?

(i) Consider a point A at 80 mol% naphthalene at 320 K. What are the percentage amounts of liquid and solid phases corresponding to this composition?

(ii) Consider a melt having 70 mol% biphenyl at 345 K. Explain the changes taking place on cooling. Show the temperature time plot relevant to this cooling.

5. From the following data pertaining to the magnesium–zinc system, draw the phase diagram and label all the regions. Melting point of Mg is 924 K; melting point of Zn is 693 K. A compound having the composition $MgZn_2$ and a congruent melting point of 848 K is also observed. The system has two eutectic points (i) having eutectic temperature 653 K (composition 15 mol% Mg) and (ii) another with a eutectic temperature of 620 K and overall composition 75 mol% Mg. Explain the significance of eutectic points and the maximum in the diagram. Consider the overall composition of 60 mol% Mg (point L in diagram) and follow the isoplieth as the melt is cooled and describe the sequence of events occurring.

6. The copper nitrate–water system has the following features. It has a eutectic point with the temperature being 249 K and 35 wt% $Cu(NO_3)_2$ composition. There are two incongruent melting points, one at 253 K with a copper nitrate composition of 38 wt% and another at a temperature of 300 K with a copper nitrate composition of 62 wt%. The solid phases undergoing incongruent melting are $Cu(NO_3)_2.9H_2O$ in the former case and $Cu(No_3)_2.6H_2O$ in the latter case. Draw the phase diagram of the system which may include the extreme point of $Cu(NO_3)_2$ with the solubility being 76 wt% at a temperature of 393 K. (The temperature range may vary between 243 K and 400 K). Label all the regions. What is the value of F at the eutectic point and the points where incongruent melting occurs? Explain their significance. Specify the phases in equilibrium at these points. Consider point F in the diagram having an overall composition of 55 wt% $Cu(NO_3)_2$ at a temperature of 373 K. Follow the isoplieth and explain the sequence of changes until the temperature falls below 245 K.

7. The following table contains the solubility data relating to the three-component system Na_2SO_4–NaCl–H_2O at 298 K. Use these data to draw the phase diagram and label all the regions.

NaCl (mol%)	Na_2SO_4 (mol%)	H_2O (mol%)	Solid phase
0.00	3.37	96.63	$Na_2SO_4.10H_2O$
2.91	2.52	94.56	$Na_2SO_4.10H_2O$
5.76	2.37	91.86	Na_2SO_4+$Na_2SO_4.10H_2O$
7.24	1.68	91.08	Na_2SO_4
8.82	1.14	90.04	Na_2SO_4 +NaCl
9.47	0.54	90.00	NaCl
10.03	0.00	89.97	NaCl

Convert the above data into wt% (for better clarity) and label the regions and calculate F. What is the composition of the system at points H and I (in the diagram)?

8. The liquid–vapour composition data and the corresponding boiling points at 1 atm. are given for water–HNO_3 mixtures in the table.

Boiling point (K)	375.5	381.5	386.0	390.5	393.4	393.8	393.8	394.0	393.8
HNO_3 in liquid phase (mol%)	4.23	11.73	18.5	25.07	32.39	35.33	37.16	37.77	39.33
HNO_3 in vapour phase (mol%)	0.132	1.45	4.75	12.59	25.46	31.90	35.65	37.76	40.72
Boiling point (K)	392.5	385.0	380.0	370.0	364.5	357.0	356.8		
HNO_3 in liquid phase (mol%)	43.92	55.75	60.64	72.06	81.56	96.74	97.34		
HNO_3 in vapour phase (mol%)	52.75	76.86	86.45	91.26	95.02	97.40	96.98		

Draw the phase diagram and label the regions. Apply the phase rule and calculate F above the curves, below the curves and in the region between the liquid and vapour phase regions. Also, answer the following questions.

(i) What is the azeotropic temperature and composition?
(ii) Do H_2O–HNO_3 mixtures obey Raoult's law? If not, why do they show deviation from Raoult's law?
(iii) How do the vapour–pressure composition curves appear (at constant temperature)?
(iv) Give an example of a liquid mixture which shows a minimum in the boiling point–composition curves.
(v) Do benzene–toluene mixtures obey Raoult's law? Give a sketch of the vapour pressure–composition (at constant temperature) and boiling point–composition (at 1 atm.) curves and label the various regions.

9. The distribution coefficient of diethylamine between chloroform and water is 2.2 at 298 K. Starting with a 0.2 M solution of diethylamine in water, calculate the volume of chloroform required to extract 90% of diethylamine into the chloroform layer.
(i) Calculate the total volume of chloroform required to extract 90% of the diethylamine in five steps.
(ii) Using 200 mL of chloroform each time, calculate the number of extractions required to obtain 90% diethylamine in the chloroform layer.

10. The composition of conjugate solutions of hexane–nitrobenzene mixtures at a constant pressure of 1 atm. are given.

Temperature (K)	Hexane-rich layer (mol%)	Nitrobenzene-rich layer (mol%)
273	90.0	93.0
290	65.0	83.0
281	80.0	90.0
285	75.0	91.0
270	95.0	95.0

The critical solution temperature of the system is 294 K and the composition at this temperature is $X_{Hexane} = 0.4$ (or 40 mol%). Draw the phase diagram and label the various regions. Calculate F above the curve, along the line and at the critical solution temperature. What is the significance of this temperature? Suppose a solution having an overall composition 40 mol% hexane at a temperature of 296 K is gradually cooled. Explain the sequence of changes taking place. What happens at 285 K and explain and calculate if necessary, the compositions of the layers.

11. (a) The distribution of *p*-hydroxy benzoic acid between water and benzene at 298 K yielded the following data.

Concentration in aqueous layer (mol dm^{-3})	Concentration in benzene layer (mol dm^{-3})
1.39×10^{-5}	6.4×10^{-7}
3.28×10^{-5}	3.40×10^{-6}
3.87×10^{-5}	4.80×10^{-6}
4.40×10^{-5}	6.30×10^{-6}

By suitable calculations, show the nature of the acid in the organic solvent.

(b) The organic compound terpinene distils freely on steam distillation at a temperature of 368 K under an atmospheric pressure of 99.2 kPa. The vapour pressure of water at this temperature is 84.52 kPa. The distillate was found to contain 57 wt% of terpinene. What is its molecular weight?

12. 20 g of aniline hydrochloride was shaken with 1 L of water and 0.1 L of benzene at 298 K. When equilibrium was established, 25 mL of the benzene layer was found to contain 0.001505 mol of aniline. Given that the partition coefficient of aniline between benzene and water is 11.5, calculate the hydrolysis constant of aniline hydrochloride.

13. The freezing points of chloroform and carbon tetrachloride are 210 K and 250 K respectively. Their enthalpies of fusion are 8.8 kJ mol^{-1} and 3.28 kJ mol^{-1} (for $CHCl_3$ and CCl_4) respectively. Calculate the melting points at compositions of $X_{CHCl_3} = 0.2, 0.4, 0.6, 0.8$ and similarly for CCl_4 at the same compositions from ideal solubility relation and plot the composition–melting point data for each. Determine eutectic temperature and composition from the intersection of the curves.

14. The experimentally determined freezing point–composition data for the $CHCl_3$–CCl_4 system is given. Use these data to draw the phase diagram of the system and determine the eutectic temperature and composition. Label the regions and calculate F along the line and at the eutectic point. Compare the results from those obtained in Problem 13. If a mixture of both components at an overall composition of 20 mol% CCl_4 at 250 K is cooled gradually, explain the sequence of changes taking place. Draw the cooling temperature–time curve.

X_{CCl_4}	0.0	0.1	0.2	0.3	0.4	0.6	0.7	0.8	1.0
Freezing point (K)	210.0	207.0	202.0	195.0	192.0	200.0	208	218	250.0

15. Draw the triangular phase diagram for the system NH_4CNS–aniline–water from the data given at 298 K.

Aniline (wt%)	H_2O (wt%)	NH_4CNS (wt%)
0.0	35.5	64.5
5.4	33.7	60.9
16.8	28.7	54.6
20.0	25.5	54.5
40.2	20.5	39.3
53.4	15.4	31.2
72.4	8.0	19.2
81.5	4.3	14.2
89.5	0.0	10.6

Composition of conjugate solutions

Water-rich layer	Salt (wt%)	Aniline-rich layer
98 wt% water + 2 wt% aniline	0.0	98 wt% aniline + 2 wt% water
95 wt% water + 2 wt% aniline + 3 wt% salt		90 wt% aniline + 5 wt% salt + 5 wt% water
80 wt% water + 14% salt + 6 wt% aniline		84 wt% aniline + 8 wt% salt + 8 wt% water
68 wt% water + 22% salt + 10 wt% aniline		80 wt% aniline + 12 wt% salt + 8 wt% water
65 wt% water + 25 wt% salt + 10 wt% anline		60 wt% aniline + 15 wt% salt +25 wt% water

(Continued)

Label the various regions and calculate F. Mark the plait point and indicate its significance. Is NH_4CNS more soluble in water or aniline?

16. For the partition of salicylic acid between benzene and water at 298 K, the following results were obtained.

Conc. in water (mol L^{-1}) (C_W)	Conc. in benzene (mol L^{-1}) (C_{org})
3.49×10^{-3}	13.92×10^{-3}
4.10×10^{-3}	17.03×10^{-3}
5.13×10^{-3}	23.67×10^{-3}
5.98×10^{-3}	32.66×10^{-3}
2.745×10^{-3}	9.097×10^{-3}

The dissociation constant of salicylic acid in water is 1.05 $\times 10^{-3}$ at 298 K. Considering ionisation of the acid in water and also that salicylic acid dimerises completely in benzene, derive an equation given by,

$$\frac{C_{org}}{N^2} = \frac{P}{N} + \text{constant where } P = \frac{\text{Concentration of single acid molecules in org phase}}{\text{Concentration of single acid molecules in aqueous phase}}$$

and $N = (C_{\text{Total}}$ of acid in water – acid dissociated at conc. $C)$

Make a plot of C_{org}/N^2 vs $1/N$ and calculate P from the slope.

17. The solubility of ammonium azide in ammonia was investigated and the resulting data are given.

Freezing point (K)	195.3	190.5	189.2	187.8	186.9	190.1	194.5	196.2
Composition of NH_4N_3 (wt%)	0.0	15.0	17.8	20.5	24.5	27.6	30.4	31.8

Freezing point (K)	197.6	201.0	205.0	206.2	218.3	231.8	240.3
Composition of NH_4N_3 (wt%)	32.9	35.7	36.0	36.4	37.3	39.2	40.7

Freezing point (K)	249.0	260.5	262.8	268.0	279.9
Composition of NH_4N_3 (wt%)	42.70	45.8	46.6	47.3	48.1

Draw the phase diagram, label the regions and explain the features of the curve.

NOTE

- The solid phases observed are as follows: up to 187 K NH_3; up to 201.6 K $NH_4N_3.5NH_3$; up to 263.6 K $NH_4N_3.2NH_3$; transition temperature of $NH_4N_3.5NH_3$ to $NH_4N_3.2NH_3$ is 202 K.

18. The distribution coefficient of bromine between CCl_4 and water is 27.0 at 298 K. Starting with 0.25 M bromine in water, calculate the volume of CCl_4 required to extract 95% of the bromine into the organic layer. Also, calculate the total volume of CCl_4 required to extract 95% of the bromine in five steps.

19. Benzene and octane (C_8H_{18}) form an ideal solution and their vapour pressures in the pure state are 74 kPa for benzene and 35 kPa for octane at 298 K. Plot the partial vapour pressure of each and the total vapour pressure curve against mole fraction at this temperature. Calculate the composition of the vapour in equilibrium with a solution containing 0.4 mole fraction of octane. Also, calculate the composition of the liquid phase in equilibrium with vapour containing 0.3 mole fraction of benzene and 0.7 mole fraction of octane.

20. The boiling point and liquid–vapour composition data of aqueous acetic acid mixtures at a constant pressure of 1 atm are given.

Boiling point (K)	373.0	375.0	377.4	380.5	386.8	391.1
Wt% $C_2H_4O_2$ in liquid phase	0.0	55.6	76.9	88.6	96.8	100.0
Wt % $C_2H_4O_2$ in vapour phase	0.0	43.1	66.6	81.9	94.3	100.0

Draw the phase diagram, label the regions and answer the following questions.
(i) Do water–acetic acid mixtures form an azeotrope?
(ii) Can pure acetic acid be realised by fractional distillation of acetic acid–water mixtures?
(iii) What approach can be adopted for efficient separation of acetic acid from its mixtures with water?
(iv) The activity coefficients of both acetic acid and water are greater than 1.0 in the region around $X_i = 0.5$. Does it indicate +ve or –ve deviation from Raoult's law?

21. (a) Aniline was steam distilled and analysis of the distillate showed it to contain 27 wt% aniline. Calculate the vapour pressure of aniline at the distillation temperature under a pressure of 101.325 kPa.
(b) In the steam distillation of an organic compound at 371 K, the wt% of the compound in the distillate was 43.75%. Given that the vapour pressure of water at this temperature is 94.94 kPa, calculate the molar mass of the organic compound.

Multiple Choice Questions

1. The total number of triple points in the sulfur system are
(a) 1 **(b)** 2
(c) 4 **(d)** 3

2. The system having both an upper and lower consolute temperature from the following are
(a) triethylamine–water **(b)** water–nicotine
(c) phenol–water **(d)** water–aniline

3. In a mixture of methanol and formic acid, the number of components are
(a) 4 **(b)** 2
(c) 1 **(d)** 3

4. In a solution containing a mixture of NaCl and $LiNO_3$ in equilibrium with the vapour at 298 K and 1 atm, the number of components and the number of degrees of freedom (in that order) are
(a) 2, 0 (b) 4, 2
(c) 3, 1 (d) 0, 3

5. The coefficient of thermal expansion of water is 2.126×10^{-4} K^{-1}. Given the density of water at 20°C as 0.9982 cm^{-3} its density at 25°C is
(a) 0.98926 (b) 0.99714
(c) 1.0178 (d) 0.97838

6. The molar volumes of ice and water at 0°C are 19.63 cm^3 mol^{-1} and 18.0 cm^3 mol^{-1} respectively. Its enthalpy of fusion at this temperature is 6000 J mol^{-1}. The rate of variation of the freezing point of water with pressure is (in K bar^{-1})
(a) 7.42×10^{-3} (b) 3.71×10^{-2}
(c) -7.42×10^{-3} (d) -3.71×10^{-2}

7. The gold–silicon system forms a simple eutectic with the eutectic temperature being 650 K. Given that the melting point of pure silicon is 1700 K, the mole fraction of silicon in the melt (liquid phase) at 1500 K is (ΔH_{fusion} of Si = 50.6 kJ mol^{-1})
(a) 0.785 (b) 0.216
(c) 0.394 (d) 0.621

8. The number of degrees of freedom at the upper consolute temperature of the phenol–water system is
(a) 0 (b) 2
(c) 1 (d) 3

9. Which of the following statements is true from the study of the phase diagram of carbon dioxide?
(a) Above 5.1 atm and a temperature of –40°C, the solid, liquid and vapour forms are in equilibrium.
(b) At normal temperatures and pressures, solid carbon dioxide melts on heating to form liquid.
(c) Above 31.1°C, carbon dioxide exhibits super critical behaviour.
(d) The solid–liquid line in the phase diagram lies slightly towards the left because $\Delta V_f = (V_1 - V_s)$ is +ve.

10. Consider a two component solid system involving the formation of a eutectic and a compound. Pick out the correct statement about this system from the following.
(a) Both the eutectic and the compound are similar in all properties as they have sharp melting points.
(b) Addition of an inert substance at the eutectic temperature at the particular composition raises the temperature, while the temperature falls in the case of addition to a compound at its melting point.
(c) Increase of pressure does not affect the composition in the case of a eutectic as well as a compound.
(d) None of the above statements truly represent the characteristics of a eutectic or a compound.

11. From the phase diagram of the ammonium sulfate–water system, at an overall composition of 30 mol% salt and 323 K, the weight % of solid and the solution, in that order, are
(a) 45.94, 54.06 (b) 77.78, 22.22
(c) 54.06, 45.94 (d) 22.22, 77.78
(**Hint:** Use the phase diagram given in the text)

12. Sodium sulfate decahydrate shows efflorescence at normal temperatures because
(a) the vapour pressure of water in air is less than the dissociation pressure of the hydrate
(b) the vapour pressure of water in air is greater than the dissociation pressure of the hydrate
(c) the salt hydrate acts as a dessicant
(d) the salt hydrate gains more water from the air

13. A mixture of 3 mol of *n*-pentane and 2 mol of benzene has a vapour pressure of 350.2 mmHg at 298 K. The vapour pressure of the mixture goes up to 376.0 mmHg on the addition of 1 mol of *n*-pentane. The vapour pressures of pure *n*-pentane and benzene (in the same order) are (assuming ideal behaviour)
(a) 200, 345 (b) 504, 119
(c) 345,200 (d) 119,504

14. Indicate the mixture which is least ideal and hence shows greatest deviation from ideal behaviour.
(a) Benzene–toluene (b) *n*-butanol–water
(c) *n*-hexane–*n*-octane (d) acetone–ethanol

15. The extensive property from the following is
(a) molar volume
(b) vapour pressure of a solution
(c) vapour pressure of a pure solvent
(d) density of a solution

16. Which of the following statements relating to an azeotrope is correct?
(a) An azeotrope can be separated into its pure components by fractional distillation.
(b) A mixture of HNO_3–H_2O is an example of a system which forms an azeotrope with a boiling point lower than either of the pure components.
(c) The number of degrees of freedom at the azeotropic composition of the system is zero.
(d) The azeotropic composition of a system does not vary with a change in external pressure.

17. In the steam distillation of bromobenzene at the boiling point of the mixture, 368 K, the ratio of the weight of water to bromobenzene in the distillate is (the vapour pressure of water = 84.5 kPa, vapour pressure of bromobenzene = 15.7 kPa, both at 368 K)
(a) 0.617 (b) 0.724
(c) 0.528 (d) 0.789

18. 13.5 g of aniline hydrochloride was shaken with 1 dm^3 of water and 100 ml of benzene until equilibrium was

established. At equilibrium, 25 ml of the benzene layer was found to contain 0.121 g of aniline. Given that the distribution coefficient of aniline between benzene and water is 11.3, the degree of dissociation of the anilinium ion in water is

(a) 0.058 **(b)** 0.075
(c)0.069 **(d)** 0.084

19. In the three-phase diagram of the water–phenol–acetone system, the addition of acetone to the phenol–water mixture shifts the tie lines to one side (rather than being parallel to the phenol–water axis) because

(a) acetone preferentially goes into the phenol layer
(b) acetone dissolves more in water than in phenol
(c) acetone does not prefer either water or phenol
(d) the observed behaviour has no connection whatsoever to the solubility of acetone in the two chemical species

20. The distribution coefficient of succinic acid between water and ether is 5.5. Starting with 0.1 M of acid in ether, the minimum volume of water required to extract 98% of the acid into the aqueous layer in one step (in litres) is

(a) 7.0 **(b)** 9.8
(c) 7.5 **(d)** 8.9

Answer Keys

Practice Problems

1. (i) $C = 2, P = 1, F = 3$

(ii) There is no equilibrium, therefore, $C = 3, P = 1, F = 4$

(iii) $C = 2, P = 2, F = 1$

(iv) $C = 2, P = 1, F = 1$

(v) Number of chemical species = 5 since $Na^+(aq)$, $NO_3^-(aq)$, $K^+(aq)$, $H_2O(l)$, $H_2(g)$. Also the condition of electroneutrality, and equilibrium condition $H_2O(l) \rightleftharpoons H_2O(g)$ exist.

$\therefore C = 5 - 2 = 3,\ P = 2,\ F = 3 - 2 = 1$

(vi) Number of chemical species = 4, equilibrium condition = 1. $\therefore C = 4 - 1 = 3,\ P = 3,\ F = 0$

(vii) $C = 4,\ P = 2,\ F = 2$

(viii) $C = 1,\ P = 2,\ F = 1$

2. (a)

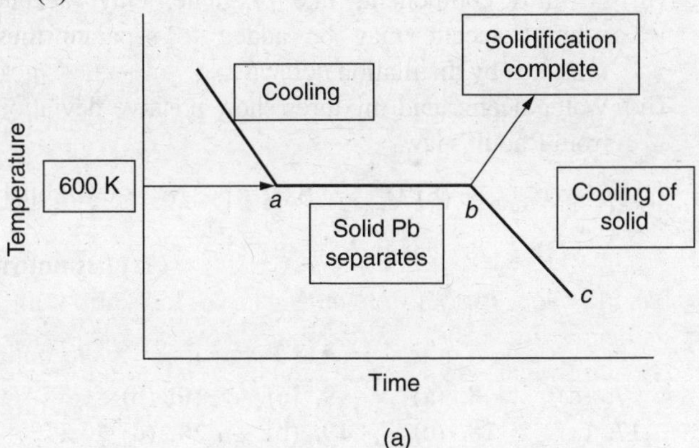

(a)

As Pb cools, at *a* it starts separating out and the temperature remains constant until *b* at which Pb solidifies completely. Further cooling occurs along *bc*.

(b)

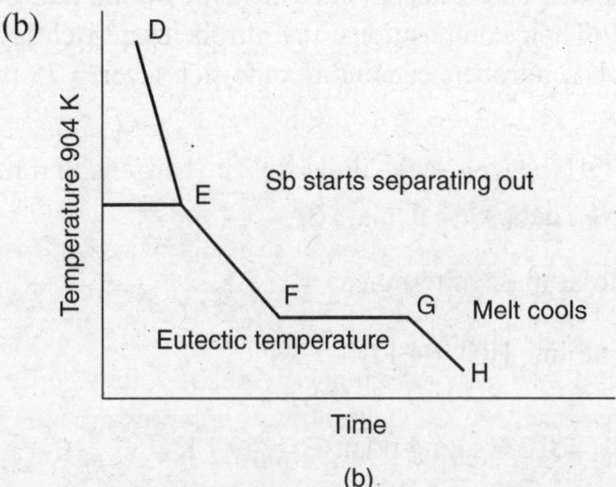

(b)

The Sb melts above 904 K, cools along DE. The addition of Pb results in separation of solid Sb at E. Thus, there is a break in the curve and along EF, antimony continues to separate. At F the eutectic composition of Sb and Pb separates out and the temperature remains constant until the melt of the eutectic composition separates out completely. Therefore, the temperature falls along GH and the solid cools.

3. (a) At 240 K and 91 kPa, the compound sublimes. No liquid phase exists above 197 K.

(b) (i) Red phosphorus sublimes.

(ii) Red phosphorus melts.

(iii) ΔG^0 is positive and conversion is not feasible.

(iv) Conversion of white to red phosphorus under high pressure is feasible.

(v) Red phosphorus sublimes at 816.9 K under a pressure of 20 atm.

4. Eutectic temperature = 310.5 K, eutectic composition = 55 mol% biphenyl.

(i) wt% of solid phase = 55.6; wt% liquid phase = 44.4

5. Theoretical question.

6. From the diagram at B, the eutectic point, $F = 0$. At the incongruent melting C and D also, $F = 0$.

7. Descriptive question.

8. (i) Above the curve, $F = 2$. Below the curve, $F = 2$. Between the curves, $F = 1$. Azeotropic composition = 38 wt% HNO_3, azeotropic temperature = 395 K.

   (ii) Deviations from Raoult's law. HNO_3–H_2O interactions are stronger leading to negative deviations from Raoult's law.

   (iii) Show a minimum in the diagram.

   (iv) Ethanol–water mixtures.

   (v) Benzene–toluene mixtures obey Raoult's law.

9. 4.095 litres (i) 1.330 litres (ii) 6 extractions.

10. Two liquid phases, appear once the temperature falls below 294 K. Their compositions: the nitrobenzene-rich layer = 88 mol%, nitrobenzene and hexane-rich layer = 75 mol% hexane.

11. (a) p-OH benzoic acid dimerises in benzene. From the given data, we calculate $C_1/\sqrt{C_2}$.

    (b) Molar mass of terpinene = 137.2.

12. K_h of aniline HCl $= 4.11 \times 10^{-4}$

13. $X_{CHCl_3} = 0.2$, melting point $T = 159.1$ K

    $X_{CHCl_3} = 0.4$, melting point $T = 177.8$ K

    $X_{CHCl_3} = 0.6$, melting point $T = 190.65$ K

    $X_{CHCl_3} = 0.8$, melting point $T = 201.1$ K

    The eutectic temperature = 182 K, eutectic composition = 53 mol% CCl_4

14. Experimental eutectic temperature = 188 K, eutectic composition = 45 mol% CCl_4 (both for $CHCl_3$–CCl_4 system).

15. Theoretical question.

16. $P = 5; K_D = \dfrac{[HA]^2}{[HA]_2} = 6.67 \times 10^{-2}$

17. Theoretical question.

18. (i) Volume of CCl_4 required to extract 95% Br_2 is to be extracted into organic layer, 0.704 litres.

    (ii) Volume of CCl_4 required to extract 95% Br_2 is to be extracted into organic layer in five steps = 0.152 litres.

19. At $X_{octane} = 0.4$ in liquid phase, the vapour phase equilibrium composition $X_{benzene} = 0.76$. When $X_{benzene} = 0.3$ in vapour phase, the liquid phase in equilibrium composition $X_{benzene} = 0.1$ or $X_{octane} = 0.9$.

20. The boiling point–composition diagram is shown.

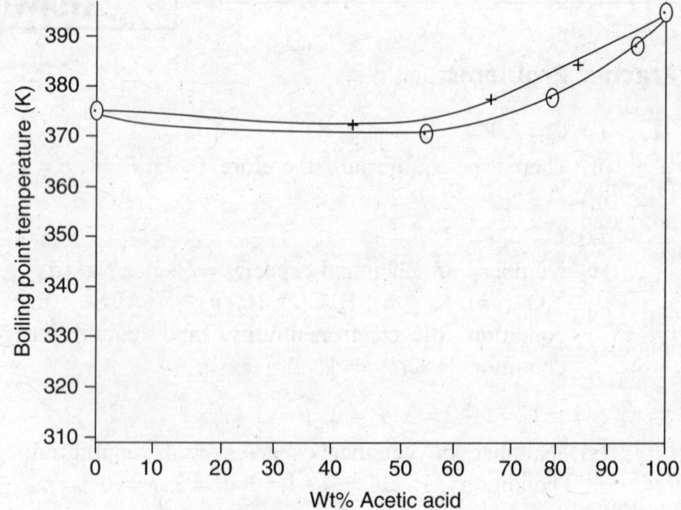

    (i) Water–acetic acid mixtures do not form an azeotrope.

    (ii) Separation is not possible by fractional distillation.

    (iii) A third component, like p-xylene, ethyl acetate, n-butyl acetate may be added to separate these mixtures by distillation.

    (iv) Water–acetic acid mixtures show positive deviations from Raoult's law.

21. (a) $p_A^0 = 6.77$ kPa (b) $M_{org} = 208.2$

Multiple Choice Questions

1. (c)	2. (b)	3. (d)	4. (a)	5. (b)	6. (c)	7. (d)	8. (a)	9. (c)	10. (b)
11. (d)	12. (a)	13. (b)	14. (d)	15. (b)	16. (c)	17. (a)	18. (c)	19. (b)	20. (d)

CHAPTER 11

POLYMER CHEMISTRY

CHAPTER OUTLINE

Types of Molecular Weights
- Number Average Molecular Weights
- Weight Average Molecular Weights
- Viscosity Average Molecular Weights

Important Concepts and Formulae
- Molecular Weight using Sedimentation and Light Scattering Methods
- Glass Transition Temperature

11.1 Types of Molecular Weights

In contrast to other fields, polymers possess different types of molecular weights depending upon the method of synthesis and extent of reaction. Hence, different approaches have been envisaged to deduce estimates of molecular weights. Thus, it is essential to denote the type of molecular weight referred to, while carrying out numerical calculations. For monodisperse polymers, the number average molecular weight is equal to the weight average molecular weight and generally, these polymers are denoted as having a polydispersity index of unity. Table 11.1 provides the different types of molecular weights and the corresponding experimental technique used to determine them.

Table 11.1 Different types of molecular weights

Molecular weight	Equation	Experimental technique
$\bar{M}_n$ (Number average)	$\dfrac{\sum N_i M_i}{\sum N_i}$	Osmotic pressure, end-group analysis
$\bar{M}_w$ (Weight average)	$\dfrac{\sum N_i M_i^2}{\sum N_i M_i}$	Light scattering sedimentation
$\bar{M}_z$ (z-average)	$\dfrac{\sum N_i M_i^3}{\sum N_i M_i^2}$	Sedimentation
$\bar{M}_v$ (Viscosity average)	$\left(\dfrac{\sum N_i M_i^{1+\alpha}}{\sum N_i M_i}\right)^{1/\alpha}$ $0.5 < \alpha < 1.0$	Viscosity measurements (α is an exponent characterising the geometry of the polymers)

From the above equations, it can be shown that the trend in molecular weights is $\bar{M}_n \leq \bar{M}_v \leq \bar{M}_w \leq \bar{M}_z$.

11.1.1 Viscosity Average Molecular Weights

A popular method of estimating molecular weights of polymers is by viscosity measurements (viscometry). The flow time of the solvent and solution is then correlated. The values deduced by viscosity methods refer to the corresponding solvent–polymer pairs.

Relative viscosity (η_{rel}) $\dfrac{\eta}{\eta_0}$ —relative to that of the solvent.

Specific viscosity (η_{sp}) $\dfrac{\eta}{\eta_0}-1$ —indicates the deviation of the relative viscosity from unity.

Reduced viscosity (η_{red}) $\dfrac{\eta_{\text{sp}}}{C}$ —specific viscosity normalised with respect to the concentration of the solution.

Inherent viscosity $\ln\dfrac{\eta_{\text{rel}}}{C}$

Intrinsic viscosity $[\eta] = \lim\limits_{c\to 0}\dfrac{\eta_{\text{sp}}}{C}$

The intrinsic viscosity refers to $\dfrac{\eta_{\text{sp}}}{C}$ as the concentration of the polymer tends to zero. Intrinsic viscosity measurements lead to the viscosity average molecular weight of polymers. If viscosity is expressed in poise, the conversion is 1 poise = 1 g cm^{-1} s^{-1}. The Mark–Howink–Sakurada equation relating intrinsic viscosity and molecular weight for a given polymer–solvent system is as follows.

$$\text{Intrinsic viscosity } [\eta] = K(\bar{M})^{\alpha}$$

According to polymer dynamics theories, $\alpha = 0$ for hard spheres, $\alpha = 2.0$ for rigid rods and $\alpha = 0.5$ for random coils. The constant K and exponent α are the Mark–Howink–Sakurada constants which are characteristic of a given polymer–solvent system. Furthermore, $\bar{M}$ can refer to either number average, weight average or z-average molecular weight. Accordingly, the values of K and α vary.

11.2 Important Concepts and Formulae

Molecular weights of polymers using light scattering: An approximate relation governing the molecular weight of polymers and light scattering characteristics is given by

$$\frac{Kc}{P(90)} = 1/M_w + A_2 \times c + A_3 \times c^2 + \ldots.$$

where K is the optical constant and $P(90)$ denotes the Rayleigh ratio measured at 90°, c being the concentration of the polymer. It is the norm to construct a plot of $\frac{Kc}{P(90)}$ vs c and therefrom calculate the weight average molecular weights.

Diffusion coefficient and viscosity relation: The relation between viscosity and diffusion coefficient can be understood by several methods, the simplest being the analysis due to Stokes–Einstein. Here, the polymer is assumed to be a three-dimensional spherical molecule and the solution is considered dilute in order to ignore polymer–polymer interactions.

Stokes-Einstein equation: The diffusion coefficient of a solute is $D = \frac{kT}{f}$ where the frictional coefficient $f = 6\pi r\eta$, r = hydrodynamic radius or Stokes–Einstein radius and η = viscosity coefficient.

Entropy change in the elongation of polymers: For an ideal rubber undergoing elongation, the entropy change can be estimated using

$$\Delta S = \frac{-Nk}{2}(\alpha^2 + 2/\alpha - 3)$$

where N = number of polymer chains in mol cm^{-3}, α = degree of elongation and k = Boltzmann constant.

Kinetics of polymerisation: Extent of polymerisation,

$$p = 1 - \frac{M_1}{\bar{M}_n}$$

where M_1 is the weight of the *monomer* and $\bar{M}_n$ is the number average molecular weight of the polymer.

$$\text{Degree of polymerisation (DP)} = \frac{1}{1-p}$$

$$\text{Weight average molecular weight } \bar{M}_w = \bar{M}_n(1+p)$$

Molecular weight from osmotic pressure measurements: Virial equation of state for the osmotic pressure of a solution,

$$\frac{\pi}{C} = \frac{RT}{\bar{M}_n} + B_2C + B_3C + \ldots$$

C denotes the concentration of the solution in grams per unit volume. $\bar{M}_n$ denotes the number average molecular weight of the polymer. B_2, B_3, B_4, are known as the virial coefficients. Their magnitudes can be related to theta, poor and good solvents. For theta solvents, the second virial coefficient B_2 is 0 and the corresponding temperature is known as the theta temperature. The positive and negative values of B_2 reflect the Flory–Huggins interaction parameter.

$$\lim_{c\to 0}\frac{\pi}{C} \text{ vs } C \text{ yields approximate } \bar{M}_n$$

The slope is related to the second virial coefficient (B_2).

Sedimentation method for estimating the molecular weights of polymers: In the sedimentation method, the centripetal force is related to the angular velocity ω (radians s^{-1}) by the following equation.

$$F_{\text{centripetal}} = mr\omega^2$$

for a particle of mass m where r refers to the distance from the axis of rotation. The sedimentation velocity is measured using an ultracentrifuge.

$$\text{Sedimentation coefficient } (s) = \frac{1}{\omega^2(t_2 - t_1)}\ln\frac{r_2}{r_1}$$

The subscripts 1 and 2 for r and t refer to the initial and final states in the moving boundary. The molecular weight of the polymer is estimated from the equation,

$$M_2 = \frac{RTs}{D_2\left(1 - \frac{\rho_1}{\rho_2}\right)}$$

where D_2 = diffusion coefficient of the polymer, s = sedimentation coefficient (svedberg; 1 svedberg = 10^{-13} s), ρ_2 = density of the polymer (g cm^{-3}), ρ_1 = density of the solvent (g cm^{-3}).

Significance of glass transition temperatures: The glass transition temperature of a polymer denotes the temperature at which it undergoes transitions from a hard, glassy state to a soft, elastic state. It is thought of as a second order phase transition and its value is of immense technological significance in the viscoelasticity behaviour. Its magnitude can often be controlled by chemical modifications and by adding plasticisers. Above T_g, the viscosity decreases by several orders of magnitude. A common method of estimating T_g is by differential scanning calorimetry.

The dependence of the viscosity η on T_g is given by the Williams–Landel–Ferry (WLF) equation.

$$\log_{10}[\eta_T/\eta_{T_g}] = -17.4(T - T_g)/[51.6 + (T - T_g)]$$

Glass transition temperatures for copolymers: The two well-known equations for estimating the new glass transition temperature of copolymers in terms of the weight fractions are as follows:

$$\frac{1}{T_g} = \frac{W_1}{T_{g1}} + \frac{W_2}{T_{g2}} \quad \text{(Fox equation)}$$

$$T_g = W_1T_{g1} + W_2T_{g2} \quad \text{(Arithmetic mean)}$$

Worked Out Problems

Molecular Weight From Compositions

Problem 11.1

A solution contains equal *number* of particles with molecular weights 8222 g mol^{-1} and 18210 g mol^{-1}, respectively. Calculate the number average and weight average molecular weights.

Solution

$$\bar{M}_n = \frac{N_1M_1 + N_2M_2}{N_1 + N_2}$$

But $N_1 = N_2 = N$

$$\text{Therefore } \bar{M}_w \frac{N(M_1 + M_2)}{2N} = 13{,}216 \text{ g}$$

$$\bar{M}_w = \frac{\sum N_i M_i^2}{\sum N_i M_i}$$

$$= \frac{8222^2 + 18210^2}{26432}$$

$$= 15103.1 \text{ g}$$

Problem 11.2

A suspension contains equal *masses* of particles with molecular weights 2250 g mol^{-1} and 9250 g mol^{-1}, respectively. Calculate the number average and weight average molecular weights.

Solution

$$\bar{M}_n = \frac{N_1M_1 + N_2M_2}{N_1 + N_2}$$

$$\text{Therefore} \quad \bar{M}_n = \frac{M(N_1 + N_2)}{N_1 + N_2}$$

$$\text{But } N_1 = \frac{W_1}{M_1} \text{ and } N_2 = \frac{W_2}{M_2}$$

$$\text{and} \quad W_1 = W_2 = W$$

$$\text{Therefore } \bar{M}_n = \frac{2W}{\frac{W}{M_1} + \frac{W}{M_2}} = 3628 \text{ g mol}^{-1}$$

$$\bar{M}_w = \frac{W_1M_1 + W_2M_2}{W_1 + W_2}$$

$$W_1 = W_2 = W$$

Therefore

$$\bar{M}_w = \frac{W(M_1 + M_2)}{2W} = \frac{2250 + 9250}{2} = 5750 \text{ g}$$

Problem 11.3

Estimate the number average and weight average molecular weights for a sample of polyaniline with the following percentages and molecular weights.

15%	2500
12%	7000
50%	140000
23%	18000

Solution

$$\bar{M}_n = \frac{(0.15)(2500) + (0.12)(7000) + (0.5)(140000) + (0.23)(18000)}{1}$$

$$= 75355 \text{ amu}$$

$$\bar{M}_w = \frac{\sum N_i M_i^2}{\sum N_i M_i}$$

$$= \frac{(0.15)(2500^2) + (0.12)(7000^2) + (0.5)(140000^2) + (0.23)(18000^2)}{75355}$$

$$= 131130 \text{ amu}$$

Problem 11.4

Estimate the number average molecular weight ($\bar{M}_n$) and weight average molecular weight ($\bar{M}_w$) if a sample is made by mixing 17 g of polystyrene ($\bar{M}_w = 90{,}000$, $\bar{M}_n = 60{,}000$) and 12 g of another polystyrene ($\bar{M}_w = 65{,}000$, $\bar{M}_n = 25{,}000$).

Solution

$$\text{Weight average molecular weight} = \frac{W_1M_1 + W_2M_2}{W_1 + W_2}$$

$$= \frac{(17)(90{,}000) + (12)(65{,}000)}{29}$$

$$= 79{,}655 \text{ g}$$

Number average molecular weight = ?

	17 g	12 g
Number of moles	$\frac{17}{60{,}000}$	$\frac{12}{25{,}000}$

$$\text{Total number of moles} = \frac{17}{60{,}000} + \frac{12}{25{,}000}$$

Therefore

$$\bar{M}_n = \frac{\frac{17}{60{,}000} \times 60{,}000 + \frac{12}{25{,}000} \times 25{,}000}{\frac{17}{60{,}000} + \frac{12}{25{,}000}} = 38007.9 \text{ g}$$

Problem 11.5

If equal weights of two polymers A and B are mixed, calculate the weight average ($\bar{M}_w$) and number average ($\bar{M}_n$) of molecular weights of the mixture using the following data.

Polymer A: $\bar{M}_n = 20{,}000$ $\bar{M}_w = 95{,}000$

Polymer B: $\bar{M}_n = 140{,}000$ $\bar{M}_w = 310{,}000$

Solution

$$\text{Weight average molecular weight} = \frac{W_1M_1 + W_2M_2}{W_1 + W_2}$$

$$= \frac{(1)(95{,}000) + (1)(310{,}000)}{2}$$

$$= 202500$$

Number average molecular weight = ?

$$\text{Number of moles } \frac{1}{20{,}000}, \frac{1}{140{,}000}$$

$$\text{Total number of moles} = \frac{1}{20{,}000} + \frac{1}{140{,}000}$$

$$\text{Therefore } \bar{M}_n = \frac{\frac{1}{20{,}000}\times 20{,}000 + \frac{1}{140{,}000}\times 140{,}000}{\frac{1}{20{,}000} + \frac{1}{140{,}000}}$$

$$= 35000$$

Problem 11.6

Find the viscosity average molecular masses for a sample of a polyvinyl alcohol that has molecules with the following percentages and molecular masses.

8.00%	2500 amu
23.00%	16,000 amu
58.00%	26,000 amu
11.00%	34,000 amu

if $\alpha = 0.75$ for polyvinyl alcohol.

Solution

$$\text{According to the equation, } (M_v) = \left(\frac{\sum N_iM_i^{1+\alpha}}{\sum N_iM_i}\right)^{1/\alpha}$$

$$= \left[\frac{(0.08\times 2500^{1.75}) + (0.23\times 16000^{1.75}) + (0.58\times 26000^{1.75}) + (0.11\times 34000^{1.75})}{(0.08\times 2500) + (0.23\times 16000) + (0.58\times 26000) + (0.11\times 34000)}\right]^{1/0.75}$$

$$= 24{,}674 \text{ amu}$$

Problem 11.7

Polymers A and B are monodisperse methyl polystyrenes. Molecular weight of polymer A is twice the molecular weight of polymer B. Polymer C is polydisperse PS with $M_w = 3.2\times 10^5$ g mol^{-1}. The molecular weight of a mixture containing 27 g of A, 46 g of B and 27 g of C polymer is estimated as 103500 g mol^{-1} using light scattering. The molecular weight obtained using osmotic pressure data is 50000 g mol^{-1}. Estimate the number average molecular weight M_n of the polymer C.

Solution

Since the polymers A and B are monodisperse, the number average and weight average molecular weights are equal, that is, $M_n = M_w$.

$$M_B = \frac{M_A}{2}$$

The light scattering measurement yields weight average molecular weight M_w as 103500 g mol^{-1}. Osmotic pressure yields the number average molecular weight for the mixture. Hence $M_n = 50000$ g mol^{-1}.

Molecular weight for the polymer A can be estimated as

$$\bar{M}_w = \frac{\sum w_iM_i}{\sum w_i}$$

$$= \frac{w_A\times M_A + w_B\times\frac{M_A}{2} + w_C\times M_{w,c}}{w_A + w_B + w_C} = 103500 \text{ g mol}^{-1}$$

$$\text{Therefore } M_A = \frac{\bar{M}_w\times(w_A + w_B + w_C) - w_C\times M_{w,c}}{w_A + \frac{w_B}{2}}$$

$$= \frac{103500\text{ g mol}^{-1}\times(27\text{ g} + 46\text{ g} + 27\text{ g}) - 27\text{ g}\times 3.2\times 10^5\text{ g mol}^{-1}}{27\text{ g} + \frac{46\text{ g}}{2}}$$

$$= 34{,}200 \text{ g mol}^{-1}$$

$$M_B = M_A/2 = 17100 \text{ g mol}^{-1}$$

$$\bar{M}_n = \frac{\sum n_iM_i}{\sum n_i} = \frac{\sum w_i}{\sum n_i}$$

$$= \frac{w_A + w_B + w_C}{\frac{w_A}{M_A} + \frac{w_B}{M_B} + \frac{w_C}{M_{n,C}}} = 50000 \text{ g mol}^{-1}$$

$$M_{n,C} = \frac{w_C}{\frac{w_A + w_B + w_C}{\bar{M}_n} - \frac{w_A}{M_A} - \frac{w_B}{M_B}}$$

$$= \frac{27\text{ g}}{\frac{27\text{ g} + 46\text{ g} + 27\text{ g}}{50000\text{ g mol}^{-1}} - \frac{27\text{ g}}{103500\text{ g mol}^{-1}} - \frac{46\text{ g}}{34200\text{ g mol}^{-1}}}$$

$$= 27027 \text{ g mol}^{-1}$$

Problem 11.8

List the following in order of increasing values— $\bar{M}_z$, $\bar{M}_n$ and $\bar{M}_w$.

Solution

$$\bar{M}_z = \frac{\sum M_i^3N_i}{\sum M_i^2N_i}$$

$$\bar{M}_w = \frac{\sum M_i^2 N_i}{\sum M_i N_i}$$

$$\bar{M}_n = \frac{\sum M_i N_i}{\sum N_i}$$

Therefore $\bar{M}_z > \bar{M}_w > \bar{M}_n$.

Problem 11.9

The number average and weight average molecular weights are denoted as $\bar{M}_n$ and $\bar{M}_w$ respectively. For a monomer whose molecular weight is 100 g mol^{-1}, estimate $\bar{M}_n$ and $\bar{M}_w$ if the degree of polymerisation is 0.89.

Solution

$$\bar{M}_n = \frac{M_1}{1-p} = 909.09$$

$$\bar{M}_w = \frac{1+p}{1-p} \times M_1 = 1718.18$$

Problem 11.10

The weight of a sample of poly(hexamethylene adipamide) having two carboxylic acid groups, is 12.20 g. It is found to contain 4×10^{-3} mol of COOH groups by end-group analysis. Calculate $\bar{M}_n$ of the polymer.

Solution

Since each molecule contains two terminal –COOH groups, the number of moles equals 2×10^{-3}. Hence

$$\bar{M}_n = \frac{12.20\text{ g}}{2\times10^{-3}\text{ mol}} = 6100\text{ g mol}^{-1}$$

Problem 11.11

Upon polymerisation of $HO(CH_2)_5$–COOH, the product formed has a number average molar mass of 15,000 g mol^{-1}. Calculate the (a) extent of reaction p and (b) degree of polymerisation.

Solution

(a) Extent of reaction $= p$

$$M_n = \frac{M_1}{1-p} \qquad 15{,}000 = \frac{132}{1-p}$$

$$p = 0.9912$$

(b) Degree of polymerisation $= \dfrac{1}{1-p} = 113.6$

Molecular Weight Using Sedimentation Methods

Problem 11.12

A sample of γ globulin gives the following experimental results at 20°C. Specific volume of the solute = 0.702 mL g^{-1}, sedimentation coefficient = 6.13×10^{-13} s, diffusion coefficient = 3.9×10^{-11} m^2 s^{-1}. If the density of the solvent is 1.012 g mL^{-1}, calculate the molecular weight of γ globulin.

Solution

$$M_2 = \frac{RTs}{D_2\left(1-\frac{\rho_1}{\rho_2}\right)}$$

But $\rho_2 = \dfrac{1}{v_2}$ where v_2 = specific volume of the polymer.

$$\text{Therefore } M_2 = \frac{RTs}{D_2\left[1-(1.012\text{ g mL}^{-1})(0.702\text{ mL g}^{-1})\right]}$$

$$= \frac{8.314\times293\times6.13\times10^{-13}\text{ s (kg m}^2\text{ s}^{-2})}{3.9\times10^{-11}\text{ m}^2\text{ s}^{-1}\times0.2896}$$

$$= 132.21\text{ kg}$$

Problem 11.13

The diffusion coefficient of myosin in water is 1.13×10^{-7} cm^2 s^{-1} at 293 K and its sedimentation coefficient is 6.38×10^{-13} s. If the density of myosin is 1.29 g cm^{-3}, estimate its molecular weight.

Solution

$$M_w = \frac{RTs}{D_2\left(1-\frac{\rho_1}{\rho_2}\right)}$$

$$= \frac{8.314\text{ kg m}^2\text{ s}^{-2}\times293\text{ K}\times6.38\times10^{-13}\text{ s}}{1.13\times10^{-9}\text{ m}^2\text{ s}^{-1}\left(1-\frac{1}{1.29}\right)}$$

$$= 611.87\text{ kg}$$

Problem 11.14

The molecular weight of a macromolecule is 3.5×10^6 g mol^{-1} and its sedimentation coefficient is 12.0×10^{-13} svedberg. The product of the specific volume of macromolecule and density of solvent is 0.60 (at 25°C). Calculate the diffusion coefficient.

Solution

$$\text{Molar mass} = M = \frac{sRT}{D(1-v_2\rho)}$$

$$D = \frac{sRT}{M(1-v_2\rho)}$$

$$= \frac{12\times10^{-13}\text{ Sv}\times8.314\text{ kg m}^2\text{ s}^{-2}\text{ K}^{-1}\text{ mol}^{-1}\times298.15\text{K}}{3.5\times10^6\text{ g mol}^{-1}\ (1-0.6)}$$

$$= 0.21\times10^{-14}\text{ m}^2\text{ s}^{-1}$$

Problem 11.15

Polystyrene is ultracentrifuged at 970 revolutions per second, the boundary moves from 5.187 cm to 5.297 cm in 120 minutes. Calculate the sedimentation constant of polystyrene.

Solution

$$s = \frac{1}{\omega^2(t_2-t_1)}\ln\frac{x_2}{x_1}$$

$$= \frac{1}{(970\times2\pi)^2\text{ s}^{-2}\times120\times60\text{ s}}\ln\frac{5.297}{5.187} = 7.85\times10^{-14}\text{ s}$$

Problem 11.16

The sedimentation coefficient is 5.30×10^{-13} s for a polypeptide. The diffusion coefficient and volume at 293 K are 4.9×10^{-7} cm^2 s^{-1} and 0.72 mL g^{-1}. Calculate the molecular weight of the polypeptide.

Solution

$$M=\frac{sRT}{D(1-v_2\rho)}$$

$$=\frac{(5.3\times10^{-13}\ \text{s})\left(8.314\times\dfrac{10^{7}\ \text{erg}}{\text{Kmol}}\right)(293.16\ \text{K})}{4.90\times10^{-7}\ \text{m}^2\ \text{s}^{-1}\left(1-\left(0.72\dfrac{\text{mL}}{\text{g}}\right)\left(0.9988\dfrac{\text{g}}{\text{mL}}\right)\right)}$$

$$= 93{,}800\ \text{g mol}^{-1}$$

Problem 11.17

In an experiment to determine the sedimentation coefficient of a solution at 20°C using an ultracentrifuge rotating at an angular frequency of 5600 s^{-1}, the movement of boundary occurred from 6.3 cm to 6.4 cm in 180 minutes. The diffusion coefficient of the solution is 1.6×10^{-7} cm^2 s^{-1} and the specific volume of the polymer as 0.6 cm^3 g^{-1}. Assuming the density of the solvent (ρ) as 1.0 g cm^{-3}, estimate the (a) sedimentation coefficient and (b) molar mass.

Solution

(a) $$s=\frac{1}{\omega^2(t_2-t_1)}\ln\frac{x_2}{x_1}$$

$$=\frac{1}{(5600\times2\pi)^2\text{s}^{-2}\times180\times60\ \text{s}}\ln\frac{6.4}{6.3}=1.18\times10^{-15}\ \text{s}$$

(b) Molar mass $$M=\frac{sRT}{D(1-v_2\rho)}$$

$$=\frac{(1.18\times10^{-15}\ \text{s})(8.314\ \text{J})(293\ \text{K})}{1.16\times10^{-7}\ \text{m}^2\ \text{s}^{-1}\ (0.4)}$$

$$=619.5\ \text{g mol}^{-1}$$

Molecular Weight Using Viscosity Measurements

Problem 11.18

The viscosity of solutions of nitrocellulose in butyl acetate at 25°C is given below:

Concentration (moles of the monomer)	Specific viscosity
0.007	0.710
0.0142	1.486
0.0269	5.163
0.0459	18.110
0.107	99.700

Determine the molecular weight of nitrocellulose given that the Mark–Houwink constant is $K = 1.4\times10^{-3}$.

$\log\dfrac{\eta_{sp}}{C}$	C
2.0060	0.007
2.0197	0.0142

(Continued)

$\log\dfrac{\eta_{sp}}{C}$	C
2.2830	0.0269
2.5960	0.0459
2.9690	0.107

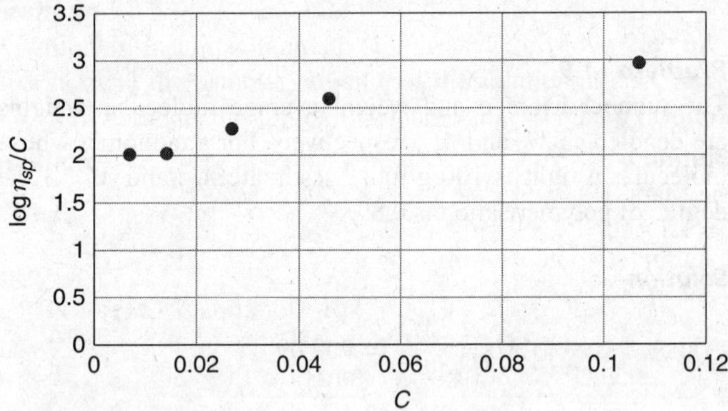

Solution

$$K=1.4\times10^{-3}$$

$$\log\frac{\eta_{sp}}{C}\ \text{vs}\ C$$

Limiting slope = 1.90

Therefore $\left(\dfrac{\eta_{sp}}{C}\right)=79.43$

$K\times$ Molecular weight = 79.43. Hence $M = 56738$

Problem 11.19

When determining the molecular weight of a fraction of polyvinyl chloride by the viscosimetric method, the following values were obtained at 298 K.

C (g/L)	90	200	498	890
$\dfrac{\eta}{\eta_0}$	2.0586	2.3341	3.5450	4.6721

Estimate the molecular weight of polyvinyl chloride.

$\log\dfrac{\eta_{sp}}{C}$	C
–1.9295	90
–2.1758	200
–2.2915	498
–2.3845	890

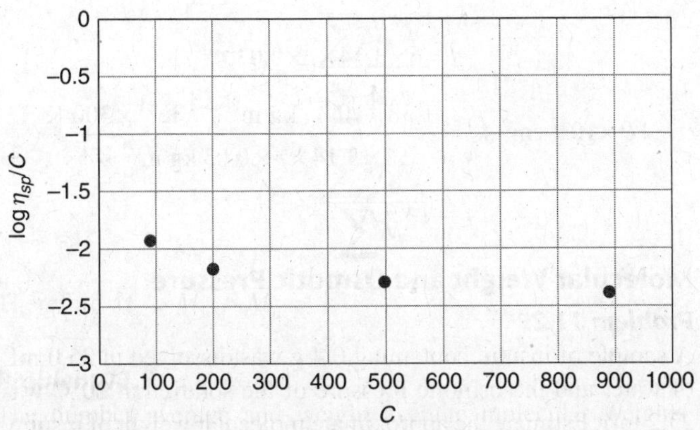

Solution

The Mark–Houwink constant K for this system is known to be $K = 1.1 \times 10^{-5}$

Since $\dfrac{\eta}{\eta_0} - 1 = \eta_{sp}$

$$\eta_{sp} = 1.0586,\ 1.3341,\ 2.5450,\ 3.6721$$

Hence $\log \dfrac{\eta_{sp}}{C}$ vs C yields the limiting slope as –1.75.

Therefore $\left(\dfrac{\eta_{sp}}{C}\right) = 0.01778$

Hence $M = 1616.6$

since $[\eta] = \lim\limits_{C \to 0} \dfrac{\eta_r}{C} = K \times M$

Problem 11.20

Determine the molecular weight of a polystyrene sample having $\alpha = 0.70$ and $K = 1.5 \times 10^{-4}$ dL g^{-1} if the intrinsic viscosity is 0.06 dL g^{-1}.

Solution

From Mark–Howink equation $[\eta] = KM^{\alpha}$

$$K = 1.5 \times 10^{-4} \text{ dL g}^{-1}$$

$$\alpha = 0.7$$

Therefore $M = 0.5 \times 10^{4}$

Problem 11.21

If the values of K and α in are 1×10^{-2} cm^3 g^{-1} and 0.6 respectively, deduce using the Mark–Howink equation, the average molecular weight of a polymer whose intrinsic viscosity is 130 cm^3 g^{-1}.

Solution

According to Mark–Howink equation; $[\eta] = KM^{\alpha}$

Hence $130 = 1 \times 10^{-2} \times M^{0.6}$

$$M = 71.9 \times 10^{5}$$

Problem 11.22

The limiting viscosity number of a polymer is 0.3 dL g^{-1}. Calculate the molecular weight of the polymer if $\alpha = 0.56$ and $K = 5 \times 10^{-5}$ dL g^{-1}.

Solution

According to Mark–Howink equation; $[\eta] = KM^{\alpha}$

Hence $0.3 \text{ dL g}^{-1} = 5 \times 10^{-5} \times M^{0.56}$

$$M = 5.6 \times 10^{6}$$

Problem 11.23

Two monodisperse polystyrenes (a and b) are mixed in equal quantities by weight. Their molecular weights are 38,000 and 2,89,000 respectively. What is the intrinsic viscosity of the blend in benzene at 25°C. The Mark–Howink constants for polystyrene/benzene at 25°C are $K = 7.18 \times 10^{-5}$ dL g^{-1} and $\alpha = 0.75$.

Solution

$$w_A = w_B = 0.5$$

$$\bar{M}_v = (\Sigma w_i M_i^{\alpha})^{1/\alpha}$$

$$= [0.5(38{,}000)^{0.75} + 0.5(289{,}000)^{0.75}]^{1/0.75}$$

$$= 149{,}198$$

$$K = 7.18 \times 10^{-5} \text{ dL g}^{-1}$$

$$\alpha = 0.75$$

$$\eta = 7.18 \times 10^{-5}\ (149198)^{0.75}$$

$$\eta = 0.55 \text{ dL g}^{-1}$$

Molecular Weight of Polymers Using Light Scattering

Problem 11.24

A sample of poly(methyl methacrylate) was subjected to light scattering studies and the Rayleigh ratio $P(90)$ was measured at different concentrations with the scattering constant as K. A plot of $Kc/P(90)$ vs concentration (c) yielded an intercept of 2×10^{-6} mol g^{-1} when $P(90)$ was 0.67. Estimate the molecular weight of the polymer.

Solution

$$1/[M_w \times P(90)] = 2 \times 10^{-6} \text{ mol g}^{-1}$$

$$M_w = 7.46 \times 10^{6} \text{ g mol}^{-1}$$

Molecular Weight and Diffusion Coefficient

Problem 11.25

The diffusion coefficient of polystyrene at 160°C is related to the weight average molecular weight by the equation:

$$D = 7.9 \times 10^{-3} (M_w)^{-2} \text{ cm}^2 \text{ s}^{-1}$$

If the welding of polystyrene (with $M_w = 10^5$) is carried out at 160°C, estimate the diffusion flux of polystyrene across the interface over a distance of 110 Å. (Density of polystyrene = 1.023 g cm^{-3})

Solution

Diffusion flux $J = -D \dfrac{dc}{dx}$ Fick's first law of diffusion

Upon substituting the molecular weight,

$$D = (7.9 \times 10^{-3})\ (10^5)^{-2} \text{ cm}^2 \text{ s}^{-1}$$

$$= 7.9 \times 10^{-13} \text{ cm}^2 \text{ s}^{-1}$$

$$\text{Concentration of the polymer} = \frac{1.023 \text{ g cm}^{-3}}{10^5 \text{ g mol}^{-1}}$$

$$= 1.023 \times 10^{-5} \text{ mol cm}^{-3}$$

$$J = \frac{(-7.9\times10^{-13}\ \text{cm}^2\ \text{s}^{-1})(-10^{-5}\ \text{mol m}^{-3})}{(110\times10^{-8}\ \text{cm})}$$

$$= 7.35\times10^{-12}\ \text{mol cm}^{-2}\ \text{s}^{-1}$$

$$= 4.427\times10^{12}\ \text{molecules cm}^{-2}\ \text{s}^{-1}$$

It is assumed that dc/dx is nearly equal to polymer concentration/distance of diffusion.

Problem 11.26

Derive the relation between the diffusion coefficient of a polymer and its molecular weight from the Stokes–Einstein equation.

Solution

Stokes–Einstein equation: $D = \frac{kT}{6\pi r\eta}$

Molecular weight, $M = \frac{4}{3}\pi r^3\rho$

where r = hydrodynamic radius and ρ = density.

Therefore $r = \left(\frac{3M}{4\pi\rho}\right)^{1/3}$

Hence $D \propto M^{-1/3}$ if the Stokes–Einstein formalism is valid.

Problem 11.27

The molar mass of hemoglobin is 72 kg mol^{-1} while its density is 1.215 g cm^{-3}. Estimate the following quantities when it is dissolved in water: (a) diameter (b) diffusion coefficient and (c) mean square displacement in 40 s. The viscosity is 1.002×10^{-3} kg m^{-1} s^{-1}.

Solution

(a) $\rho = \frac{m}{V}$

$$1.215\ \text{g cm}^{-3} = \frac{72\ \text{kg mol}^{-1}}{V}$$

$V = 9.839\times10^{-20}$ cm^3

Therefore, $\frac{4}{3}\pi r^3 = 9.839\times10^{-20}$ cm^3

Radius = 2.86×10^{-7} cm

Diameter of the molecule = 5.73×10^{-7} cm

(b) $D = \frac{kT}{f}$

$f = 6\pi r\eta$

$\eta = 1.002\times10^{-3}$ kg m^{-1} s^{-1}

$T = 293$ K

Therefore $D \approx 7.49\times10^{-11}$ m^2 s^{-1}

(c) Assuming three dimensional random walk models,

$\langle r^2\rangle \approx 6Dt \approx 1.79\times10^{-8}$ cm^2

Problem 11.28

The diffusion coefficient of a protein in water at 300 K is 4.9×10^{-7} cm^2 s^{-1}. If the viscosity of the medium is 0.03 poise, what will be the approximate hydrodynamic radius of the protein?

Solution

$$D = \frac{kT}{f}$$

$$f = 6\pi r\eta$$

$$f = 6\times3.14\times r\times0.03$$

$$4.9\times10^{-7}\ \text{cm}^2\ \text{s}^{-1} = \frac{1.38\times10^{-23}\ \text{kg m}^2\ \text{s}^{-2}\ \text{K}^{-1}\times300\ \text{K}}{6\times3.14\times r\times0.03\ \text{kg m}^{-1}\ \text{s}^{-1}}$$

$$r = 14.9\ \text{Å}$$

Molecular Weight and Osmotic Pressure

Problem 11.29

A sample of insulin containing 0.4 g was dissolved in 25.0 mL of water and the osmotic pressure of the solution at 20°C was 27.2 torr. Estimate the approximate molecular weight of insulin.

Solution

$$M_w = \frac{CRT}{\Pi}$$

1 torr = 1 L atm mol^{-1} K^{-1}

$$M_w = \frac{0.0821\ \text{L atm mol}^{-1}\ \text{K}^{-1}\times293\ \text{K}\times0.4\ \text{g}}{0.035\ \text{atm}\times25\times10^{-3}\ \text{L}}$$

$$M_w \approx 10{,}997\ \text{kg}$$

Problem 11.30

Estimate the osmotic pressure (in atm) of a polymer solution (concentration = 0.02 g mL^{-1}) at 293 K, if the molecular weight is 1×10^5 g mol^{-1}.

Solution

Concentration of polymer = 20 g in 1 litre

Therefore using van't Hoff's equation

$$\Pi = \frac{R\times\text{L atm mol}^{-1}\text{K}^{-1}T\times\text{K}\times C\times\text{g L}^{-1}}{M\times\text{g mol}^{-1}}$$

$$\Pi = \frac{(0.0821)(293)(20)}{1\times10^5}$$

$$= 4.81\times10^{-3}\ \text{atm}$$

Thermodynamic Parameters for Polymerisation Reactions

Problem 11.31

In the formation of polyisobutylene, the enthalpy change of polymerisation at 1 atm and 25°C is 118.26 kJ mol^{-1} and the entropy change is –365 J mol^{-1} K^{-1}, calculate the temperature (ceiling temperature) above which butyl rubber will depolymerise.

Solution

At constant temperature, $\Delta G = \Delta H - T\Delta S$

Since the rubber is depolymerising, $\Delta H - T\Delta S = 0$

$$(118.26\ \text{kJ}) - T(-365\ \text{J mol}^{-1}\ \text{K}^{-1}) = 0$$

$$T = 324\ \text{K}$$

At temperatures > 324 K, butyl rubber will depolymerise.

Problem 11.32

For an ideal rubber that contains 2.5×10^{-6} mol of cross-linked polymer chains, find the work done, heat and internal energy for stretching it to four times its original length at 300 K.

Solution

$$w = \frac{Nk_BT}{2}(\alpha^2 + \frac{2}{\alpha} - 3)$$

Therefore

$$w = (2.5\times10^{-6}\text{ mol})\times(6.023\times10^{23}\text{ mol}^{-1})\times \frac{(1.38\times10^{-23}\text{ JK}^{-1})\times 300\text{ K}}{2}\left(4^2 + \frac{2}{4} - 3\right)$$

$w = 0.0421$ J

$q = \Delta E - w = 0 - 0.0421$

$= -0.0421$ J (since $\Delta E = 0$ for an isothermal process)

Problem 11.33

Calculate ΔS for 100% elongation of a polymer sample if the number density of the chains is 4.3×10^{-5} mol cm^{-3} and the degree of elongation is 3.

Solution

If the original length is denoted as L_0, the new length L after elongation is given by

$$L = L_0 + \Delta L$$

If $\alpha = \frac{L}{L_0}$; $\gamma_t = \frac{\Delta L}{L_0}$

$$\alpha = 1 + \gamma_t$$

We can show that for a polymer of ν sub chains, the entropy change is

$$\Delta S = \frac{-kN\nu}{2}\left(\alpha^2 + \frac{2}{\alpha} - 3\right)$$

$= -\frac{1}{2}$ (8.314 J K^{-1} mol^{-1}) (4.3×10^{-5} mol cm^{-3}) [3^2 + (2/3) – 3]

$= -11.9\times10^{-4}$ J K^{-1} cm^{-3}

Since α is > 1, entropy change during elongation is always negative.

Kinetics of Polymerisation

Problem 11.34

Styrene monomer containing 0.04% (by wt.) was polymerised using benzoyl peroxide as the initiator till the initiator was entirely consumed, while 44% of the monomer remained unreacted. Calculate the average degree of polymerisation, assuming the absence of any side reactions.

Solution

In 100 grams, the % of the initiator = 0.04

In 1000 g, the weight of the initiator = 0.4 g

Molecular weight of benzoyl peroxide = $(C_6H_5CO)\,O_2$

= 242 g mol^{-1}

Molecular weight of styrene = (C_8H_8) = 104 g mol^{-1}

Number of moles of the initiator $= \frac{0.40\text{ g}}{242\text{ g mol}^{-1}} = 1.65\times10^{-3}$

Since the termination occurs by coupling, the number of moles of the polymer = 1.65×10^{-3}

Amount of styrene formed $= \frac{(100-44)\,(1000)}{100} = 560$ g

Number of moles of styrene $= \frac{560}{104} = 5.38$ mol

Degree of polymerisation (DP) $= \frac{5.38\text{ mol}}{1.65\times10^{-3}\text{ mol}} = 3255$

Problem 11.35

Poly(hexamethyleneadipamide) (Nylon-6,6) was synthesised by the condensation of hexamethylenediamine and adipic acid in a 1:1 mole ratio. Calculate the acid equivalent of the polymer if the degree of polymerisation is 540.

Solution

(a) Degree of polymerisation = 540

Molecular weight of hexamethyleneadipamide = 226 g

Total molecular weight of the polymer and H_2O = (540×226) + 18 = 122058 g mol^{-1}

Therefore, weight of the polymer containing one acid equivalent = 122058 g mol^{-1}

Problem 11.36

The weight average molecular weight of polystyrene is 3500 amu. What is the degree of polymerisation?

Solution

Extent of reaction = p

$$M_n = \frac{M_1}{1-p}$$

$$3500 = \frac{104.15}{1-p}$$

Hence $p = 0.9703$

Degree of polymerisation $= \frac{1}{1-p} = 33.67$

Problem 11.37

The activation energies of the initiation, propagation and termination of the radical polymerisation of acrylamide are (in kJ mol^{-1}) 0.01, 24.25 and 45.68 respectively. Estimate the overall activation energy of the reaction.

Solution

The total activation energy $= E_i/2 + E_p - E_t/2$

$= 0.01/2 + 24.25 - 45.68/2$

$= 1.415$ kJ mol^{-1}

Problem 11.38

The initial concentration of a monomer undergoing stepwise polymerisation is 0.1 mol L^{-1} and after 60 minutes, the concentration became 0.001 mol L^{-1}. Estimate the degree of polymerisation.

Solution

For stepwise polymerisation,

$$\text{Concentration }(t) = \text{Concentration }(t = 0)\times(1 - p)$$

Hence $0.1 = 0.001/(1 - p)$

$$p = 0.99$$

Chain Length and Polymerisation

Problem 11.39

Polyethylene $(–CH_2–CH_2–)_n$ has a molecular weight of 83980 g mol^{-1} and assuming that each bond is freely joined and that the length of a C–C bond is 0.153 nm, calculate the approximate chain length.

Solution

Molecular weight of polyethylene = 83980 g mol^{-1}

$$= (12 + 3\times1)\times2 + (12 + 2\times1)\times n = 30 + 14n$$

where n denotes the number of atoms.

Hence $n = 5996$

Number of C–C bonds $= \frac{n}{2} = 2998$

Number of repeating units ≈ 2998

Chain length = number of repeating units × bond length

Hence chain length ≈ 459 nm

Problem 11.40

Estimate the number of $H_2C{=}CH_2$ monomeric units in one molecule of polyethylene with a molar mass of 35,000 g. How many carbon atoms are present in this molecule?

Solution

Molar mass of ethylene = 28.052 g

Molar mass of polyethylene = 35000 g

Number of monomeric units = 1248

The total number of carbon atoms = 1248×2 = 2496

Melt Viscosity and Polymerisation

Problem 11.41

A vinyl polymer of chain length 250 has a melt viscosity of 150 poise. What will be the viscosity when the chain length becomes 850?

Solution

$$\eta = Kz^1 \text{ for vinyl polymers}$$

$$\frac{\eta_1}{\eta_2} = \left(\frac{250}{2\times850}\right)^1$$

$$\frac{1}{\eta_2} = \frac{0.1471}{150}$$

$$\eta_2 = 1020\,\text{poise}$$

Problem 11.42

A polymer with a (weight) average degree of polymerisation of 1500 and six atoms in the repeating unit has a melt viscosity of 1400 poise at 180°C. What will be the viscosity at the same temperature if its molecular weight is tripled?

Solution

$$\text{Molecular weight } (M_w) = \text{DP}\times n = 1500\times6 = 9000$$

where n denotes the number of atoms in the repeating unit. Melt viscosity measures the rate of extrusion of thermoplastics through an orifice.

The melt viscosities are $\eta_1 = K_H Z_1^{3.4}$; $\eta_2 = K_H Z_2^{3.4}$

where K_H is a constant for polymers of a given degree of polymerisation and Z is defined as the weight average number of chain atoms in the polymer molecules.

$$\text{Therefore } \frac{1400}{\eta_2} = \left(\frac{1}{3}\right)^{3.4}$$

$$\eta_2 \approx 58685 \text{ poise}$$

End-Group Analysis and Polymerisation

Problem 11.43

A polymer with one carboxylic acid functional group was prepared using chain polymerisation and 1 g of polymer was dissolved and titrated against methanolic KOH of 0.02 M concentration. The titration required 11 mL of KOH. Estimate the number average molecular weight of the sample.

Solution

Number of moles of KOH added = 11×10^{-3} L × 0.02 mol L^{-1}

$$= 22\times10^{-5}$$

Number of moles of –COOH group = 22×10^{-5}

Therefore number of moles of polymer = 22×10^{-5}

Since one gram of polymer contains 22×10^{-5} mol

$$\text{The molecular weight} = \frac{1}{22\times10^{-5}} = 4545.45 \text{ g mol}^{-1}$$

Glass Transition Temperature and Phase Transitions

Problem 11.44

How much of a low molecular weight plasticiser with $T_g = -70$°C should be added to a film of nylon-6,6, in order to decrease T_g from 60°C to 15°C? Assume the arithmetic mean to be valid.

Solution

The two well-known equations for estimating the new glass transition temperature in terms of the weight fractions are as follows:

$$T_g = W_1T_{g1} + W_2T_{g2}$$

$$15 = 60w_1 + (1 - w_1)(-70)$$

$$85 = 130w_1$$

$$w_1 = 0.62$$

Hence $w_2 = 1 - w_1 = 0.38$

where w_1 and w_2 are the weight fractions of nylon-6,6 and plasticiser respectively.

Therefore $W_2 = 0.38\%$

Problem 11.45

The glass transition temperature of an amorphous polymer is +12.5 °C. At 27 °C, it has a melt viscosity of 4×10^8 poise. Estimate its viscosity at (a) 12.5° C and (b) 50.5°C.

Solution

From the WLF equation,

(a) $\log_{10}[\eta_T/\eta_{T_g}] = -17.4(T - T_g)/[51.6 + (T - T_g)]$

$\log_{10}[4\times10^8/\eta_{T_g}] = -[17.4\times14.5]/[51.6 + 14.5]$

$\eta_{T_g} = 2.59\times10^{12}$ poise

(b) $\log_{10}[\eta_{323.5\text{ K}}/2.59\times10^{12}] = -[17.4\times38]/[51.6 + 38]$

$\eta_{323.5\text{ K}} = 1.10\times10^5$ poise

Problem 11.46

Describe the factors affecting the glass transition temperature.

Solution

(i) Sensitivity of the measurement
(ii) Structural and compositional factors
(iii) Number average molecular weight
(iv) Cross-linking
(v) Morphological effects
(vi) The effects of orientation
(vii) Conformational factors

Problem 11.47

Consider a copolymer consisting of two miscible polymers with weight fractions w_1 and w_2. Write the formulae for the glass transition temperature using (i) arithmetic (ii) geometric and (iii) harmonic mean.

Solution

(i) Using arithmetic mean

$$T_g = w_1 T_{g_1} + w_2 T_{g_2}$$

(ii) Using geometric mean

$$T_g = T_{g_1}^{w_1} \times T_{g_2}^{w_2}$$

(iii) Using harmonic mean

$$\frac{1}{T_g} = \frac{w_1}{T_{g_1}} + \frac{w_2}{T_{g_2}}$$

Note that in each of the above, if one of the weight fractions is unity, the corresponding T_g is obtained.

Problem 11.48

Depict the glass transition temperature using the temperature dependence of (i) volume, (ii) coefficient of thermal expansion, (iii) enthalpy and (iv) heat capacity at constant pressure.

Solution

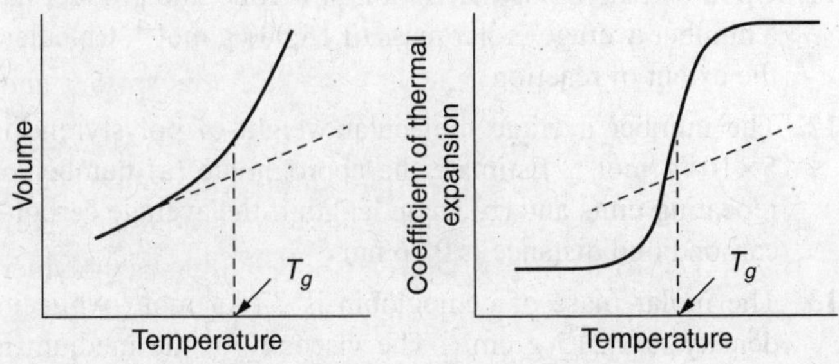

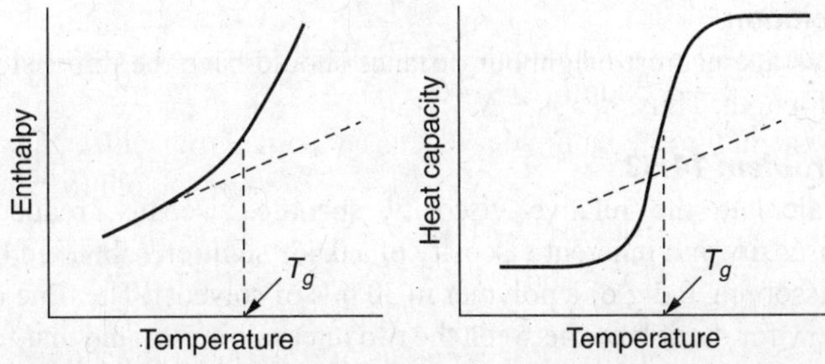

Problem 11.49

Schematically draw the phase diagram for a polymer solution and describe the significance of the upper critical solution temperature.

Solution

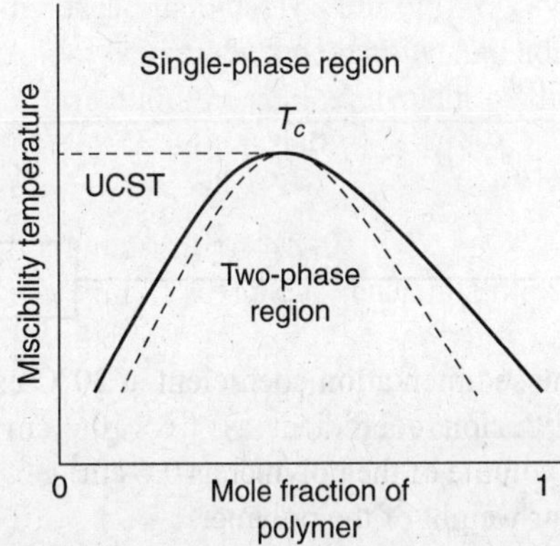

The UCST denotes the temperature above which a single phase exists.

Structures of Random and Block Copolymers

Problem 11.50

Provide a schematic representation of random and block *copolymers.*

Solution

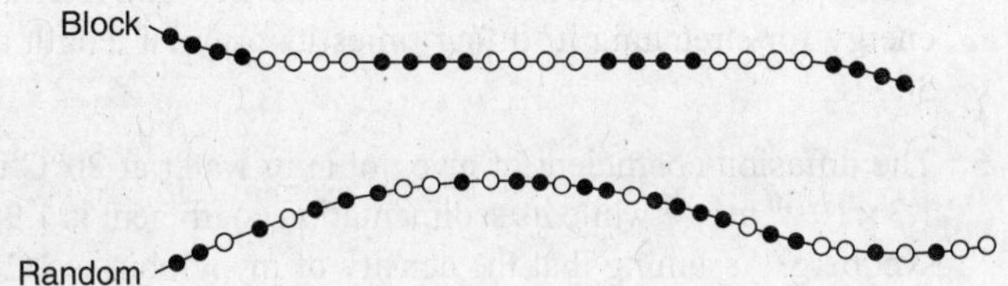

Problem 11.51

Indicate the structural features required for thermosetting polymers and elastomers.

Solution

Cross-linked polymers are designated as thermosets because they cannot be re-processed and altered into different shapes upon heating. Linear and branched polymers are known as thermoplastics since these can often be re-processed upon heating.

Miscellaneous Problems

Problem 11.52

A polymer with the number density of n is dissolved in a solvent. Assuming point molecules, provide an approximate relation for the average nearest neighbour distance if the solution is very dilute.

Solution

Average nearest neighbour distance should have the dimension of length. Thus, <r> is ~ $N^{-1/3}$.

Problem 11.53

Calculate the relative viscosity, specific viscosity, reduced viscosity and inherent viscosity of a 0.4% solution (obtained by dissolving 0.2 g of a polymer in 50 mL of solvent). The time of flow for the solvent between the two intervals is 50 s and that for the solution is 70 s.

Solution

$$\frac{t}{t_0} = \frac{\eta}{\eta_0}$$

$$\frac{\eta}{\eta_0} = \eta_{relative} = 1.4$$

$$\eta_{sp} = \frac{\eta}{\eta_0} - 1 = \frac{t}{t_0} - 1 = \frac{70}{50} - 1 = 1.4 - 1 = 0.4$$

$$\frac{\eta_{sp}}{c} = \eta_{red} = \frac{0.4}{0.4} = 1\,\text{dL g}^{-1}$$

$$\ln\frac{\eta_{rel}}{c} = \ln\frac{1.4}{0.4} = 1.25$$

EXERCISES

Practice Problems

1. Using the sedimentation coefficient at 20°C as 80.6×10^{-13} s and diffusion coefficient as 1.6×10^{-5} cm² s^{-1} with a specific volume of the polymer as 0.6 cm³ g^{-1}, calculate the molecular weight of the polymer.

2. Estimate the degree of polymerisation if the number average molecular weight of polypropylene is 10^5.

3. The number average molecular weight of a poly(styrene–butadiene) alternating copolymer is 1,400,000 g mol^{-1}, determine the average number of styrene and butadiene repeat units per molecule.

4. For an ideal rubber that contains 2.36×10^{-5} mol of cross-linked polymer chains, find the work done, heat and internal energy for stretching it to four times its original length at 300 K.

5. The diffusion coefficient of myoglobin in water at 20°C is 1.3×10^{-10} m² s^{-1} while its sedimentation coefficient is 1.98 svedberg. Assuming that the density of myoglobin is 1.21 g cm^{-3}, calculate its molar mass in kg.

6. Estimate the osmotic pressure (in atm) of a polymer solution (concentration = 0.2 g mL^{-1}) at 293 K, if the molecular weight is 1.5×10^5 g mol^{-1}.

7. The glass transition temperatures of polyvinylchloride and polyvinylacetate are respectively 87°C and 32°C. Calculate the glass transition temperature of a copolymer of 90 weight% of vinylchloride and 10 weight% of vinylacetate using the weighted arithmetic mean formula.

8. The glass transition temperature of a polymer is 100°C and the viscosity at this temperature is 1.0×10^{13} poise. Estimate the viscosity at 150°C.

9. Fill up the blanks:
   (a) Below the glass transition temperature elastomers become ________.
   (b) The number average molecular weight of polymers can be estimated using ________ measurements.
   (c) The condensation polymerisation of hexamethylene-diamine and adipic acid in a 1:1 mole ratio yields ________.
   (d) In the determination of the molecular weight of macromolecules using osmotic pressure measurements, the linear relation between $\frac{\pi}{RTC}$ and $1/C$ is studied. The slope and intercept indicate ________ and ________ respectively. (C indicates the concentration of the polymer.)
   (e) The diffusion coefficient (D) of polymers is approximately related to its molecular weight by the relation D = ________ if the Stokes–Einstein equation is valid.
   (f) For a 'good' solvent, the dimensionless Flory interaction parameter χ is always less than the magnitude of the ________.

10. The dynamic viscosity of a polymer solution is 0.01 g cm^{-1} s^{-1} and the intrinsic viscosity is 0.12 dL g^{-1}. The Mark–Houwink constants are 0.65 and 6×10^{-5} dL g^{-1}. Calculate the molecular weight of the sample. What is the kinematic viscosity of the solution if the density is 0.2 g cm^{-3}?

11. Upon oxidative polymerisation of aniline, the product has a number average molar mass of 15,000 g mol^{-1}. Calculate the extent of reaction.

12. The number average molecular weight of polystyrene is 5×10^4 g mol^{-1}. Estimate the approximate (a) number of repeating units and (b) chain length if the average carbon–carbon bond distance is 0.16 nm.

13. The molar mass of hemoglobin is 72 kg mol^{-1} while its density is 1.215 g cm^{-3}. The viscosity of the medium is 1.002×10^{-3} kg m^{-1} s^{-1}. Estimate the following quantities at

293 K when hemoglobin is dissolved in water, (a) diameter of the molecule, (b) diffusion coefficient and (c) mean square displacement in 40 seconds.

14. Fill up the blanks with appropriate phrases or equations
    (a) Upon stretching of a rubber band adiabatically, the final temperature is ________ than the initial value.
    (b) In the case of *theta* solvents, the second virial coefficient is ________.
    (c) Plasticisers are often organic liquids having ________ molecular weights and are employed to ________ the glass transition temperatures.
    (d) The Gibbs free energy of mixing of polymer solutions using the Flory interaction parameter χ as approximation can be written in terms of the appropriate volume fractions (ϕ_1 and ϕ_2) and polymer segments r_1 and r_2 as $\Delta G_{\text{mixing}}/RT =$ ________.
    (e) Thermoplastic polymers ________ upon heating and ________ when cooled.
    (f) For a polymer sample, the number average molecular weight (M_n) equals the viscosity average molecular weight (M_v). The weight average molecular weight (M_w) of the sample equals ________.
    (g) The glass transition temperature of a cross-linked poly(styrene–butadiene) *random* copolymer is –52°C. At temperatures higher than the glass transition temperature, this copolymer will function as an ideal ________.
    (h) Let the entropy of an ideal rubber of length L be denoted as S and its tensile force as f. Using the Maxwell relations, the quantity $(\partial S/\partial L)_T$ can be written as $(\partial S/\partial L)_T =$ ________.
    (i) The tensile strength of polymers depends upon the ________ average molecular weights.
    (j) The addition of a low molecular weight compound ________ the free volume of the polymer and ________ the glass transition temperature.

15. Fill up the blanks with appropriate phrases or equations.
    (a) The prerequisite for polymers to exhibit enhanced conductivity upon doping, is the presence of ________ double bonds.
    (b) The elasticity of rubber arises from ________ of molecules upon applying a force.
    (c) The addition of sulphur to rubber, followed by heating leads to ________ of rubber and the process is known as ________.
    (d) The glass transition temperature denotes the temperature at which the viscosity of polymers is approximately ________ poise.
    (e) The polymer electrolyte membrane used in methanol fuel cells is a ________ exchange membrane.
    (f) One of the precursors for the production of Nylon is adiponitrile, which is synthesised using electrochemical techniques. The electrochemical reaction involved in the synthesis of adiponitrile is ________.
    (g) Proteins consist of precise sequences of ________ linked together by covalent ________ bonds.

Multiple Choice Questions

1. Elastomers when cooled below the glass transition temperature become
   (a) soft (b) hard and brittle
   (c) thermosetting (d) unstable

2. Which of the following is the most suitable method to estimate the number average molecular weight of polymers?
   (a) Differential scanning calorimetry
   (b) Gel permeation chromatography
   (c) Viscosity data
   (d) Osmotic pressure measurements

3. The oxidative polymerisation of aniline in aqueous solutions involves
   (a) *n*-doping
   (b) *p*-doping
   (c) both *n* and *p* doping
   (d) semiconductor to insulator transition

4. Which of the following is true for the structure of thermoplastic polymers?
   (a) Linear
   (b) Cross-linked
   (c) Random
   (d) Helical

5. Among the following, identify the thermoplastic.
   (a) Natural rubber (b) Polystyrene
   (c) Polyesters (d) Polyurethane

6. Identify the polymer formed using addition polymerisation.
   (a) Nylon (b) Polycarbonate
   (c) Polyisoprene (d) polypeptide

7. Which is the most suitable method to estimate the glass transition temperature of polymers?
   (a) Nuclear magnetic resonance spectroscopy
   (b) Differential scanning calorimetry
   (c) Osmometry
   (d) Viscometry

8. At the glass transition temperature, which parameter exhibits a sharp discontinuity?
   (a) Logarithm of viscosity
   (b) Heat capacity at constant pressure
   (c) Entropy
   (d) Activation energy

9. Upon adding plasticisers, polymers become
   (a) more crystalline (b) less crystalline
   (c) amorphous (d) elastomers

10. The dimensionless Flory interaction parameter χ in the case of 'good' solvents is
(a) zero **(b)** 1/2
(c) greater than unity **(d)** $3RT/2$

11. In hydrogen–oxygen fuel cells, the function of the solid polymer electrolyte membranes such as Nafion is to
(a) eliminate the liquid junction potential
(b) provide electrical contact
(c) allow cations to migrate
(d) decrease the ionic conductivity

12. Upon complete dissolution of polymers in suitable solvents, the second virial coefficient at θ-temperature becomes
(a) zero **(b)** positive
(c) negative **(d)** π/RT

13. Assuming an one-dimensional random walk model for a polymer of N segments (each of length a), the mean square displacement $< r^2 >$ is proportional to
(a) Na^2 **(b)** N^2a^2
(c) a^2 **(d)** $(N)^{1/3}$

14. In order to determine the molecular weight of a polymer, the osmotic pressure π was measured at different concentrations (Xg per litre). A plot of π/RTX vs X yields a straight line with a slope m and an intercept c. The values of m and c can be employed to deduce respectively
(a) Weight average molecular weight and second virial coefficient
(b) Number average molecular weight and osmotic pressure
(c) Second virial coefficient and number average molecular weight
(d) Sedimentation coefficient and Flory interaction energy

15. Which of the following is true regarding the glass transition temperature (T_g) of polymers?
(a) Above the T_g, the segmental motion of the polymer chains becomes negligible.
(b) Below the T_g, polymers function as plastics.
(c) Below the T_g, polymers are useful as elastomers.
(d) The addition of plasticisers increases the T_g.

Answer Keys

Multiple Choice Questions

1. (b) **2.** (d) **3.** (b) **4.** (a) **5.** (b) **6.** (a) **7.** (b) **8.** (b) **9.** (b) **10.** (b)
11. (c) **12.** (a) **13.** (a) **14.** (c) **15.** (b)

Practice Problems

1. 306.95 g mol^{-1}
2. 2376
3. 8848
4. –0.397 J
5. 23,133 kg
6. 3.21×10^{-3} atm
7. 81.5°C
8. 2.6×10^4 poise
9. (a) hard and brittle (b) osmotic (c) nylon-6,6 (d) $1/M_n$ and B_2/RT (e) $M^{-1/3}$ (f) 0.5
10. (a) 1.19×10^5 (b) 0.05 cm^2 s^{-1}
11. Extent of reaction = 0.9938
12. (a) 480 (b) 740 nm
13. (a) 55.5 Å (b) 7.74×10^{-11} m^2 s^{-1} (c) 1.86×10^{-8} m^2
14. (a) higher (b) zero (c) low and decrease (d) $[(\phi_1)\ln(\phi_1)/r_1 + (\phi_2)\ln(\phi_2)/r_2 + \chi\,\phi_1\,\phi_2]$ (e) harder and softer (f) M_n (g) elastomer (h) $-(\partial f/\partial T)_L$ (i) number (j) increases and decreases
15. (a) conjugated (b) uncoiling (c) hardening and vulcanisation (d) 10^{12} (e) proton (f) dimerisation of acrylonitrile (g) aminoacids and peptide (or amide)

CHAPTER

12 QUANTUM CHEMISTRY

CHAPTER OUTLINE

Mathematical Preliminaries

Important Concepts and Formulae
- Postulates of Quantum Mechanics
- Quantum Mechanical Operators for Physical Quantities

Time-independent Schrödinger Equation
- Systems for which the Schrödinger Equation can be Solved Exactly

Essential Equations
- Particle in a Box
- Harmonic Oscillator
- Rigid Rotor
- Hydrogen Atom

12.1 Mathematical Preliminaries

Commutator algebra: The commutator of two *operators* Â and B̂ is represented as [Â, B̂]. If [Â, B̂] = [B̂, Â], the two operators commute. The significance is that the quantities corresponding to these operators can be measured simultaneously and precisely. For operators Â, B̂ and Ĉ, the following relations are valid.

$[\hat{A}, \hat{A}] = 0$

$[\hat{A}, \hat{B}] = -[\hat{B}, \hat{A}]$

$[\hat{A}, c] = 0$, for any scalar quantity c

$[\hat{A}, c\,\hat{B}] = c\,[\hat{A}, \hat{B}]$

$[\hat{A} + \hat{B}, \hat{C}] = [\hat{A}, \hat{C}] + [\hat{B}, \hat{C}]$

$[\hat{A}, \hat{B}\hat{C}] = [\hat{A}, \hat{B}]\,\hat{C} + \hat{B}\,[\hat{A}, \hat{C}]$

$[\hat{A}, [\hat{B}, \hat{C}]] + [\hat{B}, [\hat{C}, \hat{A}]] + [\hat{C}, [\hat{A}, \hat{B}]] = 0$

$[\hat{A}\hat{B}\hat{C}] = [\hat{B}, \hat{A}\hat{C}] + [\hat{A}\hat{B}]\hat{C}$

Hermitian operator

$$\int [f(x)]^*[\hat{X}g(x)]dx = \int [g(x)][\hat{X}f(x)]^*\,dx$$

$$= \int g(x)[\hat{X}^* f(x)^*]dx$$

Matrices

Hermitian: In this case, the transpose of the complex conjugate of the matrix is identical with the original matrix.

Example, Matrix [A] = $\begin{bmatrix} l & i \\ -i & 1 \end{bmatrix}$ → (complex conjugate of [A] $\begin{bmatrix} l & -i \\ i & 1 \end{bmatrix}$

Transpose of [B] = $\begin{bmatrix} l & i \\ -i & 1 \end{bmatrix}$ = Original Matrix [A]

Orthogonal matrix O: Here, the transpose of a matrix is identical with its inverse, namely,

$$\hat{O}\text{ transpose} = \hat{O}\text{ inverse}$$

$$2\times 2\text{ orthogonal matrix} = \begin{bmatrix} \cos\theta & \sin\theta \\ -\sin\theta & \cos\theta \end{bmatrix}$$

Standard integrals

$$\int_0^\infty e^{-a^2x^2}\,dx = \frac{1}{2a}\sqrt{\pi}$$

Table 12.1 A few standard integrals

Even integrals	Odd integrals
Integration limits: 0 to ∞	Integration limits: −∞ to + ∞
$\int_0^\infty e^{-\beta x^2}\,dx = \frac{1}{2}\sqrt{\frac{\pi}{\beta}}$	$\int_{-\infty}^\infty e^{-\beta x^2}dx = 2\int_0^\infty e^{-\beta x^2}dx$
$\int_0^\infty xe^{-\beta x^2}\,dx = \frac{1}{2\beta}$	0
$\int_0^\infty x^2e^{-\beta x^2}\,dx = \frac{1}{4\beta}\sqrt{\frac{\pi}{\beta}}$	$\int_{-\infty}^\infty x^2e^{-\beta x^2}\,dx = 2\int_0^\infty x^2e^{-\beta x^2}\,dx$
$\int_0^\infty x^3e^{-\beta x^2}\,dx = \frac{1}{2\beta^2}$	0
$\int_0^\infty x^4e^{-\beta x^2}\,dx = \frac{3}{8\beta^2}\sqrt{\frac{\pi}{\beta}}$	$\int_{-\infty}^\infty x^4e^{-\beta x^2}\,dx = 2\int_0^\infty x^4e^{-\beta x^2}\,dx$

Kronecker delta: $\delta_{ij} = 1$, if $i = j$

$= 0$ if $i \neq j$

Trigonometric identities

$\cos n\pi = 1$ (for $n = 0$ or even numbers)

$= -1$ (for $n =$ odd numbers)

$\sin n\pi = 0$ (for any integer value of n)

$\sin(\theta + 2\pi) = \sin \theta$

$\cos(\theta + 2\pi) = \cos \theta$

Definition and generating functions for Hermite polynomials

$$G(x,t) = e^{-t^2+2tx} = \sum_{n=0}^{\infty} H_n(x)\frac{t^n}{n!}$$

$$H_n(x) = (-1)^n e^{x^2} \frac{d^n}{dt^n} e^{-x^2}$$

$H_0(x) =$ zeroth order Hermite polynomial

$H_1(x) =$ first order Hermite polynomial

Time evolution operator $= e^{-i\hat{H}t/\hbar}$ where $\hat{H}$ is the Hamiltonian.

12.2 Important Concepts and Formulae

12.2.1 Essential Postulates of Quantum Mechanics

(i) Any physical system can be represented by an appropriate wave function denoted as ψ or $|\psi>$.

(ii) To every experimental observable there exists a linear Hermitian operator.

(iii) The expectation value of any observable can be obtained using the corresponding wave functions.

(iv) It is permissible to construct wave functions obeying the superposition principle if orthonormality is valid.

12.2.2 Quantum Mechanical Operators for Physical Quantities

Table 12.2 Quantum mechanical operators for various physical quantities

Property	Operator
Coordinate	r x, y, z
Momentum in x, y and z axis	$-i\hbar\frac{\partial}{\partial x}, -i\hbar\frac{\partial}{\partial y}, -i\hbar\frac{\partial}{\partial z}$

(Continued)

Property	Operator
Angular momentum	$\hat{L} = -i\hbar[r \times \nabla]$ $\widehat{L_x} = -i\hbar\left(y\frac{\partial}{\partial z} - z\frac{\partial}{\partial y}\right)$ $\widehat{L_y} = -i\hbar\left(z\frac{\partial}{\partial x} - x\frac{\partial}{\partial z}\right)$ $\widehat{L_z} = -i\hbar\left(x\frac{\partial}{\partial y} - y\frac{\partial}{\partial x}\right)$
Total energy	$\hat{H} = -\frac{\hbar^2}{2m}\nabla^2 + \hat{V}(r)$
Kinetic energy, one dimension	$\hat{K} = -\frac{\hbar}{2m}\frac{d^2}{dx^2}$
Kinetic energy in three dimensions	$\hat{K} = -\frac{\hbar}{2m}\left(\frac{\partial^2}{\partial x^2} + \frac{\partial^2}{\partial y^2} + \frac{\partial^2}{\partial z^2}\right)$
Potential energy for one dimensional harmonic oscillator	$\hat{V} = \frac{1}{2}km\omega^2x^2$ where k refers to the force constant
Potential energy between two charges q_1 and q_2 separated by a distance r in vacuum	$\hat{V} = \frac{q_1q_2}{4\pi\epsilon_0 r}$

- $\hbar$ is known as the reduced Planck constant defined as $h/2\pi$.
- $V(r)$ or $V(x)$ denotes the potential energy as a function of distance. For the one-dimensional harmonic oscillator, the radial frequency ν is related to the force constant k and mass m by the equation,
- $\nu = \frac{1}{\tau} = \frac{1}{2\pi}\sqrt{\frac{k}{m}}$
- The unit of the force constant k is kg s^{-2}. Further, the angular frequency $\omega = 2\pi\nu$.

Characteristic properties of wave functions: Any acceptable wave function in quantum mechanics should satisfy the following criteria.

Normalisation of wave functions: $\int N\psi_1\psi_2^* d\tau = 1$ where N denotes the normalisation constant to be evaluated appropriately.

Orthogonalisation: If $N\int \psi_1\psi_2 = 0$, the wave functions are orthogonal.

The probability of finding an observable between intervals a and b is given by

$$\int_a^b \psi\psi^* d\tau$$

Any expectation value can be estimated from the wave function as follows (Table 12.3):

Expectation value =

$$\frac{\int \psi(\text{operator corresponding to the property})\psi^* d\tau}{\int \psi\psi^* d\tau}$$

The angular momentum vector

$$\hat{L} = \begin{vmatrix} \hat{i} & \hat{j} & \hat{k} \\ x & y & z \\ p_x & p_y & p_z \end{vmatrix} = \hat{L}_x \hat{i} + \hat{L}_y \hat{j} + \hat{L}_z \hat{k}$$

$$\hat{L}_x = \begin{vmatrix} y & z \\ p_y & p_z \end{vmatrix}; \hat{L}_y = \begin{vmatrix} x & z \\ p_x & p_z \end{vmatrix}; \hat{L}_z = \begin{vmatrix} x & y \\ p_x & p_y \end{vmatrix}$$

Table 12.3 Expectation values of a few systems

System	<x>	<x²>	<p>	<p²>
One-dimensional harmonic oscillator	0	$\frac{\hbar}{2\sqrt{(mk)}}$ when the vibrational quantum number = 0	0	$\frac{\hbar\sqrt{mk}}{2}$
Particle-in-a-box of length *L*	*L*/2	$\left(\frac{L}{2n\pi}\right)^2\left[\frac{(2n\pi)^2}{3}-2\right]$	0	$\frac{n^2\pi^2\hbar^2}{L^2}$

Upon substituting the units for each quantity, the dimensions should match for $\langle x^2 \rangle$ and $\langle p^2 \rangle$.

12.3 Time-independent Schrödinger Equation

$$\hat{H}\psi = \hat{E}\psi$$

On account of the similarity with matrix equation $AX = \lambda X$, the above equation is commonly referred to as eigenvalue–eigenvector problem. Instead of writing $AX = \lambda X$, it is customary to write the same as $A|\psi\rangle = \lambda|\psi\rangle$ in the Dirac notation.

12.3.1 Systems for which the Schrödinger Equation can be Solved Exactly

(i) ***Particle in a box (one dimension):*** The potential energy of the free particle of mass *m* is $u(x) = 0$ within the box and $u(x) = \infty$ outside the box.

Schrödinger equation in this case is

$$-\frac{h^2}{8\pi^2 m}\frac{d^2\psi}{dx^2} = E\psi \quad (12.1)$$

(Second order ordinary linear differential equation)

Normalised wave functions for a particle in a one-dimensional box of length *a*,

$$\psi_n(x) = \sqrt{\frac{2}{a}}\sin\frac{n\pi x}{a} \quad (12.2)$$

$$n = 1, 2, 3......$$

Note that ψ_n is 0 at $x = 0$ and at $x = a$, indicating that the particle is confined within the box (Fig. 12.1).

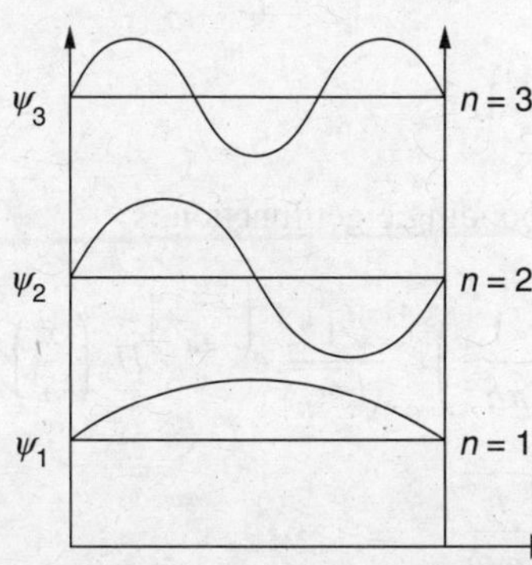

Fig. 12.1 Schematic representation of the wave function ψ_1 to ψ_3 for the free particle in a one-dimensional box

Energies in a one-dimensional box of length *a* are proportional to n^2.

$$E_n = \frac{h^2 n^2}{8ma^2} \quad (12.3)$$

The above expression is a non-trivial method of calculating the ground state ($n = 1$) and excited state ($n = 2$) energies of molecules and hence provides a method of calculating the wavelengths associated with spectroscopic transitions.

(ii) ***Particle in a box (two dimensions):*** The normalised wave function for a free particle with lengths *a* and *b* is,

$$\psi_{n_x n_y} = \frac{2}{\sqrt{ab}}\sin\frac{n_x \pi x}{a}\sin\frac{n_y \pi y}{b} \quad (12.4)$$

The corresponding energies are given by,

$$E_{n_x n_y} = \frac{\hbar^2}{2m}\left(\frac{n_x^2}{a^2}+\frac{n_y^2}{b^2}\right) \tag{12.5}$$

(iii) *Particle in a box (three dimensions):* The normalised wave function for a free particle with lengths a, b and c is,

$$\psi_{n_x n_y n_z} = \left(\frac{8}{abc}\right)^{1/2} \sin\frac{n_x \pi x}{a}\sin\frac{n_y \pi y}{b}\sin\frac{n_z \pi z}{c} \tag{12.6}$$

The corresponding energies are given by,

$$E_{n_x n_y n_z} = \frac{\hbar^2}{2m}\left(\frac{n_x^2}{a^2}+\frac{n_y^2}{b^2}+\frac{n_z^2}{c^2}\right) \tag{12.7}$$

(iv) *One-dimensional harmonic oscillator:* In an one-dimensional simple harmonic oscillator, the restoring force is directly proportional to its displacement and the potential energy is given by $V(x) = \frac{1}{2}m\omega^2 x^2$, where $\omega = 2\pi\upsilon$.

The time-independent Schrödinger equation for one-dimensional harmonic oscillators for a particle of mass m, with frequency υ is given by,

$$\hat{H}\ \psi = \left[-\frac{\hbar^2}{2m}\frac{d^2}{dx^2}+\hat{V}(x)\right]\psi(x) = E\psi(x) \tag{12.8}$$

$$E_\upsilon = \left(v+\frac{1}{2}\right)h\upsilon \tag{12.9}$$

The corresponding eigenfunction is

$$\psi_n(x) = \left(\frac{1}{\pi a^2}\right)^{1/4}\frac{1}{\sqrt{2^n n!}}e^{\left(-\frac{x^2}{2a^2}\right)}H_n\left(\frac{x}{a}\right)$$

$$a = \sqrt{\frac{\hbar}{m\omega}} \tag{12.10}$$

where $H_n(z)$ denote the Hermite polynomials, $H_0(z) = 1$, $H_1(z) = 2z$, $H_2(z) = 4z^2 - 2, \ldots$

Hence $\psi_0(x) = \left(\frac{1}{\pi a^2}\right)^{1/4} e^{\left(-\frac{x^2}{2a^2}\right)}$

$$\psi_1(x) = \left(\frac{1}{\pi a^2}\right)^{1/4}\frac{1}{\sqrt{2}}e^{\left(-\frac{x^2}{2a^2}\right)}\left(\frac{2x}{a}\right) \tag{12.11}$$

(v) *Rigid rotor:* A rigid rotor (or rotator) refers to a system in which two point masses are separated by a distance R from their centres. Denoting the two masses as m_1 and m_2, the reduced mass is given by $\mu = \frac{m_1 m_2}{m_1 + m_2}$. Hence, the moment of inertia $I = \mu R^2$. The kinetic energy operator $\hat{H}_{\text{kin}} = -\frac{\hbar^2}{2\mu}\nabla^2$ while the potential energy term $V(R)$ is zero.

The solution of the eigenvalue–eigenvector problem yields spherical harmonics denoted as $Y_l^m(\theta,\varphi)$. The quantised rotational energy is given by

$$E_l = \frac{l(l+1)\hbar^2}{2I} = Bl(l+1) \tag{12.12}$$

where the rotational constant $B = \frac{\hbar^2}{2I}$ and the unit of B is cm^{-1}.

(vi) *Hydrogen atom:* Generally, the hydrogen atom problem is called a *central force field* problem since the potential energy depends upon the distance from the origin and is spherically symmetric.

The wave function ψ can be separated into radial and angular parts as shown. This leads to all the quantum numbers in a natural manner.

$\psi = A.R(r).Y_{lm}(\theta,\phi)$

where A is a suitable normalisation constant. The wave functions for a few hydrogen-like orbitals are as follows:

$$\psi_{1s} = \frac{1}{\sqrt{\pi}}\left(\frac{Z}{a_0}\right)^{3/2} e^{-\frac{Zr}{a_0}} \tag{12.13}$$

$$\psi_{2s} = \frac{1}{4\sqrt{2\pi}}\left(\frac{Z}{a_0}\right)^{3/2}\left(2-\frac{Zr}{a_0}\right)e^{-\frac{Zr}{2a_0}} \tag{12.14}$$

Z denotes the atomic number; for the hydrogen atom, $Z=1$ and for Li^{2+}, $Z = 3$. The energy of the electrons for any Z and n is given by,

$$E_n = -\frac{2\pi^2\mu e^4}{(4\pi\varepsilon_0)^2 h^2 n^2}Z^2 \text{ (SI unit)}$$

$$E_n = \frac{-2.179\times10^{-18}}{n^2}Z^2 \text{ J atom}^{-1}$$

$$E_n = \frac{-313.6}{n^2}Z^2 \text{ kcal mol}^{-1}$$

$$E_n = \frac{-1312}{n^2}Z^2 \text{ kJ mol}^{-1}$$

$$E_n = \frac{-13.60}{n^2}Z^2 \text{ eV atom}^{-1}$$

$$r = \frac{0.529\text{Å}}{Z}$$

The value of 0.529 Å is often referred to as the Bohr radius and the above equations for energies show the explicit manner in which the principal quantum number influences each energy level. There is an inverse quadratic dependence of E_n on n.

Table 12.4 Normalised spherical harmonics $Y_{l,m}(\theta, \phi)$ $l = 0, 1, 2, \ldots$ and $m = 0, \pm1, \pm2, \ldots, \pm l$

l	m	$Y_{l,m}(\theta, \phi)$
0	0	$\frac{1}{\sqrt{4\pi}}$
1	0	$\sqrt{\frac{3}{4\pi}}\cos\theta$
1	±1	$\sqrt{\frac{3}{8\pi}}\sin\theta\begin{Bmatrix}\cos\phi\\ \sin\phi\end{Bmatrix}$

Table 12.5 Atomic orbitals for $n = 1$, 2 and 3 with the corresponding degeneracies

Principal quantum number, n	Angular quantum number, l	Orbital name	Degeneracy, g
1	0	1s	1
2	0	2s	1
2	1	2p	$3(2p_x, 2p_y, 2p_z)$
3	0	3s	1
3	1	3p	$3(3p_x, 3p_y, 3p_z)$
3	2	3d	$5(5d_{xz}, 5d_{yz}, 5d_{xy}, 5d_{z^2}, 5d_{x^2-y^2})$

Table 12.6 The dependence of p and d orbitals on the distances

Orbital	Function
p_x	x/r
p_y	y/r
p_z	z/r
d_{z^2}	$(3z^2 - r^2)/r^2$
d_{xz}	xz/r^2
d_{yz}	yz/r^2
d_{xy}	xy/r^2
$d_{x^2-y^2}$	$(x^2 - y^2)/r^2$

It is easy to decipher the corresponding orbitals by inspecting the variables in the function. For example, xz will denote the d_{xz} orbital. For clarity, the normalisation constants are omitted.

The degeneracy of the hydrogen atom for any given n can be shown to be n^2. Typically, $n = 2$ will yield 4 degenerate states—2s, $2p_x$, $2p_y$ and $2p_z$. Similarly for $n = 4$, the degenerate states are 16—4s (1), 4p (3), 4d (5) and 4f (7). The most important fact about m is that for a given l, it can take $2l + 1$ possible values and provides the degeneracy of the levels. Example, for $l = 2$, the allowed values for m are +2, +1, 0, –1, –2. The azimuthal quantum number is denoted as l and characterises the total angular momentum of the atom. When $\hat{M}^2$ operates on a wave function, the eigenvalues are $l\,(l+1)\,\hbar^2$, where l denotes the azimuthal quantum number.

The z component of the angular momentum is given in terms of m known as the magnetic quantum number.

$$M_z = m\hbar$$

$$m = -l, -(l-1), \ldots, -1, 0, 1, \ldots, (l-1), l.$$

m varies from $-l$ to $+l$ through 0.

12.4 Essential Equations

$Ax = \lambda x$ (eigenvalue–eigenvector equation)

$\left[\hat{H}\right]\psi = E\psi$ (time-independent Schrödinger equation)

$\hat{H}\psi = i\hbar\frac{\partial\psi}{\partial t}$ (time-dependent Schrödinger equation)

$\hat{x} \equiv x$ (definition of position operator)

$\hat{p}_x \equiv -i\hbar\frac{\partial}{\partial x}$ (definition of momentum operator

$\Delta x.\Delta p_x \geq \frac{\hbar}{2}$ (Heisenberg uncertainity principle)

Probability of finding any observable within the interval

x_1 and $x_2 = \int_{x_1}^{x_2}\psi^*\psi\,d\tau$ (Born definition of probability)

$N\int_{\text{all space}}\psi^*\psi\,d\tau = 1$ (normalisation requirement of wave functions)

$\int_{\text{all space}}\psi_m^*\psi_n\,d\tau = 0, \psi_m \neq \psi_n$ (orthogonality requirement of wave functions)

$\psi(x) = \sqrt{\frac{2}{a}}\sin\frac{n\pi x}{a}$ (wave function for a particle in a one-dimensional box of length a)

$E_n = \frac{n^2h^2}{8ma^2}$ (quantised energy for a particle of mass m in a one-dimensional box)

$<\hat{A}> = \int_{\text{all space}}\psi^*\hat{A}\psi\,d\tau$ (expectation value of any quantity in terms of the corresponding operator $\hat{A}$)

For example, $<x> = \int\psi x\psi^*\,d\tau$; $<\hat{P}_x> = \int_0^L\psi^*\,\hat{p}_x\psi\,dx$

$\psi = \sqrt{\frac{8}{abc}} \sin\frac{n_x \pi x}{a} . \sin\frac{n_y \pi y}{b} . \sin\frac{n_z \pi z}{c}$ (wavefunction of a particle in a three-dimensional box of lengths a, b and c)

$E_{n_x,n_y,n_z} = \frac{h^2}{8m}\left(\frac{n_x^2}{a^2} + \frac{n_y^2}{b^2} + \frac{n_z^2}{c^2}\right)$ (energy of a particle in a three-dimensional box)

Worked Out Problems

Relations Involving Commutators

Problem 12.1

Rewrite the commutator [Â, B̂ Ĉ] in terms of [Â,B̂]

Solution

$$\begin{aligned}[\hat{A}, \hat{B}\hat{C}] &= \hat{A}\hat{B}\hat{C} - \hat{B}\hat{C}\hat{A}; \text{ since } [X, Y] = XY - YX \\ &= \hat{A}\hat{B}\hat{C} - \hat{B}\hat{A}\hat{C} + \hat{B}\hat{A}\hat{C} - \hat{B}\hat{C}\hat{A} \\ &= (\hat{A}\hat{B} - \hat{B}\hat{A})\,\hat{C} + \hat{B}\,(\hat{A}\hat{C} - \hat{C}\hat{A}) \\ &= [\hat{A}, \hat{B}]\,\hat{C} + \hat{B}\,[\hat{A}, \hat{C}]\end{aligned}$$

Problem 12.2

Find the commutator for (a) $[\hat{P}_i, \hat{x}_i]$ and (b) $[\hat{P}_i, \hat{x}_j]$

Solution

$\hat{P}_i$ and $\hat{x}_i$ denote respectively the momentum operator and position operator of the ith particle. We operate this on a hypothetical wave function ψ.

(a) $[\hat{P}_i, \hat{x}_i] = ?$

$[\hat{P}_i, \hat{x}_i] = \hat{P}_i \hat{x}_i - \hat{x}_i \hat{P}_i$

$$[\hat{P}_i, \hat{x}_i]\psi = \left[-i\hbar\frac{\partial}{\partial x_i} x_i\right]\psi - x_i\left[-i\hbar\frac{\partial \psi}{\partial x_i}\right]$$

$$= -i\hbar\frac{\partial}{\partial x_i}[x_i\,\psi] + x_i\, i\hbar\frac{\partial \psi}{\partial x_i}$$

$$= -i\hbar\left(\frac{\partial x_i}{\partial x_i}\right)\psi - x_i\, i\hbar\frac{\partial \psi}{\partial x_i} + x_i\, i\hbar\frac{\partial \psi}{\partial x_i}$$

$$= -i\hbar\psi$$

(The commutator should refer to the operator alone, that is, without the function on which it operates). Hence

$[\hat{P}_i, \hat{x}_i] = -i\hbar$

(b) For an arbitrary wave function ψ

$[\hat{P}_i, \hat{x}_j]\,\psi = \hat{P}_i\, \hat{x}_j\, \psi - \hat{x}_j\, \hat{P}_i\, \psi$

$$-i\hbar\frac{\partial}{\partial x_i}[x_j \psi] - x_j\left(-i\hbar\frac{\partial \psi}{\partial x_i}\right)$$

$$= -i\hbar\, x_j\left(+\frac{\partial \psi}{\partial x_i}\right) - i\hbar\psi(0) + i\hbar x_j\left(\frac{\partial \psi}{\partial x_i}\right) = 0$$

Problem 12.3

Obtain the result when the commutator $[\hat{L}_x\, \hat{L}_y]$ operates on x.

Solution

$$\left[\hat{L}_x \hat{L}_y\right]x = \left(\hat{L}_x \hat{L}_y - \hat{L}_y \hat{L}_x\right)x$$

$$= \left(-i\hbar y\frac{\partial}{\partial z} + i\hbar z\frac{\partial}{\partial y}\right)\left(-i\hbar z\frac{\partial}{\partial x} + i\hbar x\frac{\partial}{\partial z}\right)x -$$

$$\left(-i\hbar z\frac{\partial}{\partial x} + i\hbar x\frac{\partial}{\partial z}\right)\left(-i\hbar y\frac{\partial}{\partial z} + i\hbar z\frac{\partial}{\partial y}\right)x$$

$$= \left(-i\hbar y\frac{\partial}{\partial z} + i\hbar z\frac{\partial}{\partial y}\right)(-i\hbar z) - \left(-i\hbar z\frac{\partial}{\partial x} + i\hbar x\frac{\partial}{\partial z}\right)(0)$$

$$= \left((-i\hbar)^2 y\right) = \hbar^2 y \neq 0$$

Problem 12.4

Write all the commutators for position and momentum operators.

Solution

In general,

$[\hat{x}_i, \hat{P}_j] = i\hbar\, \delta_{ij}$ where δ_{ij} denotes the Kronecker delta.

$$[\hat{x}, \hat{P}] = i\hbar$$
$$[\hat{x}_i, \hat{x}_j] = 0,$$
$$[\hat{P}_i, \hat{P}_j] = 0,$$
$$[\hat{x}_i, \hat{P}_j] = i\hbar\, \delta_{i,j},$$
$$[\hat{x}, \hat{P}_y] = [\hat{x}, \hat{P}_z] = [\hat{y}, \hat{P}_x] = [\hat{y}, \hat{P}_z] = [\hat{z}, \hat{P}_x] = [\hat{z}, \hat{P}_y] = 0,$$
$$[\hat{x}, \hat{P}_x] = [\hat{y}, \hat{P}_z] = [\hat{z}, \hat{P}_z] = i\hbar$$

Problem 12.5

Write the Heisenberg uncertainty principle in the operator notation.

Solution

Let the commuter $\hat{c}$ be defined as

$$\hat{c} = \left[\hat{A}, \hat{B}\right] = \left[\hat{A}\hat{B}\right] - \left[\hat{B}\hat{A}\right]$$

$$\Delta\hat{A}\Delta\hat{B} \geq \langle c \rangle / 2$$

Acceptable and Non-acceptable Wave Functions

Problem 12.6

Classify the following as the allowed/forbidden wave function(s) within the given interval:

(a) $\frac{1}{3x}$ with $0 \leq x \leq \infty$

(b) 5 cos (x) with $-10 \leq x \leq \infty$

(c) e^{-10x} with $-\infty \leq x \leq \infty$

(d) $x\, e^{-5x}$ in the interval $0 > x < \infty$

Solution

(a) $\frac{1}{3x}$ with $0 \le x \le \infty$

It is not an allowed wave function since it becomes infinity at $x = 0$.

(b) 5 cos (x) with $-10 \le x \le 0$

It is an allowed wave function since it is bounded and does not become infinity within the specified interval.

(c) e^{-10x} with $-\infty \le x \le \infty$

It is not an allowed wave function since it becomes infinity as x approaches $-\infty$.

(d) $x\,e^{-5x}$ in the interval $0 > x < \infty$

It is an allowed wave function since it does not become infinity within the given range.

Problem 12.7

Deduce whether the following functions are acceptable wave functions over the indicated intervals in brackets:

(a) e^{-3x}, $(0, \infty)$

(b) e^{-10x}, $(-\infty, +\infty)$

(c) $\sin^{-1} x, (-1, 1)$

(d) $(\sin x)/x$, $(0, \infty)$

Solution

(a) e^{-3x}; acceptable (in $0, \infty$)

Single valued, without discontinuity, finite and square integrable.

(b) e^{-10x} ($-\infty$ to $+\infty$); Not acceptable since it $\to \infty$ at $x \to -\infty$

(c) $\sin^{-1}x$(-1 to $+1$); Not acceptable since multiple values exist at $x = 1$,

namely, $\frac{\pi}{2}, \frac{5\pi}{2}, \frac{9\pi}{2}\ldots$

(d) $\frac{\sin x}{x}$ $(0, \infty)$; Acceptable since it is finite at both limits.

Atomic Units and Orbitals

Problem 12.8

Write the following quantities in terms of fundamental constants and any other parameter. Provide a numerical estimate of them.

(a) Bohr radius (in cm)

(b) Binding energy of hydrogen (in eV)

(c) Compton wavelengths of electrons (in Å)

Solution

(a) $a_0 = \frac{\varepsilon_0 h^2}{\pi m_e e^2}$

$$= \frac{(8.854\times10^{-12}\ \text{J}^{-1}\ \text{C}^2\ \text{m}^{-1})(6.626\times10^{-34}\ \text{J s})^2}{(3.14)(9.109\times10^{-31}\ \text{kg})(1.602\times10^{-19}\ \text{C})^2}$$

$$= 0.529\times10^{-8}\ \text{cm}$$

(b)

$$E_n = \frac{e^2}{8\pi a_0 \varepsilon_0 n^2}$$

$$= -\frac{(1.602\times10^{-19}\,\text{C})^2}{8\times3.14\times0.529\times10^{-8}\,\text{cm}\times8.854\times10^{-12}\text{J}^{-1}\text{C}^2\text{m}^{-1}\times n^2}$$

$$= -\frac{13.60\,\text{eV}}{n^2}$$

(c) λ_{Compton}

$$= \frac{h}{m_e c} = \frac{6.626\times10^{-34}\ \text{J s}}{\left(9.109\times10^{-31}\ \text{kg}\right)\times(3\times10^8\ \text{m s}^{-1})} = 0.0243\,\text{Å}$$

Problem 12.9

Write the atomic units for (a) time and (b) energy.

Solution

(a) Time $= \frac{a_0\hbar}{e^2} = 2.4\times10^{-17}\,\text{s}$

(b) Energy = Hartree $= \frac{e^2}{a_0}$

Problem 12.10

For $4d$ orbitals, calculate the number of (a) radial and (b) angular nodes.

Solution

Here $n = 4$; $l = 2$

(a) $n - l - 1 = 4 - 2 - 1 = 1$ radial nodes

(b) $l = 2$ angular nodes

Problem 12.11

Derive the mean potential energy of an electron in an element with atomic number Z, in the (a) $1s$ and (b) $2p_z$ orbital.

Solution

(a) $$\langle V \rangle = \int_0^\infty \psi_{1s}(r)\left(-\frac{Z.e.e}{4\pi\varepsilon_0 r}\right)\psi_{1s}(r)4\pi r^2 dr$$

$$= \int_0^\infty \left(-\frac{Z.e.e}{4\pi\varepsilon_0 r}\right)\frac{e^{-\frac{2r}{a0}}4\pi r^2 dr}{\pi(a_0)^3}$$

After removing the constants from the integral sign, it follows that

$$\langle V \rangle = -\frac{Z.e.e}{\pi\,\varepsilon_0 (a_0)^3}\int_0^\infty e^{-\frac{2r}{a0}}\,r dr = -\frac{Z.e.e}{4\pi\ \varepsilon_0\ (a_0)}$$

(b) For an electron in the $2p_z$ orbital, the probability density is given by

$$P(r) = Br^4 e^{-\frac{Zr}{a_0}}$$

It is possible to estimate the mean potential energy <V> by merely evaluating the expectation value of <1/r> along with the probability density given above. Thus,

$$\langle V\rangle = -\left[\frac{Ze^2}{4\pi\varepsilon_0}\right]\left\langle\frac{1}{r}\right\rangle$$

$$\left\langle\frac{1}{r}\right\rangle = \int_0^\infty \frac{1}{r}P(r)dr = \int_0^\infty \frac{1}{r}Br^4 e^{-\frac{Z_r}{a_0}}dr = B\int_0^\infty r^3 e^{-\frac{Zr}{a_0}}dr$$

$$\left\langle\frac{1}{r}\right\rangle = \frac{1}{24}\left(\frac{Z}{a_0}\right)^5 \frac{3!}{\left(\frac{Z}{a_0}\right)^4} = \frac{1}{24}\left(\frac{Z}{a_0}\right)6 = \frac{1}{4}\left(\frac{Z}{a_0}\right)$$

$$\langle V\rangle = -\frac{Ze^2}{4\pi\varepsilon_0}\left\langle\frac{1}{r}\right\rangle = -\frac{Ze^2}{4\pi\varepsilon_0}\frac{1}{4}\left(\frac{Z}{a_0}\right) = -\frac{Z^2e^2}{16\pi\varepsilon_0 a_0}$$

$$\langle V\rangle = -\frac{Z^2e^2}{16\pi\varepsilon_0 a_0}$$

Elementary Matrix Calculations

Problem 12.12

Verify whether the following is a Hermitian matrix.

$$[A] = \begin{bmatrix} 1 & +i \\ -i & 1 \end{bmatrix}$$

Solution

$$[A] = \begin{bmatrix} 1 & +i \\ -i & 1 \end{bmatrix}; \text{ Hence } [A^*] = \begin{bmatrix} 1 & -i \\ +i & 1 \end{bmatrix}$$

$$[A^*]^T = \begin{bmatrix} 1 & +i \\ -i & 1 \end{bmatrix} = [A], \text{ therefore matrix } [A] \text{ is Hermitian.}$$

Problem 12.13

Determine the eigenvalues of the following 2×2 matrix, $\begin{bmatrix} 1 & -2 \\ 1 & 4 \end{bmatrix}$

Solution

The characteristic equation for the matrix follows from the determinant.

$$\begin{bmatrix} 1-\lambda & -2 \\ 1 & 4-\lambda \end{bmatrix} = 0$$

$$\lambda^2 - 5\lambda + 6 = 0$$

Therefore, the eigenvalues λ_1 and λ_2 are 2 and 3. Note that the sum of the eigenvalues equals the trace of the given matrix.

Problem 12.14

Mention five scalar quantities associated with any square matrix.

Solution

Trace, determinant, rank, signature and eigenvalue.

Eigenfunctions and Hermitian Operators

Problem 12.15

Find the eigenvalues for the following cases:

(a) d^2/dx^2 operating on $\sin(4x)$

(b) $x\frac{d}{dx}$ operating on x^n

Solution

(a) $\{d^2/dx^2\}\ [\sin(4x)] = (d/dx)(4\cos(x)) = 4(-4)\sin(4x)$ $-16\sin(4x)$

Eigenvalue: –16

(b) $\left\{x\frac{d}{dx}\right\}$ operating on $[x^n] = x\frac{d}{dx}x^n = x\,n\,x^{n-1} = nx^n$

Eigenvalue: n

Problem 12.16

Verify whether the following operators are Hermitian.

(a) $-\frac{d}{dx}$ (b) $i\frac{d}{dx}$ (c) $ix\frac{d}{dx}$ (d) $-i\hbar\frac{\partial}{\partial x}$ (e) $\frac{d}{dx}$ and (f) $\frac{d^2}{dx^2}$

Solution

(a) An operator X is Hermitian if

$$\int [f(x)^*[Xg(x)]dx = \int [g(x)][Xf(x)]^*dx$$

$$= \int g(x)[X^* f(x)^*]dx$$

$$\int_{-\infty}^{\infty} f(x)^*\left(-\frac{d}{dx}g(x)\right)dx = I_1$$

Integrating by parts, $\int_{-\infty}^{\infty} udv = uv - \int_{-\infty}^{\infty} vdu$

$$dv = -\frac{d}{dx}g(x);\ u = f(x)^*$$

$$\int_{-\infty}^{\infty} f(x)^*\left(-\frac{d}{dx}g(x)\right)dx$$

$$= [f(x)^*(-1)g(x)]_{-\infty}^{\infty} - \int_{-\infty}^{\infty}(-1)g(x)\left[\frac{d}{dx}f(x)^*\right]dx$$

The first term is zero.

$$\int g(x)\left[\frac{d}{dx}f(x)^*\right]dx = I_2$$

The first term is zero.

Since $I_1 = I_2$, $-d/dx$ is Hermitian

(b)
$$I_1 = \int_{-\infty}^{\infty} f(x)^*\left(i\frac{d}{dx}g(x)\right)dx$$

Using the integration by parts rule,

$$[f(x)^* ig(x)]_{-\infty}^{\infty} - \int_{-\infty}^{\infty} ig(x)\left[\frac{d}{dx}f(x)^*\right]dx$$

$$I_2 = \int_{-\infty}^{\infty} g(x)\left[-i\left[\frac{d}{dx}f(x)\right]dx\right]$$

$$= \int_{-\infty}^{\infty} g(x)\left[\text{complex conjugate of}\left[i\frac{d}{dx}f(x)\right]dx\right]$$

Since $I_1 = I_2$, $i\frac{d}{dx}$ is Hermitian.

In general, a skew Hermitian operator becomes Hermitian upon multiplication by i and vice versa.

(c) $I_1 = \int_{-\infty}^{\infty} f_1(x)^*\left(ix\frac{d}{dx}\right)g(x)dx$

$$= \int_{-\infty}^{\infty} f_1(x)^* ix\frac{d}{dx}g(x)dx$$

$$\int_{-\infty}^{\infty} g(x)\left(-ix\frac{df_1(x)^*}{dx}\right)dx$$

NOTE

- $-ix\frac{df_1(x)^*}{dx}$ is the complex conjugate of $ix\frac{df_1}{dx}$

$$I_2 = \int_{-\infty}^{\infty} g(x)\left(-ix\frac{df_1(x)}{dx}\right)dx$$

Hence $ix\frac{d}{dx}$ is Hermitian.

(d) This can be identified as the linear momentum operator.

$$p_x = -i\hbar\frac{\partial}{\partial x}$$

$$\int_{-\infty}^{\infty} f(x)^*\left(-i\hbar\frac{\partial}{\partial x}g(x)\right)dx$$

$$u = -i\hbar f(x)\,;\; dv = \frac{\partial}{\partial x}g(x)$$

$$\int_{-\infty}^{\infty} udv = uv - \int_{-\infty}^{\infty} vdu$$

$$= [0] - \int_{-\infty}^{\infty} g(x)\left(-i\hbar\frac{\partial}{\partial x}f(x)\right)^* dx$$

$$= +\int_{-\infty}^{\infty} g(x)\left(i\hbar\frac{\partial}{\partial x}f(x)\right)^* dx$$

$\left(i\hbar\frac{\partial}{\partial x}\right)$ is the conjugate complex of $-i\hbar\frac{\partial}{\partial x}$

Since $\int_{-\infty}^{\infty} f(x)\hat{O}g(x)dx = \int_{-\infty}^{\infty} g(x)[\hat{O}f(x)]^*\,dx$

Hence $-i\hbar\frac{\partial}{\partial x}$ is Hermitian.

(e) $$\int_{-\infty}^{\infty} f(x)^*\frac{d}{dx}g(x)dx = I_1$$

Using integration by parts,

$$[f(x)^*g(x)]_{-\infty}^{\infty} - \int_{-\infty}^{\infty} g(x)f'(x)^*dx = I_2$$

Since $I_1 \neq I_2$, $\frac{d}{dx}$ is not Hermitian; instead it is skew Hermitian. Since integration by parts leads to change in sign, the differential operator $(f)\frac{d}{dx}$ is not Hermitian.

(f) $$\int_{-\infty}^{\infty} f(x)^*\frac{d^2}{dx^2}g(x)dx = I_1$$

$$[f(x)^*\frac{d}{dx}g(x)]_{-\infty}^{\infty} - \int_{-\infty}^{\infty}\frac{d}{dx}g(x)f'(x)^*dx$$

$$\frac{d}{dx}g(x) = dv \text{ and } f'(x)^*dx = u$$

$$I_2 = -\int_{-\infty}^{\infty}\frac{d}{dx}g(x)f'(x)^*dx$$

$$= \int_{-\infty}^{\infty} g(x)\left[\frac{d^2f}{dx^2}\right]^* dx$$

Therefore $\int_{-\infty}^{\infty} f(x)^*\frac{d^2}{dx^2}g(x)dx = \int_{-\infty}^{\infty} g(x)\left[\frac{d^2f(x)}{dx^2}\right]^* dx$

Hence, Hermitian.

Problem 12.17

Using the Dirac notation, show that the eigenvalues of a Hermitian operator are real.

Solution

Let $|a\rangle$ be an eigenstate of Hermitian operator $\hat{O}$ corresponding to eigenvalue a, then

$$\hat{O}\,|a\rangle = a|a\rangle$$

$$\Rightarrow \langle a|\hat{O}|a\rangle = \langle a|a|a\rangle = a\langle a|a\rangle$$

$$\text{and } \langle a|\widehat{O^\dagger} = a^*\langle a|$$

$$\Rightarrow \langle a|\widehat{O^\dagger}|a\rangle = a^*\langle a|a\rangle$$

$\langle a|\hat{O}|a\rangle = a^*\langle a|a\rangle$, since $\hat{O}$ is Hermitian.

Hence, $(a-a^*)<a|a\rangle = 0$

$$\Rightarrow a = a^*, \text{ since } <a|a\rangle \neq 0$$

Thus the eigenvalues are real.

Problem 12.18

Explain why the value of the magnetic quantum number cannot exceed that of the azimuthal quantum number.

Solution

From the solution of the Schrödinger equation, it is deduced that $\cos\theta = \frac{m}{l}$

For any value of θ, $\cos\theta$ ranges from –1 to +1 and hence $\frac{m}{l}$ cannot exceed unity. Thus, $m \leq l$.

Problem 12.19

Derive the normalisation constant for the wave function associated with the magnetic quantum number m.

Solution

The acceptable wave function arising from the solution of the azimuthal equation is

$$\phi_m(\phi) = Ae^{im\phi}$$

The normalisation condition implies that

$$\int_0^{2\pi} Ae^{im\phi} Ae^{-im\phi} d\phi = 1$$

Therefore, $A^2 \int_0^{2\pi} d\phi = 1$

Therefore $A = \frac{1}{\sqrt{2\pi}}$

Normalisation constant $= \frac{1}{\sqrt{2\pi}}$

The normalised wave function $= \frac{1}{\sqrt{2\pi}} e^{im\phi}$

Expectation Values and Normalisation Constants

Problem 12.20

Deduce the expectation value of the distance of the electron from the nucleus for (a) 4s orbital and (b) 3p orbital with the azimuthal quantum number $l = 1$.

Solution

(a) $<r_s>$ for any s-orbital is given by

$$<r_s> = \frac{3n^2 a_0}{2} = 24\,a_0$$

(b) $n = 3$ and $l = 1$

The expectation value for the distance $<r>$ in terms of n and l is given by $<r> = \left[\frac{a_0}{2}\right]\left[3n^2 - l(l+1)\right] = \frac{25a_0}{2}$

Problem 12.21

For g-orbitals, estimate the value of the total angular momentum.

Solution

For g-orbitals, $l = 4$; hence the angular momentum $\sqrt{l(l+1)} = \sqrt{20}\hbar$

Schrödinger Equation and Wave Functions

Problem 12.22

Derive $\hat{H}\psi = E\psi$ from the time-dependent Schrödinger equation.

Solution

The time-dependent Schrödinger equation is

$$\hat{H}\,\psi(x,t) = i\hbar\frac{\partial\psi(x,t)}{\partial t} \quad (1)$$

The time-dependent wave function can be written as

$$\psi(x,t) = \psi(x)e^{\frac{-iEt}{\hbar}}$$

Upon substituting the above in (1)

$$\hat{H}\left\{e^{\frac{-iEt}{\hbar}}\psi(x)\right\} = i\hbar\psi(x)e^{\frac{-iEt}{\hbar}}\left(-\frac{iE}{\hbar}\right) \quad (2)$$

The LHS of Eq. (2) can be written as $e^{\frac{-iEt}{\hbar}}\hat{H}\psi(x)$ while the RHS becomes $E\psi(x)e^{\frac{-iEt}{\hbar}}$. Upon cancellation of the exponential terms, it is seen that

$$\hat{H}\,\psi = E\,\psi$$

Problem 12.23

What is meant by stationary states?

Solution

If the probability density $\psi(x, t)^*\psi(x, t)$ is independent of time, the states are known as stationary states.

Problem 12.24

Draw the radial distribution functions for 1s, 2s and 3s states.

Solution

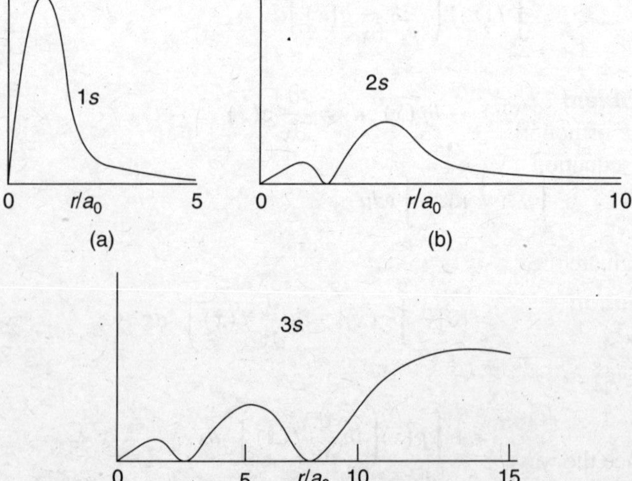

Problem 12.25

Write the angular momentum operator $\hat{L}_x$, $\hat{L}_y$ and $\hat{L}_z$ in spherical coordinates.

Solution

$$x = r\sin\theta\cos\phi$$

$$y = r\sin\theta\sin\phi$$

$$z = r\cos\theta$$

Equivalently, $r^2 = x^2 + y^2 + z^2$

$$\cos\theta = \frac{z}{\sqrt{x^2+y^2+z^2}}$$

$$\tan\phi = \frac{y}{x}$$

Therefore, $\hat{L}_x = -i\hbar\left[y\frac{\partial}{\partial z} - z\frac{\partial}{\partial y}\right]$

$$= -i\hbar\left[\begin{matrix} r\sin\theta\sin\phi\left(\frac{\partial r}{\partial z}\frac{\partial}{\partial r} + \frac{\partial\theta}{\partial y}\frac{\partial}{\partial\theta} + \frac{\partial\phi}{\partial z}\frac{\partial}{\partial\phi}\right) \\ -r\cos\theta\left(\frac{\partial r}{\partial y}\frac{\partial}{\partial r} + \frac{\partial\theta}{\partial y}\frac{\partial}{\partial\theta} + \frac{\partial\phi}{\partial y}\frac{\partial}{\partial\phi}\right)\end{matrix}\right]$$

$$= -i\hbar\left[\begin{matrix}(r\sin\theta\sin\phi\cos\theta - r\cos\theta\sin\theta\sin\phi)\frac{\partial}{\partial r} + \\ (-\sin^2\theta\sin\phi - \cos^2\theta\sin\phi)\frac{\partial}{\partial\theta} + (-\cot\theta\cos\phi)\frac{\partial}{\partial\phi}\end{matrix}\right]$$

$$= -i\hbar\left[-\sin\phi\frac{\partial}{\partial\theta} - \cot\theta\cos\phi\frac{\partial}{\partial\phi}\right]$$

$$\hat{L}_x = -i\hbar\left(-\sin\phi\frac{\partial}{\partial\theta} - \cot\theta\cos\phi\frac{\partial}{\partial\phi}\right)$$

$$\hat{L}_y = -i\hbar\left(\cos\phi\frac{\partial}{\partial\theta} - \cot\theta\sin\phi\frac{\partial}{\partial\phi}\right)$$

$$\hat{L}_z = -i\hbar\frac{\partial}{\partial\phi}$$

Problem 12.26

The orthonormal wave function of a molecule AB is given by the equation

$$|\psi_1 >= N_1\left[0.1|\psi_A\rangle + 0.8|\psi_B\rangle\right]$$

Evaluate the normalisation constant N_1.

Solution

$$\int_{-\infty}^{\infty} N_1^2\left[0.1|\psi_A > +0.8|\psi_B >\right]^2 d\tau = 1$$

Since the wavefunctions are orthonormal,

$$N_1^2\int 0.1^2\psi_A\,|\,\psi_A + 0.8^2\psi_B\,|\,\psi_B + (0.1)(0.8)\psi_A\,|\,\psi_B = 1$$

$$N_1^2\,[(0.1)^2 + (0.8)^2] = 1$$

$$(0.01 + 0.64)\,N_1^2 = 1$$

Therefore $N_1 = 1.24$

Problem 12.27

The wave function ψ is given by

$$\psi = \frac{1}{\sqrt{3}}\phi_1 + \sqrt{\frac{2}{3}}\phi_2$$

[where ϕ_1 *and* ϕ_2 are orthonormal]. Deduce the expectation value of the energy.

Solution

Any expectation value can be computed from the wave function and the appropriate operator. Energy will be obtained by operating the Hamiltonian operator $\hat{H}$ on the wave functions.

$$E = \int\psi^*\hat{H}\psi d\tau$$

$$= \int\left(\frac{1}{\sqrt{3}}\phi_1 + \frac{2}{\sqrt{3}}\phi_2\right)\hat{H}\left(\frac{1}{\sqrt{3}}\phi_1 + \frac{2}{\sqrt{3}}\phi_2\right)d\tau$$

$$\int\frac{1}{3}\left(\phi_1\hat{H}\phi_1\right)d\tau + \left(\frac{1}{\sqrt{3}}\phi_1\hat{H}\frac{2}{\sqrt{3}}\phi_2\right)d\tau + \frac{2}{3}\phi_2\hat{H}\frac{1}{\sqrt{3}}\phi_1 d\tau + \frac{4}{3}\left(\phi_2\hat{H}\phi_2\right)d\tau$$

$$= \frac{1}{3}E_1 + \frac{4}{3}E_2$$

Note that $\phi_1\hat{H}\phi_2 = 0$ and $\int\phi_1 E_1\phi_1 d\tau = E_1\int\phi_1^2 d\tau = E_1$ and so on.

Problem 12.28

For the wave function $\psi(x) = N\frac{e^{\left(\frac{ipx}{\hbar}\right)}}{\sqrt{x^2+l^2}}$; obtain the (a) normalisation constant (b) probability of finding the particle between $-l/\sqrt{3}$ and $+l/\sqrt{3}$.

Solution

(a) In order to obtain the normalisation constant, we equate the probability to unity over the entire interval. The probability is, in general given by

$$\text{Probability} = \int\psi\psi^* d\tau$$

$$\psi = N\frac{e^{\left(\frac{ipx}{\hbar}\right)}}{\sqrt{x^2+l^2}}$$

$$\psi^* = N\frac{e^{\left(\frac{-ipx}{\hbar}\right)}}{\sqrt{x^2+l^2}}$$

Since the probability is given by $\psi\psi^*$ it follows that,

$$\int_{-\infty}^{\infty} dx\psi(x)\psi^*(x) = N^2\int_{-\infty}^{\infty} dx\frac{1}{x^2+l^2} = N^2\frac{\pi}{l}$$

Therefore $N^2 \dfrac{\pi}{l} = 1$

$$N = \sqrt{\frac{l}{\pi}}$$

(b) The probability of finding any particle within the interval is obtained by evaluating the integral.

$$\frac{l}{\pi}\int_{-\frac{l}{\sqrt{3}}}^{+\frac{l}{\sqrt{3}}} \psi(x)\psi^*(x)dx = \int_{-l/\sqrt{3}}^{+l/\sqrt{3}} \frac{1}{x^2+l^2} = \frac{2}{3}$$

Hydrogen Atom

Problem 12.29

Derive the most probable radius for an electron in the 2s orbital of the hydrogen atom.

Solution

The most probable value of any quantity in quantum mechanics is obtained by (i) writing the appropriate probability which is given by the square of the wave function, and (ii) equating the first derivative of this probability to zero.

In the case of the hydrogen atom, this probability is given by the term $r^2[R_{21}(r)]^2$, where $R_{21}(r)$ denotes the radial part of the wave function.

For electrons in the 2s orbital

$$R_{21}(r) = \left(\frac{1}{3}\right)^{1/2}\left(\frac{1}{2a_0}\right)^{3/2}\left(\frac{r}{a_0}\right)e^{-r/2a_0}$$

The corresponding probability = $r^2[R_{21}(r)]^2$. The multiplication by r^2 is required due to the conversion of dV from Cartesian coordinates to spherical coordinates.

To find the most probable value of r, we solve the equation

$$\frac{d}{dr}\left\{r^2[R_{21}(r)]^2\right\} = 0$$

$$r^2[R_{21}(r)]^2 = r^2\frac{1}{3}\left(\frac{1}{2a_0}\right)^3\left(\frac{r}{a_0}\right)^2 e^{-r/a_0}$$

$$= \frac{1}{3}\left(\frac{1^5}{8a_0^5}\right)r^4e^{-r/a_0}$$

$$\frac{d}{dr}\left\{r^2[R_{21}(r)]^2\right\} = \left(\frac{1}{24a_0^5}\right)\left(4-\frac{r}{a_0}\right)r^3e^{-r/a_0}$$

$$\left(\frac{1}{24a_0^5}\right)\left(4-\frac{r}{a_0}\right)r^3e^{-r/a_0} = 0$$

Hence $r = 4a_0$

Problem 12.30

Show that the ground state of the hydrogen atom is normalised.

Solution

The ground state wave function of the hydrogen atom is given by

$$\Psi_{100} = \frac{1}{\sqrt{\pi a_0^3}}e^{\frac{-r}{a_0}}$$

The subscript 100 indicates the 1s orbital.

Method I: In spherical coordinates,

$$\int|\Psi_{100}|^2 d^3\vec{r} = \int_{r=0}^{\infty}\int_{\theta=0}^{\pi}\int_{\phi=0}^{2\pi}\frac{1}{\pi a_0^3}e^{\frac{-2r}{a_0}}r^2\sin\theta dr d\theta d\phi$$

$$= \frac{1}{\pi a_0^3}\int_{r=0}^{\infty}e^{\frac{-2r}{a_0}}r^2dr\int_{\theta=0}^{\pi}\sin\theta d\theta\int_{\phi=0}^{2\pi}d\phi$$

But $\displaystyle\int_{r=0}^{\infty}e^{\frac{-2r}{a_0}}r^2dr = \frac{a_0^3}{4}$; $\displaystyle\int_{\theta=0}^{\pi}\sin\theta d\theta = 2$; $\displaystyle\int_{\phi=0}^{2\pi}d\phi = 2\pi$

$$= \frac{1}{\pi a_0^3}\left[\frac{a_0^3}{4}\times 2\times 2\pi\right]$$

$$= \frac{1}{\pi}\times\frac{1}{4}\times 2\times 2\pi = 1$$

Method II: Since $V = (4/3)\pi r^3$, it follows that $dV = 4\pi r^2 dr$. Thus,

$$\int|\Psi_{1s}|^2 dV = \frac{1}{\pi a_0^3}\int_0^{\infty}4\pi r^2 dr e^{\frac{-2r}{a_0}}$$

$$= \frac{1}{\pi a_0^3}\int_0^{\infty}4\pi r^2 e^{\frac{-2r}{a_0}}dr \tag{1}$$

$$= \frac{4}{a_0^3}\int_0^{\infty}r^2e^{\frac{-2r}{a_0}}dr$$

Since $\displaystyle\int_0^{\infty}x^ne^{-ax}dx = \frac{n!}{a^{n+1}}$

The integral is evaluated as $\dfrac{2!}{\left(\dfrac{2}{a_0^3}\right)^3} = \dfrac{2a_0^3}{8}$

Hence Eq. (1) leads to

$$\frac{4}{a_0^3}\times\frac{2a_0^3}{8}=1$$

Free Particle in a One-Dimensional Box

Problem 12.31

Write the potential energy of a particle in a one-dimensional box if the square well potential is of finite height.

Solution

$$V(x) = V_0\theta(x-a)$$

θ (x–a) is known as the Heaviside step function with the properties

$$\theta(x-a) = 1 \text{ if } x \geq a$$
$$= 0 \text{ otherwise}$$

Problem 12.32

Determine whether the wave functions for the particle in a one-dimensional box are orthogonal.

Solution

In order to verify the orthogonality, we evaluate the integral $\int \psi_1^* \psi_2 d\tau$ over the entire space.

$$\text{Hence, } \int_{-\infty}^{+\infty} \psi_1^*\psi_2 d\tau = \frac{2}{a}\int_0^a \sin\frac{1\pi x}{a}\sin\frac{2\pi x}{a}dx$$

$$2\sin\alpha\sin\beta = \cos(\alpha-\beta)-\cos(\alpha+\beta)$$

$$\text{Hence } I = \frac{1}{a}\int_0^a\left[\cos\left(-\frac{\pi x}{a}\right)-\cos\left(\frac{3\pi x}{a}\right)\right]dx$$

$$=\frac{1}{a}\left\{\int_0^a \cos\left(-\frac{\pi x}{a}\right)dx\int_0^a -\cos\left(\frac{3\pi x}{a}\right)dx\right\}$$

$$=\frac{1}{2}\left[\frac{\sin\left(\frac{\pi x}{a}\right)}{\left(\frac{\pi}{a}\right)}\right]_0^a - \frac{1}{a}\left[\frac{\sin\left(\frac{3\pi x}{a}\right)}{\left(\frac{3\pi}{a}\right)}\right]_0^a = 0$$

Hence, the integral is zero.

Thus, the wave function is orthogonal.

Problem 12.33

Write all the wavefunctions of a cubical box having energy 26 ($h^2/8ma^2$).

Solution

In a three-dimensional box, the quantum numbers are n_x, n_y and n_z, and these have to be integers. Hence the following combinations are possible: (4,3,1), (4,1,3), (3,4,1), (3,1,4), (1,4,3) and (1,3,4) such that $n_x^2+n_y^2+n_z^2 = 26$.

$$\psi_{431} = \sqrt{\frac{8}{a^3}}\sin\frac{4\pi x}{a}\sin\frac{3\pi y}{a}\sin\frac{1\pi z}{a}$$

$$\psi_{413} = \sqrt{\frac{8}{a^3}}\sin\frac{4\pi x}{a}\sin\frac{1\pi y}{a}\sin\frac{3\pi z}{a}$$

$$\psi_{341} = \sqrt{\frac{8}{a^3}}\sin\frac{3\pi x}{a}\sin\frac{4\pi y}{a}\sin\frac{1\pi z}{a}$$

$$\psi_{314} = \sqrt{\frac{8}{a^3}}\sin\frac{3\pi x}{a}\sin\frac{1\pi y}{a}\sin\frac{4\pi z}{a}$$

$$\psi_{143} = \sqrt{\frac{8}{a^3}}\sin\frac{1\pi x}{a}\sin\frac{4\pi y}{a}\sin\frac{3\pi z}{a}$$

$$\psi_{134} = \sqrt{\frac{8}{a^3}}\sin\frac{1\pi x}{a}\sin\frac{3\pi y}{a}\sin\frac{4\pi z}{a}$$

Problem 12.34

Calculate the probability of finding a particle in a one-dimensional box of length L between 0 and a.

Solution

$$\text{Probability of finding a particle between } x_1 \text{ and } x_2 = \int_{x_1}^{x_2}\psi\psi^* dx$$

$$=\int_{x_1}^{x_2}\psi^2 dx = \frac{2}{a}\int_{x_1}^{x_2}\sin^2\frac{n\pi x}{a}dx$$

$$\cos 2x = 1-\sin^2 x$$

$$I = \frac{1}{a}\int_0^a\left(1-\cos\frac{2n\pi x}{a}\right)dx = \frac{1}{a}\left[x - \frac{\sin\frac{2n\pi x}{a}}{\left(\frac{2n\pi}{a}\right)}\right]_0^a = \frac{1}{a}[a-0] = 1$$

Problem 12.35

A particle in a box is given by the wavefunction

$$\psi_1(x) = \left(\frac{30}{3^5}\right)^{\frac{1}{2}}x(3-x) \qquad 0\leq x\leq 3$$

Verify whether (a) $\psi_1(x)$ is normalised and (b) estimate the mean kinetic energy.

Solution

$$\int_0^a[\psi_1(x)]^2 dx = \int_0^3\left(\frac{30}{3^5}\right)[x^2(3-x)^2]dx$$

$$=\left(\frac{30}{3^5}\right)\int_0^3 x^2\left(3^2+x^2-6x\right)dx$$

$$=\frac{30}{3^5}\int_0^3(3^2x^2+x^4-6x^3)dx = 1$$

Hence normalised.

(a) The mean kinetic energy is given by operating the kinetic energy operator on the wave function

$$\hat{T} = \int_0^3 \psi_1(x)\left(-\frac{\hbar^2}{2m}\frac{\partial^2}{\partial x^2}\right)\psi_1(x)dx$$

$$= -\frac{\hbar^2}{2m}\int_0^3 \psi_1(x)\left(\frac{\partial^2}{\partial x^2}\right)\psi_1^*(x)dx$$

$$T = \int_0^3 \left(\frac{30}{3^5}\right)^{\frac{1}{2}} x(3-x)\left(-\frac{\hbar^2}{2m}\right)\left(\frac{30}{3^5}\right)^{\frac{1}{2}}(-2)dx$$

$$= \left(\frac{30}{3^5}\right)^1 \frac{\hbar^2}{2m}\int_0^3 x(3-x)dx$$

$$= \frac{30}{3^5}\frac{\hbar^2}{2m}\left[\frac{3^3}{2}-\frac{3^3}{3}\right]$$

$$= 0.28\frac{\hbar^2}{m}$$

Problem 12.36

Estimate the difference in energies (in eV) between the ground and the first excited level for an electron confined to a one-dimensional box of length 10 Å?

Solution

The ground state refers to $n_1 = 1$ and the first excited state refers to $n_2 = 2$.

Hence, $\Delta E = E_2 - E_1$

Using the equation $E_n = \dfrac{n^2h^2}{8mL^2}$

$$\Delta E = \frac{h^2}{8mL^2}(n_2^2 - n_1^2) = \frac{2^2h^2}{8mL^2} - \frac{1^2h^2}{8mL^2} = \frac{3h^2}{8mL^2}$$

$$\text{Hence, } \Delta E = \frac{3(6.62618\times10^{-34}\,\text{J s})^2}{8(9.10953\times10^{-31}\,\text{kg})(10\times10^{-10}\,\text{m})^2}$$

$$\Delta E = 1.807\times10^{-19}\,\text{J} = 1.128\,\text{eV}$$

Problem 12.37

For electronic transitions from the ground state to the first excited level with a wavelength of radiation of 700 nm, estimate the length of the box assuming a one-dimensional model.

Solution

For a transition from $n = 1$ to $n = 2$, the energy difference ΔE is $\Delta E = E_2 - E_1$.

$$\Delta E = E_2 - E_1 = hc/\lambda = \frac{h^2}{8mL^2}(n_2^2 - n_1^2)$$

where h = Planck constant (6.626×10^{-34} J s) and c = velocity of light (3×10^8 m s^{-1}) and m = mass of an electron (9.109×10^{-31} kg).

Substituting,

$$\Delta E = hc/\lambda = \frac{(6.626\times10^{-34}\,\text{J s})(3\times10^8\,\text{m s}^{-1})}{(700\,\text{nm})\left(\dfrac{10^{-9}\,\text{m}}{1\,\text{nm}}\right)}$$

$$= 2.838\times10^{-19}\,\text{J}$$

But $\Delta E = \dfrac{3h^2}{8mL^2}$

Hence, the length of the box is $L = \left(\dfrac{3h^2}{8m\Delta E}\right)^{1/2}$

$$L = \left[\frac{3(6.626\times10^{-34}\,\text{J s})^2}{8(9.109\times10^{-31}\,\text{kg})(2.838\times10^{-19}\,\text{J})}\right]^{1/2} = 7.981\,\text{Å}$$

Problem 12.38

In a one-dimensional box of width L, determine the probability of finding the particle between $3L/4$ and L, if the particle is in the ground state.

Solution

$$\text{Probability} = \int_{3L/4}^{L} \psi_1^2(x)dx$$

But $\psi_1(x) = \sqrt{\dfrac{2}{L}}\sin\left(\dfrac{\pi x}{L}\right)$

$$\text{Hence the probability} = \frac{2}{L}\int_{3L/4}^{L}\sin^2\left(\frac{\pi x}{L}\right)dx$$

Let $X = \pi x/L$ and hence $dx = dX\,L/\pi$

Hence the probability can be written as

$$= \frac{2}{L}\left[\frac{L}{\pi}\right]\int_{3\pi/4}^{\pi}\sin^2(X)dX$$

$$= \frac{2}{L}\int_{3L/4}^{L}\left(\frac{1-\dfrac{\cos 2\pi x}{L}}{2}\right)dx = \frac{2}{L}\times\frac{1}{2}\int_{3L/4}^{L}\left(1-\frac{\cos 2\pi x}{L}\right)dx$$

$$= 1/4 - 1/(2\pi)$$

Problem 12.39

For a particle in a one-dimensional box of length L, estimate $\langle\Delta x\rangle$ and $\langle\Delta p_x\rangle$.

Solution

In order to compute Δx and Δp_x, we require the following equations:

$$\Delta x = \sqrt{\langle x^2\rangle - \langle x\rangle^2}$$

$$\Delta p_x = \sqrt{\langle p^2 \rangle - \langle p \rangle^2}$$

$$\langle x \rangle = \int_0^L \psi_n^*(x\psi_n)dx = \int_0^L \psi_n^* x \psi_n dx$$

But $\psi_n = \sqrt{\frac{2}{a}} \sin\frac{n\pi x}{a}$

$$\langle x \rangle = \frac{2}{L}\int_0^L x \times \sin^2\frac{n\pi x}{L}dx = \frac{L}{2}$$

Similarly, $\langle x^2 \rangle = \int_0^L \psi_n^* x^2 \psi_n dx = \frac{2}{L}\int_0^L x^2 \sin^2\frac{n\pi x}{L}dx$

$$= \left(\frac{L}{2n\pi}\right)^2\left[\frac{(2n\pi)^2}{3} - 2\right]$$

$$\Delta x = \sqrt{\left(\frac{L}{2n\pi}\right)^2\left[\frac{(2n\pi)^2}{3} - 2\right] - \frac{L^2(n\pi)^2}{4(n\pi)^2}}$$

$$= \frac{L}{2n\pi}\sqrt{\frac{(n\pi)^2}{3} - 2}$$

$$= \frac{L}{2n\pi}\sqrt{1 + \frac{n^2\pi^2 - 3^2}{3}}$$

In the case of Δp_x, we need to compute $\langle p_x^2 \rangle$ and $(\langle p_x \rangle)^2$

where $\langle p_x \rangle = \int_0^L \sqrt{\frac{2}{L}} \sin\frac{n\pi x}{L} dx \left(-i\hbar \frac{d\left[\sqrt{\frac{2}{L}} \sin\left(\frac{n\pi x}{L}\right)\right]}{dx} \right) dx$

$$= -\frac{i\hbar}{L}\sin^2\left(\frac{n\pi x}{L}\right)\Big|_0^L = 0$$

$$(\langle p_x \rangle)^2 = \int_0^L \psi_n^* p_x^2 \psi_n dx$$

$$= 2m\int_0^L \psi_n^* H \psi_n dx = 2mE_n \int_0^L \psi_n^* \psi_n dx$$

$$= 2mE_n = \frac{n^2\pi^2\hbar^2}{L^2}$$

Therefore $\Delta p_x = \sqrt{\frac{n^2\pi^2\hbar^2}{L^2}} = \frac{n\pi\hbar}{L} = \frac{nh}{2L}$

Problem 12.40

Estimate the energies of the first four levels of an electron in a box having a width of 15 Å.

Solution

$$E_1 = \frac{1^2 h^2}{8m_e a^2} = \frac{1^2(6.626\times10^{-34}\text{ J s})^2}{8(9.109\times10^{-31}\text{ kg})(1.5\times10^{-9}\text{ m})^2}$$

$$= 2.68\times10^{-20}\text{ J}$$

$$E_2 = \frac{2^2 h^2}{8m_e a^2} = \frac{2^2(6.626\times10^{-34}\text{ J s})^2}{8(9.109\times10^{-31}\text{ kg})(1.5\times10^{-9}\text{ m})^2}$$

$$= 10.72\times10^{-20}\text{ J}$$

$$E_3 = \frac{3^2 h^2}{8m_e a^2} = \frac{3^2(6.626\times10^{-34}\text{ J s})^2}{8(9.109\times10^{-31}\text{ kg})(1.5\times10^{-9}\text{ m})^2}$$

$$= 24.12\times10^{-20}\text{ J}$$

$$E_4 = \frac{4^2 h^2}{8m_e a^2} = \frac{4^2(6.626\times10^{-34}\text{ J s})^2}{8(9.109\times10^{-31}\text{ kg})(1.5\times10^{-9}\text{ m})^2}$$

$$= 42.88\times10^{-20}\text{ J}$$

Problem 12.41

For a conjugated hydrocarbon, the maximum absorption of light occurs at 480 nm. Assuming that the transition involves $n = 9$ to $n = 10$, deduce the approximate length of the molecule.

Solution

First, we obtain the energy corresponding to the wavelength of the light:

$$E_\lambda = \frac{hc}{\lambda} = \frac{(6.626\times10^{-34}\text{ J s})(3.00\times10^8\text{ m s}^{-1})}{4.80\times10^{-7}\text{ m}}$$

$$= 4.14\times10^{-19}\text{ J}$$

This is then equated to the energy difference,

$$\Delta E = E_{10} - E_9$$

$$= \frac{10^2 h^2}{8m_e a^2} - \frac{9^2 h^2}{8m_e a^2} = (100 - 81)\frac{h^2}{8m_e a^2}$$

$$= 19\frac{h^2}{8m_e a^2} = 4.14\times10^{-19}\text{ J}$$

$$\frac{19(6.626\times10^{-34}\text{ J s})^2}{8(9.109\times10^{-31}\text{ kg})a^2} = 4.14\times10^{-19}\text{ J}$$

$$a = 16.6\text{ Å}$$

Problem 12.42

If $n_x^2 + n_y^2 + n_z^2 = 14$ for a particle in a cubic box, write all the degeneracies.

Solution

$$n_x^2 + n_y^2 + n_z^2 = 14$$

$$= 3^2 + 2^2 + 1^2 = 14$$

The combinations involving 3, 2 and 1 are 321, 123, 132, 231, 213 and 312.

Rigid Rotor

Problem 12.43

Calculate the moment of inertia, the angular momentum, and the energy in the first rotational state above the ground level, $J = 1$, for O_2 using $r_0 = 121$ pm.

Solution

The moment of inertia is given by the equation $I = \mu r^2$

where $\mu = \dfrac{m_1 m_2}{m_1 + m_2}$

Therefore $\mu = \dfrac{16\times16}{16+16} = 8 \text{ g mol}^{-1}$

$$I = \frac{8\text{g}\times(121\times10^{-12}\text{m})^2}{6.023\times10^{23}}$$

$$I = 1.94\times10^{-46} \text{ kg m}^2$$

The angular momentum is given by,

$$\sqrt{J(J+1)}\hbar = \sqrt{1(1+1)}\hbar = \sqrt{2}\frac{h}{2\pi} = \sqrt{2}\frac{6.626\times10^{-34}\text{ J s}}{2\times3.14}$$

$$= 1.49\times10^{-34} \text{ kg m}^2 \text{ s}^{-1}$$

$$\text{Energy } E_J = J(J+1)\frac{\hbar^2}{2I} = 5.74\times10^{-23}\text{ J}$$

Fill Up the Blanks with Appropriate Numbers or Phrases

1. The normalisation constant for a particle in a one-dimensional box is ________.
2. The energy difference between the electrons in its ground state in the following species: H_2^+, D_2^+, HD^+ is ________ (if the electron has the same reduced mass for all the species).
3. The orbital angular momentum of an electron in the 3d orbital is ________ $\hbar$.
4. Normalised wavefunctions of the bonding and antibonding orbitals for the electron in H_2^+ molecular ion are ________ and ________ respectively.
5. If two Hermitian operators commute, they possess a complete set of common ________.
6. For any principal quantum number n, the values of l range from ________.
7. The orbital angular momentum is quantised due to ________ quantisation.
8. The classical turning points of a harmonic oscillator refer to the value of x when the kinetic energy is ________.
9. The operator $\hat{L}_z$ in spherical coordinates is ________.
10. In the case of a free particle in the one-dimensional box, the wave function with odd values of n are ________ with respect to the centre of the box.
11. The wave function for the first excited state of a simple harmonic oscillator is ________.
12. The ground state energy increases quadratically with the quantum number (n_x) for ________.
13. The total number of nodes for a 5f orbital is ________.
14. The principal (n) and azimuthal (l) quantum numbers of electrons in the 4f orbital are respectively ________ and ________.
15. The energy of 2s electronic state of Li^{2+} ions is ________ eV.
16. In the case of a free particle in the one-dimensional box, the wave functions with odd value of n are ________ with respect to the centre of the box.

EXERCISES

Practice Problems

1. Demonstrate the correspondence between the Bohr model of hydrogen atom and quantisation of the orbital angular momentum.
2. Estimate the wavelength (in Å) of the Lyman transition from 1s to 2p level in He^+.
3. Derive $<1/r>$ and $<r^2>$ for the ground state of the hydrogen atom.
4. Show that the sum of expectation value of the potential and kinetic energies of the 1s level of the hydrogen atom is consistent with the prediction from the virial theorem.
5. Estimate the distance r from the nucleus at which the probability of finding the electron equals 5% of the probability at $r = 0$.
6. The Pauli spin matrix $S_x = \dfrac{\hbar}{2}\begin{bmatrix}0 & 1\\ 1 & 0\end{bmatrix}$. Obtain S_x^2.
7. Estimate the electronic energy (in eV) for the 3s state for Li^{2+}.
8. Estimate the commutator $[p_x^2 \;\; x]$.
9. Write the quantum mechanical operator for (a) kinetic energy and (b) momentum in one dimension.

10. Among the following, identify the Hermitian operator(s).

(a) p_x (b) $\hat{L}_x$ (c) $i\frac{d}{dx}$ (d) $-\frac{d^2}{dx^2}$ (e) $\frac{d}{dx}$

11. Show that the allowed wavelengths for a particle in a one-dimensional box are $2a, a, \frac{2a}{3}, \frac{a}{2}, \ldots\ldots\ldots$

12. Verify whether the time-dependent wave function $\psi(x, t) = e^{ikx}\, e^{-iEt/h}$ is stationary.

13. Evaluate the following commutators: $[\hat{L}_x, \hat{L}_y]$ and $[\hat{L}_z, \hat{L}_2]$

14. Derive the normalisation constant for the 3s orbital of a hydrogen-like atom.

15. Show that the eigenvectors for distinct eigenvalues of a Hermitian matrix are orthogonal.

16. Write the angular momentum operator $\hat{L}_z$ in spherical polar coordinates and hence obtain the eigenvalue.

17. Deduce the value of the normalisation constant of a rigid rotator whose wave function is given by $\psi(\theta) = N \times e^{im\theta}$.

18. Estimate the total number of nodes in the following orbitals: (a) 2*p* (b) 3*p* and (c) 4*d*

19. Derive the total internal energy of *N* one-dimensional harmonic oscillators, taking into account the zero point energy.

20. Determine whether the function $f(x) = \cos(kx/l)$ is orthonormal when *x* lies in the range of 0 to 1.

21. Verify whether the matrix $\begin{bmatrix} \cos\theta & \sin\theta \\ -\sin\theta & \cos\theta \end{bmatrix}$ is orthogonal.

22. Deduce the magnitude of the angular momentum vector for $J = 3$.

23. Obtain the result when the momentum operator $\hat{p}_x$ operates on $e^{(-ikx)}$.

24. Write all the possible orbital angular momentum (*z*-component) values of an electron in the *d*-orbital.

25. Write the first two Hermite polynomials $H_0(x)$ and $H_1(x)$ and the wave functions for the ground and first excited state of a harmonic oscillator.

26. Show that eigenvalues of Hermitian operators are real.

27. Show that eigenkets corresponding to two distinct eigenvalues are orthogonal.

28. Write the determinants for the quantum mechanical operators $\hat{L}_x$, $\hat{L}_y$ and $\hat{L}_z$.

29. Plot schematically $|\psi|^2$ for various *n* values for the particle in a box model.

30. Show that for a particle in the one-dimensional potential $U = 0$, if $0 < x < a$; $U = \infty$, if $x < 0$ or $x > a$, the following relations hold:

$$\langle x \rangle = \frac{1}{2}a, \langle (x - \langle x \rangle)^2 \rangle = \frac{a^2}{12}\left(1 - \frac{6}{n^2\pi^2}\right)$$

31. Determine whether the wave function $\psi(x, t) = A \sin(t/t_m)$ is a wave function of the stationary state.

32. Consider a system wherein the number of electrons per unit volume is 2.5×10^{20} cm^{-3}. Estimate the average distance (in Å) between electrons in an approximate manner.

33. Deduce the normalisation constant *A* in the wave function, $\Psi(x) = A \sin(2\pi x/\lambda)$ where *x* ranges from 0 to 1.

34. Draw rough sketches of radial wave functions and radial distribution functions for 4*s*, 4*p*, 4*d* and 4*f* orbitals.

35. Deduce the mean position $\langle x \rangle$ and mean momentum $\langle p \rangle$ for the ground state of a simple harmonic oscillator.

Multiple Choice Questions

1. Consider the function $\Psi(r,\theta,\phi) = A\,[1 - r\sin^2(\theta/2)]\,f(r)$ where *A* is a constant. The above will be a solution of the Schrödinger equation for hydrogen atom if $f(r)$ is
   (a) $e_0^{-r/a}$ (b) $\sin(r/a)$
   (c) e^{-r} (d) $e^{-r/2}$

2. The normalised wave function $\Psi(r)$ is given by $\Psi(r) = A\,e^{-r/3}$. The value of *A* is
   (a) $2/\pi$ (b) $1/3\pi$
   (c) $1/9\pi$ (d) $(1/27\pi)^{1/2}$

3. In the ground state of hydrogen atom, the expectation value of $\langle 1/r^2 \rangle$ is
   (a) $2/a_0^2$ (b) $a_0^{1/2}$
   (c) $\sqrt{\pi}$ (d) $3/2\,a_0^2$

4. If the radial function $R(r/a_0)$ is given by $R(r/a_0) = \text{Const.}\,(r/a_0)e^{-r/2a_0}$, then this corresponds to the orbital with
   (a) $n = 2$ and $l = 0$ (b) $n = 3$ and $l = 1$
   (c) $n = 1$ and $l = 2$ (d) $n = 2$ and $l = 1$

5. The commutation relation between the angular momentum operator $\hat{L}_y$ and $\hat{L}_x$, namely, $[\hat{L}_y, \hat{L}_x]$ is
   (a) 0 (b) $i\,h\,\hat{L}_z/2\pi$
   (c) $-ih\,\hat{L}_z/2\pi$ (d) L^2

6. The following expression $Y_{lm} = \text{Const}\,(\cos\theta)$ corresponds to the values
   (a) $l = 1$ and $m = 0$ (b) $l = 1$ and $m = 1$
   (c) $m = 1$ and $l = 0$ (d) $m = 0$ and $l = 0$

7. The application of the virial theorem to the hydrogen atom in the ground state yields the relation between the kinetic

energy $<E>$ and potential energy $<V>$ as
(a) $<E> + <V> = 0$ (b) $<E> = <(V/2)>$
(c) $<E> = <V>$ (d) $<E> = <V>/2$

8. The expectation value of the kinetic energy of the hydrogen atom in the ground state may be represented in atomic units as
(a) $e^2/(2a_0)$ (b) e^2/a_0
(c) $2e^2/a_0$ (d) 0

9. Which of the following represents the quantisation condition postulated by Bohr?
(a) $E_n = (n + 1/2)\, h\nu$ (b) $mvr = nh/2\pi$
(c) $E_n = n^2h^2/8ma^2$ (d) $E_r = B_r(J)(J + 1)$

10. The coulomb potential $V(r)$ of an electron due to the charge 'Ze' in the central force field is represented by
(a) $-Z^2e^2/r^2$ (b) $-Z^2e^2/r^3$
(c) $-Ze^2/r$ (d) $-(Ze/r)e^{-ze/r}$

11. The orbits employed in Sommerfeld's model for energies of electrons are
(a) circular (b) cylindrical
(c) elliptical (d) ellipsoidal

12. The excitation energy involved in the transition from the $n = 1$ to the $n = 2$ state of the hydrogen atom is (in eV).
(a) 6.8 (b) 13.6
(c) 3.4 (d) 10.2

13. The total number of energy states E_n for the hydrogen atom in the $n = 3$ level is
(a) 9 (b) 6
(c) 24 (d) 12

14. The expectation values of the kinetic energy and potential energy of the electron in the 1s level of the hydrogen atom are respectively (in eV)
(a) 13.6 and 27.2 (b) –13.6 and –27.2
(c) 13.6 and –27.2 (d) 27.2 and 13.6

15. The function $f(t)$ required for converting the time-dependent Schrödinger equation into a time-independent equation is
(a) $e^{\frac{iEt}{\hbar}}$ (b) $\sin\left(\frac{iE}{\hbar}t\right)$
(c) $1/\left[e^{\frac{iEt}{\hbar}}\right]$ (d) $\frac{iE}{\hbar}t$

16. The *total* number of allowed values of m for $n = 5$ and $l = 4$ are
(a) 5 (b) 9
(c) 3 (d) 18

17. The uncertainty in the momentum p of an one-dimensional harmonic oscillator in the ground state is
(a) 0 (b) $mh\nu$
(c) $(mh/4\pi)^{1/2}$ (d) m

18. For the principal quantum number n, the total number of states with the energy E_n for the hydrogen atom is
(a) $n^2 + l^2$ (b) e^{-n2}
(c) n^2 (d) $n(n + 1)^2$

19. The spherical harmonics are eigen functions of the
(a) radial part of the Schrodinger equation
(b) angular momentum operators
(c) Hermitian matrix
(d) position operator

20. Which of the following was employed by Bohr to estimate Planck's constant?
(a) atomic spectra
(b) Stern–Gerlach experiment
(c) correspondence principle
(d) photoelectric effect

21. Which of the following is true regarding the angular momentum operator? The angular momentum operator commutes with the
(a) Hamiltonian (b) position
(c) momentum (d) potential energy

22. Which one of the following is linearly independent?
(a) [1,1,1] and [0,1,–2]
(b) [1,1,2] and [2,–1,2]
(c) [2,–1,1] , [3,–4,–2] and [5,–10,–8]
(d) [1,0,3] , [–2,1,–1] and [3,1,–1]

23. Identify the rotation matrix through π.
(a) $\begin{bmatrix} 1 & i \\ -i & 1 \end{bmatrix}$ (b) $\begin{bmatrix} 1/2 & 1 \\ -1 & 0 \end{bmatrix}$
(c) $\begin{bmatrix} 0 & -1/4 \\ 0 & -1/2 \end{bmatrix}$ (d) $\begin{bmatrix} -1 & 0 \\ 0 & -1 \end{bmatrix}$

24. Which one of the following is a Hermitian matrix?
(a) $\begin{bmatrix} 1 & 3-i \\ 3+i & i \end{bmatrix}$ (b) $\begin{bmatrix} 2 & i \\ -i & 5 \end{bmatrix}$
(c) $\begin{bmatrix} 0 & 3-2i \\ 3-2i & 4 \end{bmatrix}$ (d) $\begin{bmatrix} 2 & 3i \\ -i & 5 \end{bmatrix}$

25. Which of the following is a rotation matrix?
(a) $\begin{bmatrix} \sin\theta & \cos\theta \\ -\sin\theta & \cos\theta \end{bmatrix}$ (b) $\begin{bmatrix} \tan\theta & \cos\theta \\ -\sin\theta & -\cos\theta \end{bmatrix}$
(c) $\begin{bmatrix} \cos\theta & -\sin\theta \\ \sin\theta & \cos\theta \end{bmatrix}$ (d) $\begin{bmatrix} \sec\theta & \cos\theta \\ -\sin\theta & \tan\theta \end{bmatrix}$

26. Which of the following commutator relation is true?
(a) $[\partial/\partial x_1, \hat{x}_2] = 1$ (b) $[\partial/\partial x_2, \hat{x}_1] = 1$
(c) $[\partial/\partial x_1, \hat{x}_1] = 1$ (d) $[\hat{x}_2, \partial/\partial \hat{x}_1] = -1$

27. Among the following the correct relation among angular momentum operators is
(a) $[\hat{L}_1, \hat{L}_2] = i\hbar\hat{L}_3$ (b) $[\hat{L}_1, \hat{L}_1] = 1$
(c) $[\hat{L}_2, \hat{L}_2] = 1$ (d) $[\hat{L}_1, \hat{L}_3] = -i\hbar\hat{L}_2$

28. The eigenvalues of $\begin{bmatrix} -1 & 2 & 2 \\ 2 & 2 & 2 \\ -3 & -6 & -6 \end{bmatrix}$ are
(a) –6,1,2 (b) 0,–6,1
(c) 3,2,–11 (d) 0,–2,–3

29. Which of the following is correct regarding an anti-Hermitian operator?
(a) The eigenvalue of an anti-Hermitian operator is real.
(b) It becomes Hermitian upon multiplication by i.
(c) The momentum operator is an example of an anti-Hermitian operator.
(d) id/dx is an anti-Hermitian operator.

30. The spherical harmonic is denoted by $Y_{l,m}(\theta, \phi)$ in the solution of the hydrogen atom. The normalisation constant in the case of $Y_{1,-1}$ for $n = 2$, is given by
(a) $1/4(2\pi)^{1/2}$ (b) $1/2\pi$
(c) $1/4\pi$ (d) $1/\sqrt{\pi}$

31. For the $5d_{xy}$ orbital, the number of radial and angular nodes are, respectively,
(a) 2 and 2 (b) 3 and 1
(c) 4 and 0 (d) 1 and 3

32. The mean distance of an electron from the nucleus when $n = 3$ and $l = 2$ will be
(a) $2a_0$ (b) $(10.5)a_0$
(c) $4a_0$ (d) a_0

33. Which of the following is *not* an allowed magnitude for the angular momentum?
(a) $\sqrt{6}\hbar$ (b) $\sqrt{12}\hbar$
(c) $\sqrt{20}\hbar$ (d) $\sqrt{18}\hbar$

34. The potential energy of an electron in the 4th quantum state of a hydrogen atom can be represented as $V = -e^2/A_1$. The value of A_1 is
(a) $4a_0$ (b) $16a_0$
(c) $1/(4a_0)$ (d) a_0

35. For a free particle in a one-dimensional box, the wave function $\psi_n(x)$ for odd values of n is proportional to
(a) $\sinh(nx/a)$ (b) $e^{-nx/a}$
(c) $\cos(n\pi x/2a)$ (d) $\sin(n\pi x/a)$

Answer Keys

Fill Up the Blanks

1. $(2/a)^{1/2}$
2. Zero
3. $6^{½}$
4. $\frac{1}{\sqrt{2(1+S)}}(1s_A + 1s_B)$ and $\frac{1}{\sqrt{2(1+S)}}(1s_A - 1s_B)$
5. Eigenstates
6. 0 to $n - 1$
7. Spatial
8. Zero
9. $\hat{L}_z = -i\hbar\left(x\frac{\partial}{\partial y} - y\frac{\partial}{\partial x}\right) = -i\hbar\frac{\partial}{\partial \phi}$
10. Symmetric
11. $xe^{(-ax^2/2)}$
12. Free particle in a one-dimensional box
13. 6
14. 4 and 3
15. –61.2
16. Symmetric

Practice Problems

1. Theoretical question
2. 300.4 Å
3. $\frac{1}{a_0}$ and $3a_0^2$
4. Theoretical question
5. 0.794 Å
6. $\frac{\hbar^2}{4}I$ where I denotes the identity matrix
7. –13.60 eV
8. $-2i\hbar p_x$
9. (a) $-\frac{h^2}{8\pi^2 m}\frac{d^2}{dx^2}$ (b) $-i\hbar\frac{\partial}{\partial x}$
10. (a) p_x (b) L_x (c) $i\frac{d}{dx}$ (d) $-\frac{d^2}{dx^2}$
11. Theoretical question
12. Theoretical question
13. $[\hat{L}_x, \hat{L}_y] = i\hbar Lz$ and $[\hat{L}_z, L^2] = 0$
14. Theoretical question

15. Theoretical question

16. $-2\hbar$

17. $1/(2\pi)^{1/2}$

18. (a) 3 (b) 1 (c) 2

19. $\frac{Nh\nu}{2} + \frac{Nh\nu}{e^{+h\nu/kT} - 1}$

20. Theoretical question

21. Theoretical question

22. $\sqrt{12}\,\hbar$

23. $k\,\hbar$

24. $2\hbar$, $1\hbar$, 0, $-1\hbar$, and $-2\hbar$

25. $H_0(t) = 1$ and $H_1(t) = 2t$

26. Theoretical question

27. Theoretical question

28. Theoretical question

29. Theoretical question

30. Theoretical question

31. Theoretical question

32. 15.8

33. Theoretical question

34. Theoretical question

35. $\langle x \rangle = 0$ and $\langle p \rangle = 0$

Multiple Choice Questions

1. (d)	**2.** (d)	**3.** (a)	**4.** (d)	**5.** (c)	**6.** (a)	**7.** (d)	**8.** (a)	**9.** (b)	**10.** (c)
11. (c)	**12.** (d)	**13.** (a)	**14.** (c)	**15.** (c)	**16.** (b)	**17.** (c)	**18.** (c)	**19.** (b)	**20.** (c)
21. (a)	**22.** (a)	**23.** (d)	**24.** (b)	**25.** (c)	**26.** (c)	**27.** (a)	**28.** (d)	**29.** (b)	**30.** (a)
31. (a)	**32.** (b)	**33.** (d)	**34.** (b)	**35.** (c)					

CHAPTER 13

ACID–BASE AND OTHER IONIC EQUILIBRIA, BUFFERS AND BUFFER ACTION, INDICATORS

CHAPTER OUTLINE

Important Concepts and Formulae
- Arrhenius Theory of Acids and Bases
- Brönsted–Lowry Theory of Acids and Bases
- Lewis Acid–Base Theory
- Dissociation Equilibria
- Ostwald's Dilution Law
- Amphiprotic Solvents
- Levelling Effect of Solvents on Acid–Base Strength

Acids and Bases
- Hard and Soft Acids and Bases
- Relative Strengths of Acids and Bases
- Polyprotic Acids
- pH Concept
- Acidity Measurements in Concentrated Aqueous Acid Solutions and Non-aqueous Solvents
- Common Ion Effect
- Acid–Base Neutralisation Curves
- Acid–Base Indicators

Buffers and Buffer Action

Hydrolysis of Salts
- Hydrolysis of the Salt of a Weak Acid and Strong Base
- Hydrolysis of the Salt of a Weak Base and Strong Acid
- Hydrolysis of a Salt of a Weak Acid and Weak Base

Solubility Product and Solubility of Sparingly Soluble Salts
- Variation of Solubility Product with Temperature
- Effect of Complex Formation on Solubility
- Effect of Common Ion on the Solubility of Sparingly Soluble Salts
- Application of Solubility Product Principle to Qualitative Analysis

Redox Equilibria
- Balancing Equations
- Redox Potentials
- Effect of Complex Formation on Redox Potentials
- Potentiometric Titrations Involving Redox Reactions
- Potentiometric Titrations Involving Precipitation Reactions

13.1 Important Concepts and Formulae

13.1.1 Arrhenius Theory of Acids and Bases

Arrhenius developed the theory of acid–base equilibria in 1884. According to him, an acid is any compound that gives H^+ ions in an aqueous solution and a base is any compound which gives OH^- ions in an aqueous solution.

He also classified electrolytes as strong and weak. Acids such as H_2SO_4 and HCl are considered as strong acids or strong electrolytes and dissociate completely to give rise to H^+ ions while substances like NaOH give rise to OH^- ions on dissociation in water.

The main drawback of the Arrhenius theory is that it does not deal with acid–base reactions in solvents other than water. Apart from this, it does not accommodate other species such as NH_3 which neutralises acids but does not contain OH^- ions.

13.1.2 Brönsted–Lowry Theory of Acids and Bases

According to this theory, an acid (A) is any substance that can donate a proton and a base (B) is any substance that can accept a proton.

$$A \rightleftharpoons B + H^+ \tag{13.1}$$

A–B is said to be a conjugate acid–base pair. An acid–base reaction always involves two acid–base pairs such as

$$\text{Acid1} + \text{Base2} \rightleftharpoons \text{Base1} + \text{Acid2} \tag{13.2}$$

Examples of Eq. 13.2 can be dissociation of acids such as HCl and H_2SO_4 in water, dissociation of NH_3 on reacting with water to give $NH_4{}^+$ and OH^- ions. An important advantage of this theory is that an acid or base may be a neutral molecule, a cation or an anion (for example, NH_4^+, $H_2PO_4^-$)

13.1.3 Lewis Acid–Base Theory

G N Lewis proposed his theory of acids and bases in 1923. It is more broad based and defines an acid as an electron pair acceptor and a base as an electron pair donor. Examples of Lewis acids are $AlCl_3$, BF_3, which accepts electron pairs while substances like H_2O, NH_3 and OH^- can donate electron pairs. A typical example of a Lewis acid–base reaction is

$$\ddot{N}H_3(\text{base}) + H^+(\text{acid}) \rightarrow NH_4^+ \tag{13.3}$$

13.1.4 Dissociation Equilibria

The dissociation equilibrium of a weak acid such as acetic acid in water may be represented as

$$CH_3COOH + H_2O \rightleftharpoons CH_3COO^- + H_3O^+ \tag{13.4}$$

and the dissociation constant K_a may be written as

$$K_a = \frac{[CH_3COO^-][H_3O^+]}{[HA]} \tag{13.4a}$$

The strength of the acid may be conveniently represented as $pK_a = -\log K_a$. In a similar way, the strength of a base, B, may be indicated by its pK_b or $-\log K_b$ as represented by its dissociation equilibrium

$$B + H_2O \rightleftharpoons BH^+ + OH^- \tag{13.5}$$

The strength of the base may also be indicated by the dissociation constant of the conjugate acid of the base, BH^+, that is,

$$BH^+ + H_2O \rightleftharpoons B + H_3O^+ \tag{13.6}$$

$$K_{BH^+} = \frac{[B][H_3O^+]}{[BH^+]} \tag{13.7}$$

K_b and K_{BH^+} are related through the equation

$$K_{BH^+} \times K_B = K_w \tag{13.8}$$

where K_w is the equilibrium constant of the reaction, $H_2O + H_2O \rightleftharpoons H_3O^+ + OH^-$, more commonly known as ionic product of water.

13.1.5 Ostwald's Dilution Law

If α is the degree of dissociation of a weak acid (Eq. 13.4) or a weak base (Eq. 13.5) at a stoichiometric concentration C, the dissociation constant K_a or K_b is given by

$$K_a = \frac{\alpha^2 C}{1-\alpha} \approx \alpha^2 C \text{ (when } \alpha \ll 1) \tag{13.9}$$

Ostwald's law states that the degree of dissociation increases with dilution. Using the Arrhenius equation for the degree of dissociation in terms of the conductance ratio given by

$$\alpha = \Lambda/\Lambda_0 \tag{13.10}$$

The dissociation constant K_a or K_b may be determined by using the equation

$$K_a = \frac{\Lambda^2 C}{\Lambda_0(\Lambda_0 - \Lambda)} \tag{13.11}$$

There are also other methods of determining K_a or K_b, such as, by measurement of pH of the solution or by spectrophotometric measurements.

13.1.6 Amphiprotic Solvents

Amphiprotic solvents are solvents that can undergo self-ionisation. Water or other solvents such as methanol or ethanol can act both as an acid or as a base and thereby undergo self-ionisation. The self-ionisation equilibrium in case of water may be represented as

$$H_2O + H_2O \rightleftharpoons H_3O^+(aq) + OH^-(aq) \tag{13.12}$$

The auto or self-ionisation constant, K_w is given by

$$K_w = [H_3O^+][OH^-] = 1.0 \times 10^{-14} \tag{13.13}$$

at 298 K. In pure water, we thus have

$$[H_3O^+] = [OH^-] = 1.0 \times 10^{-7} \tag{13.14}$$

Analogous equations for ethanol and methanol may be written as

$$CH_3OH + CH_3OH \rightleftharpoons CH_3OH_2^+ + CH_3O^- \tag{13.15a}$$

$$C_2H_5OH + C_2H_5OH \rightleftharpoons C_2H_5OH_2^+ + C_2H_5O^- \tag{13.15b}$$

The autoprotolysis or self-ionisation constants of these solvents given (as pK_{ap}) by Eq. 13.15a for methanol is 16.56 and for ethanol by Eq. 13.15b is 18.53 at 298 K. (Note that they are pK_{ap} or $-\log K_{ap}$ values!) Thus their self-ionisation is much smaller compared to water, presumably due to their lower dielectric constants as well as weaker acidic and basic strength.

13.1.7 Levelling Effect of Solvents on Acid-Base Strength

When water is the solvent, it is sufficiently basic to capture all the protons from acids such as $HClO_4$, H_2SO_4, HCl and HNO_3 which makes it impossible to distinguish between their strengths. On the other hand, if we use a solvent like acetic acid, the extent of their equilibrium given by

$$CH_3COOH + HA \rightleftharpoons CH_3COOH_2^+ + A^- \tag{13.16}$$

depends on their inherent strength and one can distinguish the differences in their acid strength. In a similar way, it is possible to distinguish between the strengths of bases by choosing a basic solvent like NH_3.

Solvent systems

(i) ***Liquid ammonia:*** It can undergo self-ionisation according to

$$2NH_3 \rightleftharpoons NH_4^+ + NH_2^- \tag{13.17}$$

and its self-ionisation constant is about 5×10^{-27} at 298 K.

(ii) ***Liquid HF:*** The ionisation of this solvent may be represented as

$$3HF \rightleftharpoons H_2F^+ + HF_2^- \tag{13.18}$$

and its ionisation constant is about 2×10^{-12} at 0°C. The ionisation of acetic acid, which acts as a base, in this medium is given by

$$CH_3COOH + 2HF \rightleftharpoons CH_3C(OH)_2^+ + HF_2^- \tag{13.19}$$

(iii) ***Concentrated H_2SO_4:*** This has been extensively used in acid–base studies and its self-ionisation may be described by

$$2H_2SO_4 \rightleftharpoons [H_3SO_4]^+ + [HSO_4]^- \tag{13.20}$$

with the auto-protolysis constant being about 3×10^{-4} at 298 K.

(iv) *Super acids:* These acids are stronger than even concentrated H_2SO_4. Examples of such acids are a mixture of HF and SbF_5, and flourosulfonic acid (H_3SO_3F) and SbF_5. A typical reaction in the latter mixture is

$$2HSO_3F + SbF_5 \rightleftharpoons [H_2SO_3F]^+ + [F_5SbOSO_2F]^- \quad (13.21)$$

13.2 Acids and Bases

13.2.1 Hard and Soft Acids and Bases

This classification was proposed by Pearson in 1963. Hard acids and hard bases are molecules which are compact and in which the electrons are held tightly by the nucleus. They are not very polarisable. Soft acids or soft bases are large molecules or ions and are highly polarisable. Generally hard acids tend to bind to hard bases while soft acids tend to bind to soft bases.

Hard acids have high charge density. They are mostly metal ions with high positive charge and small ionic size. Soft acids on the contrary, have low charge density and greater ionic size. π-bonding can occur in these molecules or ions through their *d* orbitals.

Examples of hard acids: Alkali metal ions such as Li^+, Na^+, K^+; alkaline earth metal ions such as Be^{2+}, Mg^{2+}, Ca^{2+}; molecules such as BF_3, BCl_3, $AlCl_3$, Al^{3+}, AlH_3, Fe^{3+}, Mn^{2+}, Co^{3+}, H^+, SO_3 gas.

Examples of hard bases: H_2O, OH^-; halide ions such as F^-, Cl^-; ClO_4^-, NO_3^-, CO_3^{2-}, SO_4^{2-}, PO_4^{3-}, NH_3, amines such as RNH_2.

Examples of (borderline) soft acids: Fe^{2+}, Co^{2+}, Ni^{2+}, Cu^{2+}, Zn^{2+}, Rh^{3+}, SO_2 gas.

Examples of borderline bases: Br^-, NO^{2-}, N_2, $SO_3{}^{2-}$, aniline.

Examples of soft acids: Cu^+, Ag^+, Au^+, Cd^{2+}, Hg^{2+}, Hg_2^{2+}, Pt^{2+}, Pt^{4+}, Pd^{2+}.

Examples of soft bases: H_2S, S^{2-}, SH^-, SCN^-, CN^-, CO, $S_2O_3^{2-}$, C_6H_6.

There is no relation, whatsoever, between acid–base strength and hard–soft nature of a species. Some typical hard acid–hard base reactions are shown below.

$$\underset{\text{Hard acid}}{Ca^{2+}} + \underset{\text{Hard base}}{2OH^-} \rightarrow Ca(OH)_2 \quad (13.22)$$

$$\underset{\text{Hard acid}}{F_3B} + \underset{\text{Hard base}}{NH_3} \rightarrow BF_3 : NH_3 \quad (13.23)$$

$$\underset{\text{Soft acid}}{Ag^+} + \underset{\text{Soft base}}{I^-} \rightarrow AgI \quad (13.24)$$

$$\underset{\text{Soft acid}}{Hg^{2+}} + \underset{\text{Soft base}}{S^{2-}} \rightarrow HgS \quad (13.25)$$

$$\underset{\text{Acid}}{Zn^{2+}} + \underset{\text{Base}}{2Br^-} \rightarrow ZnBr_2 \quad (13.26)$$

Products formed from hard acid–hard base reactions tend to be more ionic in nature and are energetically favoured whereas soft acid–soft base reaction products are more covalent in nature. In considering hard acid–hard base interactions, the nature of the coordination site and oxidation numbers of the species are factors which have to be taken into account. The HSAB theory finds important applications in qualitative analysis.

13.2.2 Relative Strengths of Acids and Bases

The relative strength of an acid or a base can be gauged by the position of equilibrium in the acid–base reaction under consideration. While all acids such as $HClO_4$, H_2SO_4, HCl and HNO_3 appear equally strong in water due to the levelling effect of water on their strengths, the relative strengths of acids such as formic acid, acetic acid, and so on, can be assessed by their dissociation constants in water. Formic acid, for example, has a greater K_a than acetic acid. In a similar way, it is possible to consider the relative strengths of bases such as methylamine, ethylamine and aniline by the magnitude of their dissociation constants in water.

13.2.3 Polyprotic Acids

Acids such as H_2SO_4 and H_3PO_4 have more than one proton that can disscociate and hence have multistage equilibria. For example, in the case of H_2SO_4,

$$H_2SO_4 + H_2O \rightleftharpoons H_3O^+ + HSO_4^- \quad (13.27a)$$

and

$$HSO_4^- + H_2O \rightleftharpoons H_3O^+ + SO_4^{2-} \quad (13.27b)$$

while the first ionisation is practically complete, the second step ionisation constant is 1.3×10^{-2} at 298 K. Similarly, H_3PO_4 has three ionisable protons. While its first ionisation constant is 7.1×10^{-3}, the second and third ionisation constants are small, being 6.3×10^{-8} and 4.2×10^{-13} respectively, at 298 K.

13.2.4 pH Concept

The H^+ ion concentration in aqueous solutions varies over a wide range from 1 to 10^{-14} as seen from its ionic product value. In order to bring such widely varying values into an easily expressible manner, Sørensen introduced the pH scale which is defined by

$$pH = -\log_{10} a_{H^+} = -\log_{10} m_{H^+} \gamma_{H^+} \quad (13.28)$$

where a_{H^+} = activity of H^+ ion, m_{H^+} = molality of H^+ and γ_{H^+} = activity coefficient of H^+. In dilute solutions one may write,

$$pH = -\log_{10} m_{H^+} = -\log_{10} C_{H^+} \text{ (since } \gamma_{H^+} = 1) \quad (13.29)$$

Thus it is seen that the higher the H^+ ion concentration, the lower is the pH. The pH of water is 7 and it is neutral. One may define similarly

$$pOH = -\log_{10} [OH^-] \text{ (assuming } \gamma_{OH^-} = 1) \quad (13.29a)$$

From Eqs 13.13 and 13.14,

$$pH + pOH = pK_w = 14 \quad (13.30)$$

13.2.5 Acidity Measurements in Concentrated Aqueous Acid Solutions and Non-aqueous Solvents

A measure of acidity, which can be used in concentrated solutions of strong acids as well as in non-aqueous solvents, was proposed by Hammett in 1932. It is based on the measurement of the ratio of $[BH^+]/[B]$, that is, that of the conjugate acid of a base B and the base itself of an indicator B by a spectrophotometric measurement. If B is the base, its pK_{BH^+} may be defined as

$$pK_{BH^+} = -\log K_{BH^+} = -\log\left\{\frac{[B][H^+]}{[BH^+]} \times \frac{\gamma_B \gamma_{H^+}}{\gamma_{BH^+}}\right\} \quad (13.31)$$

Rewriting the above as activity of the H^+ ion,

$$pK_{BH^+} + \log\left\{\frac{[B]}{[BH^+]}\right\} = -\log\left\{a_{H^+}\frac{\gamma_B}{\gamma_{BH^+}}\right\} \quad (13.32)$$

a_{H^+} is the activity of proton and γs are the activity coefficients of B and BH^+. In Eq. 13.32, the right hand side has been designated as H_0, the subscript 0 referring to an uncharged base. By using indicators such as nitroanilines used by Hammett, it is possible to measure the H_0, in concentrated solutions of acids in water and also in non-aqueous and mixed solvents. By this method, it has been possible to measure acidities of even pure H_2SO_4, which has a $H_0 = -12$, and of fluorosulfuric acid which has a $H_0 = -15.0$. It overcomes the limitations of pH measurements, say by a glass electrode, which is limited only to dilute aqueous solutions. This method also provides a method of comparing acidities in different solvents.

13.2.6 Common Ion Effect

Addition of a common ion affects the position of equilibrium in an acid–base reaction. For example, consider the ionisation equilibrium of acetic acid in water (Eq. 13.4). The addition of a common ion such as acetate ions suppresses the ionisation and the equilibrium is displaced towards the left resulting in a decreased ionisation of the acid, that is, a lowered value of α. The same situation happens if H^+ ions are added in the form of, say, HCl. Similarly, the ionisation of weak bases is suppressed by the addition of a salt having a common ion.

13.2.7 Acid-Base Neutralisation Curves

The addition of a base to a solution of acid in aqueous media results in a change of pH. Although the equivalence point in such titrations may be determined by conductivity measurements or potential measurements, it is interesting to find out the pH changes during the titration which is best done by a potentiometric titration using a conventional potentimetric setup or by using a glass electrode.

(i) Titration of a strong acid vs a strong base: In this titration, the pH initially changes slowly on the addition of a base (say HCl vs NaOH) but the change is rapid and sharp when the equivalence point is reached. In fact the pH changes by as much as 7 pH units at the equivalence point enabling the determination of equivalence point sharply. Therefore any acid–base indicator which changes the colour in the pH range 3.0 to 10.0 may be used for locating the equivalence point. Indicators such as methyl orange, phenolphthalein and methyl red may be used.

(ii) Titration of a weak acid vs a strong base: Consider the titration of acetic acid solution against sodium hydroxide solution as an example. The addition of NaOH solution results in the formation of sodium acetate and the acetic acid–sodiumacetate mixture acts as a buffer. Hence the change in pH initially is relatively little but as the equivalence point is reached, there is a sharp rise in the pH but it is only about 3 pH units. Hence an indicator which changes colour in the pH range 7.0 to 10.0 is effective. Thus phenolphathalein or thymol blue are preferred indicators in this titration. In this titration, the pK_a of the acid is equal to the pH at the half-neutralisation point.

13.2.8 Acid-Base Indicators

The equivalence point in acid–base titrations can also be determined by the use of compounds known as acid–base indicators. Some examples of such indicators are methyl orange, phenolphthalein, methyl red, thymol blue, and so on. They are usually either weak acids or bases, have one colour in acid solutions and a different colour in basic solutions. They change colour over a certain pH range. Representing an indicator by HInd and its dissociation equilibrium as

$$\text{HInd} + H_2O \rightleftharpoons H_3O^+ + \text{Ind}^- \quad (13.33)$$

The indicator constant K_{HInd}, is given by

$$K_{\text{HInd}} = \frac{[H_3O^+][\text{Ind}^-]}{[\text{HInd}]} \quad (13.34)$$

$$[H_3O^+] = K_{\text{HInd}} \times \frac{[\text{unionised form of HInd}]}{[\text{ionised form of HInd}]} \quad (13.35)$$

In acid solutions, the colour will entirely be due to HInd while in alkaline solutions, the colour is entirely due to Ind^-. The actual colour at any given point of the titration depends on the ratio, $[\text{HInd}]/[\text{Ind}^-]$ and thus on the pH of the solution. Equation 13.35 may also be written as

$$\text{pH} = \text{p}K_{\text{HInd}} + \log\frac{[\text{Ind}^-]}{[\text{HInd}]} \quad (13.36)$$

The colour change interval of an indicator is $\text{pH} = pK_{\text{HInd}} \pm 1$. The indicator changes colour from one form to the other in this pH range.

13.3 Buffers and Buffer action

Buffers are solutions which resist change in their pH on the addition of small amounts of strong acids or bases. They consist of a mixture of weak acid and its salt or a weak base and its salt. Buffer solutions have a range over which they are effective and there are several buffer systems covering a wide range of pH values. The equation for the pH or pOH of a buffer consisting of a weak acid and its salt or a weak base and its salt may be obtained by considering Eq. 13.4 and 13.4a or 13.5 as

From Eqs. 13.4 and 13.4a,

$$\text{pH} = \text{p}K_a + \log\frac{[\text{Salt}]}{[\text{Acid}]} \quad (13.37)$$

From Eq. 13.5,

$$\text{pOH} = \text{p}K_b + \log\frac{[\text{Salt}]}{[\text{Base}]} \quad (13.38)$$

Equations 13.37 and 13.38 are known by the name Henderson–Hasselbach equation. Consider for example, the effect of addition of a small amount of a strong acid on an acetic acid–sodium acetate buffer, the reaction

$$H_3O^+ + CH_3COO^- \rightleftharpoons CH_3COOH + H_2O \quad (13.39)$$

prevents a significant change in pH. Similarly, the reaction

$$OH^- + CH_3COOH \rightleftharpoons CH_3COO^- + H_2O \quad (13.40)$$

prevents an appreciable change in pH on the addition of small amounts of a base like NaOH.

The quantity termed *buffer capacity* defined by

$$\beta = \frac{dZ}{dpH} \quad (13.41)$$

represents the effectiveness of a buffer. It is the ratio of the number of moles of acid or base (dZ) added to a litre of buffer to bring out a change in pH (dpH) by one unit. The larger the value of β, the more effective the buffer is. Buffers have important applications in analytical and biochemical investigations. Here, $dZ = dC_a$ or dC_b where C_a, C_b are the concentrations of acid or base added to the buffer. Since buffer capacity is always positive and in the case of added acid the pH decreases, we write

$$\beta = \frac{dC_b}{dpH} = -\frac{dC_a}{dpH} \quad (13.42)$$

13.4 Hydrolysis of Salts

A salt formed from the neutralisation of a strong acid by a strong base in aqueous solution is neutral and its pH is equal to seven, an example being sodium chloride formed by the neutralisation of HCl with NaOH. The salt solution is neutral and pH = 7.0. However, the salts formed out of the neutralisation of a weak acid by a strong base, or of a weak base by a strong acid, or of a weak acid by a weak base do not give solutions which are neutral. This is because the salts formed react with water and release acid or base, or both. Since the reaction of the salts is with water, they are known as hydrolysis reactions. These reactions are only acid–base reactions in a Brönsted–Lowry sense.

13.4.1 Hydrolysis of the Salt of a Weak Acid and Strong Base

As an example, the hydrolysis of sodium acetate will be considered. The equilibrium may be represented by

$$CH_3COO^-(Na^+) + H_2O \rightleftharpoons CH_3COOH + (Na^+) + (OH^-) \quad (13.43)$$

The hydrolysis constant, K_h, of the reaction is given by

$$K_h = \frac{[CH_3COOH][OH^-]}{[CH_3COO^-]} \quad (13.44)$$

If α is the degree of hydrolysis

$$K_h = \frac{\alpha^2 C}{(1-\alpha)} \approx \alpha^2 C \text{ (when } \alpha \ll 1) \quad (13.45)$$

and C is the stoichiometric concentration of the salt, K_h is equal to

$$K_h = \frac{K_W}{K_a} \quad (13.46)$$

where K_w = ionic product of water and K_a = dissociation constant of the weak acid. From Eq. 13.45 and 13.46,

$$\alpha = \sqrt{\frac{K_h}{C}} = \sqrt{\frac{K_W}{K_a C}} \quad (13.47)$$

The concentration of OH$^-$, that is, [OH$^-$] = αC

$$\left[OH^-\right] = \alpha C = \sqrt{\frac{K_W C}{K_a}} \quad (13.48)$$

The pH of the solution is given by

$$pH = pK_W - p_{OH^-} = \frac{1}{2}pK_W + \frac{1}{2}pK_a + \frac{1}{2}\log C \quad (13.49)$$

13.4.2 Hydrolysis of the Salt of a Weak Base and Strong Acid

As an example, we consider the hydrolysis of pyridinium chloride. The equilibrium may be represented by

$$C_5H_5NH^+ + (Cl^-) + H_2O \rightleftharpoons C_5H_5N + H_3O^+ + (Cl^-) \quad (13.50)$$

The hydrolysis constant, K_h, is given by

$$K_h = \frac{[C_5H_5N][H_3O^+]}{[C_5H_5NH^+]} = \frac{\alpha^2 C}{(1-\alpha)} \approx \alpha^2 C \quad (13.51)$$

α is the degree of hydrolysis of the salt at a stoichiometric concentration of the salt, C. K_h is then equal to

$$K_h = \frac{K_W}{K_b} \quad (13.52)$$

where K_w is the ionic product of water and K_b is the dissociation constant of the base, pyridine.

$$\alpha = \sqrt{\frac{K_h}{C}} = \sqrt{\frac{K_W}{K_b C}} \quad (13.53)$$

$$pH = -\log[H_3O^+] = -\log(\alpha C) = \frac{1}{2}pK_W - \frac{1}{2}pK_b - \frac{1}{2}\log C \quad (13.54)$$

13.4.3 Hydrolysis of a Salt of a Weak Acid and Weak Base

Consider the hydrolysis constant of anilinium acetate as an example. The hydrolysis equilibrium may be given as

$$C_6H_5NH_3^+ + CH_3COO^- + H_2O + H_2O \rightleftharpoons C_6H_5NH_2 + CH_3COOH + H_3O^+ + OH^- \quad (13.55)$$

The hydrolysis constant K_h is given by

$$K_h = \frac{[C_6H_5NH_2][CH_3COOH]}{[C_6H_5NH_3^+][CH_3COO^-]} \quad (13.56)$$

$$K_h = \frac{\alpha^2}{(1-\alpha)^2} = \frac{K_W}{K_b K_a} \quad (13.57)$$

where K_w = ionic product of water, K_a = dissociation constant of acetic acid, K_b = dissociation constant of aniline. From Eq. 13.57,

$$\alpha = \frac{\sqrt{K_h}}{\sqrt{K_h} + 1} \quad (13.58)$$

The pH of the solution may be obtained by considering that

$$K_a = \frac{[CH_3COO^-][H_3O^+]}{[CH_3COOH]}$$

$$[H_3O^+] = K_a \frac{[CH_3COOH]}{[CH_3COO^-]} \quad (13.59)$$

$$= K_a \frac{\alpha}{(1-\alpha)} = K_a \sqrt{\frac{K_W}{K_a K_b}} \quad (13.60)$$

$$pH = \frac{1}{2}[pK_W + pK_a - pK_b] \quad (13.61)$$

The pH of the solution depends on the strength of the weak acid and the weak base, that is, the values of K_a and K_b. However, α is indepenedent of salt concentration as seen from Eq. 13.58.

13.5 Solubility Product and Solubility of Sparingly Soluble Salts

Considering the solubility equilibrium of a sparingly soluble salt X_aY_b, we may write

$$X_aY_b(s) \rightleftharpoons X_aY_b(\text{saturated solution}) \rightleftharpoons aX^{z+} + bY^{z-} \quad (13.62)$$

The thermodynamic solubility product of the salt is defined by

$$K_{SP} = a^a_{(X^{z+})} \times a^b_{(Y^{z-})} \quad (13.63)$$

where $a^a_{(X^{z+})}$ and $a^b_{(Y^{z-})}$ with the corresponding superscripts represent the activities of the X and Y species. In dilute solution, the activities may be replaced by the corresponding concentrations and Eq. 13.63 may be written as

$$K'_{Sp} = C^a_{(X^{z+})} \times C^b_{(Y^{z-})} \quad (13.64)$$

where K'_{Sp} is the classical solubility product and for all practical purposes, K'_{Sp} will be considered for all further discussions. Let us express the solubility product in terms of solubility of a few salts.

Example 1: Silver phosphate, Ag_3PO_4

$$Ag_3PO_4(s) \rightleftharpoons 3Ag^+(\text{solution}) + PO_4^{3-}(\text{solution}) \quad (13.65)$$

If S is the solubility of salt in mol L^{-1},

$$[Ag^+] = 3S; [PO_4^{3-}] = S$$

$$K_{SP} = [Ag^+]^3[PO_4^{3-}]^1 = (3S)^3 \times s = 27S^4 \quad (13.66)$$

Example 2: Lanthanum oxalate, $La_2(C_2O_4)_3$

$$La_2(C_2O_4)_3(s) \rightleftharpoons 2La^{3+}(\text{solution}) + 3C_2O_4^{2-}(\text{solution}) \quad (13.67)$$

$$[La^{3+}] = 2S; [C_2O_4^{2-}] = 3S \quad (13.68)$$

$$K_{SP} = [La^{3+}]^2[C_2O_4^{2-}]^3 = (2S)^2(3S)^3 = 108S^5 \quad (13.69)$$

13.5.1 Variation of Solubility Product with Temperature

$$\left(\frac{\partial \ln K_{SP}}{\partial T}\right)_P = \frac{\Delta H^0_{Soln}}{RT^2} \quad (13.70)$$

On integration,

$$\ln\frac{(K_{SP})_2}{(K_{SP})_1} = \frac{\Delta H^0_{Soln}}{R}\left[\frac{1}{T_1} - \frac{1}{T_2}\right] \quad (13.71)$$

assuming that ΔH^0_{Soln} is independent of temperature in the temperature range T_1 to T_2. Using Eq. 13.71, either ΔH^0_{Soln} or K_{sp} at another temperature may be calculated.

13.5.2 Effect of Complex Formation on Solubility

The solubility of a sparingly soluble salt increases with the formation of complex ion. For example, the solubility of $Zn(OH)_2$ in water is almost 2.5×10^{-6} M but the solubility increases by nearly two thousand fold in 1 M NH_3. This may be attributed to the complex formation according to

$$Zn(OH)_2 + 4NH_3 \rightleftharpoons Zn(NH_3)_4^{2+} + 2OH^- \quad (13.72)$$

as it shifts the solubility equilibrium to the right.

13.5.3 Effect of Common Ion on the Solubility of Sparingly Soluble Salts

The addition of a common ion, say Cl^- ions, decreases the solubility of AgCl in water due to the shift of the equilibrium to the left.

$$AgCl(s) \rightleftharpoons Ag^+(\text{solution}) + Cl^-(\text{solution}) \quad (13.73)$$

13.5.4 Application of Solubility Product Principle to Qualitative Analysis

Let us consider the precipitation of group II metal ions such as Cu, Pb and Cd as their sulfides and those of group IV metal ions such as Mn and Zn. The group II metal ions are precipitated in the acid medium by H_2S where the sulfide ion concentration from the dissociation of

$$H_2S \rightleftharpoons 2H^+ + S^{2-}$$

is lowered due to the presence of H^+ ions from the acid. However, this low S^{2-} ion concentration is enough to exceed the solubility product of CuS, PbS, and so on. Hence they get precipitated. When one considers the precipitation of ZnS and MnS, this occurs in presence of NH_4OH which provides OH^- ions which enhances the dissociation of H_2S. Thus, a high enough concentration of S^{2-} ions is available. This enables Zn and Mn ions to be precipitated as their sulfides by H_2S.

13.6 Redox Equilibria

In considering redox reactions, the concept of oxidation number, which refers to the number of electrons gained or lost in forming an anion or a cation during a reaction must be borne in mind. Cations have positive oxidation numbers and anions have negative numbers. Oxidation numbers of an atom or ion may be positive, negative or fractional.

13.6.1 Balancing Equations

An oxidising agent is a compound which increases the oxidation number of an element of the oxidation species. In the study of redox equilibria, the relevant redox equation is important and necessitates consideration of a balanced redox reaction. The balancing of such equations is done by two methods (i) ion electron method and (ii) oxidation state method. The former method will be considered by means of an example—the oxidation of iodide ions by dichromate ions in an acid medium. The relevant partial equations are,

(i) $Cr_2O_7^{2-} \rightarrow 2Cr^{3+} + 7H_2O$

(ii) $I^- \rightarrow \frac{1}{2}I_2 + e^-$

To balance charges in (i), add $14H^+$ and $6e^-$ to the left to make the equation

(iii) $Cr_2O_7^{2-} + 14H^+ + 6e^- \rightarrow 2Cr^{3+} + 7H_2O$

And to balance electrons in (ii) with (iii), multiply it by 6

(iv) $6I^- \rightarrow 3I_2 + 6e^-$

Add (iii) and (iv) to give

$$Cr_2O_7^{2-} + 14H^+ + 6I^- \rightarrow 2Cr^{3+} + 7H_2O + 3I_2 \qquad (13.74)$$

To complete the equation, the relevant counter ions are added to give

$$K_2Cr_2O_7 + 7H_2SO_4 + 6KI \rightarrow 4K_2SO_4 + Cr_2(SO_4)_3 + 7H_2O + 3I_2 \qquad (13.75)$$

13.6.2 Redox Potentials

An oxidation–reduction reaction at an electrode is represented as

$$\text{Oxidised state} + ne^- \rightleftharpoons \text{Reduced state} \qquad (13.76)$$

And the electrode potential corresponding to this reaction is given by the Nernst equation.

$$= E^0 - \frac{RT}{nF}\ln\frac{a_{\text{reduced state}}}{a_{\text{oxidised state}}} \qquad (13.77)$$

where E = electrode potential, E^0 = standard electrode potential, n = number of electrons transferred and F = Faraday constant. The value of $\frac{RT}{nF}\ln = 2.303\frac{RT}{nF}\log = \frac{0.0591}{n}\log$ at 298 K. The a represents the activity of the relevant species. Since the electrode potentials are always represented as reduction reactions vis a vis SHE, they are referred to as reduction potentials or standard reduction potentials as the case may be. (For details of sign conventions, see the Chapter 3 on Electrochemistry.)

As an example of the information that can be obtained from the sign and magnitude of the standard oxidation–reduction potentials of various redox systems, the standard redox potentials of $MnO_4^- - Mn^{2+}$, $Cr_2O_7^{2-} - Cr^{3+}$ and $Fe^{3+} - Fe^{2+}$ will be considered.

(i) Standard oxidation–reduction potential of $MnO_4^- - Mn^{2+} / Pt = 1.51$ V

(ii) Standard oxidation–reduction potential of $Cr_2O_7^{2-} - Cr^{3+} / Pt = 1.33$ V

(iii) Standard oxidation–reduction potential of $Fe^{3+} - Fe^{2+} / Pt = 0.771$ V

Let us first consider the cell

$$Pt \mid Fe^{3+} - Fe^{2+}(\text{in } H_2SO_4 \text{ medium}) \parallel MnO_4^- - Mn^{2+}(\text{in } H_2SO_4 \text{ medium}) \mid Pt \qquad (13.78)$$

The EMF of cell under standard state conditions $E^0 = E_R^0 - E_L^0 = 1.51 - 0.771 = 0.739$ V

Since the sign of E^0 is positive, it implies that the reaction

$$2MnO_4^- + 16H^+ + 10Fe^{2+} \rightarrow 2Mn^{2+} + 10Fe^{3+} + 8H_2O \qquad (13.79)$$

(or the complete reaction $2KMnO_4 + 8H_2SO_4 + 10FeSO_4 \rightarrow 2MnSO_4 + K_2SO_4 + 5Fe_2(SO_4)_3 + 8H_2O$)

is feasible and the equilibrium constant of the reaction at 298 K is given by

$$\log K = \frac{nFE^0}{2.303RT} = \frac{10 \times 96,500 \times 0.739}{2.303 \times 8.314 \times 298} = 125$$

In the case of the $K_2Cr_2O_7$–$FeSO_4$ reaction in H_2SO_4 medium, represented by the cell,

$$Pt\ Fe^{3+} - Fe^{2+}(\text{in } H_2SO_4 \text{ medium}) \parallel Cr_2O_7^{2-} - Cr^{3+}(\text{in } H_2SO_4 \text{ medium}) \mid Pt \qquad (13.80)$$

The EMF of the cell under standard conditions, E^0_{cell} is given by

$$E^0 = E_R^0 - E_L^0 = 1.33 - 0.771 = 0.559 \text{ V}$$

The reaction

$$Cr_2O_7^{2-} + 14H^+ + 6Fe^{2+} \rightarrow 2Cr^{3+} + 6Fe^{3+} + 7H_2O \qquad (13.81)$$

(or the complete reaction $K_2Cr_2O_7 + 7H_2SO_4 + 6FeSO_4 \rightarrow K_2SO_4 + Cr_2(SO_4)_3 + 3Fe_2(SO_4)_3 + 7H_2O$)

is possible and its equilibrium constant at 298 K is given by

$$\log K = \frac{nFE^0}{2.303RT} = \frac{6 \times 96,500 \times 0.559}{2.303 \times 8.314 \times 298} = 56.7$$

It is seen that the oxidation of Fe^{2+} ions by $KMnO_4$ proceeds so much more to the right than in the case of oxidation by $K_2Cr_2O_7$. Thus, the permanganate system which has a more positive redox potential is a more powerful oxidising agent compared to the dichromate system. This analysis can be extended to other redox systems.

13.6.3 Effect of Complex Formation on Redox Potentials

The standard electrode potential of Ag^+, Ag system is 0.800 V at 298 K but this value becomes less positive in the presence of ammonia because Ag^+ ions get complexed with ammonia to form $Ag(NH_3)_2^+$. Thus, the standard potential of the reaction

$Ag(NH_3)_2^+ + e^- \rightarrow Ag + 2NH_3$ is 0.373 V. Thus, complexation has a powerful effect on the electrode potential with the potential becoming less positive or less oxidising by almost 0.43 V, in this case. In fact, these two data can be used to calculate the stability constant of the $[Ag(NH_3)_2]^+$ complex ion. The formation of a silver mirror by the reduction of silver ions by glucose in the presence of ammonia can be explained using the above data as well as the potential data of the glucose–gluconic acid system.

13.6.4 Potentiometric Titrations Involving Redox Reactions

Although acid–base titrations can be carried out by using conductivity measurements as well as potentiometric measurements, redox titrations cannot be carried by using conductivity methods because of the significant acid medium that is needed to carryout redox titrations. However potentiometric titrations of such reactions can be performed using an appropriate redox half cell in conjunction with a reference electrode such as SCE according to,

$$Pt|M^{Z+}, M||SCE \text{ (where M, } M^{z+} \text{ is the redox system)} \quad (13.82)$$

Consider the titration of ferrous ions by cerium(IV) ions being added from a burette. The reaction is

$$Fe^{2+} + Ce^{4+} \rightarrow Fe^{3+} + Ce^{3+} \quad (13.83)$$

Using the cell

$$Pt|Fe^{2+}\left(\text{a trace of } Fe^{3+} \text{ is always present}\right)||SCE \quad (13.84)$$

Before the equivalence point, the electrode potential of Fe^{2+}, Fe^{3+} half cell is given by

$$E = E_1^0 - \frac{RT}{F}\ln\frac{C_{Fe^{2+}}}{C_{Fe^{3+}}} \quad (13.85)$$

And after the equivalence point, the potential of the Ce^{4+},Ce^{3+} system becomes important and is given by

$$E = E_2^0 - \frac{RT}{F}\ln\frac{C_{Ce^{3+}}}{C_{Ce^{4+}}} \quad (13.86)$$

and at the equivalence point, the electrode potential E_{EP} will be

$$E_{EP} = \frac{1}{2}(E_1^0 + E_2^0) \quad (13.87)$$

And thus the potential changes by almost 400 mV at the equivalence point which enables it to be located accurately. Therefore, potentiometric titrations can be successfully employed to determine the equivalence point in redox titrations. Redox titrations can also be performed using suitable redox indicators. A suitable redox indicator must be found which changes colour within the potential range at which the equivalence point occurs.

13.6.5 Potentiometric Titrations Involving Precipitation Reactions

The titrations of $AgNO_3$ solution by NaCl solution may be considered as an example. The following cell, involving a silver electrode dipping into a solution of silver nitrate combined with a reference electrode like SCE, given by

$$Ag|AgCl|Cl^-||SCE \quad (13.88)$$

is often employed. Since addition of NaCl results in the formation of AgCl, the cell (Eq. 13.88) has been depicted involving the Ag, AgCl electrode. The EMF of the cell (Eq. 13.88) at 298 K is given by

$$E = E_{SCE} - E_{Ag,\,AgCl} = E_{SCE} - E^0_{Ag,\,AgCl} - 0.0591\log a_{Cl^-} \quad (13.89)$$

Since the activity of the Cl^- ions is controlled by the solubility product of AgCl, Eq. 13.89 may be written as

$$E = E_{SCE} - E^0_{Ag,\,AgCl} - 0.0591\log K_{SP} + 0.0591\log a_{Ag^+} \quad (13.90)$$

$$= \text{Constant} + 0.0591\log a_{Ag^+} \quad (13.90a)$$

Thus the EMF of the cell (Eq. 13.88) depends on the activity of the Ag^+ ions, which decreases with the addition of NaCl. The E of the cell therefore decreases initially and the change in EMF is rapid near the equivalence point.

Worked Out Problems

Brönsted Acids and Bases, Lewis Acids and Bases, Hard-Soft Acids and Bases

Problem 13.1

(a) Indicate the conjugate acid and base of the following species.

(i) $H_2PO_4^-$ (vii) HS^-
(ii) OH^- (viii) $C_2H_5O^-$
(iii) H_2O (ix) NH_3
(iv) H_2SO_4 (x) HF_2^-
(v) CH_3COOH (xi) HSO_3F
(vi) HCO_3^-

(b) Identify the Lewis acids and bases among the following. Indicate the species among them that may act as Brönsted acids or bases.

(i) NH_3, (ii) $AlCl_3$, (iii) H_2O, (iv) $SnCl_4$, (v) Cu^{2+}, (vi) $FeCl_3$
(vii) CO_2, (viii) Ag^+, (ix) Cl^-, (x) Hg^{2+}, (xi) HSO_4^-, (xii) HS^-

Solution

(a) (i) H_3PO_4 and HPO_4^{2-}

(ii) H_2O and O^{2-}

(iii) H_3O^+ and OH^-

(iv) $H_3SO_4^+$ and HSO_4^-

(v) $CH_3COOH_2^+$ and CH_3COO^-

(vi) H_2CO_3 and CO_3^{2-}

(vii) H_2S and S^{2-}

(viii) C_2H_5OH and $C_2H_5OH_2^+$

(ix) NH_4^+ and NH_2^-

(x) H_2F_2 and F_2^{2-}

(xi) $H_2SO_3F^+$ and SO_3F^-

(b) (i) Lewis base (ii) Lewis acid (iii) Lewis base and Brönsted acid and base (iv) Lewis acid (v) Lewis acid (vi) Lewis acid (vii) Lewis acid (viii) Lewis acid (ix) Lewis base (x) Lewis acid (xi) Lewis base and Brönsted acid–base (xii) Lewis base and Brönsted acid–base.

Problem 13.2

(a) Briefly explain hard acids, softacids and hard bases, soft bases, with two examples for each.

(b) Consider the following reactions and point out the soft–hard species in both reactants and products (where possible).

$$\text{ZnO (zinc oxide)} + 2LiC_4H_9 (n\text{-butyl lithium}) \rightarrow Zn(C_4H_9)_2 + Li_2O$$

$$2Al^{3+} + 3SO_4^{2-} \rightarrow Al_2(SO_4)_3$$

$$2Ag^+ + S^{2-} \rightarrow Ag_2S$$

$$Cd^{2+} + 4CN^- \rightarrow Cd(CN)_4^{2-}$$

$$Mg^{2+} + CO_3^{2-} \rightarrow MgCO_3$$

Solution

(a) Hard acids are compact molecules, small in size, have high charge density. They are mostly metal ions and the charge is held tightly to the nucleus. They are not very polarisable. Hard bases have similar properties. Soft acids and soft bases are large molecules or ions and are highly polarisable. Hard acids tend to interact with hard bases while soft acids tend to interact with soft bases. Examples of hard acids: Fe^{3+}, Al^{3+}; hard bases: CH_3COO^-, ROH; soft acids: BH_3, Tl^+; soft bases: $S_2O_3^{2-}$, I^-.

(b) $\underset{\text{Soft–hard}}{ZnO} + \underset{\text{hard–soft}}{2LiC_4H_9} \rightarrow \underset{\text{soft–soft}}{Zn(C_4H_9)_2} + \underset{\text{hard–hard}}{Li_2O}$

$\underset{\text{Hard acid}}{2Al^{3+}} + \underset{\text{Hard base}}{3SO_4^{2-}} \rightarrow \underset{\text{Hard–Hard}}{Al_2(SO_4)_3}$

$\underset{\text{soft acid}}{2Ag^+} + \underset{\text{soft base}}{S^{2-}} \rightarrow \underset{\text{soft–soft}}{Ag_2S}$

$\underset{\text{soft acid}}{Cd^{2+}} + \underset{\text{soft base}}{4CN^-} \rightarrow \underset{\text{soft–soft}}{Cd(CN)_4^{2-}}$

$\underset{\text{Hard acid}}{Mg^{2+}} + \underset{\text{Hard base}}{CO_3^{2-}} \rightarrow \underset{\text{Hard–Hard}}{MgCO_3}$

Classical and Thermodynamic Dissociation Constants of Acids and Bases, Interionic Effects

Problem 13.3

The classical dissociation constant of $HClO_2$, chlorous acid, is 1.1×10^{-2} at 298 K. (a) What is the degree of dissociation and pH of an acid solution of 0.1 M concentration? (b) If the interionic effects in the acid solution are taken into consideration, what is the thermodynamic K_a value? By how much per cent does the pH change on the basis of this value of K_a?

Solution

$$HClO_2 + H_2O \rightleftharpoons H_3O^+ + ClO_2^-$$

If C is the stoichiometric concentration of the acid (0.1 M),

$$K_a = \frac{\alpha^2 C}{1-\alpha} \text{ where } \alpha = \text{degree of dissociation}$$

$$K_a = \frac{\alpha^2 \times 0.1}{1-\alpha}; \frac{\alpha^2}{1-\alpha} = \frac{1.1 \times 10^{-2}}{0.1} = 1.1 \times 10^{-1}$$

$$\alpha^2 + 0.11\alpha - 0.11 = 0$$

$$\alpha = \frac{-0.11 \pm \sqrt{0.0121 + 0.44}}{2} = 0.281$$

$$pH = -\log[H_3^+O] = -\log(0.281 \times 0.1) = -\log(0.0281) = 1.55$$

$$\text{Thermodynamic } K_a = K_a' \times \gamma_\pm^2$$

$$\text{and } -\log \gamma_\pm = A|Z_+Z_-|\sqrt{\mu} = 0.5 \times 1 \times 1 \times \sqrt{0.0281}$$

$$\left(\text{Since } \mu = \frac{1}{2} C_i Z_i^2 = \frac{1}{2} \times 0.0281 \times 1^2 + \frac{1}{2} \times 0.0281 \times 1^2 = 0.0281 \right)$$

$$\gamma_\pm = 0.825$$

$$\text{Thermodynamic } K_a = 1.1 \times 10^{-2} \times (0.825)^2 = 7.49 \times 10^{-3}$$

$$\alpha \text{ on the basis of this } K_a \text{ is given by, } \frac{\alpha^2}{1-\alpha} = 7.49 \times 10^{-2}$$

$$\text{or } \alpha = 0.239$$

$$pH = -\log[0.239 \times 0.1] = -\log(0.0239) = 1.62$$

$$\text{Change in pH} = 1.62 - 1.55 = 0.07$$

$$\% \text{ change or increase in pH} = \frac{0.07}{1.62} \times 100 = \frac{7}{1.62} = 4.3$$

Problem 13.4

The first and second dissociation constants of carbonic acid (H_2CO_3) are 4.2×10^{-7} and 4.8×10^{-11} respectively in water at 298 K. (i) Calculate the CO_3^{2-} ion concentration in a 0.2 M solution of carbonic acid. (ii) What will be the CO_3^{2-} ion concentration in the presence of 0.1 M HCl of 0.2 M carbonic acid?

Solution

(a) $H_2CO_3 + H_2O \xrightleftharpoons{K_{a_1}} H_3O^+ + HCO_3^-$

$$HCO_3^- + H_2O \xrightleftharpoons{K_{a_2}} H_3O^+ + CO_3^{2-}$$

$$K_{a_1}\times K_{a_2} = \frac{[H_3O^+]^2[CO_3^{2-}]}{[H_2CO_3]}$$

$$= 4.2\times4.8\times10^{-18} = 2.02\times10^{-17}$$

Considering the first ionisation, $K_{a_1} = \dfrac{[H_3O^+][HCO_3^-]}{[H_2CO_3]}$

Since the second ionisation is nearly 10,000 fold smaller, it will not influence the first ionisation significantly to change the $[H_3O^+]$ concentration.

Therefore, $[H_3O^+] = [HCO_3^-]$

$$\therefore K_{a_1} = 4.2\times10^{-7} = \frac{[H_3O^+]^2}{0.2}$$

$$[H_3O^+] = \sqrt{8.4\times10^{-8}} = 2.9\times10^{-4}$$

Substituting this value in

$$\frac{[H_3O^+]^2[CO_3^{2-}]}{[H_2CO_3]} = 2.02\times10^{-17}$$

$$[CO_3^{2-}] = \frac{2.02\times10^{-17}\times0.2}{(2.9\times10^{-4})^2} = 4.76\times10^{-11}$$

(b) In this case, $[H_3O^+] = 0.1$, $[H_2CO_3] = 0.2$

$$\text{Therefore } [CO_3^{2-}] = \frac{2.02\times10^{-17}\times0.2}{(0.1)^2}$$

$$= 0.404\times10^{-15} = 4.04\times10^{-16}$$

Problem 13.5

(a) Consider the following equilibria in aqueous carbonic acid–bicarbonate buffer which regulates pH changes in blood.

$$H_3O^+(aq) + HCO_3^-(aq) \xrightleftharpoons{K_1} H_2CO_3(aq) + H_2O \xrightleftharpoons{K_2} 2H_2O(l) + CO_2(g)$$

Derive an equation for the pH of the buffer, involving a composite pK'' (comprising K_1 and K_2) and the ratio of CO_2 and bicarbonate ion concentrations.

(b) The pH of blood is 7.4. If the pH falls below 7.2 or rises above 7.6 serious complications may arise to human health. Given the composite pK'' under (a) as 6.34, calculate the ratio of $[CO_2]/[HCO_3^-]$ that needs to be maintained so that the pH does not exceed the above limits. What is this ratio at the pH of blood?

(c) Tris(hydroxymethylamino)methane $[NH_2C(CH_2OH)_3]$ and its hydrochloride is another suitable buffer for the maintenance of pH of blood. Given the K_b of the base as 1×10^{-6}, calculate the ratio of $[\text{Base}]/[\text{Conjugate acid}]$ if the pH of blood is to be maintained at 7.4.

Solution

(a) $$K_1 = \frac{[H_2CO_3]}{[H_3O^+][HCO_3^-]} \quad (1)$$

$$K_2 = \frac{[CO_2]}{[H_2CO_3]} \quad (2)$$

$$K' = \frac{1}{K_1} = \frac{[H_3O^+][HCO_3^-]}{[H_2CO_3]} \quad (3)$$

$$[H_2CO_3] = \frac{[H_3O^+][HCO_3^-]}{K'} = \frac{[CO_2]}{K_2} \quad (4)$$

Using Eq. (4),

$$[H_3O^+] = \frac{[CO_2]}{[HCO_3^-]}\frac{K'}{K_2}, \quad (5)$$

$$-\log[H_3O^+] = pH = -\log K'' - \log\frac{[CO_2]}{[HCO_3^-]}$$

$$\left(\text{where } K'' = \frac{K'}{K_2}\right)$$

$$= pK'' - \log\frac{[CO_2]}{[HCO_3^-]} \quad (6)$$

(b) If the pH is not to fall below 7.2, the ratio of $[CO_2]/[HCO_3^-]$ should be

$$\log\frac{[CO_2]}{[HCO_3^-]} = pK'' - pH = 6.34 - 7.2 = -0.86 = \bar{1}.14$$

$$\frac{[CO_2]}{[HCO_3^-]} = 0.138$$

If the pH is not to go above 7.6, the ratio of $[CO_2]/[HCO_3^-]$ should be

$$\log\frac{[CO_2]}{[HCO_3^-]} = pK'' - pH = 6.34 - 7.6 = -1.26 = \bar{2}.74$$

$$\frac{[CO_2]}{[HCO_3^-]} = 0.055$$

$[CO_2]/[HCO_3^-]$ at pH = 7.4

$$\log\frac{[CO_2]}{[HCO_3^-]} = pK'' - pH = 6.34 - 7.4 = -1.06 = \bar{2}.94$$

$$\frac{[CO_2]}{[HCO_3^-]} = 0.0871$$

(c) The pK_b of tris(hydroxymethylamino)methane $[NH_2C(CH_2OH)_3] = 6.0$, if pH = 7.4; pOH = 14 – 7.4 = 6.6

$$pOH = pK_b + \log\frac{[\text{salt}]}{[\text{base}]}; \log\frac{[\text{salt}]}{[\text{base}]} = 6.6 - 6.0 = 0.6$$

$$\frac{[\text{salt}]}{[\text{base}]} = 3.98; \text{or} \frac{[\text{base}]}{[\text{salt}]} = \frac{[\text{base}]}{[\text{conjugate acid}]} = \frac{1}{3.98} = 0.251$$

Self-ionisation Equilibria, pH in Self-ionising Solvents

Problem 13.6

The self ionisation constant of deuterium oxide (D_2O) (given as pK_D is 14.96 at 298 K and 14.62 at 308 K. (a) Calculate the enthalpy of neutralisation of DCl by NaOD. (b) Consider D_2O at 298 K and 308 K. Which of the two solutions is more acidic? (c) What is the neutral pD in D_2O at 298 K? (d) What is the pD range in D_2O at 298 K?

Solution

(a) $$\frac{\log(K_D)_2}{\log(K_D)_1} = \frac{\Delta H_{ion}}{2.303R}$$

$$\left[\frac{1}{T_1} - \frac{1}{T_2}\right]; \begin{array}{l} K_{D_1} \text{ at } 298\,\text{K} = \text{antilog}(\overline{15}.04) = 1.096\times10^{-15} \\ K_{D_2} \text{ at } 308\,\text{K} = \text{antilog}(\overline{15}.38) = 2.40\times10^{-15} \end{array}$$

$$\log\left(\frac{2.40\times10^{-15}}{1.096\times10^{-15}}\right) = \frac{\Delta H_{ion}}{2.303\times8.314}\left[\frac{308-298}{308\times298}\right]$$

$$= \frac{\Delta H_{ion}}{19.147}\times\frac{10}{91784}$$

$$\Delta H_{ion} = \frac{0.340\times19.147\times91784}{10\times1000} = 59.75 \text{ kJ mol}^{-1}$$

This value represents the enthalpy change for the reaction

$$2D_2O\ (l) \rightleftharpoons D_3O^+(\text{soln}) + OD^-(\text{soln})$$

The enthalpy of neutralisation of DCl by NaOD, therefore, will be the opposite to the above reaction.

$$\therefore \Delta H_{neut} = -59.65 \text{ kJ mol}^{-1}$$

(b) At 298 K $K_{D_1} = 1.096\times10^{-15}$

$$pD = pOD = -\log\sqrt{1.096\times10^{-15}} = 7.48$$

At 308 K $K_{D_2} = 2.40\times10^{-15}$

$$pD = pOD = -\log\sqrt{2.40\times10^{-15}} = 7.35$$

Since the self-dissociation of D_2O is higher at 308 K, and its pD being smaller than that at 298 K, D_2O at 308 K is more acidic than that at 298 K.

(c) Neutral pD in D_2O is 7.48.

(d) The pD range of D_2O at 298 K is 1 to 14.96.

Problem 13.7

The autopyrolysis constant of methanol at 298 K is 1.74×10^{-17}. Calculate the $p_{CH_3OH_2^+}$ (equivalent to pH in H_2O) of a 10^{-9} M solution of $HClO_4$ in this solvent.

Solution

The neutral point in this solvent can be calculated as

$$[CH_3OH_2^+][OCH_3^-] = 1.74\times10^{-17}$$

$$[CH_3OH_2^+] = [OCH_3^-] = \sqrt{1.74\times10^{-17}} = 4.17\times10^{-9}$$

$$p_{CH_3OH_2^+} = -\log[CH_3OH_2^+] = -\log(4.17\times10^{-9}) = 8.38$$

Any $p_{CH_3OH_2^+}$ value greater than 8.38 is basic and any value less than this is acidic. Since the given H^+ concentration is closer to the neutral point, in calculating the $p_{CH_3OH_2^+}$, one has to consider the contribution of protons from the ionisation of the solvent. Using the electroneutrality condition,

$$[CH_3OH_2^+] = [OCH_3^-] + [ClO_4^-]$$

$$[CH_3OH_2^+] = \frac{K_{CH_3OH}}{[CH_3OH_2^+]} + 1\times10^{-9}$$

Simplifying, $[CH_3OH_2^+]^2 = 1.74\times10^{-17} + 1\times10^{-9}[CH_3OH_2^+]$

From the quadratic equation,

$$[CH_3OH_2^+]^2 - 1\times10^{-9}[CH_3OH_2^+] - 1.74\times10^{-17} = 0$$

$$[CH_3OH_2^+] = \frac{1\times10^{-9} \pm \sqrt{(1\times10^{-9})^2 + 6.96\times10^{-17}}}{2}$$

$$= \frac{1\times10^{-9} \pm \sqrt{10^{-18} + 69.6\times10^{-18}}}{2}$$

$$= \frac{1\times10^{-9} \pm \sqrt{70.6\times10^{-18}}}{2}$$

$$= \frac{1\times10^{-9} \pm 8.4\times10^{-9}}{2} = \frac{9.4\times10^{-9}}{2} = 4.7\times10^{-9}$$

$$p_{CH_3OH_2^+} = -\log[4.7\times10^{-9}] = 8.33$$

This quantity being less than 8.38, is acidic and well below 9 which we normally expect.

Problem 13.8

(a) The solubility product of $Cd(OH)_2$ in water at 298 K is 7.2×10^{-15}. The stability constant for the reaction

$$Cd^{2+} + 4NH_3 \rightleftharpoons Cd(NH_3)_4^{2+} \quad K_{sta} = 1.0\times10^7$$

Calculate the solubility of Cd $(OH)_2$ in 2 M NH_3.

(b) The degree of ionisation of ethanol (in pure ethanol) is 3.1×10^{-11}. (i) Calculate the concentration of ethoxonium and ethoxide ions in the pure solvent. (ii) What is its autoprotolysis constant? Give one reason why this value is so different from that of water. (iii) What is the $p_{C_2H_5OH_2^+}$ of a solution containing 0.1 M $HClO_4$ in ethanol? (iv) 0.139 g of lithium metal is dissolved in 200 mL ethanol. What will the $p_{C_2H_5OH_2^+}$ of this solution be (assume temperature to be 298 K) (density of ethanol = 0.800 g/cm³)?

Solution

(a) Reaction of interest is

$$Cd(OH)_2 + 4NH_3 \rightleftharpoons Cd(NH_3)_4^{2+} + 2OH^- \qquad (1)$$

If b mol L^{-1} is the solubility of $Cd(OH)_2$ in 2 M ammonia,

$[NH_3]$ at equilibrium $= (2-4b)$; $[Cd(NH_3)_4^+]$
$= b$; $[OH^-] = 2b$

$$K_{\text{eq.const}} \text{ of the reaction } (1) = K_{SP(Cd(OH)_2)} k_{\text{Stab.const}}$$
$$= 7.2\times10^{-15}\times1\times10^{7} = 7.2\times10^{-8}$$

$$\frac{[b][2b]^2}{(2-4b)^4} = 7.2\times10^{-8} \text{ since } 4b \ll 2, \text{we can write}$$

$$\frac{4b^3}{16} = 7.2\times10^{-8}; b = \sqrt[3]{28.8\times10^{-8}} = 6.61\times10^{-3}\,\text{M}$$

(It is possible to refine this value of b, by iterating with the first calculated value 6.61×10^{-3}.)

(b) Reaction: $2\,C_2H_5OH \rightleftharpoons C_2H_5OH_2^+ + C_2H_5O^-$

In 1 L of ethanol, 800 g is present.
Molality of ethanol = 800/46 = 17.4

$\alpha = 3.1\times10^{-11}$, conc. of ethoxide
= conc. of ethoxonium ion

$$\alpha C = 3.1\times10^{-11}\times17.4 = 5.394\times10^{-10}$$

$$K_{\text{auto prot const}} = [5.394\times10^{-10}\times5.394\times10^{-10}]$$
$$= 2.92\times10^{-19}$$

This value is about 10^5 times smaller than that of water. One important reason for this is the lower dielectric constant of ethanol ($\approx$ 20) compared to water ($\approx$ 78).

$$[C_2H_5OH_2^+] = 0.1 \text{ or } 0.1\times[C_2H_5O^-]$$
$$= 2.92\times10^{-19};\ [C_2H_5O^-] = 2.92\times10^{-18}$$

$$p_{C_2H_5O^-} = -\log[C_2H_5O^-] = -\log[2.92\times10^{-18}] = 17.54$$

$$\text{Conc.of Li ethoxide} = \frac{0.139}{6.934}\times\frac{1000}{200}$$
$$= 0.10\text{ M};[0.10][C_2H_5OH_2^+]$$
$$= 2.92\times10^{-19}$$

$$[C_2H_5OH_2^+] = 2.92\times10^{-18}$$

$$p_{C_2H_5OH_2^+} = -\log[2.92\times10^{-18}] = 17.54$$

Since the pH range in this solvent is 0–18.54, the above solution is highly alkaline in ethanol.

Different Buffer Systems, Buffer Action and Buffer Capacity

Problem 13.9

(a) Considering the acetic acid–sodium acetate buffer, derive an equation for the buffer capacity $\frac{dC_B}{dpH}$ where C_B = number of moles of a base added to one litre of a buffer and dpH is the change in pH, in terms of K_W, $[H^+]$, K_a = dissociation constant of a acetic acid and

$$[CH_3COO^-]_T = [CH_3COO^-] + [CH_3COOH]$$

(b) Calculate the buffer capacity values at pH = 2.0 and pH = 4.0 from the derived equation using $[X]_T = 0.1$ where $[X]_T = [CH_3COO^-] + [CH_3COOH]$

Solution

(a) The dissociation equilibrium of acetic acid is given by

$$CH_3COOH + H_2O \rightleftharpoons CH_3COO^- + H_3O^+$$

When considering the buffering capacity in the presence of added NaOH, the mass balance and charge balance become important.

Mass balance equation: $[X]_T = [CH_3COO^-] + [CH_3COOH]$ (1)

Charge balance equation:

$$[H^+] + [Na^+] = [OH^-] + [X^-]\{[X^-] = CH_3COO^-\} \qquad (2)$$

If C_B is the concentration on of the base added, $[Na^+] = C_B$ (3)

$$\text{From Eq.}(2), [Na^+] = C_B = \frac{K_W}{[H^+]} + [X^-] - [H^+] \qquad (4)$$

$$\text{But} \frac{[X_T]}{[X^-]} = 1 + \frac{[CH_3COOH]}{[CH_3COO^-]} = 1 + \frac{[H^+]}{K_a} \qquad (5)$$

$$\text{From Eq. } (5),\ [X^-] = \frac{[X_T]}{1+\frac{[H^+]}{K_a}} \qquad (6)$$

Substituting for $[X^-]$ in Eq. (4),

$$C_B = \frac{K_W}{[H^+]} - [H^+] + \frac{[X_T]}{1+\frac{[H^+]}{K_a}} \qquad (7)$$

$$= \frac{K_W}{[H^+]} - [H^+] + [X_T]\times\frac{K_a}{K_a+[H^+]} \qquad (8)$$

Differentiating with respect to $[H^+]$ we get from Eq. (8),

$$\frac{dC_B}{d[H^+]} = -\frac{K_W}{[H^+]^2} - 1 + [X_T]\times\frac{-K_a\times1}{\{K_a+[H^+]\}^2} \qquad (9)$$

But $\dfrac{d[H^+]}{dpH} = -\ln 10[H^+]$ multiplying Eq. (9) by this term, we get

$$\frac{dC_B}{dpH} = +\ln 10\left[\frac{K_W}{[H^+]} + [H^+] + [X_T] \times \frac{K_a[H^+]}{\{K_a + [H^+]\}^2}\right] \quad (10)$$

$$= 2.303\log 10\left[\frac{K_W}{[H^+]} + [H^+] + [X_T] \times \frac{K_a[H^+]}{\{K_a + [H^+]\}^2}\right] \quad (11)$$

Equation (11) expresses the variation of $\dfrac{dC_B}{dpH}$ in terms of three parts.

(b) At pH = 2 with $[X]_T = 0.1$, we have

$$\frac{dC_B}{dpH} = 2.303\left[\frac{10^{-14}}{10^{-2}} + 10^{-2} + 0.1 \times \frac{1.8\times10^{-5}\times10^{-2}}{\{(1.8\times10^{-5}) + 10^{-2}\}^2}\right]$$

$$= 2.303\left[10^{-12} + 10^{-2} + 0.1\times\frac{1.8\times10^{-7}}{3.24\times10^{-10} + 10^{-4}}\right]$$

$$= 2.303\left[10^{-2} + 0.1\times\frac{1.8\times10^{-7}}{10^{-4}}\right]$$

(neglecting 10^{-12} and 3.24×10^{-10} in the denominator of last term)

$$= 2.303[0.01 + 0.1\times0.0018] = 0.0234$$

At pH = 4, we have

$$\frac{dC_B}{dpH} = 2.303\left[\frac{10^{-14}}{10^{-4}} + 10^{-4} + 0.1\times\frac{1.8\times10^{-5}\times10^{-4}}{\{(1.8\times10^{-5}) + 10^{-4}\}^2}\right]$$

$$= 2.303\left[10^{-10} + 10^{-4} + 0.1\times\frac{1.8\times10^{-9}}{3.24\times10^{-10} + 10^{-8}}\right]$$

$$= 2.303\left[10^{-4} + 0.1\times\frac{1.8\times10^{-10}}{1.0324\times10^{-8}}\right]$$

$$= 2.303\,[10^{-4} + 0.0174] = 0.040$$

From the calculation, it seems as if $\dfrac{dC_B}{dpH}$ increases as we get closer to K_a.

Problem 13.10

The methyl ester of glycine ($CH_3COOCH_2NH_2$) and its hydrochloride is a suitable buffer in buffering neutral solutions. The pK_a of its conjugate acid is 2×10^{-8}. (i) Obtain an equation relating pH of the buffer to the pK_a of the conjugate acid. (ii) To prepare two litres of buffer of pH = 7.0 containing 0.04 M of the base, calculate the amount of the hydrochloride needed to add to the solution. (iii) If 1 mL of 1 M strong acid is added to 1 L of buffer under (ii), what is the change in pH? (iv) If 1 mL of a strong base is added to the same buffer under (ii) what is the change in pH? (v) Calculate the buffer capacity in (iii) and (iv). Do they have good buffer capacity?

Solution

$$CH_3COOCH_2NH_3^+ + H_2O + (Cl^-) \rightleftharpoons CH_3COOCH_2NH_2 + H_3O^+ + (Cl^-)$$

$$K_a = \frac{[CH_3COOCH_2NH_2][H_3O^+]}{[CH_3COOCH_2NH_3^+]} = 2\times10^{-8}; pK_a = 7.7$$

$$pK_a = -\log[H_3O^+] - \log\frac{[CH_3COOCH_2NH_2]}{[CH_3COOCH_2NH_3^+]}; pH$$

$$= pK_a + \log\frac{[base]}{[salt]}$$

$$7.0 = 7.7 + \log\frac{[0.04]}{[salt]}; \log\frac{[0.04]}{[salt]} = -0.7 = \bar{1}.30$$

$$\frac{[0.04]}{[salt]} = 0.2\ [salt] = \frac{0.04}{0.2} = 0.2$$

(i) Since the molecular weight of the hydrochloride is 125.5, weight of salt needed for 2 L of H_2O = $0.2\times125.2\times2 = 50.2$ g

(ii) 1 mL of 1 M acid to 1 L of buffer = 0.001 mol

(iii) Since the equilibrium reaction is shifted to the left, [base] = 0.04 – 0.001 = 0.039

[salt] = 0.20 + 0.001 = 0.201

$$\text{pH of solution} = 7.7 + \log\frac{0.039}{0.201} = 6.988$$

Change in pH = 7.0 – 6.988 = 0.012

$$\frac{db}{dpH} = \frac{0.001}{0.012} = 0.083$$

(iv) 1 mL of 1 M base added to 1 L of buffer = 0.001 M

$$\frac{db}{dpH} = \frac{0.001}{0.01} = 0.1$$

[salt] = 0.20 – 0.001 = 0.199

[base] = 0.040 + 0.001 = 0.0410

$$\text{pH of solution} = 7.7 + \log\frac{0.041}{0.199} = 7.7 - 0.69 = 7.01$$

Change in pH = 7.01 – 7.00 = 0.01

The buffer capacity is poor in both cases.

Problem 13.11

(a) To 0.2 mol of ethylamine, 0.1 mol of $HClO_4$ is added and the resulting solution is made up to two litres. Given that K_b of ethylamine is 5.6×10^{-4}, calculate the pH of the buffer.

(b) What is the pH of the buffer after addition of 2 mL of 1 N $HClO_4$?

(c) What is the buffer capacity in the solution in (b)?

(d) The pH of a solution consisting of 0.005 M boric acid and 0.09 M sodium hydrogen borate (NaH_2BO_3) is 9.4. What is the K_a of boric acid?

Solution

$$\text{Concentration of ethylamine after addition of } HClO_4 = \frac{0.1}{2} = 0.05$$

Concentration of ethylaminiumperchlorate after addition of

$$HClO_4 = \frac{0.1}{2} = 0.05$$

$$pOH = pK_b + \log\frac{[\text{salt}]}{[\text{base}]} = 3.25 + \log\frac{0.05}{0.05} = 3.25$$

$$pH = pK_W - pOH = 14 - 3.25 = 10.75$$

(a) When $HClO_4$ is added, the reaction is

$$C_2H_5NH_2 + HClO_4 \rightarrow C_2H_5NH_3^+ \cdot ClO_4^-$$

Thus the salt concentration increases while the $C_2H_5NH_2$ concentration decreases.

Millimoles of $HClO_4$ added $= 2 \times 1$

$= 2$ or moles of $HClO_4$

$= 2 \times 10^{-3}$

$$[C_2H_5NH_3^+.ClO_4^-] = 0.05 + \frac{0.002}{2} = 0.051$$

$$[C_2H_5NH_2] = 0.05 - \frac{0.002}{2} = 0.049$$

$$pOH = 3.25 + \log\frac{0.051}{0.049} = 3.25 + 0.017 = 3.267$$

$$pH = 14 - 3.267 = 10.733$$

(b) Buffer capacity $= \dfrac{db}{dpH}$, change in pH $= -(-0.017) = 0.017$

$$\text{Conc. of } HClO_4 \text{ added} = \frac{2\times10^{-3}}{2} = 1\times10^{-3}$$

$$\frac{db}{dpH} = \frac{1\times10^{-3}}{1.7\times10^{-2}} = 0.056$$

(c) $pH = pK_a + \log\dfrac{[\text{salt}]}{[\text{acid}]}$

$$pK_a = 9.4 - \log\frac{0.09}{0.005}$$

$$= 9.4 - 1.255 = 8.145$$

$$-\log K_a = -8.145,\ K_a = \text{antilog}\,(\bar{9}.855) = 7.16\times10^{-9}$$

Solubility and Solubility Product, Application to Qualitative Analysis, Concentration of an Ion in a Mixture of Salts

Problem 13.12

The solubility product of lead arsenate is 4.5×10^{-36} at 298 K. What is its solubility in g mol L^{-1}? (i) What is its solubility in the presence of 0.1 M $Pb(NO_3)_2$ solution? (ii) Will the ionic strength contributed by $Pb(NO_3)_2$ affect the solubility?

Solution

$$Pb_3(AsO_4)_2(s) \rightleftharpoons 3Pb^{2+}(aq) + 2AsO_4^{3-}(aq)$$

If S is the solubility of the salt,

$$[Pb^{2+}] = 3s; [AsO_4^{3-}] = 2s$$

$$K_{SP} = [Pb^{2+}]^3[AsO_4^{3-}]^2 = (3s)^3(2s)^2 = 108s^5$$

$$S = \sqrt[5]{\frac{4.5\times10^{-36}}{108}} = \sqrt[5]{4.167\times10^{-38}} = 2.655\times10^{-8}\ \text{mol L}^{-1}$$

(i) Let S' be the solubility in the presence of 0.1 M $Pb(NO_3)_2$.

$$[Pb^{2+}] = (3S' + 0.1); [AsO_4^{3-}] = 2S'$$

Neglecting $3S'$ in comparision to 0.1,

$$[0.1]^3[2S']^2 = 4.5\times10^{-36}$$

$$4S'^2 = 4.5\times10^{-33}$$

$$S' = \sqrt{1.125\times10^{-33}} = 3.35\times10^{-17}\ \text{mol L}^{-1}$$

(ii) Due to the interionic forces present in the solution, this solubility will be slightly higher, especially in the presence of Pb $(NO_3)_2$. This opposes the common ion effect of the salt.

Problem 13.13

(a) The solubility products of $BaCrO_4$ and $BaSO_4$ at 298 K are 1.2×10^{-10} and 1.5×10^{-9} respectively. What is the concentration of barium ions in a mixture of saturated solutions of the two salts?

(b) The solubility product of silver iodate in water at 298 K is 3.0×10^{-8}. What is its solubility in water? If instead of water a solution of 0.25 M $HClO_4$ is used, will its solubility decrease or increase and if so by how much? ($K_{HIO3} = 1.7\times10^{-1}$)

Solution

(a) In the mixture of saturated solutions,

$$[Ba^{2+}]_T = [CrO_4^{2-}] + [SO_4^{2-}]$$

$$= \frac{K_{BaCrO_4}}{[Ba^{2+}]_{\underset{\text{salt}}{CrO_4^{2-}}}} + \frac{K_{BaSO_4}}{[Ba^{2+}]_{\underset{\text{salt}}{SO_4^{2-}}}}$$

$$= \frac{1.2\times10^{-10}}{[Ba^{2+}]_{\underset{\text{salt}}{CrO_4^{2-}}}} + \frac{1.5\times10^{-9}}{[Ba^{2+}]_{\underset{\text{salt}}{SO_4^{2-}}}}$$

$[Ba^{2+}]^2 = 16.2\times10^{-10}$

$[Ba^{2+}] = \sqrt{16.2\times10^{-10}} = 4.02\times10^{-5}$

(b) $K_{SP} = [Ag^+][IO_3^-] = S^2$ where S is the solubility in g mol L^{-1}

$S^2 = 3\times10^{-8}; S = \sqrt{3\times10^{-8}} = 1.73\times10^{-4}$

The solubility increases since IO_3^- is being removed as HIO_3.

Let α be the fraction of IO_3^- being removed as HIO_3.

$$\alpha = \frac{[IO_3^-]}{[IO_3^-]+[HIO_3]} \text{ and } K_{HIO_3} = \frac{[H^+][IO_3^-]}{[HIO_3]}$$

$$\frac{[IO_3^-]}{[HIO_3]} = \frac{K_{HIO_3}}{[H^+]}$$

$$\frac{[IO_3^-]+[HIO_3]}{[IO_3^-]} = \frac{K_{HIO_3}+[H^+]}{K_{HIO_3}}$$

$$\text{Since } \alpha = \frac{[IO_3^-]}{[HIO_3]+[IO_3^-]} = \frac{1.7\times10^{-1}}{1.7\times10^{-1}+0.25}$$

$$\alpha = \frac{1.7\times10^{-1}}{4.2\times10^{-1}} = 0.405$$

In the presence of 0.25 M $HClO_4$,

$[Ag^+] = S; [IO_3^-] = \alpha S$ or $S\times\alpha S = \alpha S^2 = 3\times10^{-8}$

$$S^2 = \frac{3\times10^{-8}}{0.405} = 7.41\times10^{-8}; S = \sqrt{7.41\times10^{-8}} = 2.72\times10^{-4}$$

The solubility has increased from 1.73×10^{-4} to 2.72×10^{-4} mol L^{-1}.

Problem 13.14

(a) The solubility products of some group II and group IV sulfides at 298 K are given below.

Group II	K_{SP}	Group IV	K_{SP}
Cu(II)S	6.0×10^{-37}	Co(II)S	4×10^{-21}
Cd(II)S	8.0×10^{-27}	Mn(II)S	2.5×10^{-13}
Hg(II)S	1.6×10^{-52}	Ni(II)S	3.0×10^{-19}

Explain using calculations, why group II sulfides and not group IV sulfides are precipitated as their sulfides from 0.1 M solution salts containing 0.3 M acid under the same conditions.

(b) Explain why group IV metal cations require ammonical medium of H_2S for their precipication as sulfides.

(c) When H_2S is passed through a mixture containing 0.15 M $Pb(NO_3)_2$ and $Hg(NO_3)_2$ in 0.1 M HNO_3 medium, calculate the ratio of $[Hg^{2+}]/[Pb^{2+}]$ at the time of precipitation of Pb. K_{SP} (PbS) = 7×10^{-27}; K_{SP} (HgS) = 1.6×10^{-52}. Also $[H_2S]$ saturated in water = 0.1 M; $[H^+]^2[S^{2-}] = 1\times10^{-22}$.

Solution

(a) K_{SP} (CuS) is given by $[Cu^{2+}][S^{2-}]$.

$[S^{2-}]$ required for precipitation from 0.1 M solution of its salt

$$= \frac{K_{SP}}{[Cu^{2+}]} = \frac{6\times10^{-37}}{0.1} = 6\times10^{-36}$$

Similarly for Cd (II)S, $[S^{2-}]$ required $= \dfrac{8.0\times10^{-27}}{0.1} = 8\times10^{-26}$

Similarly for Hg(II)S, $[S^{2-}]$ required $= \dfrac{1.6\times10^{-52}}{0.1} = 1.6\times10^{-51}$

$[S^{2-}]$ concentration in a standard solution of H_2S containing 0.3 M acid is given by

$$[S^{2-}] = \frac{K_{H_2S}[H_2S]}{[H^+]^2} = \frac{1\times10^{-22}}{(0.3)^2} = \frac{1\times10^{-22}}{0.09} = 1.11\times10^{-21}$$

Product of $[Cu^{2+}][S^{2-}] = 1.11\times10^{-21}\times0.1 = 1.11\times10^{-22}$

The products of $[Cd^{2+}][S^{2-}]$ and $[Hg^{2+}][S^{2-}]$ are also 1.11×10^{-22}.

Since these products are greater than their K_{SP}s, the group (II) sulfides are precipitated in acid medium. Similarly, the products of Mn(II)S, Ni(II)S and Co(II)S, are also 1.11×10^{-22}. In all the three cases, their products are less than their K_{SP} values. Therefore the group IV sulfides are not precipitated under these conditions.

(b) In ammonical H_2S medium, the dissociation equilibrium of H_2S is shifted to the right because of the OH^- present in NH_4OH (though to a small extent). Therefore, the $[S^{2-}]$ concentration increases under these conditions, which enables the product of $[M^{2+}][S^{2-}]$ to exceed the K_{SP}s. Therefore they are precipitated under ammonical medium of H_2S.

(c) $[S^{2-}]$ in H_2S saturated aqueous medium in the presence of 0.1 M HNO_3 is,

$$\frac{K_{H_2S}[H_2S]}{[H^+]^2} = \frac{1\times10^{-22}}{(0.1)^2} = 1.10\times10^{-20}$$

Since the sulfide concentration is the same under the given experimental conditions, we can write,

$$\frac{[Pb^{2+}]}{[Hg^{2+}]} = \frac{\sqrt{K_{SP}(PbS)}}{\sqrt{K_{SP}(HgS)}} = \frac{\sqrt{7\times10^{-27}}}{\sqrt{1.6\times10^{-52}}}$$

$$= \frac{8.32\times10^{-14}}{1.26\times10^{-26}} = 6.6\times10^{12}$$

At the time of precipitation of Pb as PbS, the ratio may be written as

$$\frac{[0.15]}{[Hg^{2+}]} = 6.6\times10^{12} \text{ or } \frac{[Hg^{2+}]}{[0.15]}$$

$$= \frac{1}{6.6\times10^{12}}, \text{ that is, } [Hg^{2+}]$$

$$= \frac{0.15}{6.6\times10^{12}} = 2.27\times10^{-14}$$

As seen above, only a trace of Hg^{2+} ions will remain in solution.

Problem 13.15

(a) The K_{SP} values of $Al(OH)_3$, $Cr(OH)_3$ and $Fe(OH)_3$ in water at 298 K are 3.0×10^{-33}, 6.3×10^{-31} and 4.0×10^{-38} respectively. Considering a solution containing 0.02 M each in Al^{3+}, Cr^{3+} and Fe^{3+} ions, (i) find out which of the ions precipitate first on addition of small amounts of solid NaOH; (ii) calculate the minimum OH^- ion concentration necessary to precipitate each of the metal hydroxides.

(b) Calculate the percentage of Fe^{3+} ions that precipitate before Al^{3+} starts precipitating out.

(c) Calculate the percentage of Al^{3+} ions that precipitate before Cr^{3+} ions start precipitating out.

Solution

(a) (i) Since $Fe(OH)_3$ has the lowest K_{SP}, it starts precipitating out first followed by $Al(OH)_3$, and $Cr(OH)_3$.

$$K_{SP}(Fe(OH)_3) = 4\times10^{-38} \text{ or } [Fe^{3+}][OH^-]^3$$

$$= 4\times10^{-38}$$

$$[OH^-]^3 = \frac{4\times10^{-38}}{0.02} = 2\times10^{-36}$$

$$[OH^-] = \sqrt[3]{2\times10^{-36}} = 1.26\times10^{-12}$$

$$K_{SP}(Al(OH)_3) = 3\times10^{-33}$$

$$[OH^-] = \sqrt[3]{\frac{3\times10^{-33}}{0.02}} = \sqrt[3]{1.5\times10^{-31}} = 5.75\times10^{-11}$$

$$K_{SP}(Cr(OH)_3) = 6.3\times10^{-31}$$

$$[OH^-] = \sqrt[3]{\frac{6.3\times10^{-31}}{0.02}} = \sqrt[3]{3.15\times10^{-29}} = 3.16\times10^{-10}$$

$$K_{SP}(Fe(OH)_3) = 4\times10^{-38} \text{ or}$$

$$[Fe^{3+}][5.75\times10^{-11}]^3 = 4\times10^{-38}$$

$$[Fe^{3+}] = \frac{4\times10^{-38}}{1.5\times10^{-31}} = 2.67\times10^{-7} \text{ mol L}^{-1} \text{ remain in solution}$$

$$\% \text{ remaining in solution} = \frac{2.67\times10^{-7}}{0.02}\times100 = 0.00134\%$$

% precipitating out = 99.991

$$K_{SP}(Al(OH)_3) = 3\times10^{-33}$$

Cr^{3+} ions start precipitating out when the concentration reaches 3.16×10^{-10}.

$$\therefore [Al^{3+}][3.16\times10^{-10}]^3 = 3.0\times10^{-33} \text{ or}$$

$$[Al^{3+}] = \frac{3.0\times10^{-33}}{3.15\times10^{-29}} = 9.52\times10^{-5} \text{ mol L}^{-1}$$

remain in solution.

% remaining in solution

$$= \frac{9.52\times10^{-5}}{0.02}\times100 = 4.76\times10^{-1} = 0.476\%$$

% precipitating out = 99.524

Problem 13.16

(a) Calculate the weight of solid ammonium chloride necessary to add to a mixture of 500 mL of 0.2 M ammonium hydroxide and 500 mL of 0.1 M $Mn(NO_3)_2$ in order to prevent precipitation of $Mn(OH)_2$. K_{SP} $Mn(OH)_2 = 2.0\times10^{-13}$.

(b) Calculate the number of moles of ammonia required to be added to 5 L of water so that it will just dissolve 0.02 mol of Cu(I)Cl. $K_{SP} = 1.9\times10^{-7}$, $K_{\text{instability constant}}$ $Cu(NH_3)_4^+$ $= 2.08\times10^{-13}$.

Solution

(a) Conc. of NH_4OH in solution = 0.1 M

Conc. of $Mn(NO_3)_2 = 0.05$ M

$$K_{SP} = [Mn^{2+}][OH^-]^2 \text{ or } [OH^-]$$

$$= \sqrt{\frac{2.0\times10^{-13}}{0.05}} = \sqrt{4.0\times10^{-13}} = 2.0\times10^{-6}$$

The $[OH^-]$ should be greater than or equal to 2.0×10^{-6} M

$$NH_4OH \rightleftharpoons NH_4^+ + OH^-;\ \left[OH^-\right] = 2.0\times10^{-6}$$

$$= \frac{K_b\left[NH_4OH\right]}{\left[NH_4^+\right]}$$

$$2.0\times10^{-6} = \frac{1.8\times10^{-5}\times0.1}{\left[NH_4^+\right]} \text{ or } \left[NH_4^+\right] = \frac{1.8\times10^{-6}}{2.0\times10^{-6}} = 0.9\text{M}$$

Weight of NH_4Cl required = 0.9×53.5 = 48.15 g (Mol. wt. of NH_4Cl = 53.5)

(b) Conc. of CuCl = 0.02/5 =0.004 M

$$\underset{4\times10^{-3}}{CuCl} + \underset{4\times4\times10^{-3}}{4NH_3} \rightleftharpoons \underset{4\times10^{-3}}{Cu(NH_3)_4^+} + \underset{4\times10^{-3}}{Cl^-}$$

$$CuCl \rightleftharpoons Cu^+ + Cl^- \quad K_{SP} = [Cu^+][Cl^-] = 1.9\times10^{-7}$$

$$\text{Equilibrium conc. of } Cu^+ = \frac{K_{SP}}{[Cl^-]} = \frac{1.9\times10^{-7}}{4\times10^{-3}} = 4.75\times10^{-4}$$

$$K_{\text{inst cont.}} \text{ of } Cu(NH_3)_4^+ = 2.08\times10^{-13} = \frac{[Cu^+][NH_3]^4}{[Cu(NH_3)_4^+]}$$

$$\frac{\left[4.75\times10^{-4}\right]\left[NH_3\right]^4}{4\times10^{-3}} = 2.08\times10^{-13}$$

$$\left[NH_3\right]^4 = \frac{2.08\times10^{-13}\times4\times10^{-3}}{4.75\times10^{-4}}$$

$$= 1.75\times10^{-12}$$

$$[NH_3]^4 = \sqrt[4]{1.75\times10^{-12}} = 1.15\times10^{-3}$$

$$[NH_3]_T \text{ required} = 1.15\times10^{-3} + (4\times4\times10^{-3})$$
$$= 17.15\times10^{-3} = 0.0172 \text{ M}$$

Moles of NH_3 of required for 5 L of H_2O = 5×0.0172 = 0.0860.

Problem 13.17

(a) The solubility of lead sulfate in water at 298 K is 48.2 mg L^{-1}. In the determination of lead as lead sulfate by gravimetry, 0.1 g of the precipitate was washed using 1 L of 0.005 N H_2SO_4, dried and weighed. What percentage of loss arises due to washing of the precipitate by the acid? (molar mass of $PbSO_4$ = 303.3)

(b) A 1×10^{-3} M solution of aniline hydrochloride in water was analysed by using a galvanic cell, Pt, $H_2(g)$1 atm | H^+(?) || SCE, which showed an EMF of 0.467 V at 298 K. What is the hydrogen ion concentration of the solution and what is the hydrolysis constant of the salt? (E_{SCE} = 0.242 V)

Solution

$$K_{SP} \text{ of lead sulfate} = [Pb^{2+}][SO_4^{2-}] = \left(\frac{0.0482}{303.3}\right)^2 = 2.5\times10^{-8}$$

Let S' be the solubility in the presence of H_2SO_4.

$$[Pb^{2+}] = S';\ [SO_4^{2-}] = (S' + 0.0025)$$

$$\text{Molarity of the acid used} = \frac{0.005}{2} = 0.0025$$

$$S'(S'+0.0025) = 2.5\times10^{-8} \text{ assuming } S' \ll 0.0025$$

$$S' = \frac{2.5\times10^{-8}}{2.5\times10^{-3}} = 1\times10^{-5}\ S' \text{ in } \frac{g}{L}$$
$$= 1\times10^{-5}\times303.3 = 3.033\times10^{-3}$$

This amount of $PbSO_4$ is lost due to washing

$$\%\text{loss} = \frac{3.033\times10^{-3}}{0.1}\times100 \approx 3.03$$

$$C_6H_5NH_3^+ + (Cl^-) + H_2O \rightleftharpoons C_6H_5NH_2 + H_3O^+$$

Cell set up: Pt, $H_2(g)$ 1 atm | H^+(?) || SCE

$$E_{cell} = E_{SCE} - E_{H_2,H^+} = 0.242 - (E^0_{H_2} + 0.0591\log a_{H^+})$$
$$= 0.242 + 0.0591\text{pH}$$

$$\text{pH} = \frac{0.467-0.242}{0.0591} = \frac{0.225}{0.0591}$$
$$= 3.81; -\log a_{H^+} = 3.81\left[H^+\right] = 1.55\times10^{-4}$$

$$K_h = \frac{[C_6H_5NH_2][H_3O^+]}{[C_6H_5NH_3^+]} = \frac{(1.55\times10^{-4})^2}{0.001}$$
$$= 2.400\times10^{-8}\times10^3 = 2.400\times10^{-5}$$

pH of Mixtures of Acids and Bases

Problem 13.18

(a) Calculate the pH of an aqueous solution consisting of a mixture of 0.05 M pyridine (K_b = 1.5×10^{-9}) and 0.01 M aniline ($K_b = 4.2\times10^{-10}$).

(b) Calculate the degree of hydrolysis and pH of a 0.1 M solution of ammonium hydrosulfide (NH_4HS), given K_b (NH_4OH) = 1.8×10^{-5} and $K_{H_2S} = 9.1\times10^{-8}$.

Solution

$$K_b \text{ of pyridine} = \frac{[Py_{H^+}][OH^-]}{[Py]};[OH^-] = \frac{K_b(py)\times[py]}{[Py_{H^+}]}$$

$$K_b \text{ of aniline} = \frac{[An_{H^+}][OH^-]}{[An]};\ [OH^-] = \frac{K_b(An)\times[An]}{[An_{H^+}]}$$

$$\text{by neutrality, } [OH^-] = [Py_{H^+}] + [An_{H^+}]$$
$$= \frac{1.5\times10^{-9}\times0.05}{[OH^-]} + \frac{4.2\times10^{-10}\times0.01}{[OH^-]}$$

$$[OH^-]^2 = 0.075\times10^{-9} + 0.042\times10^{-10}$$
$$= 75\times10^{-12} + 4.2\times10^{-12} = 79.2\times10^{-12}$$

$$[OH^-] = \sqrt{79.2\times10^{-12}} = 8.9\times10^{-6}$$
$$\text{pOH} = -\log[OH^-] = 5.05$$
$$\text{pH} = 14 - \text{pOH} = 14 - 5.05 = 8.95$$

$$K_h(NH_4HS) = \frac{K_w}{K_a\times K_b} = \frac{10^{-14}}{1.8\times10^{-5}\times9.1\times10^{-8}}$$
$$= \frac{10^{-14}}{16.38\times10^{-13}} = 6.1\times10^{-3}$$

$$\alpha = \frac{\sqrt{K_h}}{\sqrt{K_h}+1} = \frac{\sqrt{6.1\times10^{-3}}}{\sqrt{6.1\times10^{-3}}+1} = \frac{7.8\times10^{-2}}{8.8\times10^{-2}} = 0.886$$

$$\text{pH} = \frac{1}{2}\text{p}K_w + \frac{1}{2}\text{p}K_a - \frac{1}{2}\text{p}K_b = 7.0 + 3.52 - 2.37 = 8.15$$

Problem 13.19

(a) The dissociation constants of acetic and phenyl acetic acids in water at 298 K are 1.8×10^{-5} and 5.0×10^{-5} respectively. (i) Calculate the pH of a mixture of 0.1 M acetic and 0.05 M phenylacetic acid respectively. (ii) What are the concentrations of the anions in the mixture? (iii) Calculate the volume of water required to be added to 1000 mL of 0.1 M of the stronger acid so that both solutions have the same pH.

Solution

(a) (i) Using electroneutrality condition

$$[H_3O^+] = [CH_3COO^-] + [C_6H_5CH_2COO^-]$$
$$= K_a\frac{[CH_3COOH]}{[H_3O^+]} + K'_a\frac{[C_6H_5CH_2COOH]}{[H_3O^+]}$$

$$[H_3O^+]^2 = 1.8\times10^{-5}\times0.1+5.0\times10^{-5}\times0.05$$
$$= 0.18\times10^{-5}+0.25\times10^{-5}$$
$$[H_3O^+] = \sqrt{4.3\times10^{-6}} = 2.07\times10^{-3}$$
$$pH = -\log[2.07\times10^{-3}] = 2.68$$

(ii) $$[CH_3COO^-] = \frac{1.8\times10^{-5}\times0.1}{2.07\times10^{-3}}$$
$$= 8.696\times10^{-4}$$
$$[C_6H_5CH_2COO^-] = \frac{5.0\times10^{-5}\times0.05}{2.07\times10^{-3}}$$
$$= 1.208\times10^{-3}$$

(iii) H_3O^+ from acetic acid = $\alpha_1C_1\,[H_3O^+]$
From phenylacetic acid = α_2C_2

$$\alpha_1C_1 = \alpha_2C_2 \text{ but } \alpha_1C_1 = \frac{K_a}{\alpha_1} \text{ and } \alpha_2C_2 = \frac{K'_a}{\alpha_2}$$

$$\frac{C_1}{C_2} = \frac{\alpha_2}{\alpha_1} = \frac{K'_a}{K_a} = \frac{5.0\times10^{-5}}{1.8\times10^{-5}} = 2.778$$

$$\text{but } C\propto\frac{1}{V} \text{ or } \frac{C_1}{C_2} = \frac{V_2}{V_1} = 2.778 \text{ and } V_2 = V_1\times2.778$$

Thus to 1 L of 0.1 M phenylacetic acid, one has to add 1778 mL of water such that both the solutions have the same pH.

To check the above result:

The pH of solution of the weaker acid, acetic acid at 0.1 M concentration

$$K_a \approx \alpha^2C = 1.8\times10^{-5}$$
$$\alpha^2 = 1.8\times10^{-4}$$
$$\alpha = \sqrt{1.8\times10^{-4}} = 1.34\times10^{-2}$$
$$\left[H_3O^+\right] = \alpha C = 1.34\times10^{-3}$$
$$pH = -\log\left[H_3O^+\right]$$
$$= -\log\left[1.34\times10^{-3}\right] = 2.87$$

The concentration of the stronger acid will, on increasing its volume from 1 L to 2.778 L, will come down to $0.1\times1/2.778$ = 0.036 M. The degree of dissociation of the acid now is

$$\alpha^2C = \alpha^2\times0.036 = 5.0\times10^{-5}$$
$$\alpha = \sqrt{\frac{5.0\times10^{-5}}{3.6\times10^{-2}}} = \sqrt{1.389\times10^{-3}}$$
$$= 3.715\times10^{-2}$$
$$[H_3O^+] = \alpha C = 3.715\times10^{-2}\times0.036$$
$$= 13.37\times10^{-4} = 1.337\times10^{-3}$$

$pH = -\log[H_3O^+] = -\log[1.337\times10^{-3}] = 2.87$ which is the same as the pH of the weaker acid.

Problem 13.20

The thermodynamic pK_b values of a ammonium hydroxide and tris(hydroxymethylamino methane) $(CH_2OH)_3CNH_2$ in water at 298 K are 4.76 and 6.00 respectively. (i) Calculate the pH of 0.1 M solution of each in a mixture. (ii) Will the degree of dissociation of each base change in the presence of 0.1 M NaCl added to the mixture? If so, calculate the pH of the mixture in the presence of the salt.

Solution

The thermodynamic K_b of the bases is based on activities and depends only on the temperature.

$$NH_4OH \xrightleftharpoons{K_b} NH_4^+ + OH^-$$
$$(CH_2OH)_3CNH_2 + H_2O \xrightleftharpoons{K'_b} (CH_2OH)_3CNH_3^+ + OH^-$$

By electroneutrality,

$$[OH^-] = [NH_4^+]+[TrisH^+] = K_b\frac{[NH_4OH]}{[OH^-]} + K'_b\frac{[Tris]}{[OH^-]}$$
$$= \frac{1.75\times10^{-5}\times0.1}{[OH^-]} + \frac{1.0\times10^{-6}\times0.1}{[OH^-]}$$
$$[OH^-]^2 = 1.75\times10^{-6}+0.1\times10^{-6}$$
$$= 1.85\times10^{-6}$$
$$[OH^-] = \sqrt{1.85\times10^{-6}} = 1.36\times10^{-3}$$
$$pOH = -\log[OH^-]$$
$$= -[\log 1.36\times10^{-3}] = 2.87$$
$$pH = 14 - 2.87 = 11.13$$

In the presence of 0.1 M NaCl

$$K_b = \frac{m_{NH_4^+}m_{OH^-}}{m_{NH_4OH}}.\gamma_\pm^2 (\text{where } \gamma_\pm$$
$$= (\gamma_{NH_4^+}\gamma_{OH^-})^{1/2} \text{ and } \gamma_{NH_4OH}\approx1$$
$$\log\gamma_\pm = -AZ_+Z_-\sqrt{\mu} = -0.5\times1\times\sqrt{0.1}$$
$$= -0.5\times1\times0.32 = -0.16 = \bar{1}.84$$
$$\gamma_\pm \text{ (due to ions of } NH_4OH) = 0.692$$
$$K_b \approx \alpha^2m\times(0.692)^2;\ \frac{1.75\times10^{-4}}{(0.692)^2} = \alpha^2\times0.1$$
$$\alpha^2 = \frac{1.75\times10^{-4}}{0.479}$$
$$\alpha = \sqrt{3.653\times10^{-4}} = 1.911\times10^{-2}$$
$$\alpha C = [OH^-] = 1.911\times10^{-3}$$
$$pOH = -\log[OH^-] = -\log[1.911\times10^{-3}] = 2.72$$

Similarly, in the case of tris $\gamma_\pm$ = 0.692 (since, it is also a1:1 electrolyte)

$$K'_b = \alpha'^2\times m\times(0.692)^2 \text{ or } \frac{1\times10^{-6}}{(0.692)^2\times0.1} = \alpha'^2$$

$$\alpha'^2 = \frac{1\times10^{-5}}{0.479} = 2.088\times10^{-5}$$

$$\alpha' = \sqrt{20.88\times10^{-6}} = 4.57\times10^{-3}$$

$$\alpha'C = 4.57\times10^{-4} \text{ or } [OH^-] = 4.57\times10^{-4}$$

$$\text{Total } [OH^-] = 1.911\times10^{-3} + 4.57\times10^{-4}$$

$$= 2.368\times10^{-3}$$

$$pOH = -\log[OH^-]$$

$$= -\log[2.368\times10^{-3}] = 2.63$$

$$pH = 14 - 2.63 = 11.37$$

Potentiometric Acid–Base and Precipitation Titrations

Problem 13.21

(a) Draw a neat sketch of the titration curve of a 0.1 M H_3PO_4 to 0.1 M NaOH as pH vs per cent acid titrated. How many inflexions are observed and at what pHs are they seen? If visual titrations using indicators are to be performed, what indicators are used to observe these inflexions?

(b) (i) H_3PO_4 can be determined as a triprotic acid by precipitating PO_4^{3-} ions as calcium phosphate and titrating the liberated HCl using phenolphathalein or methyl orange as indicator or using a glass electrode and performing a potentiometric titration. Write the relevant reaction. (ii) 100 mL of an aqueous solution of H_3PO_4 was treated with an excess of an aqueous solution of calcium chloride. The resulting calcium phosphate was filtered off. 25 mL of the filtrate required 50 mL of 0.1 N NaOH for complete neutralisation. What is the normality of H_3PO_4? How much weight of H_3PO_4 is present in the original solution? (iii) The following galvanic cell has been set up to obtain the pH vs % reagent (acid) titrated as in Problem 13.20(a).

Pt |H_2Q : Q(1 : 1), Solution of unknown pH || SCE

where H_2Q = hydroquinone, Q = quinone, $E_0 = 0.699$ V and $E_{OCE} = 0.242$ V. What is the EMF of the above cell at the inflexion points shown in Problem 13.21(a)? What is the pH at 0.00 V?

Solution

(a) Two inflexions are observed corresponding to the neutralisation of the 1st and 2nd protons as per the reactions

$$H_3PO_4 \rightleftharpoons H_2PO_4^- + H^+$$

$$H_2PO_4 \rightleftharpoons HPO_4^{2-} + H^+$$

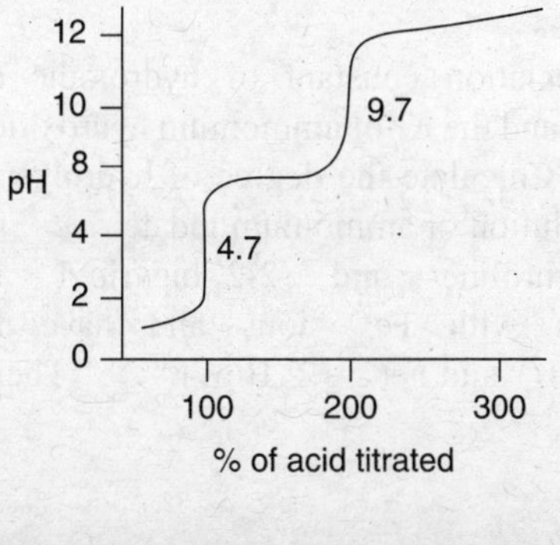

The pH indicators suitable to observe the end points at the end of each of the stages are methyl orange (1st inflexion), and thymolphthalein (2nd inflexion).

(b) (i) The normality of the filtrate is 50×0.1/25 = 0.2 N

Number of equivalents of H_3PO_4 in 100 mL

$$= \frac{100\times0.2}{1000} = 0.02$$

Weight of $H_3PO_4 = 0.02\times98/3 = 0.653$ g

(ii) Reaction corresponding to release of HCl,

$$2H_3PO_4 + 3CaCl_2 \rightarrow Ca_3(PO_4)_2\downarrow + 6HCl$$

Pt | H_2Q : Q(1 : 1), Solution of unknown pH || SCE

$$E_{cell} = E_{SCE} - E_{H_2Q,Q}$$

$$= E_{SCE} - (E^0_{H_2Q,Q} - 0.0591\log pH)$$

$$E_{cell} = E_{SCE} - E^0_{H_2Q,Q} + 0.0591\text{ pH}$$

$$\text{At pH 4.7}: E_{cell} = 0.242 - 0.699 + 0.0591\times4.7$$

$$= -0.179 \text{ V}$$

$$\text{At pH 9.7}: E_{cell} = 0.242 - 0.699 + 0.0591\times9.7$$

$$= +0.116 \text{ V}$$

when $E_{cell} = 0.00$ V, it follows from the equation that,

$$0 = 0.242 - 0.699 + 0.0591\times pH$$

$$pH = 7.73$$

Problem 13.22

Silver ions can be titrated potentiometrically with chloride ions using the galvanic cell SCE || Ag^+ | Ag ($E_{SCE} = 0.242$ V, ($E^0_{Ag^+/Ag} = 0.800$ V). Let us suppose 50 mL of a 0.02 M solution of silver nitrate is titrated against 0.2 M sodium chloride. Calculate the potential of the silver electrode (i) initially (ii) at 50% completion of the reaction (iii) at completion of the reaction, (iv) when 20% excess of NaCl is added. Trace the potential vs volume of NaCl curve.

Solution

$$\text{Volume of NaCl added for complete reaction} = \frac{50\times0.02}{0.2} = 5 \text{ ml}$$

(i) Potential of Ag electrodes initially,

$$E = E^0 + 0.0592\log\left[Ag^+\right]$$

$$E = 0.800 + 0.0592 \log 0.2$$

$$= 0.800 - 0.0592\times1.7$$

$$= 0.0800 - 0.101 = 0.699 \text{ V}$$

(ii) Potential of Ag electrodes at 50% completion of the reaction, that is, after 2.5 mL of NaCl is added.

$$[Ag^+]_{remaining} = \frac{\text{Initial moles of Ag}^+ - \text{Moles of Cl}^- \text{added}}{\text{Total volume}}$$

$$= \frac{50\times0.02 - 2.5\times0.2}{50 + 2.5} = \frac{1 - 0.5}{52.5} = 0.00952$$

$$E = 0.800 + 0.0592 \log 0.00952$$

$$= 0.800 - 0.0592\times(-2.02)$$

$$= 0.800 - 0.120 = 0.680 \text{ V}$$

(iii) When the reaction is complete,

Conc. of Ag^+ = Conc. of Cl^-

$$= \sqrt{K_{SP}} = \sqrt{1.8\times10^{-10}} = 1.35\times10^{-5}$$

$$E = 0.800 + 0.0592 \log [1.35\times10\text{–}5]$$

$$= 0.800 - 0.0592\times4.87 = 0.800 - 0.288$$

$$= 0.512 \text{ V}$$

(iv) When 20% excess of NaCl is added, that is, 6 mL of 0.2 N NaCl is added,

$$\text{Conc. of } Cl^- = \frac{\text{Moles of } Cl^- \text{ added} - \text{Initial moles of } Ag^+}{\text{Total volume}}$$

$$= \frac{6\times0.2 - 0.02\times50}{50+6}$$

$$= \frac{1.2-1.0}{56} = 3.57\times10^{-3}$$

$$[Ag^+] \text{ at this stage} = \frac{K_{SP}}{[Cl^-]}$$

$$= \frac{1.8\times10^{-10}}{3.57\times10^{-3}} = 5.04\times10^{-6}$$

$$E = 0.800 + 0.0592 \log [5.04\times10\text{–}6]$$

$$= 0.800 - 0.0592\times5.3$$

$$= 0.800 - 0.314 = 0.486 \text{ V}$$

Titration curve

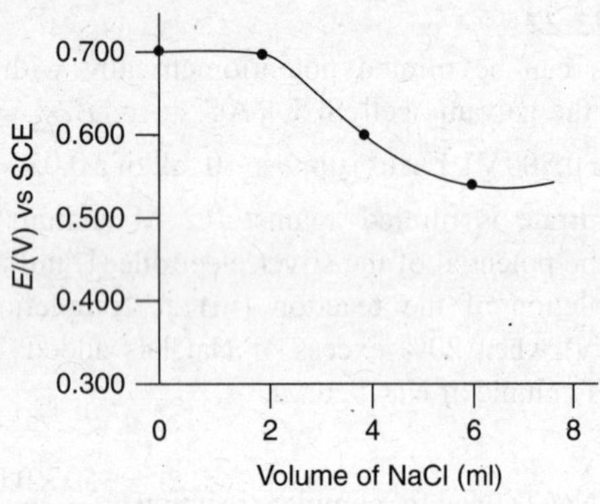

Hydrolysis Equilibria

Problem 13.23

(a) Derive an equation for the hydrolysis constant, K_h of sodium isobutyrate ($CH_3.CH.CH_3.COOH$). Also, obtain an equation for the pH of the solution. $K_W(H_2O) = 10^{-14}$; K_a of isobutyric acid = 1.44×10^{-5}, both at 298 K.

(b) Calculate the pH of the salt solution at 0.05 M and 0.1 M.

(c) β-glycerophosphoric acid has two dissociable protons of which the second one has a pK_a = 6.65 and the dissociation equilibrium may be represented as

$$C_3NaH_7O_6PH(1) + H_2O \rightleftharpoons C_3NaH_7O_6P^-(2) + H_3O^+$$

If a buffer of pH = 7.2 is to be prepared, how much weight of the species (1) salt is required to be added to 2 L of species (2) of 0.1 M concentration? Molecular weight of acid = 172.074.

Solution

(a) $$CH_3CHCH_3COO^-(Na^+)(\text{anion}) + H_2O \rightleftharpoons CH_3CHCH_3COOH(\text{acid}) + (Na^+) + (OH^-)$$

$$K_h = \frac{[\text{Acid}][OH^-]}{[\text{Anion}]}$$

If α is the degree of hydrolysis of the sodium salt at a stoichiometric concentration C,

Conc. of acid = αC

Conc. of OH^- = αC

Conc. of anion left = $(1-\alpha)C$

$$K_h = \frac{\alpha^2 C}{(1-\alpha)} \approx \alpha^2 C \text{ (assuming } \alpha \ll 1); \alpha = \sqrt{\frac{K_h}{C}}$$

$$\text{but } K_h = \frac{K_W}{K_a} = \frac{[H_3O^+][OH^-][\text{acid}]}{[\text{anion}][H_3O^+]}$$

$$[OH^-] = \alpha C = \sqrt{K_h C} = \sqrt{\frac{K_W C}{K_a}}$$

$$\text{pOH} = -\frac{1}{2}\log K_w + \frac{1}{2}\log K_a - \frac{1}{2}\log C$$

$$\text{pH} = \text{p}K_W - \text{pOH} = \frac{1}{2}\text{p}K_W + \frac{1}{2}\text{p}K_a + \frac{1}{2}\log C$$

$$\text{pH at 0.05 M : pH} = 7 + \frac{1}{2}\times4.84 + \frac{1}{2}\times(-1.3)$$

$$= 7 + 2.42 - 0.65 = 8.77$$

pH at 0.10 M: pH = 7 + 2.42 – 0.5 = 9.42 – 0.50 = 8.92

(b) $$\text{pH} = \text{p}K_a + \log\frac{[\text{salt}]}{[\text{acid}]}$$

$$= \text{p}K_a + \log\frac{[C_3NaH_7O_6P^-]}{[C_3NaH_7O_6PH]}$$

$$7.2 = 6.65 + \log\frac{[0.1]}{[\text{acid}]}$$

$$\log\frac{[0.1]}{[\text{acid}]} = 0.55$$

$$\frac{[0.1]}{[\text{acid}]} = 3.548$$

$$[\text{acid}] = \frac{0.1}{3.548} = 0.028$$

Weight of acid for 2 L = 0.028×172.074×2 = 9.64 g

Problem 13.24

(a) The dissociation constant of hydroiodic acid HIO_3 is 1.4×10^{-1} and the K_b of ammonium hydroxide is 1.80×10^{-5} at 298 K. Calculate the degree of hydrolysis and pH of a 0.02 M solution of ammonium iodate.

(b) O-phenanthroline and 2-2′-bipyridyl form strong complexes with Fe^{2+} ions and have the formulae $Fe(O-Ph)_3^{2+}$ and $Fe(2-2'\text{ Bipy})_3^{2+}$. Their standard potentials along with a few other redox systems at 298 K are given below

System	E^0 (V)
$Pt\|Fe^{2+}, Fe^{3+}$	0.770
$Pt\|Fe(O-Ph)_3^{2+}, Fe^{3+}$	1.200
$Pt\|Fe(2-2'\ Bipy)_3^{2+}, Fe^{3+}$	1.060
$Pt\|Cr_2O_7^{2-}, Cr^{3+}$	1.330
$Pt\|MnO_4^-, Mn^{2+}$	1.510

Arrange them in the order of their increasing oxidising power. Which of these redox systems will be able to oxidise H_2O to O_2 in acid medium? The E^0 for the reaction $O_2 + 4H^+ + 4e^- \rightarrow 2H_2O$ is 1.23 V.

Solution

(a) NH_4IO_3 is a salt of a strong acid HIO_3 and weak base. Its dissociation equilibrium of hydrolysis may be written as

$$NH_4^+ + IO_3^- + H_2O + H_2O \rightleftharpoons NH_3 + H_3O^+ + HIO_3 + OH^-$$

NOTE

- Although HIO_3 is a strong acid it is not completely dissociated.

$$K_h = \frac{[NH_3][HIO_3]}{[NH_4^+][IO_3^-]} = \frac{\alpha C \times \alpha C}{C(1-\alpha)C(1-\alpha)} = \frac{\alpha^2}{(1-\alpha)^2} = \frac{K_W}{K_a k_b}$$

where α is the degree of hydrolysis at a stoichiometric concentration of salt C.

$$K_h = \frac{1\times10^{-14}}{1.8\times10^{-5}\times1.41\times10^{-1}}$$

$$= \frac{1\times10^{-14}}{2.56\times10^{-6}} = 3.94\times10^{-9}$$

$$\alpha = \frac{\sqrt{K_h}}{\sqrt{K_h}+1}$$

$$= \frac{\sqrt{3.94\times10^{-9}}}{\sqrt{3.94\times10^{-9}}+1}$$

$$= \frac{6.31\times10^{-5}}{1+6.31\times10^{-5}} \approx 6.31\times10^{-5}$$

$$pH = \frac{1}{2}pK_w + \frac{1}{2}pK_a - \frac{1}{2}pK_b$$

$$= 7 + 0.425 - 2.373 = 5.052$$

(b) The more positive a standard potential is, the stronger that agent is as an oxidising agent. Thus, the oxidising power of the systems increases in the order,

$$Pt\left|Fe^{2+}, Fe^{3+} < Pt\right|Fe(2-2'\ Bipy)_3^{2+},\ Fe^{3+} < Pt\left|Fe(O-Ph)_3^{2+},\ Fe^{3+} < Pt\right|Cr_2O_7^{2-}, Cr^{3+} < Pt\left|MnO_4^-, Mn^{2+}\right.$$

Since the redox potential of $Cr_2O_7^{2-} \mid Cr^{3+}$ and $MnO_4^- \mid Mn^{2+}$ systems are more positive than that of the $O_2 \mid H_2O$ system, they only can oxidise H_2O to O_2 in acid medium.

Classification of Solvents and Acidity Function

Problem 13.25

(a) (i) Some solvents are listed below in addition to water: H_2O, glacial acetic acid, liquid ammonia, liquid hydrogen fluoride, hydrazine and ethylenediamine. Of the above, classify them as predominantly acidic, mostly basic, or those that do not belong to either of these classes.

(ii) Which of the solvents are expected to show a levelling effect on the strength of acids $HClO_4$, HCl, HNO_3, H_2SO_4?

(iii) Which of the solvents are expected to show a distinguishing effect of strengths of the acids mentioned in (i)?

(iv) Which of the solvents are expected to show a levelling effect on the bases ethylamine, aniline, methylamine and benzylamine?

(v) Which solvents given in (i) show a distinction on the strengths of the bases given in (iii)?

(b) Calculate the pH of a 0.2 M solution of H_2SO_4.

(c) Define H_0, the acidity function proposed by Hammett. What is its advantage in measuring the acidity of a solution? Given that the H_0 of pure H_2SO_4 is –12.0 and that of flourosulfuric acid is –15.0, which solution is more acidic?

(d) The K_b of *o*-nitroaniline is 5.13×10^{-15}. What is the H_0 of any solution that were to be used if the ratio of $[B]/[BH^+] = 0.1$? Calculate the proton activity of the solution.

Solution

(a) (i) Predominantly acidic solvents: glacial acetic acid, liquid HF.
Mostly basic solvents: liqiud ammonia, hydrazine, ethylenediamine.
Solvents not belonging to either of the above categories: water.

(ii) Water, liqiud ammonia, hydrazine, ethylenediamine.

(iii) Glacial acetic acid and liquid HF.

(iv) Glacial acetic acid and liquid HF.

(v) Liquid ammonia, hydrazine, ethylenediamine.

(b) The first ionisation of H_2SO_4: $H_2SO_4 \rightleftharpoons H^+ + HSO_4^-$

$\therefore$ Conc. of $[H^+] = [HSO_4^-] = 0.2$ M

For second ionisation: $HSO_4^- \rightleftharpoons H^+ + SO_4^{2-}\ K_a = 1.3\times10^{-2}$

If α is the degree of dissociation,

$$K_a = \frac{\alpha C \times \alpha C}{(1-\alpha)C} = \frac{\alpha^2 C}{(1-\alpha)} \text{ or } \frac{\alpha^2}{(1-\alpha)}$$

$$= \frac{1.3\times10^{-2}}{0.2} = 6.5\times10^{-2}$$

$$\alpha^2 + 6.5\times10^{-2}\times\alpha - 6.5\times10^{-2} = 0$$

$$\alpha = \frac{-0.065 \pm \sqrt{4.225 \times 10^{-3} + 0.26}}{2}$$

$$= \frac{-0.065 \pm \sqrt{0.514}}{2} = 0.225$$

$\alpha C = 0.225 \times 0.2 = 0.0450$. Total conc. of $(H^+) = 0.2 + 0.045 = 0.245$; $pH = -\log(0.245) = -(-0.61) = 0.61$.

In doing this calculation, the effect of H^+ already present on the dissociation of HSO_4^- is not considered. As a result, the pH may be somewhat higher than 0.61.

(c) $H_0 = pK_{BH^+} + \log\frac{C_B}{C_{BH^+}} = -\log\left(a_{H^+}\frac{\gamma_B}{\gamma_{BH^+}}\right)$

The advantage of the H_0 acidity function is that it is not restricted to aqueous solutions only, unlike the case of pH. However, it can be employed for measuring the acidity of concentrated aqueous acid solutions and also the acidity of non-aqueous solutions, such as, say, the acidity of solutions in alcohol, liquid HF, and so on. This is because it is based on using the colour changes of indicators such as nitroanilines, which can be measured accurately by means of a spectrophotometer.

The more negative a value of H_0 for a given solution, the more acidic is the solution because the C_{BH^+}/C_B ratio is larger in such a case. Thus fluorosulfuric acid is more acidic than sulfuric acid.

(d) pK_B of O – nitroaniline $= -\log\left(5.13 \times 10^{-15}\right) = 14.29$

$$pK_B + pK_{BH^+} = pK_W;\ pK_{BH^+} = 14 - 14.29 = -0.29$$

$$H_0 = pK_{BH^+} + \log\frac{[B]}{[BH^+]} = -0.29 + \log 0.1 = -1.29$$

Assuming $\frac{\gamma_B}{\gamma_{BH^+}} \approx 1, -\log a_{H^+} = -1.29; a_{H^+} = 19.5$

Redox Reactions, Nernst Equation and Related Applications

Problem 13.26

(a) Pyrolusite (MnO_2) can be reduced to manganous sulfate by arsenium oxide (As_2O_3) in the presence of conc. H_2SO_4. Any excess arsenium oxide remaining is oxidised by $KMnO_4$ to arsenic pentoxide (As_2O_5) in H_2SO_4 medium. Write the relevant ionic reactions using the ion–electron method of balancing equations.

(b) Pyrosulite can also be reduced to manganous sulfate by oxalic acid in H_2SO_4 medium and any excess oxalic acid is oxidised by $KMnO_4$ in H_2SO_4 medium. Write the relevant ionic reactions using the ion–electron method of balancing equations.

(c) Derive the Nernst expression for the reduction of $KMnO_4$ to $MnSO_4$ in H_2SO_4 medium, which is common in both reactions (a) and (b).

(d) What indicator is used in reaction (a) and (b)?

Solution

(a) $MnO_2 + 4H^+ + 2e^- \rightarrow Mn^{2+} + 2H_2O$ (1)

$As_2O_3 + 2H_2O \rightarrow As_2O_5 + 4H^+ + 4e^-$ (2)

Combining Eqs. (1) and (2) and adding the requisite molecules of H_2SO_4,

$$As_2O_3 + 2MnO_2 + 2H_2SO_4 \rightarrow 2MnSO_4 + As_2O_5 + 2H_2O \quad (3)$$

$$MnO_4^- + 8H^+ + 5e^- \rightarrow Mn^{2+} + 4H_2O \quad (4)$$

Balancing the number of e^-s and adding the required H_2SO_4 molecules,

$$4KMnO_4 + 5As_2O_3 + 6H_2SO_4 \rightarrow 5As_2O_5 + 4MnSO_4 + 6H_2O + 2K_2SO_4$$

(b) $C_2O_4^{2-} \rightarrow 2CO_2 + 2e^-$ (5)

Combining Eqs (1) and (5) and adding the necessary H_2SO_4 molecules,

$$MnO_2 + H_2C_2O_4 + H_2SO_4 \rightarrow MnSO_4 + 2CO_2 + 2H_2O$$

Combining Eqs (4) and (5),

$$2KMnO_4 + 5H_2C_2O_4 + 3H_2SO_4 \rightarrow K_2SO_4 + 2MnSO_4 + 10CO_2 + 8H_2O$$

(c) $E = E^0 - \frac{RT}{5F}\ln\left(\frac{a_{Mn^{2+}}a_{H_2O}^4}{a_{MnO_4^-}a_{H^+}^8}\right)$

(d) No external indicator is necessary and permanganate is a self-indicator in these titrations.

Problem 13.27

(a) Write the redox reaction involving the oxidation of ferrous to ferric ions by potassium dichromate in sulfuric acid medium.

(b) Derive the equation for the potential at equivalence point in the titration of ferrous ions by dichromate. What is the equivalence point potential at a pH = 0? ($E^0_{Fe^{2+}/Fe^{3+}} = 0.77$ V and $E^0_{Cr_2O_7^{2-}/Cr^{3+}} = 1.33$ V at 298 K). What is the equilibrium constant of the reaction?

(c) Write down the Nernst equation for a redox indicator. What is the useful range of potentials of the indicator?

(d) Which of the redox indicators diphenylamine $\left(E^0 = 0.75\text{ V}\right)$ and diphenylaminesulfonic acid $\left(E^0 = 0.85\text{ V}\right)$ is the better one suited for the titration in (b)?

Solution

(a) $Cr_2O_7^{2-} + 14H^+ + 6e^- \rightarrow 2Cr^{3+} + 7H_2O$ (1)

$Fe^{2+}(aq) \rightarrow Fe^{3+}(aq) + e^-$ (2)

Overall reaction:

$$Cr_2O_7^{2-} + 14H^+ + 6Fe^{2+} \rightarrow 2Cr^{3+} + 6Fe^{3+} + 7H_2O$$

(b) Nernst equation:

For reaction (1)

$$E_{Cr_2O_7^{2-}/Cr^{3+}} = E^0 - \frac{0.0592}{6}\log\frac{[Cr^{3+}]^2[H_2O]^7}{[Cr_2O_7^{2-}][H^+]^{14}} \quad (3)$$

For reaction (2)

$$E_{Fe^{2+}/Fe^{3+}} = E^0 - 0.0592 \log\frac{[Fe^{2+}]}{[Fe^{3+}]} \quad (4)$$

Multiplying Eq. (1) in (b) by 6 throughout to add the log terms,

$$6E_{Cr_2O_7^{2-}/Cr^{3+}} = 6E^0_{Cr_2O_7^{2-}/Cr^{3+}} - 0.0592\log\frac{[Cr^{3+}]^2[H_2O]^7}{[Cr_2O_7^{2-}][H^+]^{14}} \quad (5)$$

Adding Eqs (4) and (5),

$$7E = E^0_{Fe^{3+}/Fe^{2+}} + 6E^0_{Cr_2O_7^{2-}/Cr^{3+}} - 0.0592\log\frac{[Fe^{2+}][Cr^{3+}]^2[H_2O]^7}{[Fe^{3+}][Cr_2O_7^{2-}][H^+]^{14}} \quad (6)$$

At the equivalence point,

$$[Fe^{2+}] = 6[Cr_2O_7^{2-}]; [Fe^{3+}] = 3[Cr^{3+}] \quad (7)$$

Substituting Eq. (7) in Eq. (6),

$$7E_{eq\cdot pt.} = E^0_{Fe^{3+}/Fe^{2+}} + 6E^0_{Cr_2O_7^{2-}/Cr^{3+}} - 0.0592\log\frac{6\left[Cr_2O_7^{2-}\right]\left[Cr^{3+}\right]^2\left[H_2O\right]^7}{3\left[Cr^{3+}\right]^2\left[Cr_2O_7^{2-}\right]\left[H^+\right]^{14}} \quad (8)$$

$$E_{eq.pt.} = \frac{E^0_{Fe^{3+}/Fe^{2+}} + 6E^0_{Cr_2O_7^{2-}/Cr^{3+}}}{7} - \frac{0.0592}{7}\log\frac{2[H_2O]^7}{[H^+]^{14}} \quad (9)$$

(Assuming $[H_2O]^7 = 1$ because its conc. is always in excess)

$$E_{eq.pt.} = \frac{E^0_{Fe^{3+}/Fe^{2+}} + 6E^0_{Cr_2O_7^{2-}/Cr^{3+}}}{7} + \frac{0.0592\times 14}{7}\log[H^+] - \frac{0.0592}{7}\log 2$$

$$= \frac{E^0_{Fe^{3+}/Fe^{2+}} + 6E^0_{Cr_2O_7^{2-}/Cr^{3+}}}{7} - 0.1184\text{pH} - 0.0025 \quad (10)$$

At a pH = 0

$$E_{eq.pt.} = \frac{E^0_{Fe^{3+}/Fe^{2+}} + 6E^0_{Cr_2O_7^{2-}/Cr^{3+}}}{7} - 0.0025 \quad (11)$$

$$= \frac{0.77 + 6\times 1.33}{7} - 0.0025 = 1.248 \text{ V}$$

Equilibrium constant of the reaction

$$E^0 = E^0_R - E^0_L = (1.33 - 0.77) = 0.56 \text{ V}$$

$$\log K = \frac{n}{0.0592}\times E^0 = \frac{6\times 0.56}{0.0592} = 56.86; K = 7.24\times 10^{56}$$

(c) For a redox indicator,

$$\text{In(ox)} + ne^- \rightleftharpoons \text{In(red)}$$

$$E = E^0 - \frac{RT}{nF}\ln\frac{[\text{In(red)}]}{[\text{In(ox)}]} = E^0 - \frac{0.0592}{n}\log\frac{[\text{In(red)}]}{[\text{In(ox)}]} \quad (1)$$

The useful range of the indicator is obtained by considering

$$\frac{[\text{In(red)}]}{[\text{In(ox)}]} = 10; \quad \frac{[\text{In(red)}]}{[\text{In(ox)}]} = 0.1$$

Substituting these values in Eq. (1),

$$E = E^0 - \frac{0.0592}{n}\log 10 = E^0 - \frac{0.0592}{n}$$

$$\text{or } E = E^0 - \frac{0.0592}{n}\log 0.1 = E^0 + \frac{0.0592}{n}$$

$E = E^0 \pm \frac{0.0592}{n}$ is the active range for the indicator.

(d) The range of colour change of diphenylamine [reduced form (colourless) to oxidised form (violet)] is 0.74–0.76 V. The range of colour change for DPAsulfonic acid (colourless) to oxidised form (red violet) is 0.84–0.86 V. Of the two, DPAsulfonic acid is better because it is closer to the equivalence point potential in the above titration. The colour change occurs in the rising point of potential vs volume curve. Ferroin would be even better (E^0 = 1.15 V) and it would be a better indicator in this titration.

Problem 13.28

The standard potentials of $Pt|Sn^{2+}, Sn^{4+}$ and $Pt|Fe^{2+}, Fe^{3+}$ are 0.15 V and 0.77 V respectively at 298 K. A mixture of stannous and ferrous ions is titrated potentiometrically against ceric ammonium sulfate in 1 M H_2SO_4 medium. Derive the equation for the potential at the equivalence point of each reaction separately and indicate whether two inflexions will be observable in this titration. What are the equilibrium constants of the reaction with Ce^{4+} ($E^0_{Ce^{4+}/Ce^{3+}}$ = 1.70 V?

Solution

Reaction of Ce^{4+} with Sn^{2+} ions:

$$Ce^{4+} + e^- \rightleftharpoons Ce^{3+}$$

$$Sn^{4+} + 2e^- \rightarrow Sn^{2+}$$

$$2Ce^{4+} + Sn^{2+} \rightarrow 2Ce^{3+} + Sn^{4+}$$

Nernst equation for the Ce^{4+}/Ce^{3+} reaction:

$$E_{Ce^{4+}/Ce^{3+}} = E^0_{Ce^{4+}/Ce^{3+}} - 0.0592\log\frac{[Ce^{3+}]}{[Ce^{4+}]} \quad (1)$$

Nernst equation for the Sn^{4+}/Sn^{2+} reaction:

$$E_{Sn^{4+}/Sn^{2+}} = E^0_{Sn^{4+}/Sn^{2+}} - \frac{0.0592}{2}\log\frac{[Sn^{2+}]}{[Sn^{4+}]} \quad (2)$$

Adding Eqs (1) and (2) after multiplying Eq. (2) by 2,

$$3E = E^0_{Ce^{4+}/Ce^{3+}} + 2E^0_{Sn^{4+}/Sn^{2+}} - 0.0592\log\frac{[Ce^{3+}][Sn^{2+}]}{[Ce^{4+}][Sn^{4+}]} \quad (3)$$

At equivalence point, $[Sn^{2+}] = 2[Ce^{4+}]$; $[Sn^{4+}] = 2[Ce^{3+}]$ (4)

Substituting Eq. (4) in Eq. (3),

$$3E_{eq.pt.} = E^0_{Ce^{4+}/Ce^{3+}} + 2E^0_{Sn^{4+}/Sn^{2+}} - 0.0592\log\frac{[Ce^{3+}]2[Ce^{4+}]}{[Ce^{4+}]2[Ce^{3+}]}$$

$$E_{eq.pt.} = \frac{E^0_{Ce^{4+}/Ce^{3+}} + 2E^0_{Sn^{4+}/Sn^{2+}}}{3}$$

$$= \frac{1.70 + 2\times 0.15}{3} = \frac{2}{3} = 0.667\ \text{V}$$

For the reaction of Ce^{4+} with Fe^{2+} ions we have

$$Ce^{4+} + Fe^{2+} \rightarrow Ce^{3+} + Fe^{3+}$$

Repeating the same steps above, the final equation becomes,

$$2E = E^0_{Ce^{4+}/Ce^{3+}} + E^0_{Fe^{3+}/Fe^{2+}} - 0.0592\log\frac{[Ce^{3+}][Fe^{2+}]}{[Ce^{4+}][Fe^{2+}]}$$

$$E_{eq.pt.} = \frac{E^0_{Ce^{4+}/Ce^{3+}} + E^0_{Fe^{3+}/Fe^{2+}}}{2} = \frac{1.70 + 0.77}{2} = 1.235\ \text{V}$$

Since the equivalence potentials are so far away from each other (568 mV), there will be two inflexions observed in the titration of the mixture, thus allowing for their individual estimations.

Ion Selective and Glass Electrodes

Problem 13.29

(a) Briefly describe the construction of a glass electrode and explain the principle of its use for pH measurements.

(b) Derive an equation for the EMF of the cell constituting the glass electrode and a reference electrode for the measurement of pH of an unknown solution. What instruments are useful for pH measurements using a glass electrode?

(c) A cell formed of a glass electrode and reference electrode placed in a buffer of pH = 4.0, gave an emf 0.150 V at 298 K. When placed in a solution of unknown pH, the measured emf of the cell with an electrometer was 0.239 V. What is the pH of the unknown solution?

Solution

(a) The glass electrode consists of a cylindrical glass tube which ends as a thin bulb made of special glass, having a low melting point and high conductivity. This tube is filled with a solution of constant pH, usually 0.1 N HCl and is fitted with an Ag–AgCl electrode. This electrode is immersed in the solution of unknown pH along with a reference electrode such as the SCE. Because of the exchange of H^+ ions across the thin membrane of glass separating the solution of unknown H^+ ion activity and the solution containing H^+ ions of known activity (0.1 M HCl), a potential develops across the membrane. This potential varies with the H^+ ion activity in accordance with the Nernst equation, that is, by 0.0592 volts for a change of one pH unit at 298 K.

(b) The cell consists of the reference electrode (SCE) and the glass electrode placed in a solution of known pH and it is represented as

Ag, AgCl|0.1 N HCl| glass membrane placed in a solution of unknown pH|SCE

$$E_{cell} = E_{SCE} - E_G = E_{SCE} - (E_{Ag,\ AgCl} + E^0_G + 0.0592\text{pH})$$

$$= E_{SCE} - E_{Ag,\ AgCl} - E^0_G - 0.0592\text{pH}$$

$$= K - E^0_G - 0.0592\text{pH} \qquad (1)$$

where K refers to the known constant potentials of E_{SCE} and $E_{Ag,AgCl}$. To measure the pH of an unknown solution, E^0_G has to be first determined by measuring E_{cell} in solutions of known pH using the glass electrode. Once E^0_G is known, this electrode is dipped into the solution of unknown pH and E_{cell} obtained. Using Eq. (1), the pH is calculated for the unknown electrode. Because of the high resistance of the glass membrane, an ordinary potentiometer cannot be used for measuring E_{cell}. Instead, an electronic voltmeter or vacuum tube voltmeter is used. It is possible to ensure that the pH is within ±0.01 units with such instruments. The glass electrode works best in solutions whose pH does not extend beyond 10.

Equation (1) may be written for a solution of pH = 4.0 as

$$0.150 = K - E^0_G - 0.0592\times 4 = K - E^0_G - 0.2368 \qquad (1)$$

For the unknown, $0.239 = K - E^0_G - 0.0592\text{pH} \qquad (2)$

Eq. (1) – Eq. (2) gives,

$$-0.089 = K' - 0.2368 - K' + 0.0592\text{pH}$$

where $K' = \left(K - E^0_G\right)$

$$0.0592\text{pH} = 0.2368 - 0.089 = 0.1478$$

$$\text{pH} = \frac{0.1478}{0.0592} = 2.50$$

Problem 13.30

(a) What is an ion selective electrode? Draw a neat sketch of the experimental set up for the measurement of the activity of an ion (or concentration in dilute solution) using ISE. Give two examples of the application of ISE. Explain the term, selectivity of an ISE.

(b) Obtain an equation relating the activity of an ion and the measured potential.

(c) Determine the concentration of Li^+ ions in the unknown solution using the Li^+ ISE which gave an E value of 0.225 V at 298 K.

Li^+ (mol L^{-1})	*E* (V)
5.0×10^{-5}	0.200
2.5×10^{-4}	0.235
5.0×10^{-4}	0.260

Solution

(a) An ion selective electrode (ISE) is an electrode specially designed to measure the activity or concentration of an ion needed in the specific case. For example, the glass electrode is designed to measure the activity of the H^+ ions. Glass electrodes capable of measuring the activity of Na^+ ions are also available. Similarly there are ISEs capable of measuring the activity of Ca^{2+} ions and F^- ions that are commercially available.

ISEs may be classified into various categories depending upon the type of membrane material used and they function via different mechanisms during their operation. The various types are:

- Glass membrane electrode (for example, glass electrode for measuring the pH).
- Crystalline or solid state membrane electrodes: The membrane consists of an insoluble inorganic salt and the potential at the membrane arises via an exchange of ions across it. As an example, F^- ions are determined using an LaF_3 electrode.
- Liquid membrane electrode: In this case, the ion of interest is exchanged via an ion exchanger dissolved in a viscous organic liquid membrane. Ca^{2+} ions are determined using this type of electrode.
- Polymer membrane electrodes: A polymer such as polyvinyl chloride (PVC) is used as the ion exchanger. For example, ions such as Cl^-, NO_3^- and K^+ have been determined using this type of membrane.

Although ISEs are specific for determining the activity of particular ions, they are susceptible to interferences by other ions. In other words, the potential registered by the ISE may be vitiated by the interfering ions. Thus their *selectivity* may be affected. This may be accounted for by introducing a selectivity coefficient term in the equation for the potential of ISE.

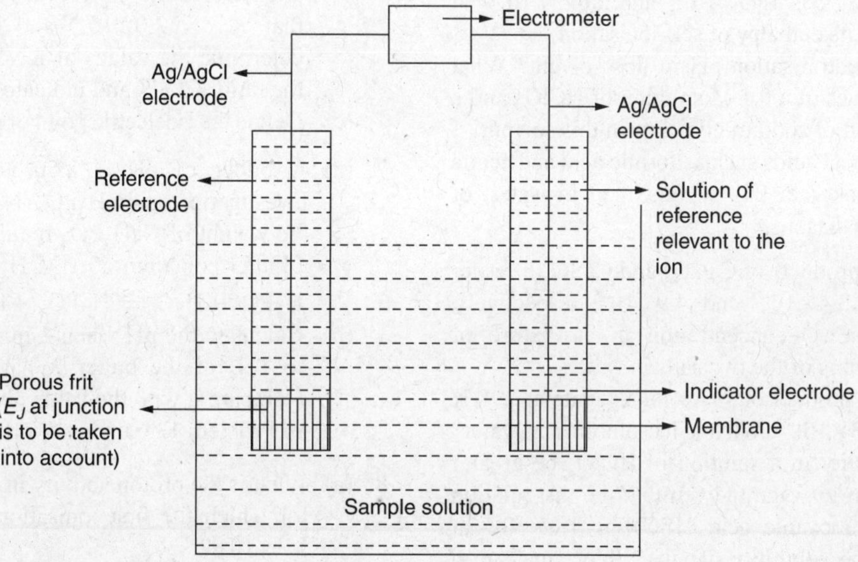

The experimental setup

(b) The potential measured at the ISE (right hand electrode in the diagram) with respect to the external reference (left hand electrode) may be given as

$$E_{ISE} = E_{outside} - E_{inside} \quad (1)$$

where $E_{outside}$ refers to the potential difference between the sample side of membrane and the reference side of the membrane. They are related via the Nernst equation.

$$E_{outside} = E^0 + \frac{0.0592}{n}\log[a]_{unknown} \quad (2)$$

$$E_{inside} = E^0 + \frac{0.0592}{n}\log[a]_{reference\ side} \quad (3)$$

E_{inside} is constant because E^0 and $[a]_{reference\ side}$ are constant.

$$\therefore E_{ISE} = K' + \frac{0.0592}{n}\log[a]_{unknown} \quad (4)$$

Using Eq. (4), the activity of an unknown ion can be determined.

(c) Using the known values of E_{ISE} at the given concentrations, K' can be calculated. Using the average K' and the obtained E_{ISE} of the unknown solution, the concentration can be calculated.

Using data (1)

$$K' = 0.200 - 0.0592\log(5\times10^{-5})$$
$$= 0.200 - (0.0592\times(-4.3)) = 0.454$$

Using data (2)

$$K' = 0.235 - 0.0592\log(2.5\times10^{-4})$$
$$= 0.235 - (0.0592\times(-3.6)) = 0.448$$

Using data (3)

$$K' = 0.260 - 0.0592\log(5\times10^{-4})$$
$$= 0.260 - (0.0592\times(-3.3)) = 0.455$$

Average $K' = 0.452$

For the unknown,

$$\log[a]_{unknown} = (E_{ISE} - K')\times\frac{n}{0.0592}$$
$$= \frac{(0.225 - 0.452)\times1}{0.0592} = -3.834$$

$$[a]_{unknown} = \text{antilog}[\bar{4}.166] = 1.466\times10^{-4}$$

EXERCISES

Practice Problems

1. (a) The specific conductivity of a 0.025 M solution of a weak organic acid is 8.26×10^{-4} ohm^{-1} cm^{-1}. If its Λ_0 is 402.5 ohm^{-1} cm^{-2} eq^{-1}, what is its degree of dissociation and what is its K_a? (Assume that the temperature is 298 K)
(b) Considering the interionic forces in the solution, what is its thermodynamic dissociation constant and what is its degree of dissociation?

2. (a) The auto-protolysis constant of ethyleneglycol (CH_2OH–CH_2OH) at 298 K is 1.45×10^{-16} and 2.69×10^{-16} at 308 K. What is its enthalpy of self-ionisation?
(b) What is the neutralisation pH in this solvent? What are the pH values of a 0.1 M solution of $HClO_4$ and a 0.01 M solution of sodium glycolate in this solvent?
(c) Would you expect acids such as formic acid and acetic acid to be weaker in this solvent than in water, or stronger? Give reasons.

3. (a) The solubility products of $CaCO_3$ and $CaSO_4$ in water at 298 K are 3.4×10^{-9} and 4.9×10^{-5} respectively. Calculate the Ca^{2+} concentration in a mixture of saturated solutions of the two salts in water.
(b) The solubility product of silver thiocyanate at 298 K in water is 1.03×10^{-12}. What is its solubility in water? Will its solubility in a solution of $HClO_4$ be greater or smaller than in water? Given that the dissociation constant of thiocyanic acid HSCN is 1.3×10^{-1} at 298 K, find the solubility of silver thiocyanate in a 0.2 M solution of perchloric acid.

4. (a) The pK_a values of propanoic and pyruvic acids at 298 K are 4.89 and 2.55 respectively. Calculate the pH of a solution consisting of 0.02 M propanoic and 0.05 M pyruvic acid.
(b) Calculate the volume of water needed to be added to 1000 mL of 0.1 M pyruvic acid so that it has the same pH as 0.1 M solution of monochloroacetic acid. K_a (pyruvic acid) = 2.82×10^{-3}; K_a ($ClCH_2COOH$) = 1.40×10^{-3}.

5. (a) Calculate the degree of hydrolysis and pH of 0.1 M solution of ammonium butanoate at 298 K. The dissociation constant K_a of butanoic acid is 1.5×10^{-5} and K_b of ammonia is 1.8×10^{-5}.
(b) The dissociation constant of dichloroacetic acid is 5.0×10^{-2} at 298 K. Assuming it to be the classical constant, calculate thermodynamic constant and the pH of a solution of the acid at 0.1 M concentration in the presence of 0.1 M LiCl.

6. (a) Calculate the OH^- ion concentration and pH of a mixture of 0.02 M hydrazine and 0.01 M of triethylamine given that their K_b values are 1.3×10^{-6} for the former and 6.4×10^{-5} for the latter at 298 K.
(b) The dissociation equilibria in glycine (aminoacetic acid) in aqueous solution may be expressed in terms of the microscopic and macroscopic equilibria as follows:
K_1 is defined as $K_1 = K + K''$ and K_2 is defined as $K_2 = \dfrac{1}{\dfrac{1}{K'}+\dfrac{1}{K'''}}$. They are experimentally determinable and have values of 4.47×10^{-3} and 1.66×10^{-10}. Given that $K'' = 2.0\times10^{-8}$ and that $K\times K' = K''\times K'''$ determine the values of K, K' and K'''. Also calculate the ratio K/K'' and indicate its significance.
(c) Define the isoelectric point of glycine. What is its value?

7. (a) A buffer of pH = 2.50 is required in a certain investigation using H_3PO_4–NaH_2PO_4 buffer. Calculate the weight of NaH_2PO_4 required for preparing 2 L of a buffer containing 0.1 M H_3PO_4. K_a for the reaction $H_3PO_4 + H_2O \rightleftharpoons H_2PO_4^- + H_3O^+$ is 7.10×10^{-3}.
(b) Calculate the pH when 2 mL of 1 N NaOH is added to 1 L of the buffer. What is the buffer capacity? Compare it with the value obtained from the equation of Problem 13.6 under 'Worked out problems'.

8. (a) What is the proton activity in a 4 M solution of H_2SO_4 for which the first ionisation is complete and K_2 is 1.3×10^{-2}?
(b) Given H_0, the Hammett acidity function, of the above solution as –2.0, what are the limits of pK_{BH^+} values of the indicator constant if the ratios of C_{BH^+}/C_B are to lie between 100 and 1, or 1 and 100?
(c) 50 mL of 0.1 M acetic acid is titrated potentiometrically with 0.5 M NaOH using the galvanic cell (an electrometer in fact) involving a glass electrode and a reference electrode given by glass electrode | acid solution || SCE.
Plot the titration curve by calculating the pH of the solution (i) at the beginning (ii) at 25% neutralisation (iii) at 50% neutralisation (iv) at the equivalence point (v) when 25% excess base is added. (pK_a of the acid = 4.75)

9. (a) The K_{SP}s of cobalt(II) sulfide, nickel(II) sulfide and zinc(II) sulfide at 298 K are 5.0×10^{-22}, 3.0×10^{-20} and 2.0×10^{-25} respectively. Considering a 0.05 M solution of each of the ions in a mixture,
(i) Calculate the minimum S^{2-} ion concentration necessary to precipitate each of the sulfides.
(ii) Find out the percentage of zinc ions that precipitate before cobalt ions start precipitating out.
(iii) Find out the percentage of Ni^{2+} ions that precipitate before cobalt (II) ions start precipitating out.

(iv) Calculate the minimum amount of OH^- ion concentration necessary to precipitate all the sulfides from the alkaline medium required for the purpose.

(b) Calculate the minimum amount of sodium cyanide required to be added to 2 L of H_2O to dissolve 5.0 g of silver iodide. The solubility product of silver iodide is 8.3×10^{-17} and the formation constant of $Ag(CN)_2^-$ complex is 7.1×10^{20} (all at 298 K). (Molar mass of AgI = 235)

10. (a) The thermodynamic K_{SP} of barium oxalate at 298 K is 2.5×10^{-8}. What is the solubility of the salt in g dm^{-3}? In the presence of 0.1 M barium nitrate, what is the solubility in mol dm^{-3}? Suppose we add $NaNO_3$ to the saturated solution of the salt such that its concentration is 0.1 M, will the solubility change or not? What is the solubility in the presence of a neutral salt, if any? (Molar mass of $BaC_2O_4 = 225.3$)

(b) A solution contains a mixture of 100 mL of 0.1 M NaOH, 200 mL of 0.2 M LiOH and 200 mL of 0.02 M $Ca(OH)_2$ (saturated solution). Assuming complete dissociation of the salts other than $Ca(OH)_2$, calculate the pH of the solution. If 500 mL of 1 M H_2SO_4 is added to this mixture, what is its pH? [K_{SP} $(Ca(OH)_2) = 5\times10^{-6}$]

11. (a) Oxalic acid is oxidised to CO_2 by $KMnO_4$ in a H_2SO_4 medium. Write down the partial equation for this oxidation and also the overall reaction.

(b) Derive the equation for the equivalence point potential in the titration of oxalic acid by $KMnO_4$ in an acid medium. Given the standard potentials $E^0_{MnO_4^-,Mn^{2+}} = 1.51$ V and $E^0_{C_2O_4^{2-},CO_2} = -0.49$ V, calculate the equivalence point potential at pH = 0. What is the equilibrium constant of the reaction? Assume the temperature is 298 K in all calculations.

12. (a) The solubility product of copper(II) hydroxide is 1.1×10^{-19} at 298 K. The stability constant for $Cu(NH_3)_4^{2+}$ is 2×10^{13} at the same temperature. Calculate the solubility of copper hydroxide in 1 M ammonia.

(b) Write down the self ionisation equilibria in anhydrous H_2SO_4. Given that the auto-protolysis constant of the acid is 3×10^{-4}, what is the concentration of the lyonium ion in this solvent?

(c) Is it possible to distinguish between the strength of $HClO_4$ and HCl in this solvent?

(d) The auto-protolysis constants of liquid ammonia and water are 5.0×10^{-23} and 1.0×10^{-14} at 298 K. Would the titration of a weak acid such as boric acid be better in an aqueous medium or in liquid ammonia? Give reasons.

13. Which of the ions gets titrated first in the potentiometric titration of a mixture of I^- and Cl^- ions given that the K_{SP} (AgI) = 1.5×10^{-16} and K_{SP} (AgCl) = 1.8×10^{-10}. Derive an equation for the potential of the silver electrode and calculate the potential when 50 mL of a mixture of equal volumes of 0.04 M sodium iodide and sodium chloride are titrated against 0.1 M silver nitrate when (i) 2 mL of $AgNO_3$ (ii) 4 mL of $AgNO_3$ (iii) 5 mL of $AgNO_3$ (iv) 7.5 mL of $AgNO_3$ (v) 10 mL of $AgNO_3$ and (vi) 15 mL of $AgNO_3$ are added. Trace the E vs volume of $AgNO_3$ curve. Assume the temperature is 298 K.

14. (a) The pK_{SP} of strontium oxalate is 7.3 at 298 K. Find out which of the following solutions: $Sr(NO_3)_2$ (I) and $K_2C_2O_4$ (II) on mixing in equal volumes can result in precipitation of strontium oxalate. (Molar masses: $Sr(NO_3)_2 = 211.6$ and $K_2C_2O_4 = 166$)

(i) 20 mL having 1.5×10^{-3} g of (I) + 20 mL having 1.5×10^{-3} g of (II)

(ii) 20 mL having 1.7×10^{-3} g of (I) + 20 mL having 1.3×10^{-3} g of (II)

(iii) 20 mL having 1.9×10^{-3} g of (I) + 20 mL having 1.9×10^{-3} g of (II)

(iv) 20 mL having 1.0×10^{-3} g of (I) + 20 mL having 2.0×10^{-3} g of (II)

(b) 50 mL of 0.02 M Na_2CO_3 is titrated with 0.05 M HCl. Construct a pH–volume of acid curve by calculating the pH at (i) the beginning (ii) 5.0 mL addition (iii) 10.0 mL addition (iv) 15.0 mL addition (v) 20 mL addition (vi) 25 mL addition of HCl (vii) 30 mL addition of HCl (viii) 35 mL addition of HCl (ix) 40 mL addition of HCl (x) 45 mL addition of HCl (K_{a1} and K_{a2} values of H_2CO_3 are 4.2×10^{-7} and 4.8×10^{-11} at 298 K).

(c) Explain your observations. Indicate which indicators are useful in this study. What indicator electrodes can be used if the titration is followed potentiometrically?

15. (a) Fluoride ion contamination is present in natural water in many countries which is a serious health hazard. Fluoride ion selective electrodes come in handy for determining this ion. Describe the method of construction of the fluoride ISE using LaF_3 crystal (doped with europium to increase conductivity). Give a neat sketch of the experimental set up and explain how the F^- ion concentration can be derived from the potential measurements using this ISE.

(b) A fluoride ISE gave the following potential values (against a standard external reference) at various F^- ion activities (= concentrations in dilute solutions) (temperature = 298 K).

F^- (mol L^{-1})	E (volts)
5.0×10^{-4}	0.260
2.5×10^{-4}	0.245
1.0×10^{-4}	0.220
5.0×10^{-5}	0.201

A solution with unknown concentration of F^- gave a potential of 0.225 V at 298 K. How much F^- (in ppm) is present in this solution?

16. (a) The specific conductance of 0.01 N solution of aniline hydrochloride is 1.267×10^{-3} S cm^{-1}. In the presence of excess of aniline it was found to be reduced to 1.12×10^{-3} S cm^{-1}. The specific conductance of 0.01 N

HCl is 4.12×10^{-3} S cm^{-1}, all data referred to are at 298 K. Calculate the hydrolysis constant of aniline hydrochloride.

(b) The following standard potentials relating to the Co^{3+}/Co^{2+} system are given along with the standard potential of the O_2 electrode.

$Pt|Co^{3+}, Co^{2+}$

$E^0 = 1.82$ V; $Pt|Co(NH_3)_4^{3+}, Co(NH_3)_4^{2+}$

$E^0 = 0.10$ V

$Pt|O_2, H^+, H_2O$

$E^0 = 1.23$ V

(i) Will dissolved O_2 oxidise Co^{2+} to Co^{3+} in the presence of NH_3?

(ii) Will the same oxidation by O_2 take place in the absence of NH_3?

(c) The following standard potential data relating to the Fe^{3+}/Fe^{2+} system are given along with that of the O_2 electrode.

$Pt|Fe^{3+}, Fe^{2+}$

$E^0 = 0.77$ V; $Pt|Fe(CN)_6^{3-}, Fe(CN)_6^{2-}$

$E^0 = 0.36$ V

$Pt|Fe(OH)_3, Fe(OH)_2$ in alkali

$E^0 = -0.56$ V; $Pt|O_2, OH^-, H_2O$

$E^0 = 0.40$ V

(i) Arrange the three electrodes in the order of their increasing oxidising power.

(ii) Will Fe^{2+} in the complexed form or combined form with OH^- ions, or in the free state, be able to reduce O_2 to OH^- ions in the alkaline medium?

17. Consider the reaction $Cr^{2+} + Ce^{4+} \rightleftharpoons Cr^{3+} + Ce^{3+}$. Derive an equation for the equivalence point potential in the oxidation of the chromous to chromic salt by cerium(IV). What is the equivalence point potential in the above reaction given that $E^0_{Cr^{3+}/Cr^{2+}} = -0.41$ V and $E^0_{Ce^{4+}/Ce^{3+}} = 1.51$ V at 298 K? Calculate the equilibrium constant of the reaction at 298 K.

18. (a) The thermodynamic solubility product of magnesium phosphate at 298 K is 1.04×10^{-24}. What is its solubility in g dm^{-3}? (Molar mass of $Mg_3(PO_4)_2 = 262.9$)

(b) Calculate its solubility in the presence of 0.1 M $Mg(NO_3)_2$ taking into consideration the interionic effects.

19. Glycine and its hydrochloride salt is a buffer system represented by the equilibrium

$$(Cl^-)H_3^+NCH_2COOH + H_2O \rightleftharpoons (Cl^-)H_3^+NCH_2COO^- + H_3O^+$$

and which has a dissociation constant of 4.47×10^{-3}. It is required to prepare 1 L of buffer of a pH = 2.0 using 0.05 M glycine hydrochloride (or glycinium chloride). How many grams of the sodium salt of glycinium chloride is required to be added to the solution above?

Succinic acid has two dissociable protons whose thermodynamic pK_1 and pK_2 values at 298 K are 4.207 and 5.636 respectively. Calculate their pK values in the presence of 0.02 M $NaNO_3$ using the extended form of the D–H limiting law given by

$$-\log\gamma\pm = \frac{A|z_+z_-|\sqrt{\mu}}{1+b\sqrt{\mu}}$$

where $A = 0.51$ and $b = 1/\left(\text{mol L}^{-1}\right)^{-1/2}$

20. (a) An acidity function $p(a_H\gamma_{Cl})$ was proposed as a measure of acidity in a buffer solution based on the measurement of emf of the cell involving the standard hydrogen electrode and silver–silver chloride electrode. The cell may be represented as

Pt, H_2(*g*, 1 atm)|acetic acid (m_1 = 0.005), sodium acetate (m_2 = 0.0048) sodium chloride (m_3 = 0.005)| AgCl, Ag

Derive an equation for $p(a_H\gamma_{Cl}) = -\log m_H(\gamma_H\gamma_{Cl})$ in terms of the measured emf of the above cell and the standard EMF of the cell. If the EMF of the above cell is 0.6390 V, and the E^0 = 0.2223 V both at 298 K, calculate the $p(a_H\gamma_{Cl})$ value. Using the Debye–Huckel extended equation, given by $\log\gamma_i = \frac{-AZ_i\sqrt{I}}{1+b\sqrt{I}}$ for calculating γ_{Cl^-}, calculate the p(a_H) of the buffer solution. Compare this value with the pH calculated for the buffer using the Henderson equation. E^0 of the above cell is 0.2223 V at 298 K.

(b) Why are alkali and alkaline earth metals ions considered as hard acids while precious metal ions such as Ag^+, Au^+, Pt^{2+}, Pd^{2+}, are considered as soft acids?

(c) Among the halide ions F^-, Cl^- and I^- and SO_4^{2-}, $S_2O_3^{2-}$, S_2^- and PO_4^{3-}, identify hard and soft bases.

Multiple Choice Questions

1. Which among the following can be considered as a Lewis base and also as a Brønsted acid too?

(a) $FeCl_3$ **(b)** NH_3
(c) HS^- **(d)** Cu_2^+

2. Find the hard acid–hard base reaction from the following:

(a) $Ca^{2+}(aq) + SO_4^{2-}(aq) \rightarrow CaSO_4$
(b) $ZnO + CO_2 \rightarrow ZnCO_3$

(c) $3\,Ba(OH)_2(aq) + 2H_3PO_4(aq) \rightarrow Ba_3(PO_4)_2 + 6H_2O$
(d) $Pb(NO_3)(aq) + H_2C_2O_4(aq) \rightarrow PbC_2O_4 + 2HNO_3$

3. The self-ionisation constant of methanol is 2.5×10^{-14} and its dielectric constant is 32 while that of water $K_W = 1\times10^{-14}$. Therefore,
(a) pure methanol is less acidic than pure water
(b) the neutral pH of methanol is 8.3
(c) the extent of dissociation of HCl is more in methanol than in water
(d) the strongest base in methanol is OH^-

4. The dissociation constant of weak organic base in water at 298 K is 6.0×10^{-4}. It is required to have a solution with $[OH^-] = 0.01$ M. The stoichiometric concentration of the base required (in M) is
(a) 0.256 (b) 0.128
(c) 0.055 (d) 0.177

5. One litre of a solution containing 25 mmol of formic acid and 16 mmol of sodium hydroxide has a pH of 4.0. The dissociation constant of the acid is
(a) 2.40×10^{-3} (b) 1.75×10^{-4}
(c) 8.0×10^{-5} (d) 6.8×10^{-3}

6. Pure H_2SO_4 and flouorosulphonic acid have H_0 values of −12.0 and −15.0 respectively, therefore,
(a) H_2SO_4 is more acidic than flouorosulfonic acid
(b) the proton activity of flouorosulfonic acid is greater than that of pure H_2SO_4
(c) both acids have the same H^+ ion concentration
(d) their proton activity can be correctly assessed by measurements with a glass electrode.

7. 10 mL of 0.1 M HCl added to one litre of an acetic acid–sodium acetate buffer changes its pH from 4.85 to 4.60. The buffer capacity of the buffer is
(a) 0.120 (b) 0.154
(c) 0.007 (d) 0.055

8. 20 mL of 0.05 N NaOH, added to a 500 mL solution of a boric acid–borate buffer, resulted in a change of pH from 9.0 to 9.25. The buffer capacity of the buffer is
(a) −0.004 (b) 0.10
(c) −0.015 (d) 0.008

9. The carbonic acid–bicarbonate system is an important buffer to control the pH of natural water. To 1 litre of such a system containing $[H_2CO_3] = 1.5\times10^{-5}$ and $[HCO_3^-] = 1.2\times10^{-4}$, 20 mL of 1 mM HCl is added. The change in pH observed on this addition is
(a) 0.092 (b) 0.056
(c) 0.085 (d) 0.079

10. The thermodynamic solubility product of lead fluoride at 298 K is 8.0×10^{-8}. Which of the following statements relating to this salt is true?
(a) Addition of F^- or Pb^{2+} to a saturated solution of the salt changes the K_{sp}
(b) The K_{sp} is a function of temperature
(c) The addition of small quantities of NaCl solution does not change the concentration of Pb^{2+} or F^- ion in a saturated solution
(d) The addition of an ion like OH^- that can complex with Pb^{2+}, alters the K_{sp}

11. Consider 1 litre of a solution containing a mixture of 0.1 mol of $Mg(NO_3)$ and $Ca(NO_3)_2$. The percentage amount of Mg^{2+} ions remaining in solution on the addition of 0.01 M NaOH by the time $Ca(OH)_2$ begins to precipitate is [K_{sp} $Ca(OH)_2 = 5.0\times10^{-6}$; K_{sp} $Mg(OH)_2 = 6.0\times10^{-12}$]
(a) 1.5×10^{-6} (b) 2.4×10^{-5}
(c) 1.2×10^{-4} (d) 3.2×10^{-3}

12. The dissociation constants K_a and K_b of acetic acid and aniline are 1.8×10^{-5} and 4.2×10^{-10} respectively at 298 K. The degree of hydrolysis of 0.02 M anilinium acetate solution is [$K_W(H_2O) = 1.0\times10^{-14}$]
(a) 0.535 (b) 0.268
(c) 0.804 (d) 0.762

13. The degree of hydrolysis of a salt formed by the neutralisation of a weak acid by a weak base
(a) decreases with an increase in the concentration of the salt
(b) increases with an increase in the concentration of the salt
(c) is not affected by a change in temperature
(d) is independent of the concentration of the salt

14. The standard reduction potentials for the reaction $Ag^+ + e^- \rightarrow Ag$ and $[Ag(NH_3)_2]^+ + e^- \rightarrow Ag + 2(NH_3)_2$ are +0.800 V and +0.3408 at 298 K respectively. From these data, the *pK(stab)* of the reaction $Ag^+ + 2NH_3 \rightarrow [Ag(NH_3)_2]^+$ is
(a) 7.87 (b) −8.77
(c) −7.78 (d) 7.95

15. The standard redox potentials of $Pt/Ce^{4+}, Ce^{3+}$, $Pt/MnO_4^-, Mn^{2+}$ and $Pt/Cr_2O_7^{2-}, Cr^{3+}$ are 1.70 V, 1.51 V and 1.33 V respectively. The oxidising powers of these oxidising agents are in the order
(a) $Pt/Ce^{4+}, Ce^{3+} > Pt/Cr_2O_7^{2-}, Cr^{3+} > Pt/MnO_4^-, Mn^{2+}$
(b) $Pt/Ce^{4+}, Ce^{3+} > Pt/MnO_4^-, Mn^{2+} > Pt/Cr_2O_7^{2-}, Cr^{3+}$
(c) $Pt/MnO_4^-, Mn^{2+} > Pt/Ce^{4+}, Ce^{3+} > Pt/Cr_2O_7^{2-}, Cr^{3+}$
(d) $Pt/Cr_2O_7^{2-}, Cr^{3+} > Pt/Ce^{4+}, Ce^{3+} > Pt/MnO_4^-, Mn^{2+}$

16. Which of the following statements relating to the dissociation constant of acetic acid is true?
(a) The pH of the solution decreases on the addition of sodium acetate.
(b) The addition of a small amount of KCl does not affect the dissociation equilibrium of the acid.
(c) The degree of dissociation of the acid increases on the addition of a small amount of KCl.
(d) The activity coefficients of all the species in the reaction have no effect on the extent of dissociation.

17. Given that the standard potential for $Cd^{2+}(aq) + 2e^- \rightarrow Cd(s)$ $E^0 = -0.400$ V and $Cd(OH)_2 + 2e^- \rightarrow Cd(s) + 2OH^-$, $E^0 = -0.880$ V, the pK_{sp} of $Cd(OH)_2$ at 298 K is
(a) 12.69 (b) 16.24
(c) 13.85 (d) 14.62

18. The $pK_{\text{self ion}}$ of ethanol is 19.1. The $p_{C_2H_5O^-}$ of an ethanolic solution containing 0.365 g of HCl in 500 mL ethanol is
(a) 17.4 (b) 18.5
(c) 17.9 (d) 18.8

19. The pK_{a1} and pK_{a2} of bromoacetic and chloroacetic acids at 298 K are 2.69 and 2.85 respectively. The concentration of bromoacetate and chloroacetate in a 0.02 M solution of the mixture are respectively
(a) 3.90×10^{-3}, 4.55×10^{-3}
(b) 6.54×10^{-3}, 5.80×10^{-3}
(c) 5.23×10^{-3}, 3.60×10^{-3}
(d) 7.42×10^{-3}, 5.23×10^{-3}

20. The pK_{sp} of barium selenate and lead selenate at 298 K are 7.46 and 6.85 respectively. Which of the following statements relating to the salts is true?
(a) The solubility of lead selenate is lower than that of barium selenate.
(b) In a mixture of saturated solutions of the salts, the concentration of $[SeO_4^{2-}]$ is 4.18×10^{-4} M.
(c) The total concentration of $[SeO_4^{2-}]$ in the presence of a neutral salt such as $NaNO_3$ is lower than that in its absence.
(d) The effect of temperature on the solubility product depends on the magnitude of the ΔG^0 of the reaction.

Answer Keys

Practice Problems

1. (a) $\alpha = 0.0820$; $K_a = 1.831\times10^{-4}$
(b) $K_{Th} = 1.649\times10^{-4}$; $\alpha = 0.078$

2. (a) $\Delta H_{\text{self ion}} = 47.8$ kJ mol^{-1}
(b) Neutralisation pH in this solvent = 7.92; pH of 0.1 M $HClO_4$ = 1.0; $p_{CH_2OH^+-CH_2OH}$ of 0.01 sodium glycollate = 13.84. The solution is alkaline.
(c) The ionisation constant of formic acid and acetic acid in EG will be expected to be smaller because EG has a smaller dielectric constant than water. Also, EG is more acidic and less basic than water.

3. (a) $[Ca^{2+}] = 7.00\times10^{-3}$. Ca^{2+} ions from the dissociation of $CaCO_3$ do not effectively contribute to the total Ca^{2+} concentration.
(b) $[Ag^+] = [CNS^-] = 1.015\times10^{-6}$ g mol L^{-1}

4. (a) pH = 1.93
(b) To 1 L of 0.1 M pyvuric acid, one has to add 1.014 L (or 1014 mL) of water so that both have the same pH.

5. (a) $\alpha = 6.06\times10^{-3}$; pH = 7.04
(b) $K_{a(Th)} = 2.048\times10^{-2}$; pH = 1.84

6. (a) $[OH^-] = 8.16\times10^{-4}$; pH = 10.91
(b) $K = 4.47\times10^{-3}$; $K' = 1.66\times10^{-10}$; $K''' = 3.72\times10^{-5}$
$$\frac{K}{K'''} = 2.24\times10^{-5}$$
(c) Isoelectric point refers to the pH at which the ratio of the dipolar ion to the neutral molecule is equal to one. pH(isoel) = 6.06

7. (a) Salt required for 2 L of water = 53.72 g
(b) pH = 2.512; buffer capacity = 0.167; $\frac{db}{dpH} \approx 0.01$
(c) $a_{H^+} = 0.793$

8. (a) The $pK_{(BH+)}$ values should be in the range 0 to –2.3.
(b) (i) pH at the beginning = 2.872
(ii) pH at 25% neutralisation = 4.27
(iii) pH at 50% neutralisation = 4.75
(iv) pH at neutralisation = 8.87
(v) pH when 25% excess base is added = 12.3

9. (a) (i) Minimum $[S^{2-}]$ for precipitating ZnS = 4×10^{-24}
Minimum $[S^{2-}]$ for precipitating NiS = 5×10^{-19} M
Minimum $[S^{2-}]$ for precipitating CoS = 1.0×10^{-20} M
(ii) Concentration of zinc ions remaining in solution at the time of precipitating $Co^{2+} = 2\times10^{-5}$ M
Percentage remaining in solution = 0.1
Percentage of zinc precipitate out = 99.9
(iii) Percentage of Co^{2+} ions remaining in solution = 5
Percentage of Co^{2+} ions precipitate out = 95
(iv) Trace of OH^- are enough to give the necessary $[S^{2-}]$ to precipitate the sulphide.
(b) For 2 L of water, 2.16 g of NaCN is required.

10. (a) $S = 3.56\times10^{-2}$ g L^{-1}; S' in the presence of 0.1 M $Ba(NO_3)_2 = 2.5\times10^{-7}$ g L^{-1}
In the presence of 0.1 M $NaNO_3$, because of the increase in ionic strength, the solubility changes.
The solubility in the presence of 0.1 M $NaNO_3$ increases from 1.58×10^{-4} to 5.43×10^{-2} M
(b) pH = 13.04. When 1 M H_2SO_4 is added, pH = 0.325
For CCl_4, r = 2.173 Å

11. (a) $2\left[MnO_4^- + 8H^+ + 5e^- \rightleftharpoons Mn^{2+} + 4H_2O\right]$

$5\left[2CO_2 + 2H^+ + 2e^- \rightleftharpoons H_2C_2O_4\right]$

Overall:

$$2MnO_4^- + +6H^+ + 5H_2C_2O_4 \rightleftharpoons 2Mn^{2+} + 8H_2O + 10CO_2$$

(b) $$E = \frac{5E^0_{MnO_4^-/Mn^{2+}} + 2E^0_{C_2O_4^{2-}/CO_2}}{7} - \frac{0.0592}{7}\log\left(\frac{\frac{1}{5}[CO_2][H_2O]^4[H_2C_2O_4]}{\frac{2}{5}[H_2C_2O_4][H^+]^{10}[CO_2]^2}\right)$$

At a pH = 0, $E = 0.9405$ V; equilibrium constant of the reaction $K = 10^{340}$

12. (a) Solubility of $Cu(OH)_2$ in 1 M $NH_3 = 8.18\times10^{-3}$
On iteration, a revised value of $S = 7.76\times10^{-3}$

(b) $H_2SO_4 + H_2SO_4 \rightleftharpoons H_3SO_4^+ + HSO_4^-$
Concentration of hydronium ion in this solvent $= \left[H_3SO_4^+\right] = 1.73\times10^{-2}$ M

(c) Yes, because both $HClO_4$ and HCl behave as weak acids in this solvent.

(d) Boric acid is a weak acid in water with a $K_a = 5.8\times10^{-10}$. However, it is a strong acid in liq. NH_3. The strongest acid in water is H_3O^+ and the strongest acid in liq. NH_3 is NH_4^+. The auto-protolysis constants play a significant role in providing the quantum of pH jump near the neutralisation point in a potentiometric titration. In liq. NH_3, the jump in pH near the neutralisation point is 18 units whereas in water the jump is only 6 units.

13. The K_{SP} of AgI is much smaller than the K_{SP} of AgCl. The first inflexion in the potentiometric titration of the mixture is due to the I- ions reacting to give AgI and the second inflexion point is due to the formation of AgCl.

(i) When 2 mL of 0.1 M $AgNO_3$ is added, $E_1 = -0.004$ V
(ii) When 4 mL of 0.1 M $AgNO_3$ is added, $E_2 = +0.025$ V
(iii) When 5 mL of 0.1 M $AgNO_3$ is added, $E_3 = +0.332$ V
(iv) When 7.5 mL of 0.1 M $AgNO_3$ is added, $E_4 = +0.370$ V
(v) When 10 mL of 0.1 M $AgNO_3$ is added, $E_5 = +0.572$ V
(vi) When 15 mL of 0.1 M $AgNO_3$ is added, $E_6 = +0.634$ V

14. (a) Composition (3): In this case $[Sr^{2+}][C_2O_4^{2-}] = 6.41\times10^{-8}$; since this product is greater than the K_{SP}, SrC_2O_4 is precipitated from this solution.

(b) (i) When 5 mL of 0.1 M HCl is added, pH = 11.18
(ii) When 10 mL of 0.1 M HCl is added, pH = 10.70
(iii) When 15 mL of 0.1 M HCl is added, pH = 10.22
(iv) When 20 mL of 0.1 M HCl is added, pH = 8.33
(v) To the solution of $NaHCO_3$ we now add 5 mL of 0.05 M HCl, pH = 6.78
(vi) When 10 mL of 0.05 M HCl is added, pH = 6.0
(vii) When 15 mL of 0.05 M HCl is added, pH = 5.12
(viii) When 20 mL of 0.05 M HCl is added, pH = 2.78
(ix) When 25 mL of 0.05 M HCl is added, pH = 2.43
(x) When 30 mL of 0.05 M HCl is added, pH = 2.26

Since the first inflexion occurs around pH = 8.3 to 10.0, thymol blue and phenolphthalein are suitable. When the second inflexion occurs around 2.8–5.1, methyl orange or bromophenol blue are suitable.

15. (a)

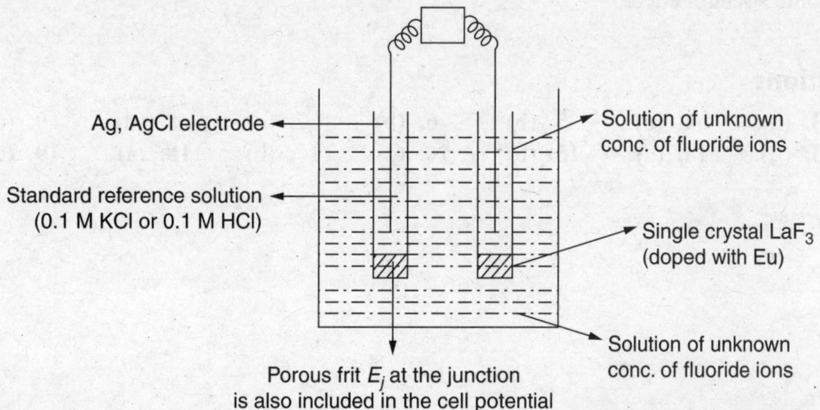

(b) a_{F^-} (unknown) = 2.13 ppm

16. (a) $K_h = 2.4\times10^{-5}$

(b) (i) For the reaction

$$4Co(NH_3)_4^{2+} + O_2 + 4H^+ \rightarrow 4Co(NH_3)_4^{3+} + 2H_2O$$

$E^0 = 1.23 - 1.102 = -1.22$ V

$\therefore O_2$ can oxidise CO^{2+} to CO^{3+} in the presence of NH_3

For the reaction

$$4Co^{2+} + O_2 + 4H^+ \rightarrow 4Co^{3+} + 2H_2O$$

$E^0 = 1.23 - 1.82 = -0.59$ V

Since E^0 is negative, O_2 cannot oxidise Co^{2+} to Co^{3+} in the absence of NH_3.
The oxidising powers lies in the order (C) < (B) < (A)

(ii) Fe^{2+} ions complexed with OH- or CN- ions can reduce O_2 to OH- ions because the E^0 for the reactions

$$2Fe(OH)_2 + \frac{1}{2}O_2(g) + H_2O \rightleftharpoons 2Fe(OH)_3 + 2OH^-$$

$E^0 = 0.40 - (-0.56) = 0.96$ V

$$2Fe(CN)_6^{4-} + \frac{1}{2}O_2(g) + H_2O \rightleftharpoons 2Fe(CN)_6^{3-} + 2OH^-$$

$E^0 = 0.40 - (+0.56) = 0.04$ V are positive

17. $E = \dfrac{E^0_{Ce^{4+}/Ce^{3+}} + E^0_{Cr^{3+}/Cr^{2+}}}{2} - \dfrac{0.0592}{2}\log\left(\dfrac{[Ce^{3+}][Ce^{4+}]}{[Ce^{3+}][Ce^{4+}]}\right)$

or $E = \dfrac{E^0_{Ce^{4+}/Ce^{3+}} + E^0_{Cr^{3+}/Cr^{2+}}}{2} = \dfrac{1.51 + (-0.41)}{2} = 0.55\,\text{V}$

Equilibrium constant of the reaction $= 2.69\times10^{32}$

18. (a) Solubility of $Mg_3(PO_4)_2 = 1.65\times10^{-3}$ g L^{-1}

(b) The solubility is reduced to 2.07×10^{-7} mol L^{-1} considering the ionic strength effect.

If ionic strength effect is not considered at all, $S' = 1.61\times10^{-11}$ mol L^{-1}

19. (a) Weight of $((HCl.NH_2CH_2COONa)M = 133.5) = 11.20\times10^{-3}\times133.5 = 1.495$ g is required for 1 L of buffer solution

(b) $K_1 = 8.30\times10^{-5}$

pK_1 (in the presence of 0.02 M $NaNO_3$) = 4.08

$K_2 = 4.13\times10^{-6}$

pK_2 (in the presence of 0.02 M $NaNO_3$) = 5.38

20. (a) $p_{(a_{H^+}\gamma_{Cl^-})} = \dfrac{E - E^0}{0.0592} + \log m_{Cl^-}$

$= \dfrac{0.6390 - 0.2223}{0.0592} + \log 0.0051 = 4.746$

Using the D–H extended equation for Cl^-, $p_{a_H} = 4.700$

From Henderson equation,

$$pH = pK_a + \log\frac{[\text{salt}]}{[\text{acid}]} = 4.756 + \log\frac{0.0048}{0.0050} = 4.738$$

(b) Alkali and alkaline earth metal ions are small in size, have a high charge density and are not very polarisable. Hence, they are considered as hard acids while the opposite is true for Ag^+, Au^+, Pt^{2+}, Pd^{2+} ions. These are large cations, have lower charge density and have filled or nearly filled *d* orbitals.

(c) Among the halide ions, Cl^- and F^- ions are hard bases while I^- is a soft base. The SO_4^{2-} ion is a hard base and the $S_2O_3^{2-}$ ion is a soft base. PO_4^{3-} is a hard base and S^{2-} is a soft base. These conclusions are arrived at on the basis of definitions of hard and soft bases.

Multiple Choice Questions

1. (c)	2. (a)	3. (b)	4. (d)	5. (b)	6. (b)	7. (c)	8. (d)	9. (d)	10. (b)
11. (c)	12. (a)	13. (d)	14. (c)	15. (b)	16. (c)	17. (b)	18. (a)	19. (c)	20. (b)

CHAPTER

14 NUCLEAR CHEMISTRY

CHAPTER OUTLINE

14.1 Phenomenon of Radioactivity

Radioactivity refers to the spontaneous emission of radiation from compounds of uranium or minerals of uranium such as pitchblende. The emitted radiations have the capability of affecting photographic plates and form charged particles such as gaseous ions or electrons when passing through air. Radioactivity is a property of the nucleus rather than of the atom and external parameters such as temperature or pressure have no effect on the rate of radioactive transformation.

14.1.1 Types of Radiation and their Properties

Three types of radiation are emitted from radioactive materials which are known as α-, β- and γ-rays. Table 14.1 summarises their characteristic properties.

Three methods are used to detect all these radiations, (i) Wilson cloud chamber, (ii) bubble chamber, and (iii) using special nuclear track emulsions. Recent devices include Geiger–Muller counters for detecting these radiations.

14.1.2 Differences between Chemical and Nuclear Reactions

Some important differences may be noted between these two reactions.

- In chemical reactions, there is rearrangement of external electrons (outside the nucleus) but nuclei are not affected. In nuclear reactions, there is a change in the number of protons or neutrons in the nucleus, leading to the formation of new species.
- The energy changes in chemical reactions are of the order of a few kJ, whereas in nuclear reactions the energy changes are a million times higher.
- Temperature or pressure changes affect the rates of chemical reactions but not those of nuclear reactions.

Table 14.1 Characteristics of nuclear radiation

Name	Nature of particles	Mass and charge	Energies and other particulars
α-particles	Helium nuclei	Mass of 4 on conventional atomic weight scale and two unit +ve charges	In the order of 4–10 MeV and velocities of about 2×10^7 m s^{-1}. Ionise gases. Cannot penetrate a thin sheet of Al foil.
β-particles	Electrons	5.48×10^{-4} g on the conventional atomic scale (1/1837 of the mass of H atom) and unit –ve charge (4.80×10^{-10} esu)	In the order of 1 MeV and velocities of about 3×10^8 m s^{-1} (equal to that of light). Can ionise atoms of gases. More penetrating than α-particles. Can be stopped by a 0.2 cm Al foil.
γ-rays	Electromagnetic radiation of very short wavelength between 10^{-1} nm and 10^{-4} nm (similar to x-rays)	No mass and no charge	In the order of 10^{-2} to 10 MeV. Can penetrate dense matter to a considerable length and can ionise gases.

14.2 Kinetics of Radioactive Disintegration

The radioactive decay of a species follows a rate law similar to that of a first order chemical reaction. The rate of decay is directly proportional to the number, N, of the atoms present at any given time. Mathematically, it may be written as

$$-\frac{dN}{dt} = \lambda_N \tag{14.1}$$

where λ = disintegration or decay constant. On integration and evaluation of the integration constant using $N = N_0$ at time $t = 0$, the relation,

$$-\ln\frac{N}{N_0} = \lambda t; \quad \frac{N}{N_0} = e^{-\lambda t} \tag{14.2}$$

is obtained. λ is evaluated from the slope of the linear plot of ln N vs time. The quantity $\tau = \frac{1}{\lambda}$ is known as the average life of a radioactive species.

Half-life period, $T_{1/2}$: It is defined as the time required for the activity (or amount) of a radioactive species to be reduced to one half of its original value. Thus, when $\frac{N}{N_0} = \frac{1}{2}$

$$\frac{1}{2} = e^{-\lambda T_{1/2}} \text{ from which } T_{1/2} = \frac{0.693}{\lambda} \tag{14.3}$$

is obtained. For example, the $T_{1/2}$ of uranium = 4.4×10^9 years and $T_{1/2}$ of radon = 3.8 days.

The relation between the number of half-life periods (n) and the number of atoms remaining after this time interval (N) is,

$$N = N_0(1/2)^n \tag{14.4}$$

Radioactive equilibrium: If a radioactive element (referred to as the parent) decays to form another radioactive element (referred to as the daughter), a state of equilibrium is said to be reached when the daughter element also decays at the same rate as the parent element. The condition of equilibrium is thus given by

$$\lambda_A N_A = \lambda_B N_B \quad \left(\text{since } -\frac{dN_A}{dt} = -\frac{dN_B}{dt}\right)$$

For a series of decay stages, we can write in general,

$$\lambda_A N_A = \lambda_B N_B = \lambda_C N_C = \ldots \text{Constant} \tag{14.5}$$

Note that A, B are the parent and daughter elements. For the decay B to C, B is the parent and C is the daughter element. Ns are the number of atoms of different species. Substituting for λ in terms of $T_{1/2}$ (from Eq. 14.3), we may write,

$$\frac{N_A}{(T_{1/2})_A} = \frac{N_B}{(T_{1/2})_B} = \frac{N_C}{(T_{1/2})_C} \quad \left(\text{that is, } N \propto T_{1/2}\right) \tag{14.6}$$

Equation 14.6 is useful for determining the $T_{1/2}$ values, especially in cases where a given radioactive element has very slow rates of decay.

14.3 Important Definitions

Fajan's group displacement law and radioactive series: According to this law, when a radioactive atom loses an α-particle, its position changes two places to the left in the periodic table because its atomic number decreases by two. Further its atomic mass decreases by 4 units. When a β-particle is emitted, its atomic number increases by one unit (shifting its place by one unit to the right in the periodic table) but there is no change of mass. There are three radioactive series—uranium series, thorium series and actinium series. All these involve emission of α- or β-particles and ultimately lead to the formation of a non-radioactive product which is a form of lead. An artificial radioactive series, namely, the neptunium series also is known.

Units of radioactivity

Curie (Ci): It is defined as equal to 3.7×10^{10} disintegrations (or disintegrating atoms) per second (dps). A smaller unit is the millicurie, which is equal to 3.7×10^7 dps. A rutherford (Rd) is equal to 10^6 dps.

The specific activity of radioactive material is given as millicuries per gram (D_g) or millicuries per ml (D_{ml}).

It is possible to convert the activity of a radioactive material in millicuries (or curies) into its weight in mg using the relation,

1 mC = $8.864 \times 10^{-14} \times$ atomic mass $\times T_{1/2}$ mg

where $T_{1/2}$ is given in seconds.

Isotopes: Two elements having the same atomic number but different masses (or mass numbers) are known as isotopes. They occupy the same position in the periodic table and have similar chemical properties.

Example 1: Radium A (atomic no. 84, atomic mass 218); radium C′ (atomic no. 84, atomic mass 214) and radium F (atomic no. 84, atomic mass 210) are isotopes.

Example 2: Uranium-I (atomic no. 92, atomic mass 238), uranium-II (atomic no. 92, atomic mass 234).

Isobars: Substances having the same mass numbers but different atomic numbers are known as isobars. For example, ^{40}Ar, ^{40}K, and ^{40}Ca are isobars. Their mass number is 40 but their atomic numbers are 18, 19 and 20 respectively. That is, the total number of protons and neutrons (nucleons) is always the same but their individual numbers differ.

Isotones: Different substances have different atomic numbers and different mass numbers, but have the same number of neutrons. They are known as isotones. For example, $_{14}Sc^{30}$ and $_{15}P^{31}$ have 16 neutrons each. It is also necessary to mention the existence of nuclear isomers, that is, those nuclei having the same mass number and same atomic number. Thus they have the same composition but they differ in their energy states. An example is the pair $_{91}U^{234}$ and $_{91}UX2^{234}$

14.4 Mass Spectrometric Analysis

The end product of the disintegration series of uranium and thorium is lead with an atomic number of 82, but it was found that the lead obtained in the two series has different mass numbers. These observations lead to the search of isotopic forms of other elements. The mass spectrometer in this context is a powerful tool to determine the masses of the isotopes of various elements and also their relative abundance.

In a mass spectrometer, a beam of positively charged ions from an ion source is generated. This is done by bombarding the element under consideration with electrons and the +ve ions so formed are passed through two slits. The ions that come out are next subjected to a powerful field of a strong electromagnet which results in the bending of the ions into circular paths of nearly same radius (R) which is equal to

$$R = \frac{mv}{Bq} \qquad (14.7)$$

where m = mass of the ion, v = velocity of the ion, B = magnetic induction and q = charge on the ion, all in their proper units.

Substituting Eq. 14.7 in the expression

$$\frac{1}{2}mv^2 = \omega q \qquad (14.8)$$

we get $$\frac{m}{q} = \frac{B^2R^2}{2\omega} \qquad (14.9)$$

Equation 14.8 equates the kinetic energy of the ions to the electric energy due to the electric potential and charge of the ion.

14.4.1 Relation between Physical and Chemical Atomic Weights

Let us consider sulfur as an example. It has four isotopes of masses 31.9822, 32.9819, 33.9786 and 35.9784 with their abundances being 95.018, 0.750, 4.215 and 0.017 per cent respectively. The physical atomic weight is therefore is given by $(0.95018 \times 31.9822) + (0.00750 \times 32.9819) + (0.04215 \times 33.9786) + (0.00017 \times 35.9784) = 32.07455$, whereas the chemical atomic weight is 32.066.

14.4.2 Separation of Isotopes

This is achieved by using several methods, (i) gaseous diffusion, (ii) thermal diffusion, (iii) electromagnetic method, and (iv) chemical exchange method.

Gaseous diffusion method: This method takes advantage of different rates of diffusion of gaseous isotopic molecules. As mentioned in Chapter 1, the rate of diffusion is inversely proportional to the square root of density and also to the square root of the molar mass. An important application in this method is in the separation of isotopes of U^{235} and U^{238} by converting them into the gaseous form as uranium hexafluoride.

Thermal diffusion method: This method is based on a combination of diffusion and convection. It results in the concentration of the lighter isotope at the top and of the higher one at the bottom of a long tube which has a central wire heated electrically to a temperature of 773 K. A temperature gradient is maintained between the hot wire and the colder wall of the tube. Separation of the isotopes of chlorine, $^{35}_{17}Cl$ and $^{37}_{17}Cl$, was achieved by this method.

Electromagnetic method: This method employs a mass spectrometer with split collectors for collecting the higher and lighter ions separately. Isotopes of lighter elements such as boron and heavy elements such as uranium have been separated by this method.

Chemical exchange method: This method is based on the fact that the chemical reactivity of different isotopic forms of an element varies significantly. This is due to the differences in activation energies of the reactions of compounds containing different isotopes of a given element. For example, consider the exchange reaction between ammonia gas and ammonium ion in solution given by

$$^{15}NH_3(g) + {}^{14}NH_4^+(soln) \rightleftharpoons {}^{14}NH_3(g) + {}^{15}NH_4^+(soln) \qquad (14.10)$$

$$K_{eq,\,298\,K} = 1.033$$

Thus

$$\frac{\dfrac{^{14}NH_3(g)}{^{15}NH_3(g)}}{\dfrac{^{14}NH_4^+(soln)}{^{15}NH_4^+(soln)}} = 1.033$$

which means that the ^{14}N to ^{15}N ratio is greater in the gas phase than in solution. Using this approach, a higher concentration of ^{15}N in solution as $^{15}NH_4^+(NO_3)$ at the bottom of a fractionating column was achieved.

Consider the isotopes of hydrogen, with atomic masses 1 and 2. Heavy hydrogen (also known as deuterium) is in the ratio H_2:D_2 = 4500:1. Fractional distillation of liquid hydrogen by H C Urey resulted in the partial separation of the isotopes. Electrolysis of water is found to result in the accumulation of water with the heavier isotope (D_2O) while the electrolytic gas contains a higher proportion of the lighter gas. Thus prolonged electrolysis of water leads to appreciable quantities of water with the heavier isotope (heavy water D_2O). The physical properties of ordinary (H_2O) and heavy (D_2O) water such as density, freezing point and boiling point are markedly different from each other.

14.5 Uses of Isotopes

Isotopes have been used as tracers in analysing many different phenomena.

- In unravelling the mechanism of the photosynthesis reaction using the ^{18}O isotope. For example, when water enriched with ^{18}O is used in photosynthesis, it was observed that the O_2 evolved in the reaction came entirely from water while the O_2 of carbon dioxide was retained in the organic compound. The overall reaction is given by

$$CO_2^{16} + H_2\,^{18}O \rightleftharpoons \frac{1}{6}(C_6H_{12}\,^{18}O_6) + {}^{18}O_2 \qquad (14.11)$$

- The mechanism of the hydrolysis of an ester was elucidated by using the ^{18}O isotope. The reaction in the case of ethyl acetate may be written as,

$$H_3C-C=O-OC_2H_5 + H_2\,^{18}O \rightarrow$$
$$H_3C-C=O-{}^{18}OH + C_2H_5OH \qquad (14.12)$$

The reaction proceeds via the breaking of the C–OC_2H_5 bond.

- Solubilities of sparingly soluble salts such as lead sulfate have been determined by using a lead isotope. Isotopic dilution analysis is another method employed for determination of total amino acid content in the hydrolysis of a protein.
- Radiocarbon dating: Using the carbon isotope $_6C^{14}$, the ages of fossils, rocks and old trees have been determined. An example for determining the age of an old wood sample using $_6C^{14}$ is illustrated. Radioactive carbon $_6C^{14}$ is produced in air by the action of neutrons of cosmic rays on $_7N^{14}$ in the upper atmosphere. The reaction is,

$$_7N^{14} \text{ (upper atmosphere)} + {}_0n^1 \rightarrow {}_6C^{14} + {}_1H^1 \qquad (14.13)$$

As a result, the CO_2 present in air as $_6C^{12}O_2$ contains a definite proportion of $_6C^{14}$ due to mixing and it is absorbed by the wood. Since the $_6C^{14}$ decays continuously according to

$$_6C^{14} \rightarrow {}_7N^{14} + {}_{-1}e^0 \quad (T_{1/2} = 5600 \text{ years}) \qquad (14.14)$$

the live wood (say from living plants) replenishes the $_6C^{14}$ [lost due to the reaction (Eq. 14.14)] from the atmosphere and keeps the ratio of $_6C^{14}/_6C^{12}$ constant. This ratio diminishes in the 'dead' wood since there is no exchange of radioactive carbon.

Using the relation,

$$\text{Age of wood} = \frac{2.303}{\lambda}\log\frac{N_0}{N}$$

$$= \frac{2.303 \times T_{1/2}}{0.693}\log\frac{\dfrac{_6C^{14}}{_6C^{12}} \text{ ratio in live wood}}{\dfrac{_6C^{14}}{_6C^{12}} \text{ ratio in dead wood}} \qquad (14.15)$$

the age of wood can be determined.

- Isotopes have extensive medical applications in cancer treatment, CT scan therapy, thyroid gland related issues, and so on.

14.6 Artificial Nuclear Reactions

(i) When α-particles having a kinetic energy of about 4.8 MeV are passed though nitrogen gas, the reaction

$$_7N^{14} + {}_2He^4 \rightarrow {}_8O^{17} + {}_1H^1 \qquad (14.16)$$

occurs resulting in the formation of protons. This was the first artificial nuclear reaction reported by E Rutherford in 1919.

(ii) ***Acceleration of particles:*** Since the discovery of the reaction (14.16), attempts were made to develop apparatus that could bring about reactions with nuclei such as hydrogen ($_1H^1$), deuterium ($_1D^2$) and so on. Devices such as the cyclotron (E O Lawrence, 1932) and the linear accelerator were developed as they can produce high velocities and energies.

When the high energy particle is a proton, the reaction may be such that the released particle is a neutron, a γ-ray or an α-particle. If the high energy projectile is a deuteron, the released particle may be a proton, neutron or an α-particle.

14.7 Nuclear Reactions Involving the Generation of Neutrons, Positrons and Mesons

14.7.1 Neutrons

James Chadwick (1932) proposed the existence of a new particle which had a mass comparable to that of a proton but carrying no charge and it was named the *neutron*. This particle is formed as a result of a (α, n) reaction with some lighter nuclei such as like $_4Be^9$. The reaction is

$$_4Be^9 + {_2He^4} \rightarrow {_6C^{12}} + {_0n^1} \quad (14.17)$$

Other reactions such as

$$_3Li^7 + {_1H^1} \rightarrow {_4Be^7} + {_0n^1} \quad (14.18)$$

result in the formation of neutrons. The half-life of a neutron, $T_{1/2}$, is 127.5 minutes and it decays according to

$$_0n^1 \rightarrow {_1p^1} + e^- + 0.783 \text{ MeV} \quad (14.19)$$

giving rise to a proton and an electron. The neutron has a mass of 1.00871 on the chemical atomic weight scale.

14.7.2 Positrons

This particle was discovered in 1932 by C D Anderson and a typical reaction involving its formation is

$$_5B^{10} + {_2He^4} \rightarrow {_7N^{13}} + {_0n^1} \text{ and}$$

$$_7N^{13} \rightarrow {_6C^{13}} + {_1e^0} \quad (14.20)$$

14.7.3 Muons or μ-Mesons

A theory for the structure of the nucleus led H Yukawa (in 1935) to postulate the existence of a new kind of particle called the μ-meson. In a totally different approach, Anderson working at Caltech, showed that the particles produced by cosmic rays have properties similar to that predicted by Yukawa. The μ-mesons or muons have a mass of 209 and a $T_{1/2}$ of 2.2 μs.

14.8 Artificial Radioactivity

The induction of radioactivity was observed during the study of the reaction of boron, magnesium or aluminium with α-particles. For example, the reaction of aluminium with α-particles is given by

$$_{13}Al^{27} + {_2He^4} \rightarrow {_{15}P^{30}} + {_0n^1} \quad (14.21)$$

The $_{15}P^{30}$ nucleide formed is unstable and decays with the emission of positrons according to

$$_{15}P^{30} \rightarrow {_{14}Si^{30}} + {_1e^0} \quad (14.22)$$

$_{15}P^{30}$ is an artificial radionuclide which is unstable unlike the stable nucleus, $_{15}P^{31}$. Another example of an artificial radionucleide is $_{94}Pu^{239}$ obtained by the reaction of $_{92}U^{238}$ via a (n, γ) reaction which is followed by two β-ray decays according to

$$_{92}U^{238} + {_0n^1} \rightarrow {_{92}U^{239}} + \gamma \rightarrow$$

$$_{93}Np^{239} + {_{-1}e^0} \rightarrow {_{94}Pu^{239}} + {_{-1}e^0} \quad (14.23)$$

14.8.1 Transuranium Elements

The two nuclides $_{93}Np^{239}$ and $_{94}Pu^{239}$ are examples of transuranium elements which have $T_{1/2}$ values of 2.3 days and 24,400 years respectively. Other transuranium elements have been produced by a variety of reactions involving (α, n), $(\alpha, 2n)$ and (n, γ). The formation of californium ($_{98}Cf^{246}$), einsteinium ($_{99}Es^{247}$) and fermium ($_{100}Fm^{250}$) has been established from reactions such as,

$$_{92}U^{238} + {_6C^{12}} \rightarrow {_{98}Cf^{246}} + 4\,{_0n^1} \quad (14.24a)$$

$$_{92}U^{238} + {_7N^{14}} \rightarrow {_{99}Es^{247}} + 5\,{_0n^1} \quad (14.24b)$$

$$_{92}U^{238} + {_8O^{16}} \rightarrow {_{100}Fm^{250}} + 4\,{_0n^1} \quad (14.24c)$$

In recent years, transuranium elements with atomic numbers beyond 100 have been synthesised using different target elements and projectiles.

14.9 Nuclear Chemistry

14.9.1 Magnitude of Energy Changes in Nuclear Reactions

As mentioned earlier, the energy changes associated with nuclear reactions is a million times more than those of chemical reactions. Einstein's mass–energy relationship

$$E = mc^2 \quad (14.25)$$

can be used to convert the change in mass in nuclear reactions into its equivalent of change in energy. Consider the reaction,

$$_7N^{14} + {_2He^4} \rightarrow {_8O^{17}} + {_1H^1}$$

The change in mass between reactants and products = (14.00752 + 4.0026) – (17.00453 + 1.00782) = 18.01012 – 18.01235 = –0.00223 amu

(All atomic masses are on the physical scale.)

Since 1 amu = 931.5 MeV, the change in mass is associated with an energy change of 0.00233×931.5 = 2.077 MeV = $\frac{2.077 \times 1.6 \times 10^{-13}}{1000} = 3.32 \times 10^{-16}$ kJ per atom of $_7N^{14}$.

14.9.2 Mass Defect, Packing Fraction and Nuclear Binding Energy

Mass defect: The difference between the sum of the masses of protons, neutrons, electrons and the isotopic atomic mass of the atom is known as the mass defect; it is represented by ΔX. Let m_p, m_n and m_e be the masses of the proton, neutron and electron respectively, and let A and Z be the mass number and atomic number of the nuclide. The mass defect is then given by

$$\text{Mass defect} = \Delta X = Z(m_p + m_e) + (A - Z)m_n - M$$

$$= Zm_H + (A - Z)m_n - M \quad (14.26)$$

where m_p+m_e is replaced by m_H (mass of a proton).

ΔX is expected to be positive for stable nuclei because a stable nucleus will have less energy than its constituent particles.

Binding energy: It is defined as the energy released in the formation of a nucleus from its constituent protons and neutrons (or nucleons). It is also numerically equal to the energy required to break up the nucleus into its constituent particles. For example, the binding energy (BE) of $_8O^{17}$ can be calculated as

$$BE = 8\times1.008145 + 9\times1.008986 - 17.00453$$
$$= 8.06516 + 9.08087 - 17.00453$$
$$= 0.1415 \text{ amu} = 0.1415\times931.5 = 131.8 \text{ MeV}$$

More often, the mean binding energy per nucleon is considered, which in this case is, 131.8/17 = 7.75 MeV.

Larger values of binding energy indicate a greater stability of the nucleus. A plot of binding energy per nucleon vs mass number reveals that over a range of mass numbers, the mean binding energy per nucleon is about 8 MeV. This value represents the amount of energy required to break the nucleus to obtain a proton or neutron.

Packing fraction: It is defined as,

$$PF = \frac{\text{Isotopic mass number} - \text{Mass number}}{\text{Mass number}} \quad (14.27)$$

It is generally expressed as parts of 10,000. For example, the packing fraction of ${}_8O^{17}$ is given by

$$PF\,({}_8O^{17}) = \frac{17.00453 - 17.0000}{17.0000}\times10^4$$
$$= 2.66 \text{ parts per } 10,000$$

It may be mentioned that a positive *PF* indicates instability, while a negative value is associated with stability.

Nuclear stability: This can be understood by considering a plot of the number of neutrons (*N*) vs the number of protons (or *Z*, the atomic number) (*x*-axis). The graph shows that $\frac{N}{Z} = 1$ up to ${}_{20}Ca^{40}$. When the nuclear charge increases (at higher values of atomic numbers) beyond $Z = 20$, $\frac{N}{Z} > 1$ indicating greater instability. This is due to the electrostatic repulsion of protons as the nucleus becomes larger. The forces of electrostatic repulsions are long range forces compared to the short range attractive forces between protons, thus more neutrons need to be added to the nucleus to balance this.

Nuclei having $N = P = 2$, 8, 20, 28, 50, 82 and 126 are very stable and these numbers are referred to as magic numbers. There are two ways of bringing $\frac{N}{Z}$ closer to unity. In cases where $\frac{N}{Z} > 1$, a loss of a β-particle enables the $\frac{N}{Z}$ values close to unity. Two examples are illustrated.

(i) ${}_{11}Na^{23}$ has $\frac{N}{Z} = \frac{12}{11} = 1.09$. When it loses a β-particle, the reaction is

${}_{11}Na^{23} \rightarrow {}_{12}Mg^{23} + {}_{-1}e^0$. Therefore, the $\frac{N}{Z}$ of the resulting nuclide $= \frac{11}{12} = 0.92$

(ii) ${}_{12}Mg^{27}$: $\frac{N}{Z} = \frac{15}{12} = 1.25$; ${}_{12}Mg^{27} \rightarrow {}_{13}Al^{27} + {}_{-1}e^0$. Here, the $\frac{N}{Z}$ of ${}_{13}Al^{27}$ is $\frac{N}{Z} = \frac{14}{13} = 1.08$

In cases where $\frac{N}{Z} < 1$, it becomes closer to unity by a positron (${}_1e^0$) emission or *K*-electron capture. The positron emission may be visualised as a break of a proton $p^+ \rightarrow {}_0n^1 + {}_1e^0$ which results in an increase of the neutrons in the nucleus, thereby increasing the $\frac{N}{Z}$ ratio. For example, in the reaction ${}_9F^{17} \rightarrow {}_8O^{16} + {}_1H^1 (\rightarrow n + {}_0e^1)$ the $\frac{N}{Z}$ ratio of the flourine nuclide is $\frac{8}{9} = 0.9$ while the $\frac{N}{Z}$ ratio of the oxygen nuclide is $\frac{8}{8} = 1.0$. Sometimes a nuclide captures a *K*-shell electron and the vacancy caused thereby is filled by electrons from higher levels. The change in the nucleus as a result of the *K*-electron capture may be given as

$$p^+ + {}_{-1}e^0 \rightarrow {}_0n^1$$

The above reaction increases the number of neutrons in the nucleus thereby increasing the $\frac{N}{Z}$ ratio.

14.9.3 Nuclear Fission Reactions

When a heavy nucleus such as ${}_{92}U^{235}$ is bombarded with neutrons, its nucleus splits approximately into two nearly equal parts of lower mass with lower atomic numbers. Also, a large amount of energy is released. Such reactions are known as fission reactions. The fission reaction of the uranium nucleus with neutrons, studied by Hahn and Strassman, may be represented as

$${}_{92}U^{235} + {}_0n^1 \rightarrow {}_{56}Ba^{144} + {}_{36}Kr^{90} + 2\,{}_0n^1 \quad (14.28)$$

which results in a release of 211.5 MeV of energy. The importance of fission reactions can be seen from Eq. 14.28 where two neutrons are produced and this leads to the possibility of a fission chain reaction. Hence the process goes on continuously once initiated, and energy is produced. The large amount of fission energy is used in the generation of electric power and in the manufacture of atom bombs.

Controlled release of fission energy is obtained in nuclear reactors. The active core of the reactor consists of the nuclear fuel (a fissionable nuclide such as ${}_{92}U^{238}$ or ${}_{92}Th^{232}$) which can be transformed into fissionable material such as Pu^{239} or U^{233} by reacting with neutrons. A moderator such as heavy water or graphite rods is used when the neutrons have to be slowed down.

14.9.4 Breeder Reactors

A breeder reactor is a nuclear reactor in which more fissionable material is generated than is consumed. In these reactors, the economy of neutrons is sufficiently high and this makes it possible to breed more fissile material than amount of fertile material (such as U^{238} or Th^{232}) used. Breeder reactors have the possibility of extracting all the energy contained in the fuels uranium and thorium.

Breeder reactors may be divided into two categories based on the variation of the neutron spectrum—(i) fast breeder reactor (FBR) and (ii) thermal breeder reactor (TBR). The former

(FBR) uses fast neutrons to breed the fissile material such as plutonium and other higher transuranium elements using U^{238}. The latter (TBR) uses moderated thermal neutrons to breed the fissile material U^{233} from thorium.

The fuel core of FBRs use 20% plutonium dioxide (PuO_2) and 80% uranium dioxide (UO_2). Another type of fuel used in FBRs is a metal alloy consisting of a blend of uranium, plutonium and zirconium.

Among the TBRs, the advanced heavy water reactor (AHWR) is gaining its importance because it uses thorium as fuel, which is abundantly available in India. India has a programme of constructing thermal breeder reactors using fertile Th^{232} as fuel.

14.9.5 Fusion Reactions

Fusion reactions may be made to occur by targeting nuclei of light elements with fast protons or deuterons. However, only small amounts of energy are released. By raising the temperature of the system, it is possible to permit the nuclear reaction to occur. Nuclear reactions taking place under these conditions are known as thermonuclear reactions. Thermonuclear reactions occur in stars and one such reaction is,

$$4\,{}_1H^1 \rightarrow {}_2He^4 + 2\,{}_1e^0 + 24.7\text{ MeV} \quad (14.29)$$

Another reaction is,

$${}_1H^1 + {}_1D^2 \rightarrow {}_2He^3 + \text{energy} \quad (14.30)$$

Some fusion reactions carried out under controlled conditions include,

$${}_1D^2 + {}_1D^2 \rightarrow {}_2He^3 + {}_0n^1 + 3.2\text{ MeV}$$

$${}_1D^2 + {}_1D^2 \rightarrow {}_1H^3 + {}_1H^1 + 4.0\text{ MeV}$$

14.9.6 Energy of Stars

The energy of the stars is derived from the highly exothermic reactions occurring in them. Stellar interiors can produce high temperatures of the order of 10^7 K (as in the sun) and under such conditions, nuclei are stripped off electrons and their velocities attain high values. The kinetic energy of an α-particle attains a high value of 10^4 eV, at the high temperature of the stars (example, the sun). The reactions such as those shown below, occur.

$${}_2He^3 + {}_2He^3 \rightarrow {}_2He^4 + 2\,{}_1H^1 + 13.0\text{ MeV} \quad (14.31)$$

14.10 Nuclear Structure

Two models were proposed towards describing the structure of the nucleus—(i) the shell model and (ii) the liquid drop model.

Shell model: When a nucleon approaches the nucleus, it is attracted by the other nucleons and the potential becomes negative. However, the potential becomes constant once the nucleon is sufficiently below the surface. The energy levels of the nucleons under these conditions have been specified by the quantum numbers n and l. Following the *Pauli exclusion principle*, each level contains $(2l+1)$ protons and $(2l+1)$ neutrons. The lower magic numbers 2, 8, 20 are thus explained in a simple manner. The larger magic numbers, which are also associated with high nuclear stability, can be explained on the basis of the same model.

Liquid drop model: Several properties of a liquid (or a liquid drop) can be correlated with that of the nucleus. For example, their densities are constant irrespective of their size. The latent heat of vaporisation of a liquid is comparable to the binding energy of nucleons in the nucleus. The evaporation of a liquid drop is likened to radioactive emission from a nucleus while condensation of a liquid is analogous to the formation of a compound nucleus formed by the absorption of a particle projectile such as an α-particle or neutron. Wheeler and Bohr were able to apply the liquid drop model to calculate the optimum neutron energy required to break a given nucleus. This is in agreement with experimental values. Niels Bohr's son, A Bohr, considered the nucleus as a rotating and vibrating drop of fluid. Nuclei are considered to be ellipsoids based on their quadrupole moments. An ellipsoidal nucleus is analogous to a rotating diatomic molecule. Both the theories have their successes as well as drawbacks and further developments are required in this area.

Nuclear cross-section rates: This term refers to the rate at which intensity the projectile (or product) particles are transmitted through a thin foil of a target (Fig. 14.1).

Area covered, A, by the beam

P_0

Fig. 14.1 Illustration of nuclear cross-section

Let P_0 be the intensity of projectile particles per second incident on a thin foil of the target nuclei. Let the transmitted intensity after reaction in the foil be P particles/s. Assuming that the projectile beam intersects an area A of the foil, and that the foil contains N atoms cm^{-2}, then the area a covered by the nuclei is N (atoms cm^{-2})$\times A$ (cm^2 atom^{-1})$\times$effective area covered by one atom. The last term, the effective area covered by the atom, is called the cross-section for the reaction under consideration. Thus, the fraction of the area that is covered is a/A which is equal to N atoms cm$^{-2}\times\sigma$ cm^2 atom^{-1}. If the number of projectile particles absorbed by the target is ΔP in unit time,

$$\Delta p = -PN\sigma \quad (14.32)$$

Note that N has the unit atoms cm^{-2}, or also, thickness $\Delta X\times$density n (atom cm^{-3}). Expressing Eq. 14.32 as a differential equation,

$$dp = -pN\sigma \quad (14.33)$$

$$\int_{p_{initial}}^{p_{trans}} \frac{dp}{p} = -n\sigma \int_0^x dx \quad (14.34)$$

$$ln\frac{P_{trans}}{P_{initial}} = -\sigma x \times n \quad (14.35)$$

$$P_{trans} = P_{initial}e^{-n\sigma x} \quad (14.36)$$

where P_{trans} = number of nuclei transmitted through the foil. This discussion refers to the attenuation of the incident beam and thus covers all reactions that may occur. It is possible to calculate the number of product nuclei during an irradiation, N, in a time t. Assuming that the product nuclei are stable,

$$N = P \times t(1 - e^{-np dx}) \qquad (14.37)$$

For a thin target (ΔX is small), expansion of $(1 - e^{-n\sigma dx})$ gives

$$N = Pn\sigma\Delta xt \qquad (14.38)$$

In the case where the product nuclei are radioactive,

$$N = \frac{n\sigma\Delta xp}{\lambda}(1 - e^{-\lambda t}) \qquad (14.39)$$

Worked Out Problems

Energy Changes in Nuclear Reactions

Problem 14.1

(a) Define the term atomic mass unit (amu). What is its value (i) in MeV (ii) in kJ mol^{-1}?

(b) Calculate the energy change associated with the reaction

$${}_7N^{14} + {}_2He^4 \rightarrow {}_8O^{17} + {}_1H^1$$

given the isotopic masses of the various species (in amu) as N^{14} = 14.00752; O^{17} = 17.00453; He^4 = 4.00387 and H^1 = 1.00783.

Solution

(a) $1 \text{ amu} = \frac{1}{12^{th}}$ mass of a ${}_6C^{12}$ atom

$$= \frac{1}{12} \times \frac{12}{6.023 \times 10^{23}}$$

$$= 1.66 \times 10^{-24} \text{ g} = 1.66 \times 10^{-27} \text{ kg}$$

From Einstein's mass–energy relation, $E = mc^2$ (where m = mass, c = velocity of light).

$$E = 1.66 \times 10^{-27} \times (2.998 \times 10^8)^2 \text{ J} = 14.92 \times 10^{-11} \text{ J}$$

Since 1 eV = $1.602 \times 10^{-19} \times 1$ V C or J; 1 amu = $\frac{14.92 \times 10^{-11}}{1.602 \times 10^{-19}} = 9.313 \times 10^8$ eV = 931.3 MeV

1 amu = 14.92×10^{-11} J = 14.92×10^{-14} kJ

(b)

Masses of species on the reactant side	Masses of species on the product side
N^{14} = 14.00752	O^{17} = 17.00453
He^4 = 04.00387	H^1 = 01.00783
18.01139	18.01236

Increase in mass = 18.01236 – 18.01139 = 0.00097 amu

Increase in energy = 0.00097×931.3 = 0.9034 MeV = $0.9034 \times 1.602 \times 10^{-13} = 1.45 \times 10^{-13}$ J

This amount of energy must be consumed in the reaction, presumably as kinetic energy supplied to the bombarding helium nuclei.

Problem 14.2

(a) Calculate the maximum velocity that a proton with an energy of 5 MeV can impart on collision with a (i) deuteron (ii) N^{14} atom (rest mass of deuteron = 3.343×10^{-27} kg; rest mass of N^{14} = 23.252×10^{-27} kg; velocity of light = 3×10^8 m s^{-1}).

(b) If a deuteron is accelerated under a potential of 500 MeV, what is its mass in kg?

Solution

(a) Using the Einstein relation $E = mc^2$

(i) $5 \times 1.602 \times 10^{-13}$ (J) $= 3.343 \times 10^{-27} \times c^2$

$$c^2 = \frac{5 \times 1.602 \times 10^{-13}}{3.343 \times 10^{-27}} = 2.396 \times 10^{14}$$

$$\text{or, } c = \sqrt{2.396 \times 10^{14}}$$

$$= 1.548 \times 10^7 \text{ m s}^{-1}$$

(ii) $5 \times 1.602 \times 10^{-13}$(J) $= 23.252 \times 10^{-27} \times c^2$

$$c^2 = \frac{5 \times 1.602 \times 10^{-13}}{23.252 \times 10^{-27}} = 0.344 \times 10^{14}$$

$$\text{or, } c = \sqrt{0.344 \times 10^{14}}$$

$$= 0.586 \times 10^7 \text{ m s}^{-1}$$

(b) $E = mc^2$; $500 \times 1.602 \times 10^{-13}$ (J) $= m \times (3 \times 10^8)^2$

$$m = \frac{500 \times 1.602 \times 10^{-13}}{(3 \times 10^8)^2} = \frac{801 \times 10^{-13}}{9 \times 10^{16}}$$

$$= 89 \times 10^{-29} \text{ kg} = 8.9 \times 10^{-28} \text{ kg}$$

Problem 14.3

(a) The energy released in the Li^7 (p, α) reaction is 17.3 MeV. Calculate the isotopic mass of Li^7. (Atomic mass of ${}_1H^1$ = 1.00783 amu; atomic mass of ${}_2He^4$ = 4.00387 amu.)

(b) How many α-particles and β-particles are emitted in the following decays?

(i) ${}_{90}Th^{232} \rightarrow {}_{86}Rn^{220}$

(ii) ${}_{92}U^{238} \rightarrow {}_{82}Pb^{210}$

(iii) ${}_{88}Ra^{226} \rightarrow {}_{82}RaB^{214}$

Solution

(a) Energy released in terms of amu $= \dfrac{17.3}{931.5} = 0.01857$

Atomic masses of reactants	Atomic masses of products
$Li^7 = ?$	$2 \times {}_2He^4 = 2 \times 4.00387$
	$= 8.00774$
${}_1H^1 = 1.00783$	

(Atomic mass of Li^7, $x + 1.00783) - (8.00774) = 0.01857$
Or, x = Atomic mass of $Li^7 = 0.01857 - 1.00783 + 8.00774$
$= 7.01828$ amu

(b) (i) Balancing mass numbers

${}_{90}Th^{232} \rightarrow {}_{86}Rn^{220} + a\,{}_2He^4 + b\,{}_{-1}e^0$ (*a* and *b* to be found)

$232 = 220 + 4a + 0 \times b = 220 + 4a$

Therefore $a = \dfrac{232-220}{4} = \dfrac{12}{4} = 3$

Balancing atomic numbers

$90 = 86 + 3 \times 2 - 1 \times b = 92 - b$

Hence $b = 2$

(ii) Following the same procedure for mass numbers

$238 = 210 + 4a + b \times 0 = 210 + 4a$

$a = \dfrac{238-210}{4} = 7$

Balancing atomic numbers,

$92 = 82 + 7 \times 2 - b \times 1 = 96 - b$.

or $b = 4$

(iii) Balancing mass numbers

$226 = 214 + 4a - 0 \times b$

or $a = 12/4 = 3$

Balancing atomic numbers

$88 = 82 + 3 \times 2 - b \times 1 = 82 + 6 - b$

or $b = 0$

Problem 14.4

The fission reaction ${}_{92}U^{235} + {}_0n^1 \rightarrow {}_{56}Ba^{141} + {}_{36}Kr^{92} + 3\,{}_0n^1$ takes place giving rise to barium and krypton nuclides. But they decay through a β-emission according to

$${}_{56}Ba^{141} \rightarrow {}_{59}Pr^{141} + 3\,{}_{-1}e^0$$

$${}_{36}Kr^{92} \rightarrow {}_{40}Zr^{92} + 4\,{}_{-1}e^0$$

so that the overall reaction is

$${}_{92}U^{235} + {}_0n^1 \rightarrow {}_{59}Pr^{141} + {}_{40}Zr^{92} + 3\,{}_0n^1 + 7\,{}_{-1}e^0$$

Calculate the energy released in kJ when 100 g of ${}_{92}U^{235}$ undergoes fission. The atomic masses of ${}_{92}U^{235} = 235.1316$; ${}_{59}Pr^{141} = 140.951$; ${}_{40}Zr^{92} = 91.934$; ${}_0n^1 = 1.00867$; ${}_{-1}e^0 = 5.487 \times 10^{-4}$, all in amu.

Solution

Masses of reactants (amu)	Products (amu)
$U^{235} = 235.1316$	$Pr^{141} = 140.951$
${}_0n^1 = 1.00867$	$Zr^{92} = 91.934$
Total mass = 236.1403	$3\,{}_0n^1 = 3 \times 1.00867 = 3.02601$
	$7\,{}_{-1}e^0 = 7 \times 5.487 \times 10^{-4} = 0.00384$
	235.9148

Difference = 236.1403 – 235.9148 = 0.2255 amu

Amount of energy released per atom of ${}_{92}U^{235}$ $= 0.2255 \times 931.5 = 210.1$ MeV

Amount of energy released per gram atom, 235.1 g of ${}_{92}U^{235}$ $= 210.1 \times 6.023 \times 10^{23} \times 1.602 \times 10^{-13} = 2027.2 \times 10^{10}$ J

Energy released for 100 g of ${}_{92}U^{235}$

$$= \frac{2027.2 \times 10^{10} \times 10^{-3} \times 100}{235.1} = 8.62 \times 10^9 \text{ kJ}$$

Problem 14.5

(a) The energy of stars arises out of high the exothermicity of nuclear reactions occurring in them. Write down the various steps in the so-called carbon cycle in very hot stars at high temperatures of the order of 10^8 K or higher.

(b) The energy of the stars at lower temperatures which is one order of magnitude lower than hot stars is due to the proton–proton cycle. Write down the various steps associated with this cycle.

Solution

(a) The energy of the hot stars arises due to a series of nuclear reactions involving C^{12} and N^{14} as catalysts. The various steps are

(i) $C^{12} + H^1 \rightarrow N^{13} + h\nu$

(ii) $N^{13} \rightarrow C^{13} + {}_0\overset{+}{e} + \nu$

(iii) $C^{13} + H^1 \rightarrow N^{14} + h\nu$

(iv) $N^{14} + H^1 \rightarrow O^{15} + h\nu$

(v) $O^{15} \rightarrow N^{15} + e^+ + \nu$ and

(vi) $N^{15} + H^1 \rightarrow C^{12} + He^4$

The overall reaction is thus the conversion of four ${}_1H^1$ into one He^4 with an energy release of the order of 481×10^{-19} kJ (or 0.03 MeV).

(b) The various steps involved in the energy release in cooler stars through the proton–proton cycle are

(i) ${}_1H^1 + {}_1H^1 \rightarrow {}_1D^2 + {}_0e^+ + \gamma$

(ii) ${}_1H^1 + {}_1D^2 \rightarrow {}_2He^3 + \gamma$

(iii) ${}_2He^3 + {}_2He^3 \rightarrow {}_2He^4 + 2\,{}_1H^1$

The overall reaction involves the conversion of four protons to one helium nucleus. The energy release is of the order of 40×10^{-19} kJ (or 0.025 MeV).

Problem 14.6

(a) The energy released in the fusion reaction

$_1D^2 + {}_1H^3 \rightarrow {}_2He^4 + {}_0n^1$

is 1.7×10^{12} J mol^{-1} of deuterium. Given the atomic masses of $_1D^2 = 2.01410$, $_2He^4 = 4.0026$, $_0n^1 = 1.00867$ all in amu, calculate the atomic mass of the $_1H^3$ atom.

(b) Calculate the $\frac{N}{P}$ ratio of (i) $_{92}U^{238}$ (ii) $_{94}Pu^{239}$ (iv) $_9F^{17}$ (v) $_{29}Cu^{64}$. How do they attain stability? $_{29}Cu^{64}$ is also known to emit positrons and can also capture a *K*-electron. What are the $\frac{N}{P}$ ratios in these cases and what are the products?

Solution

(a) Energy released per atom $= \dfrac{1.7 \times 10^{12}}{6.023 \times 10^{23}} = 0.282 \times 10^{-11}$ J

$= \dfrac{0.282 \times 10^{-11}}{1.602 \times 10^{-13}} = 0.176 \times 10^2 = 17.6$ MeV $= 0.0189$ amu

Atomic masses of reactants	Atomic masses of products
$_1D^2 = 2.01410$ amu	$_2He^4 = 4.0026$ amu
$_1H^3 = ?\ x$ amu	$_0n^1 = 1.00867$ amu

Change in mass $= (2.014 + x) - (4.0026 + 1.00867) = 0.0189$

$x = 0.0189 + 5.01127 - 2.01410 = 3.0161$ amu

(b) (i) $\frac{N}{P} = \frac{146}{92} = 1.59$ loss of a β-particle will bring down the $\frac{N}{P}$ ratio. For example,

$_{92}U^{238} \rightarrow {}_{93}Np^{238} + {}_{-1}e^0$

$\frac{N}{P}$ of $_{93}Np^{238} = \frac{145}{93} = 1.56$

Further emission of β-particles will bring down the $\frac{N}{P}$ ratio further.

(ii) For $_{94}Pu^{239}$, $\frac{N}{P} = \frac{145}{93} = 1.54$. When a β-particle is lost, the product

$_{94}Pu^{239} \rightarrow {}_{95}Am^{239} + {}_{-1}e^0$, $\frac{N}{P} = \frac{144}{95} = 1.52$

Further emission of β-particles will lower the $\frac{N}{P}$ ratio further.

(iii) For $_9F^{17}$, $\frac{N}{P} = \frac{8}{9} = 0.89$

Since $\frac{N}{P} < 1$, positron emission or *K*-electron capture increases the $\frac{N}{P}$ ratio.

When a positron is emitted, the reaction is

$_9F^{17} \rightarrow {}_8O^{17} + {}_1e^0$

In this case $\frac{N}{P} = \frac{9}{8} = 1.12$

$_9F^{17} + {}_{-1}e^0 \rightarrow {}_8O^{17}$

Thus, $\frac{N}{P} = \frac{9}{8} = 1.12$ (in the case of *K*-electron capture), and the $\frac{N}{P}$ ratio increases.

(iv) $_{29}Cu^{64}$ when a positron is emitted, the reaction is

$_{29}Cu^{64} \rightarrow {}_{28}Ni^{64} + {}_1e^0$

$\frac{N}{P}$ of $_{29}Cu^{64} = \frac{35}{29} = 1.21$ when a positron is emitted.

$\frac{N}{P}$ of $_{28}Ni^{64} = \frac{36}{28} = 1.28$

When a *K*-electron is captured, the reaction is

$_{29}Cu^{64} + {}_{-1}e^0 \rightarrow {}_{28}Cu^{64}$

$\frac{N}{P} = \frac{36}{28} = 1.28$

Problem 14.7

One of the nuclear fusion reactions of interest for deriving electrical power is

$$_1D^2 + {}_1D^2 \rightarrow {}_2He^3 + {}_0n^1$$

If the electrical output required is 60 MW and the energy of the above reaction is used with 50% efficiency, how many grams of deuterium fuel is needed in a month?

(Atomic masses of reactants and products: $_1D^2 = 2.0140$; $_2He^3 = 3.01603$; $_0n^1 = 1.00867$)

Solution

Total mass of reactants – total mass of products

$= (2 \times 2.01410) - (3.01603 + 1.00867)$

$= 4.02820 - 4.02470 = 0.0035$ amu

0.0035 amu $= 0.0035 \times 1.66 \times 10^{-27}$ kg $= 0.00581 \times 10^{-27}$ kg

$\Delta E = mc^2 = 0.00581 \times 10^{-27}$ (kg) $\times (3 \times 10^8$ m/s$)^2 = 0.053 \times 10^{-11}$ J (for 2 atoms)

$= \dfrac{0.053 \times 10^{-11}}{2} \times 6.023 \times 10^{23} = 1.6 \times 10^{11}$ J mol^{-1}

Since the efficiency is 50%, used energy $= \dfrac{1.6 \times 10^{11} \times 50}{100}$

$= 0.8 \times 10^{11}$ J mol^{-1}

Output required = 60 MW $= 60 \times 10^6$W $= 60 \times 10^6$ J s^{-1}

$= 60 \times 10^6$ Js$^{-1} \times 3600$ s h$^{-1} \times 24$ h day$^{-1} \times 30$ days $= 15552 \times 10^{10}$ J (in 30 days)

Number of moles of deuterium required $= \dfrac{155.52\times10^{12}}{0.8\times10^{11}}$

$= 194.4\times10$ mol per month

$1944\times2 = 3888$ g (for 30 days) $= 3.888$ kg

Decay of Radioactive Atoms

Problem 14.8

(a) The decay constant of ${}_{53}I^{131}$ is 6.016×10^{-5} min^{-1}. Starting with 2 g of the isotope, find the amount of the material left after a day. How much of the isotope is left at the end of five half-lives?

(b) What is the average life of its atoms in years?

(c) How many milligrams of ${}_{53}I^{131}$ are present if the activity is 5 curie?

Solution

(a) Using the first order rate equation $\lambda = \dfrac{2.303}{t}\log\dfrac{N_0}{N}$

$$= \frac{2.303}{t}\log\frac{a}{a-x}$$

where a and $(a-x)$ refer to the initial amount of isotope and that left at time t.

$$6.016\times10^{-5} = \frac{2.303}{24\times60}\log\frac{2}{a-x}$$

$$\log\frac{2}{(a-x)} = \frac{6.016\times10^{-5}\times24\times60}{2.303}$$

$$\frac{2}{(a-x)} = 1.0905$$

$$(a-x) = \text{amount left behind} = \frac{2}{1.0905} = 1.834 \text{ g}$$

$$N = N_0\left(\frac{1}{2}\right)^n \text{ or } (a-x) = a\left(\frac{1}{2}\right)^5 \text{ or } (a-x) = 2\left(\frac{1}{2}\right)^5$$

$\log(a-x) = \log 2 + 5\log 0.5 = 0.3010 + (5\times(-0.3010))$
$= -1.204 = \bar{2}.796$

$(a-x) =$ amount left behind $= 6.252\times10^{-2}$ g

It is possible to calculate $(a-x)$ by using the first order equation.

(b) $$\tau = \frac{1}{\lambda} = \frac{1}{6.016\times10^{-5}\ \text{min}^{-1}\times60\ \text{min h}^{-1}\times24\ \text{h day}^{-1}\times365\ \text{day y}^{-1}}$$

$$= \frac{1}{31.62} = 0.0316 \text{ y}$$

(c) 5 curies $= 5\times1000 = 5000$ millicuries $= 5000\times3.7\times10^7$ $= 18.5\times10^{10}$ dps

$$= \frac{18.5\times10^{10}}{6.023\times10^{23}}\text{ g atoms}^{-1} = \frac{18.5\times10^{10}}{6.023\times10^{23}}\times131\times\frac{T_{1/2}}{0.693}\text{ g s}^{-1}$$

$$\frac{18.5\times10^{10}}{6.023\times10^{23}}\times131\times\frac{0.693}{\dfrac{6.016\times10^{-5}\times1\text{ min}^{-1}\times\text{s}^{-1}\times\text{min}}{60\text{ min}}}$$

$$\frac{1}{0.693}\text{ g s}^{-1}$$

$= 4013.04\times10^{-5}$ g

$= 4013.04\times10^{-5}\times10^3 = 40.13$ mg

Problem 14.9

(a) The decay constant of ${}_{96}Cm^{247}$ is 1.21×10^{-10} day^{-1}. If its specific activity is 9.3×10^{-5} Ci (curie) at a certain time, what is its specific activity at the end of 15×10^7 years in dps?

(b) Uranium II ${}_{92}U^{234}$ is an old mineral with a $T_{1/2} = 2.7\times10^5$ y. RaB(Pb) which is a product of its disintegration is in radioactive equilibrium. Under the equilibrium, 1 mg of ${}_{92}U^{234}$ contains 1.9×10^{-10} mg of RaB. Use this data to calculate the $T_{1/2}$ of RaB in minutes.

Solution

(a) $T_{1/2}$ of ${}_{96}Cm^{247} = \dfrac{0.693}{1.21\times10^{-10}} = 0.573\times10^{10}$ days

15×10^7 y $= 15\times10^7\times365 = 5475\times10^7$ days

$$\text{No. of half-lives} = \frac{5475\times10^7}{0.573\times10^{10}} = \frac{5475\times10^7}{573\times10^7} = 9.55$$

Specific activity at the end of 9.55 half-lives =

$$N = N_0\left(\frac{1}{2}\right)^{9.55} = 9.3\times10^{-5}\left(\frac{1}{2}\right)^{9.55}$$

$\log N = \log 9.3\times10^{-5} + 9.55\log 0.5 = -4.032 + (-2.875)$
$= -6.907 = \bar{7}.093$

$N = 1.24\times10^{-7}\ C_i = 1.24\times10^{-7}\times3.7\times10^{10} = 4.59\times10^3$ dps

$$\frac{T_{1/2}(U^{234})}{T_{1/2}(\text{RaB})} = \frac{N(U^{234})}{N(\text{RaB})} = \frac{W(U^{234})}{W(\text{RaB})}$$

$$\frac{2.7\times10^5}{T_{1/2}(\text{RaB})} = \frac{1\text{ mg}(U^{234})}{1.9\times10^{-10}(\text{RaB})}$$

$$T_{1/2}(\text{RaB}) = \frac{2.7\times10^5\times1.9\times10^{-10}(\text{mg})}{1\text{ mg}}$$

$= 5.13\times10^{-5}$ y $= 5.13\times10^{-5}\times365 = 0.018725$ days

$= 0.018725\times24\times60 = 26.96$ min

Problem 14.10

(a) The $T_{1/2}$ of an isotope of copper is 4.68×10^4 s. It decays by three parallel paths, one by β-particle emission (up to 40%), a second one by K-electron capture (up to 40%) and a third one by positron emission (up to 20%). What are the rate constants for each of the paths?

(b) The $_6C^{14}$ isotope decays by β-ray emission with a $T_{1/2}$ $= 5.73\times 10^3$ y. A charcoal sample collected during an archeological excavation was found to disintegrate at the rate of 0.15 counts per gram per minute. The disintegration rate of carbon from a freshly cut tree is 15.5 counts per gram per minute. What is the age of the carbon from the archeological sample?

Solution

(a) Overall rate constant $k_{Cu} = \dfrac{0.693}{T_{1/2}}$

$$= \frac{0.693}{4.68\times 10^4} = 0.148\times 10^{-4}\,s^{-1}$$

The overall k_{Cu} is equal to the sum of the separate rate constants of each of the steps. Thus

$$k_{Cu} = k_{1(\beta\text{-particle emission})} + k_{2(K\text{-electron capture})} + k_{3(\text{positron emission})}$$

and $k_{1(\beta\text{-particle emission})}$ = (fractional yield of product)$\times k_{Cu}$
$= 0.4\times 0.148\times 10^{-4}$

$= 0.0592\times 10^{-4}\,s^{-1}$

$k_{2(K\text{-electron capture})}$ = (fractional yield of product)$\times k_{Cu}$ =
$0.4\times 0.148\times 10^{-4} = 0.0592\times 10^{-4}\,s^{-1}$

$k_{3(\text{positron emission})}$ = (fractional yield of product)$\times k_{Cu}$
$= 0.2\times 0.148\times 10^{-4} = 0.0296\times 10^{-4}\,s^{-1}$

(b) $\lambda = \dfrac{0.693}{5.73\times 10^3} = 0.121\times 10^{-3}\,y^{-1}$

Using $T = \dfrac{2.303}{\lambda}\log\dfrac{N_0}{N}$

$$T = \frac{2.303}{0.121\times 10^{-3}}\log\frac{15.5}{0.15} = \frac{2.303}{0.121\times 10^{-3}}\log 103.3$$

$= 19.033\times 10^3\times 2.014 = 38.34\times 10^3 = 38340$ y

Problem 14.11

Ba^{140} has a $T_{1/2} = 12.75$ days. What is the mass of this nuclide that has an activity of 5 curies?

Solution

$$\lambda = \frac{0.693}{T_{1/2}} = \frac{0.693}{12.75\times 24\times 3600} = 6.29\times 10^{-7}\,s^{-1}$$

The given activity, that is, 5 Ci = $5\times 3.7\times 10^{10}$ dps is an instantaneous rate measured under constant concentration conditions of the nuclide.

$$\text{Rate} = \frac{-d[A]}{dt} = \lambda[A] \text{ or, } [A] = \frac{-d[A]}{dt}\times\frac{1}{\lambda}$$

$$= \frac{5\times 3.7\times 10^{10}\,s^{-1}}{6.29\times 10^{-7}\,s^{-1}} = 2.94\times 10^{17}\text{ atoms}$$

$$\text{g of } Ba^{140} = \frac{2.94\times 10^{17}}{6.023\times 10^{23}}\times 140 = 68.34\times 10^{-6}\,g = 68.34\,\mu g$$

Problem 14.12

(a) Water contains one H^3 atom per 10^{18} H atoms. What is its specific activity in Ci L^{-1} ($T_{1/2}$ of $H^3 = 12.43$ y)?

(b) An explosion of one ton of TNT (trinitrotoluene) releases 4.2×10^9 J of energy. A certain country detonated a nuclear device whose capacity was 57 megatons of TNT. Calculate the mass of the device that released the stated energy.

Solution

(a) $A = \lambda_N;\ \lambda = \dfrac{0.693}{T_{1/2}} = \dfrac{0.693}{12.43\times 365\times 3600} = 4.243\times 10^{-8}\,s^{-1}$

$$1\text{ litre of water} = \frac{1000}{18} = 55.55\text{ mol}$$

55.55 mol of water contain $55.55\times 2 = 111.0$ mol of H atoms

Since water contains $^3_1H;\,^1_1H = 1\times 10^{-18}:1$, no. of mol of H^3 atoms $= 111\times 10^{-18} = 11.1\times 10^{-17}$

$$A = \lambda_N = 4.243\times 10^{-8}\,s^{-1}\times 11.1\times 10^{-17}\times 6.023\times 10^{23}$$

$$= 283.7\times 10^{-2}\text{ dps }L^{-1}$$

$$= \frac{283.7\times 10^{-2}}{3.7\times 10^{10}}$$

$$= 76.7\times 10^{-12}\text{ Ci} = 76.7\text{ pCi L}$$

(b) 57 mega tons of TNT = 57×10^6 tons of TNT
Energy released = $57\times 10^6\times 4.2\times 10^9$ J

Using the relation $E = mC^2, m = \dfrac{E}{C^2} = \dfrac{57\times 10^6\times 4.2\times 10^9}{(3.0\times 10^8)^2}$

$$= \frac{239.4\times 10^{15}}{9\times 10^{16}} = 26.6\times 10^{-1} = 2.66\text{ kg}$$

Mass Defect and Binding Energy in Nuclides

Problem 14.13

(a) The total binding energy of the chlorine isotope $_{17}Cl$ is 285.3 MeV. What is the atomic mass of this isotope? (Mass of proton = 1.00783 amu, mass of neutron = 1.00867 amu)

(b) The volume of helium given out by a certain mass of $_{88}Ra^{226}$ at STP is found to be 4.3×10^{-3} dm^3 over a year. How many α-particles are released per second under these conditions?

Solution

(a) Total binding energy of the isotope = 285.3 MeV

$$\text{Mass defect in amu, } (\Delta X) = \frac{285.3}{931.5} = 0.306$$

$$\Delta X = 0.306 = Z\ m_H + (A - Z\ )m_n - M$$

$$0.306 = 17\times 1.00783 + 20\times 1.00867 - M = 17.133 + 20.173 - M = 37.306 - M$$

Or, $M = 37.306 - 0.306 = 37.00$ amu

(b) Moles of He at STP $= \dfrac{4.3\times10^{-3}}{22.4} = 1.92\times10^{-4}$

Number of α-particles released in a year $= 1.92\times 10^{-4}\times 6.023\times 10^{23} = 11.56\times 10^{19}$

$$\alpha\text{-particles released per s} = \frac{11.56\times10^{19}}{365\times24\times3600} = 3.66\times10^{12}$$

Problem 14.14

(a) Calculate the mass defect and binding energy per nucleon of $_{93}Np^{239}$. The masses of the proton and neutron may be taken from Problem 14.13. (b) Neutrons used in nuclear reactors to study fission reactions have energies of 0.5 MeV. What is their wavelength in (i) Å (ii) nm?

Solution

(a) $\Delta X = Z\ m_H + (A - Z\ )m_n - M$

$= 93\times 1.00783 + 146\times 1.00867 - 239$

$= 93.7282 + 147.2658 - 239$

$= 240.994 - 239.00 = 1.994$ amu

Binding energy $= 1.994\times 931.5 = 1857.4$ MeV

(b) $E = mc^2$ (Einstein relation); $\lambda = \dfrac{h}{mc}$ or, $mc = \dfrac{E}{c}$

Therefore

$$\lambda = \frac{hXc}{E} = \frac{6.627\times10^{-34}\ \text{Js}\times3\times10^{8}\ \text{ms}^{-1}}{0.500\times1.602\times10^{-13}\ \text{J}} = 24.82\times10^{-13}\ \text{m}$$

Or $\lambda = 24.82\times10^{-11}$ cm

$\lambda = 24.8\times10^{-8}\times10^{-3} = 24.82\times10^{-3}$ Å

and $\lambda = 24.82\times10^{-4}$ nm

Problem 14.15

Calculate the binding energy per nucleon in the reaction

$$_{74}W^{184} \rightarrow 74\,_1H^1 + 110\,_0n^1$$

The atomic mass of $W^{184} = 183.9510$ amu, $_1H^1 = 1.00783$ amu, $_0n^1 = 1.00867$ amu.

Solution

There are 74 protons and 110 $_0n^1$ in the tungsten nucleus.

Δm = mass of W^{184} – (mass of $74\,_1H^1$ + mass of $110\,_0n^1$)

$= 183.9510 - (74\times1.00783 + 110\times1.00867)$

$= 183.9510 - (74.5794 + 110.9537)$

$= -1.5821$ amu

$\Delta E = \Delta mC^2$

$= -1.5821\times(3\times10^8)^2\times1.661\times10^{-27}\ \text{kg}^2\ \text{m}^2\ \text{s}^{-1}$

$= -23.65\times10^{-11}$ J

Binding energy per nucleon

$$= \left|\frac{23.65\times10^{-11}}{184}\right| = \frac{0.128\times10^{-11}\ \text{J}}{\text{nucleon}} = 1.28\times10^{-12}\ \text{J / nucleon}$$

Separation Energy in Nuclei

Problem 14.16

(a) Consider a sample of material containing 5 mg of potassium. The atom ratio of $\dfrac{K^{40}}{K^{39}} = 1.263\times10^{-4}$ in this sample.

(i) Calculate the number of K^{40} atoms present. ($T_{1/2}$ of $K^{40} = 1.3\times 10^9$ y)

(ii) Calculate the disintegration rate of K^{40} for this sample.

(b) Define the term separation energy of a particle (say neutron or proton) from a nucleus. Calculate the separation energy of a neutron from N^{14} nuclide and S^{33} nuclide. (Mass of proton = 1.00783 amu, mass of neutron = 1.00867 amu, mass of $N^{13} = 13.00574$, mass of $N^{14} = 14.00752$ amu, mass of $S^{32} = 31.97207$ amu, mass of $S^{33} = 32.97144$ amu.)

Solution

(a) (i) $\dfrac{K^{40}}{K^{39}} = \dfrac{0.0118}{93.1} = 1.263\times10^{-4}$

$$\frac{5\times10^{-3}\ \text{g}}{39.1\dfrac{\text{g}}{\text{g atom}}}\times6.023\times10^{23}$$

$= 0.0256\times10^{-3}\times6.023\times10^{23}$

$= 0.154\times10^{20}$ atoms

Number of K^{40} atoms $= 1.263\times 10^{-4}\times 0.154\times 10^{20}$

$= 0.195\times 10^{16}$ atoms

(ii) $A = \lambda N = \dfrac{0.693}{T_{1/2}}\times N$

$$= \frac{0.693}{1.3\times10^9\ \text{y}\times365\ \text{day}^{+1}\ \text{y}^{-1}\times24\ \text{h day}^{-1}\times3600\ \text{s h}^{-1}}\times N$$

$$= \frac{0.693}{1.3\times10^9\times3.15\times10^7}\times0.195\times10^6$$

$= 0.033$ dps $= 3.3\times 10^{-2}$

(b) Separation energy is defined as the energy required to separate a particle such as a neutron, a proton or an α-particle from a nucleus. In the case of a neutron, the separation energy S_n is given by

$S_n = [M(A-1)(Z) + M(n) - M(A,Z)]$ amu

For the N^{14} nuclide, $S_n = [M({}_7N^{13}) + M(n) - M({}_7N^{14})$

$= 13.00574 + 1.00867 - 14.00752$

$= 14.01441 - 14.00752 = 0.00689$ amu

$= 0.00689 \times 931.5 = 6.42$ MeV

For the S^{32} nuclide,

$S_n = [M({}_{16}S^{32}) + M(n) - M({}_{16}S^{33})$

$= 31.97207 + 1.00867 - 32.97144$

$= 32.98074 - 32.97144 = 0.0093$ amu

$= 0.0093 \times 931.5 = 8.66$ MeV

Applications and Other Aspects of Isotopes

Problem 14.17

${}_{53}I^{131}$ isotope has important medical applications in the treatment of thyroid related diseases. Its $T_{\frac{1}{2}} = 8.0$ days. Calculate the amount of this isotope to be purchased so that it has 10 curies of dps in a day.

Solution

$$T_{\frac{1}{2}} \text{ of } {}_{53}I^{131} = 8.0 \text{ days}$$

$$\lambda = \frac{0.693}{8.0} = 8.662\times10^{-2}\ \text{day}^{-1}$$

Rate required = 10 curies in a day

$=10\times3.7\times10^{10}$

$= 3.7\times10^{11}$ dps in a day

Using the relation $\lambda = \dfrac{2.303}{t}\log\dfrac{N_0}{N} = \dfrac{2.303}{1}\log\dfrac{N_0}{3.7\times10^{11}}$

$$\log\left(\frac{N_0}{N}\right) = \log\frac{N_0}{3.7\times10^{11}} = \frac{8.662\times10^{-2}}{2.303} = 0.0376$$

$$\frac{N_0}{3.7\times10^{11}} = 1.0904 \text{ or, } N_0 = 1.0904\times3.7\times10^{11}$$

$= 4.034\times10^{11}$ dps in a day

$$\frac{4.034\times10^{11}}{6.023\times10^{23}}\times131.0\times\frac{8.0\times24\times3600}{0.693}\times$$

$$1000 \text{ mg} = 8.75\times10^{-2} \text{ mg}$$

The amount of ${}_{53}I^{131}$ isotope required is 0.0875 mg.

Problem 14.18

${}_{33}As^{72}$ has a half life of 26.0 hours. A well contaminated with this nuclide is found to give a disintegration rate of 3.1×10^{12} dps L^{-1}. What is its concentration in ppm? What is its molarity?

Solution

mg of As^{72} is given by

$$\frac{3.1\times10^{12}}{6.023\times10^{23}}\times72\times\frac{26.0\times3600}{0.693}\times1000 = 50052.3\times10^{17}\times10^{-23}$$

$= 50052.3\times10^{-6}$ mg (present in 1 L of well water)

10^3 parts of well water contains $= 50052.3\times10^{-6}$ mg

$= 5.0\times10^{-2}$ mg

Thus, the concentration in ppm = 0.05

$$\text{Molarity} = \frac{50052.3\times10^{-6}}{10^3\times72} = 6.95\times10^{-7}$$

Problem 14.19

(a) A tracer technique was used to determine the volume of blood in a laboratory animal. A 0.5 mL of the sample solution having an activity of 1×10^{-5} Ci containing the Fe^{59} isotope was injected into the blood stream of the animal. After a long time for establishment of equilibrium of the circulating blood, 0.1 mL of the sample was withdrawn and found to have a count of 18.5 dps. Use this data to arrive at volume of blood of the animal.

(b) It is proposed to estimate the potassium content of a person with a body weight of 60 kg. For this purpose, the radiotracer K^{40} was used and it constitutes 0.0118% of natural potassium. K^{40} has a half life of 8.8×10^8 years and emits β-rays with an activity of 9.324×10^{11} dps over a certain interval of time. What per cent of natural potassium is present in his body?

Solution

(a) The activity of the injected sample $= 1\times10^{-5}\times3.7\times10^{10}$
$= 3.7\times10^5$ dps

The withdrawn sample has an activity of 18.5 dps

The activity of the injected sample is also the activity of the total volume of the blood. If V is the total volume of blood in mL,

$$\frac{V+0.5}{0.1} = \frac{3.7\times10^5}{18.5}; (V+0.5)\times18.5 = 3.7\times10^5\times10^{-1}$$

$= 3.7\times10^4$

$18.5V = 3.7\times10^4 - 9.25 = 37000 - 9.25 = 36990.75$

$$V = \frac{36990.75}{18.5} = 1999.5 \text{ mL} \approx 2.0 \text{ L}$$

(b) Weight of K^{40} in the body

$$= \frac{9.324\times10^{11}}{6.023\times10^{23}}\times\frac{40\times8.8\times10^8\times365\times24\times3600\times1000}{0.693}$$

$= 2.48\times10^9$ mg $= 2.48\times10^6$ g

For 60 kg = 60, 000 g body $\rightarrow 2.48\times10^6$ g

$$\text{For } 100 \text{ g} \rightarrow \frac{2.48\times10^6\times100}{60{,}000} = 0.413\times10^4 \text{ g}$$

$$\%\text{ amount of natural K} = \frac{0.413\times10^4\times0.0118}{100} = 0.487$$

Problem 14.20

Sr^{90} accumulates in the body via the milk taken in areas close to nuclear test sites. Assuming that a maximum of one μCi is ingested inadvertently, how many disintegrations can be expected over a ten year period ($T_{1/2}$ of Sr^{90} = 28.6 years)?

Solution

$$A = \lambda N; \lambda = \frac{0.693}{T_{1/2}} = \frac{0.693}{28.6} = 2.42\times10^{-2}\ y^{-1}$$

$$1\times10^{-6}\ Ci = 2.42\times10^{-2}\ y^{-1}$$

$$1\times10^{-6}\times3.7\times10^{10} = 3.7\times10^{4}\ dps$$

$$3.7\times10^{4}\ dps = 2.42\times10^{-2}\ y^{-1}\ N$$

$$N = \frac{3.7\times10^{4}\ dps}{2.42\times10^{-2}\ y^{-1}\times\frac{1}{365}\ y\ day^{-1}\times\frac{1}{24}\ day\ h^{-1}\times\frac{1}{3600}\ h\ s^{-1}}$$

$$= 1.53\times10^{6}\times365\times24\times3600$$

$$= 4.825\times10^{13}\ atoms$$

Using the relation,

$$\lambda = \frac{2.303}{t}\log\frac{A_0}{A} = \frac{2.303}{t}\log\frac{N_0}{N}$$

$$2.42\times10^{-2} = \frac{2.303}{t}\log\frac{4.825\times10^{13}}{N}\quad \text{or}$$

$$\log\frac{4.825\times10^{13}}{N} = \frac{2.42\times10^{-2}\ y^{-1}\times10\ y}{2.303} = 0.105$$

$$\frac{4.825\times10^{13}}{N} = \text{antilog}(0.105)$$

$$= 1.274$$

$$N = \frac{4.825\times10^{13}}{1.274}$$

$$= 3.81\times10^{13}\ \text{atoms or disintegrations}$$

Problem 14.21

(a) Name at least one application of the following isotopes.
(i) Cs^{137} (ii) Cr^{51} (iii) Cf^{252} (iv) Co^{60} (v) C^{14} (vi) Fe^{55} (vii) I^{131} (viii) P^{32} (ix) Tc^{99} (x) Se^{75}

(b) Tc^{99} has a $T_{1/2}$ =6.0 h. Use this data to find the mass of Tc^{99} having an activity of 1 gigaBq.

Solution

(i) Cs^{137}—used in the treatment of cancerous tumours.
(ii) Cr^{51}—used in the study of survival of red blood cells.
(iii) Cf^{252}—used in the testing of airline luggage for hidden explosives, also to check the moisture content of soil.
(iv) Co^{60}—used in food irradiation, cancer treatment.
(v) C^{14}—used in biological research, pollution control, archaeology, testing of metabolisation of new drugs.
(vi) Fe^{55}—used in the analysis of electroplating solutions, to detect the presence of sulfur in air.
(vii) I^{131}—used in the treatment of thyroid disorders.
(viii) P^{32}—used in genetic and molecular biology research.
(ix) Tc^{99}—used widely in diagnostic studies in nuclear medicine as an imaging agent for brain, bone, liver, and so on.
(x) Se^{75}—used in protein studies in the life sciences.

(b) $$\lambda = \frac{0.693}{T_{1/2}} = \frac{0.693}{6x3600} = 3.21\times10^{-5}\ s^{-1}$$

$$A = 1\ gigaBq = 1\times10^{12}\ dps = 3.21\times10^{-5}\times N$$

$$N = \frac{1\times10^{12}}{3.21\times10^{-5}} = 0.312\times10^{17}\ atoms$$

$$= \frac{0.312\times10^{17}}{6.032\times10^{23}}\times99$$

$$= 5.13\times10^{-6}\ g = 0.00513\ mg$$

Problem 14.22

The $T_{1/2}$ Of $_{92}U^{238}$ is 4.5×10^{9} y. Using this nuclide, a sample of a rock was found to contain 1.500 g of U^{238} and 0.300 g of Pb^{206}. Use the data to calculate the age of the rock.

Solution

$$\lambda = \frac{0.693}{t_{1/2}} = \frac{0.693}{4.5\times10^{9}} = 0.154\times10^{-9}\ y^{-1}$$

$$\text{Initial amount of } U^{238} = U^{238}_{final} + U^{238}_{reacted}$$

$$= \frac{1.5}{238} + \frac{0.300}{206} = 0.0063 + 0.00146$$

$$= 0.00776\ mol$$

$$\text{Final amount of } U^{238} = 0.0063\ mol$$

$$t = \frac{2.303}{1.54\times10^{-10}}\log\frac{0.00776}{0.0063}$$

$$= 1.495\times10^{10}\times0.091$$

$$= 0.135\times10^{8}\ y = 1.35\times10^{9}\ y$$

Problem 14.23

(a) The $T_{1/2}$ of Rn^{222} is 3.8 days. The Rn^{222} concentration in air is about 1×10^{-12} Ci L^{-1}. We breathe in 0.5 L of air in one breath and exhale it after 5 s. What is the chance of Rn^{222} disintegrating in our lungs?

(b) Assuming that a fissionable nuclide gives 200 MeV of energy per fission, how many fissions occur per second for the generation of one watt of power? If a nuclear plant operates with a 10 MW energy output on U^{235} as fuel, how many grams of the fuel is required for a week?

Solution

(a) In 0.5 L of the air we breathe in, there are 5×10^{-13} Ci of Rn^{222}. Using the relation

$$\lambda = \frac{2.303}{t}\log\frac{A_0}{A};\ \log\frac{(5\times10^{-13})}{A} = \frac{\lambda\times t}{2.303}$$

$$\log\frac{(5\times10^{-13})}{A}=\frac{0.693\times5.0}{3.8\times24\times3600\times2.303}=4.60\times10^{-6}$$

$$\frac{(5\times10^{-13})}{A}=\text{antilog}(4.6\times10^{-6})\approx1$$

$A = 5\times10^{-13}$. This means that in 5 s, when we breathe out, all Rn^{222} nuclei will come out. There is thus no possibility of disintegration of the Rn^{222} nuclei in the lungs.

(b) $1\text{ W} = 1\text{ J s}^{-1}$; $1.602\times10^{-13}\text{ J} = 1\text{ MeV}$

$$1\text{ W}=\frac{1\text{ J s}^{-1}}{\dfrac{1.602\times10^{-13}\text{ J}}{\text{MeV}}}=0.624\times10^{13}\,\frac{\text{MeV}}{\text{s}}$$

For 200 $\frac{\text{MeV}}{\text{fission}}$, we get 1 W

$$=0.624\times10^{13}\,\frac{\text{Mev}}{\text{s}}\times\frac{1}{200\,\text{MeV}/\text{fission}}$$

$$=3.12\times10^{10}\text{ fissions / s}$$

$$10\text{ MW}=3.12\times10^{10}\times10^{6}\times10\frac{\text{fissions}}{\text{s}}\times$$

$$7\times24\times3600\frac{\text{s}}{7\text{ day}}\times\frac{235}{6.023\times10^{23}}$$

$$=\frac{4434393.6\times10^{19}}{6.023\times10^{23}}$$

$$=736243\times10^{-4}\text{ g}=73.6\text{ g}$$

Nuclear Reaction Intensities and Rates

Problem 14.24

Calculate the ratio $\frac{I}{I_0}$ (where I_0 = initial radiation intensity, I = intensity of radiation after passing through a certain thickness) when neutrons of cross-section 13.4 barns (1 barn = 10^{-24} cm^2) impinge on the manganese isotope $_{25}Mn^{55}$ whose particle density is 2×10^{22} atoms cm^{-3} after a distance of 1 cm.

Solution

The relation $\frac{I}{I_0}=e^{-N\sigma x}$

where N = particle density, σ = neutron cross-section and x = distance.

$$\frac{I}{I_0}=e^{-2.0\times10^{22}\times13.4\times10^{-24}\times1}(\text{cm}^{-3}\times\text{cm}^2\times\text{cm})$$

$$\ln\frac{I}{I_0}=-2.0\times10^{22}\times13.4\times10^{-24}\times1$$

$$=-26.8\times10^{-2}=-0.268$$

$$\log\frac{I}{I_0}=-\frac{0.268}{2.303}=-0.116=\bar{1}.884\text{ or, }\frac{I}{I_0}=0.765$$

The intensity I decreases by 0.235 (or 23.5%) when the neutrons pass through 1 cm of the isotope $_{25}Mn^{55}$ with a density of 2×10^{22} atoms cm^{-3}.

Problem 14.25

The $T_{1/2}$ of $_{79}Au^{197}$ is 64.8 h. The cross-section for the reaction $_{79}Au^{197}(n,\gamma)_{79}Au^{198}$ is 98 barns (1 barn = 10^{-24} cm^2). If 19.7 g of $_{79}Au^{197}$ is placed in a reactor in which the neutron flux is 5.0×10^{10} neutrons cm^{-2} s^{-1} and is acted on for 30 minutes, what is its activity? What is the activity one hour after stopping the bombardment?

Solution

The relation $A_{\text{product}} = A_{\text{Au}^{197}} = N_i\sigma f\left(1-10^{-\lambda_{\text{product}}\times\frac{t}{2.303}}\right)$ is used, where N_i = number of atoms, σ = cross-section for the reaction, f = neutron flux, λ_{product} = decay constant and t = time.

Each of the terms is calculated as follows.

$$N_i=\frac{19.7}{197}\times6.0\times10^{23}=6.023\times10^{22}\text{ atoms}$$

$$\sigma=98\text{ barns}=98\times10^{-24}\text{ cm}^2$$

$$f=5\times10^{10}\text{ cm}^{-2}\text{ s}^{-1}$$

$$\lambda_{\text{product}}=\frac{0.693}{64.8\times60}=1.78\times10^{-4}\text{ min}^{-1};t=30\text{ min}$$

The term in brackets is given by

$$(1-10^{-1.78\times10^{-4}\times\frac{30}{2.303}})=\left(1-10^{-2.32\times10^{-3}}\right)$$

($10^{-0.00232}$ may be solved as $x = 10^{-0.00232}$

$$\log x=-0.00232=\bar{1}.99768$$

or $x = 9.947\times10^{-1} = 0.9947$)

Activity after 30 min of bombardment is,

$$A_{\text{product}}=6.023\times10^{22}\times98\times10^{-24}\text{ cm}^2\times5\times10^{10}\text{ cm}^{-2}\text{ s}^{-1}\times0.9947$$

$$=2935.6\times10^{8}\text{ s}^{-1}=2.936\times10^{11}\text{ s}^{-1}$$

Activity after 60 min (from stoppage of bombardment in the beginning)

$$\lambda=\frac{2.303}{t}\log\frac{A_0}{A}\text{ or }1.78\times10^{-4}=\frac{2.303}{60}\log\left(\frac{2.936\times10^{11}}{A}\right)$$

$$\log\left(\frac{2.936\times10^{11}}{A}\right)=\frac{1.78\times10^{-4}\times60}{2.303}=0.004637$$

$$\frac{2.936\times10^{11}}{A}=1.01073\text{ or, }A=\frac{2.936\times10^{11}}{1.01073}=2.905\times10^{11}\text{s}^{-1}$$

Problem 14.26

2.75 g of $_{25}Mn^{55}$ is taken in a nuclear reactor where the neutron flux is 2×10^{12} neutrons cm^{-2} s^{-1} and was bombarded for 40 min. Two hours after the bombardment stopped, the activity was found to be 8.63×10^{10} dps. The $T_{1/2}$ of the product is 1.56×10^2 min. Calculate the cross-section for the reaction $_{25}Mn^{55}(n,\gamma)$ in barns.

Solution

$$N_i=\frac{2.75}{55}\times6.023\times10^{23}=3.01\times10^{22}\text{ atoms}$$

$$f=2\times10^{12}\text{ neutrons cm}^{-2}\text{ s}^{-1}$$

$$\lambda=\frac{0.693}{1.56\times10^2}=4.442\times10^{-3}\text{ min}^{-1}$$

$$\lambda=\frac{2.303}{t}\log\frac{A_0}{A}$$

2 hours after bombardment, $A=8.63\times10^{10}$ dps

$$4.442\times10^{-3}=\frac{2.303}{120}\log\frac{A_0}{8.63\times10^{10}}$$

$$\log\left(\frac{A_0}{A}\right)=\log\left(\frac{A_0}{8.63\times10^{10}}\right)$$

$$=\frac{4.442\times10^{-3}\times120}{2.303}=0.2314$$

$$\frac{A_0}{8.63\times10^{10}}=\text{antilog}(0.2314)=1.704$$

$$A_0=A_{\text{product}}=8.63\times10^{10}\times1.704=14.705\times10^{10}\text{s}^{-1}$$

Cross-section

$$\sigma=\frac{A_{\text{product}}}{N_i f\left(1-10^{-\lambda_{\text{prod}}\times\frac{t}{2.303}}\right)}$$

$$=\frac{14.705\times10^{10}\text{s}^{-1}}{3.01\times10^{22}\times2\times10^{12}\text{ cm}^{-2}\text{s}^{-1}\left(1-\frac{10^{-4.442\times10^{-3}\times40}}{2.303}\right)}$$

$$=\frac{14.705\times10^{10}\text{s}^{-1}}{6.02\times10^{34}\text{cm}^{-2}\text{s}^{-1}\left(1-10^{-0.0772}\right)}$$

[Evaluation of $10^{-0.0772}$, let $x=10^{-0.0772}$; $\log x=-0.0772$ $=\bar{1}.9228$; $x=\text{antilog}(\bar{1}.9228)=0.8371$]

Therefore,

$$\sigma=\frac{14.705\times10^{10}\text{ s}^{-1}}{6.02\times10^{34}\text{ cm}^{-2}\text{ s}^{-1}\times0.1629)}$$

$$=\frac{14.705\times10^{10}}{0.9806\times10^{34}\text{ cm}^{-2}}$$

$$=15\,\text{barns}\left(15\times10^{-24}\text{ cm}^{-2}\right)$$

Problem 14.27

(a) For fast neutrons, the scattering cross-section of lead is 5 barns. Calculate the thickness of lead which reduces the intensity of the neutron beam to 1% of its original value. Assume the N value (per cm^3, number of target nuclei) to be 6.0×10^{22} atoms cm^{-3}.

(b) For slow neutrons, the scattering cross-section of Ag^{109} is 44 barns. Calculate the thickness of silver which reduces the intensity to 10% of its original value. Assume the N value (no. of target nuclei per cm^3) to be 5×10^{21} atoms cm^{-3}.

Solution

(a) The equation $\frac{I}{I_0}=e^{-N\sigma x}$

where N = number of target nuclei cm^{-3}, σ = neutron scattering cross-section and x = thickness of the material.

$$\frac{I}{I_0}=\frac{1}{100}=0.01$$

$$2.303\log0.01=-6.0\times10^{22}\times5\times10^{-24}\times x$$

$$-4.606=-30.0\times10^{-2}x;\ x=\frac{4.606}{0.03}=15.3\text{ cm}$$

(b) $$\frac{I}{I_0}=\frac{10}{100}=e^{-5\times10^{21}\times44\times10^{-24}\times x}$$

$$2.303\log0.1=-220\times10^{-3}x$$

$$-2.303=-2.20\times10^{-1}x$$

$$x=\frac{2.303}{0.22}=10.5\text{ cm}$$

Problem 14.28

(a) The linear absorption coefficient of nickel oxide is 460 cm^{-1}. What is the thickness of the oxide required that reduces the γ-ray emission by Ni^{59} to 0.05%?

(b) Evaluate the Q value for the reaction $B^{11}(d,\alpha)Be^9$.

Solution

(a) $I=I_0e^{-kl}$ where k is the linear absorption coefficient.

$$\ln\frac{I}{I_0}=-kl\text{ or, }2.303\log\frac{0.05}{100}$$

$$=-460\times l\,;2.303\log0.0005=-460l$$

$$2.303\times(-3.301)$$

$$=-460l\text{ or, }l=\frac{7.602}{460}=0.0165\text{ cm}=0.165\text{ mm}$$

(b) The reaction is $_5B^{11}+{}_1D^2\rightarrow{}_4Be^9+{}_2He^4$

Atomic masses of reactants	Atomic masses of products
B^{11} = 11.0280	Be^9 = 9.07686
D^2 = 02.0141	He^4 = 4.00387
13.0421	13.08073

Increase in mass = 13.08073 – 13.04210 = 0.03263 amu

or $Q = 0.03263 \times 931.5 = 30.4$ MeV

This amount of energy must be consumed presumably as kinetic energy of bombarding deuterons.

Problem 14.29

Calculate the activity of No^{254} formed after 2 min exposure of Pb^{208} by Ca^{48}. The thickness of the lead target is 1 mg cm^{-2} and the Ca^{48} beam current is 0.5 μA. The $T_{1/2}$ of $No^{254} = 55$ s. Write down the reaction before calculation. The reaction cross-section is 3×10^{-6} barns.

Solution

$$_{82}Pb^{208} + {}_{20}Ca^{48} \rightarrow {}_{102}No^{254} + 2{}_0n^1$$

$$A = N\sigma\phi\left(1-10^{\frac{-\lambda t}{2.303}}\right)$$

$$N = 1\times10^{-3}\left(\frac{g}{cm^2}\right)\times\frac{6.023\times10^{23}\ \text{atom mol}^{-1}}{208\ g\ \ \ \text{mol}^{-1}}$$

$$= 2.9\times10^{18}\ \text{atom cm}^{-2}$$

$$\sigma = 3\times10^{-6}\times10^{-24} = 3\times10^{-30}\ \text{cm}^2$$

$$\phi = 0.5\times10^{-6}\ \frac{\text{coulombs}}{\text{s}}\times\frac{\text{ion / coulombs}}{1.602\times10^{-19}}$$

$$= 0.312\times10^{13} = 3.12\times10^{12}\ \text{ions/s}$$

$$A = 2.9\times10^{18}\times3\times10^{-30}\times3.12\times10^{12}\ \frac{\text{ion}}{\text{s}}\left(1-10^{\frac{\lambda t}{2.303}}\right)$$

$$\left(\lambda = \frac{0.693}{55} = 0.0126\ \text{s}^{-1}\right)$$

$$= 27.14\left(1-10^{\frac{-1.26\times10^{-2}\times2\times60}{2.303}}\right) = 27.14\,(1-10^{-0.6565})$$

$$= 27.14(1-0.221) = 27.14\times0.779 = 21.14\ \text{dps}$$

Problem 14.30

Calculate the number of Co^{60} atoms produced in a 5 mg sample of cobalt metal exposed for 5 minutes to a thermal neutron flux of 2×10^{13} n/cm^2 s in a reactor. The cross-section for the reaction is 16.0 barns. What is the disintegration rate of the sample after one year of the bombardment? Assume that the sample thickness is negligible. The $T_{1/2}$ of $Co^{60} = 5.26$ years.

Solution

The complete equation for the number of product nuclei, N, during irradiation is

$$N = \phi t\left(1-e^{-n\sigma\Delta x}\right)$$

For a thin target, we expand $e^{-n\sigma\Delta x}$ and write $N = \phi t n\sigma\Delta x$

N has the units of atoms cm^{-2} [or thickness $\Delta x\times$density n (atom cm^{-3})]

Using the given data,

$$N = 2\times10^{13}\times5\times60\times\frac{5\times10^{-3}}{58.9}\times6.023\times10^{23}\times16\times10^{-24}$$

$$= 4.91\times10^{12}\ \text{atoms}$$

The number of atoms at the end of one year is given by

$$\lambda = \frac{2.303}{t}\log\frac{N_0}{N};\log\left(\frac{4.91\times10^{12}}{N}\right)$$

$$\frac{\lambda\times t}{2.303} = \frac{0.693}{5.26}\times1 = 0.132$$

$$\frac{4.91\times10^{12}}{N} = 1.355$$

$$N = \frac{4.91\times10^{12}}{1.355}$$

$$= 3.624\times10^{12}\ \text{atoms}$$

Nuclear Fuels

Problem 14.31

Th^{230} disintegrates releasing α-particles and an energy of 4.62 MeV is released with each disintegration. Certain parts of the world like India possess abundant reserves of thorium. If this nuclide is to be considered as a possible nuclear fuel, what is the energy released per day per kilogram of the fuel? Report the result in kJ and MeV. The $T_{1/2}$ of the nuclide is 75400 y.

Solution

$$1\ \text{kg of the fuel} = \frac{1000\,g}{230}\times6.023\times10^{23} = 26.19\times10^{23}\ \text{atoms}$$

$$\text{Using the relation } A = \lambda N = \frac{0.693}{75400\times365}\times26.19\times10^{23}$$

$$= 2.52\times10^{-6}\times26.19\times10^{21}$$

$$= 6.6\times10^{17}\ \text{disintegrations day}^{-1}$$

Power output $= 6.6\times10^{17}\times4.62 = 3.05\times10^{18}$ MeV day^{-1}

Power output in kJ = 3.05×10^{18} MeV day$^{-1}\times1.6\times10^{-16}$ kJ (MeV)$^{-1}$ = 4.88×10^2 kJ day^{-1}

EXERCISES

Practice Problems

1. (a) The $T_{1/2}$ of $U^{238} = 4.47\times10^9$ years and that of Ra^{226} is 1600 years. Assuming that radioactive equilibrium is established in uranium minerals, what is the ratio of radium to uranium atoms under these conditions?
   (b) Calculate the average atomic mass of natural uranium given that the percentage of abundances of U^{238}, U^{235} and U^{234} are 99.274, 0.720 and 0.006 respectively.

2. The $T_{1/2}$ of ${}_{83}Bi^{210}$ is 5.0 days.
   (i) What is the mass of the nuclide that has an activity of 20 millicuries?
   (ii) Calculate the number of disintegrations at the end of five half lives.
   (iii) What is its $\frac{N}{P}$ ratio and how does it attain stability?

3. (a) The binding energy per nucleon of ${}_{25}Mn^{55}$ is 8.78 MeV. What is its nuclidic mass? (Atomic masses of ${}_1H^1$ = 1.00783 amu, ${}_0n^1$ = 1.00867 amu.) What is its packing fraction in parts per 10,000?
   (b) Po^{210} disintegrates by the emission of alpha particles with an energy release of 5.30 MeV. Its half life is 138.4 days. Calculate the energy released in a week from a 0.5 g sample in kJ.

4. (a) ${}_{15}P^{30}$ is artificially produced by a $Al^{27}(\alpha, n)P^{30}$ reaction. Write down this reaction. The product is a positron emitter. Represent the product of this reaction as well.
   (b) When a positron and an electron collide, two γ-photons are formed. Calculate the wavelength of photons formed in nm.

5. The permanent hardness due to the presence of chloride ions (as magnesium chloride) and sulfate ions as calcium sulfate was estimated by using Mg^{28} ($T_{1/2}$ = 21 hours) and Ca^{45} ($T_{1/2}$ = 164 days). In one set of experiments, the disintegration rate for Mg was found to be 7.78×10^{16} dps per litre. A second set of experiments gave a disintegration rate of 3.68×10^{14} dps per litre. Calculate the hardness due to the chloride ions and sulfate ions in ppm and the molarity.

6. The $T_{1/2}$ of ${}_{83}U^{238} = 4.5\times10^9$ years. In determining the age of the earth based on this nuclide and its disintegration product, it was found that the ratio $\frac{A}{A_0} = 2.01$ where A and A_0 refer to the activity of the nuclide at the present time and that at the time of the origin of earth. Using this data, calculate the age of the earth.
   The same result was sought to be verified using the K^{40} nuclide ($T_{1/2}$= 1.28×10^9 years) in which case $\frac{A}{A_0} = 11.4$. Verify the result for the age of the earth determined with this nuclide.

7. (a) The reaction ${}_1D^2 + {}_1D^2 = {}_1T^3 + {}_1H^1$ occurs with the release of 4.0 MeV. It is a possible source of thermonuclear energy on earth. Given that the atomic masses of ${}_1D^2$ = 2.0141 amu and ${}_1H^1$ = 1.00782 amu, calculate the nuclidic mass of ${}_1T^3$.
   (b) Calculate the separation energy of a neutron from ${}_6C^{16}$ and ${}_4Be^{10}$ nuclides in MeV. (Nuclidic masses: C^{16} = 16.01470, C^{15} = 15.010599, Be^{10} = 10.013535, Be^9 = 9.012183, ${}_0n^1$ = 1.00867.)

8. (a) Explain the term 'breeder reactor'. Give an example of a reaction that can be employed in such reactors.
   (b) Pu^{239} is a potential nuclear fuel for obtaining electrical energy and the reaction is
   $${}_{94}Pu^{239} + {}_0n^1 \rightarrow {}_{94}Pu^{237} + 3\,{}_0n^1$$
   Assuming that the electrical output required is 100 MW and the energy efficiency of the reaction is 60%, calculate the amount of fuel required for a week. (Nucleidic masses: ${}_{94}Pu^{239}$ = 240.99739, ${}_{94}Pu^{237}$ = 238.98004, n^1 = 1.00867, all in amu.)

9. (a) Derive the relation $N = N_0\left(\frac{1}{2}\right)^{t/T_{1/2}}$ where N = number of atoms at a time t, N_0 = number of atoms at time 0 and $T_{1/2}$ = half life of a radioactive element.
   (b) What is the specific activity of a sample of Cs^{137} ($T_{1/2}$ = 30.2 y) which originally had an activity of 1 mCi? What will be its activity after 50 years?

10. (a) The activity of 5.0 mg of ${}_{79}Au^{198}$ is 1.22×10^{-5} Ci. Calculate its disintegration constant and the time required for 75% of this isotope to disintegrate ($T_{3/4}$).
    (b) What is its activity at the end of one day and at the end of 10 years?

11. (a) How do you express nuclear reaction rates? Explain the term nuclear cross-section of a reaction and projectile particle flux.
(b) Calculate the ratio I/I_0 (where I_0 = initial radiation intensity, I = intensity of radiation after crossing through a certain thickness of the material) when thermal neutrons of 50 barn (1 barn = 10^{-24} cm^2) impinge on arsenic As^{72} isotope whose particle density is 4.0×10^{22} atom cm^{-3} after a distance of 2 cm.

12. The cross-section for the reaction $_{49}In^{115}$ (n,γ) reaction is 145 barns. The half life of the product In^{116} = 54 min. If 29.0 g of pure In^{115} placed in a nuclear reactor in which the neutron flux is 4×10^8 neutrons cm^{-2} s^{-1} and is bombarded for an hour, what is the activity of the sample after an hour of the stoppage of bombardment?

13. 10.9 g of $_{47}Ag^{109}$ was taken in a nuclear reactor in which the neutron flux is 5×10^{15} neutrons cm^{-2} s^{-1}. The bombardment by neutrons lasted one hour. Thirty minutes after the bombardment was stopped, the activity was found to be 1.574×10^{12} s^{-1}. Given the $T_{1/2}$ of the product as 2.3 min, calculate the cross reaction for the reaction $_{47}Ag^{109}(n,\gamma)Ag^{110}$ in barns.

14. Calculate the time required for the irradiation of 7.75 g of P^{31} nuclide in a reactor with a neutron flux of 5×10^5 neutrons cm^{-2} s^{-1}, to obtain an activity of P^{32} of 2×10^5 dis/min, 50 hours after stoppage of bombardment. $T_{1/2}$ of the product = 14.5 days (σ = 0.23 barn).

15. (a) A small amount of bismuth in aqueous solution is determined by extraction into chloroform and adding BiI_4^- ions labelled with Bi^{210}. In a partition experiment, 10 mL of an aqueous solution containing an unknown concentration of Bi^{3+} was treated with 5 mL of Bi^{210} in chloroform having an activity of 0.5×10^{-6} Ci. After shaking for a minute, 1 mL of the aqueous extract showed an activity of 0.025×10^{-6} Ci. If the concentration of Bi^{3+} in this extract is 1 mM, what is the concentration of Bi^{3+} in the original aqueous solution?
(b) The scattering cross-section σ of magnesium is 2 barns. Calculate the thickness of magnesium required to reduce the intensity of the beam to 2% of its initial value. The value of N may be taken as 5×10^{22} atoms cm^{-3}.

16. (a) The $T_{1/2}$ values of Co^{60}, T (tritium) and Sr^{90} are 5.3 y, 12.3 y and 28.8 y respectively. Their activities at a certain time are 5, 2, and 4 Ci respectively. Over a period of time, the activity of Co^{60} drops to 1.5 Ci. What is the percentage decrease in the activity of tritium and Sr^{90} in the same period?
(b) The great pyramids of Egypt are believed to be 4000 y old from radiocarbon dating of ash at the site. If the ash from the contemporary wood near the same site gave 20 counts per gram per minute using a $_6C^{14}$ tracer ($T_{1/2}$ = 5730 y), how many counts would the ash sample give at this time?

17. (a) Define the units employed to denote the radiation dosage. What is the maximum dosage that is fatal for humans?
(b) How are sievert and gray defined?
(c) Write down the relation governing intensity of the dose with distance.
(d) Given that the intensity of the dose of a particular radiation is 8000 rem m^{-2} s^{-1}, at what distance will it have a value of 1 rem m^{-2} s^{-1}?
(e) Fill up the blanks in the following nuclear reactions.
(i) $_{13}Al^{27} + {_0n^1} \rightarrow {_2He^4} +$ ....
(ii) $_{94}Pu^{239} + {_2He^4} \rightarrow {_0n^1} +$ ....
(iii) .... $\rightarrow {_{48}Cd^{110}} + {_{-1}e^0} + \gamma$
(iv) .... $\rightarrow {_5B^{11}} + {_1e^0}$
(v) $_{37}Rb^{81} + {_{-1}e^0} \rightarrow$ .... + x-ray photon

18. (a) In the determination of the solubility product of lead chomate, an experimenter added a solution of potassium chromate to a solution of lead nitrate containing a known amount of the isotope Pb^{120}. The precipitated lead chromate weighed 15 mg and had an activity of 2700 dps. When the precipitate was shaken with water, the activity due to the precipitate going into solution was found to be 7.2 dps per litre. Calculate the solubility product of lead chromate. Assume the temperature as 298 K. (Atomic masses: Pb = 207, Cr = 52.0, O = 16.0)
(b) What is the type of radioactive decay expected for the following nuclides? (i) $_{79}Au^{197}$ (ii) $_{11}Na^{23}$ (iii) $_{88}Ra^{226}$ (iv) $_{15}P^{31}$ (v) $_6C^{11}$. Justify your answer by giving reasons.

19. (a) $_{84}Po^{210}$ undergoes fission through α-ray emission according to $_{84}Po^{210} \rightarrow {_{82}Pb^{206}} + {_2He^4}$. Given that the nuclidic masses Po^{210} = 209.98285 amu, Pb^{206} = 205.97444 amu, He^4 = 4.0015 amu, calculate the mass defect for this decay. What is the amount of energy released for 1 g of Po^{210}?
(b) Calculate the volume of helium gas released at STP in a 20 year period decay of 0.5 g of Ra^{226} to Pb^{206} in which five α-particles are released according to

$$_{88}Ra^{226} \rightarrow {_{86}Rn^{222}} + {_2He^4} \rightarrow {_{84}Po^{218}} \rightarrow {_{82}Pb^{214}} + {_2He^4}$$

$$_{84}Po^{214} \rightarrow {_{82}Pb^{210}} + {_2He^4};\ {_{84}Po^{210}} \rightarrow {_{82}Pb^{206}} + {_2He^4}$$

(Isotopes of Po^{218} are produced which also decay as per the above reaction; $T_{1/2}$ of Ra^{226} = 1.6×10^3 y)

20. (a) The nuclide $_{29}Cu^{64}$ ($T_{1/2}$ = 13.0 h), decays by K-electron capture to the extent of 61% and by β-ray decay to the extent of 39%. Calculate the decay constants for the two types of decay and also determine their partial half lives.
(b) An organic material contains 5 mg of carbon and it has a C^{14}/C^{12} ratio = 1.5×10^{-14}. Calculate the disintegration rate of this sample in curies. If C^{14} in ancient times has a constant specific activity of 225 becquerel/kg C (1 becquerel = 1 dps), what is the age of the carbon sample ($T_{1/2}$ of C^{14} = 5700 y)?

Multiple Choice Questions

1. Nuclear reactions differ from chemical reactions in that
   (a) energy changes between the two types of reactions are of comparable magnitude.
   (b) new chemical species are formed in way that electrons in outer shells as well as protons and neutrons in both types of reactions participate in their reaction.
   (c) the rate of chemical reaction is governed by the Arrhenius equation but for nuclear reactions, the temperature change has no effect on their rate.
   (d) the above statements do not truly reflect their characteristics.

2. The penetrating power of nuclear radiations (α, β, γ) through a thin metal foil increases in the order,
   (a) $\gamma < \beta < \alpha$ (b) $\alpha < \beta < \gamma$
   (c) $\beta < \gamma < \alpha$ (d) $\alpha < \gamma < \beta$

3. Cesium-131 is used in the treatment of cancer. It has a $T_{1/2}$ = 10 days. The percentage decay of the isotope after 90 days is
   (a) 58.6 (b) 75.8
   (c) 62.4 (d) 80.5

4. The $T_{1/2}$ of Th^{234} is 600 hours. The weight in mg of 2 curies of Th^{234} is
   (a) 0.09 (b) 0.05
   (c) 0.12 (d) 0.08

5. The energy released in the nuclear reaction (in J.per g atom for nitrogen) is
   ${}_7N^{14} + {}_2He^4 \rightarrow {}_8O^{17} + {}_1H^1$
   (Masses of various species in amu: ^{14}N = 14.00307, He^4 = 4.00260, O^{17} = 16.99913, H^1 = 1.00794)
   (a) 2.6×10^{11} (b) 1.3×10^{11}
   (c) 6.2×10^{9} (d) 3.1×10^{10}

6. The binding energy per nucleon of F^{19} (in J) is (Given masses of H^1 = 1.00782, n = 1.00866, F = 18.998403)
   (a) 24.95×10^{-12} (b) 10.94×10^{-11}
   (c) 15.63×10^{-10} (d) 12.46×10^{-13}

7. Deuterium (${}_1D^2$) and tritium (${}_1T^3$) undergo fusion reaction at very high temperatures of the order of 10^8 K resulting in
   ${}_1D^2 + {}_1T^3 \rightarrow {}_2He^4 + {}_0n^1$
   Masses in 10^{-27} kg: D^2 = 3.3427, T^3 = 5.00827, He^4 = 6.64647, ${}_0n^1$ = 1.67493
   The energy released (in kJ mol^{-1}) is
   (a) 63.51 (b) −31.33
   (c) −40.65 (d) 36.24

8. Among the following radioactive elements, the group belonging to *isotones* is
   (a) ${}_{16}S^{36}$, ${}_{20}Ca^{40}$ (b) ${}_{19}K^{39}$, ${}_8O^{17}$
   (c) ${}_7N^{14}$, ${}_5Be^9$ (d) ${}_5B^{12}$, ${}_6C^{14}$

9. A geological rock sample contains 1×10^{-4} g of U-238 and 3.0×10^{-5} g of Pb^{206}. The $T_{1/2}$ of U-238 = 4.5×10^9 years. The age of the rock is (in years),
   (a) 2.88×10^{10} (b) 5.62×10^{9}
   (c) 1.94×10^{9} (d) 4.91×10^{10}

10. For radioactive elements where the $N/P < 1$ (N = neutron, P = proton), the stability of the nucleus is achieved by the
    (a) loss of a β-particle (b) loss of an α-particle
    (c) release of a positron (d) release of γ-radiation

11. Choose the correct statement relating to radioactive equilibrium.
    (a) The rates of the forward and reverse processes are equal as in a chemical equilibrium.
    (b) The rates of decay of parent and daughter elements are proportional to the numbers of the respective species.
    (c) The number of parent or daughter atoms is proportional to its half-life.
    (d) The radioactivity of both parent and daughter elements reduces to a negligible value under this condition.

12. Which of the following statements about fission, fusion and nuclear chain reactions is correct?
    (a) A fission reaction involves the bombardment of a lighter nucleus such as Al^{27} with α-particles.
    (b) A fusion reaction requires the combination of lighter nuclei to form heavier ones at ambient temperatures.
    (c) In a nuclear chain reaction, a fissionable atom on treatment with a neutron results in the formation of a large number of neutrons.
    (d) The energy requirements in fission and fusion reactions are of comparable magnitude.

13. The specific activity of Cu^{64} is 14.2×10^{10} dps g^{-1}. Its $T_{1/2}$ = 12.8 h. Its specific activity after 64 hours is
    (a) 4.44×10^{9} (b) 9.43×10^{8}
    (c) 8.68×10^{10} (d) 5.63×10^{8}

14. The binding energy per nucleon for Sn =10.1 Mev. The atomic mass of this element is (given m_H = 1.00782 amu, m_n = 1.00867 amu)
    (a) 117.8 (b) 118.7
    (c) 121.6 (d) 116.5

15. A moderator in a thermal nuclear reactor is used to
    (a) reduce the speed of fast moving neutrons
    (b) reduce the speed of α-particles
    (c) reduce the speed of γ-radiation
    (d) reduce the rate of a fission reaction

16. The material used as a moderator in nuclear reactors is
    (a) zirconium (b) heavy water
    (c) liquid sodium (d) boron

17. The separation energy of a neutron from Be^9 (in MeV) is
    (a) 1.736 (b) 2.847
    (c) 1.673 (d) 3.761
    Be^9 = 9.012182, Be^8 = 8.005305, m_n = 1.00867 (all in amu)

18. The scattering cross-section of lead for fast neutrons is 10 barns. Assuming that the N value is 5×10^{22} atoms cm^{-3}, the thickness of lead (in cm) that reduces the intensity of a neutron beam to 5% of its original value is

(a) 4.5 **(b)** 6.0
(c) 10.8 **(d)** 8.2

19. The $T_{1/2}$ of ${}_{24}Cr^{55}$ = 6.0 min. The cross-section for the formation of Cr^{56} is 0.4 barns. 5.5 g of ${}_{24}Cr^{55}$ is taken in a nuclear reactor in which the neutron flux is 2×10^{11} neutron cm^{-2} s^{-1} and is acted upon for one hour. The activity of the nuclide is (in s^{-1})

(a) 8.54×10^5 **(b)** 4.58×10^9
(c) 5.48×10^8 **(d)** 4.58×10^6

20. Indicate the correct statement regarding the C^{14} dating technique.

(a) C^{14} is an unstable nuclide and is completely lost in the atmosphere due to decay.

(b) Living plants (or wood from plants) replenish the ${}_6C^{14}$ from the atmosphere and keep the ratio of ${}_6C^{14}/{}_6C^{12}$ constant.

(c) Living plants absorb C^{12} but not C^{14}.

(d) The amount of C^{14} in living or dead tissues is always the same.

21. The element which is likely to have chemical properties similar to ${}_{94}Pu$ is

(a) radium **(b)** xenon
(c) cesium **(d)** strontium

22. Which of the following pairs of nuclides are nuclear isomers?

(a) ${}_{88}Ra^{226}$, ${}_{86}Rn^{222}$ **(b)** ${}_{91}U^{234}$, ${}_{91}UX_2{}^{234}$
(c) ${}_{19}Ar^{40}$, ${}_{20}Ca^{40}$ **(d)** ${}_{92}U^{235}$, ${}_{92}U^{238}$

23. ${}_{94}Pu^{239}$ absorbs a neutron and the Pu^{240} formed undergoes nuclear fission according to

$${}_{94}Pu^{239} + {}_0n^1 \rightarrow {}_{94}Pu^{240} \text{ fission} \rightarrow {}_{40}Zr^{103} + {}_{54}Xe^{134} + 3\ {}_0n^1$$

(Atomic masses in amu: ${}_{94}Pu^{240}$ = 240.12911, ${}_{40}Zr^{103}$ = 102.92656, ${}_{54}Xe^{134}$ = 133.90539, n^1 = 1.00867). The amount of energy released when one gram atom of ${}_{94}Pu^{240}$ undergoes fission (in kJ mol^{-1}) is

(a) -4.42×109 **(b)** 4.88×10^{8}
(c) -2.44×1010 **(d)** 8.46×109

Answer Keys

Practice Problems

1. (a) $\dfrac{N_{Ra}}{N_U} = 3.66\times10^{-7}$

(b) Average atomic mass of U = 237.978 amu

2. (i) Mass of nuclide = 160.8×10^{-9} g

(ii) 2.3×10^7 dps

(iii) $\dfrac{N}{P} = 1.53$. By the emission of β-particles, the $\dfrac{N}{P}$ ratio decreases and the nucleus derives stability. For example,

$${}_{83}Bi^{210} \rightarrow {}_{84}Po^{210} + {}_{-1}e^0$$

$$\frac{N}{P} = \frac{126}{84} = 1.5$$

3. (a) −11.38

(b) 362.7×102 kJ

4. (a) ${}_{13}Al^{27} + {}_2He^4 \rightarrow {}_{15}P^{30} + {}_0n^1$

(b) $\lambda = 0.00243$ nm

5. (a) Molarity of $Cl^{-1} = 2.82\times10^{-2}$ mol L^{-1} = 1001 ppm

(b) Molarity of $SO_4{}^{2-} = 1.24\times10^{-2}$ g L^{-1} = 1199 ppm

6. The age of the earth based on $U^{238} = 4.53\times10^9$ y. The age of earth based on $K^{40} = 4.50\times10^9$ y.

7. (a) Mass of ${}_1T^3 = 3.01609$

(b) Separation energy of ${}_6C^{14}$ = 4.2560 MeV, separation energy of Be^{10} = 6.8167 MeV

8. (a) A breeder reactor is a nuclear reactor in which more fissile material is generated than is consumed. In these reactors, this is made possible by their economy of neutrons which is sufficiently high to breed more fissile material than the use of material such as U^{238} or Th^{232}. The ratio of the number of fissile atoms produced to the number of atoms used by fission is called the conversion factor (c). If $c > 1$ it is a breeder reactor, otherwise if $c < 1$, it is referred to as a convertor. A mixture of uranium and plutonium are applied in fast breeder reactors. The fissile nuclide Pu^{239} is continuously reproduced according to the reaction

$$U^{238}(n,\gamma)U^{239} \xrightarrow{{}_0e^1(or\beta^{-1})} Np^{239} \xrightarrow{\beta^{-1}} Pu^{239}$$

U^{233} is another fuel so prepared continuously via the reaction

$$Th^{232}(n,\gamma)Th^{233} \xrightarrow{\beta^{-1}} Pa^{233} \xrightarrow{\beta^{-1}} U^{233}$$

(b) 133.8 g of Pu^{239}

9. (a) $N = N_0\left(\frac{1}{2}\right)^{\frac{t}{T_{1/2}}}$

(b) Specific activity = 32×10^{14} dps g^{-1} = 8.65×10^{-3} mCi, Activity after 50 years = 0.616×10^{-3} mCi

10. (a) $\lambda = 29.7\times10^{-15}\ s^{-1}$, $t_{3/4} = 4.7\times10^{13}$ s = 1.49×108 y

(b) At the end of one day, the activity is

1.22×10^{-5} Ci, at the end of 10 years, $A = \frac{1.22\times10^{-5}}{10^{8.88\times10^{-7}}}$

11. (a) The nuclear reaction rates are defined in terms of nuclear cross-section, the symbol for which is σ. It has the units of area (cm^2) and may be considered as the probability that a bombarding particle (say, an α-particle or a neutron) of a specified energy react with a chosen nucleus. If a projectile source having a density of n particles per cm^3 is moving with a velocity v cm s^{-1} in a given direction, the product nv (with a unit of $cm^2\ s^{-1}$) is known as projectile flux. It is also sometimes referred to as track length.

(b) $\frac{I}{I_0} = 3.98\times10^{-1} \approx 0.4$

12. $A_{product} = 2.179\times10^9\ s^{-1}$

13. 44 barns

14. Time = 119.5 h

15. (a) $C = 2.0\times10^{-3}$ (b) $l = 39.1$ cm

16. (a) % decrease in activity of tritium = 40.4, % decrease in activity of Sr^{90} = 19.9

(b) 12.33 counts g^{-1} min^{-1}

17. (a) The most commonly used unit for radiation dosage are (i) roentgen, R (ii) radiation absorbed dose, rad and (iii) Roentgen equivalent man, Rem.

(i) ***Roentgen:*** This is a unit used to measure a quantity called exposure. This unit is used to describe the amount of γ-rays or X-rays only in air. One roentgen is equal to an energy deposition of 2.58×10^{-4} C kg^{-1} in dry air. It is a measure of the ionisations of the molecules in a mass of air.

(ii) ***Rad:*** It is a unit used to measure a quantity called the radiation absorbed dose. It is the amount of energy absorbed in some material and is used for any type of radiation and any type of material. One rad of absorption = 100 erg g^{-1} of material. It does not describe the biological effects of different radiations.

(iii) ***Rem:*** This unit is used to derive a quantity called equivalent dose. It relates the absorbed dose in human tissues to the effective biological damage of radiation. It is often expressed in mrem. In humans, any dose equal to or above 500 rems is considered fatal and may cause death within a month.

(b) One gray = 100 rads and one sievert = 100 rem

(c) The intensity of a dose decreases with increase in distance. If I_1 and I_2 are two values of intensities at distances d_1 and d_2 metres respectively,

$$I_1 d_1^2 = I_2 d_2^2$$

(d) $d_2 = \sqrt{80} = 8.94$ m

(e) (i) ${}_{11}Na^{24}$ (ii) ${}_{96}Cm^{242}$ (iii) ${}_{47}Ag^{110}$ (iv) ${}_{6}C^{11}$ (v) ${}_{36}Kr^{81}$

18. (a) $K_{sp} = 3.725\times10^{-14}$

(b) (i) β-ray emission (ii) β-ray emission (iii) nuclear fission involving α-particles (iv) β-ray emission (v) proton emission or K-electron capture

19. (a) $\Delta m = 0.00691$ amu, $\Delta E = 2.95\times10^6$ kJ g^{-1}

(b) Volume of He released at STP = 2.143 mL

20. (a) Partial rate constant for K-electron capture = $3.25\times10^{-2}\ h^{-1}$
Partial rate constant for β-ray decay = $2.08\times10^{-2}\ h^{-1}$

(b) Age of the carbon = 9661 y

Multiple Choice Questions

1. (c)	**2.** (b)	**3.** (d)	**4.** (a)	**5.** (b)	**6.** (d)	**7.** (b)	**8.** (a)	**9.** (c)	**10.** (c)
11. (b)	**12.** (c)	**13.** (a)	**14.** (b)	**15.** (a)	**16.** (b)	**17.** (c)	**18.** (b)	**19.** (d)	**20.** (b)
21. (a)	**22.** (b)	**23.** (c)							

CHAPTER 15 APPLICATIONS OF STATISTICAL THERMODYNAMICS

CHAPTER OUTLINE

Preliminary Concepts and Formulae
- Partition Functions
- Thermodynamic Properties from Molecular Information

Classical and Quantum Statistics
- Validity of the Classical Regime
- Ortho and Para-hydrogen
- Heat Capacity of Solids
- Maxwell–Boltzmann Distribution
- Fermi–Dirac Statistics
- Bose–Einstein Statistics

15.1 Preliminary Concepts and Formulae

Ensemble: A collection of particles is often referred to as an *ensemble* (assembly). Depending upon the convenience, any of the following ensembles can be used for a given situation, although more insights are obtainable if an appropriate ensemble is employed. The basic premises on which statistical thermodynamics is founded are (i) ergodic hypothesis and (ii) the principle of equal *a priori* probability. Consider a system with volume V consisting of N particles at a temperature T whose internal energy is denoted as E. The concept of ensembles enables the computation of thermodynamic properties in a facile manner. The following are the most commonly employed ones:

- canonical—N, V, T constants,
- microcanonical—N, V, E constants,
- grand canonical—V, T, μ constants, and
- isothermal–isobaric—N, P, T constants

where μ denotes the chemical potential of the system.

15.1.1 Partition Functions

These are of several kinds. The *partition function* which can be thought of as a summation over energy levels, is a central quantity in statistical mechanics, wherefrom all other thermodynamic properties can be derived. The commonly employed symbols are Q or Z.

Translational partition function: The translational partition function is derived using the concepts learnt from the particle in a three-dimensional box. Thus, the partition function resembles the energy levels arising from the solution of the Schrödinger equation for a free particle in a three dimensional box. The governing equation for the translational partition function is given by

$$Q_{tr} = 1.88\times10^{26}\,T^{3/2}M^{3/2}\times V \quad (M \text{ in g, } V \text{ in m}^3)$$

Rotational partition function: While deriving the expression for the translational partition functions, the summation over all the energy levels is often replaced by an integration procedure, thus resulting in a simple expression. On the other hand, in the case of rotational motion, it is essential to consider three different approximations (Table 15.1). For high temperatures alone, the replacement of summation by integration is permitted and it leads to a simple expression. Furthermore, the concept of symmetry number arises in order to avoid double counting of the same configuration upon rotation through 360°. Table 15.1 provides the temperature regime under which each equation is valid.

The preliminary step in estimating the rotational partition function is the calculation of the moment of inertia I.

$$I = \mu r^2$$

where μ denotes the reduced mass $= m_1m_2/(m_1 + m_2)$ with m_1 and m_2 being in *atomic mass units*. It is also common to provide the data in wave numbers, that is, wave number $= 1/\lambda = \nu/c$ and c denotes the velocity of light. θ_r is the characteristic rotational temperature and is equal to $h^2/(8\pi^2 I)k$. It is related to the rotational constant B as $\theta_r = \dfrac{B}{k}$ where k is the Boltzmann constant and B is the rotational constant (in cm^{-1}). Since spectroscopic data yields the value of B, it is possible to determine θ_r experimentally for any molecule exhibiting rotational spectra.

Table 15.1 Rotational partition functions under different approximations

Equation	Condition
$q_{rot} = \frac{1}{\sigma}\sum_{J=0}(2J+1)e^{-J(J+1)\theta_r/T}$	$\frac{T}{\theta_r} \le 3$
$q_{rot} = \frac{1}{\sigma\theta_r}\left[1+\frac{1}{3}\left(\frac{\theta_r}{T}\right)+\frac{1}{15}\left(\frac{\theta_r}{T}\right)^2 ....\right]$ Mulholland approximation using Euler–Maclaurin expansion	$3 < \frac{T}{\theta_r} \le 30$
$q_{rot} = \frac{T}{\sigma\theta_r}$	$\frac{T}{\theta_r} > 30$

Since T is often 300 K, while θ_r varies from 2 to 6, the last equation of Table 15.1 is employed.

Vibrational partition function: The vibrational energies within the harmonic oscillator approximation are given by

$$E_\upsilon = \left(\upsilon+\frac{1}{2}\right)e^{-h\vartheta}$$

where υ denotes the vibrational quantum number. The vibrational partition function is

$$Q_{vib} = \sum_{\upsilon=0} e^{-\left(\upsilon+\frac{1}{2}\right)e^{-h\vartheta/kT}}$$

$$= e^{-h\vartheta/2kT}\sum_{\upsilon=0} e^{-\upsilon h\vartheta/kT}$$

$$= e^{-h\vartheta/2kT}\left[1+e^{-h\vartheta/kT}+e^{-2h\vartheta/kT}+...\right]$$

$$= \frac{e^{-h\vartheta/2kT}}{1-e^{-h\vartheta/kT}} \qquad (15.1)$$

If the zero point energy is not included, the above expression becomes

$$Q_{vib} = \frac{1}{1-e^{-h\vartheta/kT}} \qquad (15.2)$$

Equation 15.1 can also be rewritten in terms of the hyperbolic function as

$$Q_{vib} = \frac{1}{2[\sinh(h\nu/2kT)]}$$

Equations 15.1 and 15.2 correspond to the partition function for one vibrational frequency. For various frequencies, it is essential to *multiply* the terms corresponding to each. Other partition functions such as nuclear, electronic, and so on, exist using which several spectroscopic data can be rationalised.

It is conventional to represent the partition functions for vibrational, rotational, nuclear and electronic together as the *internal partition function*. Hence the total partition function becomes

$$Q = Q_{trans} \times Q_{int.}$$

Furthermore, q is also known as the molecular partition function while the system partition function is written as $Q = q^N/N!$ or q^N depending upon the nature of the particles. Table 15.2 provides the order of magnitude of each partition function.

Table 15.2 Values of partition functions at 298 K

Partition function	Order of magnitude
Translational	10^{25} to 10^{30}
Rotational (linear and non-linear molecules)	10 to10^3
Vibrational (for one normal mode)	1 to 5

It is interesting to note that while various energies are *additive* (translational, vibrational, rotational, nuclear, electronic), the corresponding partitions functions are *multiplicative*.

$$Q\ (\text{electronic}) = \Sigma g_i\, e^{-\varepsilon_{\text{electronic},i}}$$

where g_i denotes the degeneracy of the ith energy level.

15.1.2 Thermodynamic Properties from Molecular Information

Molar entropy due to translational motion: The most important contribution of statistical thermodynamics consists in estimating the entropy of various gases at a given temperature and pressure solely using molecular information. The classical Sackur–Tetrode equation is given by

$$\frac{S_{tr}}{R} = \frac{3}{2}\ln M + \frac{5}{2}\ln T - 1.1650 \text{ at 1 atm pressure}$$

Or, $$\frac{S_{tr}}{R} = \frac{3}{2}\ln M + \frac{5}{2}\ln T - \ln P\ (\text{in atm}) - 1.1650 \qquad (15.3)$$

where M denotes the atomic or molecular weight in g mol^{-1} and T is in kelvin. The classical thermodynamic formulae to compute entropy changes of isothermal or isobaric reversible processes are direct consequences of Eq. 15.3. However, the first and last term of Eq. 15.3 arise solely from statistical thermodynamic concepts and cannot be obtained from classical thermodynamics.

Molar entropy due to vibrational motion:

$$S_{vib} = -kN\ln\left(1-e^{-\theta_\vartheta/T}\right)+\frac{R}{T}\frac{\theta_\vartheta}{\left(e^{\theta_v/T}-1\right)} \qquad (15.4)$$

where θ_ϑ denotes the characteristic vibrational temperature ($h\nu/k$). For each mode, the vibrational entropy needs to be computed and the sum of all these is the total vibrational entropy.

Molar entropy due to rotational motion:

$$S_{rot} = kNlnq_{rot} + \frac{E_{rot}}{T} \quad (15.5)$$

where q_{rot} and E_{rot} need to be evaluated taking into account the temperature (Table 15.1). The dependence of a few thermodynamic quantities on the partition function is listed in Table 15.3.

Table 15.3 Thermodynamic properties from partition functions

Quantity	Translational	Rotational	Vibrational
	$q_{tr} = \left(\frac{2\pi mkT}{h^2}\right)^{3/2} V$ or, $\frac{V}{\lambda^3}$ or, $1.88\times10^{26}T^{3/2}M^{3/2}V$ (V in m^3)	$q_{rot} = \frac{kT}{\sigma hcb}$ or $\frac{T}{\sigma\theta_{rot}}$ (high temp app)	$q_{vib} = \prod_i \frac{e^{-h\vartheta_i/2kT}}{1-e^{-h\vartheta_i/kT}}$ or $\prod_i \frac{1}{1-e^{-h\vartheta_i/kT}}$
Helmholtz free energy (A) or (F)	$-kT\ln\left(\frac{q_{tr}^N}{N!}\right)$	$-kT\ln q_{rot}^N$	$-kT\ln q_{vib}^N$
Absolute entropy	$-\left(\frac{\partial F_{tr}}{\partial T}\right)_{V,N}$	$k\ln Q_{rot} + \frac{E_{rot}}{T}$	$k\ln Q_{vib} + \frac{E_{vib}}{T}$
Internal energy	$kT^2\left(\frac{\partial \ln Q_{tr}}{\partial T}\right)$; $Q_{tr} = \frac{q_{tr}^N}{N!}$	$kT^2\left(\frac{\partial \ln Q_{rot}}{\partial T}\right)$; $Q_{rot} = q_{rot}^N$	$kT^2\left(\frac{\partial \ln Q_{vib}}{\partial T}\right)$; $Q_{vib} = q^N$
Chemical potential	$\left(\frac{\partial F_{tr}}{\partial N}\right)_{T,V}$	$\left(\frac{\partial F_{rot}}{\partial N}\right)_{T,V}$	$\left(\frac{\partial F_{vib}}{\partial N}\right)_{T,V}$

It is seen from Table 15.3 that all the thermodynamic properties can be estimated solely using the information on the partition functions. Consequently, all the changes in thermodynamic quantities deduced from the classical thermodynamic laws not only arise from the statistical perspective but also can be interpreted using a molecular basis.

While all the above are important formulae, more emphasis is placed on the entropy. Table 15.4 provides the contribution to entropy from different modes.

Table 15.4 Contributions to entropy from different energies

Mode	Entropy
Rotational	$S_{rot} = kNlnq_{rot} + \frac{E_{rot}}{T}$
Translational	$S_{trans} = R\ln\left(\frac{V}{\lambda_{dB}^3}\right) - 52.25R$
Vibrational	$S_{vib} = -RN\ln\left(1-e^{-\theta_\vartheta/T}\right) + R\frac{\theta_v/T}{\left(e^{\theta_v/T}-1\right)}$

15.2 Classical and Quantum Statistics

The classical or Maxwell–Boltzmann statistics refers to distinguishable particles arranged in various energy levels without any restrictions on the occupancy. The characteristic features of the Fermi–Dirac (F–D) and Bose–Einstein (B–E) distribution are given in Table 15.5. The particles obeying F–D statistics are known as fermions and those satisfying B–E statistics are known as bosons.

Fermions: Electrons, protons, neutrons and odd numbers consisting of these. These have spins ½, 3/2 …..

Bosons: Photons, ^{4}He and even numbers consisting of these. These have integral spins.

The distribution laws governing M–B, F–D and B–E statistics involve the use of Lagrange's multipliers (constrained optimisation) in order to identify the Boltzmann constant k and chemical potential μ. The calculation of macro and microstates using these distribution laws is a fascinating exercise and can often be obtained by simple enumeration, without the use of mathematical formulae.

Another important approximation employed in the computation is the *Stirling approximation* for ln $N!$

$$\ln N! = N\ln N - N \quad (15.6)$$

The approximation is valid only for large values of N (number of particles).

Table 15.5 Characteristic features of classical and quantum statistics

Statistics	Particle nature	Occupancy
Maxwell–Boltzmann distribution	Distinguishable	Any number of particles in a single state
Fermi–Dirac distribution	Indistinguishable	Maximum of two particles in a single state; obey Pauli exclusion principle
Bose–Einstein distribution	Indistinguishable	Any number of particles in a single state are allowed

15.2.1 Validity of the Classical Regime

There exist practical criteria by which it is easy to decipher whether quantum effects will be manifested or not for a given system. If the thermal (or de Broglie) wavelength is much smaller than the mean separation between the particles $<l>$, quantum effects can be ignored under those conditions. The following are the two methods of representing the validity of the classical regime:

(i) $n_r <<1$ (occupation number of rth energy level for various translational states), and

(ii) $\frac{\lambda_{dB}}{<l>} \ll 1$

15.2.2 *Ortho* and *Para*-hydrogen

Molecular hydrogen consists of nuclear spin isomers of the *ortho* and *para* forms. While *para*-hydrogen possess symmetrical rotational states (two proton spins are anti-parallel), *ortho*-hydrogen has antisymmetrical (parallel spins) rotational states. The parallel coupling leads to a net spin of +1 with the values +1, 0 and –1 as the components. Thus, parallel coupling is three times more likely. Hence for *ortho*-hydrogen, the rotational partition function is given by

$$Z_{ortho\text{-}H_2} = 3.\Sigma_{J=1,3,...}^{\infty}(2J+1)e^{-J(J+1)\theta_r/T} \quad (15.7)$$

Analogously, the rotational partition function for *para* hydrogen is given by

$$Z_{para\text{-}H_2} = 1.\Sigma_{J=0,2,...}^{\infty}(2J+1)e^{-J(J+1)\theta_r/T} \quad (15.8)$$

We note that the pre-factors 3 and 1 in Eqs. 15.7 and 15.8 indicate the degeneracy of the nuclear spins. The total partition function can then be written as a sum of these two and the experimental evidence for these two forms of H_2 arises from heat capacity measurements and NMR spectroscopy.

15.2.3 Heat Capacity of Solids

There are two common classical theories of heat capacity of solids—(i) Einstein's theory and (ii) Debye's model. According to the Einstein's theory of heat capacity of solids, the vibrational motion that contributes to the heat capacity has identical frequencies. The theory predicts that $C_v \to 0$ as $T \to 0$ and $C_v \to 3R$ at $T \to \infty$ (Dulong and Petit's law). The most significant contribution to comprehend the magnitude of the heat capacity of solids is that due to Debye (in 1912), according to which, the vibrational motion of the atoms of the crystal lattice dictates the heat capacity. Thus, the concepts developed for vibrational motion and partition functions can be applied *mutatis mutandis* to solids. At low temperatures, the heat capacity is proportional to T^3 while the Dulong–Petit law holds good at very high temperatures. The Debye temperature, $\theta_{Debye} = h\nu_{Debye}/k$ provides a scaling for the absolute temperature and depends upon the nature of the solid.

Worked Out Problems

Translational Partition Functions and Thermodynamic Properties

Problem 15.1

Calculate the translational energy of O_2 in a cube of volume 0.01 m^3 with all the quantum numbers equal to unity.

Solution

Volume of the cube = a^3 = 0.01 m^3

Hence, a = 0.22 m

$$E_{trans} = (n_x^2 + n_y^2 + n_z^2)\frac{h^2}{8ma^2}$$

$$= \frac{(1^2+1^2+1^2)(6.62\times10^{-34}\text{ J s})^2 6.023\times10^{23}\text{ mol}^{-1}}{8\times32\text{ g mol}^{-1}\,(0.22\times0.22)\text{ m}^2}$$

$$= 6.41\times10^{-41}\text{ J}$$

Problem 15.2

Derive the ideal gas equation of state from the canonical partition function

$$q_{tr} = \frac{V}{(\lambda_{dB})^3}$$

Solution

$Q = \frac{q_{tr}^N}{N!}$, where N denotes the total number of molecules.

$$P = kT\left(\frac{\partial \ln Q}{\partial V}\right)_T$$

$$\ln Q = N\ln q_{tr} - lnN!$$

$$\left(\frac{\partial \ln q}{\partial V}\right)_T = N\frac{1}{q_{tr}}\left(\frac{\partial q_{tr}}{\partial V}\right)_T = \frac{N}{V}$$

Therefore $P = \frac{kTN}{V} = \frac{kTnN_A}{V} = \frac{nRT}{V}$ where N_A denotes the Avogadro number.

Problem 15.3

Estimate the expectation value $<p_x^2>$ for argon at 300 K using the equipartition principle.

Solution

In a one dimensional system, we can write,

$$<p_x^2> = m \times R \times T/N$$

$$<p_x^2> = 39.95 \times 8.314 \times 300/(6.023 \times 10^{23})$$

$$= 1.65 \times 10^{-19} \text{ J g}$$

Problem 15.4

Compute the translational partition function for argon at 1 atm and 298 K (molecular weight: 40).

Solution

ln (Z_{trans}) = (3/2) ln M + (5/2) ln T – ln P + 51.104 where M should be in atomic mass units, T in kelvin and P in atm. Upon substituting the atomic weight of Ar, T as 298 K and P as 1 atm, it follows that

$$\ln(Z_{trans}) = 5.5333 + 14.2427 + 51.104$$

$$\approx 70.88$$

$$Z_{trans} = 6.07 \times 10^{30}$$

Problem 15.5

Write the equations for the following quantities using the molecular translational partition function q.

(i) Helmholtz free energy (ii) internal energy (iii) heat capacity at constant volume

Solution

(i) $F = -kT\ln\left(\frac{q^N}{N!}\right)$

(ii) $E = kT^2\frac{\partial \ln Q}{\partial T}$

(iii) $C_V = \frac{\partial}{\partial T}\left[RT^2\frac{\partial \ln Q}{\partial T}\right]$ where $Q = \frac{q^N}{N!}$

Problem 15.6

Write the canonical partition function in the case of distinguishable and indistinguishable particles in terms of the molecular partition functions.

Solution

$Q = q^N/N!$ (indistinguishable) or q^N (distinguishable)

Q refers to the system partition function while q denotes the molecular partition function.

Problem 15.7

Deduce the equation for the de Broglie wavelength in terms of temperature, from the equipartition principle.

Solution

The de Broglie wavelength is defined as

$$\lambda_{dB} = \frac{h}{p} = \frac{h}{mv} \quad (1)$$

The translational energy E_{tr} for a particle of mass m is,

$$E_{tr} = \frac{1}{2}mv^2$$

$$\text{Equivalently, } v = \left(\frac{2E_{tr}}{m}\right)^{1/2} \quad (2)$$

Using Eq. (1),

$$\lambda_{dB} = \frac{h}{m\left(\frac{2E_{tr}}{m}\right)^{1/2}} \quad (3)$$

From the equipartition principle for the translational motion,

$$E_{tr} = \frac{3kT}{2} \quad (4)$$

Substituting Eq. (4) in Eq. (3),

$$\lambda_{dB} = \left(\frac{h^2}{3mkT}\right)^{1/2}$$

Recalling that 1 joule = 1 kg m^2 s^{-2}, the substitution of the Planck constant in J s, yields the dimension of the RHS as metre.

Problem 15.8

A monatomic gas of atomic weight 4 g mol^{-1} and 1 atm pressure at 273 K has nearly 6×10^{20} molecules cm^{-3}. Estimate the (i) mean distance between molecules $<l>$ and (ii) de Broglie wavelength λ_{dB}.

Solution

$$l \cong N^{-1/3}$$

$$\approx (6 \times 10^{20} \text{ cm}^{-3})^{-1/3}$$

$$= 1.18 \times 10^{-7} \text{ cm}$$

$$\lambda_{dB}{}^2 = \left[\frac{(6.62\times10^{-34}\ \text{J s})^2 10^3\ \text{g}^2\ \text{s}^2\ \text{J}\times6.023\times10^{23}}{3\times1.38\times10^{-23}\times273\times4}\right]$$

$$\simeq 7.64\times10^{-9}\ \text{cm}$$

Problem 15.9

For an ideal monatomic gas of mass m, write the equations for (i) chemical potential, (ii) Gibbs free energy and Helmholtz free energy.

Solution

For any monatomic gas, the only contribution to various quantities arises from the translational motion. Employing the definitions given for different thermodynamic quantities in terms of the translational partition function, the following equations arise.

(i) $$\mu = kT\log p - \frac{5}{2}kT\log T - kT\log\left\{\frac{(2\pi m)^{\frac{1}{2}} k^{\frac{5}{2}}}{h^3}\right\}$$

(ii) $$G = -NkT\left[\frac{5}{2}\ln T + \frac{3}{2}\ln M - \ln P - 3.6605\right]$$

$$F = -NkT\left[\frac{5}{2}\ln T + \frac{3}{2}\ln M - \ln P - 2.6605\right]$$

where M is in g mol^{-1}, P is in atm and T is in kelvin.

Problem 15.10

Show that the entropy change of 1 mol of an ideal gas follows from the Sackur–Tetrode equation for (i) isothermal and (ii) isobaric process.

Solution

The Sackur–Tetrode equation is given by

$$\frac{S_{tr}}{R} = \frac{3}{2}\ln M + \frac{5}{2}\ln T - \ln P(\text{in atm}) - 1.1650$$

(i) Since the process is isothermal, the above equation leads to

$$\Delta S = S_{\text{tr,final}} - S_{\text{tr,initial}}$$

$$= -R\,[\ln P_{\text{final}} - \ln P_{\text{initial}}]$$

$$= -R\,[\ln V_{\text{initial}} - V_{\text{final}}]$$

$$= R\ln\,[V_{\text{final}}/V_{\text{initial}}]$$

(ii) For an isobaric process,

$$\Delta S = S_{\text{tr,final}} - S_{\text{tr,initial}}$$

$$= (5R/2)\,[\ln T_{\text{final}} - \ln T_{\text{initial}}]$$

$$= (5R/2)\ln\,[T_{\text{final}}/T_{\text{initial}}]$$

$$= C_p\ln\,[T_{\text{final}}/T_{\text{initial}}]$$

Problem 15.11

Derive the canonical partition function corresponding to the van der Waals equation of state given as

$$\left(P + \frac{aN^2}{V^2}\right)(V - Nb) = NkT$$

Solution

In the canonical ensemble, pressure is related to the Helmholtz free energy as

$$P = -\left(\frac{\partial F}{\partial V}\right)_T = +\left(\frac{\partial[kT\ln Q]}{\partial V}\right)_T \tag{1}$$

$$= kT\left(\frac{\partial \ln Q}{\partial V}\right)_T \tag{2}$$

where $$Q = \frac{q^N}{N!} \tag{3}$$

From the given equation of state,

$$P = \frac{NkT}{V - Nb} - \frac{aN^2}{V^2} \tag{4}$$

Therefore, from Eq. (2),

$$\left(\frac{\partial \ln Q}{\partial V}\right)_T = \frac{N}{V - Nb} - \frac{aN^2}{V^2 kT} \tag{5}$$

Upon integrating Eq. (5),

$$\ln Q = N\ln(V - Nb) + \frac{aN^2}{VkT} + \text{integration constant} \tag{6}$$

The integration constant can be evaluated by invoking the condition that when $a = 0$ and $b = 0$, the canonical partition function should correspond to that pertaining to the ideal gas. Hence

$$\ln Q = N\ln(V - Nb) + \frac{aN^2}{VkT} + N\ln\left(\frac{2\pi mkT}{h^2}\right)^{3/2} - \ln N!$$

That is,

$$Q = \frac{\left(e^{aN^2/VkT}\right)\left(\frac{2\pi mkT}{h^2}\right)^{3N/2}(V - Nb)^N}{N!}$$

Note that when $a = 0$ and $b = 0$, $Q_{\text{ideal}} = \dfrac{q_{\text{ideal}}^N}{N!}$

where $$q_{\text{ideal}} = \left(\frac{2\pi mkT}{h^2}\right)^{3/2} V$$

Problem 15.12

Derive the Sackur–Tetrode equation for the translational entropy of ideal gases, from the corresponding partition functions.

Solution

$$F_r = E_{tr} - TS_{tr}$$

Therefore, $$TS_{tr} = E_{tr} - F_{tr} = \frac{3RT}{2} - [-kT\ln Q_{tr}]$$

$$S_{tr} = \frac{3R}{2} + k\ln\left\{\frac{q_{tr}^N}{N!}\right\}$$

$$= \frac{3R}{2} + k\left[\ln q_{tr}^N\right] - k\left[N\ln N - N\right]$$

$$= \frac{3R}{2} + R + k\ln q_{tr}^N - R\ln N$$

$$= \frac{5R}{2} + R\ln q_{tr} - R\ln N$$

$$= \frac{5R}{2} + R\ln\left[\frac{q_{tr}}{N}\right]$$

$$\frac{S_{tr}}{R} = \frac{5}{2} + \ln\left[\frac{\left(\frac{2\pi mkT}{h^2}\right)^{3/2} \times V}{N}\right]$$

$$= \frac{5}{2} + \ln\left[\frac{\left(\frac{2\pi mkT}{h^2}\right)^{3/2}}{N}\frac{RT}{P}\right]$$

$$= \frac{5}{2} + \ln\left[\left(\frac{2\pi mkT}{h^2}\right)^{3/2}\frac{T^{5/2}R}{PN}\right]$$

$$= \frac{5}{2} + \frac{3}{2}\ln M + \frac{5}{2}\ln T - \ln P + \ln\left\{\left(\frac{R}{N}\right)\left(\frac{2\pi k}{h^2}\right)^{3/2}\right\}$$

$$\frac{S_{tr}}{R} = \frac{5}{2} + \frac{3}{2}\ln M + \frac{5}{2}\ln T - \ln P + \text{constant term}$$

where M denotes the molecular weight.

The above equation, known as the Sackur–Tetrode equation, can be simplified further upon substituting the fundamental constants as

$$\frac{S_{tr}}{R} = \ln\left(\frac{T^{5/2}M^{3/2}}{P}\right) - 1.165$$

where M is in grams, T is in Kelvin and P in atm.

Rotational Partition Functions

Problem 15.13

Compute the rotational partition function for CO_2 at 300 K if the moment of inertia is 7.20×10^{-46} kg m^2.

Solution

It is preferable to first compute the characteristic rotational temperature θ_r since it will enable the correct approximation to be employed.

$$\theta_r = h^2/(8\pi^2 I)k$$

Substituting the fundamental constants and moment of inertia, $\theta_r = 0.56$ K. Here $T/\theta_r = 535.7$. Since $(T/\theta_r) >> 3$, the high temperature approximation can be employed. The point group of CO_2 is $D_{\infty h}$ and symmetry number (σ) is 2. Hence,

$$q_{rot} = \frac{T}{\sigma\theta_r} = 267.85$$

Problem 15.14

Compute the rotational partition function for a heteronuclear diatomic molecule at 298 K, using (i) the first three terms of the power series and (ii) high temperature approximation (the moment of inertia is $I = 6.29\times10^{-48}$ kg m^2).

Solution

(i) The characteristic rotational temperature is given by $\theta_r = h^2/(8\pi^2 I)\,k$ where k denotes the Boltzmann constant. Since T/θ_r lies between 3 and 30, the Mulholland approximation is employed (Table 15.1)

$$\frac{\theta_r}{T} = \frac{h^2}{8\pi^2\left(6.29\times10^{-48}\right)\times1.38\times10^{-23}\times298} = 0.215$$

$$q_{rot} = \frac{1}{1\times0.215}\left[1\times\frac{1}{3}(0.215) + \frac{1}{15}(0.215)^2 + \frac{4}{315}(0.215)^3\right]$$
$$= 5.0$$

(ii) For a heteronuclear diatomic molecule, the symmetry number σ is unity.

$$q_{rot} = \frac{T}{\sigma\theta_r} = \frac{1}{0.215} = 4.65$$

Problem 15.15

Compute the rotational contribution to entropy for CO_2 at 300 K if the characteristic rotational temperature θ_r is 0.56 K.

Solution

The characteristic rotational temperature $\theta_r = 0.56$ K. Hence,

$$T/\theta_r = 535.7$$

The symmetry number is 2. Hence, $q_{rot} = \frac{T}{\sigma\theta_r} = 267.85$ and $q_{rot} = 267.85$

$$S_{rot} = +R\ln q_{rot} + \frac{E_{rot}}{T}$$
$$= R(5.59 + 1) = 55.62 \text{ J K}^{-1}$$

Problem 15.16

The number of molecules (N_0) in the ground state of HF is 4×10^{20}. Estimate the number of molecules when $J = 4$, if $\frac{T}{\theta_r} = 10$.

Solution

$$N_J/N_0 = (2J+1)\,e^{-J(J+1)\theta_r/T} = 9\times e^{-4\times5\times0.1} = 1.218$$

Hence, $N_J = 1.218\times4\times10^{20} = 4.872\times10^{20}$

Problem 15.17

Show that the heat capacity at constant volume for (i) diatomic and (ii) polyatomic molecules obtained using the rotational partition function is identical with the prediction from the equipartition principle.

Solution

(i) For diatomic molecules, under high temperature approximation, the rotational partition function is given by,

$$q_{rot} = \frac{T}{\sigma\theta_r} \quad (1)$$

$$E_{rot} = RT^2\left(\frac{\partial \ln q_{rot}}{\partial T}\right) \quad (2)$$

Substituting q_{rot}, yields E_{rot} as RT and $C_v = R$.

(ii) For polyatomic molecules,

$Z_{rot} = \ln [T^{3/2}]$ + terms independent of temperature

Hence $E_{rot} = 3RT/2$ using Eq. (2).

Problem 15.18

The characteristic rotational temperature of a heteronuclear diatomic molecule is 0.6 K. Estimate the (i) moment of inertia in kg m^2, (ii) partition function and entropy at 100 K.

Solution

(i) $\theta_r = h^2/(8\pi^2 I)k$

$I = h^2/(8\pi^2)k\,\theta_r$

$$I = \frac{6.626^2 \times 10^{-68}}{8\pi^2(1.38\times10^{-23})0.6} = 6.69\times10^{-46} \text{ kg m}^2$$

(ii) $$Z_{rot} = \frac{T}{\sigma\theta_r} = 100/0.6 = 166.7$$

$$S_{rot}/R = \ln(q_{rot}) + 1 = 6.11$$

Problem 15.19

Estimate kT/hc in cm^{-1} at 298 K and hence rewrite the characteristic rotational temperature in terms of the constant B.

Solution

$$\frac{kT}{hc} = \frac{1.38\times10^{-23}\times298}{(6.626\times10^{-34})3\times10^{10}} = 207.08 \text{ cm}^{-1}$$

$$\theta_{rot}/T = h^2/(8\pi^2 I)kT = \frac{hc}{kT}\times B$$

Problem 15.20

Write the rotational contribution to the internal energy, Helmholtz free energy and absolute entropy of a polyatomic molecule with three moments of inertia.

Solution

$$E_{rot} = NkT^2\frac{\partial \ln q}{\partial T} = NkT^2\frac{d}{dT}\left[\frac{1}{2}\ln\frac{8\pi^2}{\sigma h^3}(I_a I_b I_c)^{1/2} + \frac{3}{2}\ln kT\right]$$

$$F_{rot} = -NkT\ln q$$

$$= -NkT\left[\frac{3}{2}\ln T + \frac{1}{2}I_a I_b I_c - \ln\sigma + \frac{1}{2}\ln\pi + \frac{3}{2}\ln\frac{8\pi^2 k}{h^2}\right]$$

$$S_{rot} = Nk\left(\frac{3}{2}\ln T + \frac{1}{2}\ln I_a I_b I_c - \ln\sigma + 134.684\right)$$

Vibrational Partition Functions

Problem 15.21

Under what conditions, is the vibrational contribution to heat capacity (C_V) of a diatomic gaseous molecule consistent with the prediction from the equipartition principle?

Solution

$$F_{Vib} = -NkT\ln q_{Vib}$$

$$= -NkT\left\{\frac{h\vartheta}{2kT} - \ln\left(1 - e^{-h\vartheta/kT}\right)\right\}$$

$$= -\frac{Nh\vartheta}{2} + NkT\ln\left(1 - e^{-h\vartheta/kT}\right)$$

$$F_{Vib} = E_{Vib} - TS_{Vib} = E_{Vib} - T\left(\frac{\partial F_{Vib}}{\partial T}\right)_V$$

$$E_{Vib} = F_{Vib} + T\left(\frac{\partial F_{Vib}}{\partial T}\right)_V$$

$$E_{Vib} = \frac{Nh\vartheta}{2} + \frac{Nh\vartheta e^{-h\vartheta/kT}}{1 - e^{-h\vartheta/kT}}$$

$$= \frac{Nh\vartheta}{2} + \frac{Nh\vartheta}{e^{+h\vartheta/kT} - 1}$$

$$C_{Vib} = \frac{Nh\vartheta k(-1)e^{h\vartheta/kT}\,h\vartheta}{\left(e^{h\vartheta/kT} - 1\right)^2 k}\left(\frac{-1}{T^2}\right)$$

$$C_{Vib} = R\left(\frac{h\vartheta}{kT}\right)^2\frac{e^{h\vartheta/kT}}{\left(e^{h\vartheta/kT} - 1\right)^2}$$

$$\frac{C_{Vib}}{R} = \left(\frac{h\vartheta}{kT}\right)^2\frac{e^{h\vartheta/kT}}{\left(e^{h\vartheta/kT} - 1\right)^2}$$

At high temperatures, the limits of the RHS of the above equation should be obtained using the *L'Hospital rule*. Upon substituting $\frac{h\vartheta}{kT}$ as x, it is seen that Limit $x \to 0$, $(x)^2\frac{e^x}{(e^x - 1)^2} \to 1$. Thus, at high temperatures, $C_{Vib} \to R$ for one vibrational mode, in agreement with the equipartition principle.

Problem 15.22

Calculate the vibrational partition function at 300 K for a system if the vibrational frequency is 6.5×10^{12} s^{-1}.

Solution

$$q_{vib} = \frac{1}{1-e^{-\beta E_{vib}}} = \frac{1}{1-e^{-E_{vib}/kT}}$$

$$\frac{h\vartheta}{kT} = \frac{6.626\times10^{-34}\ \text{J s}\times6.5\times10^{12}\ \text{s}^{-1}}{1.38\times10^{-23}\ \text{J K}^{-1}\times300\ \text{K}}$$

$$= 1.04$$

$$q_{vib} = \frac{1}{1-e^{-1.04}} \cong 1.55$$

Problem 15.23

The following are the characteristic vibrational temperatures of three normal modes of vibration of H_2O.

Vibrational mode	Characteristic vibrational temperature
ϑ_1	5296
ϑ_2	2310
ϑ_3	5439

Estimate the total vibrational partition function at (i) 300 K and (ii) 2000 K.

Solution

(i) $$q_{vib} = \prod_i \left(\frac{1}{1-e^{-\theta_{vib,i}/T}}\right)$$

$$= \left(\frac{1}{1-e^{-5296/300}}\right)\left(\frac{1}{1-e^{-2310/300}}\right)\left(\frac{1}{1-e^{-5439/300}}\right)$$

$$\approx 1$$

(ii) $$q_{vib} = \left(\frac{1}{1-e^{-5296/2000}}\right)\left(\frac{1}{1-e^{-2310/2000}}\right)\left(\frac{1}{1-e^{-5439/2000}}\right)$$

$$= (1.076)\,(1.460)\,(1.0705)$$

$$\approx 1.68$$

Problem 15.24

The characteristic vibrational temperatures (in K) of CO_2 are nearly 1890 (symmetric stretching), 3360 (asymmetric stretching) and 954 (bending, doubly degenerate). Calculate the partition function at 298 K.

Solution

Using Table 15.3, the partition function is estimated as,

$$q_{vib} = q_{\vartheta1}\times q_{\vartheta2}\times q_{\vartheta3} = 1.002\times1.00\times1.042\times1.042 = 1.085$$

Problem 15.25

The vibrational energy levels of a harmonic oscillator are equally spaced with 5×10^{-20} J. Deduce the ratio between the ground and the first excited state at 1000 K.

Solution

$$\frac{n_1}{N} = e^{-\beta E_1}$$

$$\frac{n_2}{N} = e^{-\beta E_2}$$

$$\frac{n_1}{n_2} = \frac{e^{-\beta E_1}}{e^{-\beta 2}} = e^{-\beta(E_1-E_2)}$$

$$\frac{n_2}{n_1} = e^{-5\times10^{-20}\ \text{J}/(1.38\times10^{-23}\times1000\ \text{J})} = e^{-5/1.38} = 0.027$$

$$\therefore n_2 = (.027)n_1$$

Note that the number of molecules in the first excited state is lower than that in the ground state.

Problem 15.26

The characteristic vibrational temperature for a diatomic gas is 700 K. Evaluate the vibrational entropy at 400 K.

Solution

From Table 15.4,

$$\frac{S_{Vib}}{R} = -\ln\left(1-e^{-700/400}\right) + \frac{700/400}{e^{700/400}-1}$$

$$= 0.190 + 0.368 = 0.558$$

Problem 15.27

The three characteristic vibrational temperatures of a triatomic molecule are 2592.6 K, 715.2 K and 1728.4 K. Estimate the partition function at 298 K.

Solution

$$q_{vib} = \left(\frac{1}{1-e^{-2592.6/298}}\right)\left(\frac{1}{1-e^{-715.2/298}}\right)^2\left(\frac{1}{1-e^{-1728.4}}\right)$$

$$\approx 1.21$$

Problem 15.28

The vibrational frequency of a diatomic species is 2300 cm^{-1}. Estimate the vibrational partition function at 298 K.

Solution

$$\frac{hc}{kT} = 0.00483$$

$$q_{vib} = \left(\frac{1}{1-e^{-2300\times0.00483}}\right) \approx 1.0$$

Problem 15.29

The characteristic vibrational temperature of a molecule is 208.6 K. Estimate the (i) heat capacity at constant volume and (ii) vibrational frequency (in cm^{-1}) assuming the validity of the harmonic oscillator model, at 298 K.

Solution

(i) $$\frac{C_{Vib}}{R} = \left(\frac{\theta vib}{T}\right)^2 \frac{e^{\theta vib/T}}{\left(e^{\theta vib/T}-1\right)^2}$$

$$= \left(\frac{208.6}{298}\right)^2 \frac{e^{0.7}}{\left(e^{0.7}-1\right)^2} = 0.96$$

(ii) $\theta_{vib} = hv/k$

Hence $$v = \frac{208.6 \times 1.38 \times 10^{-23}}{6.626 \times 10^{-34}} = 4.34 \times 10^{12}\, s^{-1} = 144.8\ cm^{-1}$$

Problem 15.30

Estimate the stretching frequency for (i) C–C ; (ii) C–H and (iii) C–D bonds using the force constant (in 10^6 dynes cm^{-1}) as 1.0, 0.5 and 0.5 respectively. Obtain the corresponding partition functions.

Solution

(i) $\bar{\upsilon}_{C-C} = \dfrac{1}{2\pi c}\sqrt{\dfrac{k}{\mu}}$ where μ denotes the reduced mass and k the force constant.

$$= \frac{\sqrt{10^6 \text{ dynes cm}^{-1} / (6 \times 6.023 \times 10^{23}\text{ g})}}{2 \times 3.14 \times 3 \times 10^{10}\text{ cm s}^{-1}}$$

$\approx 1684\ cm^{-1}$

(ii) $\bar{\upsilon}_{c-H} = \dfrac{1}{2\pi c}\sqrt{0.5 \times 10^6 \text{ dynes cm}^{-1}/0.923 \times 6.023 \times 10^{23}\text{ g}}$

$= 3036\ cm^{-1}$

(iii) $\bar{\upsilon}_{c-d} = \dfrac{1}{2\pi c}\sqrt{0.5 \times 10^6 \text{ dynes cm}^{-1}/1.73 \times 6.023 \times 10^{23}\text{ g}}$

$= 2218\ cm^{-1}$

In all the cases, the partition function is nearly unity.

Electronic Partition Functions

Problem 15.31

The ground state degeneracy of a molecule is 2 while that of the first excited state is 3. The excited state lies 0.1 eV above the ground state (energy = 0). Estimate the electronic partition function at 300 K.

Solution

Electronic partition function $q_{el} = \Sigma_{i=0}^{1} g_i e^{-E_i/kT}$

0.3125 eV corresponds to 5.0×10^{-20} J

$$q_{el} = \left\{ g_0 + g_1 e^{-E_1/kT} \right\}$$

$$= 2 + 3\ e^{-E_1/kT} = 2 + 3e^{-5\times10^{-20}\text{ J}/\left(1.38\times10^{-23}\text{ J}\times300\right)}$$

$= 2.0$

The result is anticipated since the ground state degeneracies and energies only contribute to the electronic partition function.

Assembly of Molecules in Different Energy Levels

Problem 15.32

An ensemble consisting of Avogadro number of particles has two energy levels separated by 0.3125 eV. Estimate the approximate temperature at which the internal energy of this system will be 1500 J?

Solution

The internal energy of the system is given by

$$E = n_1 E_1 + n_2 E_2 \qquad (1)$$

$$\frac{n_1}{N} = \frac{e^{-E_1/kT}}{q} \qquad (2)$$

$$\frac{n_2}{N} = \frac{Ne^{-E_2/kT}}{q} \qquad (3)$$

where q denotes the partition function. 0.3125 eV corresponds to 5.0×10^{-20} J.

Using Eqs (1) to (3),

$$1500\text{ J} = \frac{N}{q}\left\{ e^{-E_1/kT} E_1 + e^{-E_2/kT} E_2 \right\}$$

$$= \frac{N\left\{ E_1 e^{-E_1/kT} + E_2 e^{-E_2/kT} E_2 \right\}}{e^{-E_1/kT} + e^{-E_2/kT}}$$

$$E_2 = 5 \times 10^{-20}\text{ J}$$

Substituting $E_1 = 0$ and $E_2 = 5.0 \times 10^{-20}$ J, along with the Avogadro number N, it follows that,

$$1500\text{ J} = \frac{30.115 Je^{-3.623\times1000/T}}{1+e^{-3.623\times1000/T}} \text{ and } e^{\frac{3623}{T}} = 19.076$$

$$\frac{3623}{T} = 2.948$$

Therefore $T \approx 1229$ K

Problem 15.33

Derive the vibrational contribution to the internal energy and heat capacity using the partition functions.

Solution

The relation governing the internal energy and partition function is given by

$$E = -\frac{N}{q_{vib}}\left[\frac{\partial q_v}{\partial \beta}\right]$$

$$E = -N\left(\frac{\partial \ln q_{vvib}}{\partial \beta}\right) = -N\frac{\partial \ln q_{vib}}{\partial\left(\frac{1}{kT}\right)}$$

$$= -NkT^2\frac{\partial \ln q_{vib}}{\partial T(-1)} = NkT^2\frac{\partial \ln q_{vib}}{\partial T}$$

The vibrational partition function is

$$q_{vib} = \frac{1}{1-e^{-\beta h\vartheta}}$$

Therefore, $\frac{\partial q_{vib}}{\partial \beta} = (-1)\frac{1(-1)e^{-\beta h\vartheta}(-h\vartheta)}{\left(1-e^{-\beta h\vartheta}\right)^2} = -\frac{h\vartheta e^{-\beta h\vartheta}}{\left(1-e^{-\beta h\vartheta}\right)^2}$

$$\frac{1}{q_{vib}}\frac{\partial q_{vib}}{\partial \beta} = \frac{-h\vartheta e^{-\beta h\vartheta}\left(1-e^{-\beta h\vartheta}\right)}{\left(1-e^{-\beta h\vartheta}\right)^2(1)} = \frac{-h\vartheta e^{-\beta h\vartheta}}{1-e^{-\beta h\vartheta}}$$

$$-\frac{1}{q_{vib}}\frac{\partial q_{vib}}{\partial \beta} = +\frac{h\vartheta e^{-\beta h\vartheta}}{1-e^{-\beta h\vartheta}}$$

$$E_{vib} = \frac{Nh\vartheta e^{-\beta h\vartheta}}{1-e^{-\beta h\vartheta}} = \frac{Nh\vartheta_{vib}}{e^{\beta h\vartheta}-1}$$

Upon differentiation of the above with respect to temperature,

$$C_v, vib = Nh\vartheta_{vib}\frac{[-1]e^{\beta h\vartheta}}{\left(e^{\beta h\vartheta}-1\right)^2}h\vartheta\left(-\frac{1}{kT^2}\right)$$

$$= \frac{kNe^{\beta h\vartheta_{vib}}}{\left(e^{\beta h\vartheta}-1\right)^2}\left(\frac{h\vartheta_{vib}}{RT}\right)^2$$

$$= R\left(\frac{\theta_{vib}}{T}\right)^2\frac{e^{\theta_{vib}/T}}{\left(e^{\theta_{vib}/T}-1\right)^2}$$

where θ_{vib} denotes the characteristic vibrational temperature.

Problem 15.34

The characteristic vibrational temperature of I_2 is 823.2 K. Estimate the vibrational contribution to the internal energy at 300 K.

Solution

$$\frac{E}{kT} = \left(hv/kT\right)\frac{e^{-hv/kT}}{1-e^{-hv/kT}}$$

$$= \frac{\theta_{vib}}{T}\frac{e^{-\theta_{vib}/T}}{1-e^{-\theta_{vib}/T}}$$

$$= (2.744)\frac{0.0643}{0.9357} = 0.188$$

$$E = 0.188\times 1.38\times 10^{-23}\times 300 = 7.782\times 10^{-22}\ \text{J}$$

Problem 15.35

Evaluate the partition function at 298 K for (i) *ortho*-H_2 (ii) *para*-H_2 if the rotational constant B of H_2 is 61 cm^{-1}.

Solution

The rotational constant $B = 61\ cm^{-1}$

$$\theta_{rot}/T = \frac{hc}{kT}\times B = 0.00483\times 61 = 0.294$$

$$q_{ortho-H_2} + q_{para-H_2}$$

$$= 3.\sum_{J=1,3,\ldots}^{\infty}(2J+1)e^{-J(J+1)\theta_r/T} + 1.\sum_{J=0,2,\ldots}^{\infty}(2J+1)e^{-J(J+1)\theta_r/T}$$

$$\simeq 5.64 + 1.90 = 7.54$$

Thus, the partition functions are (i) 5.64 and (ii) 1.90 for the *ortho* and *para* forms of hydrogen.

Classical and Quantum Statistics

Problem 15.36

N molecules are distributed in three energy levels having energies 0, 1 kT and 2 kT. Estimate the mean internal energy and molar specific heat.

Solution

The partition function of the system is given by

$$q = \sum_{i=0}^{2} e^{-E_i/kT}$$

$$= 1 + e^{-1} + e^{-2}$$

Number of molecules in the state $E_0 = 0$ is $N_0 = \frac{1}{q}$

Similarly the number of molecules in the state $E_1 = 1\ kT$ is $N_1 = \frac{e^{-1}}{q}$

Number of molecules in the state $E_2 = 2\ kT$ is $N_2 = \frac{e^{-2}}{q}$

Internal energy $E = \sum N_i E_i$

$$= N_0E_0 + N_1E_1 + N_2E_2$$

$$= 0 + \frac{e^{-1}}{q}1kT + \frac{e^{-2}}{q}2kT = \frac{kT}{q}\left(\frac{1}{e}+\frac{2}{e^2}\right)$$

$$\text{Molar specific heat} = N\left(\frac{\partial E}{\partial T}\right)_V = \frac{Nk\left[\frac{1}{e}+\frac{1}{e^2}\right]}{1+\frac{1}{e}+\frac{1}{e^2}}$$

Problem 15.37

Write the equation for the molar entropy of a system incorporating translational, vibrational and rotational partition functions.

Solution

$$S/R = \left\{\ln((q_{trans}\times q_{rot}\times q_{vib})/N)\right\} + (E_{trans} + E_{rot} + E_{vib})/RT + 1$$

Problem 15.38

In the following general distribution law governing the distribution of n_j molecules in E_j energy levels with degeneracies g_j, obtain the value of A. μ denotes the chemical potential.

$$\frac{n_j}{g_i} = \frac{1}{A + e^{(E_j-\mu)/kT}}$$

for (i) classical; (ii) Fermi–Dirac and (iii) Bose–Einstein statistics.

Solution

(i) $A = 0$

(ii) $A = +1$

(iii) $A = -1$

Problem 15.39

Derive the mean occupation number of particles obeying Bose–Einstein statistics.

Solution

The number of particles vary from 0 to ∞ since there is no restriction on the occupancy. Let the energy of a level be denoted as ε and the chemical potential of the system be μ.

Let $x = (\varepsilon - \mu)/kT$

The summation of all energy levels leads to

$$\therefore \sum_{n=0}^{\infty} e^{-nx} = 1 + e^{-x} + e^{-2x} + \ldots$$

$$= \frac{1}{1-e^{-x}}$$

The mean occupation number is defined as

$$\therefore \langle n \rangle = -\frac{\partial}{\partial x}\ln\left\{(1-e^{-x})^{-1}\right\}$$

$$= \frac{\partial}{\partial x}\ln[1-e^{-x}]$$

$$\therefore \langle n \rangle = \frac{e^{-x}}{1-e^{-x}}$$

$$= \frac{1}{e^{x}-1}$$

$$= \frac{1}{e^{\beta(\varepsilon-\mu)}-1}$$

Problem 15.40

A system of 3 fermions (having a total energy of 3ε) can occupy the energy levels 0, ε, 2ε and 3ε with the respective degeneracies 1, 3, 4 and 6. Calculate all the macro and microstates.

Solution

Since more than one fermion cannot occupy a given energy state and taking into account the degeneracies, there are 2 macro states as shown below:

(i) ----- | ↑ ----- ↑ ----- ↑ ----- | ---- --- --- ---- | ----- ----- ----- ----- ----- -----

0 ε 2ε 3ε

(ii) ↑ ----- | ↑ ----- ----- ----- ↑ ---- ---- ---- ---- ----- ----- ----- ----- ----- -----

We note that there are no other arrangements which will give a total energy of 3ε and total number of particles as 3. Hence two macro states are possible.

On account of the Pauli exclusion principle which allows a maximum of one fermion in a given energy level, there is one microstate for the arrangement in (i). However, for (ii), the microstates are $1^1 \times 3^1 \times 4^1 = 12$.

Problem 15.41

Derive the mean occupation number of photons, assuming the harmonic oscillator approximation.

Solution

Since photons are considered as bosons, with no restriction on the occupancy of energy levels, the mean occupation number may be written as

$$\langle n_r \rangle = [\Sigma n_r e^{(-\Sigma ni\, h\nu i/kT)}]\,/Q_{vib}$$

where $\langle n_r \rangle$ denotes the mean number of photons in a given state and the summation is over all r states, ranging from 0 to ∞. The partition function and Q is given by

$$Q_{vib} = \pi\frac{1}{1-e^{-h\nu i/kT}}$$

Equivalently,

$$\langle n_r \rangle = \{1-e^{-h\nu r/kT}\}\Sigma\, n_r e^{(-\,nr\, h\nu r/kT)}]$$

Upon carrying out the summation,

$$\langle n_i \rangle = \frac{1}{e^{\left(\frac{h\nu_i}{kT}\right)}-1}$$

The above equation is also known as the Planck distribution law. The chemical potential term is sometimes introduced in order to bring consistency with the Fermi–Dirac and Bose–Einstein distributions. The incorporation of the chemical potential in the case of the Planck distribution law is sometimes not justifiable.

Question 15.42

Write the equation for mean occupation numbers for the distribution of particles obeying (i) Maxwell–Boltzmann (ii) Planck (iiii) Fermi–Dirac and (iv) Bose–Einstein statistics. The mean number of particles refers to the number of particles in the ith energy level.

Solution

(i) $\frac{n_i^{MB}}{N} = \frac{e^{-\beta E_i^{MB}}}{\sum e^{-\beta E_i^{MB}}}$

(ii) $n_i^{Planck} = \frac{1}{e^{\beta(E_i-\mu)}-1}$

(iii) $n_i^{FD} = \frac{1}{e^{\beta(E_i-\mu)}+1}$

(iv) $n_i^{BE} = \frac{1}{e^{\beta(E_i-\mu)}-1}$ where $\beta = \frac{1}{kT}$

Problem 15.43

Classify the following as Fermions or Bosons.

(i) D (ii) HD (iii) N (iv) Photons (v) H_2 (vi) ^3He and (vii) ^4He

Solution

(i) Deuterium: number of protons (1) + number of neutrons (1) + number of electrons (1) = 3; fermion

(ii) HD: number of protons (2) + number of neutrons (1) + number of electrons (2) = 5; fermion

(iii) N: number of protons (7) + number of neutrons (7) + number of electrons (7) = 21; fermion

(iv) Photons: integer spins; bosons

(v) H_2: number of protons (2) + number of electrons (2) = 4; boson

(vi) ^3He: number of protons (2) + number of neutrons (1) = 3; fermion

(vii) ^4He: number of protons (2) + number of neutrons (2) = 4; boson

Problem 15.44

Four indistinguishable particles are distributed among three energy levels without any restriction on the occupancy of each level. Deduce the total number of ways of distributing them.

Solution

$$W_{Bosons} = \frac{(g_i + n_i - 1)!}{n_i!\,(g_i - 1)!} = \frac{(3+4-1)!}{4!\,(3-1)!} = 15$$

Problem 15.45

Three indistinguishable particles are distributed among four energy levels. Obtain the number of arrangements possible if the particles are (i) fermions and (ii) bosons.

Solution

(i) $W_{F-D} = \frac{g!}{n!(g-n)!} = 4$

(ii) $W_{B-E} = \frac{(g+n-1)!}{n!(g-1)!} = 20$

Problem 15.46

Assume that three particles are to be accommodated in five different energy levels in the following manner,

Level 1	Level 2	Level 3	Level 4	Level 5
occupied	empty	occupied	occupied	empty

Obtain the probability of obtaining the above arrangements if (i) F–D (ii) M–B and (iii) B–E statistics is obeyed.

Solution

(i) In F–D statistics, the total number of ways of obtaining this configuration is

$$\frac{g!}{n!\,(g-n)!} = \frac{5!}{3!\,2!} = 10$$

Hence the probability is 1/10.

(ii) In M–B statistics, the total number of ways are $g^n = 5^3 = 125$. Since the particles are distinguishable, there 3! ways of obtaining the arrangement. Hence the probability is 6/125.

(iii) In B–E statistics, the total number of ways of obtaining this arrangement is

$$\frac{(g+n-1)!}{n!\,(g-1)!} = \frac{7!}{3!\,4!} = 35$$

Hence the probability is 1/35.

Problem 15.47

Consider a system made up of (i) two fermionic species distributed in four energy levels and (ii) four bosons distributed in three energy levels. Estimate the number of microstates in each case.

Solution

(i) For fermions, $W = \frac{g!}{n!(g-n!)} = \frac{4!}{2!2!} = 6$

(ii) For bosons, $W = \frac{(g+n-1)!}{n!(g-1)!} = \frac{6!}{4!2!} = 15$

Problem 15.48

Estimate the occupation number for fermions occupying an energy level of 1.6×10^{-21} J at 298 K.

Solution

Mean occupation number for fermions

$$= n_i^{FD} = \frac{1}{e^{\beta(E_i-\mu)}+1} = \frac{1}{1+e^{-1.6\times10^{-21}/kT}} = 0.60$$

Problem 15.49

Estimate the electronic contribution to molar entropy of oxygen whose ground state is triply degenerate.

$$S_{\text{electronic}} = R \ln 3 = 9.134 \text{ J K}^{-1} \text{ mol}^{-1}$$

Equilibrium Constants

Problem 15.50

Derive the expression for the equilibrium constant K_p for (i) A(g) $\rightleftharpoons$ B(g) and (ii) aA + bB $\rightleftharpoons$ cC + dD using partition functions.

Solution

(i) Let N denote the total number of molecules.

$$N = N(\text{A}) + N(\text{B})$$

If the number of molecules of A in the ith level is denoted as N_i, the distribution law is given by

$$\frac{N_i(A)}{N} = \frac{e^{-\beta E_i(A)}}{q} \quad (1)$$

where E_i denotes the energy of the molecules A in the ith level. Analogously,

$$\frac{N_j(B)}{N} = \frac{e^{-\beta E_i(B)}}{q} e^{-\beta \Delta E}$$

where ΔE denotes the difference in the ground state energies of A and B. From Eq. (1),

$$\Sigma N_i(A) = \frac{N \Sigma e^{-\beta E_i(A)}}{q} = \frac{N q_A}{q}$$

where q_A denotes the partition function of A.

$$N(A) = \frac{N q_A}{q}$$

Similarly $N(B) = \dfrac{N q_{(B)}}{q} e^{-\beta \Delta E}$

The equilibrium constant $K_{(N)}$ is given by

$$K_{(N)} = \frac{N(B)}{N(A)}$$

$$= \frac{q(B)}{q(A)} e^{-\beta \Delta E}$$

If B is a dimer, the partition function will include translational, vibrational and rotational contributions.

Similarly, if aA + bB $\rightleftharpoons$ cC + dD, the equilibrium constant is

$$K_N = \frac{[q(\text{C})]^c [q(\text{D})]^d}{[q(\text{A})]^a [q(\text{B})]^b}$$

Employing the relation between K_p and K_c,

$$K_P = \frac{(q(D)/V)^d [q(C)/V]^c}{[q(A)/V]^a [q(B)/V]^b} (kT)^{c+d-a-b} e^{-\beta \Delta E}$$

where Δn denotes the change in the number of moles between the products and reactants = $(c + d - a - b)$

Problem 15.51

Write the equation for the equilibrium constant of the dissociation

$$2\,\text{X}(g) \rightleftharpoons \text{X}_2(g)$$

incorporating the ground state energy and degeneracy of X_2.

$$K_p = \frac{[q_{tr}(X_2)][q_{rot}(X_2)][q_{vib}(X_2)][g_{el}(X_2)]e^{-\Delta E/RT}}{[q_{tr}(X)]^2 kT}$$

Miscellaneous Problem

Problem 15.52

Consider a system of solvent dipoles with the permanent dipole moments as $+p$ and $-p$ in an external electric field E. Derive the (i) canonical partition function and (ii) net polarisation. The total number of lattice sites is N_T and all sites are occupied by either of the two orientational states.

Solution

(i) The total number of dipoles = N_T

Number of dipoles with the dipole moment $-p = N_1$

Number of dipoles with dipole moment $+p = N_2$

Partition function $q = e^{pE/kT} + e^{-pE/kT}$

(ii) Mean polarisation = $p(N_1 - N_2) / N_T$

$$N_1 / N_T = e^{pE/kT} / q$$

$$N_2 / N_T = e^{-pE/kT} / q$$

$$(N_1 - N_2) / N_T = [e^{pE/kT} - e^{-pE/kT}] / [e^{pE/kT} + e^{-pE/kT}]$$

$$= \tanh [pE/kT]$$

EXERCISES

Practice Problems

1. Under identical temperatures, which of the following will be in gross agreement with the prediction from the equipartition principle for C_v?

   (i) H_2 (ii) N_2 (iii) I_2 (iv) N_3

2. Show that the vibrational contribution to the molar internal energy is given by

   $E_{vib} = \dfrac{RTh\nu}{2kT} \coth\left(\dfrac{h\nu}{2kT}\right)$ for N harmonic oscillators of vibrational frequency ν.

3. Four particles are distributed in equally spaced energy levels (0, $\varepsilon 1$, $\varepsilon 2$,.....$\varepsilon 6$) with a total energy of 6ε. Estimate the total number of (i) macro and (ii) microstates.

4. Classify the following species as fermions or bosons. (i) α-particles (ii) ^{12}C (iii) ^3He (iv) H_2

5. Show that the second virial coefficient $B_2(T)$ in terms of the van der Waals constants a and b is given by $B_2(T) = b - a/RT$.

6. A system of N particles has three different energy levels 0, $2kT$ and $3kT$ with the degeneracies 1, 1 and 2. Estimate the heat capacity of the system.

7. Estimate the (i) thermal wavelength and (ii) mean distance for electrons in Na at 300 K using $m = 9.1\times10^{-28}$ g and number density as 2.5×10^{22} molecules cm^{-3}.

8. Estimate the (i) characteristic rotational temperature and (ii) partition function at 300 K for HCl using the bond length as 1.37 Å.

9. Estimate the (i) rotational partition function and (ii) internal energy for O_2 at 100 K using the bond length as 1.21 Å.

10. In a microcanonical ensemble of N molecules (containing the total number of microstates W), the volume V is increased by an infinitesimal amount dV. Deduce the equation for NdV/V.

11. Derive the mean kinetic and potential energies of N one-dimensional harmonic oscillators.

12. Derive the Gibbs free energy per particle for an ideal monoatomic gas.

13. Among the following, which has the largest symmetry number? Which of the following will have the largest rotational partition function, assuming the high temperature approximation to be valid?
(i) CO_2 (ii) NH_3 (iii) Cyclohexane (iv) Benzene

14. Estimate the (i) de Broglie wavelength and (ii) translational entropy of one mole of CO at 300 K and 1 atm pressure.

15. The vibrational frequencies (in 10^{13} s^{-1}) of a molecule at 298 K are) as follows: 5.4, 1.5 (doubly degenerate) and 3.6. Estimate the partition function.

16. The moment of inertia of a homonuclear diatomic molecule is 1.407×10^{-43} g m^2. Estimate the rotational partition function.

17. Estimate the translational partition function for Ar at 300 K, in a volume of 1 cm^3.

18. The characteristic vibrational temperature of a diatomic molecule is 810 K. Estimate the vibrational contribution to (i) heat capacity and (ii) entropy, at 298 K.

19. Deduce the rotational heat capacity of *para* and *ortho* forms of hydrogen in the ratio 1:3 at 298 K.

20. In a two-level system of N independent harmonic oscillators, predict the entropy as $T \to \infty$.

21. Calculate the molar entropy of H_2 at 25°C assuming the translational contribution alone.

22. In a system of 3 fermions with a total energy of 3 $h\nu$, four energy states exist—0, $h\nu$, $2h\nu$, $3h\nu$... Deduce the total number of micro and macrostates.

23. Write the equation for the partition function for N harmonic oscillators in (i) two and (ii) three dimensions.

24. Calculate the ratio of translational partition functions of H_2 and D_2 molecules at 400 K in a cube of edge length of 2 cm.

25. Fill up the blanks
At ________ temperatures and ________ densities, Maxwell–Boltzmann statistics will be valid for all particles.

26. The equation of state corresponding to Langmuir adsorption isotherm may be represented as

$$P = -RT\Gamma_{max} \ln\left(1 - \frac{\Gamma}{\Gamma_{max}}\right)$$

where Γ denotes the relative surface excess and Γ_{max} is the maximum surface excess. Using the above equation, derive the corresponding canonical partition function.

27. Derive the mean occupation number for particles obeying the Maxwell–Boltzmann equation.

Multiple Choice Questions

1. The molar heat capacity at constant volume of solids, varies as T^3 at low temperatures according to
(a) Debye's theory
(b) Einstein's theory
(c) Dulong and Petit's law
(d) Rayleigh–Jeans theory

2. Consider 1 mol of helium at 1 atm pressure and at 298 K. The ratio between the number of molecules in j^{th} state (N_j) and the corresponding degeneracy (g_j), namely, N_j/g_j is approximately equal to
(a) 10^{-6} **(b)** 10^{-4}
(c) 10^{-2} **(d)** 10^{2}

3. The characteristic vibrational frequency of N quantum harmonic oscillators is 8.1×10^{12} radians/s. The fraction of molecules in the ground state is 0.87. The temperature (in K) corresponding to this fraction is nearly
(a) 1200
(b) 600
(c) 300
(d) 200

4. The number of molecules (N_1) in the ground state (of energy ε_1) is 2.74 times `the number of molecules (N_2) in the first excited state (of energy ε_2) for particles obeying Maxwell–Boltzmann statistics at 100 K. The difference in energies ($\varepsilon_1 - \varepsilon_2$) in electron volts is
(a) –10.5 (b) –0.1
(c) –8.6 (d) –0.001

5. For particles obeying Fermi–Dirac statistics, the ratio (n_j/g_j) at the Fermi level is equal to
(a) 1/2 (b) 2/3
(c) 1/4 (d) 1/8

6. The energy of one-dimensional quantum linear oscillators of frequency v in the $n = 3$ state is given by
(a) $7hv/2$ (b) $hv/2$
(c) 0 (d) $5\pi hv/2$

7. A canonical ensemble of N molecules has two energy levels designated as $-J$ and $+J$ with the number of molecules being N_1 and N_2 respectively. The mean internal energy of the system at T is
(a) 0 (b) $J \sinh (J/kT)$
(c) $J \cosh (J/kT)$ (d) $J \tanh (J/kT)$

8. Which of the following has the largest characteristic rotational temperature?
(a) H_2 (b) HCl
(c) I_2 (d) CO

9. When the energy level of bosons is equal to the chemical `potential of a system, the occupation number of the system becomes
(a) 1 (b) ∞
(c) ½ (d) 0

10. The vibrational heat capacity predicted by the equipartition principle can be deduced from the corresponding partition functions when
(a) $T \to 0$
(b) $T >>> 0$
(c) $(\theta v/T) >>> 0$
(d) de Broglie wavelength >>> 0

11. Which of the following is a characteristic feature in the Debye model for the specific heat of solids?
(a) All the oscillators have the same vibrational frequency
(b) All the oscillators are anharmonic
(c) The oscillators possess characteristic maximum frequencies
(d) The frequency of the oscillators can be ignored

12. Which of the following equations represents the distribution function for particles obeying Bose–Einstein statistics?
(a) $\varepsilon_j = \mu + kT \ln [\,1 + (g_j/N_j)]$
(b) $\varepsilon_j = kT \ln (g_j/N_j)$
(c) $\varepsilon_j = \mu + kT \ln [(g_j/N_j) - 1]$
(d) $\varepsilon_j = \mu - kT \ln [(g_j/N_j) - 1]$

13. Consider a system of four bosons distributed in three energy levels. The number of microstates is
(a) 81 (b) 64
(c) 6 (d) 15

14. Let the Lagrange multipliers be denoted as α and β. The Fermi–Dirac and Bose–Einstein distribution laws will lead to Maxwell–Boltzmann distribution when
(a) $\alpha = 0$ and $\beta = 0$
(b) $\alpha \neq 0$ and $\beta = 0$
(c) $(-\alpha + \beta\varepsilon_j) >>> 0$
(d) $\log \alpha = -\beta\varepsilon_j$

15. The Sackur–Tetrode equation for the translational partition function may be written as
$\ln (Q_{trans}) = A \ln M + B \ln T - \ln P + 51.104$
The values of A and B are respectively
(a) 2.5 and 1.5 (b) 1.33 and 4.5
(c) 1.5 and 2.5 (d) 3.5 and 4.5

16. The symmetry numbers of (i) SF_6 and (ii) CH_4 are respectively
(a) 6 and 4 (b) 8 and 12
(c) 3 and 24 (d) 24 and 12

17. The translational partition function for 1 mol of helium in a volume of 25 L at 298 K is equal to (in 10^{29})
(a) 2.0 (b) 100
(c) 0.5 (d) $25^{1/3}$

18. Among the following, which compound has all the three moments of inertia equal?
(a) NH_3 (b) CCl_4
(c) CH_2Cl_2 (d) C_6H_6

19. The residual molar entropy of 1,2-difluorobenzene is expected to be
(a) $R \ln 2$ (b) $R \ln 3$
(c) $R \ln 6$ (d) $R \ln 12$

20. Assume that three indistinguishable fermions are to be accommodated in five different energy levels in the following manner of occupancy:

Level 1	Level 2	Level 3	Level 4	Level 5
occupied	vacant	occupied	occupied	vacant

The probability of obtaining the above arrangement is
(a) 1/10 (b) 8/125
(c) 1/35 (d) 1/27

21. The difference in the molar Gibbs free energy of He and Ar, namely, $(G_{m,\text{He}} - G_{m,\text{Ar}})/RT$, at 298 K and 1 atm pressure is
(a) 1.75 (b) 2.55
(c) 3.45 (d) 6.85

22. Consider a one dimensional harmonic oscillator of energy 0.9 kT. The ratio between the classical (q_{cl}) and quantum mechanical (q_{qm}) partition function, (q_{cl}/q_{qm}) is nearly
(a) 0.45 (b) 0.36
(c) 2.75 (d) 1.20

23. The method of Lagrange multipliers in deriving various distribution laws is an example of
(a) Unconstrained optimisation
(b) Constrained optimisation
(c) Eigen value–eigenvector problem
(d) Non-linear programming

24. A macrostate of 4 distinguishable particles is distributed in 4 energy levels with one particle in each. The degeneracy of each level is 1, 2, 2 and 1 respectively. The number of microstates is
(a) 24 (b) 6
(c) 256 (d) 96

25. The statistical weights of *ortho* and *para* deuterium are respectively
(a) 3/4 and 1/4 (b) 1/2 and 1/2
(c) 2/3 and 1/3 (d) 1/4 and 3/4

26. The ground state of O_2 is triply degenerate. Its absolute entropy is
(a) Zero (b) $R \ln 3$
(c) $R \ln 3^2$ (d) $R \ln (1/3)$

27. Identify the solid with the largest Debye temperature:
(a) graphite (b) Cu
(c) Au (d) diamond

28. A completely isolated system with adiabatic and impermeable boundaries is a
(a) canonical ensemble
(b) microcanonical ensemble
(c) isothermal–isobaric ensemble
(d) grand canonical ensemble

29. The particles obeying anti-symmetric wave functions with maximum occupation number of unity are
(a) bosons (b) fermions
(c) boltzons (d) photons

30. Dulong and Petit's law for specific heat at constant volume (C_v) of solids is a special case of Debye's theory whenever the Debye temperature is
(a) much larger than T
(b) much smaller than T
(c) equal to T
(d) equal to $h\nu/k$

Answer Keys

Practice Problems

1. (iii) I_2

3. (i) 9 (ii) 84

4. Bosons: (i) and (ii); Fermions: (iii) and (iv)

6. $\dfrac{C_v}{k} = \dfrac{2e}{(1+e)^2}$

7. (i) $\lambda \simeq 43$ Å and (ii) $\langle l \rangle \simeq 3.4$ Å

8. (i) $\theta_{rot} \cong 15$ K; (ii) $Z_{rot} = \dfrac{300}{15} = 20$

9. (i) $\theta_{rot} = 2.1$ K and $Z_{rot} = 47.6$

10. dW/W

11. $N kT/2$

12. Theoretical question

13. (iv)

14. (i) 1.97×10^{-9} cm (ii) 7.727×10^{-9} cm

15. 1.213

16. 52.1

17. 2.4×10^{26}

18. (i) 4.65 J K^{-1} (ii) 2.17 J K^{-1}

19. Theoretical question

20. $R \ln 2$

21. 118.05 J K^{-1}

22. Macrostates = 2 and microstates = 13

23. For an one-dimensional harmonic oscillator, the partition function is q_1. For two and three dimensional isotropic harmonic oscillators, the partition functions are respectively $(q_1)^2$ and $(q_1)^3$.

24. 2.83

25. High temperatures and low densities

26. Theoretical question

27. Theoretical question

Multiple Choice Questions

1. (a)	2. (a)	3. (a)	4. (c)	5. (a)	6. (a)	7. (d)	8. (a)	9. (b)	10. (b)
11. (c)	12. (a)	13. (d)	14. (c)	15. (c)	16. (d)	17. (a)	18. (b)	19. (c)	20. (a)
21. (c)	22. (d)	23. (b)	24. (d)	25. (c)	26. (b)	27. (d)	28. (b)	29. (b)	30. (a)

Miscellaneous Problems

Worked Out Problems

Problem 1

A system of heat capacity $C_{p,1}$ at a lower temperature (T_{lower}) is brought into thermal contact with another system (heat capacity = $C_{p,2}$) at a higher temperature (T_{higher}). Deduce the entropy change of the process, assuming the heat capacities to be independent of temperature.

Solution

From the heat balance equation,

$$C_{p,1}(T_{final} - T_{lower}) = C_{p,2}(T_{higher} - T_{final})$$

$$T_{final} = (C_{p,1} T_{higher} + C_{p,2}\, T_{lower})/2$$

$$\Delta S_{sys} = \int_{T_{lower}}^{T\,final} C_{p,1}\frac{dT}{T} + \int_{T_{higher}}^{T\,final} C_{p,2}\frac{dT}{T}$$

$$\Delta S = C_{p,1}\ln\frac{T_{final}}{T_{lower}} + C_{p,2}\ln\frac{T_{final}}{T_{higher}}$$

$$\Delta S = (C_{p,1} + C_{p,2})\ln\frac{T_{final}}{T_{gm}}$$

where T_{gm} is the weighted geometric mean given by

$$T_{gm} = [(T_{lower})^{C_{p,1}} \times (T_{higher})^{C_{p,2}}]^{1/(C_{p,1}+C_{p,2})}$$

If $C_{p,1} = C_{p,2}$, $T_{gm} = (T_{lower} \times T_{higher})^{1/2}$

Hence the entropy change of the system in this case is given by

$$\Delta S = 2C_p\,\ln(T^{am}/T_{gm})$$

where T^{am} is the arithmetic mean of the two temperatures.

Problem 2

Explain the Clausius inequality.

Solution

One of the methods of representing the Clausius inequality is using the entropy change dS of a system, in terms of the infinitesimal heat δQ and absolute temperature T as

$$dS \geq \delta Q/T$$

the equality sign being valid only for reversible processes. In particular, the Clausius inequality implies that the entropy change of a system is either positive or zero. Entropy change of the universe (system + surroundings) is either positive or 0. Consequently, it is possible to formulate the rate of change of entropy with time, for irreversible processes as d_iS/dT where d_iS denotes the entropy change of any irreversible process. If a process is carried out reversibly, $d_iS = 0$.

Problem 3

The hydrolysis of adenosine triphosphate using glucokinase follows the Michaelis–Menten mechanism. The rate of the reaction is 0.133 mol s^{-1} if the substrate concentration is 0.01 M. Estimate the maximum velocity of the reaction when the Michaelis–Menten constant is 0.005 M.

Solution

$$v = v_{max}\,[S]/(K_M + [S])$$

$$v_{max} = 0.15 \times 0.133/0.01 \approx 0.2 \text{ mol s}^{-1}$$

Problem 4

Consider the following mechanism

$$\mathrm{S + E} \underset{k_{-1}}{\overset{k_1}{\rightleftharpoons}} \mathrm{ES} \underset{k_{-2}}{\overset{k_2}{\rightleftharpoons}} \mathrm{B + E}$$

Write the equation for d[ES]/dt.

Solution

$$d[\mathrm{ES}]/dt = k_1[\mathrm{S}][\mathrm{E}] - k_{-1}[\mathrm{ES}] - k_2[\mathrm{ES}] + k_{-2}[\mathrm{B}][\mathrm{E}]$$

Problem 5

Derive the velocity for the Michaelis–Menten mechanism, using the rapid pre-equilibrium approximation.

Solution

$$\mathrm{E + S} \underset{k_{-1}}{\overset{k_1}{\rightleftharpoons}} \mathrm{ES} \xrightarrow{k_2} \mathrm{E + P}$$

The rapid pre-equilibrium approximation becomes valid when the first-order rate constant k_2 is much smaller than the first-order constant k_{-1}. In this case, the association reaction E + S → ES is faster than the reaction ES → E + P. Hence, the assumption of a rapid equilibrium is invoked. Thus

$$k_{-1}[\mathrm{ES}] = k_1[\mathrm{E}][\mathrm{S}]$$

$$[\mathrm{ES}] = (k_1/k_{-1})[\mathrm{E}][\mathrm{S}] = (k_1/k_{-1})([\mathrm{E}]_0 - [\mathrm{ES}])[\mathrm{S}]$$

where $[\mathrm{E}]_0$ denotes the initial concentration of the enzyme.

$$[\mathrm{ES}] = (k_1[\mathrm{S}][\mathrm{E}]_0)/\{k_{-1} + k_1[\mathrm{S}]\}$$

Hence the velocity of the reaction is given by

$$v = k_2[\mathrm{ES}] = k_2\,k_1\,[\mathrm{S}][\mathrm{E}]_0/\{k_{-1} + k_1\,[\mathrm{S}]\}$$

Problem 6

The reduction of Ni^{2+} to Ni is studied on a Pt electrode. The diffusion limited current density is 0.12 A cm^{-2}. The bulk concentration of Ni^{2+} is 0.001 mol cm^{-3}. Assuming that the diffusion coefficient of Ni^{2+} is 10^{-5} cm^2 s^{-1}, estimate (i) diffusion layer thickness (ii) diffusion flux and (iii) mass transfer rate constant.

Solution

(i) Diffusion limited current density $i_{\text{lim}} = \dfrac{nFDC_{\text{bulk}}}{\delta}$

where δ denotes the diffusion layer thickness.

$$0.12\,\text{A cm}^{-2} = \frac{2 \times 96500\,\text{C mol}^{-1} \times 10^{-5}\,\text{cm}^2\,\text{s}^{-1} \times 0.001\,\text{mol cm}^{-3}}{\delta}$$

$\delta = 0.016$ cm

(ii) For one-dimensional transport, the diffusion flux is given by

$$J_{diff} = -D\frac{dc}{dx}$$

$$= \frac{-D[C_{x=0} - C_{\text{bulk}}]}{[\delta - 0]}$$

Since the surface concentration is 0, on account of the faster activation control,

$$J_{diff} = \frac{DC_{\text{bulk}}}{\delta} = \frac{10^{-5}\,\text{cm}^2\,\text{s}^{-1} \times 0.001\,\text{mol cm}^{-3}}{0.016\,\text{cm}}$$

$$= 6.25 \times 10^{-7}\,\text{mol cm}^{-2}\,\text{s}^{-1}$$

An alternate method of obtaining the diffusion flux is by writing the current density as $i_{diff} = nFJ_{diff}$. The flux indicates the number of moles of the species reaching the surface ($x = 0$) per unit area per unit time, by diffusion.

(iii) Mass transfer rate constant = $k_{\text{mass transfer}} = D/\delta = 10^{-5}$ cm^2 s^{-1}/0.016 cm
= 6.25×10^{-4} cm s^{-1}

Problem 7

The transfer coefficient for the reaction $Ni^{2+} + 2e^- \rightarrow Ni$ is 0.4 at 298 K. What will be the activation overvoltage required to enhance the current density by one order of magnitude?

Solution

Since the reduction reaction is considered, the Tafel equations to be employed are

$$\eta 1 = \left[\frac{RT}{\alpha nF}\right][-2.303\log(i1) + 2.303\log i_0]$$

$$\eta 2 = \left[\frac{RT}{\alpha nF}\right][-2.303\log(i2) + 2.303\log i_0]$$

$$(\eta 2 - \eta 1) = -2.303\left[\frac{RT}{\alpha nF}\right][\log i2 - \log i1]$$

$$= -2.303 \times \left[\frac{RT}{\alpha nF}\right]\log(10)$$

$$= -\frac{2.303 \times 8.314 \times 300}{0.4 \times 2 \times 96500} = -0.0744\text{ V}$$

The overvoltage should be made more negative by 74.4 mV.

Problem 8

The bulk electrolysis of nitrobenzene (concentration 1 M) was carried out using a graphite electrode (area 10 cm^2) in a volume of 0.1 cm^3. The diffusion coefficient is 10^{-5} cm^2 s^{-1} while the diffusion layer thickness is 10^{-6} cm. Estimate the (i) first order rate constant, (ii) concentration of the reactant after 120 s.

Solution

(i) First order rate constant (k_{first}) = $V\delta/Ad$ = $0.1\times10^{-6}/(10\times10^{-5})$
= 0.001 s^{-1}

(ii) Concentration of the reactant remaining = $C_{\text{bulk}}\,e^{(-kt)}$ = $1\,e^{(-0.001\times120)}$
= 0.89 M

Problem 9

Derive the equation connecting centimetre and farad.

Solution

There are two different units for describing the capacitance. Since the capacitance is defined as charge/potential, the natural unit is farad (coulomb/volt). Alternately, potential is proportional to charge × distance and hence capacitance can be expressed as 1/distance (or length). Hence it is logical to expect that length^{-1} should be related to farad. The conversion factor is provided by the permittivity of vacuum (ε_0). Thus,

$$4\pi\varepsilon_0 = 1 \qquad \text{where } \varepsilon_0 = 8.854\times10^{-12}\text{ F m}^{-1}$$

Hence 1 cm = 1.113×10^{-12} farad

Problem 10

Calculate $\gamma\pm$ for a 0.01 molal solution of NaBr in water at 298 K, using the Debye–Hückel law. The distance of closest approach a is 0.37 nm while the parameter B is 3.281×10^9 kg$^{1/2}$ mol$^{-1/2}$ m^{-1} for water.

Solution

$$\log(\gamma\pm) = \left[\frac{-0.509(1)(0.1)}{1 + (1.21)(0.1)}\right] = -0.0454$$

Hence $\gamma\pm = 0.9$

Problem 11

Show that the characteristic vibrational temperature (in K) can be written as 1.44ω where ω is in cm^{-1}.

Solution

$$\theta_{vib} = h\nu/k = hc\omega/k = \frac{6.626\times10^{-34}\times3\times10^{10}}{1.38\times10^{-23}}\omega = 1.44\,\omega$$

Problem 12

Estimate the weight of benzoyl peroxide required to produce 2000 g of polyethylene of molecular weight 2×10^5 g mol^{-1}. Assume the complete utilisation of the initiator and absence of any side reactions.

Solution

Number of ethylene molecules = $(6.023\times10^{23}\times2000)/28$ = 4.3×10^{25}

Degree of polymerisation = $(2\times10^5/28) \simeq 7143$

Number of molecules of the initiator = $(4.3\times10^{25})/7143 \simeq 6\times10^{21}$

Weight of the initiator required = $(242\times6\times10^{21})/(6.023\times10^{23})$ = 2.4 g

Practice Problems

1. Estimate the constants A and B in the Debye–Hückel–Onsager equation for molar conductance for 1:1 electrolytes at 298 K when the solvent is acetone (dielectric constant = 20.7 and viscosity = 0.0032 poise)
2. Estimate the liquid junction potential between 0.10 M NaBr and 0.01 M NaBr solutions when the transport number of Na^+ ions is 0.49.
3. The hydrodynamic radius of a spherical macromolecule is 31 Å in water. If the diffusion coefficient of the species is 7×10^{-7} cm^2 s^{-1}, calculate the kinematic viscosity of water.
4. Show that the standard Gibbs free energy change (ΔG^0) for the reaction

$H_2 + ½\, O_2 \rightarrow H_2O$ is -237 kJ mol^{-1}.

Multiple Choice Questions

1. The Gibbs free energy is a
   (a) concave function with respect to T
   (b) convex function with respect to P
   (c) convex function with respect to both T and P
   (d) concave function with respect to both T and P

2. For 1 mol of an ideal gas, $[(\partial F/\partial T)_P + (\partial G/\partial T)_V]$ is equal to
   (a) $-S-R$ (b) $S+R$
   (c) $-2S$ (d) $-S+R$

3. The vapour pressures of X(s) and X(l) are given by the equations
   ln P (in torr) = 23.03 – 3750/T (in K) and
   ln P (in torr) = 19.49 – 3060/T (in K)
   From these data, the approximate triple point of X (in K) can be estimated as
   (a) 195 (b) 163
   (c) 157 (d) 273

4. $[(\partial C_v/\partial V)_T = 0$ is valid only for
   (a) ideal gases
   (b) van der Waals gases
   (c) PV^γ = constant
   (d) virial equation of state

5. Which of the following represents the Gibbs–Duhem equation?

   (a) $0 = -SdT + VdP + \sum_{i=1}^{k} n_i \cdot d\mu_i$

   (b) $dE = TdS - PdV + \sum_{i=1}^{k} \mu_i \cdot dn_i$

   (c) $dG = Vdp - SdT + \sum_{i=1}^{k} \mu_i \cdot dn_i$

   (d) $dF = -PdV - SdT + \sum_{i=1}^{k} n_i \cdot d\mu_i$

6. For 1 mole of an ideal gas undergoing isentropic expansion, $(\partial P/\partial T)_V$ is
   (a) 0 (b) C_P/V
   (c) C_V/V (d) R

7. The electrolysis of HCl using graphite electrodes yields
   (a) Cl_2 at anode and H_2 at cathode
   (b) H_2 at anode and Cl_2 at cathode
   (c) H^+ at anode and OH^- at cathode
   (d) Cl^- at anode and H^+ at cathode

8. The exchange current density for hydrogen evolution reaction at 300 K at the below mentioned electrodes, follows the trend,
   (a) Fe > Zn > Pt > Ag (b) Pt > Ag > Fe > Zn
   (c) Ag > Fe > Pt > Zn (d) Pt > Fe > Zn > Ag

9. The magnitude of the liquid junction potential between 0.1 M HCl and 0.1 M KCl is (in mV) approximately
   (a) 5 (b) 0
   (c) 59 (d) 28

10. The formation constant of Cu^{2+} with NH_3 is nearly 1.3×10^8 at 298 K. The half-wave potential for the reduction of Cu^{2+} is +0.02 V vs SCE. If 1 mol dm^{-3} concentration of NH_3 is added to the solution containing Cu^{2+}, the new half wave potential (in V) will be
    (a) +0.02 (b) –0.02
    (c) –0.22 (d) –0.24

11. Given that the half-cell potential for Fe^{3+}/Fe^{2+} is 0.77 V and that for MnO_4^-/Mn^{2+} is 1.51 V when pH is 0, what will be the approximate pH of the solution at the equivalence point during the potentiometric titration of Fe^{2+} with MnO_4^-?
    (a) 1.4 (b) 2.3
    (c) 4.6 (d) 7.0

12. The formal reduction potential of a reaction will be equal to the standard electrode potential if
    (a) temperature is constant
    (b) ionic strength is kept constant
    (c) the reaction is an elementary step
    (d) all the activity coefficients are unity

13. An electrochemical reaction will be diffusion controlled if it is
    (a) faster than activation control
    (b) slower than activation control
    (c) equal to activation control
    (d) slower than tunnelling

14. The hydrogen evolution reaction on MoS_2 nanoparticles grown on graphene was studied using Tafel polarisation. A Tafel slope of ~40 mV per decade was obtained. It indicates the validity of
(a) Volmer–Tafel mechanism
(b) Temkin isotherm
(c) Frumkin isotherm
(d) Volmer–Heyrovsky mechanism

15. The standard electrode potential of $Ag/AgCl/Cl^-$ vs standard hydrogen electrode (SHE) is 0.22 V. The potential of the saturated calomel electrode (SCE) is +0.24 V vs SHE. The potential of $Ag/AgCl/Cl^-$ with respect to SCE is
(a) +0.46 V (b) –0.46 V
(c) –0.02 V (d) +0.02 V

16. The intracellular and extracellular concentrations of Ca^{2+} ions in mammalian skeletal muscle are respectively 0.0001 mM and 1.5 mM. According to the Nernst equation, the potential difference (in mV) at 310 K is nearly
(a) 130 (b) –98
(c) –28 (d) 59

17. Which of the following is true at temperatures below the triple point of water?
(a) V(solid) < V(liquid)
(b) V(solid) = V(liquid)
(c) V(solid) > V(liquid)
(d) V(gas) < V(liquid)

18. The number(s) of degrees of freedom (F) for the water–formic acid system which exsits as an azeotrope is
(a) 2 (b) 0
(c) 3 (d) 1

19. The number(s) of degrees of freedom (F) and phases (P) for the decomposition of $CaCO_3(s)$ are respectively
(a) 1 and 3 (b) 2 and 2
(c) 1 and 2 (d) 2 and 3

20. The relaxation time of the ionic atmosphere for 1:1 electrolytes is
(a) independent of concentration
(b) directly proportional to molar conductivity
(c) inversely proportional to molar conductivity
(d) directly proportional to the degree of dissociation

21. The largest magnitude of A in the Debye–Hückel–Onsager equation for the molar conductance of strong electrolytes at a given temperature will be for
(a) water (b) acetonitrile
(c) methanol (d) nitrobenzene

22. Consider the specific conductivity of 0.01 molar solutions of HCl, NaCl, KCl and LiCl at 298 K. The electrolyte with the highest specific conductivity among these is
(a) NaCl (b) HCl
(c) KCl (d) LiCl

23. Among the following types of electrolytes, the thickness of the ionic atmosphere at a given concentration and temperature, will be the largest for
(a) 1:1 (b) 1:2
(c) 1:3 (d) 2:2

24. The critical distance for the ion pair formation for 1:1 electrolytes, dissolved in a solvent of dielectric constant 10 at 298 K is (in nm)
(a) 7.3 (b) 2.8
(c) 3.5 (d) 5.3

25. The solubility of $PbCl_2$ in water is 0.01 molal at 298 K. In the presence of 0.1 molal NaCl, its solubility will
(a) decrease
(b) increase
(c) remain unaffected
(d) increase due to the common ion effect

26. The constant A in the Debye–Hückel limiting law for log ($\gamma\pm$) is –0.059 for
(a) solvents of dielectric constant 78.4 and T = 298 K
(b) all solvents at T = 298 K
(c) solvents of dielectric constant 78.4 and at all temperatures
(d) 1:1 electrolytes with solvent dielectric constant 78.4 and T as 298 K

27. Among the various concentrations (in mM) of 1:1 electrolytes given below, the thickness of the ionic atmosphere will be largest for
(a) 100 (b) 1.0
(c) 10 (d) 0.1

28. The function of the proton exchange membrane in methanol fuel cells is to
(a) enable the oxidation of methanol at the anode
(b) prevent the cross-over of methanol from the anode to the cathode
(c) allow the transfer of H^+ ions across the membrane
(d) accelerate the hydrogen evolution reaction

29. The elastic behaviour of rubber arises from
(a) uncoiling of molecules due to the application of a force
(b) random movement of molecules
(c) chemical potential gradient between the initial and final states
(d) first order phase transition

30. According to the regular solution approximation, the entropy of mixing ($\Delta S_{mixing}/k_B$) for a copolymer with equal volume fractions is (k_B denotes the Boltzmann constant) is
(a) 0.07 (b) 0.38
(c) 0.69 (d) 0.89

31. Consider a polymer solution wherein the volume fraction of the polymer is ϕ. The probability of obtaining the polymer–solvent contacts is proportional to
(a) ϕ^2 (b) $(1-\phi)$
(c) $(1-\phi)^2$ (d) $\phi(1-\phi)$

32. Among the following, which exhibits the lowest critical solution temperature?
(a) poly (N-isopropyl acrylamide)–water
(b) polystyrene–cyclohexane
(c) polymethylmethacrylate–acetonitrile
(d) polyethylene–diphenyl ether

33. For a polymer of n chains, the number average molecular weight (M_n) follows a Poisson distribution given by $n_i/n = e^{-a}a^i/i$. The polydispersity index will be always
(a) >1 (b) $=1$
(c) e^a (d) $1/a$

34. The probability of finding a quantum one-dimensional harmonic oscillator in the classically forbidden region is nearly
(a) 0 (b) 0.5
(c) 0.2 (d) 0.9

35. For the wave function represented by $\psi(x) = Ae^{\frac{-x^2}{4a^2}}$, the normalisation constant A is
(a) $(1/2\pi)^{1/2}$ (b) $(1/2\pi a^2)^{1/2}$
(c) $(1/2\pi a^2)^{1/4}$ (d) $(x^2/2\pi a^2)^{1/2}$

36. For a hydrogen-like atom, assume that the potential energy function $V(x)$ is proportional to x^n. The expectation values of the kinetic and potential energy are denoted respectively as $\langle T\rangle$ and $\langle V\rangle$. The validity of the virial theorem in this case implies that the ratio $\langle T\rangle/\langle V\rangle$ is
(a) n (b) $n/2$
(c) $1/n$ (d) n^2

37. Which of the following equations represents the one-dimensional van der Waals equation of state of gaseous species of density ρ? a and b denote the van der Waals' constants and l is the length of the molecule.
(a) $\frac{P}{\rho kT}+\frac{a^2}{l^2 kT}=\frac{1}{1-b\rho}$ (b) $\frac{P}{\rho kT}+\frac{a\rho}{kT}=\frac{1}{1-\rho l}$
(c) $\frac{P}{\rho kT}+\frac{\rho a}{kT}=1$ (d) $\frac{P}{\rho kT}+\frac{a}{v^2 kT}=1+b\rho$

38. The glass transition temperature T_g of polymers refers to the approximate Maxwell relaxation time (in seconds) of
(a) 10^3 (b) 10^2
(c) 10^{-3} (d) 10^{12}

39. The quantum mechanical treatment of a rigid rotator for diatomic molecules is not required whenever the characteristic rotational temperature is
(a) << 1 (b) = 1
(c) >> 1 (d) = 0

40. The inverse Debye length is denoted as κ_D. The excess internal energy per particles (E_{DH}/kT) predicted by the Debye–Hückel limiting law for a solution of density ρ is proportional to
(a) $-\kappa_D^3/\rho$ (b) $-\kappa_D/\rho$
(c) $-\kappa_D^2/\rho^2$ (d) $e^{-\rho\kappa_D}$

41. Which of the following refers to the radial distribution function $g(r)$ according to the Debye–Hückel theory? $1/\kappa_D$ denotes the Debye length and q is the charge on the central ion.
(a) $\frac{e^{-\kappa_D r}}{qr}$ (b) $\frac{qe^{-\kappa_D r}}{r}$
(c) $1/r$ (d) $e^{\frac{-q^2}{rkT}}e^{-\kappa_D r}$

Answer Keys

Practice Problems

1. $A = 32.8$ S cm^2 mol^{-1} M$^{-1/2}$, $B = 1.63$ M$^{-1/2}$
2. 39 mV
3. 1.78×10^{-6} m^2 s^{-1}

Multiple Choice Questions

1. (d)	**2.** (c)	**3.** (a)	**4.** (b)	**5.** (a)	**6.** (b)	**7.** (a)	**8.** (b)	**9.** (d)	**10.** (c)
11. (b)	**12.** (d)	**13.** (b)	**14.** (d)	**15.** (c)	**16.** (a)	**17.** (c)	**18.** (d)	**19.** (a)	**20.** (c)
21. (c)	**22.** (b)	**23.** (a)	**24.** (b)	**25.** (a)	**26.** (a)	**27.** (d)	**28.** (c)	**29.** (a)	**30.** (c)
31. (d)	**32.** (a)	**33.** (a)	**34.** (b)	**35.** (c)	**36.** (b)	**37.** (b)	**38.** (a)	**39.** (a)	**40.** (a)
41. (d)									

Bibliography

1. Arthur W Adamson, *Understanding Physical Chemistry*, Third Edition, The Benjamin Cummings Publishing Company, 1980.
2. Robert G Mortimer, *Physical Chemistry*, Third Edition, Elsevier, 2008.
3. Peter Atkins and Julio de Paula, *Physical Chemistry*, Ninth Edition, Oxford University Press, 2010.
4. Rob Phillips, Jane Kondev, Julie Theriot and Herman G Garcia, *Physical Biology of the cell*, Second Edition, Taylor and Francis, 2010.
5. Mandl F, *Statistical Physics*, Second Edition, John Wiley & Sons, 1988.
6. Gupta M C, *Statistical Thermodynamics*, Second Edition, New Age International, 2007.
7. Sears F W, Salinger G L, *Thermodynamics, Kinetic theory and Statistical Thermodynamics*, Third Edition, Addison Wesley, 1982.
8. Philip H Rieger, *Electrochemistry*, Second Edition, Chapman Hall, 1993.
9. Stephen Berry R, Stuart A Rice and John Ross, *Physical Chemistry*, Second Edition, Oxford University Press, 2000.
10. Manas Chanda, *Introduction to Polymer Science and Chemistry—A problem solving Approach*, CRC Press, 2006.
11. Bagotsky V S, *Fundamentals of Electrochemistry*, Second Edition, John Wiley & Sons, 2006.
12. Fowler R and Guggenheim E A, *Statistical Thermodynamics*, Cambridge University Press, 1965.
13. James H Espenson, *Chemical Kinetics and Reaction Mechanisms*, Second Edition, McGraw Hill, 1995.
14. Sidney W Benson, *The Foundations of Chemical Kinetics*, McGraw Hill, 1960.
15. Alfred Rudin, *Elements of Polymer Science and Engineering*, Second Edition, Academic Press, 1999.
16. Robert A Alberty, *Physical Chemistry*, Wiley Eastern Limited, 1991.
17. Kalidas C, *Chemical Kinetic Methods: Principles of Fast Reaction Techniques and Applications*, New Age International (P) Limited, 2005.
18. Strehlow H and Knoche W, *Fundamentals of Chemical Relaxation*, Verlag Chemie, 1977.
19. Linus Pauling and Bright Wilson E, *Introduction to Quantum Mechanics with Applications in Chemistry*, International Students Edition, 1935.
20. Leonard I Schiff, *Quantum Mechanics*, McGraw Hill, 1968.

Appendix 1

CSIR–UGC NET JRF/SET Solved Question Paper (June 2018)

1. The commutator of $\hat{x}$ with Hamiltonian $\hat{H}, [x, \hat{H}]$, is ________.

(a) 0 (b) $ih/2\pi$

(c) $\frac{-\hbar}{2m}\hat{p}_x$ (d) $\frac{i\hbar}{m}\hat{p}_x$

Solution: The Hamiltonian is the sum of the kinetic and potential energy operators, namely,

$$\hat{H} = \frac{P_x^2}{2m} + V(x)$$

in one dimension.

$$[\hat{x}, \hat{H}] = \left[\hat{x}, \frac{P_x^2}{2m}\right] + V(x)$$

$$= \left[\hat{x}, \frac{P_x^2}{2m}\right] + [\hat{x}; V(x)]$$

$$= \frac{1}{2m}\left[\hat{x}, P_x^2\right] + 0 = \frac{1}{2m} 2i\hbar\hat{p}_x = \frac{i\hbar}{m}\hat{p}_x$$

Answer: (d)

2. The lowest energy state of $1s^1\,2s^1$ electronic configuration, according to Hund's rule, is ________.

(a) $^3S_0$ (b) $^1S_0$

(c) $^3S_1$ (d) $^1S_1$

Solution: Since the electronic configuration is $1s^1\,2s^1$

$$L = l_1 + l_2 = 0 + 0$$

This yields the *S*-term.

$$S = s_1 + s_2 = \frac{1}{2} + \frac{1}{2} = 1$$

The degeneracy is $(2S + 1) = 3$

$$J = L + S = 0 + 1 = 1$$

Thus, the term symbol is, $^3S_1$

Answer: (c)

3. If the unit of the rate constant of a reaction is $L^3\ mol^{-3}\ s^{-1}$, the order of the reaction is ________.

(a) 1 (b) 2

(c) 3 (d) 4

Solution: The unit for an nth order reaction is $(mol\ L^{-1})^{1-n}\ s^{-1}$

Hence $n = 4$. The order of reaction is 4.

Answer: (d)

4. The standard cell potential for the reaction $Zn(s) + Cu^{2+}(aq) \rightarrow Zn^{2+}(aq) + Cu(s)$ is +1.10 V. The Gibbs free energy change during the reaction is ________.

($F = 96500$ coulomb mol^{-1})

(a) –21.2 kJ mol^{-1} (b) +212 kJ mol^{-1}

(c) –212 kJ mol^{-1} (d) –212 J mol^{-1}

Solution: The Gibbs free energy change is given by

$$\Delta G^0 = -nFE^0 = -2 \times 96500 \times 1.10\ J\ mol^{-1}$$

$$\Delta G = -212.3\ kJ\ mol^{-1}$$

Answer: (c)

5. The change in entropy for a reversible adiabatic process is ________.

(a) Maximum (b) Minimum

(c) Zero (d) positive

Solution: The entropy change for the reversible adiabatic process is given by $ds = \delta Q/T$. Since there is no heat change in the adiabatic process, $\delta Q = 0$.

Answer: (c)

6. The graph that represents the Langmuir adsorption isotherm is ________.

(a)

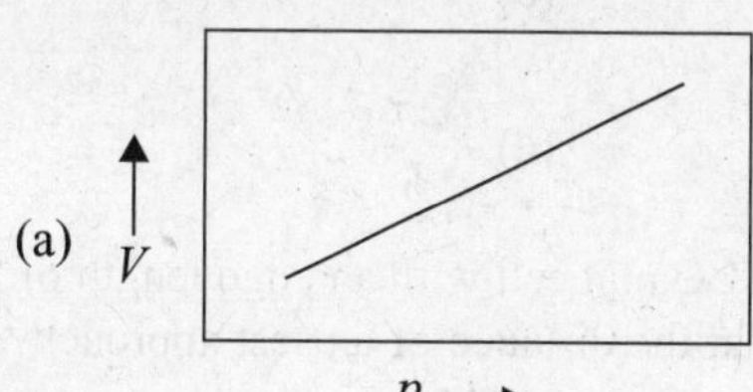

(b)

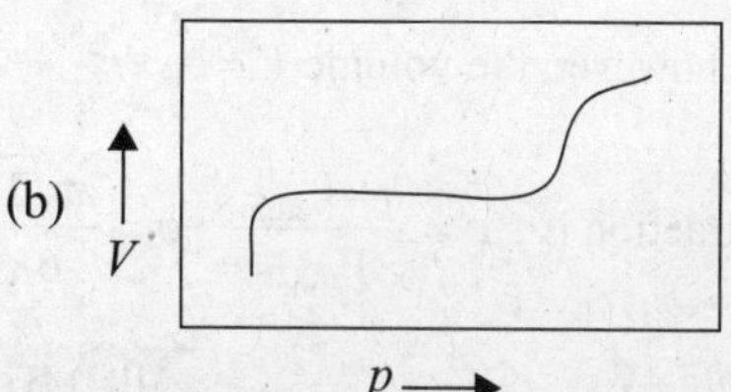

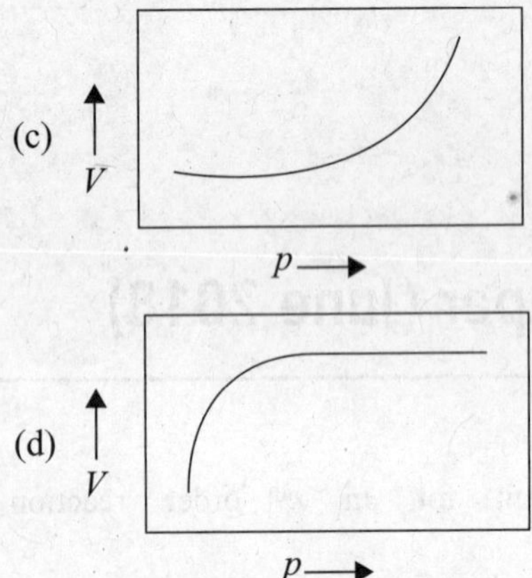

Solution: Since the Langmuir isotherm is given by $\theta/(1-\theta)$ = constant × pressure, while the surface coverage is directly proportional to the volume adsorbed, the correct dependence is given by (d).
Note that the variation depicted in (a) corresponds to Henry's adsorption isotherm.

Answer: (d)

7. Origin of the colligative properties of a dilute solution is ________.
(a) volatility of solute molecules
(b) interaction of solute–solvent molecules
(c) zero enthalpy of mixing
(d) entropy of mixing

Solution: The colligative properties are solely dictated by the number of particles in the solution. Since entropy is related to thermodynamic probability vis a vis number of arrangements, the entropy of mixing is the most appropriate answer.

Answer: (d)

8. If all the lattice points of an FCC structure are occupied by uniform hard spheres that touch each other, the fraction of volume occupied is ________.

(a) $\frac{\pi\sqrt{2}}{6}$ (b) $\frac{\pi\sqrt{3}}{6}$

(c) $\frac{\pi}{6}$ (d) $\frac{2\pi}{6}$

Solution: For an FCC unit cell with an edge length of 'a', it is well known that the distance of closest approach 'r' = face diagonal

Hence $r = \frac{a\cdot\sqrt{2}}{4}$; however, the volume $V = \frac{4}{3}\pi r^3$

Thus, the volume fraction is , $\phi = \frac{n\times V_{\text{atom}}}{V_{\text{cube}}}$; $\phi = \frac{\pi\sqrt{2}}{6}$

Answer: (a)

9. A 5 g L^{-1} polymer solution is prepared with a polymer whose molar mass is 25 kg. The osmotic pressure (in atm) of this solution at 25°C is ________.

(a) 0.002 (b) 0.05
(c) 0.005 (d) 0.008

Solution: Concentration of the polymer solution = 5 g L^{-1}
Molar mass of polymer (M) = 25 kg = 25000 g
For dilute solutions (c tending to zero), the osmotic pressure is given by

$$\frac{\pi}{c} = \frac{RT}{M} \quad \text{or} \quad \pi = \frac{c\cdot RT}{M}$$

Note that R should be substituted in L atm K^{-1} mol^{-1} =

$$\pi = \frac{5\times 2500}{25000\times 100}\cdot \text{atm} = 0.005 \text{ atm}$$

Answer: (c)

10. Which of the following statements on ground state perturbation theory, involving the zeroth order energy $E^0{}_0$, first order energy correction $E^{(1)}{}_0$ and second order energy correction $E^{(2)}{}_0$, is **false**?
(a) $E_0^{(1)}$ is the average value of the perturbation operator with respect to the ground state of the zeroth order Hamiltonian.
(b) $E_0^{(1)}$ is necessarily negative.
(c) $E_0^{(2)}$ is necessarily negative.
(d) $E_0^{(2)} + E_0^{(1)}$ is an upper bound to the exact ground state energy.

Solution: The first order energy correction can only be positive.
Hence the only false statement is (b).

Answer: (b)

11. A particle is in the state $\phi = \phi_1 + 3\phi_2$, where $\phi_1$and ϕ_2 are eigenfunctions of the Hamiltonian of the particles with eigenvalues E_1 and E_2, respectively. The average energy of the particles in the same ϕ is ________.
(a) $(E_1 + 9E_2)/10$ (b) $(E_1 + 3E_2)$
(c) $(E_1 + 9E_2)/4$ (d) $(E_1 + 3E_2)/10$

Solution: The normalised wave function is given by $\phi = A\phi_1 + 3A\phi_2$
Upon taking the complex conjugate and imposing the orthogonality condition,

$$A^2 + 9A^2 = 1, \text{ that is, } 10\,A^2 = 1$$

Hence $A = \frac{1}{\sqrt{10}}$

The expectation value of the energy is

$$\langle E\rangle = \Sigma E_n P_n$$
$$= \frac{1}{10}E_1 + \frac{9}{10}E_2$$
$$= (E_1 + 9\,E_2)/10$$

Answer: (a)

12. Consider a model system of five non-interacting fermions in a single 3-dimensional harmonic oscillator. The Hamiltonian of a single particle is

$$\hat{H} = \frac{1}{2m}\left(p_x^2 + p_y^2 + p_z^2\right) + \frac{1}{2}m\omega^2\left(x^2 + y^2 + z^2\right)$$

where m is the mass of the particles, ω is the angular frequency, $\hat{p}_x$, $\hat{p}_y$ and $\hat{p}_z$ are the momentum operators. The ground state energy of the system of 5 non-interacting fermions is

(a) $\frac{21}{2}\hbar\omega$ (b) $\frac{15}{2}\hbar\omega$
(c) $\frac{5}{2}\hbar\omega$ (d) $\frac{25}{2}\hbar\omega$

Solution:

$$H = \frac{1}{2m}\left(P_x^2 + P_y^2 + P_z^2\right) + \frac{1}{2}m\omega^2\left(x^2 + y^2 + z^2\right)$$

Total number of fermions = 5 with the quantum numbers 0 and 1 .Thus,
Energy of 3-dimensional harmonic oscillator

$$= 2\left(0 + \frac{3}{2}\right)\hbar\omega + 3\left(1 + \frac{3}{2}\right)\hbar\omega$$

$$= \frac{6}{2}\hbar\omega + \frac{15}{2}\hbar\omega$$

$$E = \frac{21}{2}\hbar\omega$$

Answer: (a)

13. The volume (cm^3) of CO adsorbed on charcoal (273 K) at two different pressures is given below.

P (kPa)	40	80
V (cm^3)	25	40

Assuming the Langmuir isotherm, the maximum possible volume (cm^3) of CO that can be adsorbed is ________.
(a) 50 (b) 100
(c) 150 (d) 200

Solution: One method of representing Langmuir adsorption isotherm is as follows

$$\frac{V_{ad}}{V_m} = \frac{K_p}{1 + K_p}$$

where V_{ad} denotes the volume of the adsorbate and V_m denotes the volume for the complete monolayer absorption. The first data yields the equation

$$\frac{V_m}{25} = \frac{1}{K_p(40)} + 1$$

Hence

$$K_p = 25/[(V_m - 25) \times 40] \quad (1)$$

The second data yields the equation

$$\Rightarrow \quad \frac{V_m}{40} = \frac{1}{K_p(80)} + 1 \quad (2)$$

Using Eqns (1) and (2),

$$V_m = 100 \text{ cm}^3$$

Answer: (b)

14. Translational partition function of D_2 molecules confined in a 100 cm^3 vessel at 25°C is ________.
($h = 6.626 \times 10^{-34}$ J s, $k = 1.381 \times 10^{-23}$ J K^{-1}
(a) 3.8×10^{22} (b) 5.8×10^{24}
(c) 7.8×10^{26} (d) 9.8×10^{28}

Solution: $q_{tr} = \left(\frac{2\pi mkT}{h^2}\right)^{3/2} V$

Substituting m = 4 g mol^{-1}, V = 100 cm^3, T = 298 K and recalling that 1 J = 1 kg m^2 s^{-2}, the value obtained is 7.8×10^{26}.

Answer: (c)

15. For an enzyme–substrate reaction, E + S ↔ ES, $ES \xrightarrow{k_2} E + P$ the slope and the intercept of the plot between $\frac{1}{r}$ and $\frac{1}{[s]}$ are 10^{-2} s and 10^2 M^{-1} s^{-1}, respectively. The forward and backward rate constants are k_1 and k_{-1}. If $E_0 = 10^{-6}$ M and $\frac{k_{-1}}{k_2} = 1000$, the value of k_1 will be close to (in units of M^{-1} s^{-1}) [r is the rate of the reaction and E_0 is the initial concentration of the enzyme]; ln (10) = 2.303.
(a) 1×10^{11} (b) 2×10^4
(c) 1×10^{-8} (d) 2×10^6

Solution: $r = r_{max}$ [S]/([S] + K_M)

$$[1/r] = K_M/(r_{max}\,[S]) + [1/r_{max}]$$

$$\text{Slope} = K_M/(r_{max})$$

$$r_{max} = 10^{-2} \text{ M s}^{-1} \text{ and } K_M = 10^{-4} \text{ M}$$

But $\quad K_M = (1 + k_{-1}/K_2)/(k_1/k_2) = 1001\, k_2/k_1$

Furthermore, the maximum rate of the reaction occurs when $r_{max} = K_2\,[E]_0$

Since $[E]_0 = 10^{-6}$ M, $r_{max} = 10^{-2}$ M s^{-1} and $k_2 = 10^4$ s^{-1}

Hence,

$$k_1 = 1.001 \times 10^{11} \text{ M}^{-1} \text{ s}^{-1}$$

Answer: (a)

16. One of the correct normalised sp^2 hybrid orbitals is ________.

(a) $\frac{1}{3}\phi_{2s} + \frac{1}{3}\phi_{2px} + \frac{1}{3}\phi_{2py}$

(b) $\frac{1}{2}\phi_{2s} + \frac{\sqrt{3}}{\sqrt{8}}\phi_{2p_x} + \frac{\sqrt{3}}{\sqrt{8}}\phi_{2p_y}$

(c) $\frac{1}{\sqrt{3}}\phi_{2s}+\frac{\sqrt{2}}{\sqrt{3}}\phi_{2p_x}$

(d) $\frac{1}{3}\phi_{2s}+\frac{2}{3}\phi_{2p_x}$

Solution: The three sp^2 hydrid orbitals arise from the hybridisation of one *s*-orbital with two *p*-orbitals whose wave function is given by $\Psi_i = a_i s + b_i p_x + c_i p_y$ with *i* ranging from 1 to 3.
The coefficients can be obtained in the following manner:

(i) $a_1^2 = a_2^2 = a_3^2 = \frac{1}{3}$; thus $a_1 = a_2 = a_3 = \frac{1}{\sqrt{3}}$

(ii) By assigning any one of the coefficients as zero and invoking the normalisation condition, as well as the constraint $c_i = 0$,

$a_i^2 + b_i^2 = 1$. Thus, $\frac{1}{3}+b_i^2 = 1$ or $bi = \sqrt{2/3}$

Hence $\Psi_1 = \frac{1}{\sqrt{3}}\cdot\phi_{2s}+\sqrt{2/3}\phi_2 p_x$

Answer: (c)

17. Assuming harmonic approximation, the energy change for the reaction $HCl + D_2 \rightarrow DCl + HD$ in cm^{-1} is (the vibrational frequency data in cm^{-1} is given in the table below)

HCl	D_2	DCl	HD
2885	2990	1990	3627

(a) −258 (b) +258
(c) −129 (d) +129

Solution: $HCl + D_2 \rightarrow DCl + HD$
The change in the internal energy taking into account the zero point energy is ΔE (cm^{-1})

$$\Delta E = \frac{1}{2}[E(\text{products}) - E(\text{reactants})]$$

$$= \frac{1}{2}[(1990+3627)-(2885+2990)] = \frac{1}{2}[5617-5875]$$

$$= -\frac{1}{2}(258) = -129\ \text{cm}^{-1}$$

Answer: (c)

18. A solution of Fe^{3+} is titrated potentiometrically using Ce^{3+} solution at 25°C. The emf (in V) of the redox system thus formed when, (i) 50% of Fe^{3+} and (ii) 80% of Fe^{3+} are titrated, would respectively be ________.

(Given $E^0_{Fe^{3+}/Fe^{2+}} = 0.77$ V, $\log_{10}2 = 0.301$)

(a) 0.734 and 0.77
(b) 0.77 and 0.385
(c) 0.77 and 0.734
(d) 0.385 and 0.367

Solution: From the Nernst equation and ignoring the volume corrections, it follows that

$$E = E^0_{Fe^{3+}/Fe^{2+}} - \frac{0.0591}{1}\log_{10}\frac{[Fe^{3+}]}{[Ce^{3+}]}$$

(i) For 50% of Fe^{3+}

$$E = 0.77\ \text{V} - 0.0591 \times \log(1) = 0.77\ \text{V}$$

(ii) 80% of Fe^{3+}

$$E = 0.77\ \text{V} - 0.0591 \times \log(80/20)$$
$$= 0.77\ \text{V} - 0.0591 \times \log(4)$$
$$= 0.77\ \text{V} - 0.0591 \times 0.602$$
$$= 0.77\ \text{V} - 0.0355$$
$$= 0.734\ \text{V}$$

Answer: (c)

19. The (002) plane of an elemental FCC crystal diffracts X-rays (λ = 0.154 nm) at Bragg angle 90°. The density of the crystal is 4×10^4 kg m^{-3}. The atomic weight of the elemental solid is ________

(a) 22 (b) 44
(c) 88 (d) 66

Solution: Wavelength of X-rays: λ = 0.154 nm, scattering angle θ = 90°, density (ρ) = 4×10^{-4} kg m^{-3}
Length of unit cell = λ/2 sin(90) = 0.154 nm/2
= 0.077×10^{-4} cm

The atomic weight can be estimated from the density as follows:

$$M = \frac{\rho \times N \times V}{n}$$

M is nearly equal to 22 g mol^{-1}.

Answer: (a)

20. The standard cell potential of cell, Pt| H_2(g) |HBr(*aq*)| AgBr(*s*)|Ag(*s*), was measured over a range of temperature, and the data was fitted as

$$E^0\ (\text{volt}) = 0.01 - 1 \times 10^{-4}(T - 298) - 2 \times 10^{-6}(T - 298)^2$$

The standard reaction entropy (J K^{-1} mol^{-1}) and enthalpy (kJ mol^{-1}) at 298 K are ________.

(a) −9.65 and −3.84 (b) −3.84 and −9.65
(c) −18.3 and −7.68 (d) −7.68 and −18.3

Solution: The standard cell potential is given as

$$E^0\ (\text{volt}) = 0.01 - 1 \times 10^{-4}(T - 298) - 2 \times 10^{-6}(T - 298)^2$$

$$\frac{\partial E^0}{\partial T} = -1\times10^{-4} - 4\times10^{-6}T + 4\times298\times10^{-6}$$

At 298 K, the temperature coefficient follows from the above as −0.0001.

Since the entropy change is related to the temperature coefficient as

$$\Delta S = nF \times 1 \times 96500 \times 10^{-4}$$

$$\Delta S = -9.65 \text{ J K}^{-1}$$

(ii) $\Delta G^0 = -nFE^0$

$$= -1 \times 96500\ (0.01)$$

$$\Delta G^0 = -965 \text{ J mol}^{-1}$$

$$\Delta H^0 = \Delta G^0 + T\Delta S^0$$

$$= -965 + 298\ (-9.65)$$

$$= -3480.7 \text{ J}$$

Answer: (a)

21. In the phase diagram of water, the solid–liquid boundary has a negative slope. The reason for this unusual behaviour can be traced to a decrease in the ________.

(a) density of the system on melting
(b) volume of the system on melting
(c) entropy of the system on melting
(d) enthalpy of the system on melting

Solution: The solid–liquid boundary of the phase diagram of water shows a negative slope. This first order phase transition is due to the decrease in the volume of system upon melting.

Answer: (b)

22. At 300 K, the thermal expansion coefficient and the isothermal compressibility of liquid water are 2×10^{-4} K^{-1} and 5×10^{-5} bar^{-1}, respectively. $\left(\frac{\partial U}{\partial V}\right)_T$ (in k bar) for water at 320 K and 1 bar will be ________.

(a) 2.4 (b) 1.2
(c) 0.6 (d) 12.0

Solution: The thermal expansion coefficient is defined as

$$\alpha = \frac{1}{v}\left(\frac{\partial V}{\partial T}\right)_p = 2 \times 10^{-4} \text{ K}^{-1} \quad (1)$$

The isothermal compressibility is given by

$$\beta = \frac{1}{v}\left(\frac{\partial V}{\partial P}\right)_T = 5 \times 10^{-5} \text{ bar}^{-1} \quad (2)$$

From Eqns (1) and (2), the ratio α/β follows as

$$= \frac{2\times10^{-4} \text{ K}^{-1}}{5\times10^{-5} \text{ bar}^{-1}}$$

Hence,

$$\frac{\alpha}{\beta} = -\left(\frac{\partial P}{\partial T}\right)$$

However,

$$\left(\frac{\partial U}{\partial V}\right)_T = T\left(\frac{\alpha}{\beta}\right)v^{-P}$$

$T = 320$ K and $P = 1$ bar. Thus, $(\partial U/\partial V)_T$ is -1281 bar

$$\left(\frac{\partial U}{\partial V}\right)_T = -1.281 \text{ k bar}$$

Answer: (b)

23. Elementary steps of a reaction are as follows

$$\text{A} + \text{B} \underset{k_{-1}}{\overset{k_1}{\rightleftarrows}} 2\text{C} \quad \text{and} \quad 2\text{C} \xrightarrow{k_2} \text{P}$$

If steady state approximation is applicable to C, the rate of product formation in the reaction is ________.

(a) proportional to [A][B]
(b) proportional to $[A]^2[B]^2$
(c) proportional to $[A]^{1/2}[B]^{1/2}$
(d) independent of [A][B]

Solution: From the given equations, $\partial P/\partial t = k_2\,[C]^2$
From the steady state approximation for C,

$$\partial C/\partial t = 0 = 2k_1\,[A]\,[B] - k_{-1}\,[C]^2$$

$$[C]^2 = 2\,k_1[A][B]/k_{-1}$$

Hence $\partial P/\partial t = k_2\,k_1[A][B]/k_{-1}$

Answer: (a)

24. A Gaussian distribution has the functional form $f(x) = \frac{2}{\sqrt{2a^2\pi}}e^{-(x-b)^2/2a^2}$. The variance of such a distribution is ________.

(a) a (b) a^2
(c) b (d) b^2

Solution: The variance of the given distribution is defined as

$$\text{Variance }(x) = \int_{-\infty}^{\infty}(x-b)^2\frac{2}{\sqrt{2\pi}}\exp\cdot\left(-\frac{(x-b)^2}{2a^2}\right)dx$$

Employing the properties of the definite integrals when the limits are between $-\infty$ and $+\infty$,

$$\text{var}(x) = \int_{-\infty}^{\infty}x^2\frac{2}{a\sqrt{2\pi}}\exp\cdot\left\{\frac{-x^2}{2a^2}\right\}dx$$

$$= a\times\frac{2}{2}\times\frac{4}{\sqrt{\pi}}\int_0^{\infty}x^2e^{-}x^2dx$$

Upon substituting $t = x^2$, the integral becomes

$$\text{Var }(x) = a^2.\frac{8}{\sqrt{\pi}}\int_0^{\infty}\left(\sqrt{t}\right)^2\left(2\sqrt{t}\right)^{-1}e^{-1}dt$$

$$= a^2\cdot\frac{8}{\sqrt{\pi}}\cdot\frac{1}{2}\int_0^{\infty}t^{3/2^1}e^{-1tdt}$$

Or, $$\text{var }(x) = a^2\times\frac{8}{\pi}\cdot\frac{1}{2}\cdot\frac{\sqrt{\pi}}{2} = 2a^2$$

Thus, the variance of the given distribution is a^2.

Answer: (b)

25. The lowest energy state of $1s^1$, $2s^1$ electronic configuration, according to Hund's rule is __________.

(a) $^3S_0$ (b) $^1S_0$
(c) $^3S_1$ (d) $^1S_1$

Solution: The lowest electronic configuration is $1s^1\ 2s^1$

$1s$	$2s$
1	1

$$L = l_1 + l_2 = 0 + 0$$

$L = 0$ corresponds to the S-term

$$S = s_1 + s_2 = \frac{1}{2} + \frac{1}{2} = 1$$

$$(2S + 1) = 3 \quad \text{and} \quad J = L + S = 0 + 1 = 1$$

Thus, the term symbol is, $= {}^3S_1$

Answer: (c)

26. In fission of ^{235}U atom, the energy released is 200 MeV. In one day, fission of 1 kg ^{235}U will give power (in MW) approximately __________.

(a) 550 (b) 650
(c) 950 (d) 1250

Solution: 1 kg of ^{235}U contains

$$= (1000/235) \times 6.023 \times 10^{23}$$

$$= 25.62 \times 10^{23} \text{ atoms}$$

The energy released during each ^{235}U fission reaction = 200 MeV

Total energy released = $25.62 \times 10^{23} \times 200 = 5124 \times 10^{23}$ MeV $= 5124 \times 10^{23} \times 1.6 \times 10^{-13}$ J

$$= 8198.4 \times 10^{10} \text{ J}$$

However, power denotes the energy released per unit time. Hence, one watt equals one J s^{-1}.

The total power released = $(8198.4 \times 10^{10}$ J$)/86400$ s is nearly equal to 950 MW

Answer: (c)

27. Difference between the activation energies of the reverse and forward steps of a reversible reaction is 9.212 RT. If the pre-exponential factor of the forward reaction is double that of the reverse reaction at the same temperature, the equilibrium constant for the reaction at the temperature will be __________.

(a) 1×10^4 (b) 2×10^4
(c) 1×10^{-4} (d) 2×10^{-4}

Solution: From the Arrhenius equation:

$$k = Ae^{-E_a/RT} \tag{1}$$

The equilibrium constant K is given by the ratio of the forward and backward rate constants.

Hence $k_f/k_b = [A_f/A_b] \exp[(1/RT)(-E_{a,1} + E_{a,2})] = 2 \exp(9.212RT/RT)$

Hence the ratio of forward and backward rate constants is 2×10^4

Answer: (b)

28. In the pure Raman rotational spectrum of $^{16}O_2$, whose electronic ground state is $^3\Sigma_g$, the transition between to/from __________.

(a) even J levels are missing
(b) odd J levels are missing
(c) all J levels appear
(d) none of the J levels appear

Solution: In the rotational Raman spectra, the transitions will be observed at $J = 1, 3, 5, 7$….(odd values of J). Hence even values of J will be missing.

Answer: (a)

29. The transition moment integral for a rotational transition between $J = 1, M_J = 0$ to $J = 2, M_J = 0$ states for a diatomic molecule along the z-axis is proportional to __________.

(a) $\int_0^\pi \cos^2\theta(3\cos^2\theta - 1)d\theta$

(b) $\int_0^\pi \cos^2\theta(3\cos^2\theta - 1)\sin\theta\, d\theta$

(c) $\int_0^\pi \cos\theta(3\cos^2\theta - 1)\sin\theta d\theta$

(d) $\int_0^\pi \cos\theta(3\cos^2\theta - 1)\sin^2 d\theta$

The transition moment integral for a rotational transition for a diatomic molecule along the z-axis is proportional to

$$\int_0^\pi \cos^2\theta(3\cos^2\theta - 1)\sin\theta d\theta$$

According to the selection rule.

Answer: (b)

Appendix 2

GATE Solved Question Paper (2019)

1. In the Freundlich isotherm, a linear relationship is obtained in the plot of (θ = surface coverage and p = partial pressure of the gas)

(a) θ vs p (b) $\ln(\theta)$ vs $\ln(p)$

(c) $\ln(\theta)$ vs p (d) θ vs $\ln(p)$

Solution: According to the Freundlich isotherm, θ is proportional to p^n

Hence $\ln\theta$ will be linear with respect to $\ln p$.

Answer: (b)

2. Micelle formation is accompanied by the

(a) decrease in overall entropy due to ordering

(b) increase in overall entropy mostly due to increase in solvent entropy

(c) increase in overall entropy mostly due to increase in solute entropy

(d) increase in overall entropy and decrease in enthalpy

Solution: Although surfactants are becoming more ordered on *micelle formation*, this loss of entropy is more compensated due to the increase in solvent entropy.

Answer: (b)

3. Consider the following phase diagram of CO_2 (not to scale). At equilibrium, the **INCORRECT** statement is:

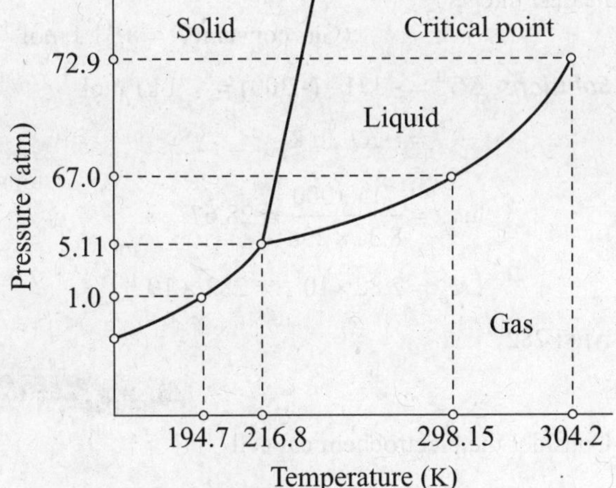

(a) At 200 K, on increasing the pressure from 1 to 50 atm, CO_2 gas condenses to liquid.

(b) It is not possible to obtain liquid CO_2 from gaseous CO_2 below 5.11 atm.

(c) Both liquid and gas phase of CO_2 coexist at 298.15 K and 67 atm.

(d) With increasing pressure, the melting point of solid CO_2 increases.

Solution: As can be seen from the figure, the condensation of CO_2 to liquid does not occur at 200 K and in the pressure given.

Answer: (a)

4. The number of CO stretching bands in the IR spectrum of trigonal bipyramidal *cis*-$M(CO)_3L_2$ is ________.

(M = metal and L = monodentate ligand)

From the point group for the *cis* isomer, the number of bands can be deduced as 3.

Answer: 3

5. On heating a sample of 25 mg hydrated compound (molecular weight = 250 g mol^{-1}) in thermogravimetric analysis, 16 mg of dehydrated compound remains. The number of water molecules lost per molecules of hydrated compound is ________.

Solution: Since 0.009 g of water was removed, this corresponds to 0.009/18 = 0.0005 moles of water removed. Number of moles of the original hydrated compound = $25 \times 0.001/250 = 0.0001$

Number of water molecules lost = 0.0005/0.0001 = 5

Answer: 5

6. The total number of α and β particles emitted in the following radioactive decay is ________.

$$^{238}_{92}\text{U} \rightarrow {}^{210}_{82}\text{Pb}$$

Solution: When ^{238}U decays to ^{210}Pb, this corresponds to the decrease of mass number by 28. Thus seven α particles are emitted. When ^{92}U decays to ^{82}Pb, there is a decrease in atomic number by 10. Since the seven α particles lead to a decrease in atomic number by 14, the number of β particles emitted is 4.

Thus, the total number of α and β particles emitted = 11

Answer: 11

7. An ideal gas occupies an unknown volume of V liters (L) at a pressure of 12 atm. The gas is expanded isothermally against a constant external pressure of 2 atm so that its final volume becomes 3 L. The work involved for this expansion process is ________ cal. (Round off to two decimal places)

(Gas constant $R = 2$ cal mol^{-1} K^{-1})

Solution: Since the expansion is isothermal and the gas is ideal, we can write

$$P_1V_1 = P_2V_2$$

$$P_1 = 12 \text{ atm}, P_2 = 2 \text{ atm}, V_2 = 3 \text{ L}$$

Hence $V_1 = 0.5$ L

$$W = -P_2(V_2 - V_1) = -5 \text{ L atm} = -121.95 \text{ cal}$$

Answer: –126 to –120

8. The entropy change for the melting of 'X' moles of ice (heat of fusion is 80 cal g^{-1}) at 273 K and 1 atm pressure is 28.80 cal K^{-1}. The value of 'X' is ________. (Round off to two decimal places)

(Molecular weight of water = 18 g mol^{-1})

Solution: Entropy change $\Delta S = \Delta H_{\text{fusion}}/T_{\text{m}}$

Hence 28.8 cal K^{-1} = 80 cal g^{-1}/273 K

Weight of ice = 98.28 g

Number of moles of ice = 98.28/18 = 5.46

Answer: 5.40 to 5.55

9. Consider a two-state system at thermal equilibrium having energies 0 and $2k_BT$ for which the degeneracies are 1 and 2, respectively. The value of the partition function at the same absolute T is ________. (Round off to two decimal places)

(k_B is the Boltzmann constant)

Solution: Partition function = $\sum g_i e^{-\epsilon_i/kT}$

$$= 1 + 2e^{-2kT/kT} = 1 + 2e^{-2} = 1.27$$

Answer: 1.25 to 1.30

10. Consider a system of three identical and distinguishable non-interacting particles and three available non-degenerate single particles energy levels having energies 0, ε and 2ε. The system is in contact with a heat bath of temperature T K. A total energy of 2ε is shared by these three particles. The number of ways the particles can be distributed is ________.

Solution: 3 distinguishable particles a, b, c

2ε	a	b	c			
ε	bc	ac	ab			
0	a	b	c	bc	ca	ab

Since the total energy should be 2ε and the number of particles is 3, the above arrangement indicates that the total number of ways is 6.

Answer: 6

11. Gas phase bond length and dipole moment of compound (MX) is 3 Å and 10.8 D, respectively. The ionic character in gas phase MX is ________%. (Round off to one decimal place)

(1 D = 3.336×10^{-30} C m)

The expected dipole moment taking into account, the charge separation is

μ (in Debye unit) = 1.60×10^{-19} C $\times 3 \times 10^{-10}$ m/(3.336×10^{-30} C m) = 14.38

% ionic character = 10.8 D × 100/14.38 = 75.06

Answer: +74 to +76

12. Consider the following two parallel irreversible first order reactions at a temperature T,

$$\text{P} \xrightarrow{k_1} \text{Q}$$

$$\text{P} \xrightarrow{k_2} \text{R}$$

where k_1 and k_2 are the rate constants and their values are 5×10^{-2} and 15×10^{-2} min^{-1} respectively, at a temperature T. If the initial concentration of the reactant 'P' is 4 mol L^{-1} then the concentration of product 'R' after 10 min of reaction is ________ mol L^{-1}. (Round off to two decimal places, assume only P is present at the beginning of the reaction.)

Solution: $[R] = \dfrac{k_2[P]_0}{(k_1+k_2)}\left[1 - e^{-(k_1+k_2)t}\right]$

$$= 15 \times 0.01 \times 4 \times [1 - e^{(-20\times0.01\times10)}]/(20 \times 0.01) = 2.60$$

Answer: 2.50 to 2.65

13. Consider the following equilibrium

$$SO_2(g) + ½\, O_2 \rightarrow SO_3(g)$$

At 298 K, the standard molar Gibbs energies of formation, $\Delta_f G^0$, of $SO_2(g)$ and $SO_3(g)$ are –300 and –371 kJ mol^{-1}, respectively. The value of the equilibrium constant, K_p, at this temperature is ________ $\times 10^{10}$. (Round off to the nearest integer)

(Gas constant $R = 8.31$ J mol^{-1} K^{-1})

Solution: $\Delta G^0 = -371 - (-300) = -71$ kJ mol^{-1}

$$= -RT \ln k_p$$

$$\ln k_p = \frac{71 \times 1000}{8.31 \times 298} = 28.67$$

$$K_p = 2.82 \times 10^{12} = 282 \times 10^{10}$$

Ans: 282

Answer: 265 to 295

14. Consider the electrochemical cell

$$M(s) \mid MI_2(s) \mid MI_2(aq) \parallel M(s)$$

where M is a metal. At 298 K, the standard reduction potentials are $E^0_{M_2+(aq)/M(s)}$ = –0.12 V, $E^0_{MI_2(s)/M(s)}$ = –0.36 V and the temperature coefficient is $\left(\dfrac{\partial E^0_{\text{cell}}}{\partial T}\right)_p$ = 1.5×10^{-4} V K^{-1}. At this temperature, the standard enthalpy change for the overall reaction, $\Delta_r H^0$, is __________ kJ mol^{-1}. (Round off to two decimal places)

(Faraday constant $F = 96500$ C mol^{-1})

$$\Delta S^0 = nF\,(\partial E/\partial T)_P$$

$$= 2\,(96500)(1.5\times 10^{-4})\text{ J K}^{-1} = 28.95\text{ J K}^{-1}$$

$$\Delta G^0 = -nFE^0 = (-2)\,(96500)\,(0.24) = -46320\text{ J mol}^{-1}$$

$$\Delta G^0 = \Delta H^0 - T\Delta S$$

$$\Delta H^0 = -46320 + (298)(28.95)$$

$$= -37.69\text{ kJ}$$

Answer: –38.00 to –37.00

15. The normal boiling point of a compound (X) is 350 K (Heat of vaporisation, $\Delta_{vap}H$, = 30 kJ mol^{-1}). The pressure required to boil X at 300 K is __________ torr. (Round off to two decimal places).
(Ignore the temperature variation of $\Delta_{vap}H$, gas constant $R = 8.31$ J mol^{-1} K^{-1} and 1 atm = 760 torr.)

Solution: $\ln\left(\dfrac{p_2}{p_1}\right) = \dfrac{\Delta H}{R}\dfrac{(T_2 - T_1)}{T_1 T_2}$

$$\ln\left(\frac{p_2}{1}\right) = \frac{30000}{8.31}\left(\frac{300-350}{300\times 350}\right)$$

$$\ln p_2 = -1.719 \text{ and } p_2 = 0.179\text{ atm}$$

$$1\text{ atm} = 760\text{ torr}$$

$$0.179\text{ atm} = 136.23\text{ torr}$$

Answer: 120.00 to 140.00

16. For the bimolecular gas phase reaction P + Q → R, the pre-exponential factor is 1×10^{13} dm^3 mol^{-1} s^{-1}. The standard entropy of activation at 25°C is __________ J K^{-1} mol^{-1}. (Round off to two decimal points.)
(The standard concentration c_0 = 1 mol dm^{-3}; Planck constant $h = 6.62 \times 10^{-34}$ J s; Boltzmann constant $k_B = 1.38 \times 10^{-23}$ J K^{-1}; Gas constant $R = 8.31$ J mol^{-1} K^{-1})

Solution: $A = e^2\left(\dfrac{kT}{hc0}\right)e^{\frac{\Delta S^{\#}}{R}}$

$$e^2\, e^{\frac{\Delta S^{\#}}{R}} = \frac{10^{13}\times 6.62\times 10^{-34}}{1.38\times 10^{-23}\times 298} = 1.6097$$

$$e^{\frac{\Delta S^{\#}}{R}} = 0.2185$$

Therefore $\dfrac{\Delta S^{\#}}{R} = -1.521$; $\Delta S^{\#} = -12.64$ J K^{-1}

Answer: –12.90 to –12.40

17. If $\langle \alpha \,|\, \hat{S}_x\hat{S}_y - \hat{S}_y\hat{S}_x\, \alpha\rangle = i\hbar^2 a$, where $\hat{S}_x$ and $\hat{S}_y$ are spin angular momentum operators and $|\alpha\rangle$ is spin up eigenfunction, then the value of a is __________.

(Round off to one decimal place.)

Solution: From the properties of Pauli spin matrices, the rhs of the given identity yields

$$(i\,\hbar/4 + i\hbar/4) = i\hbar/2$$

Hence a = ½ = 0.5

Answer: 0.5

18. A particle in a one dimensional box of length $2a$ with potential energy.

$V = \begin{cases} 0 & |x| < a \\ \infty & |x| > a \end{cases}$ is perturbed by the potential $V' = cx$ eV,

where c is a constant. The 1st order correction to 1st excited state of the system is __________ $\times\, c$ eV

Solution: Since the potential energy is zero within the box and infinity outside, the energies will not get affected and hence the correction to the energy with the altered potential is zero.

Answer: 0

19. Consider a two dimensional harmonic oscillator with angular frequency $\omega_x = 2\omega_y = 6.5 \times 10^{14}$ rad s^{-1}. The wavelength of x polarised light required for the excitation of a particle from its ground state to the next allowed excited state is __________ $\times 10^{-6}$ m. (Round off to one decimal place.)

(Speed of light $c = 3.0 \times 10^8$ m s^{-1})

Solution: Angular frequency $\omega = 2\pi\nu$
Energy of a two dimensional harmonic oscillator = $E(n_x, n_y) = (n_x + n_y + 1)\,h\nu$
Ground state energy $E(0,0) = h\nu$
First excited energy $E(1,1) = 3\,h\nu$

$$\Delta E = E(1,1) - E(0,0) = 3\,h\nu - h\nu = 2\,h\nu = 2hc/\lambda$$

Wave length $\lambda = (3 \times 10^8\text{ m s}^{-1} \times 2 \times 3.14)/(6.5 \times 10^{14}\text{ s}^{-1})$
$= 2.9 \times 10^{-6}$ m

Answer: 2.8 to 3.0

Appendix 3

Chemistry JAM Solved Question Paper (2018)

1. The value of integral $\int_{-2}^{+2} xe^{-2x^2}\,dx$ is

(a) 0 (b) ½
(c) 1 (d) 2

Solution: The integrand is an odd function of x while the limits are –2 to +2. Hence the integral becomes zero.

Answer: (a)

2. The number of crystal systems and the number of Bravais lattices are, respectively,

(a) 14 and 7 (b) 7 and 32
(c) 32 and 14 (d) 7 and 14

Solution: Number of crystal systems = 7; number of Bravais lattices = 14

Answer: (d)

3. For absorption of gas on a solid surface, the plot that represents the Freundlich isotherm is (x = mass of gas, m = mass of absorbent, P = pressure)

(a)
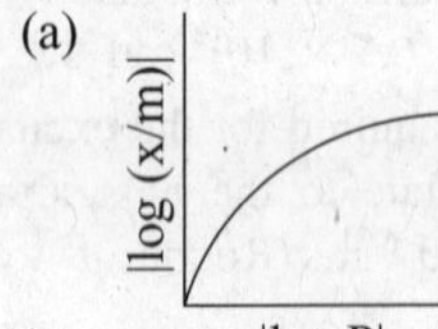

(b)
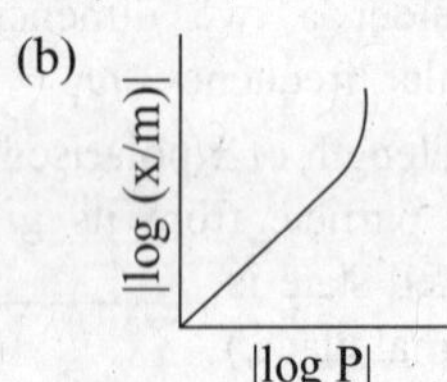

(c)
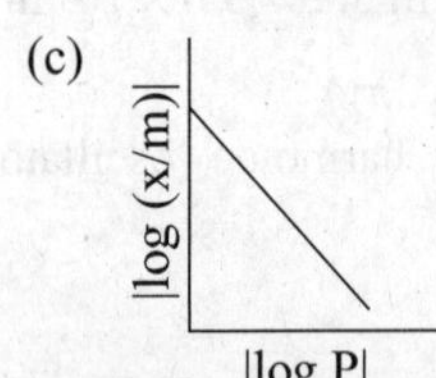

(d)
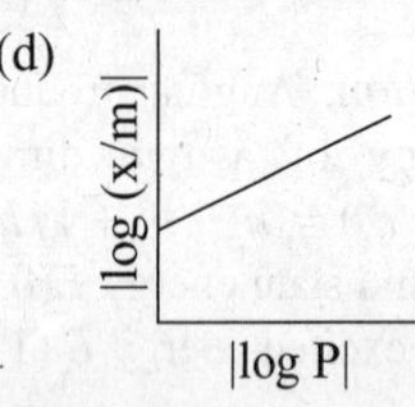

Solution: Since the amount adsorbed (x) is directly proportional to p^n in the Freundlich absorption isotherm, log x varies linearly with log p.

Answer: (d)

4. The decay modes of ^{14}C and ^{14}O are

(a) β decay
(b) positron emission
(c) β decay and positron emission, respectively
(d) positron emission and β decay, respectively

Solution: ^{14}C decays into ^{14}N through β-decay.
For ^{14}O, the number of protons /number of neutrons = 0.75, hence, it is positron emission.

Answer: (c)

5. The **CORRECT** expression that corresponds to reversible and adiabatic expansion of an ideal gas is

(a) $\Delta U = 0$ (b) $\Delta H = 0$
(c) $\Delta S = 0$ (d) $\Delta G = 0$

Solution: Entropy change of a reversible process is defined as $\Delta S_{\text{system}} = \delta q/T$
Since $\delta q = 0$ for an adiabatic process, $\Delta S_{\text{system}} = 0$

Answer: (c)

6. The reaction, A $\rightarrow$ **Products**, follows first order kinetics. If [A] represents the concentration of reaction at a time t, the **INCORRECT** variation is shown in

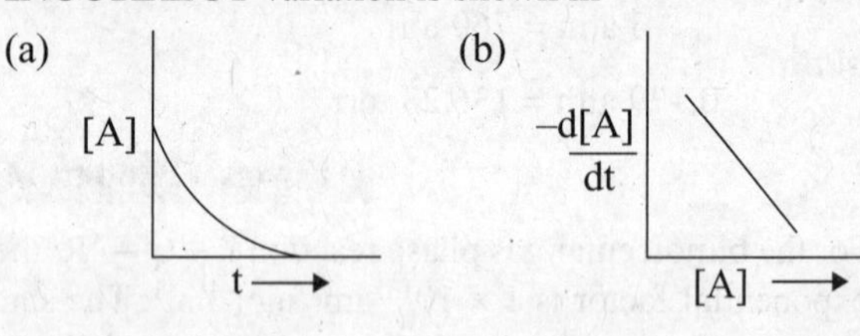

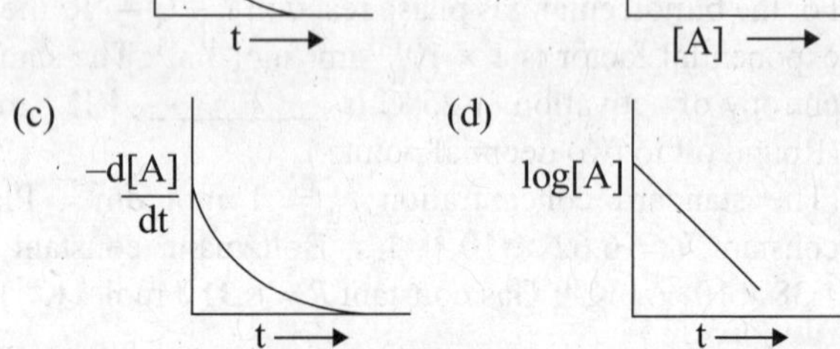

Solution: The variation shown in (a), (c) and (d) are correct for a first order reaction. Hence the incorrect variation is b.

Answer: (b)

7. The behaviour of Cl_2 is closest to ideal gas behaviour at

(a) 100°C and 10.0 atm
(b) 0°C and 0.50 atm
(c) 200°C and 0.50 atm
(d) –100°C and 10.0 atm

Solution: The behaviour of gases will be ideal at high temperatures and low pressures.

Answer: (c)

8. A vector $\vec{A}=\vec{i}+x\vec{j}+3\vec{k}$ is rotated through an angle and is also doubled in magnitude resulting in $\vec{B}=4\vec{i}+(4x-2)\vec{j}+2\vec{k}$. An acceptable value of x is

(a) 1 (b) 2
(c) 3 (d) 4/3

Solution: $2(i+xj+3k)\ 4i+(4x-2)\ j+2k$

$$2\sqrt{(1+x^2+9)}=\sqrt{16+(4x-2)^2+4}$$

Squaring on both sides,

$$4(10+x^2)=20+(4x-2)^2$$

One value for x is 2.

Answer: (b)

9. With reference to the variation of molar conductivity (Λ_m) with concentration for a strong electrolyte in an aqueous solution, the **CORRECT** statement is

(a) The asymmetry effect contributes to decrease Λ_m whereas the electrophoretic effect contributes to increase Λ_m
(b) The asymmetry effect contributes to increase Λ_m whereas the electrophoretic contributes to decrease Λ_m
(c) Both asymmetry effect and electrophoretic effect contribute to decrease Λ_m
(d) Both asymmetry effect and electrophoretic effect contribute to increase Λ_m

Solution: According to the Debye–Hückel–Onsager theory, both asymmetric and electrophoretic effects lead to a decrease in molar conductivity with increase in concentration.

Answer: (c)

10. In a saturated calomel electrode, the saturation is with respect to

(a) KCl (b) Hg_2Cl_2
(c) $HgCl_2$ (d) AgCl

Solution: In a saturated calomel electrode, the electrolyte KCl is chosen as the solution

Answer: (a)

11. Which of the following set(s) of quantum numbers is(are) **NOT** allowed?

(a) $n=3, l=2, m_l=-1$ (b) $n=4, l=0, m_l=-1$
(c) $n=3, l=3, m_l=-3$ (d) $n=5, l=3, m_l=+2$

Solution: For the principal quantum number n, the value of l ranges from $\pm n-1$ including 0.
Hence, option (c) is incorrect. For $n=4$ and $l=0$, $m_l=-1$ is not allowed. Hence, option (b) is incorrect.

Answer: (b) and (c)

12. The **CORRECT** expression(s) for isothermal expansion of 1 mol of an ideal gas is(are)

(a) $\Delta A=RT\ln\frac{V_{\text{initial}}}{V_{\text{final}}}$ (b) $\Delta G=RT\ln\frac{V_{\text{initial}}}{V_{\text{final}}}$
(c) $\Delta H=RT\ln\frac{V_{\text{final}}}{V_{\text{initial}}}$ (d) $\Delta S=R\ln\frac{V_{\text{final}}}{V_{\text{initial}}}$

Solution: $dA=-PdV-SdT$
At constant temperature, $dA=-PdV=-RT/VdV$

$$\therefore\quad \Delta A=RT\ln(V_{\text{initial}}/V_{\text{final}})$$

$$dG=VdP-SdT$$

At constant temperature

$$dG=RT/P\ dP$$

$$\therefore\quad \Delta G=RT\ln P_{\text{final}}/P_{\text{Initial}}$$

$$RT\ln V_{\text{initial}}/V_{\text{final}}$$

$$dS=\delta q_{\text{rev}}/T$$

For an isothermal process, $\Delta E=0$

$$\therefore\quad dS=\delta w/T=RTdV/VT$$

$$\therefore\quad \Delta S=R\ln V_{\text{final}}/V_{\text{initial}}$$

Answer: (a), (b) and (d)

13. The volume of 0.3 M ferrous ammonium sulphate solution required for the completion of a redox titration with 20 mL of 0.1 M potassium dichromate solution is ________ mL.

Solution: Equivalent weight of $K_2Cr_2O_7$ = molecular weight/ 6
Equivalent weight of ferrous ammonium sulphate = molecular weight

$$V_1N_1=V_2N_2$$

$$(V_1)(0.3)=(20)(0.1)(6)$$

$$V_1=40\text{ mL}$$

14. The time for 50% completion of zero order reaction is 30 min. Time for 80% completion of this reaction is ________ min.

Solution: $C_0-C_0/2=k\times 30$ min

$$C_0/2=30\ k\text{ min}$$

$$\therefore\quad k=C_0/2\times 30\text{ min}^{-1}$$

Time for 80% completion:

$$C_0-20C_0/100=k\,t$$

$$0.8\ C_0=kt$$

$$\therefore\quad t=0.8\ C_0/k$$

$$=0.8C_0\times 2\times 30/C_0\text{ min}$$

$$=48\text{ minutes}$$

15. Consider the reaction $CO(g) + \frac{1}{2} O_2(g) \rightarrow CO_2(g)$. The value of ΔU for the reaction at 300 K is –281.8 kJ mol^{-1}. The value of ΔH at same temperature is ________ kJ mol^{-1} (rounded up to the decimal place).

[$R = 8.3$ J K^{-1} mol^{-1}]

Solution: $H = U + PV$

$$\Delta H = \Delta U + \Delta(PV)$$

$$= \Delta U + (RT)(\Delta n)$$

Δn = Number of moles of gaseous products – number of moles of gaseous reactants = –1/2

$$\therefore \quad \Delta H = -281.8 + (-0.5)(8.3)(300)/1000$$

$$= -281.8 - 1.245$$

$$= -283.0 \text{ kJ mol}^{-1}$$

16. The nuclear spin quantum (I) of a nucleus is 3/2. When placed in an external magnetic field, the number of possible spin energy states it can occupy is ________.

Solution: Number of energy states = $2S + 1 = 4$

(since $S = 3/2$)

17. The value of C_v for 1 mol of N_2 gas predicted from the principle of equipartition of energy ignoring vibrational contribution, is ________ J K^{-1} mol^{-1} (rounded off to two decimal places).

[$R = 8.3$ J K^{-1} mol^{-1}]

Solution: Translational contribution to $C_v = 3R/2$
Rotational contribution to $C_v = R$. Hence, $C_v = 5\,R/2$

$$= 20.75 \text{ J K}^{-1} \text{ mol}^{-1}$$

18. Assuming ideal gas behaviour, the density of O_2 gas at 300 K and 1.0 atm is ________ g L^{-1} (rounded off to two decimal places).

[$R = 0.082$ L atm mol^{-1} K^{-1}, molar mass of $O_2 = 32$]

Solution: $PV = RT$ (1 atm) $V = (0.082$ L atm) (300)

$$\therefore \quad V = 24.6 \text{ L}$$

$$\text{Density} = 32 \text{ g}/24.6 \text{ L} = 1.30 \text{ g L}^{-1}$$

19. The emf of a standard cadmium cell is 1.02 V at 300 K. The temperature coefficient of the cell is -5.0×10^{-5} V K^{-1}. The value of ΔH^0 for the cell is ________ kJ mol^{-1} (rounded off to two decimal places).

[1 F = 96500 C mol^{-1}]

Solution: $\Delta S_0 = n\,F\,(\partial E/\partial T)_p$

$$= 2 \times 96500 \times (-5) \times 10^{-5} \text{ J K}^{-1}$$

$$= -9.65 \text{ J K}^{-1}$$

$$\Delta G_0 = -n\,FE^0$$

$$= (-2)(96500)(1.02)$$

$$= -196.86 \text{ kJ}$$

$$\Delta G^0 = \Delta H^0 - T\,\Delta S^0$$

$$\Delta H^0 = \Delta G^0 + T\,\Delta S^0$$

$$= -196.86 + (298)(-9.65)/1000$$

$$= -199.74 \text{ kJ mol}^{-1}$$

20. For the reaction $H_2(g) + \frac{1}{2} O_2(g) \rightarrow H_2O(l)$, the following information is given at $T = 300$ K.

$\overline{\Delta H}^0 = -285$ kJ mol^{-1} $\quad \overline{S}^0_{H_2O}(l) = 70$ J K^{-1} mol^{-1}

$\overline{S}^0_{O_2}(g) = 204$ J K^{-1} mol^{-1} $\quad \overline{S}^0_{H_2}(g) = 130$ J K^{-1} mol^{-1}

$\Delta\overline{S}^0_{\text{universe}}$ for the reaction is ________ J K^{-1} mol^{-1}

Solution: $\Delta S^0 = S^0_{H_2O} - S^0_{H_2} - 1/2(S^0_{O_2})$

$$= 70 - 130 - 102 = -162 \text{ J K}^{-1} \text{ mol}^{-1}$$

$$\Delta S^0_{\text{surr}} = +285 \times 1000 \text{ J}/300$$

$$= 950 \text{ J K}^{-1}$$

$$\Delta S^0_{\text{universe}} = (950 - 162) \text{ J K}^{-1} \text{ mol}^{-1}$$

$$= 788 \text{ J K}^{-1} \text{ mol}^{-1}$$

Note that $\Delta S_{\text{universe}} \geq 0$ according to the Clausius inequality.

21. For a H_2 molecule, the fundamental vibration frequency $\left(\bar{\upsilon}_e\right)$ can be taken as 4400 cm^{-1}. The zero point energy of the molecules is ________ kJ mol^{-1} (rounded up to two decimal places)

[$h = 6.6 \times 10^{-34}$ J s, $c = 3 \times 10^8$ m s^{-1}, $N_A = 6 \times 10^{23}$ mol^{-1}]

Solution: $E_0 = h\upsilon/2$

$$= 0.5 \times 6.6 \times 10^{-34} \times 4400 \text{ cm}^{-1}$$

$$= 0.5 \times 6.6 \times 10^{-34} \times 4400 \times 3 \times 10^{10} \times 6 \times 10^{23} \text{ J mol}^{-1}$$

$$= 26.14 \text{ kJ mol}^{-1}$$

22. The solubility of PbI_2 in 0.10 M KI(aq) is ________ × 10^{-7} M (rounded up to two decimal places).

[The solubility product, $K_{sp} = 7.1 \times 10^{-9}$]

Solution: $K_{sp} = [Pb^{2+}][I^-]^2 = (s)(2s + 0.1)^2$

Neglecting $2s$, in comparison with 0.1,

$$7.1 \times 10^{-9} = (s)(0.1)^2$$

$$s = 7.1 \times 10^{-7}$$

23. The electron of a hydrogen atom is in its nth Bohr orbit having a de Broglie wavelength of 13.4 Å. The value of n is ________ (rounded up to the nearest integer).

[Radius of the nth Bohr orbit = 0.53 n^2 Å, $\pi = 3.14$]

Solution: $2\pi r = n\lambda$

where r denotes the Bohr radius of the nth Bohr orbit.
Hence $2 \times 3.14 \times 0.53\,n^2 = n\,\lambda$

$$\lambda = 13.4 \text{ Å}$$

Hence $n = 4$.

Index

K

L

M

N

O

P

Q

R